HANDBOOK OF DETERGENTS
Part F: Production

SURFACTANT SCIENCE SERIES

FOUNDING EDITOR

MARTIN J. SCHICK
1918–1998

SERIES EDITOR

ARTHUR T. HUBBARD
Santa Barbara Science Project
Santa Barbara, California

ADVISORY BOARD

DANIEL BLANKSCHTEIN
Department of Chemical
  Engineering
Massachusetts Institute of
  Technology
Cambridge, Massachusetts

S. KARABORNI
Shell International Petroleum
  Company Limited
London, England

LISA B. QUENCER
The Dow Chemical Company
Midland, Michigan

JOHN F. SCAMEHORN
Institute for Applied Surfactant
  Research
University of Oklahoma
Norman, Oklahoma

P. SOMASUNDARAN
Henry Krumb School of Mines
Columbia University
New York, New York

ERIC W. KALER
Department of Chemical
  Engineering
University of Delaware
Newark, Delaware

CLARENCE MILLER
Department of Chemical
  Engineering
Rice University
Houston, Texas

DON RUBINGH
The Procter & Gamble Company
Cincinnati, Ohio

BEREND SMIT
Shell International Oil Products B.V.
Amsterdam, The Netherlands

JOHN TEXTER
Strider Research Corporation
Rochester, New York

1. Nonionic Surfactants, *edited by Martin J. Schick* (see also Volumes 19, 23, and 60)
2. Solvent Properties of Surfactant Solutions, *edited by Kozo Shinoda* (see Volume 55)
3. Surfactant Biodegradation, *R. D. Swisher* (see Volume 18)
4. Cationic Surfactants, *edited by Eric Jungermann* (see also Volumes 34, 37, and 53)
5. Detergency: Theory and Test Methods (in three parts), *edited by W. G. Cutler and R. C. Davis* (see also Volume 20)
6. Emulsions and Emulsion Technology (in three parts), *edited by Kenneth J. Lissant*
7. Anionic Surfactants (in two parts), *edited by Warner M. Linfield* (see Volume 56)
8. Anionic Surfactants: Chemical Analysis, *edited by John Cross*
9. Stabilization of Colloidal Dispersions by Polymer Adsorption, *Tatsuo Sato and Richard Ruch*
10. Anionic Surfactants: Biochemistry, Toxicology, Dermatology, *edited by Christian Gloxhuber* (see Volume 43)
11. Anionic Surfactants: Physical Chemistry of Surfactant Action, *edited by E. H. Lucassen-Reynders*
12. Amphoteric Surfactants, *edited by B. R. Bluestein and Clifford L. Hilton* (see Volume 59)
13. Demulsification: Industrial Applications, *Kenneth J. Lissant*
14. Surfactants in Textile Processing, *Arved Datyner*
15. Electrical Phenomena at Interfaces: Fundamentals, Measurements, and Applications, *edited by Ayao Kitahara and Akira Watanabe*
16. Surfactants in Cosmetics, edited by Martin M. Rieger (see Volume 68)
17. Interfacial Phenomena: Equilibrium and Dynamic Effects, *Clarence A. Miller and P. Neogi*
18. Surfactant Biodegradation: Second Edition, Revised and Expanded, *R. D. Swisher*
19. Nonionic Surfactants: Chemical Analysis, *edited by John Cross*
20. Detergency: Theory and Technology, *edited by W. Gale Cutler and Erik Kissa*
21. Interfacial Phenomena in Apolar Media, *edited by Hans-Friedrich Eicke and Geoffrey D. Parfitt*
22. Surfactant Solutions: New Methods of Investigation, *edited by Raoul Zana*
23. Nonionic Surfactants: Physical Chemistry, *edited by Martin J. Schick*
24. Microemulsion Systems, *edited by Henri L. Rosano and Marc Clausse*
25. Biosurfactants and Biotechnology, *edited by Naim Kosaric, W. L. Cairns, and Neil C. C. Gray*
26. Surfactants in Emerging Technologies, *edited by Milton J. Rosen*
27. Reagents in Mineral Technology, *edited by P. Somasundaran and Brij M. Moudgil*
28. Surfactants in Chemical/Process Engineering, *edited by Darsh T. Wasan, Martin E. Ginn, and Dinesh O. Shah*
29. Thin Liquid Films, *edited by I. B. Ivanov*
30. Microemulsions and Related Systems: Formulation, Solvency, and Physical Properties, *edited by Maurice Bourrel and Robert S. Schechter*
31. Crystallization and Polymorphism of Fats and Fatty Acids, *edited by Nissim Garti and Kiyotaka Sato*
32. Interfacial Phenomena in Coal Technology, *edited by Gregory D. Botsaris and Yuli M. Glazman*
33. Surfactant-Based Separation Processes, *edited by John F. Scamehorn and Jeffrey H. Harwell*

34. Cationic Surfactants: Organic Chemistry, *edited by James M. Richmond*
35. Alkylene Oxides and Their Polymers, *F. E. Bailey, Jr., and Joseph V. Koleske*
36. Interfacial Phenomena in Petroleum Recovery, *edited by Norman R. Morrow*
37. Cationic Surfactants: Physical Chemistry, *edited by Donn N. Rubingh and Paul M. Holland*
38. Kinetics and Catalysis in Microheterogeneous Systems, *edited by M. Grätzel and K. Kalyanasundaram*
39. Interfacial Phenomena in Biological Systems, *edited by Max Bender*
40. Analysis of Surfactants, *Thomas M. Schmitt* (see Volume 96)
41. Light Scattering by Liquid Surfaces and Complementary Techniques, *edited by Dominique Langevin*
42. Polymeric Surfactants, *Irja Piirma*
43. Anionic Surfactants: Biochemistry, Toxicology, Dermatology. Second Edition, Revised and Expanded, *edited by Christian Gloxhuber and Klaus Künstler*
44. Organized Solutions: Surfactants in Science and Technology, *edited by Stig E. Friberg and Björn Lindman*
45. Defoaming: Theory and Industrial Applications, *edited by P. R. Garrett*
46. Mixed Surfactant Systems, *edited by Keizo Ogino and Masahiko Abe*
47. Coagulation and Flocculation: Theory and Applications, *edited by Bohuslav Dobiás*
48. Biosurfactants: Production Properties Applications, edited by Naim Kosaric
49. Wettability, *edited by John C. Berg*
50. Fluorinated Surfactants: Synthesis Properties Applications, *Erik Kissa*
51. Surface and Colloid Chemistry in Advanced Ceramics Processing, *edited by Robert J. Pugh and Lennart Bergström*
52. Technological Applications of Dispersions, *edited by Robert B. McKay*
53. Cationic Surfactants: Analytical and Biological Evaluation, *edited by John Cross and Edward J. Singer*
54. Surfactants in Agrochemicals, *Tharwat F. Tadros*
55. Solubilization in Surfactant Aggregates, *edited by Sherril D. Christian and John F. Scamehorn*
56. Anionic Surfactants: Organic Chemistry, *edited by Helmut W. Stache*
57. Foams: Theory, Measurements, and Applications, *edited by Robert K. Prud'homme and Saad A. Khan*
58. The Preparation of Dispersions in Liquids, *H. N. Stein*
59. Amphoteric Surfactants: Second Edition, *edited by Eric G. Lomax*
60. Nonionic Surfactants: Polyoxyalkylene Block Copolymers, *edited by Vaughn M. Nace*
61. Emulsions and Emulsion Stability, *edited by Johan Sjöblom*
62. Vesicles, *edited by Morton Rosoff*
63. Applied Surface Thermodynamics, *edited by A. W. Neumann and Jan K. Spelt*
64. Surfactants in Solution, *edited by Arun K. Chattopadhyay and K. L. Mittal*
65. Detergents in the Environment, *edited by Milan Johann Schwuger*
66. Industrial Applications of Microemulsions, *edited by Conxita Solans and Hironobu Kunieda*
67. Liquid Detergents, *edited by Kuo-Yann Lai*
68. Surfactants in Cosmetics: Second Edition, Revised and Expanded, *edited by Martin M. Rieger and Linda D. Rhein*
69. Enzymes in Detergency, *edited by Jan H. van Ee, Onno Misset, and Erik J. Baas*

70. Structure-Performance Relationships in Surfactants, *edited by Kunio Esumi and Minoru Ueno*

71. Powdered Detergents, *edited by Michael S. Showell*

72. Nonionic Surfactants: Organic Chemistry, *edited by Nico M. van Os*

73. Anionic Surfactants: Analytical Chemistry, Second Edition, Revised and Expanded, *edited by John Cross*

74. Novel Surfactants: Preparation, Applications, and Biodegradability, *edited by Krister Holmberg*

75. Biopolymers at Interfaces, *edited by Martin Malmsten*

76. Electrical Phenomena at Interfaces: Fundamentals, Measurements, and Applications, Second Edition, Revised and Expanded, *edited by Hiroyuki Ohshima and Kunio Furusawa*

77. Polymer-Surfactant Systems, *edited by Jan C. T. Kwak*

78. Surfaces of Nanoparticles and Porous Materials, *edited by James A. Schwarz and Cristian I. Contescu*

79. Surface Chemistry and Electrochemistry of Membranes, *edited by Torben Smith Sørensen*

80. Interfacial Phenomena in Chromatography, *edited by Emile Pefferkorn*

81. Solid–Liquid Dispersions, *Bohuslav Dobiás, Xueping Qiu, and Wolfgang von Rybinski*

82. Handbook of Detergents, editor in chief: Uri Zoller Part A: Properties, *edited by Guy Broze*

83. Modern Characterization Methods of Surfactant Systems, *edited by Bernard P. Binks*

84. Dispersions: Characterization, Testing, and Measurement, *Erik Kissa*

85. Interfacial Forces and Fields: Theory and Applications, *edited by Jyh-Ping Hsu*

86. Silicone Surfactants, *edited by Randal M. Hill*

87. Surface Characterization Methods: Principles, Techniques, and Applications, *edited by Andrew J. Milling*

88. Interfacial Dynamics, *edited by Nikola Kallay*

89. Computational Methods in Surface and Colloid Science, *edited by Malgorzata Borówko*

90. Adsorption on Silica Surfaces, *edited by Eugène Papirer*

91. Nonionic Surfactants: Alkyl Polyglucosides, *edited by Dieter Balzer and Harald Lüders*

92. Fine Particles: Synthesis, Characterization, and Mechanisms of Growth, *edited by Tadao Sugimoto*

93. Thermal Behavior of Dispersed Systems, *edited by Nissim Garti*

94. Surface Characteristics of Fibers and Textiles, *edited by Christopher M. Pastore and Paul Kiekens*

95. Liquid Interfaces in Chemical, Biological, and Pharmaceutical Applications, *edited by Alexander G. Volkov*

96. Analysis of Surfactants: Second Edition, Revised and Expanded, *Thomas M. Schmitt*

97. Fluorinated Surfactants and Repellents: Second Edition, Revised and Expanded, *Erik Kissa*

98. Detergency of Specialty Surfactants, *edited by Floyd E. Friedli*

99. Physical Chemistry of Polyelectrolytes, *edited by Tsetska Radeva*

100. Reactions and Synthesis in Surfactant Systems, *edited by John Texter*

101. Protein-Based Surfactants: Synthesis, Physicochemical Properties, and Applications, *edited by Ifendu A. Nnanna and Jiding Xia*

102. Chemical Properties of Material Surfaces, *Marek Kosmulski*

103. Oxide Surfaces, *edited by James A. Wingrave*

104. Polymers in Particulate Systems: Properties and Applications, *edited by Vincent A. Hackley, P. Somasundaran, and Jennifer A. Lewis*

105. Colloid and Surface Properties of Clays and Related Minerals, *Rossman F. Giese and Carel J. van Oss*

106. Interfacial Electrokinetics and Electrophoresis, *edited by Ángel V. Delgado*

107. Adsorption: Theory, Modeling, and Analysis, *edited by József Tóth*

108. Interfacial Applications in Environmental Engineering, *edited by Mark A. Keane*

109. Adsorption and Aggregation of Surfactants in Solution, *edited by K. L. Mittal and Dinesh O. Shah*

110. Biopolymers at Interfaces: Second Edition, Revised and Expanded, *edited by Martin Malmsten*

111. Biomolecular Films: Design, Function, and Applications, *edited by James F. Rusling*

112. Structure–Performance Relationships in Surfactants: Second Edition, Revised and Expanded, *edited by Kunio Esumi and Minoru Ueno*

113. Liquid Interfacial Systems: Oscillations and Instability, *Rudolph V. Birikh, Vladimir A. Briskman, Manuel G. Velarde, and Jean-Claude Legros*

114. Novel Surfactants: Preparation, Applications, and Biodegradability: Second Edition, Revised and Expanded, *edited by Krister Holmberg*

115. Colloidal Polymers: Synthesis and Characterization, *edited by Abdelhamid Elaissari*

116. Colloidal Biomolecules, Biomaterials, and Biomedical Applications, *edited by Abdelhamid Elaissari*

117. Gemini Surfactants: Synthesis, Interfacial and Solution-Phase Behavior, and Applications, *edited by Raoul Zana and Jiding Xia*

118. Colloidal Science of Flotation, *Anh V. Nguyen and Hans Joachim Schulze*

119. Surface and Interfacial Tension: Measurement, Theory, and Applications, *edited by Stanley Hartland*

120. Microporous Media: Synthesis, Properties, and Modeling, *Freddy Romm*

121. Handbook of Detergents, editor in chief: Uri Zoller, Part B: Environmental Impact, *edited by Uri Zoller*

122. Luminous Chemical Vapor Deposition and Interface Engineering, *Hirotsugu Yasuda*

123. Handbook of Detergents, editor in chief: Uri Zoller, Part C: Analysis, *edited by Heinrich Waldhoff and Rüdiger Spilker*

124. Mixed Surfactant Systems: Second Edition, Revised and Expanded, *edited by Masahiko Abe and John F. Scamehorn*

125. Dynamics of Surfactant Self-Assemblies: Micelles, Microemulsions, Vesicles and Lyotropic Phases, *edited by Raoul Zana*

126. Coagulation and Flocculation: Second Edition, *edited by Hansjoachim Stechemesser and Bohulav Dobiás*

127. Bicontinuous Liquid Crystals, *edited by Matthew L. Lynch and Patrick T. Spicer*

128. Handbook of Detergents, editor in chief: Uri Zoller, Part D: Formulation, *edited by Michael S. Showell*

129. Liquid Detergents: Second Edition, *edited by Kuo-Yann Lai*

130. Finely Dispersed Particles: Micro-, Nano-, and Atto-Engineering, *edited by Aleksandar M. Spasic and Jyh-Ping Hsu*

131. Colloidal Silica: Fundamentals and Applications, *edited by Horacio E. Bergna and William O. Roberts*

132. Emulsions and Emulsion Stability, Second Edition, *edited by Johan Sjöblom*

133. Micellar Catalysis, *Mohammad Niyaz Khan*
134. Molecular and Colloidal Electro-Optics, *Stoyl P. Stoylov and Maria V. Stoimenova*
135. Surfactants in Personal Care Products and Decorative Cosmetics, Third Edition, *edited by Linda D. Rhein, Mitchell Schlossman, Anthony O'Lenick, and P. Somasundaran*
136. Rheology of Particulate Dispersions and Composites, *Rajinder Pal*
137. Powders and Fibers: Interfacial Science and Applications, *edited by Michel Nardin and Eugène Papirer*
138. Wetting and Spreading Dynamics, *edited by Victor Starov, Manuel G. Velarde, and Clayton Radke*
139. Interfacial Phenomena: Equilibrium and Dynamic Effects, Second Edition, *edited by Clarence A. Miller and P. Neogi*
140. Giant Micelles: Properties and Applications, *edited by Raoul Zana and Eric W. Kaler*
141. Handbook of Detergents, editor in chief: Uri Zoller, Part E: Applications, *edited by Uri Zoller*
142. Handbook of Detergents, editor in chief: Uri Zoller, Part F: Production, *edited by Uri Zoller*

HANDBOOK OF DETERGENTS

Editor-in-Chief

Uri Zoller
*University of Haifa–Oranim
Kiryat Tivon, Israel*

Part F: Production

Edited by

Uri Zoller
*University of Haifa–Oranim
Kiryat Tivon, Israel*

co-editor

Paul Sosis

CRC Press is an imprint of the
Taylor & Francis Group, an **informa** business

CRC Press
Taylor & Francis Group
6000 Broken Sound Parkway NW, Suite 300
Boca Raton, FL 33487-2742

© 2009 by Taylor & Francis Group, LLC
CRC Press is an imprint of Taylor & Francis Group, an Informa business

No claim to original U.S. Government works
Printed in the United States of America on acid-free paper
10 9 8 7 6 5 4 3 2 1

International Standard Book Number-13: 978-0-8247-0349-3 (Hardcover)

This book contains information obtained from authentic and highly regarded sources. Reasonable efforts have been made to publish reliable data and information, but the author and publisher cannot assume responsibility for the validity of all materials or the consequences of their use. The authors and publishers have attempted to trace the copyright holders of all material reproduced in this publication and apologize to copyright holders if permission to publish in this form has not been obtained. If any copyright material has not been acknowledged please write and let us know so we may rectify in any future reprint.

Except as permitted under U.S. Copyright Law, no part of this book may be reprinted, reproduced, transmitted, or utilized in any form by any electronic, mechanical, or other means, now known or hereafter invented, including photocopying, microfilming, and recording, or in any information storage or retrieval system, without written permission from the publishers.

For permission to photocopy or use material electronically from this work, please access www.copyright.com (http://www.copyright.com/) or contact the Copyright Clearance Center, Inc. (CCC), 222 Rosewood Drive, Danvers, MA 01923, 978-750-8400. CCC is a not-for-profit organization that provides licenses and registration for a variety of users. For organizations that have been granted a photocopy license by the CCC, a separate system of payment has been arranged.

Trademark Notice: Product or corporate names may be trademarks or registered trademarks, and are used only for identification and explanation without intent to infringe.

Visit the Taylor & Francis Web site at
http://www.taylorandfrancis.com

and the CRC Press Web site at
http://www.crcpress.com

Contents

Handbook Introduction xv
Handbook of Detergents Series xvii
Preface xix
Editor xxi
Co-Editor xxiii
Contributors xxv

1 Surfactant Production: Present Realities and Future Perspectives 1

  Matthew I. Levinson

2 Detergent Alkylate and Detergent Olefins Production 39

  Bipin V. Vora, Gary A. Peterson, Stephen W. Sohn, and Mark G. Riley

3 Production and Economics of Alkylphenols, Alkylphenolethoxylates, and Their Raw Materials 49

  Anson Roy Grover

4 Production of Alkyl Glucosides 69

  Jan Varvil, Patrick McCurry, and Carl Pickens

5 Production of Linear Alkylbenzene Sulfonate and α-Olefin Sulfonates 83

  Icilio Adami

6 Production of Alcohols and Alcohol Sulfates 117

  Jeffrey J. Scheibel

7 Production of Alkanesulfonates and Related Compounds (High-Molecular-Weight Sulfonates) 139

  Jean Paul Canselier

8 Production of Glyceryl Ether Sulfonates 159

  Jeffrey C. Cummins

9 Manufacture of Syndet Toilet Bars 171

  Paolo Tovaglieri

10 Phosphate Ester Surfactants 183

  David J. Tracy and Robert L. Reierson

xi

xii   Contents

11 Production of Methyl Ester Sulfonates 201

Norman C. Foster, Brian W. MacArthur, W. Brad Sheats, Michael C. Shea, and Sanjay N. Trivedi

12 Amphoteric Surfactants: Synthesis and Production 221

David J. Floyd and Mathew Jurczyk (edited by Uri Zoller)

13 Production of Alkanolamides, Alkylpolyglucosides, Alkylsulfosuccinates, and Alkylglucamides 239

Bernhard Gutsche and Ansgar Behler

14 Production of Hydrotropes 247

Robert L. Burns (edited by Uri Zoller)

15 Production of Ethylene Oxide/Propylene Oxide Block Copolymers 253

Elio Santacesaria, Martino Di Serio, and Riccardo Tesser

16 Production of Oxyethylated Fatty Acid Methyl Esters 271

Jan Szymanowski

17 Production of Silicone Surfactants and Antifoam Compounds in Detergents 285

Anthony J. O'Lenick, Jr. and Kevin A. O'Lenick

18 Production of Fluorinated Surfactants by Electrochemical Fluorination 301

Hans-Joachim Lehmler

19 Detergent Processing 323

A. E. Bayly, D. J. Smith, Nigel S. Roberts, David W. York, and S. Capeci

20 Production of Quaternary Surfactants 365

Ansgar Behler

21 Production of Detergent Builders: Phosphates, Carbonates, and Polycarboxylates 375

Olina G. Raney

22 Production of Silicates and Zeolites for Detergent Industry 387

Harald P. Bauer

23 Production of Inorganic and Organic Bleaching Ingredients 421

Noel S. Boulos

Contents

xiii

24 Inorganic Bleaches: Production of Hypochlorite 435

William L. Smith

25 Production of Key Ingredients of Detergent Personal Care Products 473

Louis Ho Tan Tai and Veronique Nardello-Rataj

26 Production of Solvents for Detergent Industry 491

Rakesh Kumar Khandal, Sapana Kaushik, Geetha Seshadri, and Dhriti Khandal

27 Production of Proteases and Other Detergent Enzymes 531

T. T. Hansen, H. Jørgensen, and M. Bundgaard-Nielsen

28 Chemistry, Production, and Application of Fluorescent Whitening Agents 547

Karla Ann Wilzer and Andress Kirsty Johnson

29 Production of Gemini Surfactants 561

Bessie N. A. Mbadugha and Jason S. Keiper

Index **579**

Handbook Introduction

The battle cry for sustainable development in our globalized world is persistent in all circles, gaining acceptance as the guiding rationale for activities or processes in the science–technology–environment–economy–society–politics interfaces, targeting at improvement and growth. Such activities are expected to result in higher standards of living leading, eventually, to a better quality of life for our increasingly technology-dependent modern society. Models of sustainable development and exemplary systems of sustainable management and applications are continually being developed and adapted and creatively applied, considering, more than before, human needs, rather than "wants" on the one hand, and long- versus short-term benefits and trade-offs on the other.

"Detergents" constitute a classic case study within this context: this is a multidimensional systemic enterprise, operating within complex sociopolitical/technoeconomical realities, locally and globally, reflecting in its development and contemporary "state of affairs," the changing dynamic equilibria and interrelationships between demands/needs, cost/benefits, gains/trade-offs, and social preferences–related policies. It is not surprising, therefore, that despite the overall maturity of the consumer market, detergents continue to advance, in the modern world and developing societies, more rapidly than population growth.

The soap and detergent industry has seen great change in recent years, requiring it to respond to the shifts in consumer preferences, requests for sustainability, the availability and cost of raw materials and energy, demographic and social trends, as well as the overall economic and political situation worldwide. Currently, detergent product design is examined against the unifying focus of delivering performance and value to the consumer, given the constraints of the economy, technological advancements, and environmental imperatives. The annual 2–3% growth of the detergent industry and the faster growth in personal care products reflect impressive developments in formulation and application. The detergent industry is thus expected to continue its steady growth in the near future in response to the ever-increasing demands of consumers for products that are more efficient, act fast, and are easier to use. For the detergent industry, the last decade of the twentieth century was one of transformation, evolution, and consolidation. On both the supplier and consumer market sides (both remain intensely competitive), the detergent industry has undergone dramatic changes, with players expanding their offerings, restructuring divisions, or abandoning the markets altogether. This has resulted in changing hands and consolidation of the market, especially in the last several years. This trend appears to be gaining momentum. Yet, the key concepts have been and still are innovation, consumer preferences, needs, multipurpose products, cost/benefit, efficiency, emerging markets, partnership/cooperation/collaboration/merging (locally, regionally, and globally), and technological advancements. Although substantial gains and meaningful rapid changes with respect to the preceding concepts have been experienced by the surfactant/detergent markets, the same cannot be said for detergent/surfactant technology itself. The $9-billion-plus detergent ingredient market and the annual global consumption of ~13 million tons of "surfactants" in 2006 have many entrenched workhorse products. This may suggest that the supply of "solutions" to most cleaning "problems" confronted by consumers in view of the increasing global demand for formulations having high performance and relatively low cost and the need for compliance with environment-related regulation are based on modifications of existing technologies.

What does all this mean for the future of the "detergents" enterprise? How will advances in research and development affect future development in detergent production, formulation, applications, marketing, consumption, and relevant human behavior as well as the short- and long-term impacts on the quality of life and the environment? Since new developments and emerging technologies are generating new issues and questions, not everything that can be done should be done; that is, there should be more response to real *needs* rather than *wants*.

Are all these aforementioned questions reflected in the available professional literature for those who are directly involved or interested, for example, engineers, scientists, technicians, developers, producers, formulators, managers, marketing people, regulators, and policy makers? A thorough examination of the literature, in this and related areas, suggests that a comprehensive series is needed to deal with the practical aspects involved in and related to the detergent industry, thus providing a perspective beyond knowledge to all those involved and interested. The *Handbook of Detergents* is an up-to-date compilation of works written by experts, each of whom is heavily engaged in his/her area of expertise, emphasizing the practical and guided by the system approach.

The aim of this six-volume handbook project (properties, environmental impact, analysis, formulation, application, and production) is to provide readers who are interested in any aspect of or relationship to surfactants and detergents, a state-of-the-art comprehensive treatise written by expert practitioners (mainly from industry) in the field. Thus, various aspects involved—properties, environmental impact, analysis/test methods, formulation application and production of detergents, marketing, environmental, and related technological aspects, as well as research problems—are dealt with, emphasizing the practical. This constitutes a shift from the traditional, mostly theoretical focus, of most of the related literature currently available.

The philosophy and rationale of the *Handbook of Detergents* series are reflected in its title and plan and the order of volumes and flow of the chapters (in each volume). The various chapters are not intended to be and should not, therefore, be considered to be mutually exclusive or conclusive. Some overlapping segments focus on the same issue(s) or topic(s) from different points of view, thus enriching and complementing various perspectives.

There are several persons involved whose help, capability, professionality, and dedication made this project possible: the volume (parts) editors, contributors, and reviewers are in the front line in this respect. Others deserve special thanks: my colleagues and friends in (or associated with) the detergent industry, whose timely help and involvement facilitated in bringing this project home. I hope that the final result will justify the tremendous effort invested by all those who contributed; you, the reader, will be the ultimate judge.

Uri Zoller
Editor-in-Chief

Handbook of Detergents Series

Editor-in-Chief
Uri Zoller

Handbook of Detergents Series Part A: Properties, edited by Guy Broze

Handbook of Detergents Series Part B: Environmental Impact, edited by Uri Zoller

Handbook of Detergents Series Part C: Analysis, edited by Heinrich Waldhoff and Rudiger Spilker

Handbook of Detergents Series Part D: Formulation, edited by Michael Showell

Handbook of Detergents Series Part E: Applications, edited by Uri Zoller

Handbook of Detergents Series Part F: Production, edited by Uri Zoller and Paul Sosis

Preface

With the annual global consumption of surface-active agents reaching 13×10^6 metric tons and more than \$9 billion worth of the detergent ingredients market, this industry embraces sustainability. Recently, the environmental impact of detergents has gone from being a fringe issue to a mainstream concern. Thus, regardless of the state of the art and affairs in the detergent industry worldwide, with respect to scientific-, technological-, economics-, safety-, and "greening"-related regulation of detergent production and formulation, the basic modes of the former will continue to be an issue of major concern. Yet, given our increasingly fast-moving world and skyrocketing oil prices, customers demand products that are more effective, energy saving, and can help to save *time* for the customers. This means demands for products that are cheaper, effective, faster acting, easier to use, more efficient, and environment friendly. This is so in view of the operating global free-market economy that is expected to ensure sustainable development, given the contemporary shifts in consumer preferences, availability and cost of basic raw materials, and energy, demographic, and social trends, as well as the overall economical/political situation worldwide.

This volume (Part F) of the six-volume series *Handbook of Detergents* deals with the production of various components of detergents—surfactants, builders, sequestering/chelating agents—as well as of other components of detergent formulations.

This volume is a comprehensive treatise on the multidimensional issues involved, and represents an international industry–academia collaborative effort of many experts and authorities, worldwide, mainly from industry. As such, *Part F—Production*, represents the state of the art concerning these multidimensional technological practices.

All of these are accompanied and supported by extensive relevant data, occasionally via specific "representative" case studies, the derived conclusions of which are transferable. Also, this resource contains several cited works and is, thus, aimed to serve as a useful and practical reference concerning the "production" aspect of surfactants—detergents—for engineers, technologists, scientists, technicians, regulators, and policy makers, associated with the detergent industry.

I thank all the contributors, reviewers, publisher's staff, and colleagues who made the realization of this and all the previous five volumes possible.

Editor

Uri Zoller is professor emeritus of chemistry and science education at Haifa University—Oranim, Kiryat Tivon, Israel. He has more than 220 published journal articles, 1 patent, and 9 books to his name, including the published five parts and the sixth part—Production, of the *Handbook of Detergents*, of which he is the editor-in-chief. He is an active member of several professional organizations, including the American Chemical Society and the Royal Society of Chemistry (United Kingdom), and is currently the chairman of the European Association for Chemical and Molecular Sciences (EuCHeMS) Committee on Education in Environmental Chemistry. His main areas of interest and research are synthetic organic chemistry, environmental chemistry, and science and environmental education and assessment. Following 10 years of research and development work in the detergent industry, Dr. Zoller received his BSc (summa cum laude) followed by an MSc in chemistry and industrial chemistry, respectively, from the Technion–Israel Institute of Technology, Haifa, Israel; an SM degree from the Massachusetts Institute of Technology in the United States; the DSc degree from the Technion–Israel Institute of Technology, Haifa, Israel; and the EdD in science education from Harvard University, Cambridge, Massachusetts. Currently, Dr. Zoller is the project coordinator of the Israeli Unified, National Infrastructural Research Project (UNIRP).

Co-Editor

Paul Sosis is currently the president of Sosis Consulting Services in Oakland, New Jersey, and vice president of Argeo Incorporated—a consulting and testing laboratory for the surfactants and detergents industry. He has served as the chairman of the Surfactants and Detergents Division of the American Oil Chemists Society (AOCS), vice chairman of the Detergents Division of the Chemical Specialties Manufacturers Association (CSMA), chairman of the Education Committee of the S&D Division of the AOCS, the Education Committee of the CSMA Detergents Division, and the Marketing Committee of the CSMA Detergents Division. Sosis was founder and chairman of the "New Horizons Conferences" since 1986 and co-chaired a technical session at the Third Detergents World Conference in Montreaux, Switzerland. He has organized and chaired several committees and technical programs with SDA, American Society for Testing and Materials (ASTM), AOCS, and CSMA.

Sosis received the Distinguished Service Award, CSMA, 1982; Award of Merit, AOCS, 2002; and The Distinguished Service Award, Surfactants and Detergents Division, AOCS, 2004. He has authored 9 patents, 18 publications, and is a co-editor and contributor to 2 books in the 52 years of his professional career.

Contributors

Icilio Adami
Desmet Ballestra S.p.A.
Milano, Italy

Harald P. Bauer
Clariant
Hürth, Germany

A. E. Bayly
Procter & Gamble
Newcastle Technical Centre
Longbenton, U.K.

Ansgar Behler
Cognis GmbH
Düsseldorf, Germany

Noel S. Boulos
Solvay Chemicals, Inc.
Houston, Texas, U.S.A.

M. Bundgaard-Nielsen
Novozymes A/S
Bagsværd, Denmark

Robert L. Burns (retired)
Rutgers-Nease Corporation
State College, Pennsylvania, U.S.A.

Jean Paul Canselier
Laboratoire de Génie
  Chimique/ENSIACET
Toulouse, France

S. Capeci
Procter & Gamble
Cincinnati, Ohio, U.S.A.

Jeffrey C. Cummins
Procter & Gamble
Cincinnati, Ohio, U.S.A.

Martino Di Serio
Dipartimento di Chimica Via Cintia
Università di Napoli "Federico II"
Complesso Universitario di Monte S. Angelo
Napoli, Italy

David J. Floyd
Goldschmidt Chemical Corporation
Hopewell, Virginia, U.S.A.

Norman C. Foster, PhD
McKinstry Company
Seattle, Washington, U.S.A.

Anson Roy Grover (retired)
Schenectady International Inc.
Schenectady, New York, U.S.A.

Bernhard Gutsche
Cognis GmbH
Düsseldorf, Germany

T. T. Hansen
Novozymes A/S
Bagsværd, Denmark

Louis Ho Tan Tai
Avenue du Maréchal Leclerc
Lambersart, France

Andress Kirsty Johnson, PhD
Ciba
Tarrytown, New York, U.S.A.

H. Jørgensen
Novozymes A/S
Bagsværd, Denmark

Mathew Jurczyk
Uniqema
Wilmington, Delaware, U.S.A.

Sapana Kaushik
Shriram Institute for Industrial Research
Delhi, India

Jason S. Keiper
Syngenta
Greensboro, North Carolina, U.S.A.

Dhriti Khandal
Shriram Institute for
  Industrial Research
Delhi, India

Rakesh Kumar Khandal
Shriram Institute for Industrial Research
Delhi, India

Hans-Joachim Lehmler
Department of Occupational and
  Environmental Health
College of Public Health
University of Iowa
Iowa City, Iowa, U.S.A.

Matthew I. Levinson, PhD
Stepan Company
Northfield, Illinois, U.S.A.

Brian W. MacArthur, PhD
The Chemithon Corporation
Seattle, Washington, U.S.A.

Patrick McCurry (retired)
Cognis GmbH
Cincinnati, Ohio, U.S.A.

Bessie N.A. Mbadugha
St. Mary's College of Maryland
St. Mary's City, Maryland, U.S.A.

Veronique Nardello-Rataj, PhD
Université de Lille,
Equipe Oxidation et Formulation
Villeneuve d'Ascq, France

Anthony J. O'Lenick, Jr.
Siltech LLC
Dacula, Georgia, U.S.A.

Kevin A. O'Lenick
SurfaTech Corporation
Dacula, Georgia, U.S.A.

Gary A. Peterson
Aromatic Derivatives & Detergents
  Process Technology & Equipment
UOP LLC
Des Plaines, Illinois, U.S.A.

Carl Pickens (deceased)
Cognis GmbH
Cincinnati, Ohio, U.S.A.

Olina G. Raney, PhD
Independent Consultant
Houston, Texas, U.S.A.

Robert L. Reierson
Rhodia Inc.
Cranbury, New Jersey, U.S.A.

Mark G. Riley
UOP LLC
Des Plaines, Illinois, U.S.A.

Nigel S. Roberts
Procter & Gamble
New Castle Technical Centre
Longbenton, U.K.

Elio Santacesaria
Dipartimento di Chimica-Università
  Federico II
Napoli, Italy

Jeffrey J. Scheibel
Procter & Gamble
The Miami Valley Innovation Center
Cincinnati, Ohio, U.S.A.

Geetha Seshadri
Shriram Institute for Industrial Research
Delhi, India

Michael C. Shea
The Chemithon Corporation
Seattle, Washington, U.S.A.

Contributors

W. Brad Sheats
The Chemithon Corporation
Seattle, Washington, U.S.A.

D. J. Smith
Procter & Gamble
Newcastle Technical Centre
Longbenton, U.K.

William L. Smith
The Clorox Company
Pleasanton, California, U.S.A.

Stephen W. Sohn
UOP LLC
Des Plaines, Illinois, U.S.A.

Jan Szymanowski (deceased)
Institute of Chemical Technology and
  Engineering, PL
Poznan University of Technology
Poznan, Poland

Riccardo Tesser
Dipartimento di Chimica Via Cintia
Università di Napoli "Federico II"
Napoli, Italy

Paolo Tovaglieri
Mazzoni LB
Varese, Italy

David J. Tracy
Tracy Consulting, L.L.C.
Amelia, Ohio, U.S.A.

Sanjay N. Trivedi
Chemithon Engineers Pvt. Ltd.
Mumbai, India

Jan Varvil (retired)
Cognis-Oleo Chemical
Cincinnati, Ohio, U.S.A.

Bipin V. Vora (retired)
Universal Oil Products
Des Plaines, Illinois, U.S.A.

Karla Ann Wilzer, PhD
Ciba Specialty Chemicals
High Point, North Carolina, U.S.A.

David W. York
Procter & Gamble
Newcastle Technical Centre
Longbenton, U.K.

1 Surfactant Production: Present Realities and Future Perspectives

Matthew I. Levinson

CONTENTS

1.1 Introduction 1
1.2 Competitive Forces Affecting the Production of Surfactants 4
1.3 Historical Perspective on Production and Feedstocks 5
1.4 Specialty Feedstocks and Surfactants 11
1.5 Basic Raw Materials 11
1.6 The Four Main Surfactant Classes and Their Production Today 12
    1.6.1 Amphoteric Surfactants 13
    1.6.2 Anionic Surfactants 16
        1.6.2.1 Sulfonates and Sulfates 17
        1.6.2.2 Phosphated Surfactants 21
        1.6.2.3 Carboxylated Synthetic Surfactants 21
    1.6.3 Cationic Surfactants 21
    1.6.4 Nonionic Surfactants 24
        1.6.4.1 Alkanolamide Nonionic Surfactants 25
        1.6.4.2 Alkoxylated Nonionic Surfactants 26
        1.6.4.3 Esterified Nonionic Surfactants 29
        1.6.4.4 Etherified Nonionic Surfactants 29
1.7 Construction and Operational Issues 31
    1.7.1 Regulatory Standards 31
1.8 Summary 35
References 35

1.1 INTRODUCTION

The annual global consumption of surface active agents or "surfactants" in 2006 was estimated to reach ~13 million metric tons,[1] with the break up of regional sales as depicted in Figure 1.1. There are arguably five major participants in the surfactant supply chain including (1) basic raw-material processors, (2) feedstock and diversified chemical producers, (3) surfactant converters, (4) product formulators, and (5) distributors/retailers,[2] some of which are listed in Figure 1.2.

Basic raw-material processors extract and refine crude oil into petrochemicals such as petroleum oil distillates including paraffins, benzene, and other basic aromatics and extract and convert natural gas into ethylene and propylene. Processors of oleochemicals extract and purify seed oils from palm, soybean, sunflower seed, palm kernel, and coconut, and render animal fats such as tallow to provide triglyceride oils with varying chain distributions.

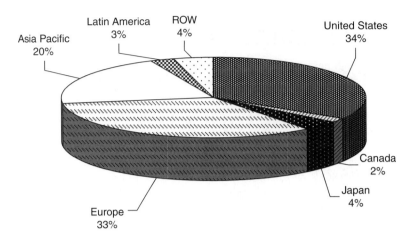

FIGURE 1.1 Estimated percentage of annual global volume sales of surface active agents for 2006 by region, based on a total of 13 million metric tons. (From Global Industry Analysts, Inc., *Surface Active Agents—A Global Strategic Business Report 08/06*, August, 2006. With permission.)

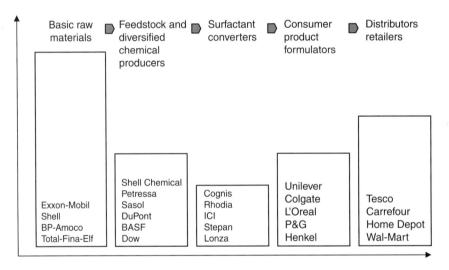

FIGURE 1.2 Consumer surfactant supply chain participants and their relative average market capitalizations.

Feedstock producers convert the aforementioned basic raw materials into numerous derivatives useful in a wide range of industries and applications, and particularly suitable for the manufacture of surfactants. These derivatives include the reaction products of paraffins with aromatics such as alkylbenzenes and alkylphenols, derivatives of ethylene and propylene such as polyalkylenes, primary alpha olefins, and their further oxidized or carbonylated derivatives such as Ziegler or oxo alcohols and their subsequent reaction products with ethylene oxide (EO) and propylene oxide (PO). Producers of petrochemical feedstocks are subsidiaries of basic raw-material processors in some cases, and so are integrated back into the key raw materials providing them economic and logistical advantages in producing their products. Examples include Shell, Sasol, ExxonMobil, Chevron, and Petressa.

Feedstocks derived from triglyceride oils include fatty acids, methyl esters, and natural alcohols through splitting, transesterification, hydrogenation, and hydrogenolysis. In several cases,

Surfactant Production: Present Realities and Future Perspectives

the natural feedstock producers are or were part of highly integrated supply strategies of consumer product companies that converted triglycerides to fatty acids for soap production, and later converted them to fatty alcohols for alcohol sulfates production, which was formulated into fabric and dish detergents, and personal wash products. Companies such as Uniqema and Cognis were originally a part of Unilever and Henkel, respectively, and the Procter & Gamble (P&G) chemicals division still supplies the consumer products division with key raw materials for internal conversion.

Diversified chemical producers are a part of the second group within the supply chain, and provide the highly reactive materials that are used by surfactant converters to affix or create a hydrophilic head group on the hydrophobic materials discussed earlier. The highly reactive reagents include sulfur trioxide (SO_3), phosphorous pentoxide (P_2O_5), EO, PO, dimethyl sulfate, hydrogen peroxide, epichlorohydrin, monochloroacetic acid, and methyl chloride. There are many materials in this category, such as alkanolamines and short chain alkyl amines, sulfur dioxide, ammonium hydroxide, sodium hydroxide, polyphosphoric acid, and volatile alcohols that are less reactive, but pose handling and safety challenges. These are provided by large, very well-known members of the chemical industry that have had some historical and continuing participation as surfactant converters in their own right, and include DuPont, Dow-Union Carbide, BASF, Bayer, Rhodia, Monsanto, FMC, and Huntsman.

Surfactant converters rely on approximately eight core chemical processes that are broadly practiced in the global manufacture of surfactants, including sulfonation, sulfation, amidation, alkoxylation, esterification, amination, phosphation, and quaternization. These process steps are used to affix or create a highly water-soluble functional group (hydrophilic head group) on a water-insoluble feedstock (hydrophobic tail group). The surfactants derived from the permutations of head groups and tail groups fall into one of the four broad categories of anionic, cationic, nonionic, and amphoteric surfactant, based on the nature of the charge that is carried by the head group. The dynamics of the surfactant market place are impacted at a fundamental level by the cost, variety, and availability of hydrophobes, and the cost and complexity of attaching or creating hydrophilic head groups.

Surfactants are consumed globally in a broad range of consumer and industrial product compositions,[3] and are formulated at active levels ranging from nearly 100% in some cleaning products down to mere parts per million levels in high-performance applications such as pharmaceutical delivery systems, precision optics coatings, and electronics manufacturing. Broad categories of applications and uses include

Laundry detergents, fabric softeners, dish washing, and household cleaning products
Personal cleansing and conditioning products, and skin creams and cosmetics
Industrial and institutional cleaning products
Emulsion polymers used in paints, coatings, and adhesives
Agricultural product formulations containing insecticides, herbicides, and fungicides
Food-grade emulsifiers
Metal-working lubrication products and metal cleaners
Pulp and paper washing, deinking, and emulsifying
Oil field and natural gas drilling, completion, and production chemicals
Plastic mold release agents, lubricants, and processing chemicals
Textile and fiber lubricants, dying aides, and scouring and finishing chemicals
Mining chemicals

The ratio between surfactants used in consumer products and commercial and industrial application is approximately 65:35 as depicted in Figure 1.3. The markets for industrial and commercial surfactants in the United States are highly segmented and range between 10 and 20% of the entire application area as shown in Figure 1.4.

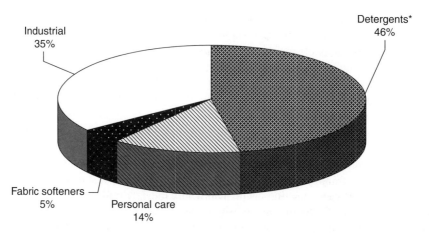

FIGURE 1.3 Percentage of global surfactant consumption by major application area for 2006 based on total sales of 13 million metric tons. (Note: * – approximately half is produced for captive use by integrated consumer companies.)

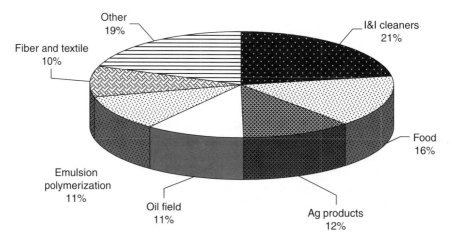

FIGURE 1.4 Industrial and commercial surfactant production, 2004 in the United States (total = 1100 t). (From Modler, R.F., Muller, S., and Ishikawa, Y., *Surfactants*, SRI Consulting; Specialty Chemical Update Program, July, 2004. With permission.)

1.2 COMPETITIVE FORCES AFFECTING THE PRODUCTION OF SURFACTANTS

There are varied and changing forces impacting the manufacture of surfactants in the world today, and the challenges producers of surfactant ingredients face are many.

1. Globalization and consolidation of surfactant users have accelerated over the past 20 years, and have affected all of the major end markets. This has driven standardization of surfactant product composition and specifications turning them into "commodity" products that command lower margins.
2. Globalization and consolidation within the retail channel, through which the preponderance of consumer products are sold, are allowing superretailers to dictate shelf space and packaging size that narrows formulation options, and to position their own house-label versions effectively even against global brands, reducing their profit margins and applying further downward pressure on surfactant margins.

3. Slowing demand and overcapacity in mature markets of North America and Europe have driven consolidation among surfactant producers attempting to achieve economies of scale, resulting in asset rationalization and product-line integration, and an ongoing need to drive out costs.
4. Medium and rapidly growing markets in Asia Pacific, India, Eastern Europe, and Latin America is creating the need for local manufacturing capacity for large volume, low-margin commodity surfactants to provide a cost-effective supply chain in the face of rising transportation costs.
5. Fluctuating and increasing raw material prices for feedstocks derived from oleochemicals and petrochemicals, respectively, as well as reagent chemicals and fuels used for manufacture and transport of intermediates and products have demanded significant price increases by surfactant converters, which until very recently were suppressed by surfactant formulators in part due to pressure by megaretailers.
6. More swings are anticipated in cost and availability of both petrochemical and oleochemical feedstocks, driving surfactant producers and formulators to develop flexible feedstock and formulation strategies.
7. Labor costs as a proportion of the total cost to produce surfactants continue to rise in mature markets, further motivating production of commodity and dilute surfactants within local markets where labor costs are low.
8. Low margin and dilute surfactants will continue to be made locally in mature markets and will not be effectively challenged by imports from developing regions due to transportation cost and service barriers such as surety of supply.
9. The cost of materials of construction and engineering services are very high in mature markets today, but are rising quickly in emerging markets as fast-paced growth in all sectors, which challenges local resources and infrastructure.
10. Ongoing concerns over the safety, health, and environmental fate of surfactants have propelled regulatory agencies in mature markets to demand extensive testing on new products, and data-gap backfilling for existing products, adding cost and slowing development of new surfactants.
11. Rapid adoption by emerging economies of regulatory standards developed in Europe or North America will drive out the use of some long-standing ingredients creating opportunities for competitive challenges.

1.3 HISTORICAL PERSPECTIVE ON PRODUCTION AND FEEDSTOCKS

The evolution of the sophisticated products and chemical-process technologies that are used today trace their origins back to the nineteenth century and the nascent chemical industry that relied on renewable oleochemical feedstocks. Synthetic surfactants prepared by the reaction of olive oil with sulfuric acid, performed by Fremy in 1831, was among the first.[4]

Some of the largest users of surfactants today originated as vertically integrated retailers of soap and candles, utilizing tallow and other animal fats obtained from the meat-processing industry, and later, vegetable oils such as palm, palm kernel, and coconut. Companies such as P&G, Lever Brothers, Colgate-Palmolive, Henkel, and others gained expertise in processing fats and oils into sodium carboxylate soaps in a variety of forms such as bars, flakes, and prills.[5]

The first "synthetic" detergents/surfactants were developed by the Germans during World War I followed by a burst of development in the late 1920s and early 1930s. Natural fats were in high demand for more important uses than soap, and this drove the search for alternatives capable of equivalent cleaning performance. The availability of coal tar as a basic raw material provided naphthalene and other polynuclear aromatics, which were alkylated using short-chain and fatty alcohols to yield feedstock alkylaromatics that were subsequently converted into surfactants by sulfonation with chlorosulfonic or sulfuric acid.[6] Although this class of surfactants delivers only moderate detergency, they were found to be good wetting agents and are still used in large quantities today as textile auxiliaries.[7]

The competitive drive for consumer products with enhanced performance and convenience, coupled with the rapid development of the chemicals industry in the 1930s, gave rise to innovations such as glyceryl ester sulfates[8] by Colgate-Palmolive-Peet Company and alcohol sulfates[9] made from fatty alcohols. Fatty alcohols were newly available feedstocks produced through catalytic hydrogenation of coconut and palm kernel oil derivatives developed in parallel by Deutsche Hydrierwerke in Germany, and E.I. DuPont in the United States in the early 1930s. P&G and Hydrierwerke pooled their U.S. interests to form American Hyalsol Corporation, which held U.S. patents for the production of alcohol sulfates. P&G was able to market and develop alcohol sulfates as synthetic detergents in household and laundry markets, and Dreft, the first household synthetic laundry detergent was launched in 1933. Finding the right builder, sodium tripolyphosphate, and formulation to maximize cleaning took another 13 years and resulted in the launch of Tide detergent in 1946.[10]

Refining of petroleum led to the separation of paraffinic alkanes, alkenes, benzene, and other aromatics that provided the feedstocks used in alkylation processes to yield alkylbenzenes. The petrochemical industry that emerged following World War II created a wide range of synthetic materials that became the alternatives to oleochemical feedstocks of the nineteenth century and the building blocks of the modern surfactant manufacturing industry of today. In the late 1940s, UOP developed a process to economically produce commercial quantities of branched alkylbenzene sulfonate (BABS), which became one of the surfactants most widely used in synthetic detergents at that time.

Even as early as 1939, the soap industry began to create laundry detergents using surfactants that were supplied to the soap manufacturers by the petrochemical industry. Because the cleaning formulations produced from these synthetic detergents were a substantial improvement over soap products in use at the time, they soon gave rise to a global surfactant industry based on branched alkyl benzene (BAB) derived from branched paraffins.

The hydrocracking of paraffins or reforming of methane gas provided the highly useful intermediates ethylene and propylene, which were used in the production of alpha olefins and polypropylenes (PP), which were used to alkylate benzene, or further converted to synthetic alcohols through Ziegler and oxo catalyst chemistry. Oxidation processes were developed to convert ethylene and propylene to their respective epoxides, EO and PO, which became building blocks for the preparation of alkoxylated alcohols and glycols, useful as nonionic surfactants and hydrophobes for further derivatization. During the 1950s and 1960s, advances in petrochemical technology provided feedstock molecules such as alkylphenol, linear and branched alpha olefins and fatty alcohols, and alkyl amines, which were suitable for derivatization and the basis for development of broad classes of synthetic surfactants as shown in Figure 1.5.

In the late 1950s, it was found that BABS had a slow rate of biodegradation that resulted in generation of large amounts of foam in surface waters such as rivers and streams.[11] Process technology was developed in the 1960s to produce linear alkylbenzene (LAB) from linear alpha olefins, as shown in Figure 1.6, or chloroparaffins. This new surfactant raw material was used to make linear alkylbenzene sulfonate (LAS), deemed to be a much more biodegradable surfactant, and grew to be the largest synthetic surfactant in use worldwide. Although it has been supplanted in some markets by alcohol ether sulfate (AES), it is still used globally in the manufacture of detergents today. The increasing use of synthetic surfactants and decline of soap sales following World War II are highlighted in Table 1.1.

Table 1.1, compiled from figures submitted by the American Soap and Detergent Association and the German firm of Henkel & Cie, shows both soap and detergent sales in the United States for various years from 1940 to 1972.[12]

In parallel with the evolving supply of petroleum raw materials, natural oil production from seed crops has increased worldwide to the volumes depicted in Figure 1.7. Palm oil has grown globally to become the single largest oil crop, comprising >35 million metric tons/year of the global production

Surfactant Production: Present Realities and Future Perspectives

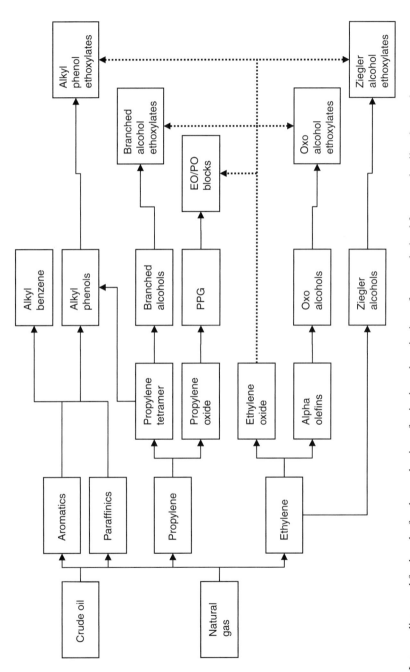

FIGURE 1.5 Intermediates and feedstocks for the production of anionic and nonionic surfactants derived from crude oil and natural gas.

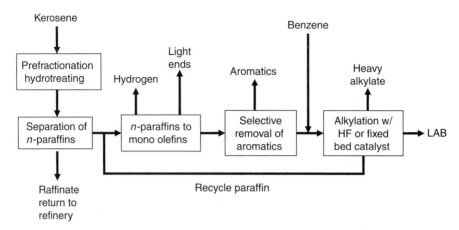

FIGURE 1.6 Integrated complex for production of alkylbenzene from normal paraffins based on UOP process technologies. (From UOP LLC, at http://www.uop.com. With permission.)

TABLE 1.1
Soap and Detergent Sales in the United States for Various Years from 1940 to 1972

Year	Soap Sales (1000 t)	Synthetic Sales (1000 t)
1940	1410	4.5
1950	1340	655
1960	583	1645
1972	587	4448

Note: Compiled from figures submitted by the American Soap and Detergent Association and the German firm of Henkel & Cie.

Source: Information available from About.com, accessible at http://www.chemistry.co.nz/deterghistory.htm.

of >110 million metric tons/year, and is used predominantly for food, and in substantially lesser quantities for derivatives and feedstocks for the chemical industry. For thousands of years, the production of soap for personal and clothes washing relied on natural triglyceride oils for preparation of fatty acids and their respective neutral sodium salts. Today palm oil, palm kernel oil, coconut oil, and tallow are converted, in significant volumes, into fatty acids, methyl esters, and alcohols, which are extensively used in the surfactant industry.[13]

However, tallow is composed of ~30% C16 and 70% C18 chains with substantial unsaturation. The choices of carbon numbers available are limited by the type of oil used as a feed material. Coconut oil is ~50% C_{12} with up to 20% C_{14} and ~15% each of C_{8-10} and C_{16-18}. Palm kernel oil has a similar distribution. However, tallow is mostly C_{16}–C_{18}. The shorter chain C12–C14 fatty acids and methyl esters derived from coconut and palm kernel oil are key starting materials for a host of surfactant derivatives in each of the major categories (anionics, cationics, nonionics, and amphoterics).

The commercial manufacture of fatty alcohols started in the late 1920s. The very first natural fatty alcohol was obtained by a simple ester cleavage of oil originating in the skull of the sperm whale. But a mere 4 years later, the first industrial-scale process had already been developed for producing a fatty alcohol from coconut fatty acid by high-pressure hydrogenation. In 1958, a route was developed from fatty acid methyl ester, which still remains the most economic method of producing

Surfactant Production: Present Realities and Future Perspectives

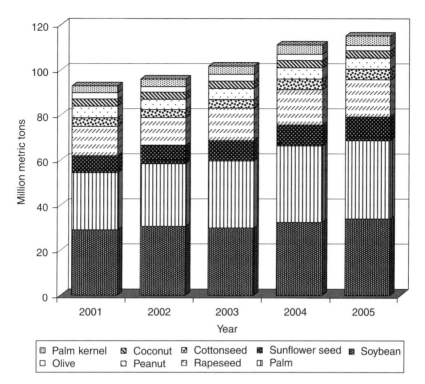

FIGURE 1.7 World vegetable oil supply and distribution between 2001–2005 showing volumes now exceed 110 million metric tons. (From Brackmann, B. and Hager, C.-D., *CESIO 6th World Surfactants Congress*, Berlin, June 20–23, 2004. With permission.)

natural fatty alcohols, and opened the door to the broad range of derivatives available from the flow depicted in Figure 1.8. Three years later, a more selective hydrogenation process allowed the preservation of the unsaturation found predominantly in the C16 and C18 chain fractions. The first unsaturated fatty alcohols became commercially available in the early 1960s, and since then, no more whales were harvested for the sake of oil.[14]

Today, natural detergent alcohols are produced using processes such as that developed by Davy Process Technology, depicted in Figure 1.9, which convert fatty acids into nonacidic intermediate methyl esters and hydrogenates these to alcohols, then separates C12–C14 and C16–C18 product streams.[15] This vapor phase process has been licensed around the world in ten ester hydrogenation plants with a total installed capacity of 350,000 t/year of alcohols. These plants have virtually no effluents; small by-product streams are recycled and consumed within the process, thus they have minimal environmental impact.

In 1963, the first petroleum-based fatty alcohols were produced based on ethylene and utilizing Ziegler's trialkylaluminum catalyst technology. This technology produces highly linear, even-numbered higher alcohols with little or no branching. The development of the oxo and modified oxo process that relies on hydroformylation of alpha olefins made mixtures of odd- and even-numbered alcohols containing around 20% methyl branching available; an example of this is depicted in Figure 1.10. Experience with slow biodegradation of BABS caused practitioners to assume that only linear alcohols would demonstrate superior biodegradability. However, over the past few years, studies sponsored by a number of groups have confirmed that if branching is properly controlled, the biodegradability of the resulting surfactant is retained and the surfactant properties are actually improved. Branched alcohols derived from Sasol's Fischer–Tropsch (FT) paraffins and alpha olefin isomerization technology developed by Shell have achieved commercial success and meet current biodegradability standards.[16]

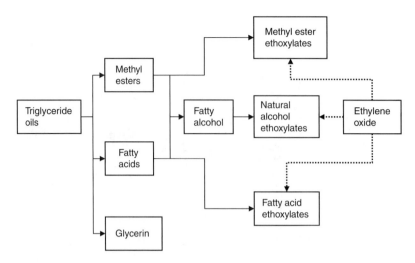

FIGURE 1.8 Intermediates and feedstocks for the production of anionic and nonionic surfactants derived from natural triglyceride oils.

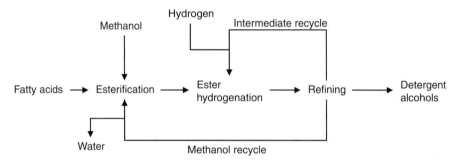

FIGURE 1.9 Process flow for the conversion of fatty acids to detergent alcohols via the Davy Process Technology Natural Detergent Alcohol process. (From Renaud, P., *CESIO 6th World Surfactants Congress*, Berlin, June 20–23, 2004. With permission.)

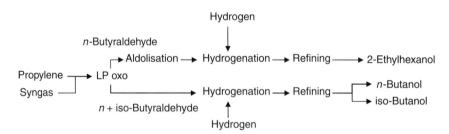

FIGURE 1.10 Davy Process Technology low-pressure hydroformylation technology was developed in collaboration with The Dow Chemical Company. The LP Oxo™ process has been applied commercially to produce detergent-grade alcohols from higher olefin cuts from FT synthesis. (From Renaud, P., *CESIO 6th World Surfactants Congress*, Berlin, June 20–23, 2004. With permission.)

The ultimate impact of FT technology is yet to be determined. Coal-to-liquid (CTL) processes are based on a technology that was developed in the early 1920s by German scientists Franz Fischer and Hans Tropsch. The FT process has been used since 1955 for CTL in South Africa where a government-sponsored plant was built by the South African Synthetic Oil Ltd corporation, now known as Sasol.[17] Currently, the U.S. Department of Energy through the National Energy

Surfactant Production: Present Realities and Future Perspectives

Technology Laboratory is supporting demonstration projects in the United States for CTL and GTL processes.[18] However, the cost to construct CTL plants is high, and the add-on capital cost to convert the branched paraffins generated from FT processes into alcohols is not likely to compete effectively with ethylene-based or natural alcohol processes.

1.4 SPECIALTY FEEDSTOCKS AND SURFACTANTS

Although the hydrophilic "head" groups of surfactants usually fit into one of the four categories described earlier, there are a number of exotic hydrophobic "tail" groups, both synthetic and natural, that populate the niches of specialty surfactants. Hydrophobes based on telomers of tetrafluoroethylene[19] or polydimethylsiloxane[20] bring unique surface-active properties to all classes of surfactants, reach extraordinarily low air/water and interfacial tensions, and enhance consumer and industrial product performance at amazingly low levels of use.[21]

Similarly, naturally derived surfactants extracted from fermentation broths or prepared by partial hydrolysis of natural extracts can contain polysaccharides, proteins, and phospholipids.[22,23] For example, rhamnolipids and sophorolipids have unique structural features that cause them to deposit on chemically similar surfaces and modify surface energy even at very low concentrations. Clearly, the emergence of biotechnology in the twenty-first century will drive the development of new surfactants from microbial fermentation, and improve the commercial viability of known surfactants from such processes.

Yet another class of niche surface-active agents includes higher molecular weight polymers based on acrylate or maleate esters, vinyl pyrolidone, and other vinyl monomers that contain, or can be modified with hydrophilic head groups. There are numerous chapters dedicated to polymeric surfactants and polymer surfactant interactions that enhance surfactant efficiency.[24,25] The use of surface-active polymers across all categories is increasing as these materials are customized and optimized to deliver enhanced product performance at very low levels.

Because of the cost, complexity to produce, and specificity of the surfactants, the development of new surfactants in these categories are conducted by highly specialized research organizations with strong technical depth in the core chemistries, often pursuing broad strategies with the same technology platform well beyond surfactants. Much of the fermentation-based surfactant development has originated from academia or federal research programs, and has been driven to commercial implementation through government seed money or private investment funds. As the large volume of surfactants described in this chapter are increasingly pushed toward commodity status, the continued development of new specialty surfactants will help to expand the limits of product performance and create value for technology-driven organizations.

1.5 BASIC RAW MATERIALS

In 2004, there were approximately 73.5 million bbl of crude oil produced per day, totaling annual production of 3600 million metric tons of oil worldwide, of which 90% was used for energy, and only 8% for chemical production.[26] The quantity of gas produced, converted into the equivalent quantity of oil, amounted to ~2400 million metric tons. A much larger portion of natural gas is consumed by the chemical industry, both for energy and as raw material feedstocks. Out of the refining, cracking, and reforming processes of these two key raw materials, ~90 million metric tons of ethylene and higher olefins were produced, and ~3 million metric tons of paraffins, of which <5% of these raw materials were consumed in the production of detergent alcohols.[13]

In the same year, a total of 5500 million metric tons of coal was produced, but only a small fraction was consumed in the production of surfactants.[27] Thus, the demand for basic petrochemicals to produce surfactant feedstocks represents a very small portion of total production, and feedstock producers find their raw material cost and supply position dictated by world energy demand and other larger chemical uses such as poly(ethylene terephthalate) (PET), polyethylene (PE), and PP plastic manufacturing.

Counterbalancing the fossil-based raw materials, in 2005, the world produced 180,000 million metric tons of biomass, of which only 67 million metric tons was used as a commercial source of energy. In the same year, the world harvested ~115 million metric tons of vegetable oils, of which ~80, 15, and 5% was consumed for food, chemical products, and animal feed, respectively.[28] A significant portion of the 15% for chemical use included 7.5 million metric tons of high-lauric content coconut and palm kernel oil, approximately half of which is converted into fatty acids for bar soaps, and the other half consumed in the manufacture of detergent grade alcohols used in surfactant production.[13]

Food and chemical uses for oils may soon be challenged by energy consumption as biodiesel production cuts into the available supplies. The 2005 global demand of 3 million metric tons of biodiesel is less than half of the 2006 demand of 6.9 million metric tons, and supply–demand is expected to increase at 133% through 2010 and projected to reach 45 million metric tons, with production capacity projected to greatly exceed supply–demand by as much as 100%.[29] Although the chain length of soybean and palm oil used for biodiesel is much longer than the optimal surfactant chain length of C12–14 found in coconut and palm kernel oils, high demand for the longer chain oils will raise their price, and drive substitution with shorter chain oils when necessary. It is not clear how global agricultural production of fats and oils can keep pace with these projections, suggesting an imminent conflict of interest between the basic need for food in the developing world and government incentives and mandates for renewable fuel use in the developed world.

Both oleo and petroleum-based surfactant feedstocks have come under supply pressure in the first decade of the twenty-first century due to rapid growth in many regional economies especially throughout Asia, and due to political factors in the Middle East, as well as poor reinvestment economics in the developed markets. Demand for petroleum, natural gas, vegetable oils, their derivatives, and other chemical intermediates used in the manufacture of surfactants is expected to continue to grow due to expanding economies in several regions of the world. At the same time, consolidations in the retail supply chain in Europe and the United States have concentrated purchasing power and allowed retailers to suppress price increases up and down the supply chain from the CPCs to the chemical manufacturers for several years.

The impact has been to force chemical manufacturers and converters to reduce their cost structure, delay or abandon capacity expansions, delay equipment upgrades, and minimize their investment in infrastructure. As a result, the chemical industry in the United States and Europe has operated in recent years near their maximum capacity, such that the occurrence of mechanical failure results in a major disruption to the supply chain. The years 2004 and 2005 saw record numbers of *force majeures* due to unplanned outages and the lack of capacity in the chemical supply chain to compensate for major supply disruptions.[30]

The future perspective for surfactant production is to expect swings in the cost and availability of feedstocks influenced by all of the factors mentioned earlier. Where possible, consumer product formulators will work toward flexible feedstock strategies, allowing them to switch between natural and synthetically based surfactants as forced by the market, or to take advantage of abundant raw materials with favorable economics.

1.6 THE FOUR MAIN SURFACTANT CLASSES AND THEIR PRODUCTION TODAY

The four main surfactant classes include anionics, cationics, nonionics, and amphoterics or zwitterionics. Within each of the general classes of surfactants, there are a broad range of variants, which are summarized in the following sections.

Anionic and nonionic surfactants represent almost equally large segments with global sales approaching 6 million metric tons in 2006 as depicted in Figure 1.11. Although cationic and

Surfactant Production: Present Realities and Future Perspectives

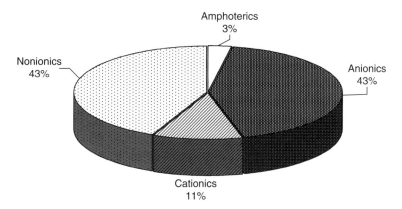

FIGURE 1.11 Percentage of global surfactant consumption by major application area for 2006 based on total sales of 13 million metric tons. (From Modler, R.F., Muller, S., and Ishikawa, Y., *Surfactants*, SRI Consulting; Specialty Chemical Update Program, July, 2004. With permission.)

amphoteric surfactants represent much smaller segments of the market, they are projected to be the fastest-growing segments in dollar terms, with compound annual growth rate (CAGR) of 5.5 and 4%, respectively.[1]

1.6.1 Amphoteric Surfactants

The amphoteric surfactants, which represent the smallest of the four categories, are identified as molecules that have the potential for a positive and negative ionic group to be present in the head group, and where the net charge may be changed by varying the pH of the system. Amphoterics, which typically have the highest raw material and production costs, are usually produced as dilute, low-active products that sell at a premium and command higher margins. Several classes of products based on chloroacetic acid or epichlorohydrin contain substantial levels of sodium chloride salt as a by-product and present long-term corrosion issues throughout manufacturing, transfer, and storage systems. Even in light of some of the barriers to producing amphoteric surfactants, their unique properties make them valuable surfactants with growing demand. They allow coupling of otherwise incompatible anionic and cationic surfactants, which greatly reduces the irritancy of primary anionic surfactants, build foam volume, increase foam stability, and reduce the total surfactant load to achieve target performance in a range of applications. Several texts are dedicated to the broad range of amphoteric surfactant chemistry.[31,32]

There are at least five main subcategories within amphoterics, the largest in use being betaines followed by amine oxides. There is a further subdivision within these two product categories based on the type of amine: Alkyl dimethyl amines (ADMAs) are used to produce alkyl betaines and amine oxides, which find substantial use in Europe. The second subdivision is derived from amides of dimethylaminopropylamine (DMAPA) where these are used to produce the alkylamidobetaines and alkylamidoamineoxides that are ubiquitous in most markets around the world. Lesser quantities of specialty amphoterics are widely used across several market segments, and these include the general product classes of amphoacetates, amphopropionates, sulfobetaines, and phosphobetaines.

The workhorse amphoteric surfactants are the alkyl- and alkylamidopropyl betaines, containing C8–C18 linear chain distributions that are derived from coconut or palm kernel oil, or ethylene-based alpha olefins. The alkyl betaines are prepared from ADMA feedstocks that are typically derived from alpha olefins and dimethyl amine through hydrohalogenation and alkylation reaction steps. The alkylamidopropyl betaines are based on tertiary amines derived from whole triglycerides or their fractionated derivative fatty acids or methyl esters reacted with DMAPA.

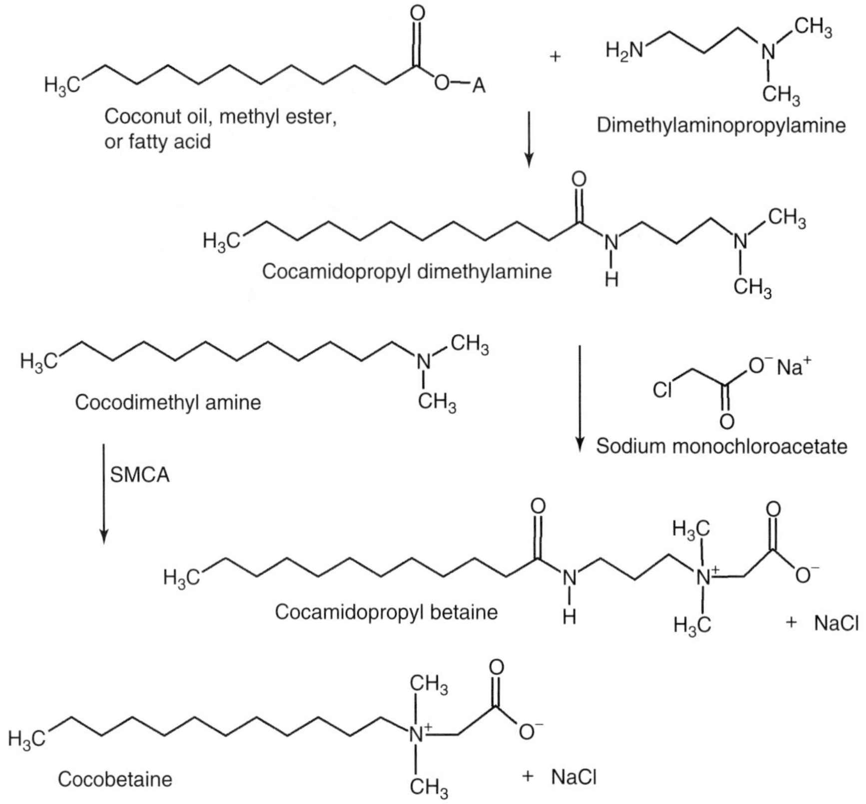

SCHEME 1.1 Preparation of alkyldimethyl betaines and alkylamidopropyl betaines.

Cocamidopropyldimethyl amine can be prepared from whole coconut oil, coconut fatty acid, or methyl esters either whole cut or stripped of the C6–C10 chain fractions. The aforementioned tertiary amines are converted to betaines by reaction with sodium monochloroacetate (SMCA) in a relatively dilute aqueous system (Scheme 1.1).

The alkylation step is conducted at temperatures between 50 and 100°C, is modestly exothermic, and requires some induction period during which a small portion of the betaine surfactant is formed and helps to emulsify the insoluble amine in the continuous aqueous phase making it available for reaction with the SMCA and accelerating the rate of conversion. Typical with two-phase surfactant forming reactions, a gel phase can be observed to develop under lower temperature conditions as conversion approaches 50%, and persists through >70% conversion at which point the viscosity drops and returns to a classic Newtonian state. Rapid conversion can occur at elevated reaction temperatures reducing the duration or virtually eliminating the appearance of a gel phase. However, the combination of a two-phase latent reaction coupled with the exothermic alkylation reaction and changing viscosity can lead to temperature excursions over 100°C and kettle foam out when operating at atmospheric pressure. The higher reaction temperatures coupled with sodium chloride levels approaching 5% constitute a corrosive environment for low-grade steel, recommending glass-lined vessels or construction from corrosion-resistant alloys.

Betaines are formulated into a wide range of personal cleansing products used on hair and skin. The inclusion of higher levels of the C16–C18 chains improves product mildness, whereas the C12–C14 chains effectively boost viscosity in the presence of additional sodium chloride, enhance

Surfactant Production: Present Realities and Future Perspectives | **15**

foam volume, stability, and quality. Betaines demonstrate excellent thermal stability and caustic and hypochlorite compatibility, and are used in aggressive formulations developed for drain cleaning and hard-surface cleaning, as well as foaming applications in oil field and other industrial products.

Amine oxides are produced by the reaction of the earlier-mentioned alkyl or alkylamidoamines with hydrogen peroxide in a two-phase system containing a large volume of water yielding dilute products typically containing <35% active (Scheme 1.2). Amine oxides have the same excellent secondary surfactant characteristics as betaines, and provide arguably better mildness and ability to mitigate skin irritation in formulations containing LAS or alcohol sulfates. They avoid corrosion problems associated with sodium chloride, and have inherent stability in the presence of hydrogen peroxide making them useful in oxygen bleaching and cleansing products. They find use in a broad range of formulations including personal cleansing products, LDLs (Light Duty Liquids–dish detergent), drain cleaners, hard surface, and fabric cleaning compositions.

The amphoacetates (also known as amphocarboxyglycinates) and amphopropionates comprise a third significant class, and are predominantly based on a mixture of amides and imidazolines derived from aminoethylethanolamine (AEEA) reacted with fatty acids or derivatives. The forcing reaction conditions and use of excess AEEA favor the imidazoline component, but when the amine mixture is reacted with aqueous SMCA solution, it undergoes partial hydrolysis to a mixture of alkylated and quaternized amidoamines and imidazolium quaternaries, yielding the product

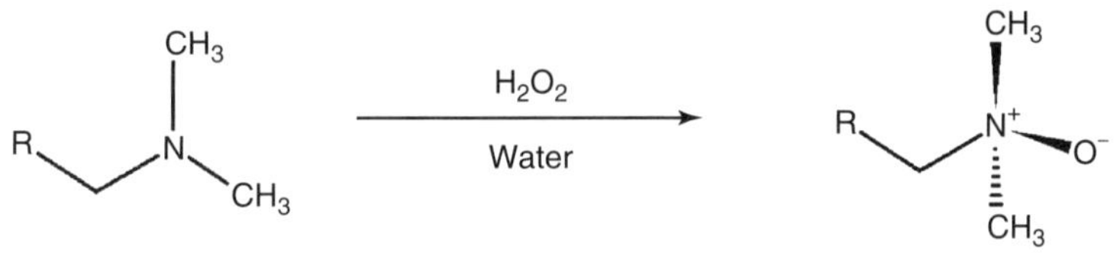

SCHEME 1.2 Oxidation of tertiary amines with hydrogen peroxide to produce amine oxides.

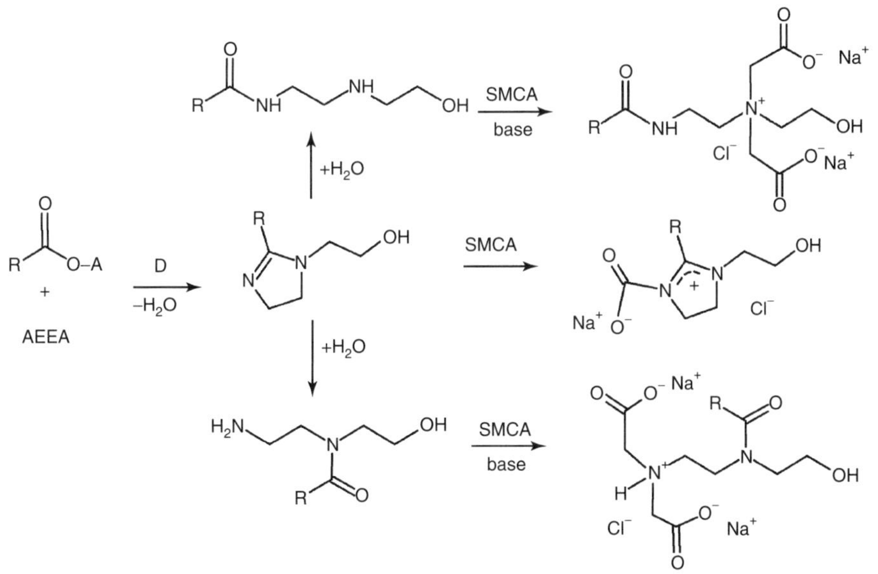

SCHEME 1.3 Preparation of amphoacetates, or amphocarboxyglycinates, through the reaction of fatty derivatives with AEEA to produce the imidazoline intermediate, then partial hydrolysis and reaction with SMCA.

mixture generically labeled amphoacetate (Scheme 1.3). The coco and lauryl amphoacetates and amphodiacetates (two moles of SMCA) find use in personal cleansing applications due to mildness and foam-building characteristics with additional sodium chloride.

Amphopropionates are prepared by the reaction of an AEEA imidazoline/amide mixture with methyl acrylate through the Michael reaction followed by mild hydrolysis of the resulting B-amino-ester giving rise to a complex mixture of components. The lauryl and cocoamphopropionates and amphodipropionates do not contain sodium chloride, but typically contain residual methanol from hydrolysis of the ester, and so are more suitable for use in hard surface cleaning formulations where chlorides may be undesirable and traces of methanol can be acceptable. The amphopropionates are stable in formulations containing high levels of inorganic salts and high levels of alkalinity.

Epichlorohydrin (ECH) is an effective linking group between tertiary amines and acid salts and is used to produce sulfonated amphoterics known as *sulfobetaines*. Sodium sulfite is reacted with epichlorohydrin in water to produce a solution of 1-chloro-2-hydroxypropane sulfonate, which is further reacted with a tertiary amine to yield a quaternary ammonium group linked to the hydroxypropane sulfonate, with sodium chloride as the primary by-product. Reaction of ECH with partially neutralized phosphorous or phosphoric acid produces an intermediate, which when reacted with tertiary amines yields the respective phosphitobetaines or phosphatobetaines.

Key issues include corrosivity of the reaction conditions to steel and proper handling of volatile alkylating agents such as epichlorohydrin. The resulting products are very stable to high temperature and aggressive pH conditions, and find use in a number of household and industrial applications.

The demand for amphoteric surfactants is expected to increase in step with the global growth of personal cleansing and home cleaning products. They provide mildness to formulations used in household applications, as well as foam stability and thickening properties. The challenges that will face amphoteric surfactant producers will include

- Cost inefficiencies associated with producing and shipping low-active materials, giving local/regional producers a competitive advantage.
- Corrosivity to ferrous reactors, piping, and storage vessels when producing amphoterics containing high levels of sodium chloride.
- Amidoamine-based amphoterics made with excess volatile amine require stripping, recovery, recycling or disposal of impure distillate streams, and process control to ensure very low levels of residual volatile amine in the intermediate.
- Use of alkylating agents including monochloroacetic acid or SMCA, ECH, and methyl acrylate require special precautions for safe use, and products must be certified to contain exceedingly low levels before shipment.
- The imidazoline-based products are very complex mixtures whose compositions are highly dependent on the precise composition of the intermediates and the reaction conditions of the quaternization, making them "products by process" requiring tight process control and often difficult to match against performance of competitive products.
- Customers look for multiple suppliers to keep pricing competitive, and usually solicit products that meet well-defined, narrow specifications.
- A number of granted patents exist that cover preferred processes for manufacture of several classes of amphoterics, and cover advantageous surfactant blends.
- Safety testing and registration costs stand as tall barriers to new amphoteric products limiting opportunities for differentiation through innovation.

1.6.2 ANIONIC SURFACTANTS

Anionic surfactants are the largest category of surfactants produced on a volume- and value basis, the bulk of which are accounted for by a relatively small number of products that are produced by sulfonation or sulfation, are used extensively in household laundry detergents and cleaning products, and also in personal cleansing products. At least two other categories of anionic surfactants are produced in substantially smaller volumes, including phosphate and polyphosphate esters, and carboxylate salts (excluding simple fatty acid soaps), and the general category is covered in several volumes.[33,34] Bar and flaked forms of traditional carboxylate soaps are still ubiquitous, and are still used globally for hand and body cleansing, and for laundering of soiled clothes in markets outside of North America and western Europe.[35]

1.6.2.1 Sulfonates and Sulfates

The broad class of products described as sulfonates results from reactions that create a carbon–sulfur bond and utilizes sulfur VI reagent SO_3 and its derivatives and adducts such as sulfuric acid. A smaller number of sulfonate products are prepared using sulfur IV reagent SO_2 as well as its derivatives and adducts such as sodium bisulfite. The preparation of sulfate esters involves the creation of carbon–oxygen–sulfur bonds, and can utilize SO_3, sulfuric acid, or chlorosulfonic acid to form alcohol sulfates that are labile and susceptible to hydrolysis in the presence of water as well as elimination reactions at elevated temperatures, and must be handled under milder conditions than sulfonates during formation and neutralization. Numerous older reviews and recent publications exist covering sulfonation and sulfation processes to produce surfactant products.[36–38]

Historically, sulfonation of alkylated aromatics for detergent was practiced with concentrated sulfuric acid, and later with fuming sulfuric acid or oleum, a mixture of SO_3 in sulfuric acid. The batch sulfonation or continuous sulfation processes utilizing these reagents are corrosive to steel, have poor productivity, and produce products with inferior color containing large amounts of sulfuric acid, which demand excess of caustic on neutralization. Resulting products are laden with sodium sulfate, and are suitable for use in low-density powdered detergents, and low concentration liquid detergents.

Chlorosulfonic acid (CSA) sulfation of alcohols or alcohol ethoxylates is usually conducted in batch reaction mode, typically in glass-lined vessels, but requires little or no excess reagent to drive the reaction to completion and can be conducted in a continuous liquid–liquid contacting process. The reaction evolves a stoichiometric quantity of HCl gas, which can be captured either as an HCl solution or with caustic to yield sodium chloride. Product color, yield, and quality can be very high, but throughput efficiency is modest in batch mode.

Liquid sulfur trioxide is the most reactive reagent used for sulfonation and sulfation, and is either obtained as a low boiling liquid or prepared at a point of use by combustion of sulfur in air and oxidation over catalyst at high temperature as depicted in Figure 1.12. The vaporized SO_3 steam

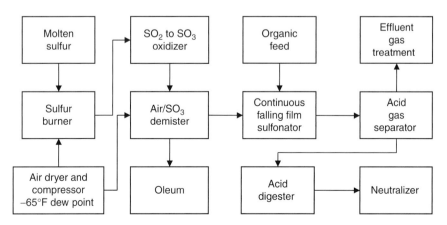

FIGURE 1.12 Continuous SO_3 falling film sulfonation process flow.

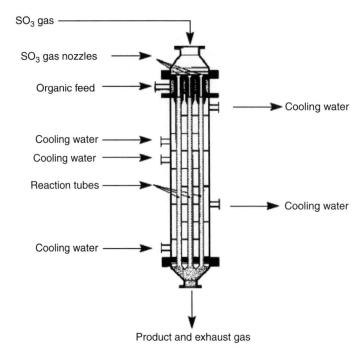

FIGURE 1.13 Ballestra multitube film sulfonation reactor. (From Ballestra, S.P.A., 2005. With permission.)

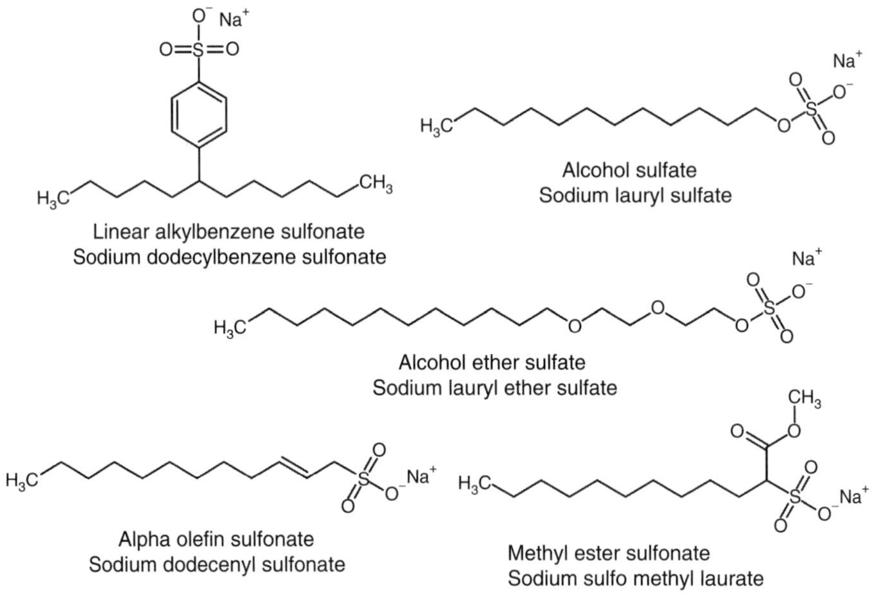

SCHEME 1.4 Anionic surfactants based on SO_3 with non-IUPAC generic and industry-specific nomenclature.

is diluted down to 2–6% in dry air and contacted with a broad variety of feedstocks in a continuous thin film reactor (Figure 1.13)[39] to produce an acid product, which is either stabilized for storage and later neutralization in the case of alkylates, or subjected to immediate neutralization in water with a variety of inorganic or organic bases as is done with fatty alcohols and fatty alcohol ethoxylates to give the range of products depicted in Scheme 1.4. In many locations around the world, LAB is sulfonated, digested, stabilized, and stored in the acid form, then later neutralized and dried along with a variety of nonionic surfactants and inorganic salts to make heavy-duty powdered detergent products.

Alpha olefin sulfonates are prepared through the falling-film sulfonation of alpha olefins, which involves the formation of a complex mixture of unsaturated sulfonic acids and cyclic sultones. The intermediate mixture is neutralized with excess base and then subjected to a rigorous high-temperature hydrolysis step during which all cyclic intermediates are converted to sodium sulfonate derivatives. The low molecular weight of alpha olefins compared to other feedstocks results in a relatively high heat of reaction per unit mass, requiring slower treatment rates to maintain product color and minimize volatilization and fouling of the reactor, separator, and emission-control equipment surfaces, and avoid subsequent volatile organic compound (VOC) emissions.

Sulfonated methyl esters are prepared through falling-film sulfonation of highly saturated methyl esters. The complex mechanism of sulfonation is discussed in several reviews and patents, but is thought to involve two molecules of SO_3 per molecule of ester, where SO_3 must first be inserted into the ester linkage. The scrambled mixture of inserted and uninserted sulfonation products requires a reesterification step before neutralization to maximize methyl ester sulfonate content. Bleaching steps and neutralization may be practiced in forward or reverse order, and can produce a low- or high-active neutralized aqueous product containing moderately low levels of impurities such as sulfonated carboxylate disalt, unreacted ester, sodium methylsulfate, and methanol. Commercially, this type of product is sold in liquid form, or can be dried, ground, and classified into uniform particles suitable for incorporation in powdered detergent products or for dissolution into liquid products.

Sulfur in oxidation state IV can be used to produce a variety of anionic sulfonates, as depicted in Scheme 1.5. Sodium bisulfite can be used to prepare sulfonates of a,b-unsaturated acids and esters, such as those prepared from maleic anhydride. The mechanism involves Michael addition to the activated double bond by the more nucleophilic sulfur atom, and is conducted in an aqueous two-phase system where, for example, a maleate half acid ester or diester is dispersed and heated under narrowly controlled pH conditions to minimize ester hydrolysis and avoid competitive hydroxide addition to the double bond. The resulting classes of surfactants include sulfosuccinates (which are in fact carboxylate sulfonate disalt surfactants) prepared from the maleic half acid esters of fatty alcohols or alcohol ethoxylates. Diesters of maleic are sulfonated by the same type of process to produce surfactants such as the ubiquitous dioctyl sulfosuccinate (DOSS) from the diester of 2-ethylhexyl alcohol and maleic anhydride.

Bisulfite or SO_2 can be reacted through free radical mechanisms with alpha olefins or with paraffins to give sulfonated alkanes. The former reaction with alpha olefins is conducted in water, usually in the presence of a cosolubilizing alcohol and a free radical initiator, and excluding all oxygen. By balancing the concentration of sulfite and bisulfite and utilizing techniques developed in the emulsion polymerization process industry, good conversion to primary alkane sulfonates can

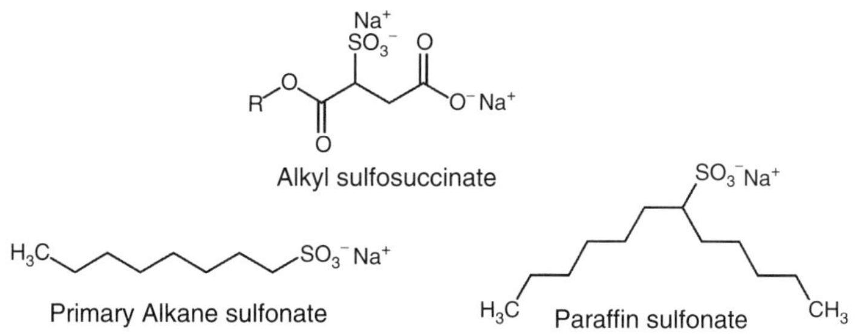

SCHEME 1.5 Examples of anionic surfactants prepared from sulfur dioxide, sodium bisulfite, or sodium sulfite.

be achieved with minimal formation of disulfonated by-products. The short-chain primary alkane sulfonates have utility as hydrotroping agents and low-foaming surfactants for use in extreme pH or strongly oxidizing formulations.

The reaction of SO_2 with paraffins is conducted in water in the presence of oxygen or chlorine, and is photoinitiated to generate oxygen or chlorine radicals that abstract hydrogen from the paraffinic backbone, creating alkane free radicals that react with gaseous SO_2 and further interact with oxygen to reach the S VI oxidation state and perpetuate the free-radical reaction. Excess SO_2 is consumed and by-products include hydrogen peroxide in the oxygen process, and hydrochloric acid in the chlorine-initiated process, as well as sulfuric acid, which must be separated at low temperatures from the alkane sulfonic acid before neutralization. The oxygen process is still practiced in several locations in Europe today, whereas the chlorine process was developed and practiced in the former German Democratic Republic, and is no longer in use due to the formation of undesirable chlorinated paraffinic by-products.

Continuous SO_3 sulfonation process equipment is available in well-designed modular packages, and involves numerous unit operations including air compressing and drying, sulfur burning, sulfur dioxide oxidation, thin-film sulfonation, gas/liquid separation, electrostatic precipitation, and final scrubbing as partially depicted in Figure 1.12. For production of alcohol and ether sulfates, immediate neutralization and deaeration are practiced to avoid degradation of color and generation of undesirable by-products such as 1,4-dioxane. A very large number of relatively small volume sulfonation units (1–5 metric tons/h) have been installed around the world, most of them relying on molten sulfur burning, several still using liquid SO_3 delivered from central production facilities.

The resulting sulfation/sulfonation overcapacity in many regions has driven down the profitability of operation, and made smaller sulfation/sulfonation units very uncompetitive. Increasingly tighter restrictions on air emissions and higher costs for disposal of scrubber wastes will present challenges for older, smaller units in mature regions, and will shape the strategy for installation of larger units in emerging countries. Since sulfation/sulfonation does not represent a significant value-added conversion for many feedstocks, it is not a highly attractive forward integration strategy for feedstock producers. Because of recent dramatic price fluctuations in petroleum and oleo feedstocks, surfactant consumers find themselves needing a flexible feedstock strategy that exceeds the simple single-flavor plant designs for making LAS or AES. Increasingly, consumer product companies are redefining themselves as marketing and technology implementation companies, and as such are abandoning backward integration strategies, divesting existing sulfation/sulfonation assets, and avoiding the construction of new facilities.

The demand for sulfated and sulfonated anionic surfactants is expected to increase in step with the global growth of consumption of laundering, personal cleansing, and home-cleaning products.[1,3] They are almost always the primary or secondary component in cleansing formulations; are responsible for reducing surface tension, forming micellar structures, which solubilize oily soil; and help suspend particulate soil by generating a significant negative zeta potential that prevents coalescence and redeposition through charge repulsion. They provide foaming and allow for thickening by manipulation of electrolyte concentration in formulations.[40] The challenges that will face anionic surfactant producers include

- Ongoing oscillations in the price and supply of both petroleum and naturally derived feedstocks due to a variety of socioeconomic forces, including significant growth in demand in east and west Asia, eastern Europe, and Latin America driven by increasing surfactant consumption as well as competing uses for fuel, food, and nonsurfactant applications
- Continued overproduction capacity in developed markets and emergence of multiple competitors in developing markets
- The need to manage a flexible feedstock strategy that supports the demand for LAS for powdered detergents as well as actives for liquid product forms based on alcohols, alcohol ethoxylates, alpha olefins, and methyl esters

Surfactant Production: Present Realities and Future Perspectives

- Continuous process control that produces products within narrow specifications, with low color, and with minimal formation of impurities such as 1,4-dioxane in alcohol ether sulfates and sultones in AOS
- Rigorous emission controls that minimize the release of SO_2 and SO_3 gas, as well as VOCs
- Ability to produce highly active neutralized products that are rheologically difficult to handle, require sophisticated storage, dispensing, shipping, and dilution infrastructure.
- Specialized assets with no general utility beyond production of sulfated or sulfonated surfactants

1.6.2.2 Phosphated Surfactants

A substantially smaller subset of anionic surfactants includes the reaction products of alcohols and alcohol ethoxylates with P_2O_5, polyphosphoric acid, or phosphoric acid, which is covered in detail in several review chapters.[33,34] The resulting products include varying mixtures of mono and diesters of phosphoric acid, which are neutralized with a variety of inorganic or organic amine bases. Phosphate esters have outstanding mildness properties and impart excellent skin feel when used in personal cleansing formulations. Depending on the ester ratio and chain length, they can have excellent emulsification properties, tolerance to electrolytes, thickening properties, and hydrolytic stability across a wide pH range making them suitable for use in acidic or basic hard surface and drain-cleaning formulations. However, the abundance of local city-, county-, and state-level legislation practically banning the use of phosphates since the 1980s in the United States in consumer and industrial cleaning products has limited this class of products to specialty applications such as agricultural emulsifiers, and plastic and fiber additives.

1.6.2.3 Carboxylated Synthetic Surfactants

In a separate category from soap are the carboxylated synthetic surfactants, which typically comprise an alcohol ethoxylate capped with a carboxymethyl group. Generally, two processes are practiced, one involving the reaction of sodium monochloroacetate and a base to form a carboxymethyl ether, with sodium chloride as a by-product. The second process involves oxidation of the terminal hydroxyl ethyl group of an alcohol ethoxylate using nitric acid or other strong oxidizing systems. These products are mild, foaming surfactants that find use in liquid and solid-form personal care products. The cost, corrosivity, and salt-removal issues of the SMCA process make it less competitive than the oxidation process.[38]

1.6.3 CATIONIC SURFACTANTS

Cationics represent the third-largest group of surfactants by volume. By virtue of a positively charged head connected to a hydrophobic tail, cationics are attracted to negatively charged molecules and surfaces, and their physical properties have been reviewed extensively.[41,42] Ion pairing and deposition can change the net surface charge and impart benefits related to dimensions of the hydrophobic chains: linear saturated hydrocarbons containing >16 carbons impart a high degree of lubricity and water repellency. Unsaturated carbon chains provide less lubrication, but can attract and hold water molecules, improving wicking properties of textiles. When used on fabrics, long-chain cationics deposit well beyond the formation of a Langmuir monolayer, creating a lubricating layer that results in fiber smoothing and fabric softening. In the dying process, cationic surfactants form highly insoluble complexes with anionic-charged dye molecules, affixing the dyes and inhibiting their loss through subsequent washings. The ability of cationic surfactants to associate with metal surfaces provides corrosion inhibition and lubricity. Biocidal properties are attributable to interaction with anionically charged surfaces of simple biological systems building sufficient local concentrations to disrupt cell wall structures of single-celled organisms and to inactivate viruses.

Cationic surfactants are produced by three general process routes that attach a nitrogen atom or nitrogen-containing fragment to a hydrophobic group. The three routes include (1) conversion of a carboxylic acid–containing group into a nitrile intermediate, followed by hydrogenation to the amine; (2) condensation products based on fatty acids and their derivatives, but without nitrile intermediates; and (3) from alpha olefins or fatty alcohols. These process routes can yield a range of product types, and some individual products may be derived through several routes.[43]

The nitrile route is exemplified by the process to manufacture one of the earliest commercial fabric softener materials, dihydrogenated tallow dimethyl ammonium chloride (DHTDMAC), developed by Armour & Company in the 1950s as a means to utilize tallow and lard by-products of their meat-packing business. The product is prepared by the process detailed in Scheme 1.6: Tallow is steam-split by the Colgate–Emery process and the glycerin is recovered as a dilute aqueous "sweet water" stream. The fatty acids are distilled, then reacted at high temperature with ammonia over a bauxite catalyst with elimination of two equivalents of water to generate fatty nitriles, which are subsequently reduced by batch hydrogenation over Rainey nickel catalyst to yield a mixture of fatty primary amines. As originally practiced, the catalyst and reaction conditions reduced the olefins in the tallow chain yielding a mixture of saturated C16 and C18 alkyl amines. New catalysts allowed selective reduction of the nitrile while preserving unsaturation in the chain. Two equivalents of fatty amine are disproportionated over the same catalyst with elimination of ammonia to yield a mixture of mono-, di-, and tritallowamines, which are purified by distillation. The di(hydrogenated)tallow amines are subjected to reductive alkylation with formaldehyde and hydrogen over nickel catalyst to yield the dialkylmethyl amine, which is quaternized with methyl chloride or other alkylating agents.

The saturated and unsaturated primary amines from this process are converted through a series of derivatization steps into tertiary amines utilizing reagents such as EO, PO, a,b-unsaturated esters, formaldehyde, or acrylolnitrile followed by reduction to the respective alkyldimethyl- and alkylaminopropylamines, all suitable for quaternization with aliphatic and aromatic alkylating agents. Alternatively, the tertiary amines may be neutralized with acids to form the respective trialkylammonium salts, which have many of the attributes of the quaternary ammonium salts when maintained at acidic pH.

The sequence of multiple processing steps including operating conditions at high temperatures and pressures is both capital- and energy-intensive.

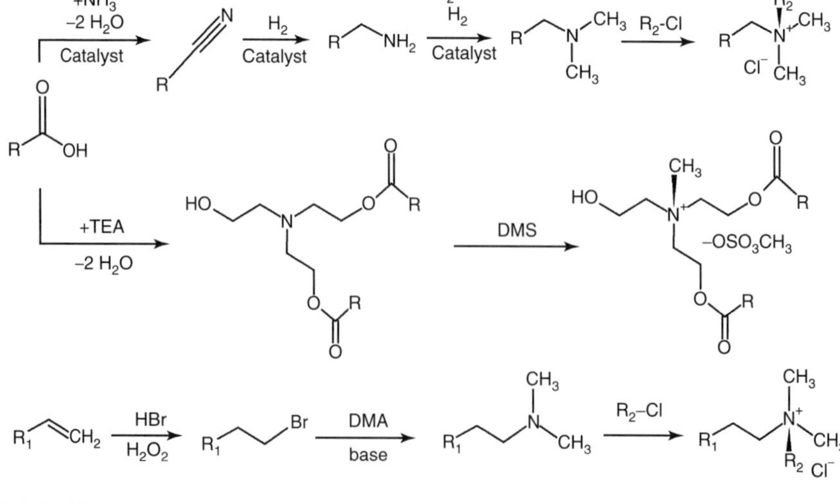

SCHEME 1.6 Three routes to cationic derivatives including fatty acid to nitrile to alkyl amine (*top*), fatty acid to alkanolamine ester (*center*), and olefin to alkyl bromide to alkyl dimethyl amine (*bottom*).

Surfactant Production: Present Realities and Future Perspectives 23

Condensation products based on fatty acids and derivatives include reaction products with alkanolamines such as triethanolamine (TEA) as in Scheme 1.6 and methyl diethanolamine (MDEA) and dimethylmonoethanolamine (DMMEA) to give estersamines, which are quaternized with methyl chloride or dimethyl sulfate (DMS). Hoechst and BASF filed patent applications, in 1970 and 1971, respectively, disclosing diester-based quaternaries of MDEA with methyl chloride, DMS, or benzyl chloride (BC) for use as fabric-softening agents.[44,45] Esterquats based on TEA were first patented for use as fabric softeners in 1974 by Stepan Company,[46] and comprise a thermodynamically controlled statistical distribution of mono-, di-, and triesters prepared from fatty acid or methyl esters that are commercially quaternzied with DMS. Various versions of esterquat products are now used globally in fabric softener formulations and have displaced DHTDMAC in many regions of the world by virtue of their biodegradability and lower capital cost to produce.[47] Cationics based on coconut fatty acid esters of DMMEA quaternized with DMS or methyl chloride are used as additives in laundry detergents and claimed to enhance the cleaning performance of LAS in numerous patents.[48]

A wide range of derivatives are prepared from alkyleneamines and poly(ethyleneamines) such as diethylene triamine (DETA), triethylenetetramine (TETA), imidazolines, amidoamines, and their ethoxylated and propoxylated derivatives as depicted in Scheme 1.7.[49] Reaction products of fatty acids and their derivatives with DMAPA (see Section 1.6.1) can be quaternized with various alkylating agents to produce amidopropyl ammonium quaternaries, many of which have been patented and commercialized in a variety of applications, including personal care, fabric softening, and industrial/oil-field applications.

Alpha olefins are reacted with HBr or HCl in the presence of peroxides to give the antimarkovnikov addition product primary alkyl halides, which are further reacted with alkylamines such as dimethylamine to give the respective alkyldimethylamines. Addition of inorganic bases such as sodium hydroxide to break the resulting amine salts generate highly purified NaBr which has significant market value in electronic and other specialty chemical applications.

Fatty alcohols are less commonly converted into intermediate alkyl halides through a reaction with $POCl_3$, PCl_3, or $SOCl_2$, and then reacted with dimethylamine to form the tertiary amine. The resulting by-products include mixtures of phosphorous and phosphoric acids and their partially chlorinated forms that present handling and disposal challenges.

Alternatively, aliphatic alcohols may be converted directly to the respective dimethyl alkylamines by catalytic amination in the presence of dimethylamine and low-pressure hydrogen over copper catalyst. The mechanism is believed to involve catalytic dehydrogenation of the alcohol to an aldehyde, addition of DMA with concomitant water elimination to form the enamine, and then subsequent reduction to the alkyldimethylamine. This route is particularly favored with longer-chain alcohols, which are derived through hydrogenation of tallow, or palm fatty acids, or methyl esters

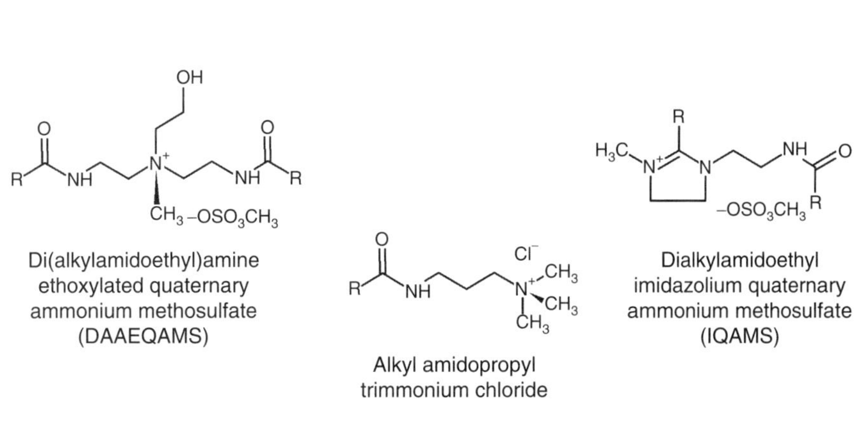

Di(alkylamidoethyl)amine
ethoxylated quaternary
ammonium methosulfate
(DAAEQAMS)

Alkyl amidopropyl
trimmonium chloride

Dialkylamidoethyl
imidazolium quaternary
ammonium methosulfate
(IQAMS)

SCHEME 1.7 Examples of quaternary ammonium surfactants based on DETA and DMAPA.

and are more plentiful than alpha olefins in the C16–C18 chain range. The hydrogenation of imines derived from the condensation of aliphatic aldehydes and primary amines is also used to produce secondary amines, which may be exhaustively alkylated to produce quaternary surfactants.

The global demand for cationic surfactants will be driven primarily by consumer fabric-softening applications, and to a lesser degree by personal care-product applications. The proliferation of washing machines coupled with increasing disposable income in developing regions and relatively low-current market penetration will create opportunities for significant growth even where hand washing still predominates. Cationic surfactants are uniquely functional in a broad range of consumer and industrial applications, and will continue to grow in use across the diverse range of application areas, from biocides to agricultural products, to oil-field chemicals. Challenges facing manufacturers of cationic surfactants include

- Increasing price and tightening supply of both petroleum and naturally derived fatty amines, PE amines, and alkanolamines due to increasing demand across the broad range of application areas, and the high capital cost of building new capacity
- Limitations on feedstock substitution options for fatty amines due to the unique performance features associated with amines produced from olefins versus fatty alcohols
- Cost of building and operating batch-processing facilities to produce customized and made-to-order (MTO) products to support consumer product differentiation strategies, as well as the diversity of small volume specialty products used in nonconsumer applications
- Ability to safely operate pressurized reactions with methyl chloride, a flammable gas reactant, or work with highly hazardous liquid alkylating agents such as dimethyl sulfate, diethylsulfate, or BC
- Rigorous emission controls that minimize the release of MeCl gas as well as VOCs

1.6.4 NONIONIC SURFACTANTS

Nonionic surfactants have found strong utility as compliments or alternatives to anionic surfactants in household cleaning and industrial applications due to their tolerance of hard water, lesser foaming characteristics, and efficiency at removing greasy and oily soils.[50–52] They can act synergistically with anionic surfactants and improve packing in mixed micelles, providing charge separation and mitigating the effects of repulsion between head groups of anionic surfactants, as well as maintaining micelle stability in the presence of polyvalent cations found in hard water. Nonionic surfactants have been tailored to produce little or no foam, and can act as defoaming agents in optimized formulations. When treated to remove trace levels of metal salts used as manufacturing catalysts, nonionic surfactants can be used effectively in high-performance cleaning applications where electrolyte residues are unacceptable, such as metalworking and electronics and optics manufacturing. Several classes of nonionic surfactants are used as foam boosting and skin-feel enhancing agents in personal cleansing products, and as emulsifying and conditioning agents in leave-on products such as creams and lotions.

The three broad classes of nonionic surfactants consist of esters, amides, and ethers, where the class name characterizes the essential bond-making process involved in linking the hydrophobic tail to the hydrophilic head. Some of the earliest nonionic surfactants were derived through esterification, typically by reaction of triglycerides or fatty acids and glycerin to yield mixtures of mono-, di-, and triesters. Esters, as a class, represent a small but significant volume of nonionic surfactants in commercial use today. Alkanolamide nonionic surfactants evolved based on the availability of alkanolamines, from which Kritchevsky developed diethanolamine (DEA) and monoethanolamine (MEA) amides of coconut fatty acid and triglycerides,[53] and these products found their way into synthetic liquid dish detergents and personal care products in the 1940s and 1950s. Ethers derived from the reaction of active hydroxyl groups with ethylene, propylene, or butylene oxide constitute the largest general class of nonionics by volume, and accounting for at least half of the 2 million

Surfactant Production: Present Realities and Future Perspectives 25

metric tons of nonionics produced annually. Ethers based on polyhydroxy compounds such as glycerin, sorbitol, and glucose or dextrose, have achieved some significant commercial volumes in consumer applications involving skin contact where product mildness justifies their higher cost of manufacture.

1.6.4.1 Alkanolamide Nonionic Surfactants

Alkanolamides are the reaction products of mono- or dialkanolamines with fatty acids or their esters to form the respective secondary or tertiary amides, examples of which are depicted in Scheme 1.8.[54] Monoalkanolamides can be prepared by the reaction of monoalkanolamines with triglycerides such as coconut or palm kernel oil as well as from fatty acids or their methyl esters. The reaction of MEA with triglyceride at temperatures of 140–160°C liberates glycerin as a by-product, and because of the stability of the secondary amide formed, the reaction can be driven to a very low ester content with only a slightly super-stoichiometric amount of MEA.

Alkanolamides based on DEA are tertiary amides and are not as stable as MEA amides such that a significant amount of esters can remain in equilibrium with the amide. The ester amines and esteramides of DEA have undesirable performance properties, but these components can be reduced by utilizing an excess of DEA to drive equilibrium toward the amide form. The most common version is prepared by the reaction of 2 moles of DEA with 1 mole of coconut fatty acid or ester to give the Kritchevsky or Ninol-type DEA amide, which is liquid at room temperature. Products made with slightly more than a 1:1 molar ratio of DEA to fatty acid or ester are referred to as *superamides*; and at reaction temperatures of 140–160°C, the mixture contains high level of ester components and free amine. However, with sufficient time at storage temperatures <50°C, the composition will increase in amide and decrease in ester and free-amine components and thus can be "aged" into specification for free DEA and ester content.

Solid amides from MEA and DEA can be chill flaked, but during formulation the mixture must be heated to 50–70°C for effective incorporation, whereas liquid DEA amides are stored and transported at <50°C, and are easier to incorporate at lower formulation temperatures. A variety of blends have been developed where high-melting amides are incorporated into aqueous anionic surfactant products such as the typical mixture of neutral alcohol sulfate and alcohol ether sulfates used in shampoo products, providing improved ease of transport, storage, and dilution, even at near-ambient temperature.

Alkanolamides that are liquid at room temperature have been commercialized based on increased content of unsaturated C18 acids, use of branched alkanolamines such as monoisopropanolamine (MIPA), or propoxylation of MEA amides, as well as blends of all these with PEG esters to improve thickening performance of the products. The higher cost of the earlier-mentioned feedstocks

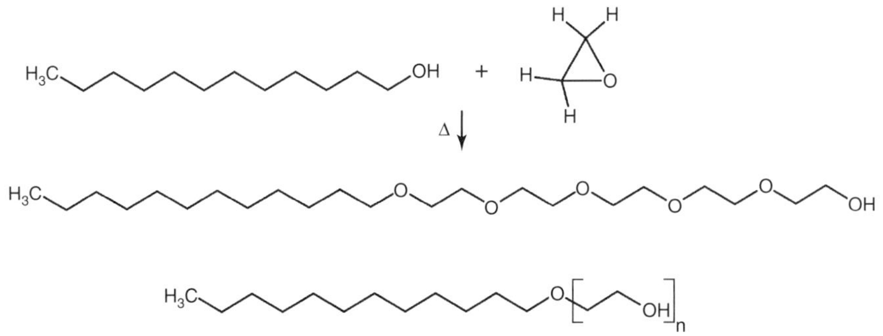

SCHEME 1.8 Reaction of dodecyl alcohol with EO to prepare the nominal 5-mole ethoxylate, also referred to as Dodeceth-5. The product is a mixture containing the Gaussian distribution of dodecyl polyoxyethylene ethers units from $n = 0$ to > 10.

1.6.4.2 Alkoxylated Nonionic Surfactants

The production and properties of alkoxylated nonionic surfactants have been reviewed extensively.[55,56] They are produced by the reaction of hydrophobic feedstocks containing hydroxyl or amine groups with alkylene oxides under pressures of 30–90 psia and at temperatures of 100–140°C in the presence of catalytic quantities of basic or transition metal elements. The degree and type of alkylene oxide appended and the blocking or random copolymerization that is allowed to occur drastically affects the physical and performance properties of the resulting nonionic surfactant.

Because of the competing reactions of water or short-chain alcohols with target functional groups, the stepwise process involves pretreatment of the hydrophobe with a catalytic quantity of base, typically NaOH or KOH, applying heat and vacuum to eliminate water and drive the composition to the respective metal alkoxylate or carboxylate salt of the substrate, optionally releasing the vacuum with nitrogen to provide an inert diluent for operation with EO, then introducing EO or PO as a gas or liquid into the catalyzed substrate. Although alkoxylated amines are often exploited for their cationic character when formulated under basic conditions, this class of surfactants functions comparably with alkoxylated alcohols with some potential advantages due to the presence of two independent PEG chains in the case of alkoxylated primary amines.

The process of alkoxylation is exothermic and involves a balance of working reactor pressure with batch temperature to achieve acceptable product color and reasonable cycle times. For reactions involving EO, there is a well-defined partial pressure of EO that should be controlled by the use of nitrogen dilution gas, batch temperature, and headspace volume to exclude the accidental inclusion of oxygen and avoid conditions that could result in a spontaneous vapor-phase reaction such as decomposition, isomerization, or polymerization, any of which generates between 2000 and 3000 kJ/kg heat of reaction and associated pressure spike. Batch alkoxylation processes have been conducted safely and efficiently at numerous plant locations since the 1950s with only a small number of serious safety issues. A consortium of EO producers actively promotes safe handling, process design, and operation and provides supporting information to this end.[57]

The rate of reaction of EO in typical ethoxylation reactions is mass transfer and often heat-transfer limited, where the low vapor pressure of EO coupled with reaction temperatures from 70 to 140°C allows a large quantity to remain in the gaseous phase occupying the headspace of a traditional batch reactor. Several reactor designs have been introduced over the past 25 years that specifically enhance gas–liquid mixing to the extent that very low partial pressures of EO exist during the course of the reaction, and the time to complete the addition is dramatically reduced. Designs such as the buss loop reactor[58] educt the headspace gas into the reaction mixture through recirculation loops comprising passive or active gas–liquid mixing devices. The rate of ethoxylation is accelerated to the extent that production of large volume commodity ethoxylates can be conducted in a semibatch mode. A supply of catalyzed feed thoroughly stripped of water and low molecular weight alcohols is prepared in the first pretreatment stage, whereas the second stage loop reactor completes the alkoxylation step, and the third finishing stage reacts or strips away trace levels of EO, where the catalyst can be neutralized, or removed by treatment through an ion-exchange bed (Figure 1.14). This arrangement is particularly suitable for production of low-mole ethoxylated alcohols used as feedstocks for sulfonation to produce ether sulfates.

Ethoxylation involves a random reaction that results in a Poisson distribution of products with varying amounts of alkylene oxide groups appended to the substrate. For conventional catalysis with sodium or potassium alkoxylates, the distribution is particularly broad and consistent. When applying <6 moles of EO, the amount of unethoxylated substrate can be significant. A large body of development work over the course of the past 20 years has identified a range of alternative catalysts

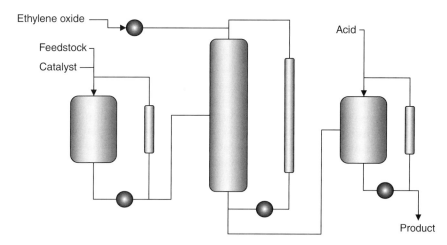

FIGURE 1.14 Commercial process for alkoxylation with precatalysis reactor, alkoxylator, and posttreatment reactor. (From Buss Process Technology, U.S. Patent 3,915,867, 1975; U.S. Patent 6,017,874, 2000. With permission.)

that can reduce the amount of unreacted substrate and narrow the distribution of alkoxylated products for reactions where <10 moles of EO or PO are added. The catalysts typically have a lower "basicity" and comprise simple or complex salts of calcium, magnesium, or a host of transition or rare earth metals that are added as powders, and usually require a longer induction or reaction period to conduct the alkoxylation reaction. Many of these same catalysts have functionality for catalyzing insertion alkoxylation reactions of methyl esters, yielding methyl-PEG or methyl-PPG esters.

Propoxylation is typically a slower reaction due to the steric hindrance between the propagating secondary hydroxide ion and the approaching PO. Attack of the hydroxide occurs essentially completely at the primary carbon of PO, regenerating a secondary hydroxide ion that is more basic than a primary hydroxide ion. In the presence of a primary alcohol substrate, complete capping with 1 mole of PO will occur before substantial polypropoxylation begins due to both steric constraints and the equilibrium concentration of alkoxide favoring the primary alcohol. This phenomenon allows for controlled capping of primary alcohols or block ethoxylates with a small amount of PO. Conversely, this effect makes it difficult to "cap" a secondary alcohol with a small amount of EO. Addition of EO first occurs on a small portion of the substrate and creates primary alcohols that are preferential sites for subsequent EO addition, driving the formation of a low mole fraction of a product with long EO blocks, and leaving a high mole fraction of the substrate unethoxylated. The production of "random" EO/PO block alkoxylate products by premixing EO and PO in a feed tank and pumping into the reactor thus tends to result in products containing blocks of EO separated by very short sectors of PO, and terminating in a large block of PO as the excess is consumed at the end of the reaction. Similarly, before adding an EO block to a propoxylate block, it is necessary to react away or strip off residual PO to a very low level to avoid random insertion of PO into the EO block, or PO capping at the end of the ethoxylation reaction that can dramatically change the performance characteristics of the product.

Longer reaction times are usually required for propoxylation reactions due to the preference for lower reaction temperatures necessary to maintain product color and avoid troublesome side reactions such as elimination of a terminal alcohol to generate allyl ethers. Propoxylation reactions can be selectively accelerated at lower temperatures with the use of a double metal cyanide catalyst. The catalyst is extremely sensitive and easily deactivated by moisture and conventional groups I and II catalysts, and will not effectively catalyze ethoxylation reactions.

The most common substrates for alkoxylated nonionics are listed as follows:

Natural C12–C16 fatty alcohols or C10–C15 synthetic fatty alcohols with 1–4 moles of EO. These nonionics are typically used as feedstocks for sulfonation to produce a wide range of ether sulfates used in household, personal care, and industrial applications (see Section 1.6.2.1).

Natural C12–C16 fatty alcohols or C10–C15 synthetic fatty alcohols with >4 moles of EO. Alcohol ethoxylates with 5–11 moles of EO are used in liquid and powdered laundry detergents as coactives with LAS and fatty alcohol ethers sulfate (FAES) in hard surface–cleaning formulations and in a host of industrial applications. Alcohol ethoxylates with 12–50 moles of EO and their respective sulfated and phosphated derivatives find use in emulsion polymerization and other select applications.

Alkylated phenol. The most commonly used alkylated phenol ethoxylates (APE) have included octyl phenol ethoxylate and nonylphenol ethoxylates with 3–11 moles of EO, which were produced by alkylation of butylene dimer or propylene trimer onto phenol and subsequent ethoxylation. They had been used extensively in laundry and hard surface–cleaning applications in the nonionic form, and as the sulfated and phosphated derivatives of the low-mole ethoxylates in a variety of industrial applications. Concerns over the environmental impact of the partial metabolites generated during the waste treatment of these surfactants has prompted their elimination from European consumer product formulations, and their deformulation from most consumer products in North America and elsewhere in the world.

Block propylene oxide. Several broad subfamilies of these surfactants were originally developed by Wyandotte Corp, later proliferated by BASF and are today produced through a process where PPG of varying molecular weight (MW) is used as the substrate for ethoxylation, or where PEG or alkyl PEG ethers of varying MW are block propoxylated. Concerns in the European Union (EU) over the anaerobic biodegradability of the block PO/EO surfactants will affect their ongoing breadth of use in consumer applications.

Secondary alcohols. Secondary alcohols derived from oxidation of primary alfa olefins are first selectively monoethoxylated under acid-catalyzed conditions before being polyethoxylated under basic conditions to avoid the EO blocking phenomenon described in the propoxylation discussed earlier. They offer the advantages of a highly branched surfactant such as low bulk-handling temperature and cloud point in formulations without the disadvantages of reduced biodegradability. The two-step process to ethoxylate represents a cost and efficiency penalty, and when ethoxylated even once, they are less stable substrates for sulfation, more susceptible to elimination reactions as a pathway to decomposition.

Glyceryl esters. Glyceryl esters such as glyceryl monostearate (GMS), including triglycerides such as castor oil are ethoxylated with classic group I catalysts and undergo insertion ethoxylation as well as ester interchange leading to highly complex mixtures of esters of ethoxylated glycerin where PEG ether chains are capped with fatty acid groups. As the degree of ethoxylation increases, the amount of remaining glyceryl esters decreases proportionately.

Polyhydric alcohol esters. Reactions of fatty acids with polyhydric alcohols derived from reduction of sugars such as sorbitol results in a mixture of esters of both the sorbitol and its dehydrated ethers sorbide and isosorbide. Ethoxylation of this complex mixture of esters involves insertion of ethoxylation and transesterification ethoxylation as described earlier, resulting in a highly complex composition with performance properties that are dependent on the precise conditions for both the esterification and the ethoxylation steps. This is described classically as a product-by-process, which is difficult to reproduce due to the affect of the commercial-scale manufacturing kit on the precise conditions of each step of the reaction.

Surfactant Production: Present Realities and Future Perspectives   **29**

Fatty acids. Fatty acids can be ethoxylated in what amounts to a two-step process where the first mole of EO adds slowly to the dry, precatalyzed acid to yield the hydroxyethyl ester. Because of the presence of carboxylic acid, the reaction runs slowly under general acid catalysis and does not produce substantial quantities of polyethoxylated product. Once the free carboxylic acids are completely capped, the pH of the system becomes alkaline and the reaction proceeds in a comparable manner to any primary alcohol ethoxylation. Some competing transesterification occurs at temperatures above 120°C, leading to a product distribution of PEG monoesters, diesters, and free PEG.

Amide ethoxylates. Mono- or dialkanolamides are alkoxylated to produce amide ethoxylates, which are noted to have some cationic character. At temperatures of alkoxylation, even monoalkanolamides will undergo ethoxylation on the amide nitrogen at a low rate, creating a low level of dialkanolamides. Dialkanolamides at temperatures of alkoxylation undergo an equilibrium 1,4 shift to form the respective ester amine, which is immediately alkoxylated to form an ester of a trialkanolamine. Thus, the concentration of esters of trialkanolamines increases during the course of and proportionately with the degree of alkoxylation.

1.6.4.3 Esterified Nonionic Surfactants

Fatty acids or their derivatives are connected to a variety of hydrophilic head groups through esterification of fatty acids or transesterification using fatty acid methyl esters or triglycerides[52] as depicted in Scheme 1.8. Hydrophilic head groups that are used to prepare nonionic ester surfactants fall into several of the following subcategories.

Nonether containing polyhydric alcohols derived from petroleum feedstocks include ethylene glycol, propylene glycol, butylene glycol, trimethylol propane, pentaerythritol, and several other feedstocks derived from epoxidation of olefins or condensation reactions based on formaldehyde. The exhaustively esterified derivatives are not surface active, but are used as emollients and lubricants. The partially esterified derivatives are typically used as emulsifying agents, and contain the distribution of mono- to polyesterified components. The monoester version can be produced at >70% yield by using more than three equivalent excess of hydroxyl to fatty acid residue where practical, and for the simple glycols the excess can be removed *in vacuo* using a wiped film distillation unit.

Nonether containing polyhydric alcohols derived from oleochemical sources include glycerin and reduced sugars consisting of erythritol, sorbitol, and manitol. The reactions and derivatives of glycerin are long known and well described,[59] and recent work on monoesters and polyesters of erythritol show them to be useful in several food-related applications. Esterification of polyhydric alcohols of more than four carbons involves competing ring-closing etherification reactions that generate anhydrides such as sorbides and isosorbides from sorbitol at reaction temperatures between 170 and 240°C with a basic catalyst. The esterification reaction rate exceeds the etherification rate, and open-chain polyol derivatives predominate as 60–80% of the product mixture depending on reaction time, temperature, and catalyst level.

Esterified nonionic surfactants are prepared from ether-containing polyhydric alcohols that include derivatives of EO and PO such as polyethylene glycol (PEG), polypropylene glycol (PPG), EO/PO block copolymers, polyglycerin that is prepared by thermal dehydration with basic catalyst, and ethoxylated glycerin (Scheme 1.9).

1.6.4.4 Etherified Nonionic Surfactants

The etherification of monosacharides with linear alcohols to produce alkyl polyglucosides has been known since Fischer developed this chemistry in 1893, but has only been practiced commercially since the early 1990s.[60] Variations and improvements on the process have been laid open in patents

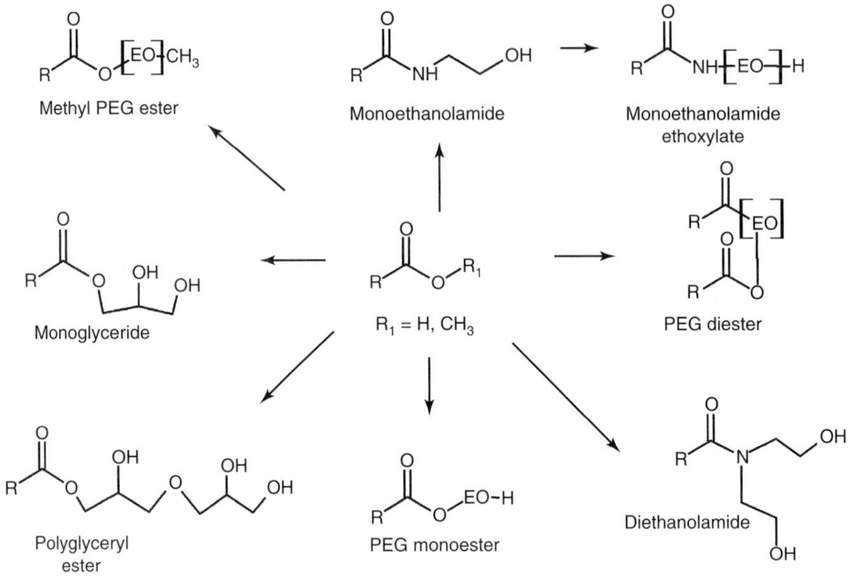

SCHEME 1.9 Nonionic surfactants derived from fatty acids or esters.

issued to Henkel, Union Carbide, Akzo, Huls, and BASF, but fundamentally it involves a continuous plug-flow etherification reaction between a heterogeneous mixture of dextrose and a 3–10 molar excess of linear alcohol in the presence of an acid catalyst at elevated temperature. The excess amount of alcohol and reaction conditions affect the degree of polymerization (DP) of glucose, with the most useful products having a DP of <2, preferably between 1.1 and 1.3 and based on C10, C12, or coconut alcohol distributions. The excess alcohol must be stripped off for performance reasons and recycled for cost control. The alcohol stripping is best performed by subjecting the neutralized product to thin film distillation *in vacuo*.

The challenges facing producers of nonionic surfactants are multiple, and vary based on the type of surfactants they seek to produce.

- Alkoxylated surfactants will require construction of manufacturing facilities adjacent to EO production facilities to receive EO "over the fence" to achieve competitive economics and guarantee surety of supply.
- The cost of capital to construct alkoxylation facilities requires a large volume base load such as detergent-grade midrange alcohol ethoxylates and sulfonation grade low-mole ethoxylates, and the scale of such equipment is not effective at servicing small volume, specialty ethoxylate products.
- The operation of alkoxylation processes involves handling of volatile, flammable, highly reactive ingredients at elevated pressures and temperatures requiring a very high degree of attention to mechanical integrity and human safety.
- Nonalkoxylated nonionics based on amide or ester chemistry are easily prepared in small-volume batch-manufacturing facilities, making them low barrier-to-entry products, inviting competition and driving down margins and profits.
- Products that involve both esterification and alkoxylation steps comprise extremely complex mixtures that are best described as *products by process* and are often difficult to replicate.

Surfactant Production: Present Realities and Future Perspectives

1.7 CONSTRUCTION AND OPERATIONAL ISSUES

The history of surfactant manufacturing is related to its position in the value chain. Some of the largest original practitioners of surfactant manufacturing were, and still are the end-formulators, consumer product companies who sought control over the cost, quality, and availability of the key anionic surfactant ingredients for powdered laundry detergents. Their goal was to build and operate the most capital-efficient operation that produced essentially one type of product on a continuous or semicontinuous basis year round. Continuous sulfonation processing, as a key example, is usually designed and implemented with efficient logistics in mind: raw materials are received in bulk, transferred through hard piping to dedicated storage, and pumped into highly automated processes requiring minimal operator intervention, with finished product piped to dedicated tanks.

The second tier of surfactant manufacturers consists of nonintegrated toll or merchant producers who started their participation in the business by making one or two families of secondary surfactants such as amides, nonionics, or amphoterics. In most cases, the original product specifications were broad, their customers' formulated performance was very forgiving, and there was a lot of latitude in manufacturing procedures that would produce an acceptable product. Batch production operations were, and can still be run today in a highly manual mode, with operators receiving and handling many or all raw materials in drum quantities, transferring feeds, intermediates, and finished products through hoses and other temporary equipment, and operating through manual control of valves and pumps located on or near the reactors.

In locations where labor costs are modest, this can be the most cost efficient and flexible means of producing a wide range of customized products that are capable of meeting demanding performance specifications. In developed countries where labor costs are the highest, there is a natural drive toward increased efficiency through automation. Automation achieves its greatest cost benefit when it allows operators to multitask efficiently and operates multiple production units simultaneously, and without needing to make constant adjustments to process variables such as heating, cooling vacuum, mixing, or level control. When designed with appropriate raw material and finished-product storage infrastructure, a single person can operate several production batches in different reactors simultaneously, and help manage the receipt of inbound raw materials and the loading and certification of outbound products.

The cost and complexity of designing, building, and maintaining facilities with this high level of automation are most easily justified when producing a single version of a product that requires efficient production while operating within tight specifications, and especially when handling hazardous raw materials. The capital cost begins to increase significantly when trying to serve a family of products based on a broader range of feedstocks. The cost associated with installing and maintaining multiple feedstock and product storage facilities can easily rival or far exceed the cost of the primary reaction vessel. Since there are a limited number of processing hours, once a vessel is "sold out," demanding additional product variants can result in a shift to MTO production, where reactors end up as temporary hold tanks awaiting product transfer to tank trucks or drums, reducing throughput efficiency dramatically.

1.7.1 REGULATORY STANDARDS

Surfactants are used in a broad range of products that enter the environment through consumer use, and agricultural and industrial products. In recent years, there has emerged a consensus view about what tests should be conducted to evaluate both new and existing chemicals, particularly those that are manufactured in large quantities.

Legislation on chemicals was first adopted and implemented in Europe in 1967. However, only after 1979, the legislation systematically addressed the generation of a minimum set of information for chemicals. Under a 1979 EEC directive, an inventory was established comprising existing chemicals, with more than 100,000 substances. New chemicals not listed in the inventory and produced in annual volumes of more than 1 t had to be notified. The notification dossier included a base set of data covering certain environmental and toxicological properties. At annual production volumes more than 100 or 1000 t, additional data focusing on the effects of long-term exposure had to be provided. Since 1981, more than 3700 new chemicals have been notified in the EU, constituting ~3% of all marketed substances. In 1993, Council Regulation 93/793/EEC was adopted requiring the submission of available data for existing chemicals placed on the market in volumes of more than 10 t/year. Based on this information, authorities identified 141 high-priority chemicals for risk assessment. By 2004, the assessment had been completed for only 32 of these chemicals.

Owing to the slow progress, a critical review of the EU chemicals legislation was initiated in 1997, and in 1999, the council adopted conclusions requesting the EU Commission to review the existing legislation and to develop appropriate proposals for a new chemicals policy in the EU. Following an extensive discussion with all the involved stakeholders and the adoption of a white paper on a strategy for a future chemicals policy in 2001, the European Commission adopted a proposal for an European Parliament and Council Regulation on the Registration, Evaluation, Authorization and Restrictions of Chemicals (REACH) in October 2003.*

According to this proposal, after the new regulation comes into force, the industry would be obliged to register all marketed chemicals above 1 t annually within 11 years. A minimum dataset would be required for chemicals between 1 and 10 t and additional information—the extent depending on the tonnage—would have to be provided for chemicals exceeding 10, 100, and 1000 t. For chemicals exceeding 10 t, industry would in addition have to draw up a chemical safety report, which includes a risk assessment. The results of this assessment would be forwarded to the downstream users transformed into the traditional safety data sheet to enable them to meet their responsibilities.

The program was codified under the EU Existing Chemicals Regulation, which covered all European Inventory of Existing Chemical Substances (EINECS)-listed existing substances manufactured or imported at >10 t/annum. The available test data have been reported for 1408 substances supplied at >1000 t/annum. Of these, only 110 have been selected so far for complete testing and risk assessment.

REACH also applies to all chemicals that are considered to be of very high concern to health or the environment—regardless of volume. Depending on the substance in question and its use, producers and importers may be obliged to investigate its effects on human health and the environment. REACH applies to all chemicals imported or produced in the EU. The European Chemicals Agency will manage the technical, scientific, and administrative aspects of the REACH system. Highlights include the following:

- Establish one system for registration, evaluation, authorization, and restrictions of chemicals that will become immediately effective in all member states.
- Establish a central agency for the management of the system.
- Focus competent authorities on high-tonnage substances.
- Register 30,000 chemicals over 11 years.
- Scope is limited to industrial chemicals.
- Pharmaceuticals, pesticides, and biocides are covered by separate regulations and testing requirements.
- Polymers and nonisolated intermediates are excluded.
- At least 1 million more animal tests are expected to be conducted.
- Estimated costs of approximately 5 billion euros for business over 11 years.
- REACH legislation contains 1,000 pages of text, rising potentially to 15,000.
- By October 2006, there had been 1,000 amendments voted by the EU parliament.

---

* AISE, the International Association for Soaps, Detergents, and Maintenance Products at www.aise-net.org.

Surfactant Production: Present Realities and Future Perspectives 33

The U.S. Toxic Substances Control Act (TSCA) was first enacted in 1976 and has been amended significantly three times. TSCA gives the U.S. Environmental Protection Agency (USEPA) broad authority to regulate the manufacture, use, distribution in commerce, and disposal of chemical substances. TSCA is a federally managed law and is not delegated to states. The law is overseen by the USEPA Office of Pollution Prevention and Toxics (OPPT).*

A major objective of TSCA is to characterize and evaluate the risks posed by a chemical to humans and the environment before the chemical is introduced into commerce. TSCA accomplishes this through the requirement that manufacturers perform various kinds of health and environmental testing, use quality control in their production processes, and notify EPA of information they gain on possible adverse health effects from use of their products. Under TSCA, "manufacturing" is defined to include "importing," and thus, all requirements applicable to manufacturers apply to importers as well.

TSCA requires manufacturers, importers, and processors of certain chemical substances and mixtures to conduct testing on the health and environmental effects of chemical substances and mixtures, unless they qualify for an exemption. Testing requirements cover existing chemicals (but not new chemicals, because these are addressed in the EPA premanufacturing notice process) and mixtures as well as individual substances. EPA has established a Master Testing List that lays out testing priorities based on risk and exposure potential.

EPA has the authority to ban the manufacture or distribution in commerce, limit the use, require labeling, or place other restrictions on chemicals that pose unreasonable risks. Among the chemicals EPA regulates under TSCA are asbestos, chlorofluorocarbons (CFCs), and polychlorinated biphenyls (PCBs).

By the early 1990s, it was evident that many of the existing major industrial chemicals had not been evaluated completely, although in 1989 the Organization for Economic Cooperation and Development (OECD) countries had initiated their voluntary testing program aimed at providing this data by developing Screening Information Data Sets (SIDS) on high production volume (HPV) chemicals.

Later in the decade, concern over chemical safety prompted the USEPA, in collaboration with the American Chemical Council (ACC) and the U.S.-based nongovernmental organization the Environmental Defense Fund, to launch the Chemical Right-to-Know initiative. A key element to this was the HPV challenge program, which was designed to obtain SIDS information on ca. 2800 HPV chemicals supplied in the United States at >1 million pound annually according to the 1990 TSCA Inventory Update. More than 300 companies and consortia volunteered to develop and supply data on a completion schedule targeted at the end of 2004.

In March 2005, the ACC, in cooperation with U.S.–based Soap and Detergents Association (SDA) and Synthetic Organic Chemical Manufacturers Association (SOCMA), announced a joint initiative to extend the industry's work on HPV chemicals—those produced in the United States or imported in quantities greater than 1 million pound annually. The Extended HPV (EHPV) Program was designed to publish health and environmental information on 574 newly designated HPV chemicals. These are substances that did not qualify as HPV chemicals at the start of the original program, but now meet the volume threshold according to EPA's 2002 inventory. In addition to gathering health and environmental information, companies will be asked to provide information on use and exposure for both the "extended" HPV as well as the original Challenge Program substances. In this way, the EHPV Program will provide EPA and the public with an extensive source of chemical safety information on HPV chemicals.

Collaboration between national and regional industry organizations representing surfactant chemical manufacturers have attempted to avoid redundant testing through consortia organized to support specific groups or families of products.

---

* ChemAlliance, a resource organization funded by USEPA and private industry to provide current information on environmental regulations to the chemical industry, accessible at www.ChemAlliance.org.

Cost sharing and data ownership issues have been negotiated in attempts to achieve equitable distribution of burden. As national and regional chemical inventories are translated from developed regions to developing regions, the expenditures borne by the consortia will create opportunity for the local production of surfactants.

As a result of regional testing and registration, the use of certain surfactants have been restricted to specific areas of the world, such as the EU, although these products continue to find acceptance and use in other regions. A well-established example is DHTDMAC.

Regulatory challenges that face production of surfactants include

1. The primary surfactant ingredient (and possibly coproducts) must appear on the regional chemical inventories of approved chemicals such as TSCA in the United States, Canadian Designated Substance List (DSL) or Canadian Non-Designated Substance List (NDSL), and EINECS in Europe.
2. When the product is ultimately discharged to the environment, it must meet biodegradability requirements, the results of which vary by test method and the acceptable levels of which vary from region to region.
3. The presence of some unreacted starting materials, by-products, or impurities may cause concerns around chronic exposure due to results generated for the concentrated isolate, and by modes of exposure that are irrelevant to the use of the product, for example, concerns for the carcinogenicity of nitrosamines from unreacted DEA found in shampoos and body washes based on results of oral testing.

In the CEFIC Pan European Survey "Image of the Chemical Industry 2006,"* the summary report notes several expected trends.

> At pan European level, the development shows a trend back towards a more business-critical climate, namely publics' declining belief in the benefits of modern technology and publics' increasing feeling of companies' profits being too high.
>
> Publics' belief in a sacrosanct value of nature is, despite the measurable decline, still very strong and a potential barrier to the implementation of new technologies such as genetically modified plants or nano-technology.*

It is clear that the chemical industry up and down the supply chain will continue to be scrutinized and judged, and our "right to practice" will continue to be challenged in various quarters, but can be maintained by complying with the evolving and ever harmonizing standards for safety, health, and environmental (SHE) performance. More than compliance, there is an expectation by stakeholders surrounding the industry that the tenets of responsible care and product stewardship will be embraced and advanced at more than a symbolic level, and that real, tangible, continuous improvement will be the standard to which we are held and against which we are judged.

The good news is that the chemical industry in Europe as a whole continues to gain positive ground in the public's eyes.

> Most detailed ratings of the chemical industry on qualitative aspects have improved too, slightly on items related to "Importance & Benefits" and "SHE-Performance", more significantly on "Social Responsibility".
>
> Practically unchanged from 2004, publics' perceptions of "chemical products" are more positive than negative and focus on cleaners, detergents, toiletries, pharmaceuticals, cosmetics, pesticides, paints.
>
> There is no measurable effect from media reports or NGO-campaigns related to REACH.*

Clearly, consumers recognize they have a high level of interaction with chemicals in the form of laundry detergents, household cleaning, personal cleansing, and health and beauty products. Thus,

---

* CEFIC, the European Chemical Industry Council, from the Cefic Pan-European Survey, Image of the Chemical Industry 2006, accessible at www.cefic.org.

Surfactant Production: Present Realities and Future Perspectives   **35**

the surfactant manufacturing and formulating industries have borne and continue to bear especially important roles in creating and preserving public trust through responsible and ethical operation, paying close attention to worker and consumer health and safety issues, and acting as responsible environmental stewards by minimizing acute releases and reducing chronic emissions.

The European perception is of particular importance since the European governments and competent authorities have proven to be the most aggressive at creating and implementing continually higher regulatory standards for the industry to meet. Governments and local regulatory bodies in countries around the world are choosing to adopt European or U.S. standards rather than negotiate and legislate internally to lower standards. Thus, as Europe does, so will ultimately the rest of the world.

1.8 SUMMARY

The pathway to survival and prosperity in the surfactant-manufacturing business will be based on the following strategies:

1. *Drive to low-cost operations.* The inevitable drive toward commoditization of large-volume work-horse surfactants will require manufacturers to relentlessly drive cost out of their systems and increasingly shift the burden associated with sales, general operations and administration (SG&A), and research and development (R&D) onto higher-value, smaller-volume specialty products.
2. *Create a diversified product range.* R&D will be needed to add value by leveraging core technologies to create value-added specialty products for niche applications. This will require capabilities to design, synthesize, screen, and test for performance alone, and in collaboration with customers across the spectrum of sophistication.
3. *Add value downstream.* By providing blends, ready-to-dilute formulas, and innovative logistic solutions and inventory management, surfactant manufacturers will gain value in a spectrum of partnerships with customers that can approach virtual vertical integration (VVI).
4. *Selective backward integration.* It is clear that commodity surfactant converters occupy the lowest-value creation step in the supply chain. Back integration to control some strategic raw materials allows access to higher value, diversification of risk, and surety of supply for customers. Feedstock suppliers are often unmotivated to move down the supply chain because the converting steps offer lower returns, require an order of magnitude, more diverse product line, are technical service wise and logistically more intensive, and can disrupt their relationships with key customers.

In brief, those who focus purely on commodity products will not survive.

REFERENCES

1. Global Industry Analysts, Inc., Surface Active Agents—*A Global Strategic Business Report 08/06*, August 2006.
2. Bragg, C. D., *The Future of Detergents*, World Oleochemical Conference, Athens, Greece, April 2005.
3. Modler, R. F., Muller, S., and Ishikawa, Y., *Surfactants*, SRI Consulting; Specialty Chemical Update Program, July 2004.
4. Levitt, B., *Oil, Fat and Soap*, Chemical Publishing Co., New York, 1951, p. 33.
5. Spitz, F. and Spitz, L., *The Evolution of Clean, A Visual Journey through the History of Soaps and Detergents*, AOCS Press, Champaign, IL, 2006, pp. 43–69.
6. U.S. Patent No. 1,686,837, 1928; U.S. Patent No. 1,708,103, 1929; and U.S. Patent No. 1,872,736, 1932.
7. Vaughan, J., *The Future of Surfactant Feedstocks*, CESIO 6th World Surfactants Congress, Berlin, June 20–23, 2004.

8. U.S. Patent No. 2,212,521, 1940.
9. U.S. Patent No. 1,968,796, 1934, and U.S. Patent No. 2,141,245, 1939.
10. U.S. Patent No. 2,486,921, 1949.
11. UOP LLC, accessible at http://www.uop.com, higher olefins web page.
12. About.com accessible at http://www.chemistry.co.nz/deterghistory.htm.
13. Brackmann, B. and Hager, C.-D., *The Statistical World of Raw Materials, Fatty Alcohols and Surfactants*, CESIO 6th World Surfactants Congress, Berlin, June 20–23, 2004.
14. Renaud, P., *Natural Based Fatty Alcohols: Completely in Line With the Future?* CESIO 6th World Surfactants Congress, Berlin, June 20–23, 2004.
15. Davy Process Technology, accessible at www.davyprotech.com, 2006.
16. Brent, J., *Standing Out From the Crowd: New Surfactant Molecules Create Competitive Opportunities*, CESIO 6th World Surfactants Congress, Berlin, June 20–23, 2004.
17. CAIA, The Chemical and Allied Industries' Association of South Africa, accessible at http://www.caia.co.za/chsahs05.htm.
18. Official publication available from the U.S. National Energy Technology Laboratory Web site accessible at http://www.netl.doe.gov, *NETL Report 2-06*, April 25, 2006.
19. Kissa, E. (Ed.), *Fluorinated Surfactants: Synthesis Properties Applications*, Surfactant Science Series Vol. 50, Marcel Dekker, New York, 1993.
20. Hill, R. M. (Ed.), *Silicone Surfactants*, Surfactant Science Series Vol. 86, Marcel Dekker, New York, 1999.
21. Farn, R. J. (Ed.), *Chemistry and Technology of Surfactants*, Blackwell Publishing, Oxford, 2006, chapter 6.
22. Kosaric, N., Cairns, W. L., and Gray, N. C. C., *Biosurfactants and Biotechnology*, Surfactant Science Series Vol. 25, Marcel Dekker, New York, 1987.
23. Kosaric, N. (Ed.), *Biosurfactants: Production Properties Applications*, Surfactant Science Series Vol. 48, Marcel Dekker, New York, 1993.
24. Piirma, I. (Ed.), *Polymeric Surfactants*, Surfactant Science Series Vol. 42, Marcel Dekker, New York, 1992.
25. Kwak, J. C. T. (Ed.), *Polymer-Surfactant Systems*, Surfactant Science Series Vol. 77, Marcel Dekker, New York, 1998.
26. Official statistics available from the U.S. Department of Energy, Energy Information Agency (EIA), accessible at http://www.eia.doe.gov/.
27. Published statistics by the U.S. Department of Energy, Energy Efficiency and Renewable Energy, accessible at http://www.eere.energy.gov.
28. Published statistics by the U.S. Department of Agriculture, Foreign Agricultural Service, USDA Economics and Statistics System accessible at http://usda.mannlib.cornell.edu/MannUsda/homepage.do.
29. Gabler, R., *Biodiesel, CEH Marketing Research Report*, SRI Consulting, Menlo Park, CA, November 2006.
30. Force majeur reference, *Insight: Charting the Operating Challenge*, ISIS News, January 17, 2006.
31. Bluestein, B. R. and Hilton, C. L. (Eds.), *Amphoteric Surfactants*, Surfactant Science Series Vol. 16, Marcel Dekker, New York, 1982.
32. Lomax, E. G. (Ed.), *Amphoteric Surfactants*, 2nd Ed., Surfactant Science Series Vol. 59, Marcel Dekker, New York, 1996.
33. Linfield, W. M. (Ed.), *Anionic Surfactants*, Surfactant Science Series Vol. 7, Marcel Dekker, New York, 1990.
34. Stache, H. W. (Ed.), *Anionic Surfactants*, Surfactant Science Series Vol. 56, Marcel Dekker, New York, 1995.
35. Spitz, L. (Ed.), *Soaps and Detergents: A Theoretical and Practical Review*, AOCS Press, Champaign, IL, 1996.
36. Dado, G. P., Knaggs, E. A., and Nepras, M. J., Sulfonation and sulfation, In *Kirk-Othmer Encyclopedia of Chemical Technology*, 5th Ed., Vol. 23, Wiley, New York, 2007.
37. De Groot, W. H. (Ed.), *Sulfonation Technology in the Detergent Industry*, Kluwer, Dordrecht, 1991.
38. Hibbs, J., Anionic surfactants, In *Chemistry and Technology of Surfactants*, Farn, R. J. (Ed.), Chapter 4, Blackwell Publishing, Oxford, 2006.
39. Ballestra, S.P.A., 2005.

40. Lucassen-Reynders, E. H. (Ed.), *Anionic Surfactants: Physical Chemistry of Surfactant Action*, Surfactant Science Series Vol. 11, Marcel Dekker, New York, 1990.
41. Jungermann, E. (Ed.), *Cationic Surfactants*, Surfactant Science Series Vol. 4, Marcel Dekker, New York, 1970.
42. Rubingh, D. N. and Holland, P. M. (Eds.), *Cationic Surfactants: Physical Chemistry*, Surfactant Science Series Vol. 37, Marcel Dekker, New York, 1990.
43. Richmond, J. N. (Ed.), *Cationic Surfactants: Organic Chemistry*, Surfactant Science Series Vol. 34, Marcel Dekker, New York, 1990.
44. French Patent Application FR 1,593,921, 1970.
45. German Patent Application DE 1,935,499, 1971.
46. U.S. Patent 3,915,867, 1975.
47. Levinson, M. I., Rinse-added fabric softener technology at the close of the twentieth century, *J. Surf. Det.*, 2(2), 223, 1999.
48. U.S. Patent 6,017,874, 2000.
49. Friedli, F. E., Amidoamine surfactants, In *Cationic Surfactants, Organic Chemistry*, Richmond, J. M. (Ed.), Surfactant Science Series Vol. 34, Marcel Dekker, New York, 1990, p. 51.
50. Schick, M. J. (Ed.), *Nonionic Surfactants*, Surfactant Science Series Vol. 1, Marcel Dekker, New York, 1967.
51. Schick, M. J. (Ed.), *Nonionic Surfactants, Physical Chemistry*, Surfactant Science Series Vol. 23, Marcel Dekker, New York, 1987.
52. Hepworth, P., Non-ionic surfactants, In *Chemistry and Technology of Surfactants*, Farn, R. J. (Ed.), Chapter 5, Blackwell Publishing, Oxford, 2006, pp. 145–148.
53. Kritchevsky, J., The Alkylolamides, *J. Am. Oil Chem. Soc.*, 34, 178, 1957.
54. Jungerman, E. and Tabor, D., In *Nonionic Surfactants*, Schick, M. (Ed.), Surfactant Science Series Vol. 1, Marcel Dekker, New York, 1967.
55. Nace, V. M. (Ed.), *Nonionic Surfactants: Polyoxyalkylene Block Copolymers*, Surfactant Science Series Vol. 60, Marcel Dekker, New York, 1996.
56. Van Os, N. M. (Ed.), *Nonionic Surfactants: Organic Chemistry*, Surfactant Science Series Vol. 72, Marcel Dekker, New York, 1997.
57. *Ethylene Oxide Users* Guide (Second Edition), accessible at the American Chemistry Council www.americanchemisty.com, December 21, 2006.
58. Buss process technology, accessible at http://www.buss-ct.com/.
59. Stauffer, C. E., Emulsifiers for the food industry, In *Bailey's Industrial Oil and Fat Products, Edible Oil and Fat Products*, Fereidoon, S. (Ed.), Vol. 4, 6th Ed., Wiley, New York, 2005, chapter 8.
60. Balzer, D. and Luders, H. (Ed.), *Nonionic Surfactants: Alkyl Polyglucosides*, Surfactant Science Series Vol. 91, Marcel Dekker, New York, 2001.

2 Detergent Alkylate and Detergent Olefins Production[*]

Bipin V. Vora, Gary A. Peterson,
Stephen W. Sohn, and Mark G. Riley

CONTENTS

2.1 The Production of Alkylbenzenes 39
    2.1.1 General 39
    2.1.2 Alkylbenzene Consumption 39
    2.1.3 Branched Alkylbenzenes Production 40
    2.1.4 Linear Alkylbenzenes Production 41
    2.1.5 Alkylation 41
        2.1.5.1 Hydrofluoric Alkylation 41
        2.1.5.2 $AlCl_3$ Alkylation 42
        2.1.5.3 UOP/CEPSA Solid Bed Catalyst Alkylation 43
    2.1.6 Alkylbenzene Quality 44
2.2 Higher *n*-Olefins for Surfactant Production 45
    2.2.1 Applications 45
    2.2.2 Internal Olefins Production 45
    2.2.3 α-Olefin Production 46
    2.2.4 Market Demand 46
2.3 *n*-Paraffins for Lab Production 46
    2.3.1 Linear Alkylbenzenes Production 46
    2.3.2 Liquid- and Vapor-Phase Extraction 47
    2.3.3 Fisher–Tropsch Process 47
2.4 Future Trends and Technology Development 47
References 48

2.1 THE PRODUCTION OF ALKYLBENZENES

2.1.1 GENERAL

Of all the various surfactants used in detergent formulations, alkylbenzene sulfonates are by far the most widely used. As a result, alkylbenzenes are the most widely used surfactant raw materials. There are two basic types of alkylbenzenes used in surfactant manufacturing: linear alkylbenzenes (LAB) and branched alkylbenzenes (BAB). Both types of alkylbenzenes are homolog mixtures. Typical homolog representatives of LAB and BAB are depicted in Figures 2.1 and 2.2, respectively.

2.1.2 ALKYLBENZENE CONSUMPTION

Since the introduction of light-duty liquid detergents in the 1960s and subsequent need for the solubility of detergent components, which include surfactants, there has been a steady conversion in

---

[*] © retained by authors

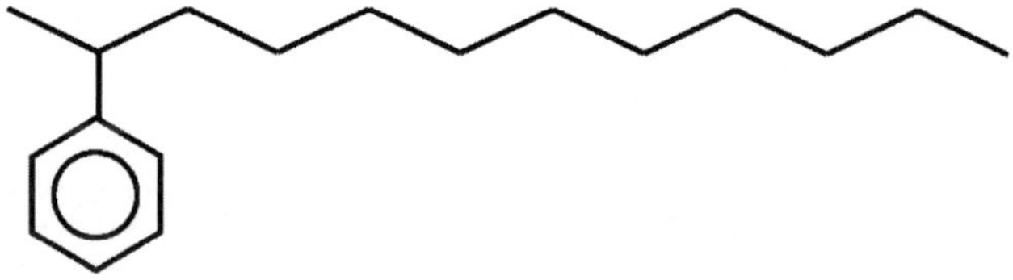

FIGURE 2.1 Structure of a simple linear alkylbenzene (2-phenyldodecane).

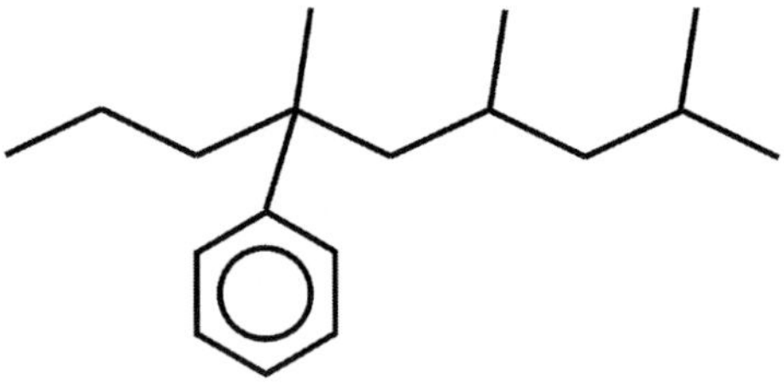

FIGURE 2.2 Structure of a branched alkylbenzene (4-phenyl-4,6,8-trimethylnonane).

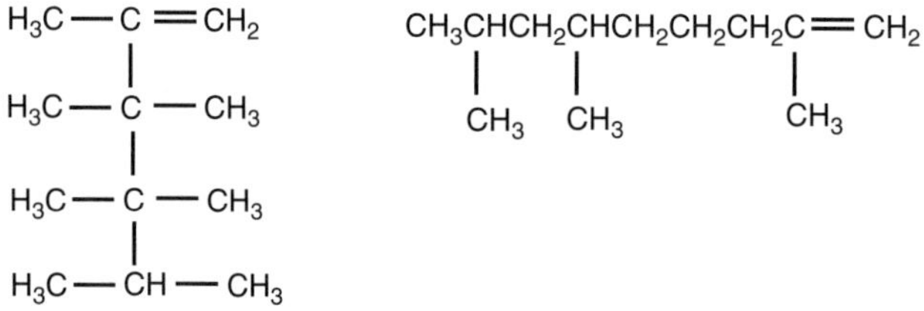

FIGURE 2.3 Various structures for propylene tetramer ($C_{12}H_{24}$).

alkylbenzene usage from BAB to LAB because of nonbiodegradability issues involving BAB. The use of BAB has been discontinued in all of the economically developed regions of the world, and its use in developing regions has declined substantially. As a result, LAB usage has increased at the expense of BAB and also due to the increase in overall detergent consumption. This is especially true in the developing regions of the world. LAB production is projected to grow at an average annual rate of about 4%. At the end of 2006, the world production of LAB was little over 3.5 million metric tons per annum. BAB production has fallen from 246,000 metric tons per annum in 1995 to 196,000 metric tons per annum in 1999. BAB demand is expected to drop to 78,000 metric tons per annum by 2010.

2.1.3 BRANCHED ALKYLBENZENES PRODUCTION

In many instances, BAB has been produced at the same locations that also produce LAB. However, the principal raw material is different. To produce BAB, propylene tetramer (tetrapropylene) is used as the principal olefinic raw material to alkylate with benzene. Propylene tetramer is produced at two types of plants: those dedicated to propylene oligomerization and those plants which have flexible oligomerization capability for mixtures of propylenes, butylenes, and amylenes. Propylene tetramer is a highly branched material and has many structures, some of which are shown in Figure 2.3. The alkylation catalyst used to produce BAB has exclusively been hydrofluoric (HF) acid.

2.1.4 Linear Alkylbenzenes Production

There are various processes for the alkylation of benzene to produce LAB, depending on the feedstock that is used as the alkylation agent. In almost all cases, much like that for the production of BAB, benzene is alkylated with a higher-molecular-weight linear-olefin in the carbon range of C_{10}–C_{13}. The olefin feedstock consists of α-olefins, where the double bond of the olefin is in the *alpha* or primary position between the first and the second carbon atoms in the linear chain. Alternatively, it can be composed of an equilibrium distribution of both alpha and *internal* olefins, where the double bond can be located anywhere along the carbon chain. Paraffin chlorination followed by benzene alkylation is another process for LAB production in which C_{10}–C_{13} carbon range normal paraffins (*n*-paraffins) are the feedstock. During the early days of LAB production, paraffin chlorination followed by alkylation gained some prominence. However, with the emergence of *n*-paraffin dehydrogenation technology and its superior economics and product quality in the late 1960s, the use of paraffin chlorination method has since fallen to the point where only one plant based on this method is still operating worldwide.

2.1.5 Alkylation

Regardless of the type of linear olefin used as the feedstock, there are three different types of alkylation processes currently available to produce LAB. The differences between the alkylation processes are determined by the type of catalyst used. Until 1990, the most widely used alkylation process employed HF acid as the Lewis-type catalyst. In this process, liquid HF acid is used to catalyze the alkylation of benzene with the linear olefins. A second process involves the use of an aluminum chloride ($AlCl_3$) slurry as the catalyst. The presence of water with both HF acid and $AlCl_3$ acts as a co-catalyst to transform each of these catalysts into active species. The third and newest commercial alkylation process to produce LAB uses a solid bed catalyst, which eliminates the safety, handling, and disposal issues that are otherwise inherent with the other two processes.

2.1.5.1 Hydrofluoric Alkylation

As of 2006, ~80% of the LAB manufactured throughout the world was produced using HF alkylation. A typical HF alkylation flow scheme is shown in Figure 2.4. This figure shows a two-reactor system. The first-stage reactor completes the major part of the alkylation reaction. In the second-stage reactor, the last traces of unsaturated hydrocarbons react, and a large portion of the soluble polyaromatics produced as by-products is removed. Units with lower diene-containing feedstocks can utilize a single reactor design.

The olefinic feedstock and benzene are charged to the first-stage reactor. The reaction occurs in the liquid phase at atmospheric pressure and 50°C. Acid strength is maintained between 80 and 90% of HF acid. The first-stage reactor effluent flows to the first-stage settler where the acid and hydrocarbon phases are separated. Most of the acid phase from the first-stage settler is returned to the first-stage reactor inlet. A portion of the acid from the first-stage settler is directed to the HF regenerator where the acid is taken overhead and undesirable heavy alkylate components (polymer), which are soluble in the HF regenerator feed acid, are removed as a bottom product. The regenerator also removes excess water that may have entered the process with one of the hydrocarbon feedstocks. The water forms a constant boiling azeotropic mixture (CBM) that is removed with the polymer in the bottom product. After phase separation of the bottom product, the CBM is removed and neutralized for safe disposal. The polymer phase is also washed with lime or caustic solution to remove residual HF acid and is typically used as a fuel in either a specially designed burner in one of the furnaces used in the HF process or another process unit located nearby.

The hydrocarbon phase from the first-stage settler is sent to the second-stage reactor where it is mixed with more acid. The effluent of this reactor flows to the second-stage settler where the acid

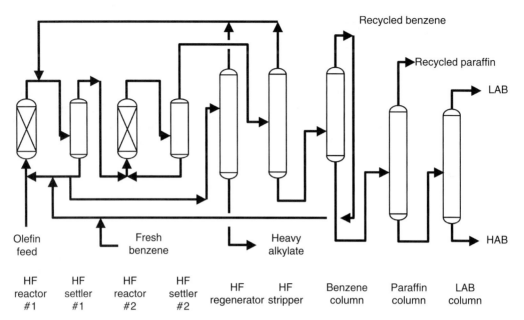

FIGURE 2.4 Typical HF alkylation process flow scheme.

and hydrocarbon phases are again separated. The acid phase is then returned to the second-stage reactor inlet.

Hydrocarbon from the second-stage settler is sent to the HF stripper where essentially all the HF and some of the benzene are taken overhead. These streams are then returned to the reactor section. The acid-free bottom is charged to the benzene column where all the benzene is removed overhead. A recycle benzene stream is taken from the benzene column overhead system and returned to the first-stage reactor. If required, a bleed or drag stream can be withdrawn from the recycle stream to maintain benzene purity. However, in most cases, the recycled benzene is of sufficient quality so that a drag stream is not required.

Depending on the feedstock used, there may be additional fractionation columns included within the HF alkylation unit. For example, when linear paraffins are fed to a dehydrogenation unit upstream of the HF alkylation unit, there are unreacted linear paraffins contained within the mixture of linear olefins. To fractionate and recover these linear paraffins, the benzene column bottom is sent to a paraffin column where the unreacted paraffins go overhead and returned to the dehydrogenation unit. The alkylated products in the paraffin column bottom are then charged to a rerun column where the LAB is recovered from the overhead system. A heavy alkylate by-product (HAB) is removed as bottom product.

2.1.5.2 AlCl$_3$ Alkylation

As of 2006, ~2% of LAB manufactured throughout the world was produced using AlCl$_3$ alkylation. A typical AlCl$_3$ alkylation flow scheme is shown in Figure 2.5.

Benzene, olefinic feedstock, and AlCl$_3$ catalyst are all added to the alkylation reactor. The alkylation reactor effluent passes to a settler, where catalyst sludge is separated out in a settler and is largely recycled. Makeup AlCl$_3$ is added to maintain a high activity level in the heterogeneous catalytic phase of the reactor. This heterogeneous catalytic phase causes any polyalkylate formed to be transalkylated, raising the yield of LAB and decreasing the amount of heavy alkylate produced. Hydrochloric acid, which is formed during the alkylation reaction, is recovered as an aqueous solution and either sold, if possible, as a by-product or neutralized and disposed off by standard methods.

Detergent Alkylate and Detergent Olefins Production

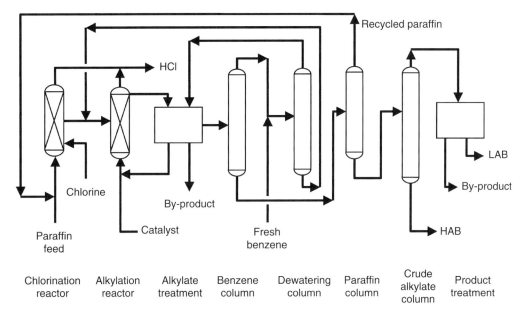

FIGURE 2.5 Typical AlCl$_3$ alkylation process flow scheme.

The LAB product is recovered by distillation. Typically, the distillation scheme is similar to the fractionation section of an HF alkylation unit. Benzene is recovered and recycled along with n-paraffins if n-paraffins are used as the feedstock. The HAB is separated in the third column and can be sold. Alternatively, the heavy alkylate can be recycled to the LAB reactor, where it undergoes the transalkylation reaction to LAB. The spent catalyst, typically as a 25% AlCl$_3$ solution, can be sold for use typically in water treatment or neutralized as necessary and sent for disposal.

2.1.5.3 UOP/CEPSA Solid Bed Catalyst Alkylation

The Detal™ process, jointly developed by UOP and Compania Espanola de Petroleos, S.A., uses a solid heterogeneous acid catalyst. Since its commercial introduction in 1995, 75% of all new unit capacity additions have been based on this process technology. As of 2006, the installed capacity of Detal process units represents ~18% of the 3.5 million metric tons per annum total world production of LAB; and by 2011, it is expected to surpass 22% of the world production capacity. A typical flow scheme for the Detal process is shown in Figure 2.6.

In the Detal process, a solid catalyst is used instead of liquid HF acid catalyst or an AlCl$_3$ catalyst in slurry. The feedstock combines with makeup and recycled benzene and flows through a reactor, which contains a fixed bed of solid catalyst. Reactor effluent then flows directly to a fractionation section that is essentially the same as that used in the HF or AlCl$_3$ alkylation processes. As is the case for the other alkylation processes, the number of distillation columns in the fractionation section for the Detal process is determined in large part by the feedstock used. The standard fractionation scheme consists of columns to (i) separate and recycle benzene, (ii) separate and recycle paraffins, and (iii) finally separate the LAB product from the HAB.

The Detal process is much simpler in design and construction because by-product neutralization and waste-stream treatment are not required. The elimination of equipment otherwise needed as part of the other two processes and use of much simpler metallurgy due to the lack of a corrosive environment results in a plant that is safe and easy to operate, as well as having much lower maintenance costs.

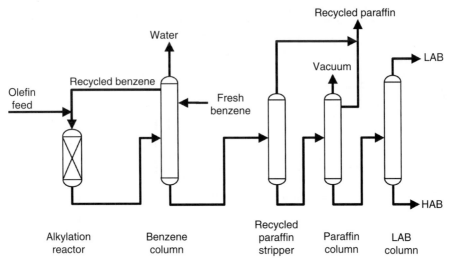

FIGURE 2.6 Detal process flow scheme.

2.1.6 ALKYLBENZENE QUALITY

Because alkylbenzene product is typically sulfonated to produce a surfactant used in detergent formulations, the quality of the alkylbenzene product must be compatible with the sulfonation processes that are currently practiced. Depending on whether LAB or BAB is produced, the primary measure of the alkylbenzene quality is the extent of either linearity or branching, respectively, as a component of the homolog mixture of the product. Another significant measure of the alkylbenzene quality is a lack of color. The lack of color allows the detergent formulator to use the alkylbenzene without the concern of adversely affecting the color of the final formulation. Whether the detergent product is either a powder or a liquid detergent, the formulator desires a colorless alkylate with excellent color stability. Otherwise, a costly bleaching step of the sulfonated alkylate is required.

In addition, there are principally four other quality measures which are significant for LAB as currently used in detergent formulations. The first additional measure of LAB quality is the average molecular weight (AMW). Increasing the AMW increases the surface activity and solution viscosity of the sulfonated LAB surfactant, which is called linear alkylbenzene sulfonate (LAS); however, increasing AMW also decreases water solubility and water hardness tolerance. The reverse is true when AMW is decreased. Optimal AMW depends largely on conditions of the final surfactant used in the detergent formulation such as washing temperature, water hardness, and detergent concentration. The most typical molecular weight range for LAB is between 230 and 250.

The second additional measure of LAB quality is the concentration of 1,4-dialkyltetralin. The presence of the dialkyltetralin by-product can be either positive or negative in terms of LAB quality, depending on what other formulation properties are desired of the LAS. If high solubility is a desired property, then a modest amount of dialkyltetralin is desired, because less or no additional hydrotrope formulation additive is required to achieve desired solubility and viscosity. However, if maximum biodegradability is preferred, then little or no concentration of dialkyltetralin is desirable. The amount of dialkyltetralin contained in the LAB product is largely determined by its production process. The LAB produced through the Detal process or HF alkylation process typically contains <1 wt% dialkyltetralin, whereas those produced through the $AlCl_3$ process may contain between 5 and 10 wt% dialkyltetralin.

The third measurement of the quality of the LAB product is the bromine index (BI). BI is a measure of the extent of double-bond unsaturation in the alkyl chain of the LAB molecule. Unsaturated double bonds tend to cause poor color during the sulfonation and thus correlate with the sulfonation

Detergent Alkylate and Detergent Olefins Production

TABLE 2.1
LAB Properties for Various Alkylation Processes

	Detal Process	HF Alkylation	AlCl₃ Alkylation
1,4-Dialkyltetralins (wt%)	< 0.5	0.5	9
BI	< 5	5	10
2-Phenyl LAB content (wt%)	26	18	29

color of the LAS product. A very low BI of <30, and preferably <10, is desired for the LAB product.

The fourth LAB quality measure is the distribution of phenyl substitution on the alkyl chain, which can influence solubility and viscosity. A higher concentration of the 2-phenyl isomer generally improves the solubility of the resultant LAS surfactant. Much like the concentration of dialkyltetralin present in the LAB, the concentration of the 2-phenyl isomer in the LAB is largely determined by the particular LAB production process. Table 2.1 compares these three additional LAB quality measures as they typically result from the different alkylation processes.

2.2 HIGHER *n*-OLEFINS FOR SURFACTANT PRODUCTION

2.2.1 APPLICATIONS

Higher *n*-olefins of C_8–C_{14} are used as intermediates in the manufacture of several types of surfactant materials. Linear internal olefins are used in the production of linear alkylbenzene; alkylphenol; detergent alcohols, which in turn is used to produce alcohol sulfates, alcohol ethoxylates, and alcohol ether sulfates; and synthetic lubricants. α-Olefins are used in the production of detergent alcohols, linear alkylbenzene, synthetic lubricants, and α-olefin sulfonates (another ionic surfactant).

2.2.2 INTERNAL OLEFINS PRODUCTION

Because of its cost effectiveness, internal olefins are the preferred feedstock for the alkylation of benzene to produce LAB. Most internal olefins are produced by the dehydrogenation of *n*-paraffins, which are extracted from kerosene by a selective adsorption process.

A kerosene fraction with an initial and a final boiling point of 190 and 250°C, respectively, along with high paraffin content will typically contain 20% or more of *n*-paraffins in the C_{11}–C_{16} chain length. The kerosene feedstock is generally hydrotreated to remove polar compounds, which would poison the downstream adsorbents used in the separation process. The commercially important separation processes all use molecular sieve adsorbents to separate *n*-paraffins from branched or cyclic paraffins through size exclusion. The separation process can be done in the vapor phase or liquid phase. The *n*-paraffins enter the molecular sieve channels, and the branched and cyclic hydrocarbons are excluded from the channels. The desorption and regeneration of the molecular sieve adsorbent is accomplished by using a competing desorbent. The desired *n*-paraffins are separated and isolated by fractionation and catalytically dehydrogenated to their corresponding *n*-olefins. The principal process employed involves passing diluted feedstock over a proprietary catalyst to affect an equilibrium mixture of *n*-paraffins, *n*-olefins, and hydrogen. By-product light ends and hydrogen are removed by fractionation, and the *n*-olefin product is then alkylated. The unreacted *n*-paraffins are recycled and eventually processed to extinction to *n*-olefins on continuous recycling.

When high-purity linear internal olefins are desired, as would be the case for the production of detergent alcohols or synthetic lubes, one takes the dehydrogenation product after the removal of light-cracked hydrocarbon fraction to an adsorptive separation process. In this case, olefins are separated from the mixture of paraffins and olefins, the paraffins are recycled for dehydrogenation to its ultimate conversion to olefins. Linear olefins of 95% purity can be obtained. This process is known as the VOP Olex™ process and its operation is similar to that of the VOP Molex™ process described in Section 2.3.2.

2.2.3 α-Olefin Production

With the exception of butene-1, which is derived from refinery feedstocks, linear α-olefins (LAO) are primarily derived by the oligomerization of ethylene feeds. By this method, a wide spectrum of LAO products having even-numbered carbon chain lengths is produced. The production of α-olefins dates back to the 1960s when several producers introduced technology based on Ziegler chemistry using ethylene and a trialkyl aluminum catalyst to build up molecules of α-olefins having only an even number of carbon atoms. α-Olefins based on Ziegler chemistry are produced employing proprietary processes by Chevron Phillips and BP.

Another process, the Shell Higher Olefins Process (SHOP), differs from the trialkyl aluminum process in that it uses a nickel-phosphine ligand and provides for the separation of the C_6–C_{18} α-olefins for downstream use. The high- and low-end fractions are subject to further processing to yield internal olefins. The process is carried out at moderate temperatures, but high pressures.

Sasol produces C_5–C_6 α-olefins from coal-derived synthesis gas processing followed by olefin extraction by fractionation. Sasol has announced that they will extend the process to also recover higher carbon number α-olefins. The extraction process has the ability to further recover C_8 α-olefins and there is the potential to also produce C_{12}–C_{18} α-olefins by the same process.

2.2.4 Market Demand

The higher α-olefin market is driven by the demand for C_8 α-olefins for use as co-monomers in polyolefins production, including high- and low-density polyethylene. There is also a growing demand for the C_{10} fraction for use in poly-α-olefin materials used in synthetic lube formulations. In 1994, only 29,000 MTA of 1.6 million MTA of LAO produced was used for LAB production. This compares with 2 million MTA of n-paraffins used in LAB production.

2.3 n-PARAFFINS FOR LAB PRODUCTION

2.3.1 Linear Alkylbenzenes Production

As discussed earlier, the AMW of LAS has a significant impact on LAS surfactant performance. The AMW of the LAS is determined by the molecular weight of its precursor, LAB, which in turn is dictated by the molecular weight of its components: n-paraffins and benzene. The AMW of LAB that is typically used in producing detergent LAS ranges from 230 to 250. Over this AMW range, surface activity and solution viscosity of the corresponding LAS increase with increasing molecular weight, but decrease with respect to its water solubility and water hardness tolerance.

Only a narrow molecular weight range of n-paraffins is suitable to meet the LAS performance targets. The desirable n-paraffins, C_{10}–C_{13}, are commonly referred to as detergent-range n-paraffins. The C_{10}–C_{13} n-paraffin molecular weight range is controlled between 154 and 174 and, consequently, gives the LAB product of the desired molecular weight. Currently, ~96% of LAB produced worldwide is derived from n-paraffins, which are dehydrogenated and then used to alkylate benzene. The remaining 4% of LAB produced is derived from α-olefins or internal olefins based on ethylene. There are three major processes for detergent-range n-paraffin production: liquid-phase extraction from kerosene, vapor-phase extraction from kerosene, and the Fisher–Tropsch process. The following sections explore the various production processes for n-paraffins.

Detergent Alkylate and Detergent Olefins Production 47

2.3.2 LIQUID- AND VAPOR-PHASE EXTRACTION

The liquid- and vapor-phase processes used in the extraction of detergent-range n-paraffins from kerosene have many similarities and significant differences. Both processes utilize calcium A zeolite with a controlled pore size to perform a molecular sieving operation on kerosene feed. Kerosene consists of n-paraffins and non-normals such as isoparaffins, cyclic paraffins, and aromatics. The adsorbent pores are sized to allow the straight-chained n-paraffins to pass through the pores and enter internal crystal cavities. Non-normal hydrocarbons have larger molecular diameters than the pore openings, which exclude them from entering the pores.

In both the liquid- and vapor-phase extraction processes, the kerosene feed is typically prefractionated to narrow the feed to the desired four-carbon number range (either C_{10}–C_{13} or C_{11}–C_{14}). This heartcut is hydrotreated to remove the majority of kerosene contaminants that may compromise the performance of life of the adsorbent or subsequent quality of the LAB or LAS properties. In some process flow schemes, the fractionation into the discrete n-paraffin cuts may be deferred until after the extraction process.

Liquid-phase extraction is an isothermal and continuous or steady-state operation. In the liquid-phase extraction process, known as the VOP Molex process, feed and product enter and leave the adsorbent chamber at essentially constant flows and compositions. In contrast, vapor-phase extraction is a cyclic process consisting of the following three steps: (i) vapor-phase adsorption of the n-paraffins, (ii) copurge with vapor phase to purge interstitial voids, and (iii) n-paraffin desorption with a vapor-phase desorbent. This operating sequence is integrated into a continuous process by cyclic use of the multiple adsorbent beds. At any given time, one or more beds may be in the adsorption step, whereas other beds are in purge and desorption steps.

Both processes utilize a desorbent to remove the adsorbed n-paraffins from the pores of the adsorbent. Once displaced from the adsorbent, both processes use some form of fractionation to separate the desorbent from the n-paraffin product. In both cases, this desorbent is recycled back to the process. In the case of liquid-phase extraction, the desorbent consists of n-pentane; whereas in vapor-phase extraction, hexane or ammonia desorbents are employed.

The liquid-phase extraction process operates at 177°C and 24.6 kg/cm^2 (g), whereas the vapor-phase extraction process operates at 310–350°C and 2.2 kg/cm^2 (a). Most importantly, the n-paraffin product from either process can surpass 99% purity and attain paraffin recoveries exceeding 95%.

Approximately 62% of n-paraffin production worldwide is from liquid-phase extraction, whereas another 33% is from vapor-phase extraction.

2.3.3 FISHER–TROPSCH PROCESS

The remaining 5% or so detergent-range n-paraffin produced worldwide are derived from natural gas reacted with syngas (a mixture of carbon monoxide and hydrogen) to create a wide range of heavy paraffins. This product is further processed and fractionated into the desired detergent-range n-paraffin.

2.4 FUTURE TRENDS AND TECHNOLOGY DEVELOPMENT

As discussed earlier, the predominant technology for the production of LAB is n-paraffin dehydrogenation followed by HF alkylation. However, with the recent commercialization of the solid bed alkylation process and its inherent advantages, it is anticipated that all future LAB complexes will utilize the solid bed alkylation process.

The average age of the HF and $AlCl_3$-based alkylation units is increasing since the first units were built in the 1960s. It is anticipated that some of these older units will be eventually converted to the solid bed alkylation process.

Of all the surfactants used in the detergent industry, LAS continue to be affordable at the lowest cost. As such, it is anticipated that LAB will continue to see a healthy worldwide growth rate.

The growth rate is the highest (4–5%) in the developing regions of Asia Pacific and is expected to continue well beyond 2010. Meanwhile, BAB (except for in some parts of Africa) has been phased out.

Technology development for the production of alkylates has been largely focused on LAB with most of the development in processes that utilize kerosene as the raw material to produce *n*-paraffin feedstock. Research and development efforts have been focusing on the process technology, as well as the catalysts and adsorbents used in the processes.

Efforts have been made to maximize the yield of LAB from its feedstocks, improve the quality of LAB, and minimize the production of its by-products such as heavy alkylates and acidic polymeric materials. This has been accomplished by eliminating the diolefins and aromatics from the feed to the alkylation unit. The removal of diolefins has been practiced widely in the industry, whereas the technology for the removal of aromatics has only become available recently and its use is increasing.

Every few years, a new catalyst for *n*-paraffin dehydrogenation or new adsorbent for the extraction of *n*-paraffins from kerosene is developed. Generally, an improvement in the dehydrogenation catalyst focuses on catalyst stability and selectivity at higher conversion, whereas improvement in the adsorbent for the extraction of *n*-paraffins focuses on increasing the ability of the adsorbent to recover more *n*-paraffins at a faster rate.

Although the technology for BAB is dormant due to its continually declining use, the technology for LAB production continues to evolve with the goal of making LAB an even more economical raw material. Since the early 1990s, various naturally derived alcohol-based surfactants have been threatening LAS market share. Some substitution has taken place in certain markets, but this number has been small. The future of LAB depends on its relative pricing versus detergent alcohols, as well as relative cost–benefit considerations. With the low cost of LAB maintained through better technology, it is unlikely that large-scale substitution of LAB will take place.

REFERENCES

Berna, J.L. et al. Growth and developments in LAB technologies: 30 years of innovation and more to come. *World Surfactant Conference*, Montreaux, September 23, 1993.

Bozzano, A. *Handbook of Petroleum Refining Processes*. 3rd Edition, McGraw-Hill, New York, pp. 1.57–1.67, 2004.

Chem Systems, Inc., PERP Report No. 93510—Linear Alkylbenzenes, Chem Systems, Inc., August 1995.

Cox, M.F. and Smith, D.L. Effect of LAB composition on LAS performance. *INFORM*, 8(1), 19, 1997.

Erickson, L.C., et al. New solid-bed alkylation technology for LAB. *CESIO—4th World Surfactants Conference*, Barcelona, June 3–7, 1996.

Greer, D. et al. Advances in the manufacture of linear alkylbenzene. *CESIO—6th World Surfactants Conference*, Berlin, June 20–23, 2004.

Imai, T., Kocal, J.A. and Vora, B.V. Solid bed alkylation process for LAB production. *Science and Technology in Catalysis*, pp. 339–342, 1994.

3 Production and Economics of Alkylphenols, Alkylphenolethoxylates, and Their Raw Materials

Anson Roy Grover

CONTENTS

3.1 Introduction 49
3.2 Nomenclature of Alkylphenols 50
3.3 Major Alkylphenols Used in Surfactant Applications 50
    3.3.1 Introduction 50
3.4 Processes of Alkylphenol Production 52
    3.4.1 The Reaction Mechanism 52
    3.4.2 Reaction Steps and Reactor Configurations 54
    3.4.3 Product Isolation Steps 57
    3.4.4 Physical Properties 58
3.5 Economics of Alkylphenol Manufacture 60
3.6 Alkylphenol Ethoxylates and Their Raw Materials 61
    3.6.1 Introduction 61
    3.6.2 Raw Materials for Alkylphenol Ethoxylates 63
        3.6.2.1 Phenol 63
        3.6.2.2 The Olefins 65
        3.6.2.3 Discussion 65
3.7 Environmental Considerations, Product Research, and the Future 66
    3.7.1 Product Research 67
References 67

3.1 INTRODUCTION

An alkylphenol is a derivative of phenol wherein one or more of the aromatic ring hydrogens have been replaced by (substituted with) an alkyl group. The alkylphenols of greatest commercial importance have alkyl groups ranging in size from 1 to 12 carbons. Those alkylphenols having alkyl groups containing four or more carbons are produced in a straightforward manner by the direct reaction (i.e., alkylation) of phenol with the corresponding alkene (olefin) by acid or aluminum catalysis followed by a number of purification steps to isolate the desired product. Depending on the olefin used, specific reaction conditions, mole ratio of phenol to olefin, and catalyst employed, a wide variety of mono-, di-, and trisubstituted alkylphenols may be produced in this general manner [1, pp. 113–114]. Annual worldwide production of all alkyphenols exceeds

49

1 billion pounds [1, pp. 135–136]. Of this, monosubstituted alkylphenols make up ~80 to 85%. It is this type, especially those having alkyl groups of 8 to 12 carbon atoms and predominant in the manufacture of nonionic surfactants, which is the focus of this chapter.

3.2 NOMENCLATURE OF ALKYLPHENOLS

In the Chemical Abstracts Service (CAS) indexing system, phenol is the heading parent and the appropriate names of alkylphenols can thus be derived by using common practices.

For monosubstituted alkylphenols, the position of the alkyl group is designated either with a numerical locant (i.e., 2, 3, or 4) relative to the aromatic ring hydroxyl function or by use of the terms *ortho*, *meta*, or *para* and the alkyl side chain typically retains a trivial name. Although the names of the alkylphenols, which are generated in this manner, are unambiguous and refer to specific compounds, they may become lengthy and cumbersome to use. Therefore, common names are more often used, especially for those alkylphenols, which have gained commercial importance. Thus, each of the names 4-(1,1,3,3-tetramethylbutyl) phenol, 4-*tert*-octylphenol, or *p-tert*-octylphenol refer to the following monosubstituted structure with either of the last two being the most commonly used in commercial industry (Structure 3.1).

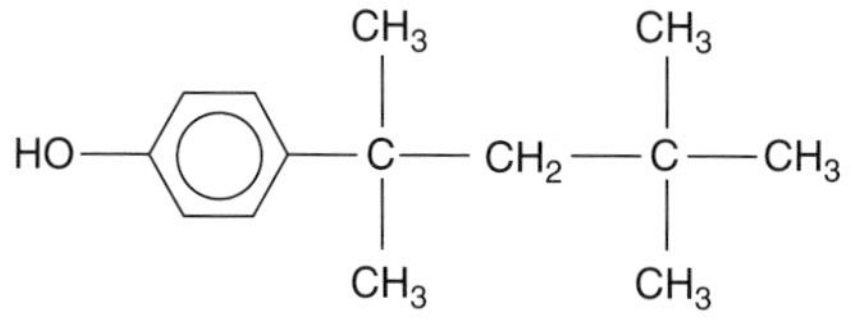

STRUCTURE 3.1

Disubstituted alkylphenols most commonly employ two numerical locants to designate the positions of each alkyl group on the aromatic ring. Thus 2,6-*bis* (1-methylpropyl) phenol or 2,6-di-*sec*-butylphenol (2,6-DSBP) refer to the following disubstituted structure (Structure 3.2):

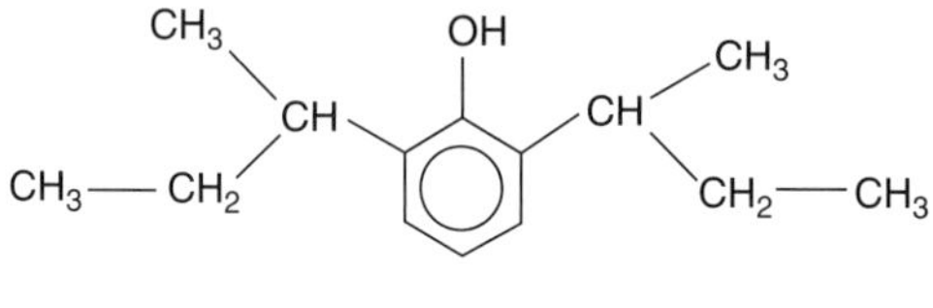

STRUCTURE 3.2

3.3 MAJOR ALKYLPHENOLS USED IN SURFACTANT APPLICATIONS

3.3.1 INTRODUCTION

The worldwide annual production of all alkylphenols exceeds 1 billion pounds. The direct use of alkylphenols is, however, limited to only a few minor applications. The vast majority of alkylphenols are used as intermediates in the synthesis of derivatives, which (in turn) have a wide range of applications ranging from surfactants to pharmaceuticals. The principal markets for alkylphenols are nonionic surfactants, lube-oil additives, phenolic resins, polymer additives, and agrochemicals. The alkylphenols that are most significant in surfactant and applications, along with their general mode of manufacture and other major commercial applications are

1. Alkylation of phenol with nonene (propylene trimer and, therefore, it has a branched-chain alkyl group) produces 4-nonylphenol. It is principally used in the manufacture of a wide variety of surfactants but also of compounds used as antioxidants for rubber, other

Production and Economics of Alkylphenols and Alkylphenolethoxylates   **51**

polymers, and flame retardants; it is also used in the manufacture of resins for the production of carbonless copy paper. Nonylphenol accounts for ~80% of the total alkylphenol usage in the manufacture of nonionic surfactants. It is available as a homologic mixture in a liquid form and generally sold in bulk shipments of 5000 gallon tank wagons or up to 25,000 gallon railcars [1, pp. 135–136;2] (Structure 3.3).

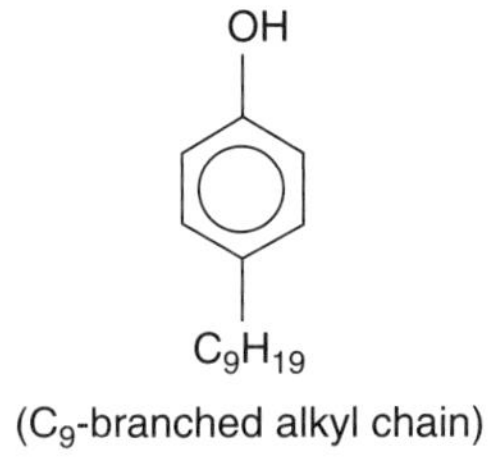

C_9H_{19}
(C₉-branched alkyl chain)

STRUCTURE 3.3

2. Alkylation of phenol with diisolbutylene produces 4-*tert*-octylphenol. It is used in the manufacture of specialty surfactants and resins and compounds also for rubber industry tackifiers and ultraviolet (UV) stabilizers. It is available in either flaked or molten form and generally sold in 25 kg bags, bulk shipments of 5000 gallon tank wagons, or up to 25,000 gallon railcars [1, pp. 136–138;2] (Structure 3.4).

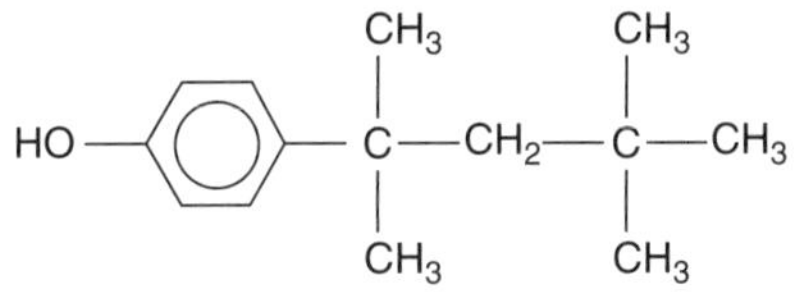

STRUCTURE 3.4

3. Alkylation of phenol with dodecene (propylene tetramer) produces 4-dodecylphenol. It is most commonly used in the manufacture of compounds for a wide variety of lube-oil additives but also of specialty surfactants and demulsifiers. It is available in liquid form and generally sold in 55 gallon drums, bulk shipments of 5000 gallon tank wagons, or up to 25,000 gallon railcars [1, p. 133;2] (Structure 3.5).

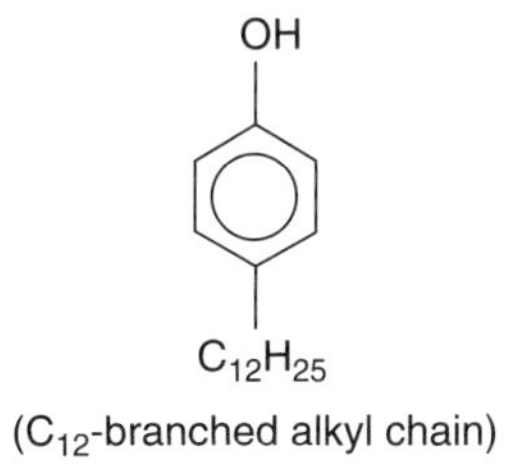

$C_{12}H_{25}$
(C₁₂-branched alkyl chain)

STRUCTURE 3.5

4. Dialkylation of phenol with butene produces DSBP. It is used in the manufacture of specialty surfactants used as wetting agents for agrochemicals (i.e., Structures 3.6 and 3.7). It is available in liquid form and is generally sold in 55 gallon drums or bulk shipments of 5000 gallon tank wagons [1, p. 140;2].

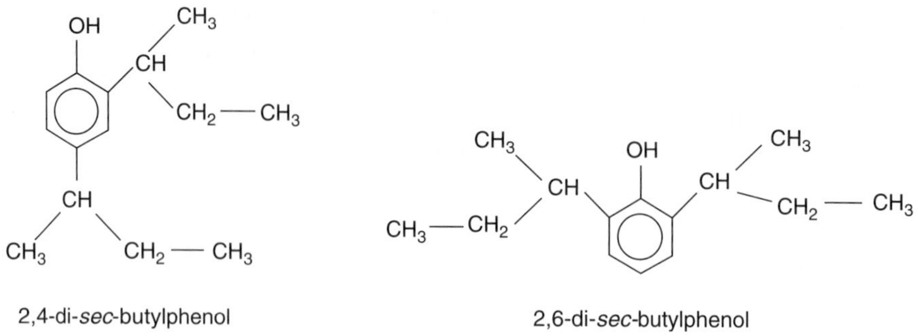

STRUCTURES 3.6 AND 3.7

2,4-di-*sec*-butylphenol

2,6-di-*sec*-butylphenol

5. Dialkylation of phenol with nonene (propylene trimer) produces di-nonylphenol. It is used in the manufacture of specialty surfactants. The volume of this product is extremely limited. The latter (i.e., Structures 3.8 and 3.9) has been extensively used within surfactant mixtures (Structures 3.8 and 3.9) aimed for laundry detergents in the late 1960s.

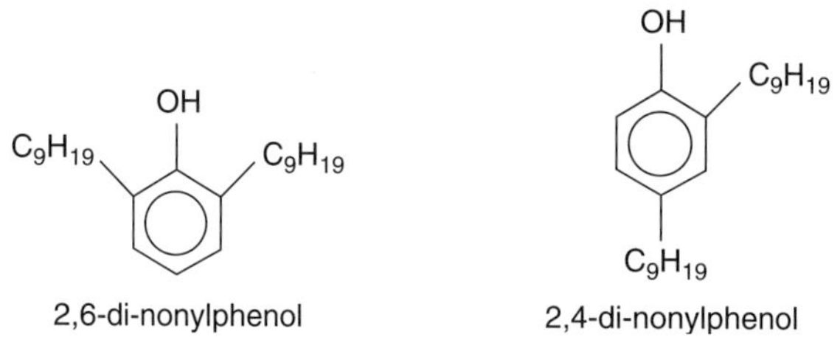

2,6-di-nonylphenol

2,4-di-nonylphenol

STRUCTURES 3.8 AND 3.9

3.4 PROCESSES OF ALKYLPHENOL PRODUCTION

3.4.1 THE REACTION MECHANISM

The approach used to synthesize most of the commercially available alkylphenols is Friedel–Crafts alkylation. The specific procedure typically uses an alkene as the alkylating agent and an acid catalyst, generally a sulfonic acid. Scheme 3.1 shows the interaction of the alkene and catalyst to form a carbonium ion (a), which interacts with phenol to form a Π complex (b), which in turn rearranges resulting in the desired monoalkylation of phenol (c).

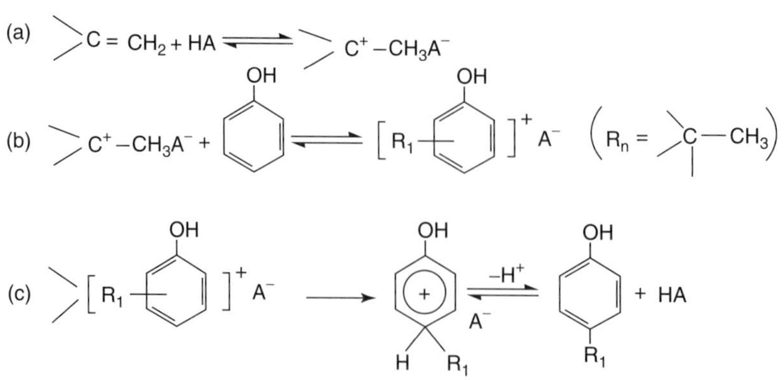

SCHEME 3.1

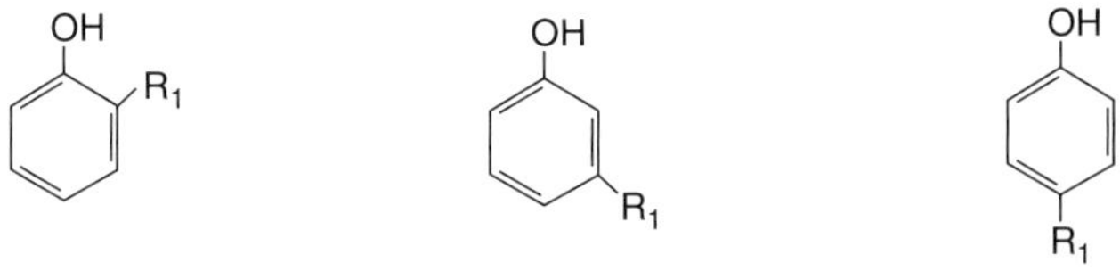

Production and Economics of Alkylphenols and Alkylphenolethoxylates

SCHEME 3.2

Several by-products may be formed in the course of the alkylation reaction. First, the alkylation will yield three different ring-positioned isomers, that is, ortho (a), meta (b), and para (c) as shown in Scheme 3.2.

Under normal, relatively mild reaction conditions and with a tertiary olefin as the alkylating agent, the formation of the para-substituted structure is favored over ortho substitution by approximately 10:1 ratio. This is mainly due to steric effects. Under relatively the same conditions but using a primary or secondary olefin, ortho substitution is favored over a para substitution by approximately 2:1 ratio. The incidence of meta-substitution-deactivated position is low in either case and the resulting meta-substituted product is generally present in the reaction mass at very low concentrations. Second, if the driving force for the formation of the Π complex is thought of as electrostatic interaction (i.e., an electron-deficient species being attracted to an electron-rich species), this type of interaction will explain some of the other by-products that are formed. For example, phenylalkyl ethers (Equation 3.1) are the result of an interaction between the unshared electron pairs of phenol oxygen and carbonium ion. Also, an alkene oligomer may form when the carbonium ion complexes with the filled Π orbital of another alkene. The resulting alkene oligomer can in turn react with phenol to yield a correspondingly higher-molecular-weight alkylphenol (Equation 3.2).

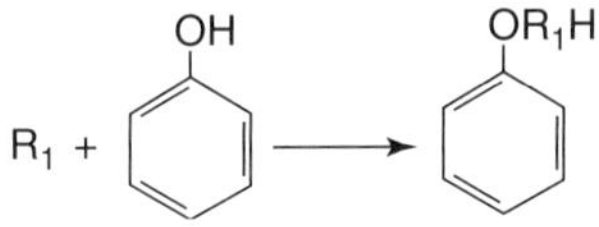

$$\text{(3.1)}$$

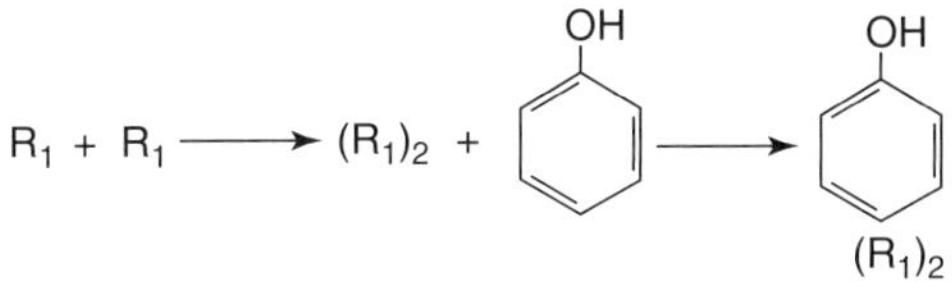

$$\text{(3.2)}$$

Finally, multisubstitution products may also be formed where localized high concentrations of the alkene relative to phenol exist (Equation 3.3).

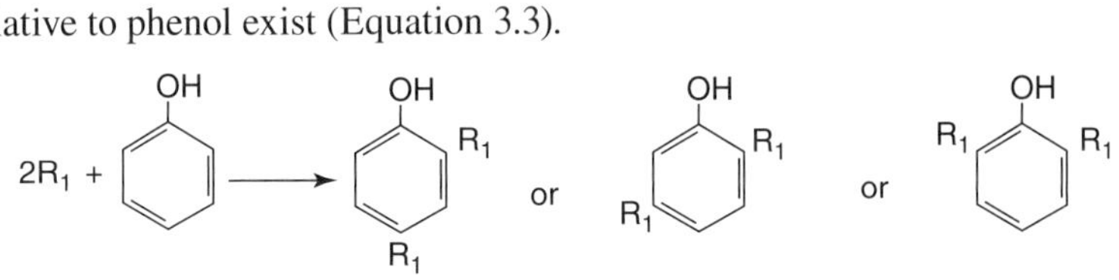

$$\text{(3.3)}$$

Thus, the alkylation reaction step may yield some of each of the aforementioned by-products, and the resulting mixture of desired product and undesired by-products presents a challenge in the isolation of the desired product at the required purity.

Although the choice of catalyst, reaction conditions, and reactor design all allow control over by-product formation and isomer selectivity in an acid-catalyzed alkylation reaction; the type of the alkylating agent is also a major contributor in isomer selectivity as already discussed. A para-substituted, monoalkylated final product accounts for nearly 85% of all commercial alkylphenol manufacture.

The development of the means to selectively alkylate phenol in the ortho position with primary, secondary, or even tertiary alkenes using aluminum catalysis was of great significance to

54   Handbook of Detergents/Part F: Production

the alkylphenol industry. This catalyst type has now made various di-substituted, ortho–ortho type dialkylphenols available. Although these type of products have recently assumed a great commercial importance, they are not generally utilized in the subsequent manufacture of alkylphenol surfactants and hence is not elaborated further in this chapter [2].

3.4.2 REACTION STEPS AND REACTOR CONFIGURATIONS

A wide choice of reactor design exists in the manufacture of alkylphenols. Several factors to be considered before selection of the specific reactor design are

1. Type of catalyst to be utilized
2. Size of the plant and type of downstream equipment for product recovery
3. Desired versatility of the plant, including range of olefins to be processed

A brief discussion of the impact of these factors on reactor type along with a description of various reactor designs used in industry is given in the following. The two factors inherent in all alkylation reactor designs, however, are good mixing characteristics and excellent heat removal capability.

Good mixing is required for contacting the alkene and catalyst with phenol as well as minimizing locally high alkene concentrations. Since alkene–alkene reactions do compete with the desired phenol–alkene reactions, good mixing will favor the desired reaction over undesired alkene oligimer or disubstitution by-product formation.

Excellent heat removal capability is needed to maintain a controlled reaction temperature as the alkylation reaction is highly exothermic. For example, the reaction between 1 mole of phenol and 1 mole of isobutylene to yield 1 mole of $p\text{-}tert$-butyl phenol liberates ~19.1 kcal [1, p. 125]. In an adiabatic system, this reaction if initiated at 40°C would liberate sufficient heat to raise the temperature of the reaction mass to ~250°C. Temperatures above 200°C are considered to be unacceptably high since the control of undesired by-product formation (reaction selectively) is greatly affected by temperature and, in the case of a heterogeneous catalyst system, the protection of the catalyst integrity is difficult under such circumstances.

Broadly speaking, there are two classifications of acid-type-alkylation catalysts—homogeneous (i.e., soluble in the reaction medium) and heterogeneous (i.e., insoluble in the reaction medium). Examples of the former are used in sulphonic acids, boron triflouride, aluminiumphenoxide, methanesulfonic acid, and toluene-xylene sulfonic acid. Examples of the latter are cationic-exchange resins, acidic clays, and modified zeolites. Although the homogeneous catalysts are relatively inexpensive, neutralization of the alkylate is required as an additional processing step before product purification, thereby leading to an additional capital investment plus an additional operating cost to recover and dispose of the products of the neutralization in an environmentally friendly manner. By using a heterogeneous catalyst, therefore, the neutralization step and most of the waste disposal costs are avoided. Heterogeneous catalysts are often used in a continuous flow, fixed-bed-type reactor. A simple batch-type reactor with adequate internal agitation and a means of discharging the raw alkylate while still retaining the catalyst within the reactor is, however, a viable alternative with the use of a heterogeneous catalyst.

The simple batch reactor generally consists of a cooled, agitated mixing tank designed for operation at ambient pressure. Figure 3.1 shows a type of a simple batch reactor arrangement. There are four basic operating steps for this type of reactor used in alkylphenol service.

1. The desired quantity of phenol is loaded.
2. The homogeneous-type catalyst is loaded (if a heterogeneous-type catalyst is used, the catalyst remains in the tank from the previous reaction and hence this step is avoided).
3. The alkene is loaded at such a rate that the desired reaction temperature is maintained by not exceeding the reactor's heat removal capability.

Production and Economics of Alkylphenols and Alkylphenolethoxylates

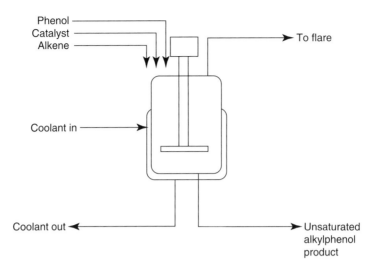

FIGURE 3.1 Flow sheet of simple batch reactor.

4. The alkylate is discharged at the completion of the desired alkylation. If a homogeneous catalyst is used, neutralization of the raw alkylate will follow before further processing. In the case of a heterogeneous catalyst, separation of the catalyst from the raw alkylate through some method of filtration is required.

The simple batch reactor has a relatively low capital cost and, of course, fits well where the downstream equipment for product recovery already exists in batch design. It also provides relatively wide versatility regarding the range of capacity of plant operation. The controlled feed of alkene into an environment of a molar excess of phenol generally leads to a relatively low by-product formation as long as temperature control is adequate. However, the inability to operate the simple tank-type reactor under pressure limits its use to the higher-molecular-weight alkenes, that is, C8 and above. These reactors are generally operated at a final mole ratio of phenol to alkene of approximately 1:1, since this results in the maximum production of alkylphenol for a given reactor volume.

The complex batch reactor is a specialized pressure vessel with both excellent heat transfer and gas–liquid contacting capability. As a result, the lower-molecular-weight olefins (i.e., C4) may be processed, higher concentrations of olefins may be utilized, and the more complex di- and tri-alkylations may be carried out in addition to more efficient monoalkylations. Despite a significantly higher investment as compared to the simple batch reactor, the more varied and powerful capabilities of the complex batch reactor are causing them to become more common in the industry. Figure 3.2 illustrates a flow sheet of the complex batch reactor.

The operating steps are the same as for the simple batch reactor. However, a pressure letdown step to a flare may be included at the completion of the reaction to burn off unreacted olefin or trace hydrocarbons. These reactors can operate at phenol to alkene mole ratios as low as 0.3:1 by designing for positive pressures in the range 30–300 psig. The use of more highly selective catalyst systems for ortho alkylations is also ideal with this reactor type.

Among continuous reactors, the most widely used for production of para-substituted monoalkylphenols is a fixed-bed reactor holding a solid acid catalyst, most generally the cation form of an ion-exchange resin. Figure 3.3 depicts an example of this type of reactor.

A continuous flow of phenol and alkene in the desired mole ratio is premixed and heated to the desired feed temperature before being fed to the reactor. Within the reactor, the reactants contact the solid but porous acid-bearing catalyst so that the desired alkylation takes place at the fixed-acid sites. The raw reaction mixture exits the continuous flow reactor in a neutral condition; hence, no additional neutralization step is required before further processing. A key design consideration for

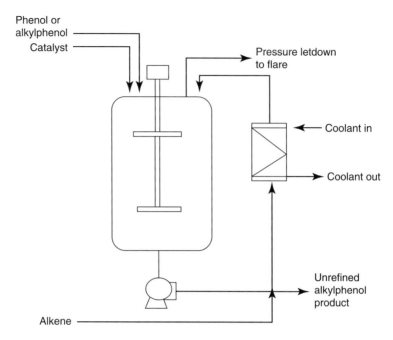

FIGURE 3.2 Flow sheet of complex batch reactor.

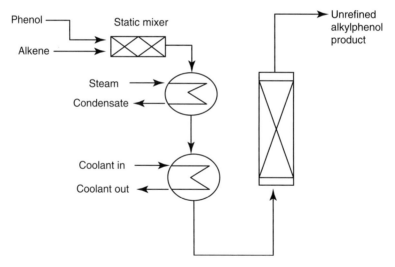

FIGURE 3.3 Flow sheet of continuous reactor.

this type of reactor is the removal of the heat of reaction. This may be accomplished by means of internal (e.g., shell- and tube-type design) or external cooling by rapid circulation of the catalyst-free reaction mass through an external heat exchanger or, alternatively, a diluent. Another important design consideration is the configuration and support of the fixed-bed catalyst from the standpoints of avoiding excessive pressure drop or channeling through it and eliminating solid catalyst carryover with the exiting reaction stream. Although this reactor type is generally restricted to para-substituted or monoalkylated products, it can process a wide variety of olefins from the C4 type to the long, branched-chain type (i.e., C12 and higher). It is an excellent choice for the economies of large-scale, multimillion pound per year continuous plants. Once designed, the versatility of a wide range of operating parameters is, however, limited from the standpoints of flow rate through the fixed-bed catalyst and heat removal. Also, the capabilities of the downstream product purification equipment are generally designed to "fit" the reactor capacity design so that it may limit production expansion efforts. A further consideration in the choice of this reactor type is the necessary allowance for the downtime required in spent catalyst removal and commensurate costs for proper environmental disposal of the same.

3.4.3 PRODUCT ISOLATION STEPS

Although there are few downstream applications (e.g., the production of alkylphenol-formaldehyde-type resins) that require little (if any) refining of the raw alkylate; generally, the majority of commercial alkylphenols are manufactured and sold at purities ranging from 90 to >99.5%, depending on the product and end-use application to which the product is being offered. Achieving these levels of purities require a series of postreaction refining steps. Table 3.1 summarizes some commercially available typical product purities.

The most common method of product isolation and refinement is the removal of impurities through a series of consecutive distillation steps. The final product may be recovered as an overhead stream or from the bottom of the distillation tower depending on the specific alkylphenol or market needs. Figure 3.4 illustrates a typical, continuous flow, monoalkylphenol distillation train.

TABLE 3.1
Typical Final Product Purities of Alkylphenols

Product Name	Minimum Purity (%)	Comment
p-tert-Octylphenol	99	
p-Nonylphenol	90–95	Two grades offered
p-Dodecylphenol	90–97	Two grades offered
2,6-DSBP	95	
Di-nonylphenol	90	

Source: Data taken from Schenectady International Inc., *Technical Data Sheets and/or Catalog (Alkylatated Phenols and Derivatives)*, Schenectady International Inc., New York.

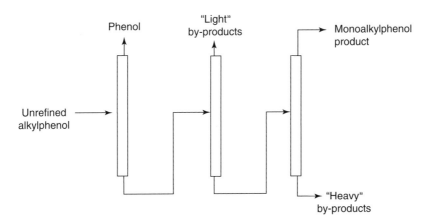

FIGURE 3.4 Monoalkylphenol distillation train.

In the earlier illustration, raw alkylate is fed to the first tower wherein the excess phenol is removed overhead for recycling to the reaction system. The by-products that are less volatile than phenol, but more volatile than the desired product (e.g., separation of the ortho-substituted isomer from the para), are removed as overhead from the second tower. The product is then removed as overhead from the third tower whereas less volatile by-products (e.g., di-substituted by-product and other by-products formed from the oligomerization of the alkene and subsequent alkylation with phenol) are removed as tower bottoms. A batch distillation of raw alkylate is similar in processing sequence with the various separations described earlier, which are taken as separate "cuts."

Proper design of these distillation trains is vital to maintain a proper balance between the operating conditions necessary to achieve the desired product quality and yet avoid thermal dealkylation of the contents of the reboiler. Hence, essentially all alkylphenol distillation systems operate under vacuum to maintain reasonable reboiler temperatures (200–350°C). Design for low-pressure drop within the distillation column operating under high vacuum conditions is especially vital for the longer chain or di-substituted alkylphenols wherein the volatility of the product is low requiring high operating temperatures. There are some products of such low volatility (generally those produced from C20 alkenes and above) in which the product cannot be economically taken as a distillation column overhead and is recovered as a bottoms.

Finished product is stored in stainless steel or phenolic-lined carbon steel tanks, generally blanketed with nitrogen, all in an effort to avoid product discoloration associated with iron-catalyzed contact and oxidation. Narrow-range temperature control of the contents of the tank is vital to maintain the product in the molten state in case of those products with high freeze point (e.g., *p-tert*-butyl phenol) or a pumpable viscosity in the case of a long, branched-chain product (e.g., nonylphenol) while preventing thermal dealkylation.

Alkylphenols are shipped in large volume in liquid form by railcar, tank wagon, or export tank-container, all of which must be constructed of stainless steel or phenolic-lined carbon steel and capable of being heated for product discharge at the customer end. For smaller volumes of low freeze-point products, shipment is made in drums (usually 55 gallon) or specially designed tote tanks. For smaller volumes of high freeze-point products, shipment as a flaked material in bags (usually 25 kg) or larger volume supersacks is common.

3.4.4 PHYSICAL PROPERTIES

The physical properties of alkylphenols, although comparable to phenol, are most strongly influenced by the type of alkyl substituent, its branching, their number and position on the aromatic ring. The melting points of para-substituted products go through a maximum for *tert*-butyl phenol and then decrease. Longer, branched-chain alkylphenols (e.g., nonyl or dodecyl phenols) are viscous liquids at 25°C. As the carbon chain of the alkyl group surpasses 20, the resulting phenol takes on a waxy form. Alkylphenols, especially when di- or tri-substituted tend to supercool. Alkylphenols have similar sensitivity to oxidation as phenol with the presence of trace amounts of metals or alkaline impurities accelerating the reaction. The solubility of alkylphenols in water falls off dramatically as the number of carbons attached to the ring increases. However, all are generally soluble in common organic solvents. The ortho-substituted isomer has a higher vapor pressure (lower boiling point) and lower melting point than the para-substituted isomer of the same product. As a class, alkylphenols have a toxicity range of moderate to very mild and are generally regarded as corrosive to skin. In all cases, precautions should be taken during handling, including wearing of appropriate protective clothing. Material safety data sheets (MSDSs) for all alkylphenols are available from the respective manufacturers [2].

The physical propertics and health and safety data for the alkylphenols that are most often used to produce detergents are found in Tables 3.2 and 3.3.

Production and Economics of Alkylphenols and Alkylphenolethoxylates

TABLE 3.2
Physical Properties of Selected Alkylphenols

Name	CAS Registry Number	Molecular Formula	Molecular Weight	Physical Form (25°C)	Boiling Point (°C[a])	Freezing Point (°C)
p-tert-Octylphenol	[140-66-9]	$C_{14}H_{22}O$	206.3	Solid	290	81.0
p-Nonylphenol	[84852-15-3][c]	$C_{15}H_{24}O$	220.3	Liquid	310	<20.0
p-Dodecylphenol	[27193-86-8]	$C_{18}H_{30}O$	262.0	Liquid	334	<20.0
DSBP	[31291-60-8]	$C_{14}H_{22}O$	206.3	Liquid	254–263	<20.0
Di-nonylphenol	[1323-65-5]	$C_{24}H_{42}O$	346.3	Liquid	>350	<20.0

Name	CAS Registry Number	Molecular Formula	Density[b] (g/mL)	Typical Assay	Flash Point (°C[a])	Molten Color (APHA)
p-tert-Octylphenol	[140-66-9]	$C_{14}H_{22}O$	0.940^{25}	90–98	132	200
p-Nonylphenol	[84852-15-3]c	$C_{15}H_{24}O$	0.933^{43}	90–95	146	100
p-Dodecylphenol	[27193-86-8]	$C_{18}H_{30}O$	0.914^{20}	89–95	>100	500
DSBP	[31291-60-8]	$C_{14}H_{22}O$	0.902^{66}	90	127	500
Di-nonylphenol	[1323-65-5]	$C_{24}H_{42}O$	0.913^{66}	90	>150	500

[a] At 101.3 kPa = 1 atm.

[b] At the temperature indicated by the superscript °C.

[c] Mixture, branched chains.

Source: Data taken from *Kirk-Othmer Encyclopedia of Chemical Technology*, Fourth Edition, Vol. 2, Wiley, New York, p. 115, 1992; Schenectady International Inc., *Technical Data Sheets and/or Catalog (Alkylatated Phenols and Derivatives)*, Schenectady International Inc., New York.

TABLE 3.3
Health and Safety Data

Compound Name	RTECS Number[a]	Irritation Data		Oral (rat) LD$_{50}$ (mg/kg)	SLOM (rabbit) LD$_{50}$ (mg/kg)	Dot Skin Corrosive
		Skin (rabbit)	Eye (rabbit)			
p-tert-Octylphenol	SM9625000	Moderate (20 mg/24 h)	Severe (50 µg/24 h)	2160	NA	Yes
p-Nonylphenol	SM563000	Severe (10 mg/24 h)	Severe (50 µg/24 h)	1620	2140	Yes
p-Dodecylphenol	SL367500	Severe (8.0–8.0 mg/24 h)	Moderate (33.3 µg/110)	2140	5000	Yes
2,6-DSBP	SK8225000	Severe (500 mg/24 h)	Severe (50 µg/24 h)	1320	NA	Yes
Di-nonylphenol	NA	NA	NA	NA	NA	Yes

The Toxicity Data columns consist of Oral (rat) LD$_{50}$ (mg/kg), SLOM (rabbit) LD$_{50}$ (mg/kg), and Dot Skin Corrosive.

[a] RTECS = Registry of Toxic Effects of Chemical Substances.

Source: Data taken from *Kirk-Othmer Encyclopedia of Chemical Technology*, Fourth Edition, Vol. 2, Wiley, New York, p. 130, 1992; Schenectady International Inc., *Technical Data Sheets and/or Catalog (Alkylatated Phenols and Derivatives)*, Schenectady International Inc., New York; Registry of Toxic Effects of Chemical Substances National Institute for Occupational Safety and Health, Cincinnati, OH, July 1990.

TABLE 3.4
Stoichiometric Raw Material Usage

Product Type	Phenol Usage (lb/lb)	Olefin Usage (lb/lb)	Olefin Type
Octylphenol[a]	0.4556	0.5444	C8
Nonylphenol	0.4267	0.5733	C9
Dodecylphenol	0.3582	0.6418	C12
Di-butylphenol	0.4556	0.5444	C4
Di-nonylphenol	0.2717	0.7283	C9

[a] Thus, phenol cost ¢/lb × 0.4556 + di-isobutylene cost ¢11b × 0.5444 = raw material cost 11b of octylphenol.

3.5 ECONOMICS OF ALKYLPHENOL MANUFACTURE

The manufacturing cost of alkylphenols includes raw-material cost and non-raw-material variable costs, fixed costs, and plant depreciation. Of these, raw materials are by far the largest single contributor as they typically make up 60–80% of the total manufacturing cost. The price of phenol, of course, is a major factor in alkylphenol manufacture, but the price of the specific olefin involved in each different product assumes an increasingly greater importance as the hydrocarbon chain length increases or as the number of aromatic ring substitutions increases. This is shown in Table 3.4 concerning the products used in detergent formulations.

Because of the importance of raw-material cost and relatively slim product sales margins involved in the more mature market areas (such as detergents), it is necessary to continually seek improvements in raw-material efficiencies. This is accomplished by gaining a greater understanding of the effects of reaction conditions and catalyst type on by-product formation and "homologic" distribution. Furthermore, judicious recovery and reuse of all by-product streams assumes great importance in product cost control. As a result, raw material efficiencies generally exceed 95% in the manufacture of most monoalkylphenols and exceed 90% in the manufacture of the more complicated di- and tri-alkylphenols. The establishment of strong and long-term relationships with major phenol and alkene suppliers is imperative to raw material price and supply stability. Finally, the placement of the alkylphenol plant near the source of the phenol or olefin producers is important to minimize freight costs to the purchaser. Alternatively, the alkylphenol plant can be situated near the consumption location to reduce freight costs for the alkylphenol.

Non-raw-material variable manufacturing costs (i.e., those costs which vary with volume produced) include most significantly utilities and environmental waste control. These costs generally account for ~5% of the total product cost. In this discussion, direct manufacturing labor is treated as a fixed cost as the size of the labor force is considered fixed and hence independent of volume.

Fixed manufacturing costs include manufacturing labor, plant maintenance, and other plant-related costs that do not vary with production volume. These costs, plus corporate services and associated overhead, generally contribute 10–20% of total product cost.

The contribution to total product cost played by depreciation depends on the relationship among original plant investment, plant capacity, and the largeness of the plant. The investment required for a multiproduct alkylphenol plant of moderate size (50–75 million pounds per year) is generally in the range of 25–30 million dollars (as of 1997). Grass roots installation will increase this investment by at least 50%.

Between 1985 and 1996, there has been a considerable consolidation in the alkylphenol industry with the shutting down of older, smaller, less efficient, and environmentally deficient plants. The current trend is to larger size, continuous, single-product plants for low-cost operation. Hence, there are currently fewer alkylphenol producers compared to 10 years ago. Overall capacity versus demand remains well balanced and an overall growth of 3–5% is foreseen over the next several years. The approximate worldwide and U.S. demands for alkylphenols are shown in Table 3.5.

TABLE 3.5
Worldwide and U.S. Demand For Alkylphenols

Product	Worldwide Demand (million lb/year)	U.S. Demand (million lb/year)
p-Octylphenol	55	27
p-Nonylphenol	425	210
p-Dodecylphenol	275	175
DSBP	2	2
Di-nonylphenol	9	3

Source: Data taken from *Kirk-Othmer Encyclopedia of Chemical Technology*, Fourth Edition, Vol. 2, Wiley, New York, p. 131, 1992.

3.6 ALKYLPHENOL ETHOXYLATES AND THEIR RAW MATERIALS

3.6.1 INTRODUCTION

A surface-active agent is a chemical compound, which when dissolved or dispersed in a liquid is preferentially absorbed at a liquid–gas interface giving rise to "lowering" in the surface tension of the liquid and consequently to a number of physicochemical properties of practical value. The molecule of the compound includes at least one group, which has little affinity for water. The designations most commonly used to describe the two dissimilar "ends" of the molecule are hydrophilic (water seeking) and hydrophobic (water repelling). The term *surfactant* is a contraction of the term *surface-active agent*. Surfactants may be classified by ionic type (i.e., anionic, cationic, nonionic, or amphoteric) and structure (i.e., benzenoid or nonbenzenoid).

In nonionic surfactants, the water-solubilizing contribution (i.e., the hydrophilic "end") is supplied by a chain of ethylene oxide groups. Since this polyoxyethylene group can be obtained by the base-catalyzed reaction of ethylene oxide with an organic molecule containing an active hydrogen atom, a wide variety of structures can be easily water-solubilized by ethoxylation.

Generally, the hydrophobic "end" consists of a hydrocarbon chain containing several carbon atoms. In the case of those surfactants based on an alkylphenol, the hydrophobic "end" typically contains 8 to 12 carbons in a branched chain, which provides the best cleansing action (i.e., detergency) to the surfactant system and is provided for by the corresponding alkene attached to the aromatic ring in the para position by alkylation of phenol. Thus the phenol nucleus serves as an excellent intermediate link for easy attachment of both hydrophobic and hydrophilic "ends" as well as for enhancing oil solubility through the attachment of the hydrophilic acid (the branched alkyl group) at the p-position (mainly) to the hydroxyl group in the aromatic nucleus.

The two reactions that generally describes the preceding reactions are shown in the following, with the alkylation of phenol to obtain the hydrophobic "end" shown first (Structure 3.10).

STRUCTURE 3.10

Thus the hydrophilic "end" is obtained through polyethoxylation of the ring hydroxyl by its reaction with ethylene oxide under mild basic conditions. Thus the nonionic surfactants derived from alkyl phenols have the generalized structure as shown in the following (Structure 3.11).

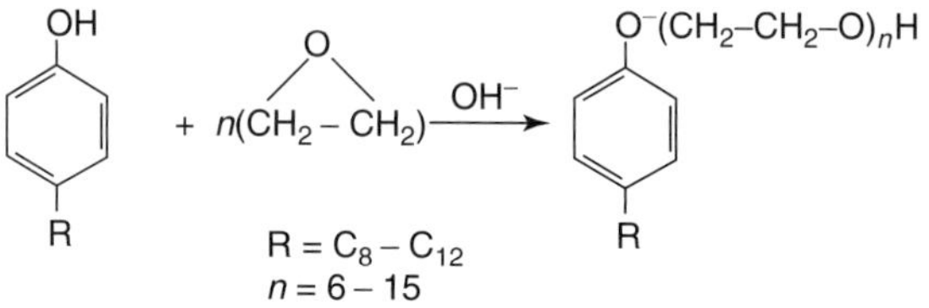

STRUCTURE 3.11

In Structure 3.11, R is the alkyl group containing 8 to 12 carbons and n the number of ethylene oxide units ranging from 1 to 30, most typically 6 to 15. A surfactant based on an ethoxylated alkylphenol falls into the general classification of a nonionic (benzenoid) type. The alkylphenol ethoxylate is most commonly referred to as APE in the industry.

APEs are chemically stable and highly versatile surfactants that find wide application in a large variety of products. The major APE markets are industrial processing (55%), which include emulsion polymerization enhancers, metal surface cleansing formulations, paper pulping dispersants, various functions in the textile manufacture, agricultural chemical emulsifiers or wetting agents for liquid herbicides or pesticides, and oil-well drilling fluids; industrial and institutional cleaners and detergents (30%); and household and personal care products (15%).

The most widely used alkylphenols in the manufacture of nonionic surfactants are described as follows in the order of their importance. APEs derived from p-nonylphenol account approximately 80% of the total market whereas those derived from octyl phenol account for 15–20%. Dodecyl phenol, di-nonylphenol, and DSBP ethoxylates run a poor third at <5%.

1. The alkylphenol p-nonylphenol is represented by a product consisting of a mixture of branched chain, nine-carbon-containing olefins, and phenol yielding the general structure shown in Structure 3.12. The p-PNP-based surfactants are the most versatile, economical, and widely used of the APEs, the most common being that of the 8 mole ethoxylan (i.e., 9 units of EO attached to the PNP molecule). PNP can be converted to phosphate esters or sulfonated to the corresponding sulfate yielding higher performing surfactants. Consumption of PNP-based surfactants is estimated to exceed 400 million lb per year.

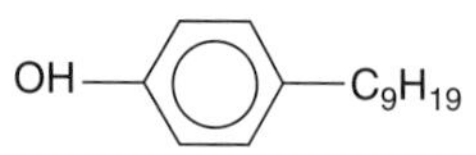

STRUCTURE 3.12

2. The alkylphenol p-$tert$-octyl phenol (PTOP) is the product of a branched-chain, eight-carbon-containing olefins, and phenol yielding the general structure as shown in Structure 3.13. The PTOP-based surfactants are used mainly in the same applications as PNP-based surfactants. In addition, a major use is as a surfactant in the emulsion polymerization of acrylic and vinyl polymers. Consumption of PTOP-based surfactants is believed to exceed 60 million lb per year. These surfactants are considered to be more of a "specialty" type because of the higher raw material cost of the diisobutylene as compared to nonene.

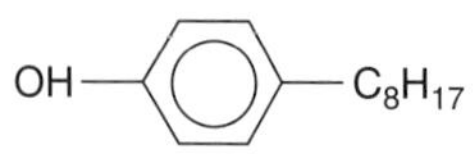

STRUCTURE 3.13

Production and Economics of Alkylphenols and Alkylphenolethoxylates 63

3. Alkylphenol *p*-dodecyl phenol is the product of a mixture of branched-chain, 12-carbon-containing olefins, and phenol yielding the general structure as shown in Structure 3.14. Although the major use of PDDP is in lube-oil additives (primarily through its conversion to a calcium phenolate), a minor, higher purity portion is ethoxylated to produce specialty surfactants.

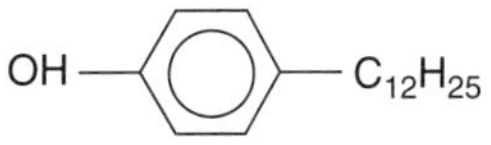

$$ OH - \langle \bigcirc \rangle - C_{12}H_{25} $$

STRUCTURE 3.14

4. Alkylphenols di-nonylphenol and 2,6-DSBP are both building blocks for specialty nonionic surfactants through polyethoxylation for use in specialty niche applications. The only significant usage for DSBP, for example, is in emulsifying liquid agrochemicals (see Structures 3.7 and 3.9 found on p. 52).

In all the cases, the specific alkylphenol used and number of ethylene oxide units (moles) added to it depends on the desired application of the surfactant being produced.

3.6.2 RAW MATERIALS FOR ALKYLPHENOL ETHOXYLATES

The immediate parent raw materials of the APEs are phenol, the corresponding olefin (e.g., diisobutylene, nonene, or dodecene), and ethylene oxide. All are derived from benzene, the hydrocarbon family, and ethylene and propylene, respectively. A short description of each is as follows:

3.6.2.1 Phenol

Worldwide phenol capacity currently exceeds 12.5 billion pounds of which >95% is produced by cumene peroxidation, or more simply put, the cumene process. To produce the precursor cumene, benzene is reacted with propylene, usually in the vapor phase at ~425°F and 400 psi over a supported bed of phosphoric acid catalyst with a 99.5% purity (technical grade) cumene product recovered from unreactive paraffin, excess benzene and undesired by-products by fractional distillation [3,4].

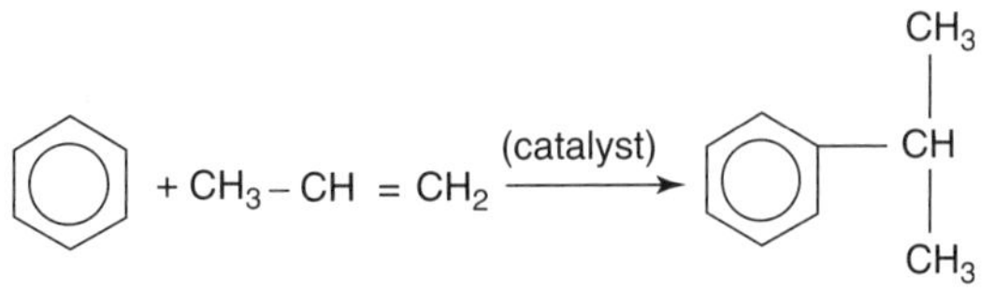

$$ \langle \bigcirc \rangle + CH_3 - CH = CH_2 \xrightarrow{\text{(catalyst)}} \langle \bigcirc \rangle - \underset{\underset{CH_3}{|}}{\overset{\overset{CH_3}{|}}{CH}} \qquad (3.4) $$

By far, the major use for cumene is currently in the manufacture of phenol. The two-step reaction process for manufacture of phenol from cumene first involves air oxidation of an aqueous cumene emulsion at ~250°F. Conversion to the hydroperoxide is controlled at the 25–30% range to maintain the highly exothermic reaction under control. Following the reaction, nitrogen and unreacted oxygen are removed, unreacted cumene is recovered for recycling, and the concentrated cumene hydroperoxide is isolated for the second reaction step.

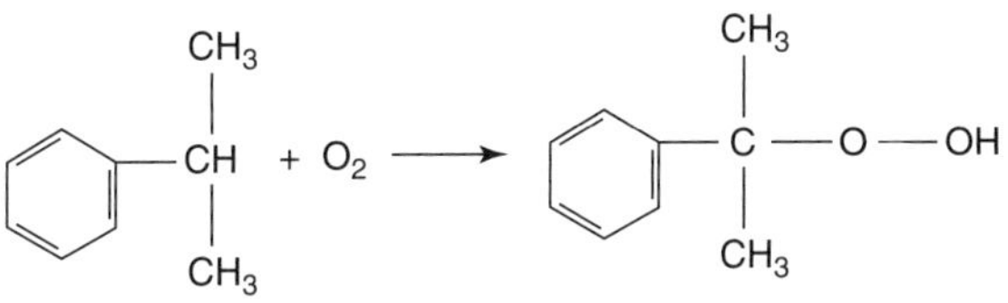

$$ \langle \bigcirc \rangle - \underset{\underset{CH_3}{|}}{\overset{\overset{CH_3}{|}}{CH}} + O_2 \longrightarrow \langle \bigcirc \rangle - \underset{\underset{CH_3}{|}}{\overset{\overset{CH_3}{|}}{C}} - O - OH \qquad (3.5) $$

The hydroperoxide concentrate is fed to a second reactor where a selective decomposition takes place in the presence of dilute sulfuric acid and well-controlled reaction conditions yielding primarily phenol and acetone. The reactor effluent is passed through a separator to remove the water, acid, and salts followed by a series of fractionation steps to isolate both the phenol and acetone from undesired by-products.

$$\underset{\underset{CH_3}{|}}{\overset{\overset{CH_3}{|}}{C_6H_5-C-O-OH^-}} \xrightarrow{H^+} C_6H_5OH + CH_3-\underset{\overset{\|}{O}}{C}-CH_3 \tag{3.6}$$

Two commercially important by-products, which are generally isolated in a "pure" form from the preceding process are α-methyl styrene and acetophenone. Overall process yields are shown in Table 3.6.

Overall, the cumene process for phenol is an economic one, the major drawback being the coproduction of acetone. When the acetone market is down and phenol demand is strong, this presents a supply–demand problem. The reverse situation is rare.

There are 17 major phenol producers worldwide, including PhenolChemie and Enichem in Europe; Allied, General Electric, Aristech, Shell, Georgia Gulf, and Dow in the United States; and Mitsui and Mitsubishi in the far east.

Major end uses for phenol are given in Table 3.7.

Since the manufacture of alkylphenols is of relatively minor importance in the overall phenol supply picture, it is incumbent on the alkylphenol producer to develop firm vendor–customer relationships with major phenol producers to ensure relative price stability and an uninterrupted supply chain during times of phenol shortages. In 1998, the worldwide phenol capacity and demand were in good balance with capacity utilization at about 87%. The U.S. market at about 4.7 billion pounds, however, is currently tighter with capacity utilization approaching 100% and price increasing with the cost of oil from which benzene is derived. A growth rate of 3–4% per year is forecasted over the next several years, driven mostly by growth in bisphenol A usage in polycarbonate resins.

TABLE 3.6
Phenol Cumene Production, the Process, and Material Balance

Step One	Quantity	Yield (%)
Feed Stream[a]		
Benzene	681	—
Propylene	367	—
Product[b]		
Cumene	1000	95%
Step Two		
Feed Stream[a]		
Cumene	1000	—
Oxygen (net)	265	—
Products[b]		
Phenol	725	57
Acetone	442	35

[a] benzene/phenol (lb) = 0.94–0.95.
[b] propylene/phenol (lb) = 0.50–0.51.

TABLE 3.7
World Phenol Demand by End Use

End Use	Percent of Total
Biphenol A	33
Phenolic resins	30
Caprolactam	14
Alkylphenols	5
Analine	4
Other (miscellaneous)	14

3.6.2.2 The Olefins

Both nonene and dodecene are branched-chain olefins obtained by the oligomerization of a propylene-containing feedstock. Nonene (C_9H_{18}) and dodecene ($C_{12}H_{24}$) are also referred to as propylene trimer and propylene tetramer, respectively. Refinery-generated propylene (50–70% propylene in propane) is generally of sufficient quality to be used as feedstock and the most common process, developed by UOP, which involves initiation by a supported phosphoric acid catalyst bed at temperatures ranging from 120 to 225°C. Reaction temperature and feedstock composition determine the range of olefins obtained in the product stream. Subsequent fractional distillation yields the desired product fractions. Six major U.S. companies currently produce either the trimer or the tetramer of both. Nameplate capacity for both chemicals was 1.290 billion pounds per year in 1994 with all the producers supplying the merchant market as well as, in some cases, for captive use [4,5].

The chemical composition of either olefins is an extremely complex mixture of predominately 9 or 12 carbon atom olefins, respectively. Indeed, high-resolution gas chromatographic analysis of *p*-nonylphenol has achieved resolution of 22 distinct para isomers, each alkene moiety identified as being a distinct and different configuration of the compound described for simplicity as propylene trimer or *nonene*. Although such an exhaustive study has not yet been conducted on *p*-dodecyl phenol, results would be expected to be at least as, if not more, complex.

Diisobutylene obtained from the C4 hydrocarbon family is a mixture of 2,4,4-trimethyl-1-pentene and 2,4,4-trimethyl-2-pentene. There is currently only one U.S. producer, two producers in the far east, and two Europe an manufacturers. The cost of diisobutylene is 50–75% higher than the more simply obtained trimer (nonene) or tetramer (dodecene).

3.6.2.3 Discussion

The very basic key starting materials used in APE synthesis are liquefied petroleum gas and crude oil to yield the commodities benzene, propylene, ethylene, and the hydrocarbons. Surfactants represent, however, a relatively minor usage of them. Although approximately 21% of benzene production goes to the manufacture of phenol, only 5% of phenol production finds its way into the production of alkylphenols. The value of benzene, and hence phenol, is therefore set by factors and markets (including the gasoline market) well outside of the surfactant industry. The uses for propylene are several, including approximately one-third for the manufacture of polypropylene. Only ~7 and 4% of propylene goes into the production of cumene (a phenol precursor) and C9 and C12 olefins, respectively. Of the multibillion pound ethylene production, well over one-half is consumed by manufacture of polyethylene with only about 14% finding its way into the manufacture of ethylene oxide. The ethylene oxide is used primarily for the production of ethylene glycol or polyethoxylation of alkylphenols or alcohols in the manufacture of surfactants.

3.7 ENVIRONMENTAL CONSIDERATIONS, PRODUCT RESEARCH, AND THE FUTURE

APEs play an important role in a number of industrial processes—pulp and paper, textiles, coatings, agricultural pesticides, lube oils and fuels, and metals and plastics. APEs are the most effective compounds in nonionic surfactants for industrial, institutional, and commercial cleaning products. In 1990 alone, more than 450 million pounds of APE surfactants were produced in the United States for domestic sale. Industrial applications consists of industrial and institutional cleaning products (30%), household cleaning products (15%), and other uses (<1%).

The most important and commonly used APEs are nonylphenol ethoxylates (NPEs), which account for 80% of the APE market. The remaining 20% of the market consists of octylphenol ethoxylates (OPE) and other lesser-known alkylphenols.

Over the last few years, concerns have been raised about the safety of several industrial chemicals in the environment and their impact on the health of humans and wildlife. Specifically, scientists are studying whether small amounts of APE-based chemicals in the environment are interfering with the endocrine systems of humans and wildlife. The central assumption to this hypothesis is that some industrial chemicals persist or bioaccumulate in the environment and therefore are available for direct or indirect contact with humans and wildlife. It was postulated that exposure to these compounds may be a risk or hazard to the health of humans and wildlife.

During the 1980s, questions arose when a number of peculiar trends were observed in certain wildlife populations. For example, some alligators in Lake Apopka in Florida were born with genital deformities and other reproductive tract problems. It was determined that these deformities may have been the result of a massive pesticide spill as no other deformed alligators were found in the surrounding areas. Other observations included the strange nesting behaviors of female gulls in the Great Lakes and inability of male panthers to reproduce in the Everglades. In trying to explain all of these phenomena, theories (linking industrial chemicals to unexplained animal behavior) began to emerge. Many European governments embraced this issue to explain their environmental problems. The Swedish and Danish governments even went so far as to ban the domestic use of all alkylphenol-based products based on general concerns rather than specific data.

At the same time, others were investigating developmental and reproductive trends in human health including the perceived global fall in sperm counts and an increase in global reproductive cancers. When this group of researchers compared notes with their colleagues observing wildlife populations, some suggested that there might be a connection.

It was during the Wingspread Conference in 1991 that this connection was first introduced. Researchers prepared a list of suspected industrial chemicals based on a set of criteria including bioaccumulaton and persistence, biodegradation, and the potential to affect the endocrine system. Certain alkylphenols and APEs were included on this list, in particular, nonylphenol (PNP) and NPEs, because it was known that these did not readily biodegrade (1993–1997). This assumption was not based on actual test data.

In 1987, U.S. industry, in response to commercial pressures from Europe and growing U.S. Environmental Protection Agency (U.S. EPA) concerns over PNP's aquatic toxicity, formed a panel under the Chemstar program of the Chemical Manufacturer's Association to study and develop an accurate and scientifically sound understanding of the human health and environmental profile of alkylphenols and APEs. This panel, known as the Alkylphenol and Ethoxylates Panel (APE Panel), is an association of manufacturers, processors, users, and raw material suppliers.

Although the debate continued in academic circles, *Assault on the Male*, a British television documentary about the endocrine disruption issue in Europe. As a result of this single act, this issue was transformed from a scientific hypothesis to a media blitz on the general public. Coverage on this grew more slowly, but steadily increased culminating with the significant press-related release of *Our Stolen Future*. This book was authored by Theo Colburn, the World Wildlife Fund biologist, who organized the Wingspread Conference. As a result of the massive publicity efforts surrounding

Production and Economics of Alkylphenols and Alkylphenolethoxylates 67

the book launch, Colburn introduced the American public to this issue. In fact, *Our Stolen Future* has the same effect on the United States as the BBC documentary had on Europe.

Ironically, the APE panel and other researchers had found data to prove many of Colburn's conclusions incorrect at the same time to the book's release, including the following findings:

- There is no scientific evidence that alkylphenols affect the human endocrine system.
- There is no positive correlation between observations in laboratories and alkylphenol concentrations found in the environment.
- Alkylphenols do not persist in the environment. Recent studies of ground water indicate that the "hard" APEs do persist.

These statements are the result of several and varied testing programs, many of which were jointly designed with the help of the U.S. EPA. One important fact remains—industry, government, or academia agree that more research is needed to answer a number of important questions relevant to this and closely related issues (see, e.g., Ref. 7).

3.7.1 PRODUCT RESEARCH

Since the alkylphenol products used in the surfactants industry border on the "commodity" type, margins are slim and reductions in manufacturing costs of the parent alkylphenol are a must to stave off competition, not only from other producers, but also from other nonionic surfactant types. For this reason, process development has become a fruitful area of research for the alkylphenol industry, particularly those areas that improved catalysis or reaction conditions and recovery for recycle (by isomerization or transalkylation) of by-product streams.

Some research has been devoted to find a suitable, low-cost, versatile, and efficient replacement for APEs. The wide range of product benefits, however, makes it difficult to find economically viable alternatives to APE-based surfactants. Furthermore, little is known about the environmental safety and health impact of these alternatives. Thus, no fix economically viable substitute has been found, and in light of the current NPE data, most "worst-case scenario research" has been discontinued.

REFERENCES

1. *Kirk-Othmer Encyclopedia of Chemical Technology*, Fourth Edition, Vol. 2, Wiley, New York, pp. 113–114, 115, 116–118, 125, 130, 131, 133, 135–136, 136–138, 140, 1992.
2. Schenectady International Inc., *Technical Data Sheets and/or Catalog (Alkylatated Phenols and Derivatives)*, Schenectady International Inc., New York.
3. Registry of Toxic Effects of Chemical Substances National Institute for Occupational Safety and Health, Cincinnati, OH, July 1990.
4. Burdick, D. L. and William L., *Petrochemicals in Non-Technical Language*, Leffler, 1990.
5. *Chemical Economics Handbook*, SRI International, Menlo Park, CA.
6. Gigger et al., *Water Res.*, 1999.
7. Toller, U. (Ed.), *Groundwater Contamination and Control*, Marcel Dekker, New York, 1994.

4 Production of Alkyl Glucosides

Jan Varvil, Patrick McCurry, and Carl Pickens

CONTENTS

4.1 Introduction...69
4.2 Raw Materials..69
4.3 Chemistry...71
4.4 Process and Technology..77
4.5 Research and Development...79
4.6 Summary...80
References..80

4.1 INTRODUCTION

Alkyl glycosides or alkyl glucosides are unique nonionic surfactants. Alkyl glycosides are the acetal products from the reactions of alcohols and saccharides, which contain an aldehyde functionality. Alkyl glucosides are the products from the reactions of alcohols and, specifically, the saccharide D-glucose or other glucose sources. The commercial production of alkyl glycosides is primarily the production of alkyl glucosides from a glucose source. This chapter discusses the raw materials, chemistry, process and technology, and research and development of alkyl glucosides.

4.2 RAW MATERIALS

Alkyl glucosides are the acetal products from the reactions of a glucose source and long carbon chain primary alcohols or alcohol mixtures. A strong acid catalyst is also necessary for their production.

Alkyl glucosides can be produced from a number of possible glucose sources. From a commercial standpoint, the initial feedstock for most of the possible glucose sources is cornstarch. Cornstarch is a high-molecular-weight polymer of D-glucose. The chemical bonds between the D-glucose subunits in starch are in fact acetal linkages. Refined cornstarch is obtained from corn by the wet milling process in which corn kernels are separated by a series of water soaking, milling, grinding, screening, centrifugation, and washing procedures into corn oil, fiber, gluten (protein), and starch. If a final saccharide product such as maltodextrins, corn syrups, or D-glucose is desired, the cornstarch is pasted in water at temperatures from ~90 to 160°C and hydrolyzed (acetal hydrolysis) at increasing temperatures by the addition of strong acids or enzymes. The aqueous starch paste may be further hydrolyzed by enzymes at 50–65°C into products containing higher levels of D-glucose monosaccharide. The aqueous saccharide products may be roll dried or spray dried to produce maltodextrins or corn-syrup solids. Anhydrous glucose and glucose monohydrate are obtained by crystallization from high-dextrose corn syrups. A number of possible glucose sources for alkyl glucosides are listed in the following table:[1]

Sources of Glucose

Glucose Product or Source Basis	Water (%)	Composition on a Dry Substance
Anhydrous glucose	<1	>99% D-glucose, remainders are disaccharides, etc.
Glucose monohydrate	9	>99% D-glucose, remainders are disaccharides, etc.
High-dextrose corn-syrup solids	<1	>95% D-glucose, remainders are di-, trisaccharides, etc.
High-dextrose corn syrup	29	>95% D-glucose, remainders are di-, trisaccharides, etc.
Corn-syrup solids	<3	<50% D-glucose, remainders are di-, tri-, tetrasaccharides, etc.
Corn syrups	15–25	<50% D-glucose, remainders are di-, tri-, tetrasaccharides, etc.
Maltodextrins	<5	<5% D-glucose, remainders are di-, tri-, tetrasaccharides, etc.
Cornstarch	11	High-molecular-weight glucose polymer, polysaccharide of D-glucose

The percentage of D-glucose monosaccharide decreases in the order listed in the preceding table. Generally, the price of the glucose sources also decreases in the order listed in the preceding table. The ease of conversion of the glucose sources into alkyl glucosides products also decreases in the order listed in the preceding table due to the decreasing solubility, and the reactivity of the higher saccharides with higher alcohols and the necessity of removing water are greater than stoichiometric amounts.

The alcohols used for the production of commercial alkyl glucoside products are mainly long carbon chains, primary alcohols, typically with about 8–18 carbon atoms. Secondary alcohols are less reactive, and tertiary alcohols undergo dehydration under conditions currently used to produce alkyl glucosides products. Because of economics and an enhanced synergistic effect on the functionality of the alkyl glucosides products, mixtures of alcohols with a range of the number of carbon atoms are commonly used.

There are several possible sources of the alcohols used for alkyl glucosides products. The alcohols may be derived from the triglycerides or tri-fatty acid esters of glycerol in natural oils such as coconut and palm kernel oils. The naturally occurring mixtures of fatty acid esters in triglycerides are converted into fatty acid methyl esters or free fatty acids. If they are converted into free fatty acids, these react with fatty alcohols to form fatty acid esters of fatty alcohols. The fatty acid methyl or fatty alcohol esters are converted into the corresponding alcohols by high-temperature and high-pressure hydrogenolysis in the presence of a copper-containing catalyst. The resulting mixtures of alcohols are separated into the desired final alcohols or alcohol mixtures by fractional distillation. The final products are saturated, linear primary alcohols. They contain even numbers (>99%) of carbon atoms, typically from about 8 to 18 carbon atoms. The alcohols may be synthetically produced from ethylene by the Ziegler process. Initially, triethylaluminum is produced by the reaction of ethylene, hydrogen, and aluminum. Triethylaluminum is reacted with additional ethylene to produce aluminum trialkyls with mixtures of longer, saturated, alkyl groups. The aluminum trialkyls are oxidized with air to form aluminum alkoxides. The aluminum alkoxides are hydrolyzed to produce the corresponding alcohols, which are fractionally distilled. The final products from the Ziegler process are saturated primary alcohols that are essentially linear, although they contain a small percentage of branching, typically <5%. They contain an even number of carbon atoms, typically from about 6 to 20 carbon atoms. The alcohols may be synthetically produced by the oxo process from carbon monoxide and olefins. A mixture of carbon monoxide, an appropriate olefin or olefin mixture, and hydrogen are reacted at high temperature and high pressure in the presence of a cobalt-containing catalyst to produce a mixture of aldehydes having one more carbon atom than the

Production of Alkyl Glucosides

olefin feedstock. The aldehydes will have more branching than the olefin feedstock. The aldehydes are hydrogenated to the corresponding alcohols in the presence of a catalyst in a high-temperature and high-pressure reaction. The alcohols are fractionally distilled. The final products from the oxo process are saturated primary alcohols. The amount of branching and the mix of even and odd carbon atoms in the alcohols depend on the olefin feedstock and specific process. The number of carbon atoms in the products varies from about 7 to 16.[2]

The production of alkyl glucoside products from alcohols and a glucose source also requires the use of a strong acid catalyst. The catalysts used may include sulfuric-, *para*-toluenesulfonic-, sulfosuccinic-, dodecylbenzenesulfonic-, or dinonyinaphthalenesulfonic acid. Sulfuric acid is typically produced by burning sulfur in air to produce sulfur dioxide, catalytically oxidizing the sulfur dioxide to sulfur trioxide at a high temperature, and absorbing the sulfur trioxide in water to form sulfuric acid. Sulfuric acid is typically sold at concentrations of ~93%, the remainder being water. *Para*-toluenesulfonic acid is produced by the sulfonation of toluene with sulfuric acid. The product is typically sold as a 65% aqueous solution or as a crystalline monohydrate. Sulfosuccinic acid is produced by the addition of sodium bisulfite to maleic acid followed by the conversion of the sodium salt into the free acid. The product is sold as an aqueous solution. Dodecylbenzenesulfonic acid is produced by the alkylation of benzene with a mixture of linear, primary alkylchlorides ($AlCl_3$ catalyst) or α-olefins (HF catalyst) that have an average number of about 12 carbon atoms.

The dodecylbenzene mixtures are sulfonated with sulfur trioxide or air or oleum (sulfur trioxide in sulfuric acid) to produce the dodecylbenzenesulfonic acid products. Dodecylbenzenesulfonic acid is sold in concentrated form. Other single or multiple alkylated benzene or naphthalene sulfonic acids are similarly manufactured.

The efficiency of strong acid catalysts for the production of alkyl glucosides products increases as their hydrophobicity or lipophilicity increases. For example, on an equimolar basis, dodecylbenzenesulfonic acid is a more efficient catalyst than *para*-toluenesulfonic acid in that it produces a higher yield of alkyl glucosides with fewer by-products.

4.3 CHEMISTRY

Commercial alkyl glycosides or, more specifically, alkyl glucosides are complex mixtures of acetals. Acetals, in general, are formed by the reaction of aldehydes and alcohols in the presence of a strong acid catalyst. The reaction involves an intermediate known as a hemiacetal. When acetaldehyde is dissolved in an excess of anhydrous ethyl alcohol in the presence of anhydrous hydrogen chloride, the following reactions occur:

$$CH_3C(=O)H + HOCH_2CH_3 \rightleftharpoons CH_3CH(OH)OCH_2CH_3$$

Acetaldehyde   Ethyl alcohol   Ethyl hemiacetal of acetaldehyde

$$CH_3CH(OH)OCH_2CH_3 + HOCH_2CH_3 \underset{}{\overset{H^+}{\rightleftharpoons}} CH_3CH(OCH_2CH_3)_2 + H_2O$$

Ethyl hemiacetal of acetaldehyde   Ethyl alcohol   Diethyl acetal of acetaldehyde   Water

The formation of the hemiacetal can occur with or without the presence of a strong acid catalyst. The reaction is a reversible, equilibrium reaction, and the relative amount of hemiacetal can be increased by increasing the amount of ethyl alcohol. Hemiacetals contain the chemical grouping H–O–CHO–C–. Generally, hemiacetals are not stable enough to be isolated; however, hemiacetals may be quite stable if they are the result of an intramolecular reaction of a compound containing both aldehyde and alcohol functionality that forms a five- or six-membered ring. The formation of a hemiacetal from an aldehyde requires 1 mole of alcohol and forms no water. The subsequent formation of the acetal from the hemiacetal requires the presence of a strong acid catalyst, and forms 1 mole of water. Once again, the reaction is a reversible, equilibrium reaction, and the amount of acetal can be increased by increasing the amount of ethyl alcohol or by removing the water formed. Acetals contain the chemical grouping −CO−CHO−C. Generally, acetals are stable and can be isolated once the acid catalyst has been neutralized. The formation of an acetal from a hemiacetal requires 1 mole of alcohol and forms 1 mole of water. Acetals (especially furanosides) tend to be unstable in aqueous media if a strong acid is present due to their reversion to the alcohols and aldehydes or hemiacetals from which they were formed; however, in the presence of an aqueous base, they are quite stable.[3]

The commercial production of alkyl glycosides is primarily the production of alkyl glucosides from a glucose source. The structure and chemistry of D-glucose is somewhat complex. D-glucose $C_6H_{12}O_6$ is an aldohexose or six-carbon monosaccharide that has both aldehyde and alcohol functionality. It can be structurally represented in several different, but increasingly informative ways, such as the Fischer projection of the free aldehyde form in Figure 4.1; the Fischer projections of the two cyclic, hemiacetal, six-membered ring or pyranose forms in Figure 4.2.

The conformational structures of the two anomers of the pyranose forms of α-D-glucopyranose in Figure 4.3 and β-D glucopyranose in Figure 4.4 are also shown.

In Figure 4.1, D-glucose is represented in the free aldehyde form with the aldehyde group on the C-1 carbon atom and with five free hydroxyl groups on the C-2, C-3, C-4, C-5, and C-6 carbon atoms. The C-2, C-3, C-4, and C-5 carbon atoms are chiral or optically active. Both in solid form and in aqueous solution, D-glucose exists almost totally in the stable, hemiacetal, six-membered ring, or pyranose forms in Figure 4.2, which are formed by the reaction of the aldehyde group and the C-5 hydroxyl group. The formation of the hemiacetal by cyclization creates a new chiral carbon atom at C-1 and the possibility of the two α- and β-D-glucopyranose forms. As represented in the conformational structures of α-D-glucopyranose in Figure 4.3 and β-D-glucopyranose in Figure 4.4, there are two possible isomers or anomers where the hydroxyl group formed at the C-1 carbon atom in Figure 4.2 are respectively axial or equatorial to the six-membered ring. In addition to the α-D-glucopyranose and β-D-glucopyranose forms of D-glucose, the two anomers of the

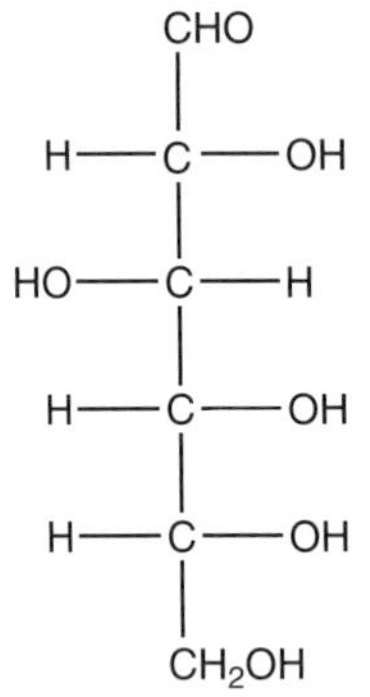

FIGURE 4.1 D-glucose.

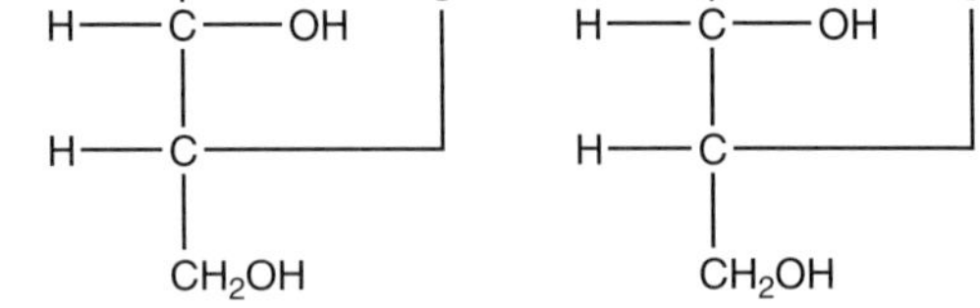

FIGURE 4.2 D-glucose cyclic hemiacetals.

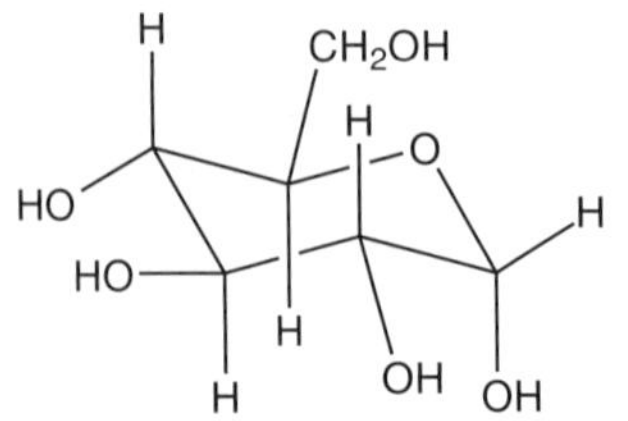

FIGURE 4.3 α-D-glucopyranose.

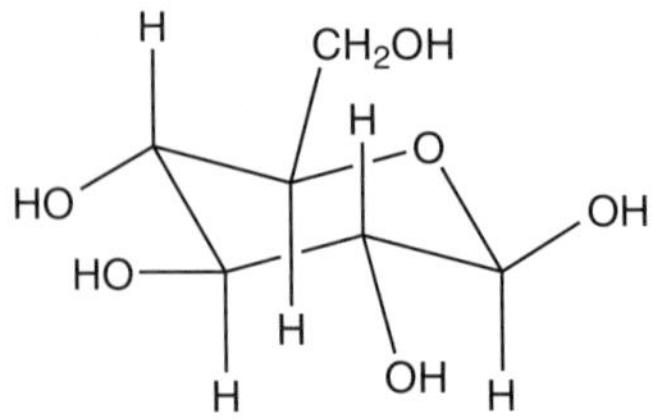

FIGURE 4.4 β-D-glucopyranose.

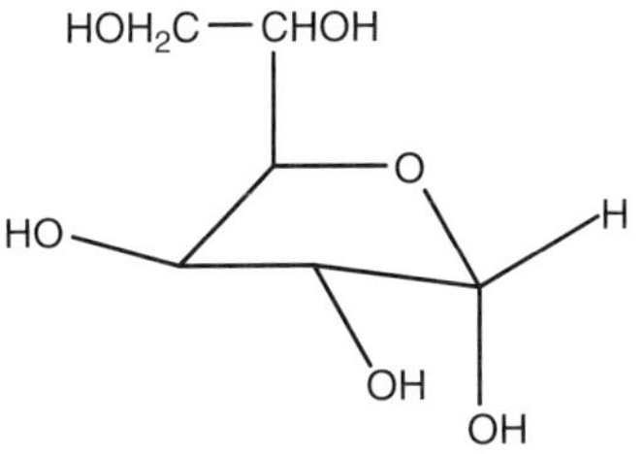

Production of Alkyl Glucosides

73

FIGURE 4.5 α-D-glucofuranose.

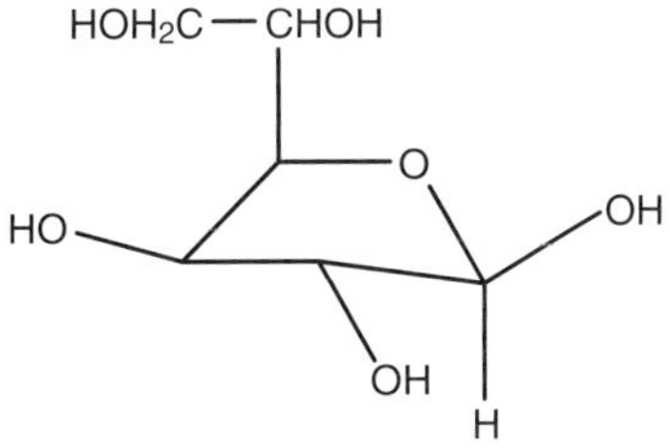

FIGURE 4.6 β-D-glucofuranose.

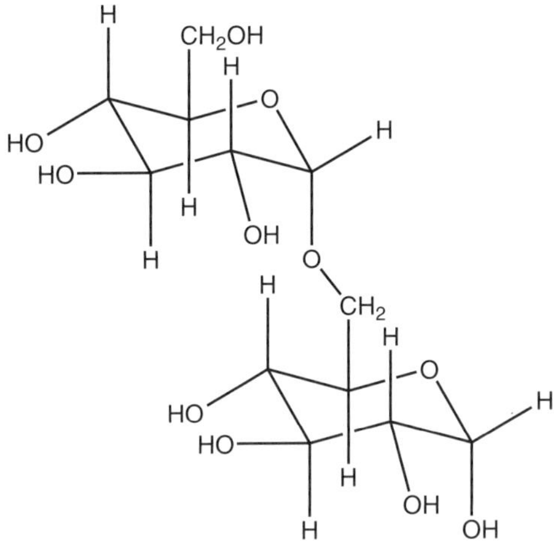

FIGURE 4.7 6-0-(α-D-glucopyranosyl)-α-D-glucopyranose.

cyclic, hemiacetal, five-membered ring, or furanose forms of α-D-glucofuranose in Figure 4.5 and β-D-glucofuranose in Figure 4.6 are also formed by the reaction of the aldehyde group and the C-4 hydroxyl group.

D-glucose in solid form and in aqueous solution contains only very small amounts of the furanoses as shown in Figures 4.5 and 4.6; however, the corresponding alkyl furanosides play a significant role in the chemistry of alkyl glucosides.

Monosaccharides such as D-glucose can also form disaccharides, trisaccharides, etc., by forming acetal linkages between D-glucose molecules. Isomaltose, shown as the 6-0-(α-D-glucopyranosyl)-α-D-glucopyranose anomer in Figure 4.7, is the acetal formed between the C-1 carbon atom of one D-glucose molecule and the hydroxyl of the C-6 carbon atom of a second D-glucose molecule.

To form the acetal, the first D-glucose molecule functioned as an aldehyde or hemiacetal, whereas the second D-glucose molecule functioned as an alcohol. Although the primary hydroxyl of the C-6 carbon atom of the second D-glucose molecule is the most reactive of the four hydroxyls that are not part of its hemiacetal grouping, the secondary hydroxyls on the C-2, C-3, and C-4 carbon atoms have relative reactivities equal to about one-third to one-half of the C-6 hydroxyl. Considering only the pyranoside and pyranose forms of D-glucose, there are four possible disaccharide isomers of D-glucose if the α and β forms of their anomeric carbons are not specified. There are 16 possible disaccharide isomers if the α and β forms of both anomeric carbon atoms are specified. There is also the possibility of each D-glucose existing in the five-membered ring, furanoside or furanose form, which brings the total number of disaccharide isomers of D-glucose to 64. Similarly, trisaccharides, tetrasaccharides, etc., may be formed from D-glucose. It is obvious that the total number of possible isomers for the higher saccharides of D-glucose becomes very large (4 for

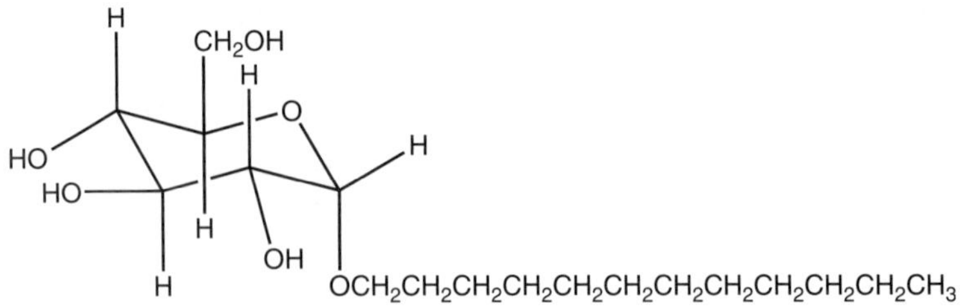

FIGURE 4.8 *n*-Dodecyl α-D-glucopyranoside.

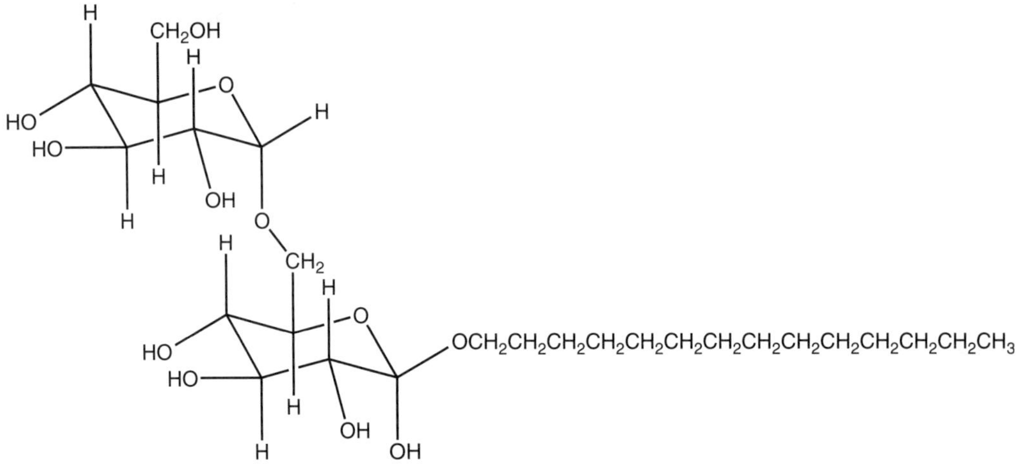

FIGURE 4.9 *n*-Tetradecyl 4-0-(α-D-glucopyranosyl)-β-D-glucopyranoside.

D-glucose, i.e., α-D-glucopyranose, β-D-glucopyranose, α-D-glucofuranose, and β-D-glucofuranose; 64 for the disaccharides; 1,408 for the trisaccharides; 35,840 for the tetrasaccharides; 992,956 for the pentasaccharides; etc.).

The α- and β-D-glucopyranose and α- and β-D-glucofuranose forms of the D-glucose monosaccharides, disaccharides, etc., are present as alkyl acetals or the corresponding α- and β-D-glucopyranosides and α- and β-D-glucofuranosides in alkyl glucosides products. *n*-Dodecyl α-D-glucopyranoside and *n*-tetradecyl 4-0-(α-D-glucopyranosyl)-β-D-glucopyranoside are shown in Figures 4.8 and 4.9, respectively.

n-Dodecyl α-D-glucopyranoside is an acetal from single molecules of D-glucose and *n*-dodecyl alcohol. The degree of polymerization (DP) of D-glucose in *n*-dodecyl α-D-glucopyranoside is one. *n*-Tetradecyl 4-0-(α-D-glucopyranosyl)-β-D-glucopyranoside is the product of two molecules of D-glucose and a single molecule of *n*-tetradecyl alcohol. The DP of D-glucose in *n*-tetradecyl 4-0-(α-D-glucopyranosyl)-β-D-glucopyranoside is two. Similarly, the product from three molecules of D-glucose or a trisaccharide with a single molecule of alcohol would have a DP of three, and so on.

The relative lipophilicity (hydrophobicity) and hydrophilicity (lipophobicity) of commercial alkyl glucosides products are determined primarily by two factors: the number of carbon atoms of the alkyl group or the average number of carbon atoms of the alkyl groups of the glucosides present, the lipophilicity, and the average DP of the glucosides present, the hydrophilicity. The number of carbon atoms of the alkyl group or the average number of carbon atoms of the alkyl groups of the glucosides is normally the same as that of the primary alcohol or primary alcohol mixture that is used for their production. The average DP of the glucosides can be mathematically represented as follows:

$$\text{Average DP} = \frac{1}{100} \sum_{i=1}^{i=\infty} p_i * i$$

In the preceding equation, p_i is the mole percent of the ith alkyl glucoside x-mer, where i is 1 to ∞; p_1 is the mole percent of the total alkyl monoglucoside isomers based on the total alkyl glucosides present, p_2 is the mole percent of the total alkyl diglucoside isomers based on the total alkyl glucosides present, and so on. The term *x-mer*, where x is 1 to ∞, includes monomers, oligomers, and polymers. A 1-mer is a monomer, a 2-mer is a dimer, and so on. Total isomers refer to all of the α and β anomers of the glucopyranosides and glucofuranosides: total alkyl monoglucoside isomers include the α- and β-D-monoglucopyranosides and the α- and β-D-monoglucofuranosides, etc. In the equation, i is the integer value of the DP of D-glucose for each alkyl glucosides x-mer, where i = 1 to ∞; for all alkyl monoglucoside isomers, i = 1; for all alkyl diglucoside isomers, i = 2; and so on.

Typically, the average DP of commercial alkyl glucosides products is between 1.20 and 1.80. Alkyl glucosides have a distribution of x-mers that is determined by probability or statistics. It has been demonstrated by P. M. McCurry that a mathematical description presented by P. J. Flory and modified by S. Erlanger and D. French can be applied to the mole percent distribution of alkyl glucosides x-mers.[4-6] For a given average DP, the mole percent distribution of the x-mers can be determined to produce the results shown in Figure 4.10.

As the average DP increases, the mole percent of monomers decreases. Regardless of the average DP, the monomer is the most prevalent x-mer. For each average DP, the distribution of the mole percent for increasing x-mers monotonically decreases. If the molecular weight of the alcohol or the average molecular weight of the alcohol mixture used to produce the alkyl glucosides product is known, the weight-percent distribution of the alkyl glucosides x-mers can be calculated from the mole percent distribution of the x-mers to produce the results shown in Figure 4.11.

For each average DP and for each alcohol or alcohol mixture and corresponding alkyl group molecular weight or average alkyl group molecular weight, the monomer is the most prevalent x-mer and the weight percent for increasing x-mers monotonically decreases. If the alcohol or alcohol mixture used for a particular alkyl glucosides product and the weight percent of the total alkyl monoglucosides isomers based on the total alkyl glucosides present are known, then the average DP

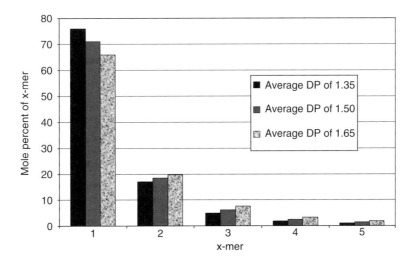

FIGURE 4.10 Mole distribution of x-mers of alkyl glucosides.

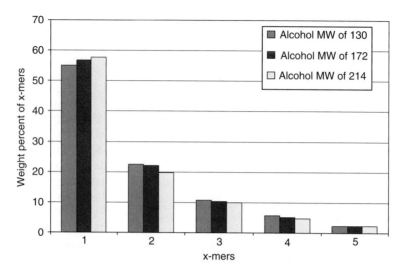

FIGURE 4.11 Weight distribution of x-mers of alkyl glucosides with average DP of 1.50.

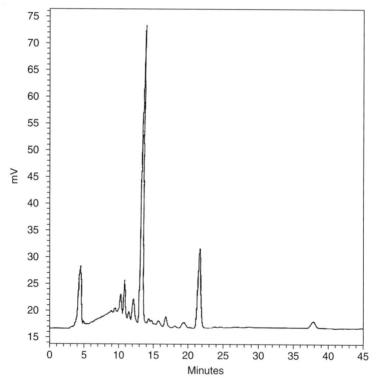

FIGURE 4.12 HPLC chromatogram of dodecyl, tetradecyl, hexadecyl alkyl glucosides product.

is known. The weight percent of the total alkyl monoglucoside isomers of the total alkyl glucosides present in an alkyl glucosides product can be determined by high-performance liquid chromatography (HPLC) or by high-temperature gas chromatography (HTGC). An example of an HPLC chromatogram of an alkyl glucosides product made from a mixture of dodecyl, tetradecyl, and hexadecyl alcohols (70, 25, and 5% by weight, respectively) is shown in Figure 4.12.

Production of Alkyl Glucosides

The largest peaks at ~13, 22, and 38 min are the dodecyl α- and β-D-monoglucopyranosides combined, the tetradecyl α- and β-D-monoglucopyranosides combined, and the hexadecyl α- and β-D-monoglucopyranosides combined.

4.4 PROCESS AND TECHNOLOGY

The commercial production of alkyl glycosides is primarily the production of alkyl glucosides from a primary alcohol or primary alcohol mixture containing about 8–18 carbon atoms and a glucose source. Two of the most common processes for the reaction of an alcohol or alcohol mixture and a glucose source are described as follows:

In the *direct* (or *one-step*) *reaction*, the glucose source reacts directly with an excess of the alcohol or alcohol mixture. The reaction is carried out at ~95 to 115°C under relatively high vacuum (relatively low pressure) in the presence of a strong acid catalyst. The water formed during the acetalization reaction is removed as a distillate. When the reaction is complete, the catalyst is neutralized and the product is recovered from the excess alcohol.

In the *transacetalization* (or *two-step*) *reaction*, the glucose source reacts first with an excess of a lower molecular weight, more polar alcohol than that of the final alkyl glucoside product. A low-molecular-weight alcohol such as butyl alcohol is typically used. The reaction is carried out at ~95 to >140°C under relatively low vacuum to pressures greater than atmospheric pressurein the presence of a strong acid catalyst. The bulk of any free water initially present in the glucose source or formed during acetalization is removed as a distillate. Glucose sources initially containing increasingly higher levels of glucose oligo or polysaccharides require increasingly higher reaction temperatures and pressures. Under appropriate conditions, these glucose sources can be converted by transacetalization into low-DP butyl glucosides. The butyl glucosides, by transacetalization with an excess of a higher-molecular-weight alcohol or alcohol mixture, can then be converted into the final alkyl glucoside product. The reaction conditions for this transacetalization are similar to those of the direct or one-step reaction. During this reaction, the bulk of the low-molecular-weight alcohol initially present as either free or reacted alcohol is removed by distillation. When the reaction is complete, the catalyst is neutralized and the product is recovered from the excess, high-molecular-weight alcohol.

The possible ways of converting the glucose sources discussed earlier into alkyl glucosides products include

Glucose Product or Source	Possible Reaction Processes
Anhydrous glucose	Direct reaction, transacetalization reaction
Glucose monohydrate	Direct reaction (an initial dehydration of the glucose monohydrate is required at an elevated temperature and high vacuum before the direct reaction), transacetalization reaction
High-dextrose corn-syrup solids	Direct reaction, transacetalization reaction
High-dextrose corn syrup	Direct reaction (possible on a continuous basis with lower-molecular-weight alcohols such as mixtures of primary octyl, nonyl, or decyl alcohols), transacetalization reaction
Corn-syrup solids	Direct reaction (possible on a continuous basis with lower-molecular-weight alcohols such as mixtures of primary octyl, nonyl, or decyl alcohols), tranacetalization reaction
Corn syrups	Transacetalization reaction
Maltodextrins	Transacetalization reaction
Cornstarch	Transacetalization reaction

The main factors in determining the way an alkyl glucosides product can be commercially produced are the saccharide distribution, water content of the glucose source, and number of carbon atoms in the alcohol. The higher the D-glucose monosaccharide content and the lower the water content that the glucose source has, the easier it can be converted into alkyl glucosides products. The lower the number of carbon atoms of the alcohol, the easier it is to convert higher saccharides into alkyl glucosides products. In those cases where both direct and transacetalization reactions are possible, the direct reaction is less complicated and requires less time and equipment; but the transacetalization reaction may be more efficient in the conversion of the glucose source, especially the less expensive ones containing higher saccharides, into alkyl glucosides products.

An example of a laboratory direct reaction was done using particulate, anhydrous α-D-glucose and an excess of n-dodecyl alcohol at 110°C and 30 mm of mercury pressure with *para*-toluenesulfonic acid catalyst. The reaction mixture was rapidly stirred to maintain a uniform suspension of the α-D-glucose particles and promote dissolution. Rapid stirring also prevented the accumulation of a portion of the reaction mixture at the bottom of the reactor where the pressure was necessarily higher than the pressure at the top due to the liquid depth of the reaction mixture, which leads to higher, undesirable moisture levels. The water formed during the reaction was removed by distillation. Samples were periodically removed from the reaction mixture, neutralized, and analyzed by gas chromatography for n-dodecyl α-D-monoglucopyranoside, n-dodecyl β-D-monoglucopyranoside, total n-dodecyl (α- and β-) D-monoglucofuranosides, total n-dodecyl diglucosides, total n-dodecyl triglucosides, and total n-dodecyl tetraglucosides. The reaction was terminated by neutralization of the catalyst after 350 min. The results are shown in Figure 4.13.

As the anhydrous α-D-glucose gradually dissolved and reacted in the dodecyl alcohol reaction mixture, monoglucosides began to be formed. Both experimental evidence and mechanistic theory support the fact that among the first kinetic products to be formed are the monoglucofuranosides and that they are then transformed into the more thermodynamically stable α- and β-D-monoglucopyranosides. The monoglucofuranosides reached their highest concentration at ~140 min and then decreased as the α- and β-D-monoglucopyranosides continued to increase.

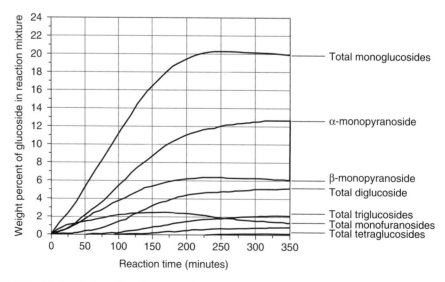

FIGURE 4.13 Dodecyl glucosides direct reaction.

As α-D-glucose continued to dissolve and react, the total monoglucosides increased, and measurable amounts of the diglucosides, followed by the triglucosides and tetraglucosides were formed. The total monoglucosides reached their highest concentration at ~250 min and then decreased as the diglucosides, triglucosides, and tetraglucosides continued to increase. During the reaction, water distillate was initially collected at a slow rate that gradually increased to a maximum and then decreased as there was little undissolved and unreacted glucose remaining. At ~350 min, all the glucose had essentially dissolved and reacted. The reaction was terminated by neutralizing the *para*-toluenesulfonic acid catalyst with sodium hydroxide. As the reaction progressed, the average DP of the total alkyl glucosides, including the higher oligo and polyglucosides (DP > 5) that were not measured, increased. If the reaction mixture had not been neutralized at 350 min, polar by-products and the average DP would continue to increase as the mono and lower oligoglucosides combined to form higher oligo and polyglucosides. At the end of the reaction, the ratio of α-D-monoglucopyranosides to β-D-monoglucopyranosides was slightly greater than 2:1, and the ratio of total monoglucopyranosides to total monoglucofuranosides was about 19:1. These ratios were expected to be representatives of the ratios of α- to β-D-glucopyranosides and of total glucopyranosides to total glucofuranosides in all oligo and polyglucosides. Ideally, the reaction would be terminated by neutralization when a desired residual, unreacted glucose level, and a desired DP have been reached. The average DP of the final alkyl glucosides reaction mixture is sensitive to, and can be controlled by, changing the alcohol or alcohol mixture, the saccharide and particle size distribution of the glucose source, the relative amount of alcohol to glucose source, the strong acid catalyst, the agitation, the reaction temperature, the reaction vacuum, and the reaction time.

The generalized, commercial production of the final alkyl glucosides products from α-D-glucose monohydrate is described in the following text.

Particulate α-D-glucose monohydrate in an excess of a primary alcohol or alcohol mixture containing octyl to octadecyl alcohols is vigorously agitated. The initial mixture of glucose monohydrate in the alcohol is dehydrated at ~10 to 100 mm of mercury pressure at ~50 to 115°C to form an anhydrous glucose mixture. The dehydrated mixture is heated at ~80 to 115°C, and a strong acid catalyst is added. The reaction is continued at ~95 to 115°C under relatively high vacuum until essentially all of the glucose has dissolved and reacted and there is no further water distillate. The catalyst is neutralized with a metal hydroxide or metal oxide. The residual, unreacted alcohol in the mixture may be partially removed from the alkyl glucosides in an evaporator at ~10 to 40 mm of mercury pressure at a temperature of ~130 to 180°C. Essentially, all the unreacted alcohol in the mixture is removed from the alkyl glucosides in a thin-film evaporator at a pressure of ~2 to 20 mm of mercury pressure at a temperature of ~180 to 230°C. The alkyl glucosides residue from the thin-film evaporator is dispersed in water at ~50 to 80% dry substance. The slightly dark product may be adjusted to a high pH with sodium hydroxide and bleached with hydrogen peroxide at ~80 to 110°C to produce a light-colored product. Final pH and dry substance adjustments are made to the alkyl glucosides product.

The alcohol or alcohol mixture used is the largest factor in determining the appropriate temperatures and pressures for the direct reaction and evaporation steps. There is a general positive relationship between temperature and pressure for a given alcohol or alcohol mixture: higher or lower temperatures allow or even require higher or lower pressures.

4.5 RESEARCH AND DEVELOPMENT

Alkyl glucosides products may be represented as $H[C_6H_{10}O_5]_nOC_mH_{2m+1}$, where n is the average DP of the glucosides and m is the number or average number of carbon atoms in their alkyl groups and in the alcohols used for their production. Typical, commercial alkyl glucosides products include the following:

Alcohol Mixture Used for Production (Percentage by Weight)	Average DP (n)
40–48% octyl alcohol (m = 8)	1.50–1.55, 1.60–1.75
51–59% decyl alcohol (m = 10)	
13–23% nonyl alcohol (m = 9)	1.55–1.70
37–47% decyl alcohol (m = 10)	
33–43% undecyl alcohol (m = 11)	
65–75% dodecyl alcohol (m = 12)	1.39–1.45, 1.50–1.60
22–28% tetradecyl alcohol (m = 14)	
4–8% hexadecyl alcohol (m = 16)	

Future studies of alkyl glucosides products will include their production with different mixtures and types of alcohols or with different average DPs for new applications.

By taking advantage of the reactivities of the free, glucosidic hydroxyl groups, chemical derivatives of the basic alkyl glucosides products (AG-OH) have been prepared for new applications. These include the following:

Nonionic alkyl glucoside derivatives
    Glycerol ethers, AG–OCH$_2$CH(OH)CH$_2$OH
    Butyl ethers, AG–OCH$_2$CH$_2$CH$_2$CH$_3$
    Benzyl ethers, AG–OCH$_2$C$_6$H$_5$
    Ethoxylates, AG–O(CH$_2$CH$_2$O)$_x$H
    Carboxylic acid esters, AG–OC(=O)(CH$_2$)$_x$H
    Ethyl carbonates, AG–OC(=O)OC$_2$H$_5$
Anionic alkyl glucoside derivatives
    Sulfates, AG–OSO$_3$–
    Sulfosuccinates, AG–OC(=O)CH$_2$CH(SO$_3$–)CO$_2$–
    Alkyl sulfosuccinates, AG–OC(=O)CH$_2$CH(SO$_3$–)CO(CH$_2$)$_x$H
    Isethionates, AG–OCH$_2$CH$_2$SO$_3$–
    Ether carboxylates, AG–O(CH$_2$)$_x$CO$_2$–
Cationic alkyl glucosides derivatives
    Hydroxyethers, AG–OCH$_2$CH(OH)CH$_2$N ((CH$_2$)$_x$H))$_3$, where x is about 1–20

4.6 SUMMARY

Commercial alkyl glucoside products are unique nonionic surfactants. They can be produced from renewable resources such as corn and natural oils. Alkyl glucosides products are produced by well-defined commercial processes to a high-quality standard. Additional alkyl glucoside products are potentially available with properties for new applications.

REFERENCES

1. Lloyd, N.E., W.J. Nelson. Glucose- and fructose-containing sweeteners from starch. In: R.L. Whistler, J.N. Bemiller, E.F. Paschall, eds. *Starch: Chemistry and Technology*. Academic Press, New York, 1984, pp. 611–660.

Production of Alkyl Glucosides

2. Wagner, J.D., G.R. Lappin, J.R. Zeitz. Alcohols, higher aliphatic-synthetic processes. In: J.I. Kroschwitz, M. Howe-Grant, eds. *Kirk-Othmer Encyclopedia of Chemical Technology*, Volume 1, 4th ed. Wiley, New York, 1991, pp. 893–913.
3. Solomons, T.W.G. *Organic Chemistry*, 2nd ed. Wiley, New York, 1980, pp. 705–708.
4. Flory, P.J. Molecular size distribution in three dimensional polymers. VI. Branched polymers containing A-R-Bf-, type units. *J. Am. Chem. Soc.* 74:2718–2723, 1952.
5. Erlanger, S., D. French. A statistical model for amylopectin and glycogen. The condensation of A-R-Bf-, units. *J. Polymer Sci.* 20:7–28, 1956.
6. McCurry, P.M., N. Lauer. *The Gospel According to Flory*. Henkel Corporation, Dusseldorf, 1987.

5 Production of Linear Alkylbenzene Sulfonate and α-Olefin Sulfonates

Icilio Adami

CONTENTS

5.1 Historical Importance of Linear Alkylbenzene Sulfonate.................................83
5.2 Linear Alkylbenzene Sources and Specifications 84
5.3 Linear Alkylbenzene-Sulfonation Technology...86
    5.3.1 The Production of α-Olefin Sulfonate by SO_3 Sulfonation 102
5.4 Linear Alkyl Benzene Sulfonation Economics...109
5.5 Perspective Development 112
References...115

5.1 HISTORICAL IMPORTANCE OF LINEAR ALKYLBENZENE SULFONATE

Soap was considered and used as the universal anionic surfactant in both home-laundry and personal-care detergents till the 1940s.

The shortage of natural oils and fats during the Second World War has triggered the development of synthetic surfactants, mainly due to the better performance of the latter in hard water. However, the progressive replacement of soap by petrochemical-based surfactants began in the 1950s [1].

The success of this replacement, on an industrial scale, has particularly affected the production of linear alkylbenzene sulfonate (LAS) since it has resulted in an outstanding improvement of the production economics and product performances.

LAS is considered the ideal "all-purpose" surfactant suitable for the widest range of detergent applications due to its optimized performances in terms of solubility, foaming, wetting, detergency, and compatibility with other detergent components.

The general formula of LAS is indicated in Figure 5.1.

LAS results from the linear alkylbenzene (LAB) sulfonation reaction and its specific chemical composition, as well as application performance, depend on two main factors—the purity of the LAB raw material and sulfonation technique used.

Both these important topics are dealt with in Sections 5.2 and 5.3, but the close correlation between the LAB characteristics and the application properties of the obtainable LAS constitutes the key point in maintaining LAS as the predominant important surfactant capable of coping with the most severe demand of the detergent industry and the related market.

LAB is still the most commonly used raw material for surfactant production in the world and nowadays its consumption in the detergent industry alone accounts for some 2700 kilo/tons/year. The worldwide consumption shown in Figure 5.2 confirms the importance of LAB in the detergent world scenario.

83

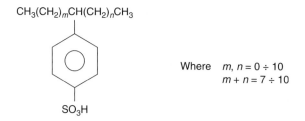

FIGURE 5.1 LAS structure.

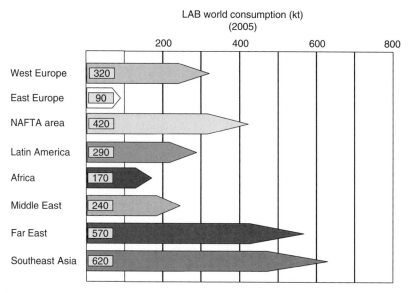

FIGURE 5.2 LAB world consumption.

Historically, LAS has progressively substituted the branched-chain alkyl benzene sulfonate (ABS) since the mid-1960s, mainly because of its substantially higher biodegradability due to its dominant straight alkyl chain; thus resulting in both higher and more valuable application performance and much improved environmental acceptability.

Today, LAS is the only type of alkylbenzene sulfonate used by the detergent industry and its applications have been extended to a whole range of formulated products.

5.2 LINEAR ALKYLBENZENE SOURCES AND SPECIFICATIONS

The hydrophobic portion of the LAS molecule is provided by the LAB, which is the main determinant of the modified surface-tension property of LAS.

Today, LAB is commercially produced by benzene alkylation reaction, catalyzed by Lewis-type acid catalysts (AlCl3, HF, and Detal) [2].

The use of chloroparaffins as an alkylating agent has declined in the last two decades; today, linear paraffins in the C10–C13 range are the hydrocarbons of choice for alkylation.

LAB, having the structure formerly indicated in Figure 5.1, is a mixture of isomers and homologues whose quantity and type depend on the alkylation process adopted for its production.

The basic steps of the commercially used benzene alkylation process are summarized in Figure 5.3.

The linear paraffins used for LAB production are obtained by liquid- or vapor-phase extraction from kerosene followed by fractionation to obtain the C10–C13 cut or narrower C-chain

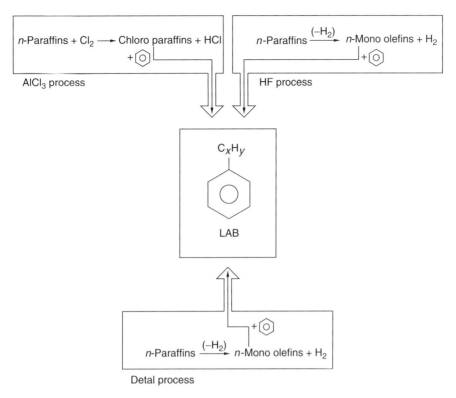

FIGURE 5.3 Benzene alkylation processes.

TABLE 5.1
Typical LAB Characteristics

LAB Type	Alkyl-Chain Range	Average Molecular Weight	C10	C11	C12	C13	C14
C11-LAB	C10–C13	231–236	18–19	48–49	31–32	1–2	—
C12-LAB	C10–C13	239–245	11–15	26–38	32–34	18–25	0–1
C13-LAB	C10–C14	252–256	0–1	5–8	28–31	54–58	6–8

length range and thus producing alkylates with different molecular weights, as indicated in Table 5.1.

The C-chain length of the n-paraffins used for the LAB production directly affect the molecular weight of the resulting alkylate, whereas the different catalysts are responsible for the structure and isomeric composition of the final LAB.

Particularly, 2-phenyl isomers and dialkyltetralins (DAT) are the components that can be considered as more representative of the adopted alkylation process, as shown in Table 5.2.

Independent of its manufacturing process, the most widely used type of LAB has an average molecular weight of 240–242 corresponding to an average alkyl chain of C12. However, C11-LAB type (MW = 235) and C13-LAB (MW = 255) are also being produced in response to specific request from the detergent formulators.

The most typical specifications of C12-LAB produced by the two most adopted alkylation processes (HF and Detal) are shown in Table 5.3.

TABLE 5.2
LAB Characteristics vs. Alkylation Process

	2-Phenyl Isomers (Maximum %)	DAT (Maximum %)	Alkyl-Chain Linearity (%)	Bromine Index (Maximum)
AlCl$_3$	25	5	90–92	5
HF	17	0.5	93–94	2
Detal	30	0.5	93–95	1

TABLE 5.3
Typical Properties of Commercial C12-Type LAB

Property	Typical Figures
Bromine index	2–3
Specific gravity	0.86–0.87
Carbon distribution (wt%)	
C10	11–15
C11	26–40
C12	30–35
C13	18–25
C14	0–1
2-Phenyl isomers	17–30
Flash point (°C)	140–145
Average molecular weight	235–242

The specificity of LAB required to be used for the production of LAS by sulfonation is based on its several important quality characteristics; particularly

- Molecular weight/C-chain distribution (requiring the correct setting of the sulfonation conditions)
- Sulfonability (to ensure the highest conversion of LAB into LAS and minimize, particularly, by-products and the presence of unsulfonated matter in the LAS product)
- Bromine index (low values indicate low unsaturation in the alkyl chain and respectively, the light color of the LAS product, entailing the achievement of sulfonation completeness under mild conditions)
- 2-Phenyl isomers content (high values ensure high solubility in water and high viscosity of LAS)
- DAT content (low value means higher purity, better performance, and higher biodegradability of the LAS product)

5.3 LINEAR ALKYLBENZENE-SULFONATION TECHNOLOGY

LAS is obtained by reacting LAB with the SO$_3$ functional group in a one-to-one ratio, thus resulting in a molecule capable of reducing the surface tension between two nonmiscible phases.

There are many sources of SO$_3$: H$_2$SO$_4$, oleum, pure gaseous SO$_3$, chlorosulfonic acid, and sulfamic acid; however, the basic path leading to LAS formation is essentially the same as outlined in Figure 5.4.

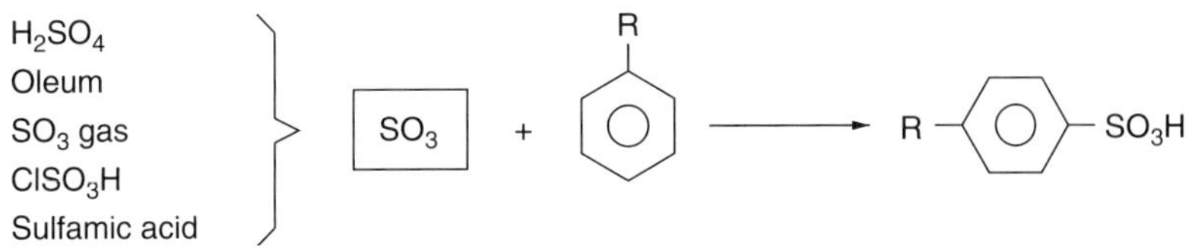

Production of Linear Alkylbenzene Sulfonate and α-Olefin Sulfonates

FIGURE 5.4 LAB–SO₃ reaction.

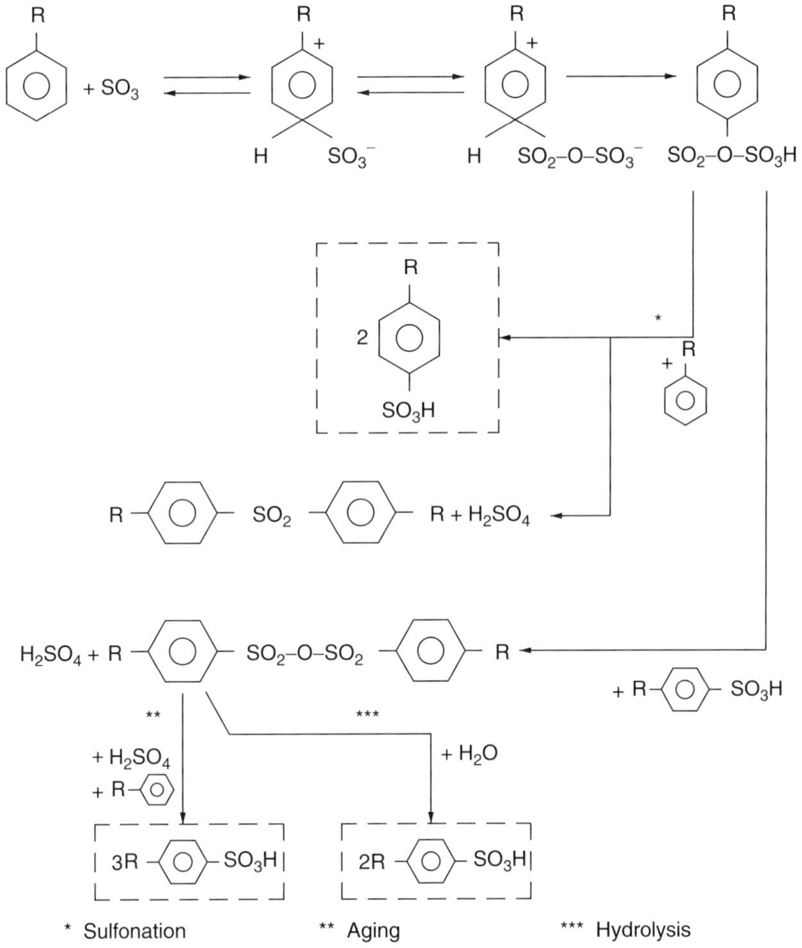

FIGURE 5.5 LAB–SO₃ reaction mechanism.

The reaction between SO_3 and the aromatic substrate is an electrophilic substitution reaction of the second order, and in the specific case of LAB, this reaction proceeds in accordance with the mechanism shown in Figure 5.5 [3].

When sulfonating LAB, the reaction is extended to the by-products in the LAB raw material; consequently, branched alkylates, DATs, and diphenylalkanes undergo sulfonation too, although each with a different speed (see Figure 5.6).

The result of the preceding reactions is a *sulfonic acid* product characterized by the chemical and physical specifications indicated in Table 5.4.

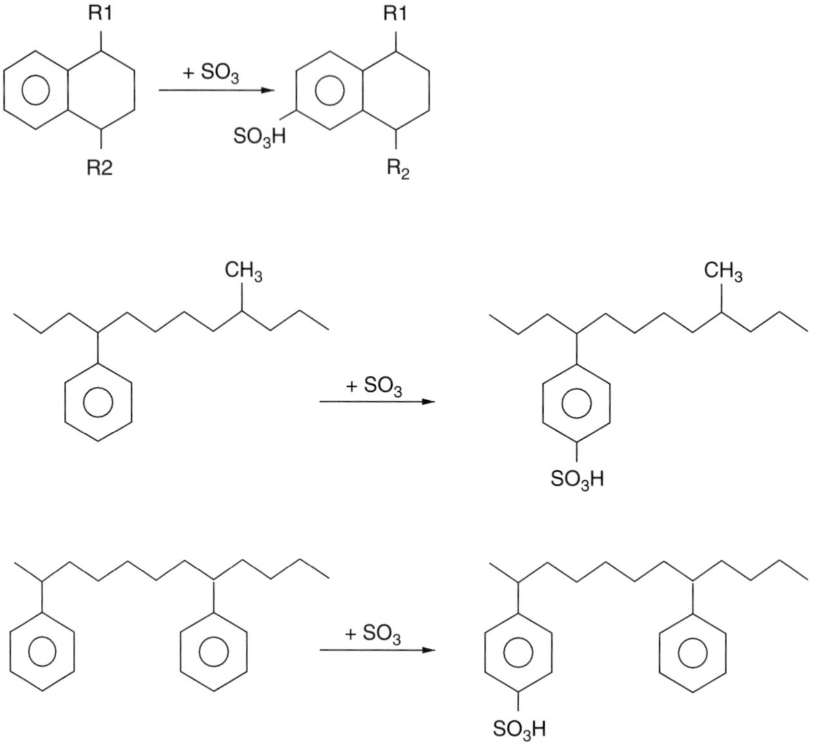

FIGURE 5.6 SO$_3$ reaction with LAB by-products.

TABLE 5.4
Chemical and Physical Characteristics of LAS

	Range	Typical
Active content (%)	96.7–97.5	97.2
Unsulfonated matter (%)	0.8–1.5	1.3
Free H$_2$SO$_4$ (%)	0.9–1.2	1.0
Water (%)	0.3–0.6	0.5
Pour point (°C)	(-10)–(-12)	-10
Specific heat at 20°C (kJ/kg · °C)	1.6	1.6
Viscosity at 20°C (mPa·s)	1500–1850	1500
Density at 20°C (g/cm^3)	1.050–1.055	1.052

The control of the preceding outlined reactions is the key point to ensure the production of LAS having the best quality. Therefore, specific operating conditions should be properly set up to minimize the side reactions negatively affecting both the conversion yield and the product quality.

The latter can also be affected by the main characteristics of the processed LAB raw material, and the specific influence detectable on the parameters commonly adopted for the quality evaluation of the LAS product are summarized in Table 5.5.

The continuous development of the process design focused on appropriate sulfonation equipment that is necessary to transfer all the findings from the parallel laboratory studies on the LAB–SO$_3$ reaction kinetics and mechanisms to an industrial scale. Although these efforts have been initiated from the beginning of the use of synthetic surfactants, new doubts concerning sulfonation chemistry

TABLE 5.5
LAB Specs versus LAS Characteristics

		Solubility	Detergency	Color	Viscosity	Biodegradability
Molecular weight	↑	↓	↑	↑	–	–
	↓	↑	↑	↓	–	–
2-Phenilisomer	↑	↑	↑	–	–	–
	↓	↑	↑	–	–	–
DAT	↑	↑	↑	–	–	↓
	↓	↑	↑	–	–	↑
Bromine index	↑	↑	–	–	↑	–
	↓	↑	–	–	↓	–
Alkyl-chain linearity	↑	↑	–	–	↑	↓
	↓	↑	–	–	↑	↓

are still arising, which is being followed by efforts to give a fully correct explanation to the involved reaction mechanism.

Historically, the source of the SO_3 for LAB sulfonation was sulfuric acid (H_2SO_4) and, subsequently, the so-called oleum or fuming sulfuric acid made by enriching the concentrated sulfuric acid with SO_3 (~20% excess in most cases) [4].

At the very beginning of the synthetic detergent production, most of the sulfonation process based on sulfuric acid and oleum were of the discontinuous-batch type. Their evolution into a continuous process was fast as a result of the effort to optimize the rate of production capacity and quality as demanded by the outstanding and ever-increasing development of the detergent manufacturing industry.

Both sulfuric acid and oleum when used as a sulfonating agent involve the stoichiometric formation of water as a by-product from the reaction with an organic substrate (see Figure 5.4).

This water, due to the dilution effect on the still unreacted sulfuric acid, causes the progressive loss of the latter's reactivity. This loss implies the necessity of continuous removal of the formed water or operating the process with excess of the sulfonating agent and eventually to separate, by physical settling, the weak-spent sulfuric acid that is not capable to comply with the desired sulfonation reaction kinetics anymore.

Regarding the involved process equipment and plant configuration, the sulfuric acid and oleum-based sulfonation plants have demonstrated evident limits to the possibility of their further improvement and development due to the foregoing chemical constraint.

The process block diagram relevant to continuous oleum and SO_3-based sulfonation processes are indicated in Figures 5.7 and 5.8, respectively.

When oleum is the sulfonating agent, the reaction with the LAB raw material is carried out in liquid homogeneous phase and it requires a "digestion" time for achieving its completeness; this time demand is a direct function of the excess oleum added with respect to the stoichiometric quantity.

After digestion, the reaction mass has to be diluted with H_2O to separate (by settling) the "spent acid;" the sulfonic acid mixture can be sent for storage or neutralization, whereas the spent acid

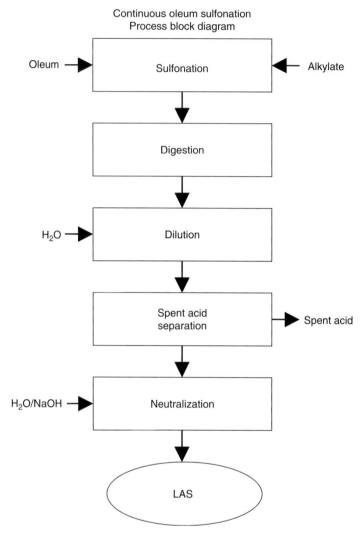

FIGURE 5.7 Continuous oleum sulfonation.

may undergo storage or neutralization process as well, or it is collected (as a waste) and reprocessed to recover its SO_2/SO_3 content. This operation is usually done in the frame of a H_2SO_4 production plant; therefore, the logistic of this operation is a key factor in determining the possibility of reusing the spent acid.

The process block diagram of the SO_3-based sulfonation plant clearly shows the total absence of liquid effluents and wastes, and it is based on a substantially stoichiometric reaction of SO_3 and the organic feedstock. This important point reflects the extreme "sulfonating strength" of the SO_3, particularly when compared to oleum and H_2SO_4. The SO_3 sulfonation does not involve the formation of water as a by-product, with consequent possibility to use the whole amount of SO_3 for the main reaction. After sulfonation, the aging and stabilizing steps are required to allow the "rearrangement" of sulfoanhydrides directly to sulfonic acid; thus maximizing, over a short time, the conversion of the feedstock into the final sulfonate.

The aged and stabilized sulfonic acid is sent for storage or neutralization, without the necessity to dispose off any other effluent or by-product.

Production of Linear Alkylbenzene Sulfonate and α-Olefin Sulfonates

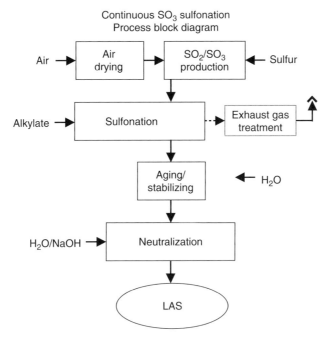

FIGURE 5.8 Continuous SO₃ sulfonation.

The technological switch to the use of gaseous SO₃ as the sulfonating agent for the synthetic detergent production has therefore been the most characteristic factor of the development that the surfactant and detergent industry have experienced since the early 1960s [4,5].

The SO₃ gas, directly produced from the combustion of elemental sulfur and subsequent oxidation of the formed sulfur dioxide (SO_2) with dry air to yield sulfur trioxide (SO_3), is nowadays the most widely used sulfonating agent, therefore, it deserves a deeper description.

The most updated LAB sulfonation plants are designed and operated in accordance to the configuration shown in Figure 5.9 and they should cope with the industry demand in terms of maximum yield, highest product quality, and minimized operation cost and environmental impact.

The gaseous SO_3 is generated by oxidation of the SO_2 formed by direct combustion of elemental sulfur with dry air having a dew point of at least −60°C.

Figures 5.10 and 5.11 illustrate the typical scheme of these process sections.

The SO_3 gas directly produced from the combustion of elemental sulfur and subsequent SO_2 oxidation has been a milestone in the design of continuous sulfonation processes, and has led to a fast development of the design and efficiency of the process equipment employed for reacting the SO_3 gas with organic substrates.

The control over the LAB–SO_3 reaction is totally dependent on the design approach adopted for the reactor where the two reactants are contacted; therefore, the sulfonation reactor is indeed the heart of a sulfonation plant and its concept should be based on the deepest possible knowledge on what exactly takes place when LAB and gaseous SO_3 react.

The "cascade-type" sulfonation reactors [6] have been adopted in a large number of plants and considered as the "state-of-the-art" reactors until the mid-1970s, when new reactors based on the principle of the falling film were introduced [7].

In such type of reactors, the combination of the LAB–SO_3 reaction exothermicity (40.6 kcal/mol) and the almost instantaneous increase in the viscosity of the organic undergoing sulfonation indicate that the control of the reaction temperature in the organic phase is the most difficult target that the reactor should comply with.

92 Handbook of Detergents/Part F: Production

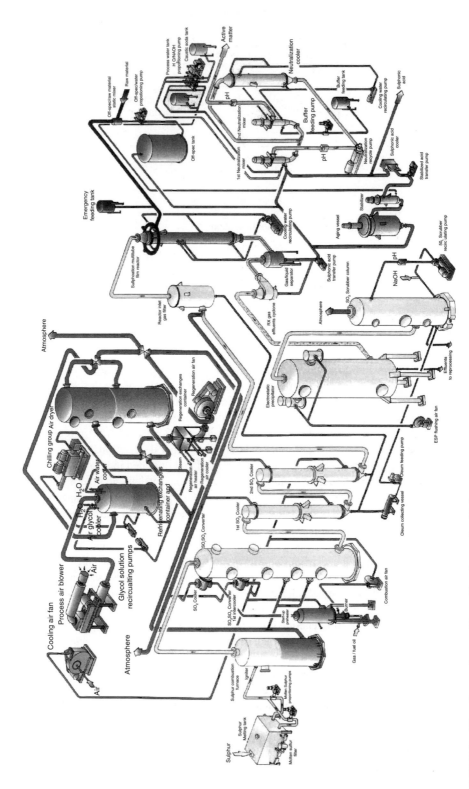

FIGURE 5.9 Plant for continuous film sulfonation and sulfation.

Production of Linear Alkylbenzene Sulfonate and α-Olefin Sulfonates

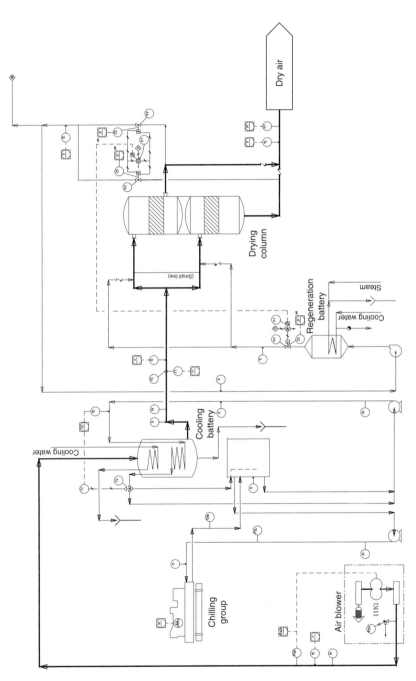

FIGURE 5.10 Air drying unit.

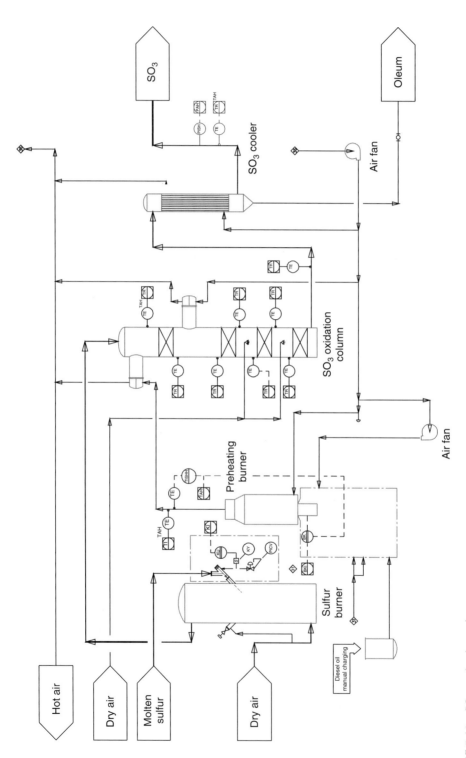

FIGURE 5.11 SO$_3$ production unit.

Production of Linear Alkylbenzene Sulfonate and α-Olefin Sulfonates

In a falling-film reactor, the evolved reaction heat can be controlled and balanced by diluting the SO_3 with dry air, thus reducing its partial pressure and flow to the gas–liquid interface. The latter is of utmost importance as the rate of the reaction is controlled by the transport of SO_3 through the gas phase and the distribution of the reactants in the reactor tubes is consequently a crucial point for a complete control over the reaction thermodynamics [8–10].

A typical example of a falling-film reactor (i.e., Ballestra multitube type film reactor) is shown in Figure 5.12, whereas Figure 5.13 refers to a process scheme of the LAB–SO_3 sulfonation section.

In this type of film reactor, the combination of cooling efficiency, reaction tube geometry, and reactants distribution allows to achieve a practically complete absorption of the stoichiometric amount of SO_3 necessary to convert LAB into LAS, as indicated in Figure 5.14 where the variation of temperature versus the reaction completeness along the reactor length is shown.

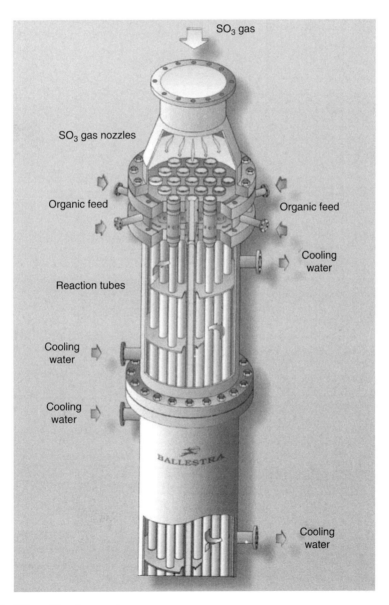

FIGURE 5.12 Film sulfonation reactor.

96   Handbook of Detergents/Part F: Production

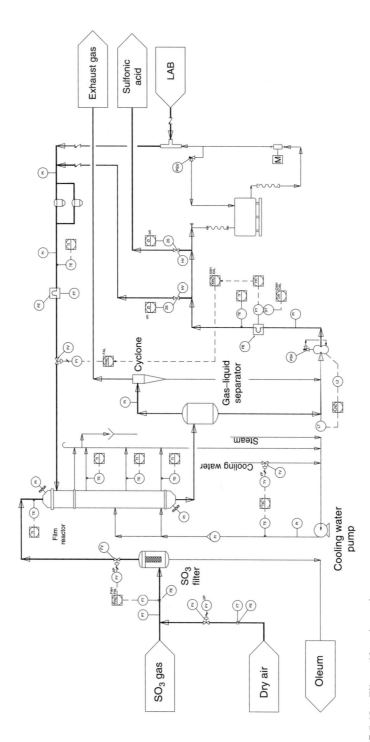

FIGURE 5.13 Film sulfonation unit.

Production of Linear Alkylbenzene Sulfonate and α-Olefin Sulfonates

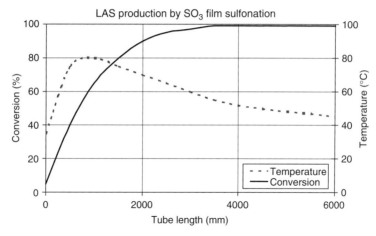

FIGURE 5.14 LAS production by SO$_3$ film sulfonation—reaction completeness in multitube film reactor.

In commercial-scale plants based on multitube film reactor, the following operating parameters are typically adopted:

SO$_3$/LAB mole ratio fed to the reactor	1.01–1.03
Reactor cooling water temperature (°C)	25–30
SO$_3$ gas concentration (% volume)	5–7
SO$_3$ gas infeed temperature (°C)	48–50
LAB infeed temperature (°C)	20–25
LAB–sulfonic acid temperature (°C)	45–50

The LAB–sulfonic acid has to be then aged and residual traces of anhydrides have to be hydrolyzed; standard condition for this step of the LAS production is to operate at a temperature range 45–50°C to allow the complete conversion of LAB into LAS over ~45 to 60 min, whereas the amount of H$_2$O added for anhydride hydrolysis accounts for ~0.5% with respect to the mass of the acid.

Recent trends in liquid detergent formulation have resulted in the demand of LAS characterized by very low content of inorganic sulfates, and this demand has been satisfied by the surfactant industry with the implementation of the so-called "extended aging" where the sulfonic acid is allowed to reach complete conversion over a time length of 2–4 h at a temperature not above 40°C.

In Figure 5.15, the effect of the extended aging over the most outstanding and detectable LAS characteristic (i.e., color) is shown; whereas Figure 5.16 illustrates the scheme of the aging and anhydride hydrolysis ("stabilizing") section.

The typical chemical characteristics of the LAB–sulfonic acid obtainable by processing international standard LAB by updated sulfonation plants are as follows:

Unsulfonated matter (%)	0.8–1.2
H$_2$SO$_4$ (%)	0.5–1.0
Active matter (%)	96.7–97.3
H$_2$O (%)	0.3–0.5

The last step in LAS production is to convert the LAB–sulfonic acid obtained by sulfonation and postsulfonation treatment into a chemically neutral and stable form, and this is typically

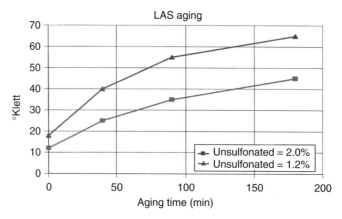

FIGURE 5.15 LAS aging—time versus color variation.

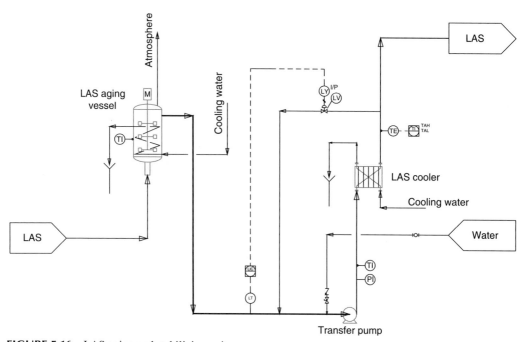

FIGURE 5.16 LAS aging and stabilizing unit.

accomplished by neutralizing it with caustic soda solution to yield a neutral paste with active content (LASNa) in the range 50–60%.

LAS acid neutralization can be performed by means of a process unit based on the principle of the *forced loop* [11] in which the reactants are continuously dosed and the heat of reaction is immediately dispersed over the mass of neutralized product kept under recycling.

Figure 5.17 shows the scheme of a typical LAS neutralization.

The neutralization reaction is exothermic (~25 kcal/mol) as shown in Figure 5.18 in which a typical heat and material balance is observed.

Commercially, the relevance of LAS in the form of neutral water solution of its sodium salt is overtaken by the corresponding sulfonic acid, due to its easy handling and use in the detergent production process where the LAS acid can be directly neutralized.

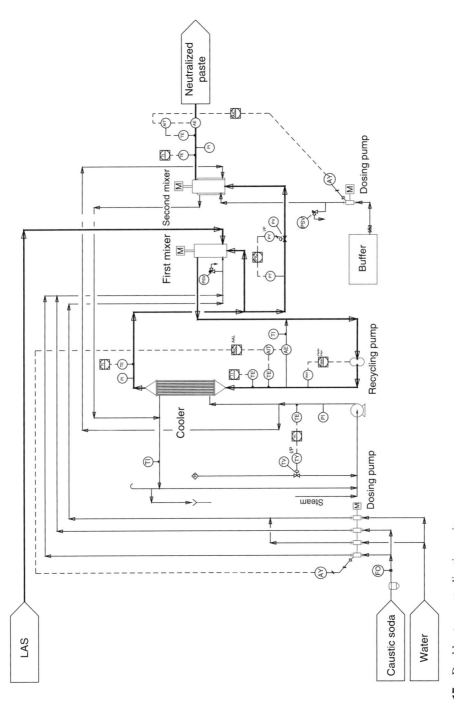

FIGURE 5.17 Double-step neutralization unit.

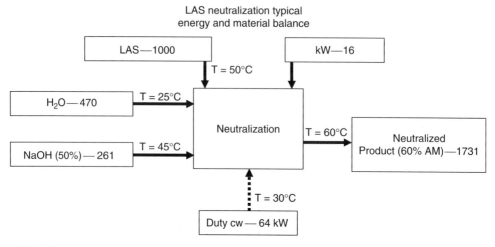

FIGURE 5.18 LAS neutralization—energy and material balance.

The critical point connected with the use of LAS-sulfonic acid is represented by its tendency to undergo color increase during storage [12] and this sensitivity may depend on factors such as the LAB quality (i.e., iron, heavy alkylate, and DAT/indane content) as well as operating conditions adopted for its production (i.e., SO_3/LAB mole ratio, aging time, temperature, and hydrolysis conditions). Storage of the LAS–sulfonic acid should be preferably made in stainless steel tanks or carbon steel containers lined with epoxy resins and kept at a temperature not exceeding 30°C.

In the frame of the sulfonation plant for LAS production, the exhaust gas leaving the sulfonation reactor (~1.8 to 2.0 kg/kg of processed LAB) contains a pollutant load of a few thousand parts per million mainly constituted by SO_2 (derived by the incomplete conversion of SO_2 to SO_3), unreacted SO_3, and organic mist composed of partially sulfonated material physically entrained by the gas.

This gaseous stream is conventionally treated in a gas-cleaning unit based on physical scrubbing of the "polar" and ionizable pollutant (i.e., organic mist and SO_3) by means of a dedicated electrostatic precipitator (ESP), whereas the SO_2 is scrubbed off by chemical reaction with an alkaline water solution countercurrently contacted with the gas stream.

Figure 5.19 illustrates the scheme of this unit, which enables the gas leaving the sulfonation plant to be released into the atmosphere with the following maximum pollutant contents:

SO_2	5 ppmv
SO_3	15 ppmv
Organic mist	25 mg/m^3

The effluents from the exhaust gas scrubbing unit are constituted by acidic drippings from the ESP (~1.8 to 2.0 kg/t of LAS) and Na_2SO_3/water solution (9–10% dry matter content).

The acidic dripping needs to be selectively collected and sent for incineration (with consequent cost for their disposal), whereas the sulfite solution can be sent to a conventional collecting and treatment system before being discharged to the sewage network.

Recently, a new exhaust gas treatment based on the chemical scrubbing of the unreacted SO_3 (by reaction with freshly added LAB) has been developed and retrofitted in several industrial plants located in areas where the acidic dripping disposal indeed has a high cost and poor feasibility.

This system, illustrated in Figure 5.20, operates by spraying fresh LAB where the concentration of unreacted SO_3 is high (i.e., at the outlet of the film reactor) and the contact between the same SO_3

Production of Linear Alkylbenzene Sulfonate and α-Olefin Sulfonates

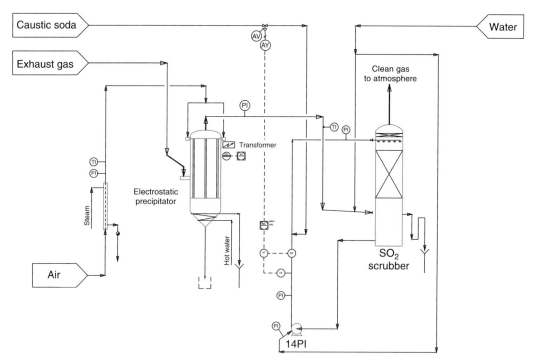

FIGURE 5.19 Exhaust gas treatment unit.

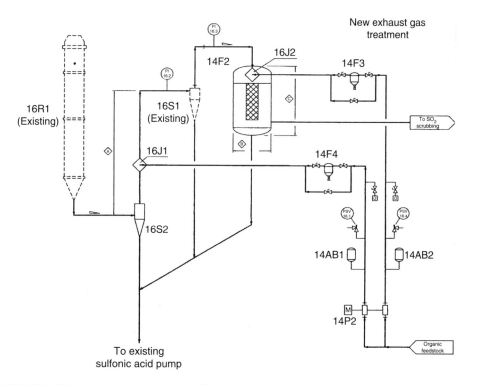

FIGURE 5.20 Exhaust gas treatment by wet filter.

and fine droplets of organic feedstock is made in a "climbing tube" that is positioned upstream of a cartridge filter operating in a wet pattern [13].

The droplets of partially reacted LAB are filtered on the surface of the high-efficiency wet filter where an additional quantity of LAB is sprayed on purpose to keep the filtering surface wet and free, thus preventing the progressive product charring that might cause the parallel clogging of the filtering elements.

The advantages of this system are the absence of acidic dripping to be disposed off and the consequent cost of the (heavy) maintenance required for the ESP.

Moreover, the new system entails an increase in the overall production yield (~1%) obtained by recovering the partially sulfonated materials generated by contacting the exhaust gas with the sprayed amount of LAB. The impact of this recovered stream on the quality of the final LAS–sulfonic acid is limited to a minor increase in the unsulfonated matter content and color. The overall gas scrubbing efficiency has proven to be the same ensured by the standard ESP-based gas treatment units.

The validity of the SO_3-based sulfonation technology is confirmed when its technical potential is compared to the continuously increasing request of the detergent industry for good performance and cost-effective surfactants.

This request can be complied with by the surfactant manufacturers only by offering a wide variety of anionic surfactants to cover all the necessities of detergent formulators.

Consequently, the validity and operation profitability of a sulfonation plant depend mainly on the capability and suitability of the critical key equipment to process strongly different demanding feedstocks.

5.3.1 THE PRODUCTION OF α-OLEFIN SULFONATE BY SO_3 SULFONATION

An example of the flexibility of the preceding sulfonation technology making use of gaseous SO_3 and film sulfonation reactor is represented by the production of α-olefin sulfonates (AOS).

AOS are the only anionic surfactants whose production has no alternative feedstock to linear olefins. The latter are olefins with the double bond positioned in the terminal part of the carbon chain (α-olefin [AO]) and, by sulfonation, they generate a molecule having surfactant properties.

Historically, the olefins of such type have been obtained by both alcohol dehydration and paraffin wax cracking, but these processes have been progressively abandoned in favor of processes based on ethylene oligomerization.

Presently, ethylene oligomerization for the production of AO is industrially operated according to the three main processes.

- Ziegler process
- Modified Ziegler process
- SHOP process

These three technologies differ from one another in their operating conditions, number of process steps, and catalyst of choice and therefore they give rise to blends of olefins characterized by a different C-chain length distribution and, to some extent, also the C-chain structure.

Ethylene oligomerization proceeds in accordance with two mechanisms of C-chain growth, namely, catalytic or stoichiometric [14–16].

Both the reaction mechanisms generate at the same time only even C-chain number so that it is not possible to produce a specific carbon number.

The C-chain length of interest for detergent application ranges between C14 and C18 and the most typical AO-cut composition used for AOS production are those indicated in Table 5.6.

The impurities contained in the AO obtained by these processes consist of vinylidene olefins (branched isomers), internal olefins, paraffins, and diolefins.

Production of Linear Alkylbenzene Sulfonate and α-Olefin Sulfonates

TABLE 5.6
AO-Cut Composition

AO Cut	Composition (wt%)		
	C14	C16	C18
C14	99	1	—
C14–C16	65	35	—
C14–C18	15	50	35
C16–C18	—	55	45

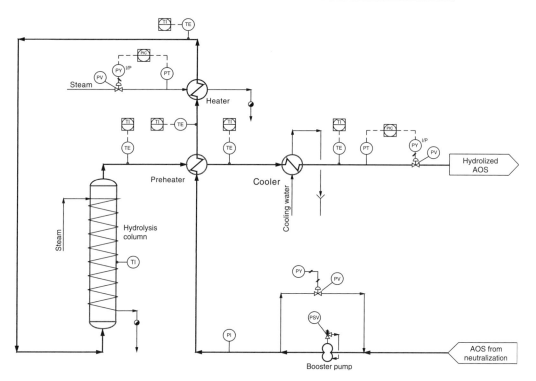

FIGURE 5.21 AOS hydrolysis unit.

The production of AOS matching the present market requirements can be based on the standard process configuration of an existing sulfonation plant, where in addition to the fitting of the AOS hydrolysis unit (see Figure 5.21), the sulfonation section (based on multitube film reactor) should be completed with cooling devices for the infeed organic stream and with independent cooling circuit for the top part of the reactor, as shown in Figure 5.22.

Both linear (terminal and internal types) and branched olefins show high reactivity with sulfonating agents, and this is particularly evident when the latter is gaseous SO_3 in dry air.

This reaction has been studied since the beginning of the twentieth century [17] and its first industrial application for surfactant production dates to the 1970s [18].

For the three main types of AO components (namely, linear alpha, linear internal, and branched olefins), the sulfonation reaction chemistry can be summarized as shown in Figure 5.23.

In all the cases, the first step of the reaction leads to the formation of cyclic intermediates (β-sultones), which rapidly undergo isomerization to unsaturated sulfonic acids and (in the case of linear and internal, but not vinylidene olefins) the more stable γ- and δ-sultones. These sultones can be hydrolyzed to hydroxyalkane sulfonates.

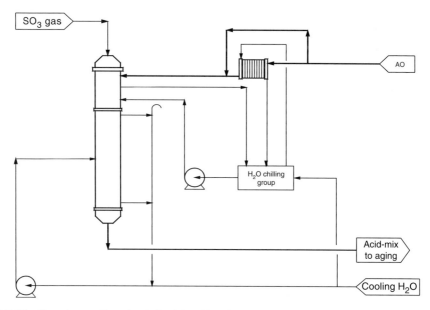

FIGURE 5.22 Reaction cooling circuit for AO sulfonation.

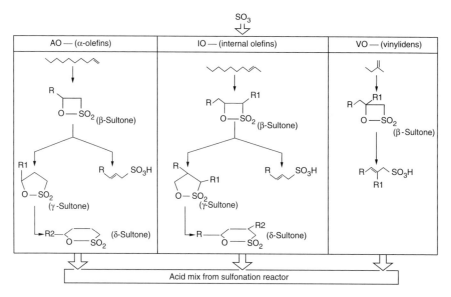

FIGURE 5.23 SO$_3$/olefin reaction mechanism.

Thus, the acidic mass immediately after sulfonation consists mainly of a mixture of alkene sulfonic acids and γ- and δ-sultones, with small amounts of β-sultones.

The transformation of AO feedstock into AOS therefore involves, after the sulfonation and aging steps, the neutralization and hydrolysis steps where the reactions necessary to convert AO into AOS follow the scheme indicated in Figure 5.24.

When the content of internal olefins in the sulfonation feedstock is significant, these isomers, despite their high reactivity toward SO$_3$, are less highly converted than the corresponding α-isomers. The reaction mechanisms have been extensively studied [3,19–21], including kinetic experiments carried out on reaction products obtained by SO$_3$/olefin contacting in a falling-film

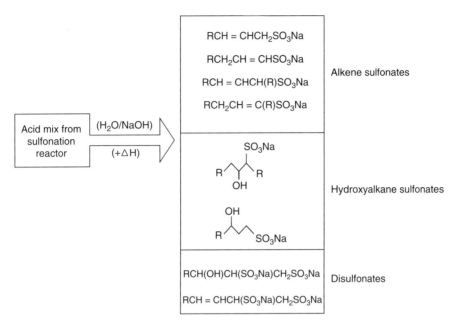

FIGURE 5.24 Main components of AOS.

reactor [22]. It is thought that the difference in sulfonation behavior of alpha versus internal olefins is due to the greater resistance to isomerization of the β-sultones derived from the latter olefins and to the ability of these internal β-sultones to act as sulfonating agents toward alkene sulfonic acids, forming disulfonated products and at the same time regenerating internal olefins. Therefore, optimum conversion of internal olefins requires higher excess of SO_3 and stronger cooling in the reactor than what was required as optimal for α-olefin sulfonation. A further consequence of the greater resistance to isomerization of internal β-sultones is that the AOS generated by AO feedstocks rich in internal isomers show lower ratios of alkene sulfonates versus hydroxyalkane sulfonates.

The reaction completeness and the color of the sulfonated product are the most important parameters considered in the design of a sulfonation reactor suitable to process AO feedstocks.

When a "latest-generation" multitube falling-film reactor [23] is adopted for AO sulfonation, the reaction heat is almost completely evolved over a few seconds after the contact between the reactants, and to remove the reaction heat it is necessary to ensure, particularly in the first part of the reactor, an efficient cooling capability, without negatively affecting the viscosity profile of the mass undergoing sulfonation.

To cope with the processing of AO, the overall production capacity of a sulfonation plant for AOS is comparatively lower than with other feedstocks, mainly for the loading conditions of the reactor necessary to cope with the kinetics and thermodynamics of the reaction between SO_3 and AO. Figure 5.25 shows a comparison of the "tube load" of falling-film reactor for the processing of AO and other feedstocks.

Typically, for each reactor tube (loaded with 0.07–0.09 kmol/h of AO), some 3500–4500 kcal/h are removed by the circulating cooling water whose delta-T (in/out reactor) is kept at ~0.5°C.

The high sensitivity of the AO sulfonation also requires particular care in controlling the viscosity variation before and during the reaction itself. The extremely low viscosity of the AO feedstock, and the relatively low organic flow through each reaction tube, may enhance random liquid film breaking, with consequent charring; cooling of AO fed to the reactor is recommended to increase the viscosity so as to avoid such a possibility.

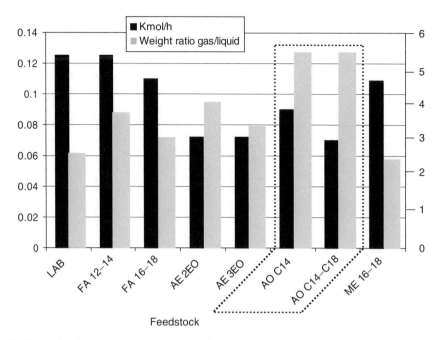

FIGURE 5.25 Multitube film reactor—tube load figures (FA—fatty alcohol; ME—methylester).

As a result of the preceding reactions, the AOS products (after sulfonation, aging, neutralization, and hydrolysis) show the following typical chemical composition:

Alkene sulfonates	50–65% (typically 60%)
Hydroxyalkane sulfonates	30–40% (typically 33%)
Disulfonates	3–10% (typically 7%)

The disulfonates formation mechanism is complicated and not yet completely clear. Disulfonates in AOS are considered to arise from further sulfonation of alkene sulfonic acids.

The AOS composition and quality are also linked to the structure and characteristics of the processed AO, particularly, the content of paraffins and internal isomer are the feedstock characteristics that adversely affect the completeness of AO to AOS conversion as well as the composition and color of the final product.

Independent of the feedstock characteristics and C-chain length, the sulfonation of AO requires accurate control of

- Overall process mass balance (see typical data for C14 AOS production in Figure 5.26)
- SO_3/AO molar ratio (to be kept at about 1.05–1.07)
- Cooling water temperature and flow to the reactor (to ensure color preservation without sultone crystallization)
- Profile of product temperature in sulfonation, neutralization, and hydrolysis (see typical data in Figure 5.27)

By processing an AO feedstock at a commercial scale with characteristics within those listed in Table 5.6, the AOS quality figures shown in Table 5.7 are obtained and accepted for their use in detergent products.

It must be noted that the indicated figures of active matter content correspond to the physical form of liquid and pumpable solution. Higher active matter content (i.e., 70% and higher) although

Production of Linear Alkylbenzene Sulfonate and α-Olefin Sulfonates

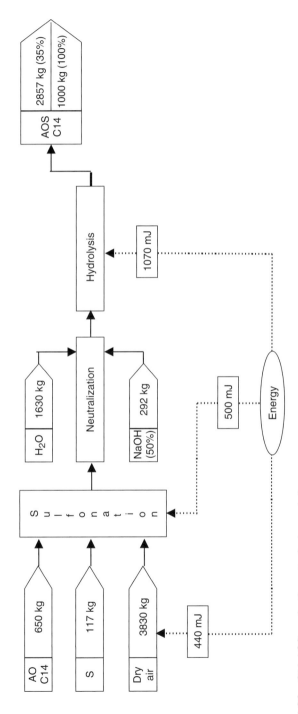

FIGURE 5.26 C14 AOS production—typical mass balance.

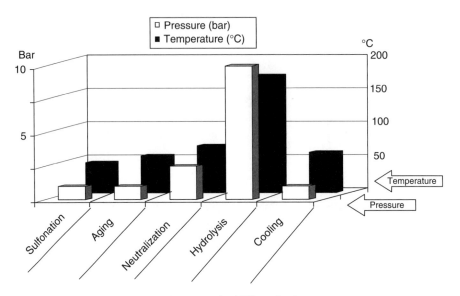

FIGURE 5.27 Profile of temperature and pressure in AOS production.

TABLE 5.7
AOS Specifications

	AOS C14	AOS C14–016	AOS C14–18
Active matter (AM) (%)	35–38	34–36	34–36
Na_2SO_4 (maximum % on 100% AM)	1.0	1.0	1.5
Free oil (maximum % on 100% AM)	1.5	1.5	2.0
Total sultones (maximum ppm on 100% AM)	30–50	30–50	30–50
Color (maximum °Klett on 5% AM solution)	40–70	40–60	40–70

achievable does not represent a commercial specification, as in this form the product is a viscous paste quite sensitive to temperature and pH and its handling and reuse is not easy.

The production of dry AOS represents the solution to the technical and economic factors that have limited the possibility and convenience of large-scale use of AOS in detergent production.

The most updated dry AOS manufacturing technology [23] is therefore focused to generate a product characterized by

- Optimized chemical specification (i.e., low content of free oil, inorganic sulfate, and moisture)
- Regular physical shape
- Easy handling, storage, and reusing

Industrially, this target is accomplished by the following operations:

1. AO sulfonation in multitube film reactor and aging/digestion of acid mixture
2. Neutralization (as 70% active content) of the acid mixture and subsequent hydrolysis

Production of Linear Alkylbenzene Sulfonate and α-Olefin Sulfonates   **109**

3. Preconcentration of the 70% active AOS paste
4. Drying of the preconcentrated AOS paste up to the maximum achievable level

Presently, dry AOS powder is produced with an average granulometry of 150–200 μ and bulk density at ~500 g/L. The commercially available powders are manufactured starting from both C14–C16 and C14–C18 feedstocks and they are used for detergent powder production as well as for other industrial application (i.e., textile and concrete industries).

5.4   LINEAR ALKYL BENZENE SULFONATION ECONOMICS

The economics of the LAB-sulfonation process can be defined as the balance between the overall operating costs and the value of the product output.

As the quantification of the LAB-sulfonation process is quite difficult and significantly affected by several factors such as the plant capacity, production mix, integration of the plant with up- and downstream facilities, and the plant features, a few simplifications in establishing the economy evaluation of this process appear necessary.

This is confirmed, for instance, when the obtainable product quality, energy and utilities demand, and raw material consumption for LAS production are comparatively taken into account for oleum- and SO_3-based processes, as shown in Table 5.8. For the oleum-based sulfonation plant, the spent acid represents an important variable (either for its neutralization or simple disposal for further SO_2 recovery) with heavy impact on the process cost. Moreover, the availability and cost of oleum derived from "local" situation that, case by case, might contribute to make this sulfonating agent comparatively cheaper than the more modern, clean, efficient, and versatile gaseous SO_3.

The only (apparent) advantage of the oleum-based process versus the gaseous SO_3-based process is the overall lower energy demand, which is derived from the liquid–liquid reaction without the necessity of the energy-intensive step of air drying (as strictly necessary for the SO_3 process).

TABLE 5.8
Comparative Process Performances

Oleum-based process			SO_3-based process
		Product quality	
88–90	←	LAB-SO_3H (%) →	97 min
1–1.5	←	LAB (%) →	0.8–1.2
5–6	←	H_2SO_4 (%) →	0.8–1.0
5–6	←	H_2O (%) →	0.3–0.5
40–50	←	Color (°Klett) →	20–35
		Energy and utility demand	
40	←	Electric power (kW/t AM) →	160
20	←	Cooling water (m³/h t AM) →	100
		Raw material consumption	
720	←	LAB (kg/t AM) →	718
274[a]	←	Sulfur →	99
160	←	Process H_2O →	—
159[b]	←	NaOH (100%) →	131

[a] Corresponding to 792 kg oleum (20% SO_3).
[b] Without neutralization of spent acid.

On the basis of the present market demand for higher product quality, for LAS production it seems more logical to focus the economic analysis on the SO_3-based process, where the main factors accounting for the overall process cost are shown in Figure 5.28.

By increasing the plant capacity, the contribution of the two main fixed accounting costs (energy and personnel) decreases with benefit for the global economy; the latter indeed can be clearly stated only by the quantification of the income derived from the product sales. It is therefore necessary to take into account the current market value of the raw materials and corresponding sulfonates to get an idea of their effect on the global process economy.

As the manufacturing of the LAB feedstock is closely integrated in the transformation chain of crude oil, it is understandable that its price trend is not only variable, but also extremely difficult to be predicted or controlled.

With the aim to correlate the sulfonation profitability and cost ratio of feedstocks and sulfonates, it is more appropriate to evaluate how the "upgrading of feedstock value" (UFV) obtainable by sulfonating LAB feedstock has changed over the past five years.

The UFV is a parameter based on factors such as the ratio of molecular weight of feedstock and sulfonate and relevant process conversion to yield the sulfonate; and it quantifies the percentage increase of value of the "unit weight" of feedstock when undergoing sulfonation [24].

The UFV trend for LAB (see Figure 5.29), when compared to the variation of feedstock price in the same period, confirms that the sulfonation ensures an average increase of the LAB value in the range 13–15%, despite the instability of LAS market price (in turn deriving from rising manufacturing costs and commercial contingencies).

Out of the situation resulting from the preceding figures, it is important to note that, as mentioned earlier, the detergent industry requires a variety of sulfonates that is well exceeding the field covered by LAS; the use of feedstocks such as fatty alcohols and ethoxylated fatty alcohols are consolidated and the market value of their corresponding sulfated/sulfonated products as well as their application importance in detergent manufacturing are increasing too; therefore, an example of overall process economy for a modern sulfonation plant operating for a production volume of 65/35 of LAB versus ethoxysulfates (AES) seems appropriate to state the convenience of operation.

The figures indicated in Table 5.9 confirm the global convenience of the LAS production by SO_3-based plant, particularly in case of plants operated by surfactant manufacturers, where

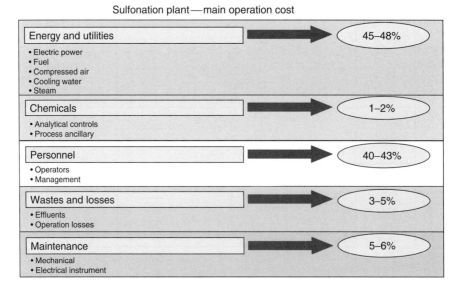

FIGURE 5.28 Operating costs of sulfonation.

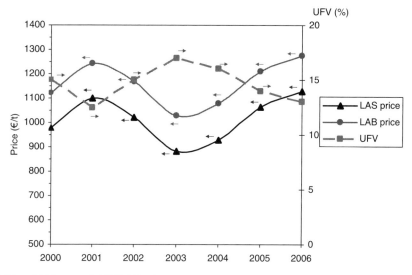

FIGURE 5.29 UFV versus LAB/LAS prices.

TABLE 5.9
LAB and AE Process Economy

	Quantity (Metric Tons/Year)	Cost (€) Unit	Cost (€) Total
Costs			
I. Raw materials			
LAB	11.970	1,100/t	13,167.000
Ethoxylated alcohol-2EO	4.287	1,200/t	5,144.400
Sulfur	2.115	95/t	200.925
NaOH (100%)	1.484	170/t	252.280
Process water	3.737	0.400/m^3	1.495
Total raw material cost			18,766.100
II. Manufacturing cost			
Utilities			
Fuel oil (start-up and steam production)	102	1,200/t	122.400
Electric energy (kWh year)	2,888	0.05/kWh	144.400
Total utilities cost			
Personnel (excluding administration, sales, logistic, etc.)			
N. 1 plant manager (€70,000/year)			
N. 6 operators (€240,000 /year)			
Total personnel cost			310.000
Maintenance and spare parts			70.000
Total manufacturing cost			646.800
Total cost (I + II)			19,412.900
Return (product sales)			
LAS (97% AM)	15,990	960/t	15,350.400
AES-2EO (70% AM)	8,000	950/t	7,600.000
Total return			22,950.400
Margin			3,537.500

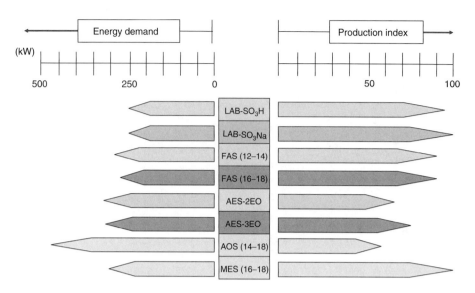

FIGURE 5.30 Production energy demand and yield of sulfonates.

the enlargement of the range of processed feedstocks contributes significantly to the achievable economical margin.

The figures in Table 5.9 can be confirmed and corroborated by considering the high flexibility afforded by modern sulfonation plants. The production yield and relevant energy demand when processing strongly different feedstocks comparative to LAB sulfonation are outlined in Figure 5.30 [25].

5.5 PERSPECTIVE DEVELOPMENT

The future of LAS and its development is closely linked to the technological trend of LAB production and SO_3 sulfonation process.

The production safety, product purity, and commercial price characterize all the efforts of the petrochemical industry to make more and more competitive and appealing LAB for the detergent industry.

The continuous increase in the cost of the raw materials for LAB production indicates that the *cost versus benefit* ratio for this product will necessarily have to be reset, and new and more specific application range should be investigated to obtain the proper criterion for the most fruitful application of LAS in modern detergents.

Independent of economic factors, the perspective development of the SO_3 sulfonation technology will be based on the following targets:

- Reduction of the overall energy demand
- Minimizing operation losses, by-products, and effluents
- Increase of operation flexibility
- Improvement of the obtainable sulfonates quality
- Diversification of the range of commercial product's physical shape

As already discussed, the structure and demand of the surfactant and detergent industries nowadays require the highest production flexibility coupled with the maximum product quality and consistency; these characteristics can only be derived by operating sulfonation in which the technical innovations already available will be a part of the process configuration.

Process units such as the vacuum neutralization (where the heat of reaction is removed by vacuum evaporation of the water formed as by-product) specifically designed for high-quality AES production (see Figure 5.31) and surfactant drying (see Figure 5.32) designed to obtain sulfonates

Production of Linear Alkylbenzene Sulfonate and α-Olefin Sulfonates

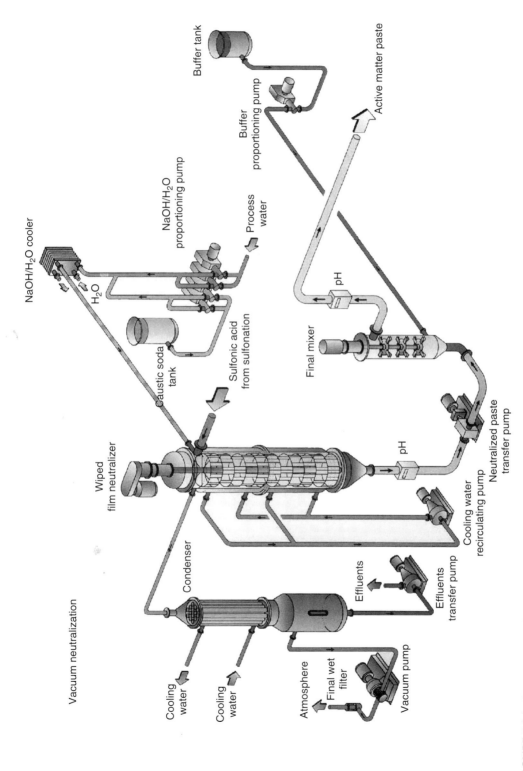

FIGURE 5.31 Vacuum neutralization unit.

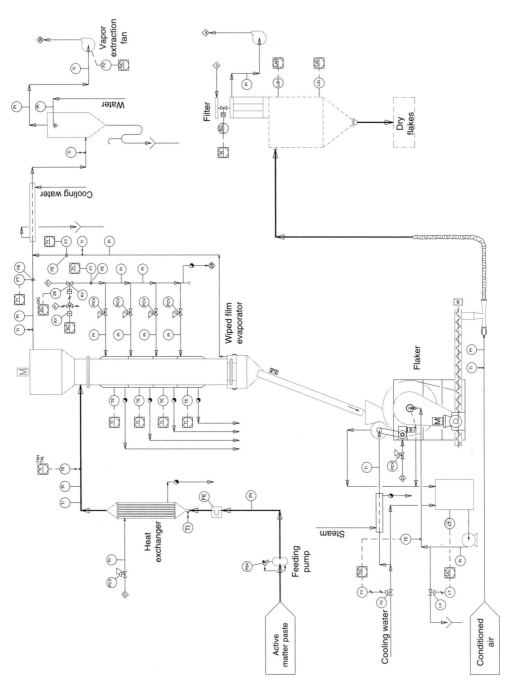

FIGURE 5.32 Anionic surfactant drying unit.

from LAS, AOS, fatty alcohol sulfate (FAS), and methylester sulfonate (MES) in dry shape without changes in their chemical specifications will represent the tool to improve the technical importance of sulfonation (and relevant obtainable products) in the detergent-world scenario.

The continuous improvement in the design and implementation of specific technical solution aimed to reduce the energy demand in the SO_3 gas section of the sulfonation process, such as the partial reblowing of the exhaust gas from the sulfonation, the recovery of the SO_3 generation reaction, and the use of highly efficient moisture absorber and oxidation catalyst are further examples of the extent in which the modern chemical engineering is committed to the development of the SO_3 sulfonation technology.

When the impact of the preceding technical growth is examined under the economic point of view, it is reasonable to expect that the strong technological roots of the SO_3-based process and its continuous development will for sure guarantee to maintain the level of profitability that makes sulfonate production similar to the so-called fine chemicals.

REFERENCES

1. Smulders, E. *Laundry Detergents*, Wiley-VCH, Weinham, pp. 45–52, 2002.
2. Berna, J. L. Raw Material for Sulphonation and Sulphation: Production, Characteristics and Uses. *Proceedings of 1st SODEOPEC Conference*, AOCS, pp. 242–248, 2002.
3. Roberts, D. W. Sulphonation technology for anionic surfactant manufacture. *Org. Proc. Res. Dev.* 2(3), 194–202, 1998.
4. Degroot, W. H. *Sulphonation Technology in the Detergent Industry*, Kluwer Academic, Norwell, MA, pp. 10–11, 1991.
5. Ballestra, M. and Triberti, D. US Patent 713401.
6. Degroot, W. H. *Sulphonation Technology in the Detergent Industry*, Kluwer Academic, Norwell, MA, pp. 132–134, 1991.
7. Johnson, G. and Crynes, B. L. Modelling of a thin film sulphur trioxide sulphonation reactor. *Ind. Eng. Chem. Process Des. Dev.* 13(1), 6–14, 1974.
8. Moretti, G. *NOE' Sergio*. UK Patent GB 204306 7B, 1983.
9. Moretti, G. F., Adami, I., Nava, F., and Molteni, E. The Multitube Film Sulphonation Reactor for the 21st Century. *Proceedings of 5th World Surfactant Congress*, CESIO, vol. 1, pp. 161–170, 2000.
10. Adami, I. Design criteria, mechanical features, advantage and performance of multitube falling film sulphonation reactor. *Tenside Surfact. Det.* 41(5), 240–245, 2004.
11. Moretti, G. F. US Patent 4311650.
12. Berna, J. L., Moreno, A., and Bengoechea, C. LAB Sulphonation: Factors Affecting the Color of Final Sulphonic Acid. *Proceedings of ICSD Conference on Surfactants and Detergents (China – 2002)*, 2002.
13. Adami, I. *New Exhaust Gas Treatment System for Sulphonation Plants*. 7th International Conference on Surfactants and Detergents (paper), Shensen, 2002.
14. Herron, S. J. Alpha olefins and alpha olefinsulphonates. *Chim. Oggi*, 7/8, 1993.
15. Ziegler, K. et al. *Ann. Chem.*, 629, 53, 121, 172, 198, 1960.
16. Fernald, H. B., Gwynn, H. B., and Kresge, A. N. US Patent N. 3.482.000, 1969.
17. Guenter, H. and Hansmann, G. F. US Patent N. 2.094.451, 1937.
18. Yamane, I. *Syn. Org. Chem. Jpn.*, 38, 593, 1980.
19. Roberts, D. W. and Williams, D. L. Why internal olefins are difficult to sulphonate. *Tenside Surfact. Det.* 22, 193D95, 1985.
20. Roberts, D. W. and Williams, D. L. Formation of sultones in olefin sulphonation. *J. Am. Oil Chem. Soc.*, 67, 1020D7, 1990.
21. Roberts, D. W. Kinetics and mechanism in olefin sulphonation. *Riv. Ital. Sostanze Grasse*, 74(12), 567–570, 1997.
22. Roberts, D. W. and Jackson, P. S. Sulphonation of internal olefins. *J. Com. Esp. Deterg.*, 22, 21D35, 1991.
23. Adami, I. Dry-Olefinsulphonates: Cost-Effective and High-Performance Multipurpose Surfactants. *Proceedings of 6th World Surfactant Congress (CESIO)*, 2004.
24. Adami, I. The Role of Sulphonation in the Detergent Industry: Status and Perspectives. *Proceedings of IX Giornate CID*, 2001.
25. Adami, I. Dry-Anionic Surfactants: A Valuable Tool for the Detergent Industry. *Proceedings of XI Giornate CID*, 2005.

6 Production of Alcohols and Alcohol Sulfates

Jeffrey J. Scheibel

CONTENTS

6.1 Introduction to the Development Surfactants: Alcohol Sulfates 117
6.2 Past and Current Alcohols and Alcohol Sulfates 118
    6.2.1 Oleochemical-Derived Alcohols 118
        6.2.1.1 Raw Materials 118
        6.2.1.2 Processes 119
    6.2.2 Petroleum-Derived Linear and Branched Alcohols 120
        6.2.2.1 Raw Materials, Compositions, and Olefin Chemistry 120
        6.2.2.2 Processes: Ethylene Derived 122
        6.2.2.3 Processes: Paraffin-Derived 125
    6.2.3 Sulfation of Natural and Synthetic Alcohols.. 126
6.3 Alternative Alcohol Sulfate Production Technology 128
    6.3.1 Overview of New Technology in Petroleum and Naturally Derived
        Alcohol Sulfates 128
    6.3.2 Petroleum-Derived Alcohols and Alcohol Sulfates 129
    6.3.3 Oleochemical-Derived Alcohols and Alcohol Sulfates... 133
Acknowledgments.. 135
References 135

6.1 INTRODUCTION TO THE DEVELOPMENT SURFACTANTS: ALCOHOL SULFATES

Surfactants, particularly anionic surfactants, have been important to society for many millennia. Surfactants are the most important component of several consumer goods such as laundry detergents, hair shampoos, hard surface cleaners, and industrial cleaners. Laundry detergents are the largest market for anionic surfactants. The needs of the laundry cleaning market have clearly played a key role in shaping the changes that have occurred over the past century in anionic surfactant technology including alcohol sulfates, which are half esters of sulfuric acid derived from long-chain alcohols. These alcohols are produced from both natural and synthetic raw materials today. It is easy to understand why laundry detergents have such a strong influence on the surfactant industry when considering the total cleaning product market is a $60 billion global business. Laundry cleaning products command the largest portion of this market globally. Anionic surfactants are the largest component in laundry detergent formulations. As a result, surfactant worldwide consumption in all markets was a staggering 22+ billion pounds by the year 2000[1] with a large portion attributed to laundry and cleaning products for the home. Detergent alcohols accounted for over 6 billion pounds a year by 2005.[2] The four big alcohol producers'—Shell, Procter & Gamble (P&G), Sasol,

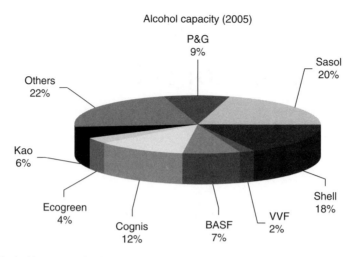

FIGURE 6.1 Alcohol by companies in 2005 as a percentage of the total global capacity.

and Cognis—combined supply is more than half the world volume of alcohol capacity as shown in Figure 6.1.

This large market is completely dependent on the needs of the ever-changing detergent market, which is influenced by laundry consumer habit change and environmental pressures such as reduced energy use in washing machines. Other factors that influence the market and development of alcohol sulfates are raw material costs. Cost can be an agent of change energizing new developments in process technology to reduce production cost and improve efficiency. Historically, petroleum or plant fats and plant-derived oils have been the two main raw material sources for detergent alcohols. New process developments and catalyst technology in the petroleum or fats and oils industry are often reapplied into the surfactant industry since the linkage is strong between these raw materials and the surfactant industry. This chapter explains the raw materials, processes, and the resulting products, which are often complex molecular mixtures and the key drivers for past, current, and future alcohol and alcohol sulfate surfactants process technology development. This chapter also focuses only on the detergent-range alcohols and sulfation of the alcohols, which are defined as alcohols containing 12 or more carbon atoms per molecule and are used mainly in detergent applications. Thus, plasticizer alcohols C_{4-11} are not discussed.

6.2 PAST AND CURRENT ALCOHOLS AND ALCOHOL SULFATES

6.2.1 Oleochemical-Derived Alcohols

6.2.1.1 Raw Materials

The introduction of natural alcohol sulfates dates to around 1932.[3] Natural alcohol sulfates such as soap are derived from natural oils. One of the first laundry products to use the natural alcohol sulfates was Dreft® in 1933, and later Tide® in 1947.[3] Today's natural alcohol sulfates are derived mainly from coconut oil, palm kernel oil, and tallow fat (a by-product of the rendering industry). The compositions of the fatty acids present in tryglycerides that comprise the major raw materials vary substantially with coconut and palm kernel containing high levels of C_{12-14} chain lengths and tallow high in C_{16-18} chain lengths[4] as shown in Table 6.1. As a result, tallow alcohols are used mainly in specialty applications instead of high-volume laundry detergents due to their poor solubility in the presence of water hardness.

TABLE 6.1
Comparison of Fatty Acid Compositions from Natural Sourced Feedstocks

	Coconut	Palm Kernel	Tallow
Saturated Chain Length			
6	0–0.8	0–1.5	—
8	5–9	3–5	—
10	6–10	3–7	—
12	44–52	40–52	0–0.2
14	13–19	14–17	2–8
16	8–11	7–9	24–37
18	1–3	1–3	14–29
20+	0–0.4	0–1	0–1.2
Range (%)	88.5–95	78–86.5	42–57
Unsaturated Chain Length			
14	—	—	0.4–0.6
16	0–1	0–1	1.9–2.7
18	5–10.5	13.4–21	41–55
Range (%)	5–11.5	13.5–22	43–58

Source: Adapted from Swern, D., in *Bailey's Industrial Oil and Fat Products*, 3rd ed., Interscience Publishers, New York, 1964, 176 and 192.

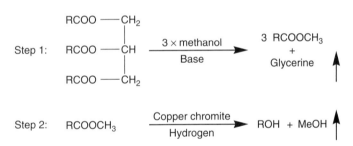

FIGURE 6.2 Natural alcohol process.

6.2.1.2 Processes

The process of converting natural oils and fats into natural alcohols is a two-step process as shown in Figure 6.2. The first step in process is by the reaction of the triglyceride with base catalyst in the presence of methanol at molar ratios of slightly higher than 1:3 triglycerides to methanol. This reaction is called *transesterification* because one ester is exchanged with another. The reaction takes place within minutes of moderate heating. The driving mechanism for the speed of the reaction is the formation of glycerin, a very hydrophilic by-product, which phase separates during reaction from the fatty methyl ester product driving reaction to rapid completion with yields of fatty methyl ester >99%. The glycerin has a higher density than the fatty methyl ester and forms a bottom phase. Minor components such as the base catalyst and other hydrophilic impurities are also retained in the glycerin phase. The top phase, the fatty methyl ester phase, is quite clean and light in color. The ester is usually distilled in preparation for conversion to fatty alcohols. Fractionation can be by individual chain lengths, but majority of the production is cut into three main components—a light fraction (C_{6-10}), a middle fraction (C_{12-14}), and a high fraction (C_{16-18}). The middle fraction or mid cut is the main detergent fraction. Higher cuts find their way into softener actives and other beauty care applications, whereas the light fraction often ends up in the production of plasticizer alcohols for polymers production. The by-product, crude glycerin, must be refined and distilled for sale in other markets. The economics of the processing of fats and oils to surfactants is dependent on the additional value of purified glycerin because glycerin makes up 7–13% of the total production. The second step of fatty alcohol production involves hydrogenation of the methyl esters in a continuous process at pressures above 3000 psi in the presence of a copper chromite catalyst and two equivalents of hydrogen at temperatures above 200°C forming methyl ester and methanol. This process also hydrogenates the unsaturated chain lengths during the process eliminating the need for a prehydrogenation step. Most of the methanol from the process is recycled for use in the transesterification step. The alcohol product (95% overall yield from triglyceride) is fractionated to eliminate small amounts of high boiling impurities and, if desired, to further fractionate the fatty alcohol into various chain lengths for applications such as hair shampoo, which mainly uses C_{12} alcohol for high-foaming characteristics. The ester reduction process is employed by most manufacturers of detergents such as P&G, Cognis, Sasol, and Kao as well as dedicated producers from Indonesia and Malaysia.[5]

The Philippines also produces some natural alcohol but in lower volumes. Fats and oils may be hydrolyzed to the corresponding fatty acids and subsequently reduced catalytically to alcohols. Variations in catalysts and operating conditions can permit retention of double bonds, if desired. Until 1991, the former Sherex Chemical Company, Inc. (now part of Degussa Corporation) used its own variation of this method to produce oleyl and tallow alcohols.

Sasol Olefins & Surfactants GmbH of Germany and United Coconut Chemicals, Inc., in the Philippines produce saturated alcohols by direct reduction of acids.[5]

6.2.2 PETROLEUM-DERIVED LINEAR AND BRANCHED ALCOHOLS

6.2.2.1 Raw Materials, Compositions, and Olefin Chemistry

A number of synthetic routes have been developed for producing detergent range alcohols from petroleum-derived raw materials. The basic raw materials have been ethylene, propylene, and kerosene. Propylene-derived alcohols are still in use today. However, insignificant amounts are used in the detergent alcohol market due to the highly branched nature of the alcohols and, thus, slower to biodegrade making them less suitable for the detergent market. I will briefly explain propylene-derived alcohols in Section 6.3 as it relates to comparison of future technologies.

Ethylene comes from numerous sources including natural gas and wax cracking as a by-product from refining to produce fuel. Kerosene is the source of paraffins for surfactant production. Ethylene and paraffins together are the largest-volume raw materials for production of long-chain olefins for synthetic detergent alcohols. Ethylene is used to produce long-chain olefins for detergent alcohol production by oligomerization using homogeneous transition metal complex catalysts.[6,7] An example of how the coordination-catalyzed process works is shown in Figure 6.3.

The oligomerization process produces homologous, even-numbered, linear alpha olefins. The oligomerization is performed at 80–120°C and at pressures of 1000–2000 psig. A solvent like 1, 4-butanediol is used, in which only the catalyst and the ethylene are soluble but not the formed higher molecular weight olefins. This enables ease of separation of the product from the catalyst by simple phase separation.

The coordination-catalyzed oligomerization of ethylene leads to a mixture of homologous terminal olefins having a broad molecular weight distribution and high purity of 96–98% terminal.

Shell employs the oligomerization process at their Geismar plant in Louisiana and Stanlow plant in England. This process is part of the Shell Higher Olefins Process® (SHOP). The linear terminal olefins can then be used to produce detergent alcohols. The quality of the Shell olefins is shown in Table 6.2.[8] The olefins are extremely low in aromatics, dienes, paraffins, internal olefins, and branched olefins.

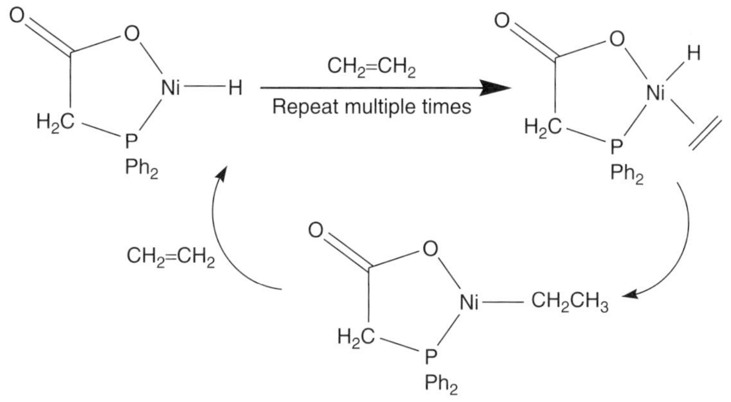

Production of Alcohols and Alcohol Sulfates

FIGURE 6.3 Coordination-catalyzed ethylene oligomerization into *n*-alpha olefins.

TABLE 6.2

Composition of Shell Olefins Compositions from the Shop® Process

Product	C_6	C_8	C_{10}	C_{12}	C_{14}	C_{16}	C_{18}
n-Alphaolefins	98.5	98.5	97.5	96.0	95.0	94.5	94.0
Branched olefins	0.5	0.5	1.0	2.0	3.0	3.5	4.0
n-Omegaolefins	1.0	1.0	1.5	1.5	1.5	1.5	1.5
Paraffins	0.1	0.1	0.1	0.1	0.1	0.1	0.1
Conjugated dienes	0.1	0.1	0.1	0.1	0.1	0.1	0.1
Aromatics	0.1	0.1	0.1	0.1	0.1	0.1	0.1

Source: Adapted from Fell, B., in *Anionic Surfactants Organic Chemistry*, Surfactant Science Series Vol. 56, Marcel Dekker Inc., New York, 1996, 15.

Another part of the Shell process is the isomerization and disproportionation unit. This unit allows Shell to take short-chain olefins of low value such as the excess olefins of C10 or less and mix them with the long-chain low-value olefins of C20+ to give C12–C16 detergent range, higher-value olefins. A result of the process is the formation of internal olefins as well as odd chain lengths due to disproportionation. These internal olefins are then used to produce detergent alcohols through Shell's unique oxo process to be discussed in later sections. The overall SHOP[9] involves an integrated process of steps from the cracking unit producing ethylene all the way to the end–result-desired higher olefins for a wide variety of chemical processes as shown in Figure 6.4.

The other raw materials for production of linear olefins are paraffins. These are used due to low cost of the main feedstock kerosene. Long-chain linear paraffins are produced in a modern production facility by sieving kerosene. Two main processes used today are, gas-phase separation, and the other by UOP, liquid-phase separation. Gas-phase separations are practiced by a number of companies.[10] They supply linear paraffins to the detergent manufacturers directly. Most modern integrated detergent plants use the liquid-phase separation process of UOP called MOLEX®.[11–13] Pretreatment of the kerosene is required before sieving. This involves a hydrogenation step to saturate any aromatics or other components, which could interfere with the sieving process. The molecular sieves adsorb the linear fraction of kerosene and reject the branched and cyclic fractions. Typical sieve sizes are 5 A. A solvent carrier is required to desorb the long-chain linear paraffin from the sieve and is usually a C5–C7 linear hydrocarbon, which is also readily available from a commercial refinery. The second step of the process is catalytic dehydrogenation of the paraffin in a continuous process.

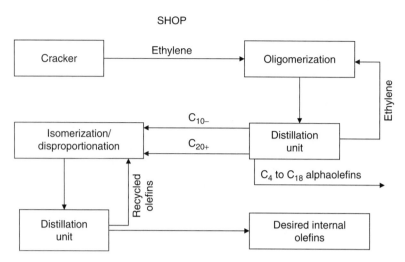

FIGURE 6.4 Illustrative block diagram of SHOP.

Past catalyst technology employed resulted in substantial formation of by-products such as dienes, aromatics and cracking, and isomerization productions. Typically, conversion to olefin would be low—in the range of 10–25%—to minimize formation of these impurities. UOP has developed particularly efficient high activity, selective catalysts such as DeH9,[11] which also has reduced residence times for higher throughput. Typical reaction conditions for a UOP dehydrogenation catalyst are 470–490°C with hydrogen/paraffin ratio of 8:1. Pressures are typically low (30 psig) and conversion can be as high as 30%.[14] The resulting paraffin–olefin mixture formed mostly of internal olefins is used for alkylation of benzene, but is typically not used for alcohol production. A third process step is necessary to concentrate the olefin for hydroformylation. One such process is called OLEX® by UOP.[15] The integrated process for making olefins for detergent alcohol production requires four steps: (1) kerosene hydrogenation, (2) sieving to obtain linear paraffins, and (3) dehydrogenation to a mixture of internal olefins and paraffins, and (4) sieving again to obtain a concentrated internal olefin stream. Figure 6.5 illustrates this process.

6.2.2.2 Processes: Ethylene Derived

Ethylene-derived alcohols were first produced in the early 1960s by two main processes. The hydroformylation process and the Ziegler chemistry are the most important commercial routes today. The Ziegler process involves three basic steps: (1) addition of ethylene to triethylaluminum to build the higher molecular weight trialkylaluminum called the ethylene growth product, (2) oxidation of the growth product with air to form the corresponding aluminum alkoxide, and (3) hydrolysis with water forming a mixture of mostly linear primary alcohols, which have the same number of carbon atoms as the alkyl groups in the trialkylaluminum growth product. Ziegler process is illustrated in Figure 6.6.

Three variations on the Ziegler process are reported. Sasol uses the Alfol® process[16] developed by the former Conoco Chemical Company. This first step chain growth is conducted at relatively moderate reaction temperatures of around 100°C, which minimizes competing displacement reactions that produce by-product olefins. The growth reaction proceeds until an average length of 10 carbon atoms. The distribution of chain lengths in the growth product and derived alcohols conform to a statistical distribution predicted by a Poisson curve yielding a wide spectrum of alcohols from C2 to C28 as shown in Table 6.3.[17]

After the oxidation and hydrolysis steps, the alcohols from this process are highly linear, nearly 100%, resembling oleochemical alcohols since they contain an even number of carbon atoms.

Production of Alcohols and Alcohol Sulfates

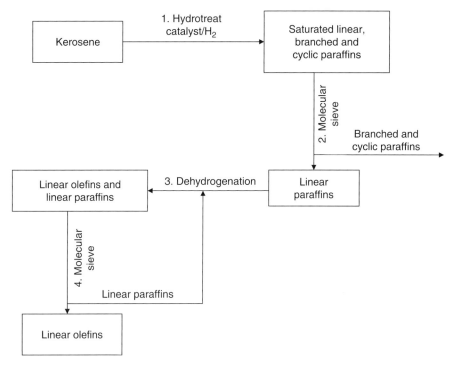

FIGURE 6.5 Olefin process for alcohol production from kerosene.

a. Synthesis of triethyl aluminum

$$Al + 2Al(C_2H_5)_3 + 1.5 H_2 \xrightarrow[70 \text{ bar}]{130°C} 3HAl(C_2H_5)_2$$

$$3HAl(C_2H_5)_2 + 3CH_2=CH_2 \xrightarrow[20 \text{ bar}]{120°C} 3Al(C_2H_5)_3$$

b. Growth reaction

$$Al(C_2H_5)_3 + 3nCH_2=CH_2 \xrightarrow[120 \text{ bar}]{120°C} Al[(CH_2CH_2)_n-C_2H_5]_3$$

c. Oxidation of the trialkylaluminum and hydrolysis of aluminum alcoholates

$$Al[(CH_2CH_2)_n-C_2H_5]_3 + 1.5O_2 \xrightarrow[5 \text{ bar}]{50-100°C} Al[-O-(CH_2CH_2)_n-C_2H_5]_3$$

$$Al[-O-(CH_2CH_2)_n-C_2H_5]_3 + H_2O \xrightarrow{90°C} 3C_2H_5-(CH_2CH_2)_n-OH + Al(OH)_3$$

FIGURE 6.6 Ziegler-alfol process.

TABLE 6.3
Poisson Distribution of Chain Lengths from the Ziegler Process

Chain	C_2	C_4	C_6	C_8	C_{10}	C_{12}	C_{14}	C_{16}	C_{18}	C_{20}	C_{22}	C_{24+}
Percent	1	3	9	18	19	18	14	8.5	5	2.5	1	1

Source: Adapted from Modler, R.F., Gulber, R., and Inoguchi, Y., in *Chemical Economics Handbook*, 2004.

TABLE 6.4
Olefin Chain Composition from the Ethyl Process

Chain	C_2	C_4	C_6	C_8	C_{10}	C_{12}	C_{14}	C_{16}	C_{18}	C_{20}	C_{22}	C_{24+}
Percent	0	1	2	3	7.5	32	24	15	7	3	2	3

Source: Adapted from Modler, R.F., Gulber, R., and Inoguchi, Y., in *Chemical Economics Handbook*, 2004.

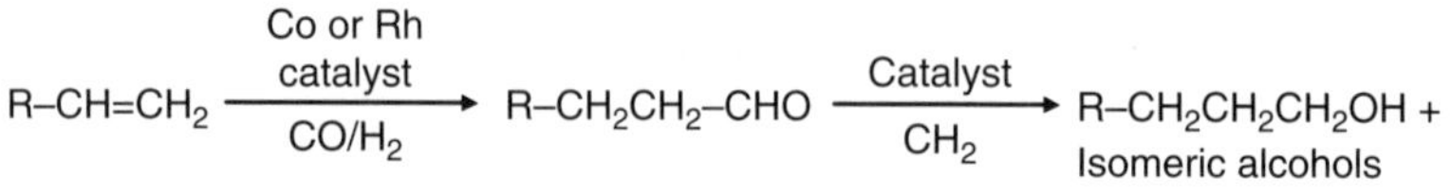

FIGURE 6.7 Hydroforymation of olefins.

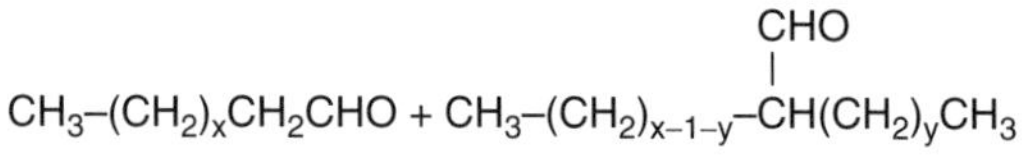

FIGURE 6.8 Linear aldehydes and branched aldehyde isomers.

One drawback of this simple process is lack of flexibility to target chain lengths to match market dynamics.

The Ethyl process[18] maximized the production of C12 and C14 alcohols by a peaking process. Here, multiple tranalkylation reactions are used with lower-chain growth product mixed with higher-chain growth product resulting in a rearrangement between shorter and longer chains giving a peaked chain distribution. Short-chain material is recycled to be reblended and reacted with the long-chain growth product to continue the peaking process. Thus, about 85% of the products are C12 and higher with about 95% linearity as shown in Table 6.4.[17]

The Ethyl process was costly relative to the simple one-step process, which is likely the reason for Ethyl selling their plant to BP that was later shut down in 2002 because it could not compete with the other process routes such as the simpler single-step and oxo processes.

The hydroformylation of olefins into aldehydes is called the oxo process, which was discovered in 1938 by O. Roelen and was later applied to synthesis of longer detergent-range alcohols by Shell.[19] The aldehydes can then be easily hydrogenated to give alcohols. The oxo reaction, as applied by various producers today in the synthesis of detergent-range alcohols, is currently employed in multiple modifications commercially. All involve reaction of olefins with synthesis gas, carbon monoxide and hydrogen, in the presence of an oxo catalyst to yield a product that is a one carbon higher alcohol. The major differences in processes are type of olefin, catalysts, cocatalysts, stoichiometry, process conditions, catalyst recovery, handling of intermediates, and product composition. Most processes, with the exception of Shell, isolate the intermediate aldehydes, then purify, and finally hydrogenate the aldehyde in a second reactor to make detergent alcohols as in Figure 6.7.

Catalyst and olefin feed are the key to what type of alcohols are formed. The industrially relevant catalysts employed are cobalt carbonyl[20] or cobalt carbonyl/*tert*-phosphine complexes.[21–24] In the hydroformylation of an olefin, linear terminal aldehydes are formed along with other isomeric 2-alkyl-branched aldehydes[25,26] as noted in the structures in Figure 6.8.

The formation of isomeric 2-alkyl-branched aldehydes is caused by the cobalt organic intermediates, which are formed by the reaction of the olefin with the cobalt carbonyl catalyst. These intermediates isomerize rapidly into a mixture of isomer positions on the cobalt catalyst–olefin compound.

Production of Alcohols and Alcohol Sulfates 125

The primary cobalt organic compound carrying a terminally fixed metal atom is thermodynamically more stable than the isomeric internal secondary cobalt organic compounds. Thus, the terminal cobalt organic compound, which is less sterically hindered, will further react in the catalytic cycle favorably generating the terminal or unbranched alcohol product independent of the position of the double bond in the olefin reactant ("contrathermodynamic" olefin isomerization).[25,26] The ratio of terminal linear alcohol to internal alcohol can further be enhanced by additional steric hindrance to the metal through phosphine ligands. Thus, if one desires to produce more linear alcohols from internal olefins of the type generated by the kerosene feedstock described earlier, then the use of a catalyst with phosphine ligands is required. If substantial 2-alkyl alcohols are desired, then cobalt carbonyl is the catalyst of choice. One advantage of the phosphine-containing catalyst is its lack of volatility making it possible to separate the catalyst from the alcohol product by distillation. For production of Shell's Neodol alcohols® by conversion of internal olefins from the SHOP, shown earlier, Shell is believed to employ a modified cobalt catalyst with sterically hindered phosphine ligands. Compared with others, Shell's oxo reaction is also unique in another aspect. Shell's process combines both hydroformylation and reduction in one step, eliminating the need for isolation of the aldehyde intermediate and further conversion to alcohol. Thus, Shell has the highest terminal alcohol content of all catalyst systems at 80% or greater.[27]

Other large-volume producers of alcohols using the oxo process are BASF and Noroxo S. A. in western Europe. BASF and Noroxo S. A. convert both linear alpha olefins and other olefins using nonselective oxo catalysts resulting in much higher isomeric products than Shell's alcohols.

6.2.2.3 Processes: Paraffin-Derived

The largest volume production of detergent-range alcohols derived from kerosene feed today is from Sasol's Augusta plant in Italy. Sasol's integrated facility also produces large volumes of alkyl benzenes from kerosene feed.[28] Sasol uses a nonselective oxo catalyst (no ligands). The result is a substantial formation of 2-alkyl branched alcohols. The trade name for their alcohol is LIAL® Alcohol. They produce both LIAL 123 alcohol (C12/13) and LIAL 145 alcohol (C14/15). Compositions of the LIAL 123 alcohol and LIAL 145 alcohol are shown in Table 6.5.[29] The main application of the resulting product alcohol sulfates is in granular laundry products and dishwashing liquid detergents.

Other alcohols from the Augusta facility include some production of ALCHEM® 123 and 145 and ISALCHEM® 123 and 145 alcohols.[30] These alcohols are obtained from freeze fractionation of some of the LIAL 123 and LIAL 145 alcohols. This process is similar to the one practiced in the fats and oils industry called *winterization*. The neat alcohol is cooled with or without a solvent. The linear terminal alcohols have fairly high melting points and crystallize from the liquid.

After cold filtration, ISALCHEM is the remaining liquid. Analyses of these alcohols are shown in Table 6.6.[30] The properties of the ISALCHEM alcohols make them suitable for use in numerous products, but are used only in specialty applications due to the limited commercial volume available today. The linear crystalline fraction is designated as ALCHEM alcohol and is limited in supply.

TABLE 6.5

Sasol's Lial Alcohol® Chain Compositions and Properties

	<C12	C12	C13	>C13	Linear	Isomeric	Pour Point (°C)	Flash Point (°C)
LIAL 123A	0.5	42	56	1.5	46	54	5	126
	<C14	C14	C15	>C15				
LIAL 145A	1	62	36	1	40	60	15	140

Source: Adapted from Author Unknown, *LIAL® Alcohols for North America*, Sasol North America Inc.

TABLE 6.6

Sasol's Isalchem Alcohol® Chain Compositions and Properties

	<C12	C12	C13	>C13	Linear	Isomeric	Pour Point (°C)	Flash Point (°C)
ISALCHEM 123A	0.5	41	55	2.5	6	94	−45	137
	<C14	C14	C15	>C15				
ISALCHEM 145A	1.5	59	39	1	5	95	−31	146

Source: Adapted from Author Unknown, *ISALCHEM® Alcohols for North America*, Sasol North America Inc.

$$R-CH_2-O-H + SO_3(g) \longrightarrow R-CH_2-O-SO_3H$$

FIGURE 6.9 Sulfation with SO_3.

Some smaller amounts of alcohols are also produced from paraffins by direct oxidation in the liquid phase. This causes considerable scission of the chain and gives a wide variety of oxygenated products including peroxides, alcohols, acids, ketones, and esters. The resulting product is a secondary alcohol with random position on the chain except for the terminal position. This chemistry is a free-radical oxidation and, thus, only secondary radicals are sufficiently stable to react further with air (see Refs 31 and 32 [Chapter 4]).[31,32] The producer today is Nippon Shokubai Company, Ltd., but the volume is very limited. Another alcohol of limited commercial volume is called guerbet. This alcohol is produced by condensation of two shorter-chain alcohols. The process proceeds through formation of an aldehyde intermediate, which then proceeds through an aldol condensation resulting in dehydration followed by hydrogenation giving the resulting guerbet alcohol with 2-alkyl-specific branching. 2-Butyl-1-octyl alcohol is an example of a guerbet reaction of a 1-butyl alcohol and 1-octyl alcohol. Cross condensation also occurs forming the 2-butyl-1-butanol and 2-octyl-1-octanol.

6.2.3 Sulfation of Natural and Synthetic Alcohols

Sulfation involves a reaction where a –COS– linkage is formed by the action of a sulfating agent on, in this case, an alcohol as shown in Figure 6.9.

Sulfation of alcohols is extremely exothermic releasing 35.8 kcal/mole of energy.[33] The mechanism of sulfation is believed to go through a metastable pyrosulfate species that decomposes rapidly to alkyl hydrogen pyrosulfate, $ROSO_2OSO_3H$, which reacts with a second alcohol to produce an alcohol sulfate product mixture.

Unlike sulfonates, such as linear alkyl benzene sulfonates, sulfates of alcohols are readily susceptible to hydrolysis in acidic media. Alcohol sulfates and alkoxylated alcohol sulfates can be produced from the corresponding alcohols and alkoxylated alcohols using a wide variety of reagents including chlorosulfuric acid, sulfur trioxide, sulfuric acid, and sulfamic acid as shown in Table 6.7.[34] The resulting surfactants from all these reagents give similar properties with the exception of chlorosulfuric acid. The product is slightly better in color, but requires disposal of HCl. The production of alcohol sulfates was first performed on natural alcohols with concentrated sulfuric acid and was used mainly before the early 1940s. This reaction is reversible and, thus, requires molar excess of the acid shown in Figure 6.10.

Relatively low conversions are achieved, no matter what operating conditions are used. Besides the reversibility of the reaction, the low yield is also a result of several side reactions involving

TABLE 6.7
Alcohol Sulfonation Reagents

Sulfonation Reagent	Chemical Formula	Reaction with SO_3
Sulfur trioxide	SO_3	—
Sulfuric acid	H_2SO_4	$H_2O + SO_3$
Oleum	$H_2SO_4 \ nSO_3$	$H_2O + (n + 1)SO_3$
Chlorosulfuric acid	$ClSO_3H$	$HCl + SO_3$
Amidosulfonic acid (sulfamic acid)	H_2NSO_3H	$NH_3 + SO_3$
Sulfur trioxide complex	Base SO_3	Base $+ SO_3$

Source: Adapted from Schaurich, K., *J. Prakt. Chem.*, 15, 332, 1962.

$$R–CH_2–O–H + excess \ H_2SO_4 \longrightarrow R–CH_2–O–SO_3H + Water$$

FIGURE 6.10 Sulfation with sulfuric acid.

formation of dialkyl sulfates, dehydration to olefins, formation of aldehydes, and decomposition to short-chain hydrocarbons.[35] If the alcohols contain unsaturation, then reaction of the double bond can occur forming other products such as ethers and alkylene disulfates.[36] Sulfation with theoretical amounts of sulfuric acid at 20–40°C gives yields of 25–55%. Theoretically, if 100% excess sulfuric acid is used, the yield rises to 83% for sulfation of primary alcohols.[37] Chlorosulfuric acid was widely used in the past for sulfation of detergent alcohols.[38] For batch processes, using a glass-lined stirred reactor with gradual addition of the $ClSO_3H$ to the alcohol over a period of several hours at temperatures of 26–32°C with evolution and removal of HCl aided by a slow continuous nitrogen purge is a general procedure. Stepan utilized such a process from 1952 to 1962.

Another approach with better results involves rapid mixing of acid and alcohol, combined with vigorous air sparging to remove the generated HCl, which improves the process. In continuous sulfation with chlorosulfuric acid, the reactants flow through concentric tubes.[39] The reaction mixture is quenched in recirculated product and effluent gas is further separated. A continuous $ClSO_3H$ sulfation process has been patented and used by Henkel for production of fatty alcohol sulfates.[40,41] Fatty alcohol and $ClSO_3H$ are continuously injected into the bottom of a 1 cm annular space within a concentrically cooled, vertically tapered spiral reactor. The reaction is conducted at 30°C with reaction mixture propelled upwardly, owing to the HCl gas generated by the reaction. Product residence times are around 1–2 min. This process produces excellent quality products, but requires refrigeration. The lower cost of liquid SO_3, *in situ* sulfur burner generated gaseous SO_3 reagents and problems, and additional cost associated with handling and disposal of HCl has largely displaced the use of chlorosulfuric acid for sulfation of fatty alcohols.

Oleum, a mixture of sulfuric acid and SO_3 that is obtained when sulfuric acid is concentrated to 98–100%, has also been employed in sulfation of alcohols. The formation of large quantities of sodium sulfate as by-products of neutralization is a consequence of oleum sulfationare formed. This has been advantageous for making granular detergent products. However, this is a disadvantage for liquid laundry products and dishwashing products, since salts limit the solubility and ability to concentrate the surfactant formulations. Thus, as liquid detergents have grown in market, particularly in North America, and heavy-duty granular detergents have declined, the use of oleum for sulfation has also declined.

Today, by far, the most widely used process for sulfation of alcohols is surfur trioxide. Numerous process variations are commercial around the globe. The key feature of all is continuous processing.

Stepan was the first to develop and commercialize a continuous falling film SO$_3$ sulfonation process. The design is a multitubular unit.[42] The company operates about 12 falling film SO$_3$ sulfonation units in the United States, not only for the production of linear alkyl benzene sulfonates, but substantial amounts of fatty alcohol and fatty alcohol ethoxylates are also sulfated. Other key commercial reactor designs are by Chemithon, Ballestra SpA, Lion, Mazzoni SpA, and Meccaniche Moderne.[32,43–49] Several features are common to all falling-film systems. Fatty alcohol and alcohol ethoxylates are reacted at a rate of about 0.3 kg/h/mm with SO$_3$ concentration at about 2–3%. Liquid residence times are estimated at 10–30 s and most units operate with gas velocities in the range of hurricane wind velocities (121–322 km/h).[50] Linear alcohols and linear alcohol ethoxylates are by far the easiest to sulfate. Caution is required with branched alcohols as color and conversion can suffer.

6.3 ALTERNATIVE ALCOHOL SULFATE PRODUCTION TECHNOLOGY

6.3.1 Overview of New Technology in Petroleum and Naturally Derived Alcohol Sulfates

Most of the technologies discussed in the past had one theme; they focused mainly on utilizing existing feedstocks from either the oleochemical industry or the petroleum industry. Since the advent of Katrina and the rising price of oil, alternative feedstock sources have begun to take on new relevance for chemical industry including feedstock for detergent alcohols. Predicting the future of both feedstocks that will be available and the technology to reapply into production of detergent alcohols and surfactants is challenging. However, many choices are emerging as possible alternatives. A few of the possible new sources of raw materials are shown in Figure 6.11.[51]

There is activity to develop a variety of plant, animal, gas, and even coal. For years to come, we completely expect that surfactants will be derived from a combination of green- and petroleum-based sources. Moreover, as natural and bioderived sources become cheaper, and as such attract more investment, natural sourced feedstocks will compete head to head with synthetics and perhaps, someday, dominate the mix as a supply source for the surfactant industry.[52] New process technology may also begin to play a role to change how we apply even natural oils to production of surfactant alcohols. New research in the fuels industry by Neste Oil has now demonstrated an ability to convert fats and oils into paraffins, making potential feedstock for synthetic surfactants

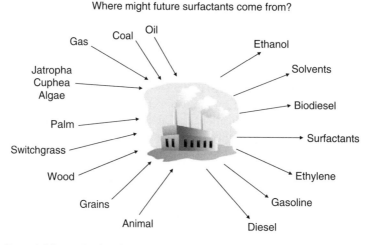

FIGURE 6.11 Potential future feedstock sources for fuels and chemicals.

Production of Alcohols and Alcohol Sulfates

derived from naturally sourced materials.[53] Neste's process called NexBTL is in the commercialization stage. Even chicken fat will not be converted into a product resembling fuel by a recent partnership between Tyson Chicken and Syntroleum Corporation.[54] This then opens up the potential for reapplication of natural oils and fats to make synthetic-type detergent alcohols through the paraffins produced using processes as described earlier in this chapter.

Other factors affecting the detergent industry include colder wash temperatures, less efficient builders, liquid products without calcium control, and the push for reduced surfactant use overall due to the perceived environmental impact of surfactants. Many of today's mainframe linear surfactant technologies such as linear alcohol sulfates (LAS) described earlier cannot address all of these challenges. The poor solubility of LAS forces formulators to use more and more linear alcohol ethoxylated sulfates, which have higher solubility. However, the ethoxylated sulfates have lower surface activity as well as lower mass efficiency than the LAS. LAS is also not a solution, because cost is now comparable to detergent alcohol sulfates. Thus, a challenge for the present and future surfactants including fatty alcohol sulfates is to provide high solubility with maximum hydrophobicity for high efficiency and cost-effective formulation. Surfactants must work under reduced temperature, reduced total water including the rinse water, or reduced wash time to meet the demands of today's fast pace society. Thus, a need exists now and in the future for alcohol sulfate surfactants, which do not suffer from the compromise between cold, hard-water solubility and high surface activity, while having a favorable environmental profile and comparable economics versus current LAS, alcohol sulfates (AS), and alcohol ethoxylated sulfates (AES). Compaction is also driving product form and requires even higher formulatibility than earlier surfactants. New research in this direction related to alkyl sulfates has been reported and are discussed.

6.3.2 PETROLEUM-DERIVED ALCOHOLS AND ALCOHOL SULFATES

Recently, a number of companies have been exploring alternative alcohol technologies. One such company is P&G. P&G sought new surfactant technology that is highly surface-active and weight-efficient to compete economically versus current commodity surfactants, although maintaining a good environmental profile. A surprising finding was that simple substitution of a short alkyl group, as shown in Figure 6.12, in the middle of the hydrophobe severely disrupted the crystallinity of the surfactant in solution.[55]

A specific example, using model surfactant structures, was studied in the P&G laboratories. Snoody found that the Krafft boundary could be altered dramatically by addition of a small methyl substituent on the chain of a methyl-substituted octadecylsulfate. P&G[56-58] has published some work on mixtures of midchain methyl-substituted alcohol sulfate surfactants. The resulting increased solubility of these alcohol sulfates called highly soluble alcohol sulfates (HSAS) allows use of long-chain alcohols, such as C_{14-17}, in laundry detergents under today's cold and hard water environments. The solubility effect is maximized when the methyl substitution is away from the end of the surfactant chain as shown in Figure 6.13.

Unlike C_{14-15} AS, HSAS is extremely tolerant of calcium even with a chain length of C_{16-17}. HSAS, with its surface activity and solubility, is weight effective and can be used by the formulator to deliver superior cleaning performance versus traditional surfactants. HSAS can also be used to optimize surfactant level in laundry detergent products. P&G partnered with Shell to develop HSAS.

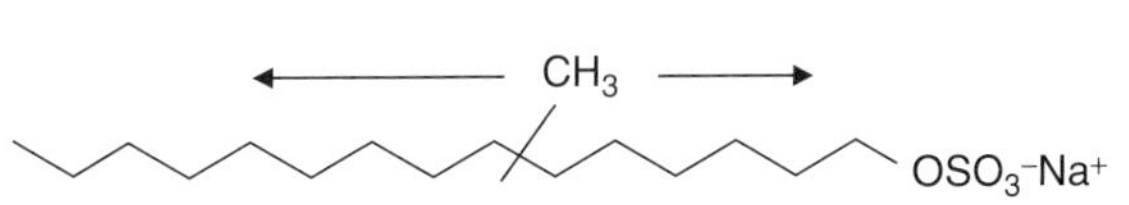

FIGURE 6.12 Illustrative structure of the HSAS molecule.

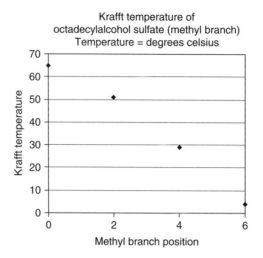

FIGURE 6.13 Effect of methyl position on Krafft temperature for pure isomers of isooctadecylalcohol sulfate.

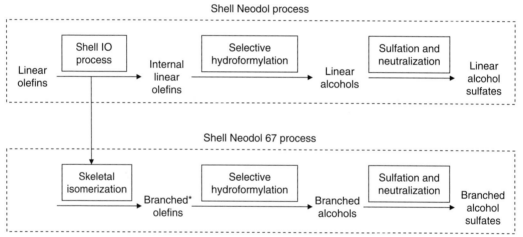

* Branches are mainly methyl and ethyl substitutions.

FIGURE 6.14 Comparison between Shell Neodol process and new Neodol 67 process.

Shell[59–64] published information on the biodegradability of methyl-substituted surfactants. The Shell and P&G research paved the way for new anionic surfactant design with controlled alkyl substitution on the hydrophobe and acceptable biodegradability. Selective isomerization of the linear olefins using recent advances in zeolite catalytic sieve technology[65] are the key technology advancements for these new surfactant hydrophobes, which require only one additional step in the current Shell linear alcohol process as shown in Figure 6.14.

Catalytic sieves make viable the controlled incorporation of methyl substitution on the parent linear olefin.[66–68] The HSAS' added efficiency and performance compensate for the outcome of an additional process step in the standard alcohol processing.

Another recent breakthrough is the commercialization of alcohols from coal using Fischer–Tropsch (FT) technology practiced by the synthetic fuels expert, Sasol. The technology produces a

Production of Alcohols and Alcohol Sulfates

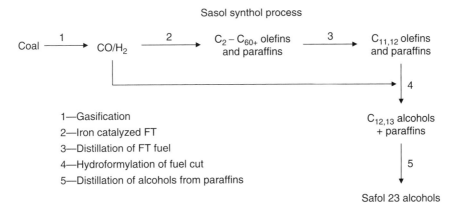

FIGURE 6.15 Production of Safol alcohols by the Synthol process.

mixture of linear and nonlinear surfactant alcohols.[69,70] The raw materials are derived from Sasol's Synthol® process. Sasol has made several announcements on the technology, which is marketed under the trade name Safol®. The material produced to date in the alcohol plant is a mixture of C_{12-13} alcohols. The production of Safol alcohols by the Synthol process, which produces linear and nonlinear alcohols in five steps from coal, is shown in Figure 6.15. This stands in some contrast to the Shell Neodol® 67 process, which also requires six steps: cracking of petroleum, separation of ethylene, ethylene oligomerization, isomerization/disproportionation, distillation, and finally hydroformylation, as well as uses the more expensive petroleum feedstocks. Thus, the Safol process should be a lower-cost process based on processing costs and raw materials costs. Sasol has the capability to also produce C_{14-15}-type Safol alcohols through the Safol process.[71]

This trend toward new uses of low-cost feeds for adding value to surfactants is continuing. Recently, BASF[72] reported specific C_{13} detergent alcohols from hydroformylation of a C_{12} olefin derived from dimerization of 3-hexene, a by-product of butane metathesis to propylene.[73,74] Since 3-hexene appears to have no direct market value compared to ethylene, expectations are that the cost of the alcohol will be competitive with the ethylene-based alcohols.

The surfactant industry of the future will continue to face many challenges. Commodity surfactants will continue to be driven by cost and environmental safety. The market will demand reliable low-cost supply and environmental acceptance. New surfactant technology will be a portion of the future revolution in surfactants. This innovation will likely come from gas to liquids (GTL[75,76]) technology and catalyst/process breakthrough such as those already demonstrated by Sasol. GTL is the general process name for conversion of natural gas, coal, biomass, or other carbon-containing raw materials to higher liquid hydrocarbons, and more specifically to naphtha, jet/diesel, and diesel fuel.

A more detailed description of the GTL process is (1) the production of synthesis gas (syngas) from carbon-based feedstocks, (2) the FT reaction for conversion of syngas to higher liquid hydrocarbons and long-chain waxy paraffins, and (3) the conversion of the solid long-chain waxy paraffins through hydrocracking and hydroisomerization back to liquid fuels.

There are a number of suitable carbon sources for syngas including natural gas, coal, and heavy petroleum to name a few. High sulfur coal requires the use of iron catalysts. The original high temperature process of Sasol in South Africa uses coal, an iron catalyst, and high temperatures to convert syngas to fuel as described earlier. This process produces substantial olefins content along with paraffins. The new cobalt processes for natural gas conversion produce high-purity paraffins using lower temperatures and lower pressures. Low-temperature iron processes are also being developed and produces the highest olefin content with the paraffin formed although details of the composition are not widely published. Wax-cracking processes are critical to a GTL plant because wax is a major product of the FT process. New processes[77–82] have been developed with state-of-the-art catalytic zeolitic sieves, which can control the cracking and isomerization process of the GTL wax to produce linear and lightly branched paraffins for fuel or other use such as surfactant feedstocks as shown in Table 6.8.[83]

Many announcements have been made for new GTL plants from all the large oil producers. The key players BP, Chevron, ConocoPhilips, ExxonMobil, Sasol, and Shell all have plans for large-size plants. The volume of GTL-based fuels, by the year 2020, could be over 833,000 bbl/day or about 175 million pounds a day.[84] A medium size plant alone will produce 22 million pounds a day. Thus, a medium-sized GTL plant could produce enough paraffin in about 2 weeks to run a large-scale fatty alcohol plant. Conceptually, the number of process steps from feedstocks to key surfactant precursors like fatty alcohols could be lowered from five, such as that practiced in Sasol's Augusta plant or South Africa coal conversion process to four if the FT wax were processed through the old steam-cracking process of the wax versus hydrocracking as shown in Figure 6.16.

GTL plants will also make lightly branched paraffins from wax cracking. The branched paraffins can then be used to produce the lightly branched alcohol technologies resembling the Shell Neodol 67 from coal instead of crude oil. More cost-effective and efficient ways to produce surfactants from alternative feedstocks will likely continue to be pursued in the industry to remain

TABLE 6.8
500–555°C Distillation Fraction from Hydroisomerization and Hydrocracking of an FT Wax

Carbon No.	n-Paraffin Weight (%)	Monomethyl Paraffins (Wt. %)	Other IsoParaffins (Wt. %)[a]
14	9	2	1
15	27	10	4
16	14	12	12
17	0	2	7
Total	50	26	24

[a] Other isoparaffins include both multimethyl branched species (e.g., dimethyl and trimethyl) and higher carbon number branches (e.g., ethyl and propyl).

Source: Adapted from Wittenbrink, R.J., Mart, C.J., Ryan, D.F., and Cook, B.R., *High Performance Environmentally Friendly, Low Temperature, Drilling Fluids*, US Patent 6,455,474, 2002.

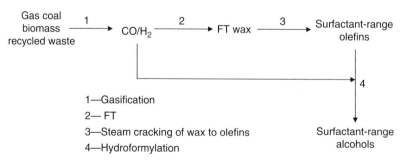

FIGURE 6.16 Conceptual diagram of future alcohol production process from FT.

competitive in the marketplace. Thus, GTL is one of the keys to the future of surfactants to meet consumer needs using clean natural gas to make clean surfactants. Several articles have stated that economics of large GTL facilities based on today's technology will require crude oil to remain above $40 a barrel.[85] If today's trend toward $60+ a barrel holds, then GTL is indeed a profitable venture and a key low-cost source of feedstock paraffins, isoparaffins, olefins, and isoolefins for the detergent alcohol markets of the future.

6.3.3 Oleochemical-Derived Alcohols and Alcohol Sulfates

The majority of the world's oleochemical-derived alcohol sulfates comes from four main sources today—coconut, palm, palm kernel, and tallow. The majority of surfactant alcohol volume comes from coconut and palm kernel oil due to the high content of C_{12-14} for fatty-alcohol production as discussed earlier. However, several new initiatives are now underway to broaden the available existing sources of plant oils.

Cuphea is a new plant crop being explored as a potential U.S. replacement for coconut and palm kernel; see Figure 6.17. *Cuphea* is a genus of about 260 species of annual and perennial flowering plants growing in warm temperate to tropical regions and is indigenous to the Americas. The species range from low-growing herbs to semiwoody shrubs up to 6 ft tall.[86]

The tall species have potential for use as oilseed crops.[87] Recently, there has been a growing interest in *Cuphea* by the U.S. Department of Agriculture (USDA) because many of the species have oilseeds, which contain medium chain length ($C_{12,14}$) triglycerides resembling coconut and palm kernel oil as shown in Table 6.9.[88] *C. painteri*, for example, is rich in caprylic acid (73% C_{10}), whereas *C. carthegenensis* oil consists of 81% lauric acid (C_{12}). *C. koehneana* oil may be the richest source of a single fatty acid, with 95% of its content consisting of capric acid (C_8).[88] Several varieties, which can be grown in the plains states, are being developed by the USDA.[89]

Currently, key issues for crops being explored for plains states are yield per acre, increasing C_{12} content of varieties. Full commercialization may still be many years away.

Jatropha curcus is another plant being developed today for its oil-bearing fruit. Biodiesel and fuel oil are key reasons for its development today. Both India and Africa are pursuing this crop as it can grow on sparse soil and in dry climates. *Jatropha*, shown in Figure 6.18, from a surfactant fatty alcohol feedstock is much less interesting, since it mainly contains C_{18} unsaturated fatty acids with limited C_{12} and C_{14} fatty acid content and, thus, bears no further discussion.

FIGURE 6.17 Close-up of *Cuphea* plant with seeds and harvesting of experimental crop.

TABLE 6.9
Comparison of *Coconut* Tryglyceride Chain Compositions to Various *Cuphea* Species

Species	C8 (%)	C10 (%)	C12 (%)	C14 (%)	Other (%)
C. painteri	73.0	20.4	0.2	0.3	6.1
C. hookeriana	65.1	23.7	0.1	0.2	10.9
C. koehneana	0.2	95.3	1.0	0.3	3.2
C. lanceolata		87.5	2.1	1.4	9.0
C. viscosissima	9.1	75.5	3.0	1.3	11.1
C. carthagenensis		5.3	81.4	4.7	8.6
C. laminuligera		17.1	62.6	9.5	10.8
C. wrightii		29.4	53.9	5.1	11.6
C. lutea	0.4	29.4	37.7	11.1	21.4
C. epilobiifolia		0.3	19.6	67.9	12.2
C. stigulosa	0.9	18.3	13.8	45.2	21.8
Coconut	8.0	7.0	48.0	18.0	19.0

Source: Adapted from Kleiman, R., *Advances in New Crops*, Timber Press, Portland, OR, 1990, 196–203.

(a)

(b)

FIGURE 6.18 *Jatropha* plants growing in India.

The most recent and interesting new development is the exploration of microalgae as future biodiesel fuel source. Microalgae are small single-celled organisms with more than 60,000 species. Algae are usually found in damp places or bodies of water and are thus, common in terrestrial as well as aquatic environments. Like plants, algae require three primary components to grow: sunlight, CO_2, and water. Prehistoric algae are believed to be the source of a substantial number of large petroleum oil reserves and exploration for new reserves today often start with a mapping of potential past microalgae sites. Microalgae in the oceans are responsible today for a major portion of the oxygen generated by all plants on earth and can generate massive blooms in the ocean under the right conditions as shown in Figure 6.19. This algae bloom recorded by National Aeronautics and Space Administration (NASA) is the size of Wisconsin. Algae contain 1–70% triglycerides, 20–70% protein, and 10–60% carbohydrates.[90] A substantial number of algae can be manipulated to consistently produce over 50% oil by weight. Compositionally, most microalgae

Production of Alcohols and Alcohol Sulfates 135

FIGURE 6.19 NASA Modis satellite image of algae bloom of the size of Wisconsin.

resemble common biodiesel sources today, such as palm oil, soybean oil, and canola oil. They are high in C_{18} unsaturated fatty acids with some C_{16} fatty acid content. However, no real effort has been undertaken to seek algae species for shorter chain-length composition and it is believed that the potential to find such species exists.

Earlier studies[91] on microalgae took place during 1978–1996 with projects funded by the National Renewable Energy Lab (NREL). The major conclusion of the study in a closeout report by NREL in 1998 showed that algae oil-derived biodiesel is, even with all the improvements, at least two times higher in cost than crude oil-derived diesel. Other results of the NREL report (Ref. 91, pp. 28–31) indicate that some species contain substantial amounts of unusual components not found in other plants such as C13-0, C14-1, C15-0, C16-1, and even some with methyl-branched fatty acid content in concentrations of 3–5% to as much as 15–20%.

Only time will tell if these new oil sources will bear fruit in the world of low cost and new alternatives for the fatty alcohol surfactant market.

ACKNOWLEDGMENTS

I thank the Procter & Gamble Company for allowing me to write this chapter and all my industry and academic colleagues who shared their wisdom in surfactant technology and their assistance on data presented herein.

REFERENCES

1. Challenger, C., Soaps and detergents—industry overview in chemical market, *Reporter*, 263(4), 2003.
2. Unknown Author, Soap and Detergents History, Surfactant and Detergents Association, www.sdahq.org/sdalatest/html/soaphistory1.htm.
3. Dyer, D., Dalzell, F.F. and Olegario, R., *Rising Tide*, Harvard Business School Press, Boston, MA, Chapter 4, pp. 70–79, SRI Consulting, Menlo Park, CA, 2004.
4. Swern, D., *Bailey's Industrial Oil and Fat Products*, 3rd ed., Interscience Publishers, New York, pp. 176 and 192, 1964.
5. Modler, R.F., Gulber, R. and Inoguchi, Y., *Chemical Economics Handbook*, Detergent Alcohols, Section 609.5000A-609.5002S, SRI Consulting, Menlo Park, CA, 2004.
6. Freitas, E.R. and Gum, R.C., *Chem. Eng. Progress*, 75(1), 73, 1979.
7. Spitzer, E.L.T.M., *Seifen-Ole-Fette-Wachse*, 107, 141, 1981.

8. Fell, B., *Anionic Surfactants Organic* Chemistry, Surfactant Science Series Vol. 56, Chapter 1, Table 8, Marcel Dekker, Inc., New York, p. 15, 1996.
9. Hons, G., *Anionic Surfactants Organic Chemistry,* Surfactant Science Series Vol. 56, Chapter 1, Figure 5, Marcel Dekker, Inc., New York, p. 50, 1996.
10. Hons, G., *Anionic Surfactants Organic Chemistry,* Surfactant Science Series Vol. 56, Chapter 1, Table 2, Marcel Dekker, Inc., New York, p. 47, 1996.
11. Sterba, M.J., *Hydrocarbon Proc. Petrol. Ref.*, 44(6), 151, 1965.
12. Broughton, D.B. and Carson, D.B., *Hydrocarbon Proc. Petrol. Ref.*, 47(9), 238, 1968.
13. Ritzer, H. and Cremer, G., *Erdol und Kohle-Erdgas-Petrochemie*, 22(3), 132, 1969.
14. Vora, B.V., Pujado, P.R. and Spinner, J.B., *Hydrocarbon Proc.*, 86, 1984.
15. Vora, B.V., U.S. Patent 5,300,715, 1994.
16. Modler, R.F., Gulber, R. and Inoguchi, Y., *Chemical Economics Handbook*, Detergent Alcohols, Section 609.5000 M-609.5000 N, SRI Consulting, Menlo Park, CA, 2004.
17. Modler, R.F., Gulber, R. and Inoguchi, Y., *Chemical Economics Handbook*, Detergent Alcohols, Chart, Section 609.5000 N, SRI Consulting, Menlo Park, CA, 2004.
18. Modler, R.F., Gulber, R. and Inoguchi, Y., *Chemical Economics Handbook*, Detergent Alcohols, Section 609.5000 N, SRI Consulting, Menlo Park, CA, 2004.
19. Verbrugge, P.A., U.S. Patent 3,776,975, 1973.
20. Radici, P., Cavalli, E., Veluscek, E. and Braca, G., *Proceedings of the 2nd CESIO World Surfactants Congress*, Vol. 1, Paris, pp. 164–176, 1988.
21. Slaugh, S.H. and Mullineaux, R.D., *J. Organomet. Chem.*, 13, 469, 1968.
22. Spooncer, J., *J. Organomet. Chem.*, 13, 327, 1969.
23. Slaugh, S.H. and Mullineaux, R.D., Belg. Patent 606,408, 1962.
24. U.S. Patent 3,448,157, 1969.
25. Asinger, F. and Berg, O., *Chem. Ber.*, 88, 445, 1955.
26. Keulemanns, A.J.M., Kwantes, A. and Van Bevel, Th., *Rec. Trav. Chim. Pays-Bas*, 67, 298, 1948.
27. Modler, R.F., Gulber, R. and Inoguchi, Y., *Chemical Economics Handbook*, Detergent Alcohols, Section 609.5000 L, SRI Consulting, Mento Park, CA.
28. Scheibel, J.J., *Industrial Colloquium Program*, University of Kansas, MI, February 8, 2007.
29. Author Unknown, *LIAL® Alcohols for North America*, Sasol North America Inc., Houston, TX.
30. Author Unknown, *ISALCHEM® Alcohols for North America,* Sasol North America Inc., Houston, TX.
31. Modler, R.F., Gulber, R. and Inoguchi, Y., *Chemical Economics Handbook*, Detergent Alcohols, Chart, Section 609.5000 N-O, SRI Consulting, Menlo Park, CA, 2004.
32. DeGroot, W.H., *Sulphonation Technology in the Detergent Industry*, Kluwer Academic Publishers, Dorrecht, Netherlands, 1991.
33. Gilbert, E.E. and Weldhuis, B., *J. Am. Oil Chem. Soc.*, 36, 208, 1959.
34. Schaurich, K., *J. Prakt. Chem.*, 15, 332, 1962.
35. Varlamov, V.S. and Ivanova, T.M., *Maslob.-Zhir. Prom.*, 28(12), 19, 1962.
36. Karnaukh, A.M. and Deinekhovskaya, Z.P., *Maslob.-Zhir. Prom.*, 27(5), 28, 1961.
37. Sosis, P. and Dringloi, J., *J. Am. Oil Chem. Soc.*, 47, 229, 1970.
38. Brooks, R. and Brooks, B., U.S. Patent 3,069,242, 1962.
39. Tischbirek, G., U.S. Patent 2,931,822, 1960.
40. Knaggs, E.A. and Nussbaum, M.L., U.S. Patent 3,169,142, 1965.
41. Knaggs, E.A., *Chem.Tech.*, 436–445, 1992.
42. Brooks, R.J. et al., U.S. Patent 3,427,342, 1969.
43. Vander Mey, J.E., U.S. Patent 3,328,460, 1967.
44. Pisoni, C., EU Patent Application, 0570844 A1, 1993.
45. Lanteri, A., U.S. Patent 3,931,273, 1976.
46. Ballestra, M. and Moretti, G., GB Patent 2043067B, 1983.
47. Ballestra, M. and Moretti, G., *High Quality Surfactants from SO3 Suphonation with Multi-tube Film Reactors*, Chimica Oggi, Luglio, Italy, p. 41, 1984.
48. Falk, K. and Taplin, W., U.S. Patent 2,923,728, 1960.
49. Toyoda, S. and Ogoshi, T., U.S. Patent 3,925,441, 1975.
50. Knaggs, E.A. and Nepras, M.J., *Sulfonation and Sulfation, Kirk-Othmer Encyclopedia of Chemical Technology*, Vol. 23, 4th ed., Wiley, New York, p. 180, 1997.
51. Scheibel, J.J., *The Impact of Feedstocks on Future Innovation in Surfactant Technology for the Detergent Market*, Samuel Rosen Memorial Award Lecture, 98th AOCS Annual Meeting and Expo, Quebec, 2007.

Production of Alcohols and Alcohol Sulfates

52. Kubickova, I. et al., Hydrocarbons for diesel fuel via decarboxylation of vegetable oils, *Catal. Today*, 106(1–4), 197–200, 2005.

53. Linnaila, R., *Status of Neste Oil's Biobased NExBTL Diesel Production for 2007*, SYNBIOS Conference, Stockholm, May 18–20, 2005.

54. Hess, G., Tyson and Syntroleum partner for fat-based fuel, *Chemical and Engineering News*, June 26, 2007.

55. Cripe, T.A., *Future Surfactants for Laundry Products*, New Horizons, Williamsburg, VA, 1998.

56. Cripe, T.A. et al., U.S. Patent 6,020,303, 2000.

57. Cripe, T.A. et al., U.S. Patent 6,060,443, 2000.

58. Cripe, T.A. et al., U.S. Patent 6,008,181, 2000.

59. Battersby, N., Kravetz, L. and Salanitro, J.P., *Effect of Branching on the Biodegradability of Alcohol Based Surfactants*, World Surfactants Congress, 5th, Firenze, May 29–June 2, 2000.

60. Shi, J. et al., *Biodegradable High Solubility Alkyl Sulfate Surfactants—Environmental Safety Profiles*, World Surfactants Congress, 5th, Firenze, May 29–June 2, 2000.

61. Singleton, D.M., U.S. Patent 5,780,694, 1998.

62. Singleton, D.M., Kravetz, L. and Murray, B.D., U.S. Patent 5,849,960, 1998.

63. Singleton, D.M., Kravetz, L. and Murray, B.D., U.S. Patent 6,150,322, 2000.

64. Singleton, D.M., U.S. Patent 6,222,087, 2000.

65. Szostak, R., *Molecular Sieves*, 2nd ed., Blackie, London, 1997.

66. Murray, B.D, Wingquist, B.H.C., Powers, D.H., Wise, J.B. and Halsey, R.B., U.S. Patent 5,648,585, 1997.

67. Murray, B.D., U.S. Patent 5,510,306, 1996.

68. Giaccobe, T.J. and G.A. Ksenic, U.S. Patent 5,112,519, 1992.

69. SASOL Alpha Olefins, SASOL Detergent Alcohols, *R&D Technical Bulletin*, SASOL Alpha Olefins, pp. 1–12, October 1, 1996.

70. Lawrence, M. and Potgieter, I., *Res. Disclosure*, 427, 1432–1453, November 1999.

71. Ong, R., Reed Business Information Limited, *Chemical News and Intelligence*, January 23, 2003.

72. Maas, H. et al., Patent Application WO 0,039,058, 2000.

73. Schwab, P. et al., U.S. Patent 6,166,279, 2000.

74. Schwab, P. et al., U.S. Patent 6,433,240, 2002.

75. Saunders, B., Coming next: natural gas refineries, *Oil Gas Investor*, 18(8), 48–51, 1998.

76. Taffe, P., Gas to liquids—an expanding market? *European Chemical News*, April 17–23, 2000.

77. Scherzer, J. and Gruia, A.J. (Eds.) *Hydrocracking Science and Technology*, Marcel Dekker, Inc., New York, pp. 174–199, 1996.

78. Wittenbrink, R.J., Ryan, D.F., Berlowitz, P.J. and Habeedb, J.J., U.S. Patent 6,332,974, 2001.

79. O'Rear, D.J., *Pat, Appl.*, U.S. Patent 2002/0,111,521, 2002.

80. Kalnes, T.N., U.S. Patent 6,361,683, 2002.

81. Moore, R.O., Van Gelder, R.D., Hilton, G.C. and Jones, C., Patent Application U.S. 2002/0,115,732, 2002.

82. Marcilly, C., Benazzi, E. and George-Marchal, N., U.S. Patent 6,198,015, 2001.

83. Wittenbrink, R.J., Mart, C.J., Ryan, D.F. and Cook, B.R., *High Performance Environmentally Friendly, Low Temperature, Drilling Fluids*, U.S. Patent 6,455,474, 2002.

84. Houston, C.A., *The Lab Market Report*, Colin A. Houston & Associates, Inc, Brewster, New York, January, 2001.

85. Rockwell, J. and Phillips, C. *California Alternative Diesel Symposium*, Sacramento, CA, August 19, 2003.

86. Kleiman, R., Chemistry of new industrial oilseed crops. In *Advances in New Crops*, Janick, J. and Simon, J.E. (Eds.), Timber Press, Portland, OR, pp. 196–203, 1990.

87. McCormick, B., *New Crops Update: Cuphea 2004, First Steps to Market*, IOP 3 Session, AOCS Annual Meeting and Expo, Cincinnati, OH, 2004.

88. Forcella, F., Gesch, R.W. and Isbell, T.A., *Crop Sci.*, 45, 2195–2202, 2005.

89. Forcella, F., Gesch, R.W. and Isbell, T.A., DOI: 10.2135/cropsci 2004.0593, Crop Ecology, Management & Quality, Seed Yield, Oil, and Fatty Acids of *Cuphea* in the Northwestern Corn Belt, USDA-ARS, North Central Soil Conservation Research Lab, Morris, MN, 2004.

90. Spolaore, P., Joannis-Cassan, C., Duran, E. and Isambert, A., *J. of Biosci. Bioeng.*, 101(2), 87–96, 2006.

91. Sheehan, J., Dunahay, T., Benemann, J. and Roessler, P., A Look Back at the U.S. Department of Energy's Aquatic Species Program-Biodiesel from Algae, *Close Out Report*, U.S. Department of Energy's Office of Fuels Development, NREL/TP-580-24190, pp. 28–31, Washington, DC, July, 1998.

7 Production of Alkanesulfonates and Related Compounds (High-Molecular-Weight Sulfonates)

Jean Paul Canselier

CONTENTS

7.1 Introduction 139
    7.1.1 Main Commercial Anionic Surfactants 139
    7.1.2 Manufacturing Processes of Sulfates and Sulfonates 140
7.2 Primary Alkanesulfonates 142
7.3 Secondary Alkanesulfonates 142
    7.3.1 Historical Considerations 142
    7.3.2 The Manufacture of Secondary Alkanesulfonates: Chemistry ... 143
        7.3.2.1 Sulfochlorination 143
        7.3.2.2 Sulfoxidation 144
    7.3.3 The Manufacture of Secondary Alkanesulfonates: Industrial Developments 146
        7.3.3.1 Separation of *n*-Alkanes from Petroleum Fractions 146
        7.3.3.2 Sulfochlorination with Chlorine and Sulfur Dioxide 147
        7.3.3.3 Sulfoxidation 148
    7.3.4 Production and Consumption 151
    7.3.5 Uses 151
7.4 Petroleum Sulfonates 151
    7.4.1 Manufacture 151
    7.4.2 Consumption and Uses 152
7.5 Lignosulfonates 152
    7.5.1 Formation 152
    7.5.2 Consumption and Uses 154
References 155

7.1 INTRODUCTION

7.1.1 MAIN COMMERCIAL ANIONIC SURFACTANTS

With the exception of soaps, commercial anionic surfactants mainly belong to the sulfonate and sulfate families. Sulfonated species include alkylbenzenesulfonates (linear alkylbenzene sulfonate [LAS], with linear alkyl chain for detergence and still, to a much lesser extent, long, branched-chain compounds as lubricating oil additives), secondary alkanesulfonates (SAS), α-olefin and internal olefin sulfonates (AOS and IOS, respectively), α-sulfo fatty acid esters (FAES, principally methyl esters), sulfosuccinic acid alkyl esters (e.g., bis-2-ethylhexylsulfosuccinate, sodium salt, a.k.a. AOT, one of the best wetting agents known so far), fatty acid isethionates (acyl oxyalkanesulfonates), petroleum sulfonates, and lignosulfonates. Alkyl sulfates (AS) are derived by direct sulfation of long-chain alcohols, either natural oil-based ones (fatty alcohols) or synthesized by the Oxo process; alkyl ether sulfates (AES) are obtained from slightly ethoxylated (most commonly 2–3 ethylene oxide units) alcohols [1]. This chapter only deals with alkanesulfonates, petroleum sulfonates, and lignosulfonates, all of them reviewed thoroughly earlier in this series by Bluestein and Bluestein [2]. Alkylbenzenesulfonates, AOS, alcohol and alcohol ether sulfates, soaps, alkyl phosphates, sulfosuccinates, sarcosinates, and taurates are dealt with in other chapters of this book. α-Sulfomonocarboxylic acids and derivatives were reviewed in this series by Stirton and Weil [3].

7.1.2 Manufacturing Processes of Sulfates and Sulfonates

The various sulfonation and sulfation processes of hydrocarbons and alcohols and the corresponding agents are summarized in Figure 7.1 [1]. Among the pure reagents or mixtures, sulfuric acid is the cheapest but is not very efficient, and its use is practically limited to making benzenoid hydrotropes, requiring azeotropic water removal; sulfur trioxide (also rather cheap when made *in situ* by burning elemental sulfur and catalytically converting the resulting SO_2 into SO_3) is the most reactive, the most versatile, and, consequently, the most used in large chemical plants, more often gaseous; it is diluted to 3–8% in air so that its aggressivity can be controlled. Chlorosulfonic acid, which is more expensive, reacts readily on hydroxyl groups and ethylenic bonds making it applicable for commercial sulfation of AS and AES. Unfortunately, none of the preceding reagents can be employed to functionalize alkanes by introducing a sulfonic group. Thus, photosulfochlorination (with sulfur dioxide and chlorine) and photosulfoxidation (with sulfur dioxide and oxygen) are the only ways for manufacturing alkanesulfonic acids on a large scale [4].

Figure 7.2 illustrates the main routes for sulfonated or sulfated anionic surfactant manufacture [5]. It appears that paraffin sulfonates or alkanesulfonates (i.e., practically only SAS) are made from linear paraffins produced from kerosene or gas oil.

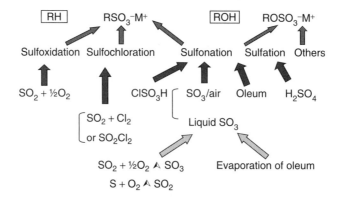

FIGURE 7.1 Sulfonation and sulfation agents for the manufacture of sulfur-containing anionic surfactants. (After Falbe, J., ed., *Surfactants in Consumer Products: Theory, Technology and Application*, Springer, Berlin, 1986. With permission.)

Production of Alkanesulfonates and Related Compounds

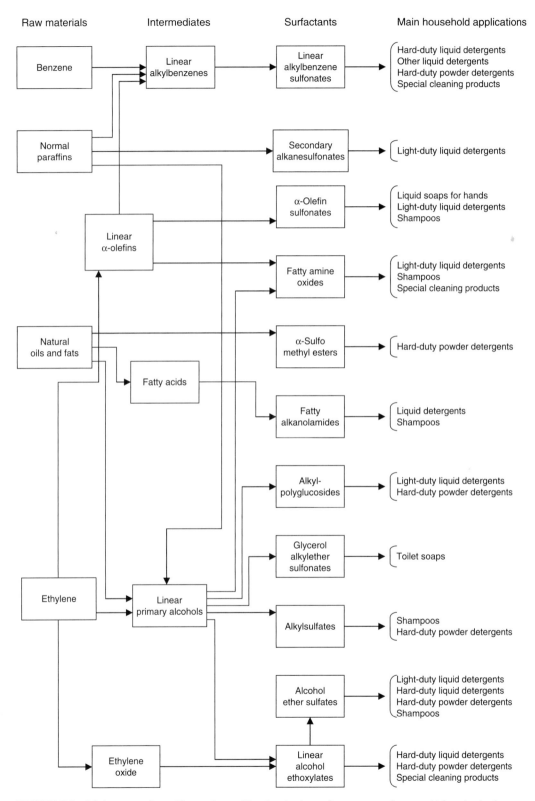

FIGURE 7.2 Main routes for sulfonated or sulfated anionic surfactant manufacture. (After Agriculture et Agroalimentaire Canada. Profils sectoriels–AC–25 June 2002 [report].)

7.2 PRIMARY ALKANESULFONATES

Although, historically, not made from fossil fuel material, contrary to Günther's naphthalenesulfonates [6], the C_{16} primary alkanesulfonate (PAS) can be considered as the first synthetic surfactant, obtained by Reychler in 1914 through the oxidation of hexadecanethiol [7]. Later, Reed and Tartar [8,9] employed the Strecker reaction to prepare PAS in 70% yield.

$$RBr + Na_2SO_3 \xrightarrow{200°C,\ 9\ h} RSO_3Na + Na_2SO_4$$

But the only possibly commercial way of manufacturing PAS is the sulfitation of α-olefins in the presence of peroxide according to an anti-Markownikov mechanism [10].

$$RCH = CH_2 + NaHSO_3 \xrightarrow{Peroxide} RCH_2CH_2SO_3Na$$

The preparation of PAS by this way has been thoroughly studied, especially in the United States, Germany [10], and the Soviet Union [11,12]; but PAS had failed to gain commercial success, may be because, apart from their behavior independent of pH (strong electrolytes), they resemble common carboxylic acid soaps which themselves are very inexpensive—they are completely biodegradable and show no hard water tolerance; ammonium salts are rather water-soluble [13]. However, for molecules longer than sodium dodecanesulfonate, the PAS possess a Krafft point (Tk) higher than room temperature [14], which significantly limits their applicability for household and personal care products. Accordingly, compared to SAS, PAS are less surface active and less water-soluble [13], but are excellent detergents above Tk. As colorless products, obtained in high yield [12] and in a state of high purity at low cost, they could compete in price with heavy-duty surfactants [10], should the Krafft point limitation be overcome.

In fact, the use of PAS in toilet bars [15], textile and fur industry [1], and analytical chemistry (medium-chain-length compounds for premicellar, ion-pairing liquid chromatography [16]) has been reported. It remains quite marginal.

7.3 SECONDARY ALKANESULFONATES

7.3.1 HISTORICAL CONSIDERATIONS

Sulfochlorination and sulfoxidation processes—primarily used for the production of paraffin sulfonates, more precisely SAS—have been thoroughly reviewed till the 1960s by Asinger [17]. The first patents on sulfochlorination were taken out by Reed [18–20] in the United States, and the DuPont company was intensely interested in such a reaction [21,22], but the process was mainly employed by IG Farbenindustrie that manufactured the Mersolate products in Germany during World War II (up to 80,000 t per annum) [23] from Fischer–Tropsch hydrocarbon sources, with coal as a raw material [17]. Eastern European countries (Volgograd, Soviet Union; Leuna Werke, East Germany) and China have been the principal users of this technology [24,25], although a few Western companies patented related processes [26,27].

The first successful photosulfoxidation was carried out under ultraviolet (UV) light by C. Platz of the IG Farbenindustrie at Hoechster Farbwerken in 1943 [28]. Shortly afterward, the "light-water" and "acetic anhydride" processes were used in small units at Hoechst and Leuna, respectively [17,29]. Then, the conditions for improving the sulfoxidation of straight-chain paraffins were studied since the 1960s, but the first production units for alkanesulfonate manufacture were operated only in the 1970s and continuous processes were developed, principally by the Hoechst Company in Europe (Germany, the Netherland, and France). Sulfoxidation studies were still of interest in the 1990s [30]. Making SAS on a commercial scale has been boosted by environmental concern—their biodegradability is easier than that of LAS [31]. Nevertheless, despite this advantage and other qualities (good detergent properties and aqueous solubility), SAS could not compete with LAS because of their higher cost due to their special manufacturing processes [4,32].

Production of Alkanesulfonates and Related Compounds **143**

7.3.2 The Manufacture of Secondary Alkanesulfonates: Chemistry

Let us first examine the chemical reactions, their stoichiometry, thermochemistry, mechanism, and kinetics. The ancient name paraffins meaning alkanes are poor chemical reagents and activation of the C—H bond must first occur to generate free radicals. This can be effected with UV radiation [33] or even visible light, X- or γ-rays [34,35], or peroxides.

7.3.2.1 Sulfochlorination

7.3.2.1.1 Chlorine and Sulfur Dioxide

The universal sulfochlorination reaction consisting of the substitution of a sulfochloride group, SO_2Cl, for a H atom, thus leading to a mixture of sulfonyl chlorides, is as follows:

$$RH + SO_2 + Cl_2 \xrightarrow{\ h\nu\ } RSO_2Cl + HCl \quad \Delta H^\varnothing = -163\,\text{kJ/mol at } 298\,\text{K}$$

This reaction enthalpy value has been measured by Geiseler and Nagel [36] for dodecane monosubstitution. A likely mechanism of this free-radical process, first involving chlorine dissociation into its atoms (free radicals), more often with UV light from mercury vapor lamps, is as follows [4]:

$$Cl_2 \xrightarrow{\ h\nu\ } 2Cl^\bullet$$
$$RH + Cl^\bullet \rightarrow R^\bullet + HCl$$
$$R^\bullet + SO_2 \rightarrow RSO_2^\bullet$$
$$RSO_2^\bullet + Cl_2 \rightarrow RSO_2Cl + Cl^\bullet$$

The major products are secondary alkanesulfonyl chlorides with a statistical distribution of sulfonic groups on the internal carbon atoms of the linear chain. In contrast, the presence of the SO_2Cl group at the end of the chain is unlikely because of the much lower stability of primary radicals $RCH_2^\bullet$. Besides, di- and polysulfochlorides, as well as chloroalkanes are also formed as by-products and a large amount of HCl is generated. Selectivity reaches 85–90% with excess SO_2 and low paraffin conversion [32,37]. The quantum yield of a photochemical reaction at a wavelength λ is given by:

$$\Phi_\lambda = \Delta n/(P_{0,\lambda} \cdot t)$$

where $P_{0,\lambda}$ is the incident photonic flux and Δn the number of molecules of reagent transformed during the period of time t. The quantum yield of sulfochlorination with SO_2 and Cl_2 can reach 40,000 in laboratory conditions, but remains lower than 5,000 in plant conditions (no solvent, less pure starting material) [32]. Earlier kinetic and selectivity studies have been summarized by Joschek [38].

In the second step, sulfonyl chlorides are neutralized with a strong base, more often sodium hydroxide, yielding the SAS isomeric mixture:

$$RSO_2Cl + 2Na^+OH^-(aq) \rightarrow RSO_3^-Na^+ + NaCl + H_2O \quad \Delta H^\varnothing = -83\,\text{kJ/mol at } 298\,\text{K}$$

7.3.2.1.2 Sulfuryl Chloride

An alternative process to classical sulfochlorination with the gas mixture uses liquid sulfuryl chloride; the reaction is operated under UV or visible light ($\lambda = 300\text{–}500$ nm).

$$RH + SO_2Cl_2 \xrightarrow{\ h\nu\ } RSO_2Cl + HCl \quad \Delta H^\varnothing = -72.6\,\text{kJ/mol at } 298\,\text{K}$$
$$RH + SO_2Cl_2 \rightarrow RCl + HCl + SO_2 \quad \Delta H^\varnothing = -57.5\,\text{kJ/mol at } 298\,\text{K}$$

The reaction enthalpies have been calculated from Geiseler and Nagel's [36] result and enthalpy of formation of sulfuryl chloride. Their absolute values are lower than those of the corresponding reactions involving sulfur dioxide and chlorine because the dissociation of SO_2Cl_2 is endothermic [39].

This reaction scheme was first observed by Kharasch and Read [40], but no industrial application has yet been developed because of poor yield and selectivity with linear paraffins. The yield of alkyl chlorides may be higher than that of sulfochlorides [23]. However, some recent works have shown that in the presence of an adequate catalyst (e.g., pyridine) for rather low conversion rates (e.g., 15%), sulfuryl chloride can lead to acceptable RSO_2Cl/RCl ratio (about 1 in pure phase and higher in benzene solvent) and the quasi absence of di- and polysubstituted compounds [39–45]. Without doubt, the main drawback of the process is still the large amount of alkyl chlorides produced; it is then necessary to consider the possible recovery and transformation of the chloroalkane mixtures through quaternization into alkyltrimethylammonium chlorides (cationic surfactants) [46,47]:

$$RCl + (CH_3)_3N \rightarrow RN^+(CH_3)_3, Cl^-$$

Chloroalkanes are also used, although less commonly than 1-alkenes, in Friedel–Crafts alkylation of benzenoid hydrocarbons.

Kharasch et al. [48] had proposed a mechanism for the sulfochlorination with sulfuryl chloride in the presence of pyridine, simply assuming the dissociation of SO_2Cl_2 into chlorine and sulfur dioxide, but the different secondary/primary selectivity with respect to that of the SO_2-Cl_2 mixture [41] suggests three more possible steps [42]:

$$SO_2Cl_2 \rightarrow SO_2Cl^\bullet + Cl^\bullet$$

$$SO_2Cl^\bullet \rightarrow SO_2 + Cl^\bullet$$

$$RH + SO_2Cl^\bullet \rightarrow R^\bullet + SO_2 + HCl$$

To the best of our knowledge, quantum yields have not been determined for the sulfochlorination with SO_2Cl_2, always carried out under polychromatic light. The only related data is a value of quantum efficiency (η = number of molecules of product/total number of absorbed photons) of the order of 10^3. There seems to be no real kinetic order for this complex set of elementary reactions; between 30 and 40°C, an attempt of modeling at low hydrocarbon conversion rate gave apparent reaction orders of 2.2 and 2.8 for SO_2Cl_2 and activation energies of 114.5 and 126.5 kJ/mol for sulfochlorination and chlorination, respectively [39].

Rather than a distillation of labile sulfochlorides or their direct saponification, the extraction of sulfochlorides with acetonitrile appears to be the best way of separation of the products from the reaction medium [17,39,47].

A variation of sulfochlorination with sulfuryl chloride involves a highly selective porphyrin catalyst that is active in the absence of light, but giving favorable results only in benzene as a solvent [49].

7.3.2.2 Sulfoxidation

The other way to functionalize n-paraffins by introducing a sulfonic group is photosulfoxidation. The free-radical mechanism involves the direct replacement of a hydrogen atom by a sulfonic acid group, SO_3H. The global sulfoxidation reaction is strongly exothermic.

$$RH + SO_2 + \frac{1}{2}O_2 \xrightarrow{h\nu} RSO_3H \quad \Delta H^\varnothing = -246\,kJ/mol \text{ at } 298\,K$$

Production of Alkanesulfonates and Related Compounds

All the reagents and products are supposed to be in the gaseous state. The value of $\Delta H^{\varnothing}$ is calculated according to Bares et al. [50], Cox and Pilcher [51], and Joshi [52].

As in sulfochlorination, a free-radical mechanism is also considered—one was proposed by Graf [53]:

$$RH \xrightarrow{\;hv\;} R^{\bullet}$$

$$R^{\bullet} + SO_2 \rightarrow RSO_2^{\bullet}$$

$$RSO_2^{\bullet} + O_2 \xrightarrow{\;k_a\;} RSO_2-O-O^{\bullet}$$

$$RSO_2-O-O^{\bullet} + RH \rightarrow RSO_2-O-OH + R^{\bullet}$$

$$RSO_2-O-OH \rightarrow RSO_2-O^{\bullet} + {}^{\bullet}OH$$

$$RSO_2-O^{\bullet} + RH \rightarrow RSO_2-OH + R^{\bullet}$$

$${}^{\bullet}OH + RH \rightarrow H_2O + R^{\bullet}$$

However, it is likely that C–H bond breaking is made possible after SO_2 excitation [32,54]:

$$SO_2 \rightarrow SO_2^{*} \rightarrow {}^{3}SO_2^{*}$$

$$RH + {}^{3}SO_2^{*} \rightarrow R^{\bullet} + H^{\bullet} + SO_2$$

or

$$RH + {}^{3}SO_2^{*} \rightarrow R^{\bullet} + HSO_2^{\bullet}$$

The reaction is self-catalyzed and irradiation can be cut off after the process has started. Free radicals can be produced by UV light (with optional water), ozone, X- or γ-ray, and acetic anhydride or chlorine. The main intermediates are thus sulfur dioxide in its triplet state (${}^{3}SO_2^{*}$); alkyl, $R^{\bullet}$; hydrogenosulfonyl, $HSO_2^{\bullet}$; alkanesulfonyl, $RSO_2^{\bullet}$; alkanepersulfonyl, $RSO_2-O-O^{\bullet}$; alkanesulfonyloxy, $RSO_2-O^{\bullet}$; and hydroxyl, ${}^{\bullet}OH$ radicals and alkaneperoxysulfonic acid. Since $RSO_2^{\bullet}$ is the most stable free radical, the most probable chain rupture reaction is:

$$2RSO_2^{\bullet} \xrightarrow{\;k_b\;} RSO_2-SO_2R \rightarrow RSO_2-R + SO_2$$

In the presence of water, alkaneperoxysulfonic acid (RSO_2-O-OH) reacts with water and dissolved SO_2 (sulfurous acid) whose oxidation forms H_2SO_4:

$$RSO_2-O-OH + H_2O \rightarrow RSO_2-OH + H_2O_2$$

$$RSO_2-O-OH + SO_2 + H_2O \rightarrow RSO_2-OH + H_2SO_4$$

The global reaction, much more strongly exothermic, is then:

$$RH + 2SO_2 + O_2 + H_2O \xrightarrow{\;hv\;} RSO_3H + H_2SO_4 \quad \Delta H^{\varnothing} = -472\,kJ/mol \text{ at } 298\,K$$

In the preceding reaction, all the reagents and products are supposed to be in the gaseous state, except H_2O and H_2SO_4.

According to this stoichiometry, 1 mol of sulfuric acid is produced for each mole of sulfonic acid. Moreover, water stops the chain reaction so that continuous irradiation is needed.

In this case again, almost only secondary alkanesulfonic acids are produced—the selectivity of the substitution of a CH_2 hydrogen atom versus a CH_3 one may reach 30/1 at room temperature [55]. The distribution of sulfonic groups on the internal carbon atoms of the linear chain is statistical. This reaction should not be restricted to linear paraffins and could be applied to substituted hydrocarbons (alkyl chlorides, alcohols, carboxylic acids, etc.) as well. However, special interest is lacking in the related processes because the resulting compounds that will make effective detergents are not yet known. However, branched alkanes and, above all, alkenes and aromatics react in the same conditions, but interfere seriously with the photosulfoxidation, leading to different products [54].

The rate of formation of sulfonic acids may be expressed as a function of the concentration of dissolved oxygen, $[O_2]_\alpha$; the light intensity absorbed by SO_2, I_a; the rate of formation of alkyl free radicals, $\phi R^\bullet$; and the rate constants of two of the preceding reactions.

$$\frac{d[RSO_3H]}{dt} = [O_2]_\alpha (I_a \phi_{R^\bullet})^{0.5} \frac{k_a}{(k_b)^{0.5}}$$

The sulfoxidation reaction is followed by a neutralization process, usually with aqueous sodium hydroxide.

$$RSO_3H + Na^+OH^-(aq) \rightarrow RSO_3^-Na^+ + H_2O \quad \Delta H^\emptyset \sim -56.9\,kJ/mol \text{ at } 298\,K$$

7.3.3 THE MANUFACTURE OF SECONDARY ALKANESULFONATES: INDUSTRIAL DEVELOPMENTS

Apart from sulfur oxidation, the general scheme of the manufacture of SAS (Figure 7.3) includes at least four steps: extraction of paraffins, sulfochlorination or sulfoxidation, neutralization, and one or several separation and purification operations. Sulfochlorination and sulfoxidation are photochemical reactions, each requiring special equipment. With both types of processes, undesirable polysulfonate generation greatly limits the extent of sulfonation and, consequently, such processing involves recovery and recycling of large quantities of unreacted feedstocks. Regarding the remarks made earlier in this chapter, it is this recovery and recycling that must be optimized to make SAS compete more cost effectively with the world's principally used sulfonates and sulfates—LAS, AS, and AES.

7.3.3.1 Separation of *n*-Alkanes from Petroleum Fractions

Mixtures of linear C8–C22 and, more specifically, C12–C18, paraffins are desirable raw materials to obtain sulfonates with good surfactant properties. Nowadays, petroleum cuts have replaced the coal-derived (Fischer–Tropsch) mixture "Kogasin II" (boiling range 230–320°C) fully hydrogenated to "Mepasin" before treatment. (Ironically, however, the present state of the world's energy industry may give way to the reemergence of gas-to-liquid and coal-to-liquid paraffin supply.) Regarding petroleum-cut paraffins, after distillation of a Koweit crude, the kerosene fraction contains 25% straight-chain paraffins, 65% branched paraffins and naphthenes, and 10% aromatics. The selective extraction of straight-chain alkanes may be carried out in two main ways:

- Clathrate formation with urea
- Adsorption–desorption on molecular sieves (zeolites)

Production of Alkanesulfonates and Related Compounds

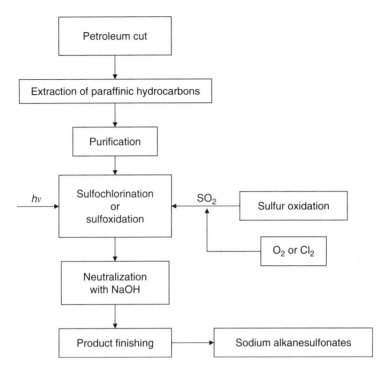

FIGURE 7.3 General scheme of the manufacture of SAS. (From Braun, A.M., Maurette, M.T. and Oliveros, E., *Technologie Photochimique*, Presses Polytechniques Romandes, Lausanne, 1986. With permission.)

Since the former process does not give more than 90% pure *n*-alkanes, the latter is, by far, preferred, leading to a maximum 99.5% purity. Liquid–solid and gas–solid adsorption processes are used according to the chain length of the feed. A 5 Å molecular sieve retains straight-chain compounds and rejects branched ones. The longer the chain, the stronger the adsorption; therefore, besides thermal or low-pressure regeneration of zeolites, energy- and time-consuming or not very efficient adsorption of a less-adsorbed alkane, a shorter-chain one (e.g., *n*-heptane), is recommended to desorb a C10–C14 petroleum cut [17,56,57].

7.3.3.2 Sulfochlorination with Chlorine and Sulfur Dioxide

The simplest photochemical reactors are certainly those of the immersion type, where the lamps are immersed in the reaction medium that they irradiate directly, the gases being injected through perforated tubings at the bottom of the reactor. However, these reactors rapidly get dirty because of by-product deposits and they are difficult to cool and favor the production of chloroalkanes due to higher absorption of chlorine. Residence times range between 2 and 10 h. For example, with a residence time of 2 h and a 34% sulfochlorination rate, the production of sulfochlorides containing ca. 12% di- and polysulfochlorides can reach 80 kg/h m^3 [37].

Various other continuous processes have been patented. Many of them use tubular reactors—high-speed injection of gases ensures good fluid contacting and flow, short residence times, easy cooling, and therefore, good reaction control and high conversion rate. Unfortunately, they are rather difficult to operate and maintain. Figure 7.4 shows one of the proposed process schemes with cocurrent feeding of a tubular photoreactor [58].

Nevertheless, a countercurrent process may be more efficient, that is, the gases are fed into the bottom of the reactor, whereas the hydrocarbons flow from the top. The other parameters are $\lambda > 400$ nm (visible light, able to dissociate Cl$_2$), *t* ranging between 20 and 40°C, and maximum conversion rate of alkanes is ~30% [1].

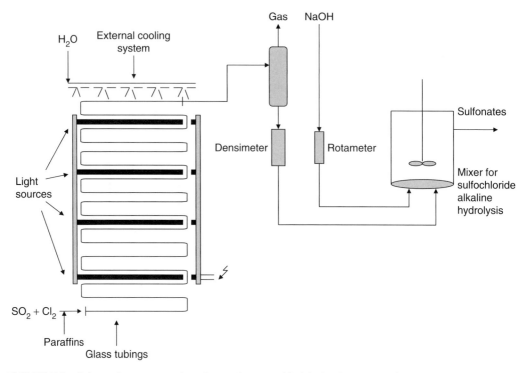

FIGURE 7.4 Schematic representation of a continuous sulfochlorination process in a tubular reactor. (From Braun, A.M., Maurette, M.T. and Oliveros, E., *Technologie Photochimique*, Presses Polytechniques Romandes, Lausanne, 1986. With permission.)

The final product mixture contains unreacted alkanes and mono- and disulfochlorides in the approximate ratio of 15:1. This mixture is saponified with 10% aqueous sodium hydroxide at 80°C. Higher temperatures should be avoided so that sulfochlorides do not undergo desulfonation [1]. Phase separation occurs in a settler—the upper and lower phases contain almost all the alkanes and chloroalkanes and the water-soluble salts (sulfonates and sodium chloride), respectively. After cooling, a second-phase separation allows to eliminate NaCl almost totally. The 20% aqueous solution of sulfonates is then fed to an evaporator working at 300–350°C (melting range of the SAS) from which a nearly dry product is recovered. Several patents, issued in the early 1990s, are related to the "saponification" of sulfochlorides [59–63].

7.3.3.3 Sulfoxidation

In addition to the most classical simple immersion reactor, three main types of photochemical reactors can be used in plants [32]. Their respective advantages and drawbacks are presented in Table 7.1 [57].

The most recent processes of SAS manufacture are based on the continuous reaction of a sulfur dioxide/oxygen gaseous mixture with C13–C17 linear paraffins, possibly in the presence of water and UV irradiation [4,64]. In principle, it is also possible to maintain or even initiate sulfoxidation without irradiation so that several modes of initiation can be considered—UV light (namely, in the "light-water" process), γ-radiation, acetic anhydride, ozonized oxygen, chlorine, or peroxycompounds [54].

The large-scale, always continuous, process, involving circulation of gas and liquid, appears as rather complex despite the absence of solvents and solid catalysts and whatever the type of the initiator. First, sulfur dioxide and oxygen flow rates must be controlled carefully since only a constant

Production of Alkanesulfonates and Related Compounds   **149**

TABLE 7.1

Comparison of Industrial Photochemical Reactors

Type	Advantages
Falling-film cylindrical reactor	Avoids deposits on the wall of the reactor irradiated from outside
	Lower energy cost with a mercury arc lamp
	No need to stop the reactor for cleaning
	Better reactor type for gas–liquid reactions
	Lower but homogeneous illumination, even toward the ends of the tube
Continuous-flow annular reactor	Reduced reactor size allowing easier scaling-up
	Adjustable flow rates
	Higher turbulence
	No need to stop the reactor for cleaning
Bubble reactor	Higher turbulence due to gas injection
	Adaptable to any type of vertical, cylindrical reactor
	No need for cleaning (absence of deposit)

Source: After Student report, ENSIGC, Toulouse, 1993 (unpublished).

mole ratio at the reactor inlet allows an optimal yield; theoretically, 2 mol of SO_2 for 1 mol of O_2 are required, but an excess of sulfur dioxide is strongly recommended for a good yield. The gaseous mixture must be fed from below and thoroughly dispersed into the liquid phase to ensure a quick and efficient reaction. In addition, because of sulfoxidation and hydration strong exothermicities, reactor cooling is needed—temperature regulation at 40°C at the maximum prevents by-product formation (namely, through oxidation) and yield decreases. Optimization of the residence times avoids the formation of undesirable by-products. Alkanesulfonic acids must be continuously conveyed to the separation zone. In fact, another issue of the commercial process lies in the separation–purification problems—unreacted materials (paraffins, SO_2, H_2O) and by-products (di- and polysulfonic acids, H_2SO_4) are to be recovered and recycled or used in another part of the plant or disposed off properly [1].

Let us now consider the UV (light-water) process, so far the most frequently used, namely, by the Clariant company (formerly Hoechst), which is currently operating several substantial-sized sulfoxidation plants in Europe for the production of C13–C17 SAS (e.g., <1% of chains shorter than C13, 58% C13–C15, 39% C16–C17, and <1% of chains longer than C17) [65]. The corresponding flowchart is illustrated in Figure 7.5 [1,54,65].

Since, in the presence of water, the reaction is not self-catalyzed, continuous irradiation is necessary. It is produced by several immersed quartz glass high-pressure mercury vapor lamps. The liquid (alkane (11) and water) and finely dispersed gaseous (SO_2 and O_2) reactants are fed into the reactor (1) at the optimum temperature (30–38°C) to induce an efficient mass transfer near the surface of the lamps. The multilamp system ensures a large irradiated surface area. A quantum yield higher than 12 equivalents of alkanesulfonic acids per molar quantum can thus be achieved. The ratio of monosulfonated products to disulfonated ones is ca. 9:1, higher than with the other sulfoxidation processes (e.g., down to 1.5 with γ rays). There are virtually no other by-products. The respective advantages of the light-water and dry UV processes are summarized in Table 7.2.

In (2), the heavier sulfonic acid–rich phase is separated from part of the residual hydrocarbon, which is recycled (13). The remaining paraffin is solubilized in sulfonic acid aggregates. In (3), the sulfur dioxide dissolved in the lower phase is expelled (10) by means of air (16). Water is then distilled off *in vacuo* (4), together with some paraffin, and on such concentration of the aqueous

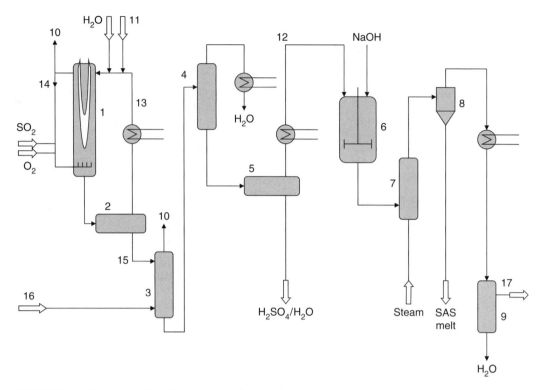

FIGURE 7.5 Flowchart of the light-water sulfoxidation process. Photoreactor (1), separators (2,5,8,9), SO$_2$ stripper (3), evaporating column (4), neutralization reactor (6), evaporator (7), exhaust gas (10), fresh paraffin (11), sulfonic acid (12), recycle paraffin (13), recycle gas (14), acid phase (15), air (O$_2$) (16), back paraffin (17). (After Falbe, J., ed., *Surfactants in Consumer Products: Theory, Technology and Application*, Springer, Berlin, 1986. With permission.)

TABLE 7.2
Respective Advantages of the Light-Water and Dry UV Sulfoxidation Processes

Process	Advantages
Light water	Avoids the settling of colored materials on the light source (which would reduce the photochemical energy available) [23]
	Colorless product
	Low ratio of disulfonated compounds
Dry (no water)	Irradiation only necessary to initiate the reaction
	Longer-chain reactions
	Low amount of H$_2$SO$_4$, avoiding extraction step

Source: After Student report, ENSIGC, Toulouse, 1993 (unpublished).

medium and addition of concentrated sulfuric acid and medium chain-length alcohol (C5–C12, derived from linear or branched-chain paraffins or cycloparaffins) [66], a lower phase consisting of most of the sulfuric acid separates as a 20 wt% solution (5). Alkanesulfonic acids are then fed to the neutralization reactor (6) where the rest of the sulfuric acid is precipitated as Na$_2$SO$_4$. The remainder of the hydrocarbon is removed in (7) by high-temperature steam distillation. The final product, water white, comes out of the separator (8) as an alkanesulfonate melt, further made up for sale as flakes or concentrated (30 or 60%) aqueous solution. Paraffin and water are separated in (9) [54,65]. Electric power required ranges between 0.005 and 0.020 kWh/kg sulfonates [32,38,55].

7.3.4 PRODUCTION AND CONSUMPTION

SAS are sold as waxy flakes (93% active matter), 60% paste, or 30% aqueous solution. Flakes must be stored in the absence of moisture. The aqueous paste, which tends to undergo segregation on standing, must be homogenized before use. The solution remains homogeneous and can be easily pumped [67]. In western Europe, the sales and captive uses of alkanesulfonates were 75,000 t in 1996 and around 85,000 t in 2005, placing them far behind alkylbenzenesulfonates and alcohol ether sulfates and just after AS [68]. The main European manufacturer is Clariant (all employ sulfoxidation) [69], whereas Stepan and Witco share the market in the United States.

7.3.5 USES

The industrial and household uses of SAS are numerous and various. Their advantages are higher aqueous solubility, lower solution viscosity, and better skin compatibility compared to those of LAS [70], but also faster biodegradability (more than 99% after a few days) [31]. These qualities are appreciated as well as their good wetting, foaming (although inferior to those of LAS), and emulsifying properties [71], for instance, in the polymerization of vinyl monomers [70]. These features make SAS very suitable for dishwashing liquids (light-duty detergents) and all-purpose cleaners, showing strong defatting properties. In Japan, for instance, they are preferably used in completely made detergents. They allow flotation of oxide ores, apatite, fluorite, baryte, and so on [1] and are potentially useful for enhanced oil recovery (EOR) [72,73]. They have found a lot of uses in the textile and fiber industries, especially in textile finishing—emulsifiers for spin finish, scouring, mercerizing and carbonizing auxiliaries, and wetting agents, for example, in exhaust dyeing or padding processes, resist agents to dyeing [74]. Other fields of application of SAS are the photographic industry [1], polymer processing (antistatic agents for polyvinyl chloride and polystyrene) [70], and even cosmetics (shampoos) [75].

7.4 PETROLEUM SULFONATES

7.4.1 MANUFACTURE

Originally, the sulfonation of lateral cuts of vacuum distillation units (namely, mixtures of C30–C40 paraffinic, naphthenic, and aromatic hydrocarbons) was carried out for the removal of aromatics in those cuts dedicated to the manufacture of lubricating oil [4]—because of their reactivity at high temperature and their inappropriate viscosity index, aromatics are indeed unwanted in lubricant formulations. The sulfonic acids so produced allowed separation of the aromatic fraction of the oil through liquid–liquid extraction with an aqueous alkaline solution. "Mahogany" alkylarylsulfonates (so-called because of their reddish color) were thus obtained as valuable by-products, contrary to green acids, present in a lower aqueous layer containing high amounts of polysulfonic acids of no economic value. Nowadays, aromatic hydrocarbons are extracted directly with solvents such as furfural or phenol [64,76]. Besides, due to the shortage of these traditional (or "natural") petroleum sulfonates and the need for mixtures with more regular properties, "synthetic" petroleum sulfonates are obtained from alkylates (e.g., detergent alkylate bottoms). Still obtained as coproducts in the manufacture of white mineral oils, lube oils, and process oils (e.g., for rubber compounding), petroleum sulfonates have also become primary petrochemicals [4].

The sulfonation with sulfuric acid or oleum is currently carried out either on an appropriate petroleum cut of the aromatic stream [76] or on selected petroleum fractions containing benzenoid, cycloaliphatic, and even paraffinic hydrocarbons, the latter operation often requiring several steps using different sulfonating agents [69]. The presence of polycyclic aromatics makes it difficult to avoid polysulfonation. In addition, a lot of side reactions occur such as oxidation polymerization, isomerization, cyclization, disproportionation, and dehydrogenation. The substitution of SO_3 for oleum yields more monosulfonic acids and less total sludge. However, oil sulfonation for dearomatization has been declining since the past few decades and synthetic alkylated aromatic sulfonates, made by continuous SO_3 sulfonation processes, tend to replace the so-called natural petroleum sulfonates. Some processes, utilizing improved technologies, have been discovered, which give products directly suitable for EOR.

Alkaline earth salts are manufactured by direct neutralization of sulfonic acids with calcium or magnesium hydroxide or by a cation exchange reaction on the sodium salt. With natural petroleum sulfonates, the presence of polysulfonates, which destabilize lubricant formulations, often requires a desludging procedure involving the use of a hydrocarbon solvent followed by centrifugation or water or sulfuric acid wash, clay percolation, and filtration. However, with high-quality synthetic alkylated aromatic feedstocks of molecular weight (MW) 332 or higher (yielding oil-soluble monosulfonates), it is possible to suppress the solvent desludging step. Overbased alkaline earth sulfonates are produced by direct carbonation of excess hydroxide with CO_2 [4].

Two drawbacks of petroleum sulfonates are that they are highly colored and always contain rather large amounts of residual (unsulfonated) oil [70].

7.4.2 CONSUMPTION AND USES

The main current U.S. producers of petroleum sulfonates are Penreco, EXXON, Shell (natural), Pilot Chemical (synthetic), and Witco [4,64,69,77]. Note that like lignosulfonates, petroleum sulfonates are particularly inexpensive surfactants with good performances. Their production amounts to ~10% of that of the total sulfonates. Since in EOR applications they enable ultra-low interfacial tensions (down to ~0.1 μN/m) to be obtained [1,4,64,69,76], their manufacture is likely to be boosted by the strong current increase in crude oil market prices.

The MW of sodium petroleum sulfonates ranges approximately from 400 to 550 Da; some uses of lower MW salts (up to ca. 460) are as oil-in-water emulsifying agents in cutting oils, frothers in ore flotation, and ingredients in dry-cleaning formulations. Higher MW salts find use in rust prevention and pigment dispersion in organic media. Ammonium salts are also corrosion inhibitors and soluble dispersants in hydrocarbon mixtures (gasoline, fuel oil) [64,70].

Mg, Ca, and Ba sulfonates are also used as sludge dispersants for fuel oils and, above all, as lube additives used for sludge dispersion, detergency, corrosion inhibition, and micellar solubilization of water [4,64,70,76]. Overbased oil-soluble lube sulfonates may consist of up to 12–20 mol MCO_3, in suspension in the clear oil, associated to 1 mol $(ArSO_3^-)_2$, M^{2+}—they are added to such formulations to neutralize the organic acids formed in engines at high operation temperatures [4]. A crystalline form of overbased calcium sulfonate, formulated with alkyd resins, is particularly suitable for corrosion prevention on metallic parts [78], for instance, automotive vehicles and bridges.

Petroleum sulfonates are also used as demulsifiers of petroleum (oil-well) emulsions and in epoxy resin elastomers where they improve anticorrosion properties of coatings and sealants [69].

7.5 LIGNOSULFONATES

7.5.1 FORMATION

With cellulose and hemicellulose (both accounting for 70–80 wt%), lignin is one of the three major components of wood, representing 20–30% of its mass. Lignins consist of linked phenylpropane

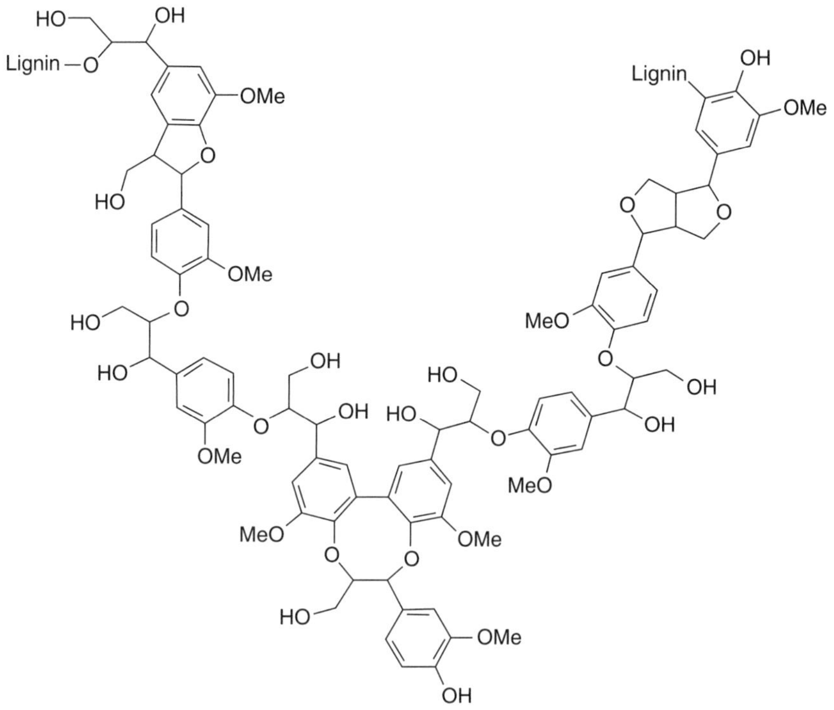

Production of Alkanesulfonates and Related Compounds

FIGURE 7.6   Simplified structural formula of lignin (after http://en-wikipedia.org/wiki/Image:LigninStructure.jpg).

units possibly containing ethylenic unsaturations and always oxygenated functional groups (ethers, e.g., as methoxy and phenoxy groups, or five-membered heterocycles, alcohols, phenols, aldehydes, and lactones). A very partial formula is shown in Figure 7.6 [79].

Many types of lignins exist according to the kind of wood (softwood and hardwood). More complete formulas can be found in the literature [80–82]. Lignins can be selectively isolated before reacting with Na_2SO_3. In the Kraft process, delignification is first carried out with a 10–20% mixture of NaOH and Na_2S and then, after filtration and acidification, submitted to sulfonation at relatively high temperatures and pressures [4]. But, most of the time, lignosulfonates are obtained through the acid sulfite process—they are complex mixtures typically produced by the sulfonation/neutralization of lignin separated from cellulose, that is, as by-products of sulfite wood pulping [69]. They are one of the biomass-derived chemicals by chemical processes. The production capacity of lignin in the Western world was estimated to be approximately 8.10^5 t/year in the mid-1990s [81,82].

Sulfonation yields of 95–98% are obtained with $NaHSO_3$ and NaOH between 120 and 150°C and the sulfonate content can amount to almost 2% (as sulfur) [4]. Bisulfite sulfonation, with SO_2, $NaHSO_3$, $Ca(HSO_3)_2$, or magnesium or ammonium bisulfite, occurs on the aliphatic side chain of the guaiacyl (4-hydroxy-3-methoxyphenylpropane) or similar units (almost exclusively at the α-carbon), next to the benzenoid ring by a nucleophilic substitution, which displaces either an OH or an OR group ($HC–SO_3^-$ replacing HC–OH or HC–OR at benzyl alcohol, benzyl aryl ether, and benzyl alkyl ether linkages). At neutral pH, sulfonation is favored by phenolic groups para to the propane side chain. One of the earliest and most efficient ways for isolating lignin products from spent pulping liquors is the Howard process involving the precipitation of calcium lignosulfonates with excess lime [80–82]. Numerous improvements in the Kraft pulping lignosulfonate process have been developed. In one of these, sulfonation occurs on the benzenoid ring due to the use of $FeSO_4$ or $FeCl_3/Na_2S_2O_8–Na_2SO_3$ combination, allowing solubilization of Kraft lignin [83]. Typical formulas of "monomer" lignosulfonates from soft- and hardwood are $C_9H_{8.5}O_{2.5}(OCH_3)_{0.85}(SO_3H)_{0.4}$

and $C_9H_{7.5}O_{2.5}(OCH_3)_{1.39}(SO_3H)_{0.8}$, respectively, and the molar masses of these very disperse macromolecules range between 1,000 (more often 4,000) and 140,000 Da [81,82].

Lignosulfonates are soluble in water, but are insoluble in most common organic solvents. They exhibit surface activity but neither do they reduce surface tension to the extent that true surfactants do nor do they form oriented monolayers or micelles [64,81,82]. However, they are able to stabilize emulsions.

7.5.2 CONSUMPTION AND USES

Lignosulfonates are produced in large volume as aluminum, ammonium, calcium, chromium, ferrochromium, magnesium, potassium, sodium, and amine salts. In the early 1990s, their production in the United States was approximately 300,000 t/year for the Ca salt (sold at ~$110/t), the cheapest of all the sulfonates, and 90,000 t/year for the Na salt (sold at ~$340/t), representing ca. 38% of the total U.S. production of sulfonic acids and sulfonates, but only 18% of their commercial value [4,69]. The largest U.S. manufacturers are DuPont de Nemours, EXXON, Georgia-Pacific, Ligno Tech, and Westvaco [69]. Lignosulfonates are almost the sole outlet for lignin, even if further desulfonation occurs in some cases. Their consumption is divided into extremely various uses [4]. First, lignosulfonates are intermediates in the manufacture of chemicals such as vanillin, often involving calcium salt precipitation [80] and oxidative desulfonation [81,82], and dimethylsulfoxide through reaction with sulfide or elemental sulfur followed by oxidation of the dimethyl sulfide produced [81,82]. Second, sulfonated lignins are useful additives in diverse formulations—they are ore flotation [64], inexpensive wetting, dispersing, and sequestering agents [69] used, for instance, in pesticides and herbicides [1,81,82], in oil-well drilling fluid formulations for deflocculating clay [1,64], in "black liquor soap acidulation" (tall oil recovery and processing), and in textile finishing industry (protective colloids for dissolving or dispersing dyestuffs [74,81,82]); sulfonated kraft lignins show a strong dispersing power to kaolin [83]. Dry forms of the materials are used as road binders [69]. Pelletizing animal feed products and holding ruminant feed pellets together are made easier due to lignosulfonate incorporation [64]. Lignosulfonate materials can be modified to serve as water-reducing agents and superplasticizers for cement. Regarding drilling muds, it should be noted that, in addition to heavy minerals, many formulations contain Cr(III) lignosulfonates, which is derived from the reduction of sodium dichromate (as in a tanning formula) by lignosulfonate waste from sulfite pulp mills. This use amounts to ~4% of the total Cr compound consumption [84–86].

The approximate distribution of some of the main uses of lignosulfonates is presented in Table 7.3.

TABLE 7.3

Distribution of the Main Uses of Lignosulfonates

Use	Consumption(%)
Road dust control	19
Animal feed pellets	15
Concrete additives	14
Oil-well drilling muds	4
Pesticide dispersant	3
Others	45

Source: After Knaggs, E.A. and Nepras, M.J. in *Kirk-Othmer Encyclopedia of Chemical Technology*, Wiley, New York, 1997, 146–193.

REFERENCES

1. Falbe, J. ed., *Surfactants in Consumer Products: Theory, Technology and Application*, Springer, Berlin, 1986.
2. Bluestein, C. and Bluestein, B.R., Petroleum sulfonates, in *Anionic Surfactants*, Surfactant Science Series, vol. 7, part 2, W.M. Linfield, ed., pp. 316–343, Marcel Dekker, New York, 1976.
3. Stirton, A.J. and Weil, J.K., Alpha-sulfomonocarboxylic acids and derivatives, in *Anionic Surfactants*, Surfactant Science Series, vol. 7, part 2, W.M. Linfield, ed., pp. 381–404, Marcel Dekker, New York, 1976.
4. Knaggs, E.A. and Nepras, M.J., Sulfonation and sulfation, in *Kirk-Othmer Encyclopedia of Chemical Technology*, vol. 23, 4th edn., J.I. Kroschwitz, ed., pp. 146–193, Wiley New York, 1997.
5. Agriculture et Agroalimentaire Canada. Profils sectoriels–AC–25 June 2002 (report).
6. www.spd-lu.de/aktuelles.
7. Reychler, A., Cetylsulfonic acid, *Bulletin des Sociétés Chimiques Belges.*, 28, 227–229, 1914.
8. Reed, R.M. and Tartar, H.V., The preparation of sodium alkyl sulfonates, *J. Am. Chem. Soc.*, 57, 570–571, 1935.
9. Reed, R.M. and Tartar, H.V., A study of salts of higher alkyl sulfonic acids, *J. Am. Chem. Soc.*, 58, 322–332, 1936.
10. Weschler, J.R. and Koberda, A.M., Primary alkane sulfonates, *J. Am. Oil Chem. Soc.*, 60(12), 2012–2014, 1983.
11. Tyutyunnikov, B.N., Kovalev, V.M. and Bukhshtab, Z.I., Factors affecting the preparation of primary alkanesulfonates from α-olefins and sodium bisulfite, *Nefterabotka i neftekhimiya*, 12, 48–50, 1973.
12. Tyutyunnikov, B.N., Bukhshtab, Z.I., Kovalev, V.M. and Goryushko, V.E., Optimization of the preparation of primary alkanesulfonates by an experiment planning method, *Khimi. Teknol.*, 3, 48–50, 1974.
13. Möhle, L., Opitz, S. and Ohlerich, U., Zum Grenzflächenverhalten von Alkansulfonaten, *Tenside Surfact. Det.*, 30(2), 104–109, 1993.
14. Saito, M., Moroi, Y. and Matuura, R., Dissolution and micellization of sodium n-alkylsulfonates in water, *J. Colloid Interf. Sci.*, 88(2), 578–583, 1982.
15. Prince, L.M., Toilet soaps of synthetic surfactants, German Patent DE 2236727, 1974.
16. Thomsen, J.K. and Cox, R.P. Alkanesulfonates as effluents for the determination of nitrate and nitrite by ion chromatography with direct UV detection, *J. Chromatogr.*, 521(1), 53–61, 1990.
17. Asinger, F., *Paraffins—Chemistry and Technology*, Pergamon Press, Oxford, pp. 483–571, 645–668, 1968.
18. Reed, C.F., Forming substitution products by treating hydrocarbons with a halogen and a dioxide of sulfur, selenium or tellurium, US Patent 20968, 1939.
19. Reed, C.F., Alkanesulfonyl chlorides, US Patent 2,174,492, 1939.
20. Reed, C.F., Surface-active agents formed by the action of sulfur dioxide and chlorine on petroleum oils, US Patent 2,263,312, 1941.
21. Lockwood, W.H. and Richmond, J.L., Countercurrent reaction of hydrocarbons with sulfur dioxide and chlorine, US Patent 2,193,824, March 19, 1940.
22. Dupont de Nemours, Continuous photochemical preparation of aliphatic sulfonyl chlorides, US Patent 2,528,320, July 1946.
23. Gilbert, E.E., *Sulfonation and Related Reactions*, Wiley, New York, 1965.
24. Balakirev, E.S., Gershenovich, A.I., Genin, L.S., Kats, M.B., Babenko, V.E., Bisekenov, M.A., Pevnev, V.A., Kutyanin, L.I. and Lukmanov, A.S. et al., Photochemical sulfochlorination of paraffins, German (East) DD Patent 0,160,830, April 11, 1984.
25. Kats, M.B., Sidorov, V.G., Serov, V.A., Chernyshev, G.N. and Sheshenev, A.A. (to Volg Khimprom Production Association), Process for preparing alkyl sulfochlorides, SU Patent 2,007,392, February 15, 1994.
26. Lubrizol Corporation, Preparation of hydrocarbon sulfonyl chlorides, US Patent 2,836,093, 1960.
27. Continental Oil, Preparation of biodegradable alkanesulfonamides, US Patent 3,808,272, 1974.
28. Platz, C. and Schimmelschmidt, K. (IG Farbenindustrie), German Patent 735,096, 1943.
29. Orthner, L., The introduction of the sulfo group into alkanes by means of sulfur dioxide and oxygen (sulfoxidation), *Angew. Chem.*, 62A, 302–305, 1950.
30. Stanciu, I., Papahagi, L., Avram, R., Cristecu, C., Duta, I. and Savu, D., Sulfoxidation of *n*-alkanes, *Rev. Chim. (Bucharest)*, 41(11–12), 885, 1990.
31. Trautmann, M. and Jürges P., Secondary alkanesulfonate—An ecological and economical alternative, *Tenside Surfact. Det.*, 21(2), 57–61, 1984.

32. Braun, A.M., Maurette, M.T. and Oliveros, E., *Technologie Photochimique*, Presses Polytechniques Romandes, Lausanne, 1986.
33. Hagemeyer, A. and Kuhlein, K., Photoinitiation of sulfochloration or sulfoxidation of alkanes using UV-excimer radiation for the manufacture of alkylsulfonyl chlorides and alkylsulfonates, European Patent Application EP 1,028,107, 2000.
34. Schneider, A. and Chu, J.C., Sulfochlorination of cyclohexane induced by gamma radiation, *Ind. Eng. Chem. Proc. Des. Dev.*, 3(2), 164–169, 1964.
35. Schneider, A. and Chu, J.C., Kinetics of sulfochlorination of cyclohexane in carbon tetrachloride induced by gamma radiation, *AICHE J.*, 10(6), 930–934, 1964.
36. Geiseler, G. and Nagel, H.D., Über die Reaktionswärme der Sulfochlorierung höhermolekularer geradkettiger Alkane, *Chem. Ber.*, 91, 204–211, 1958.
37. VEB Leuna Werke, Verfahren zur Herstellung von Alkansulfochloriden, German (East) DD 147,844, 1965.
38. Joschek, H.I., Die sulfochlorierung, *Chem. Ztg.*, 93(17), 655–663, 1969.
39. Ferreira de la Salles, W., Sulfochloration par le chlorure de sulfuryle: un procédé alternatif pour la fabrication d'alcanesulfonates secondaires, Thesis, Institut National Polytechnique de Toulouse, 217, 2004.
40. Kharasch, M.S. and Read, A.T., Sulfonation reactions with sulfuryl chloride, *J. Am. Chem. Soc.*, 61, 3089, 1939.
41. Tazerouti, A., Rahal, S. and Soumillion, J.P., The photochemical chlorosulfonation of heptane by sulfuryl chloride: the role of solvent and catalyst—a reinvestigation, *J. Chem. Res.*, 1, 1101–1119, 1994.
42. Tazerouti, A., Etude systématique de la chlorosulfonation des *n*-paraffines par voie photochimique, Ph. D. thesis, Université des Sciences et de la Technologie Houari Boumediene, Alger, 1994.
43. Azira, H., Assassi, N. and Tazerouti, A., Synthesis of long-chain alkanesulfonates by sulfochlorination using sulfuryl chloride, *J. Surf. Det.*, 6(1), 55, 2003.
44. Ferreira de la Salles, W., Canselier, J.P., Gourdon, C. and Tazerouti, A., An alternative process for the manufacture of alkanesulfonates, *J. Com. Esp. Det.*, 34, 339, 2004.
45. Ferreira de la Salles, W., Canselier, J.P. and Gourdon, C., An approach to the manufacture of secondary alkanesulfonates without gaseous reagents, *Household and Personal Care Today* (suppl. to *Chimica Oggi*), June, 70–73, 2004.
46. Westphal, O. and Jerchel, D., Reaction of higher 1-chloroparaffins with ammonia, primary, secondary and tertiary amines, *Chem. Ber.*, 73, 1002–1011, 1940.
47. Student report, ENSIACET, Toulouse, France, 2003 (unpublished).
48. Kharasch, M.S., Chao, T.H. and Brown, H.C., Sulfonation reactions with sulfuryl chloride. II. The photochemical of aliphatic acids with sulfuryl chloride, *J. Am. Chem. Soc.*, 62, 2393–2397, 1940.
49. Khanna, V., Tamilselvan, P., Kalra, S.J.S. and Iqbal, J., Cobalt (II) porphyrin catalyzed selective functionalization of alkanes with sulfuryl chloride: A remarkable substituent effect, *Tetrahedron Lett.*, 35(32), 5935–5938, 1994.
50. Bares, J., Cerny, C., Fried, V. and Pick, J., *Recueil De Problèmes De Chimie Physique*, Gauthier-Villars, Paris, 1966.
51. Cox, J.D. and Pilcher, G., *Thermochemistry of Organic and Organometallic Compounds*, Academic Press, London, 1970.
52. Joshi, R.M., Bond energy scheme for estimating heats of formation of monomers and polymers. VI. Sulfur compounds, *J. Macromol. Sci. Chem.*, A13(7), 1015–1044, 1979.
53. Graf, R., The mechanism of the sulfoxidation, *Justus Liebigs Ann. Chem.*, 578, 50–82, 1952.
54. Ramloch, H. and Taüber, G., Modern large-scale chemical processes: Sulphoxidation, *Chem. unserer Zeit.*, 13(5), 28–36, 1979.
55. Quack, J.M. and Trautmann, M., Sekundäre Alkansulfonate, *Tenside Surfact. Det.*, 22(6), 281–289, 1985.
56. Cooper, D.E., Griswold, H.E., Lewis, R.M. and Stokeld, R.W., Improved desorption route to normal paraffins, *Chem. Eng. Prog.*, 62(4), 69–73, 1966.
57. Student report, ENSIGC, Toulouse, France, 1993 (unpublished).
58. Roberts, J.B., Gage, H.B. and Brautcheck, C.H. (to E.I. Du Pont de Nemours), Aliphatic sulfonyl chlorides by photochemical reaction, US Patent 2,528,320, October 31, 1950.
59. Berthold, H., Haase, B., Mörlein, J., Rockstuhl, R. and Wirth, D., Preparation of sodium alkanesulfonates, German (East) DD 283,729, 1990.

Production of Alkanesulfonates and Related Compounds

60. Berthold, H., Haase, B., Müller, W., Rockstuhl, R., and Wirth, D., Preparation of alkanesulfonates by saponification of sulfochlorinated alkanes, German (East) DD 284,790, 1990.

61. Berthold, H., Haase, B., Lipfert, G., Rockstuhl, R., Werner, W. and Wirth, D., Preparation of solid sodium alkanesulfonates from aqueous solutions, German (East) DD 290,775, 1991.

62. Berthold, H., Matte, E., Mörlein, J., Rose, F., Scholz, J., Werner, W. and Wirth, D., Preparation of alkanesulfonate solutions with high viscosity from sulfochlorinated alkanes, German (East) DD 292,448; DD 292,449, 1991.

63. Ludwig, W., Janisch, I., Senge, F., Moritz, R.J., Tegtmeyer, D. and Koll, J., Continuous process for the manufacture of alkali metal alkanesulfonates, *Ger. Offen.* DE 4,212,086, 1993.

64. Lynn Jr., J.L. and Bory, B.H., Surfactants, in *Kirk-Othmer Encyclopedia of Chemical Technology*, vol. 23, 4th edn., J.I. Kroschwitz, ed., pp. 478–541, Wiley, New York, 1997.

65. Gaillard, R. and Borenfreund, E., *Les industries de la Chimie, Pour la Science*, Belin, Paris, 1992.

66. Boy, A., Brard, R. and Passedroit, H., Paraffin-sulfonates process boasts new extraction step, *Chem. Eng.*, 82(22), 84–85, 1975.

67. Kosswig, K., Surfactants, in *Ullmann's Encyclopedia of Industrial Chemistry*, vol., 7th edn., Wiley-VCH, New York, 2002 (http//:www.mrw.interscience.wiley.com/ueic/articles/a25_747/toc.html).

68. Karsa, D.R., Bailey, R.M., Shelmerdine, B. and McCann, S.A., Overview: A decade of change in the surfactant market, in *Industrial Applications of Surfactants IV*, D.R. Karsa, ed., Royal Society of Chemistry, Cambridge, 1999.

69. Tully, P.S., Sulfonic acids, in *Kirk-Othmer Encyclopedia of Chemical Technology*, vol. 23, 4th edn., J.I. Kroschwitz, ed., pp. 194–217, Wiley, New York, 1997.

70. Rosen, M.J., *Surfactants and Interfacial Phenomena*, 3rd edn., Wiley, New York, 2004.

71. Bogach, E.V. and Uskach Y.L., Manufacture of alkanesulfonates emulsifiers, Russian Patent SU 1,768,589, 1992.

72. Barakat, Y., Fortney, L.N., LaLanne-Cassou, C., Schechter, R.S., Wade, W.H., Weerasooriya, U. and Yiv, S., The phase behavior of simple salt-tolerant sulfonates, *Soc. Petrol. Eng. J.*, 23(6), 913–918, December, 1983.

73. Schechter, R.S., Wade, W.H., Weerasooriya, U., Weerasooriya, V. and Yiv, S., *J. Disp. Sci. Technol.*, 6(2), 223–235, 1985.

74. OECD series on emission scenario documents, 7: Emission scenario document on textile finishing industry (ENV/JM/MONO(2004)12, June 24, 2004.

75. www.stepan.com.

76. Salager, J.L., FIRP booklet # E300-A, *Surfactants: Types and Uses*, FIRP, version 2, Merida, Venezuela, 2002.

77. www.penreco.com.

78. Christhilf, H.H., Morrison, P.E. and Niemczura, P.W., Pipe varnish compositions, US Patent 4, 631, 083, 1986.

79. http://en.wikipedia.org/wiki/Image:LigninStructure.jpg.

80. Goheen, D.W. and Hoyt, C.H., Lignin, in *Kirk-Othmer Encyclopedia of Chemical Technology*, vol. 14, 3rd edn., M. Grayson, ed., pp. 294–312, Wiley, New York, 1981.

81. Lin, S.Y. and Lebo Jr., S.E., Lignin, in *Kirk-Othmer Encyclopedia of Chemical Technology*, vol. 15, 4th ed., J.I. Kroschwitz, ed., pp. 268–289, Wiley, New York, 1995.

82. Lebo Jr., S.E., Gargulak, J.D. and McNally, T.J., *Kirk-Othmer Encyclopedia of Chemical Technology*, 5th edn., Wiley, New York, 2001.

83. Meshitsuka, G. and Nakano, J., Studies on lignin. 113. Studies on water solubilization of lignin. 2. *Kami Pa Gikyoshi*, 34(11), 7443–7749, 1980.

84. Rogers, W.F., *Composition and Properties of Oil Well Drilling Fluids*, 3rd edn., Gulf Publishing Co., Houston, TX, pp. 420–422, 1963.

85. Skelly, W.G. and Dieball, D.E., Behavior of chromate in drilling fluids containing chrome lignosulfonate, *J. Soc. Petrol. Eng.*, 10(2), 140–144, 1970.

86. Page, B.J. and Loar, G.W., Chromium compounds, in *Kirk-Othmer Encyclopedia of Chemical Technology*, vol. 6, 4th edn., J.I. Kroschwitz, ed., pp. 263–311, Wiley, New York, 1993.

8 Production of Glyceryl Ether Sulfonates

Jeffrey C. Cummins

CONTENTS

8.1 Introduction 159
8.2 Step 1: Production of Glyceryl Ether 159
    8.2.1 Chemistry 160
    8.2.2 Process 160
8.3 Step 2: Production of Glycidyl Ether 161
    8.3.1 Chemistry 162
    8.3.2 Process 162
8.4 Step 3: Production of Glyceryl Ether Sulfonate 163
    8.4.1 Chemistry 163
    8.4.2 Process 164
8.5 Alternative Chemical Process 165
8.6 Benefits and Uses of Glyceryl Ether Sulfonates 166
8.7 Conclusion 168
References 168

8.1 INTRODUCTION

Glyceryl ether sulfonates are surface-active anionic surfactants, which utilize a glyceryl backbone in the sulfonated group of the molecule. On the basis of patent literature, the first production of these ethers began with Alfred Kirstahler[1] and Richard Hueter[2] in the 1930s. Since this time, the process for the production of these surfactants has been significantly improved and optimized.

These surfactants have been clinically proven to be very mild with good lather abilities. An overview of the chemistry, production, and current uses of these surfactants are given in the following sections. Literature sources from journals or reference books are rare; most research on this group of surfactants is based on patent references.

8.2 STEP 1: PRODUCTION OF GLYCERYL ETHER

The first step in the production of glyceryl ether sulfonates is the production of the intermediate glyceryl ether. There is a multitude of potential reactants that can be used as discussed in Section 8.2.1. Additionally, this reaction is an oligomer reaction, and hence, the intermediate is characterized by multiple chemical species, not just one.

This step is the most important step in the production of the final surfactant. The reactants chosen and the degree of oligomerization will affect the structure and composition of the surfactant and have a direct effect on the final properties. The chemistry and production covered herein are in general terms, and specific reaction conditions may vary depending on the type of reactants and catalyst used.

159

8.2.1 CHEMISTRY

Glyceryl ethers are formed by the reaction of an epoxide compound with an organic hydroxyl compound as seen in Figure 8.1. The epoxide compounds, which can be used for this reaction, include alkylene oxides, epihalohydrins, nitro epoxide compounds, epoxide ethers, and epoxide thioethers. There are many types of hydroxyl compounds that can be used; the most used in industry includes primary alcohols, ethoxylated alcohols, and branched alcohols. The types of catalyst used for this reaction include acid acting compounds and metal halides such as sulfuric acid and stannic chloride, respectively.[3]

The general overview of the reactants and catalysts in Figure 8.1 shows a vast array of potential intermediates; few of which have made it to production primarily due to the economics of the reactants. The most used epoxide and hydroxyl compound in industry are α-epichlorohydrin and either an alkyl alcohol or alkyl ethoxylated alcohol, respectively.

The reaction begins immediately once the reactants come in contact with the catalyst and is highly exothermic. As shown in Figure 8.1, the final product is chloro-glyceryl ether, which is in itself a hydroxyl-containing compound. As a result, an oligomerization reaction occurs. Some literature sources do mention the additional oligomers and refer to them as monomers, dimers, trimers, and tetramers. The literature does not mention ethers with more than four repeating units.

8.2.2 PROCESS

This reaction is capable of being performed in batch, semibatch, and continuous operation. Owing to the heat rise during the production, the most used process in industry is a batch system, which allows better control over the exotherm as well as allowing trimming procedures to ensure proper reaction completion. That is, if after the specified reaction time, there is still an undesirable amount of unreacted alcohol, more epichlorohydrin or catalyst can be added to the vessel.

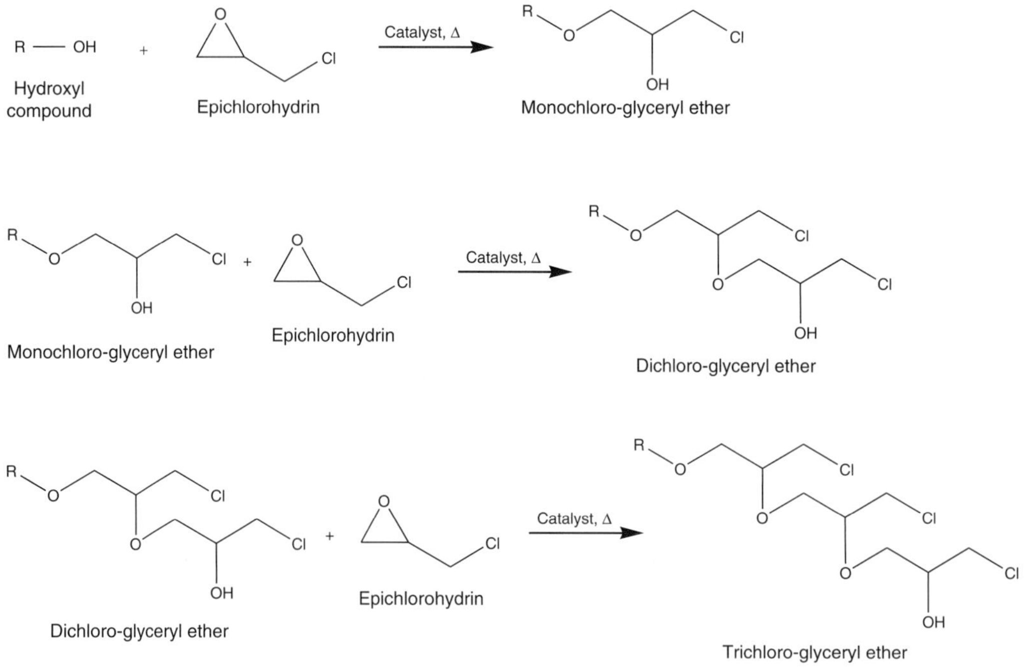

FIGURE 8.1 Chemistry for the production of glyceryl ethers.

Production of Glyceryl Ether Sulfonates

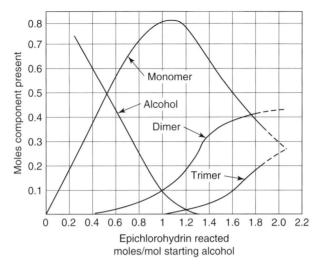

FIGURE 8.2 Graph of reactants and products as a function of the amount of epichlorohydrin. (Redrawn from Whyte, D.D. et al., U.S. Patent 3,024,273, 1962.)

Earlier work recommends a molecular excess of the hydroxyl compound to favor the formation of the monomers.[4] However, the excess hydroxyl compounds can be a negative during the final sulfonation and would require removal by distillation or other separation means. It has been discovered that small amounts of higher oligomers actually have some benefits in the sulfonated product and is discussed in more detail in Section 8.5. An excess of ~5 molar percent of the epichlorohydrin is recommended to maximize the amount of monomer as seen in Figure 8.2.[5] If a purified monomer is desired, the oligomers can be separated through distillation but with increased production costs.

A typical procedure involves adding the desired amount of the hydroxyl compound to the reactor and heating to the desired reaction temperature. The catalyst is then added in quantities of 0.001–0.1 mol/mol of the hydroxyl compound.[6] It is not recommended to add the catalysts directly to the epichlorohydrin as the epichlorohydrin will react with itself before the addition of the alcohol. The temperature of the reaction is typically controlled between 70 and 120°C.[6] To control the exotherm, the epichlorohydrin is metered into the reaction vessel. Fast addition of the epichlorohydrin can cause the reaction mixture to boil, building up pressure and even vaporizing the epichlorohydrin before it has a chance to react.

Other than temperature control, the most important aspect of this reaction is to prevent addition of moisture. Any presence of water will hydrolyze the metal halide catalyst rendering it inactive. The presence of water will require more catalyst to be used and can cause hydrogen halides to be liberated and react with the epoxide compound to form various impurities.[3]

8.3 STEP 2: PRODUCTION OF GLYCIDYL ETHER

The chloro-glyceryl ethers described earlier can be converted into the final glyceryl ether sulfonate by a Streckerization reaction, that is, use of sodium sulfite to perform sulfonation. However, the resulting composition of the surfactant will have an increased amount of sodium chloride by-product, which is very difficult to remove from the product. Therefore, it is recommended that an additional reaction step be performed to dehalogenate the product into a glycidyl ether.

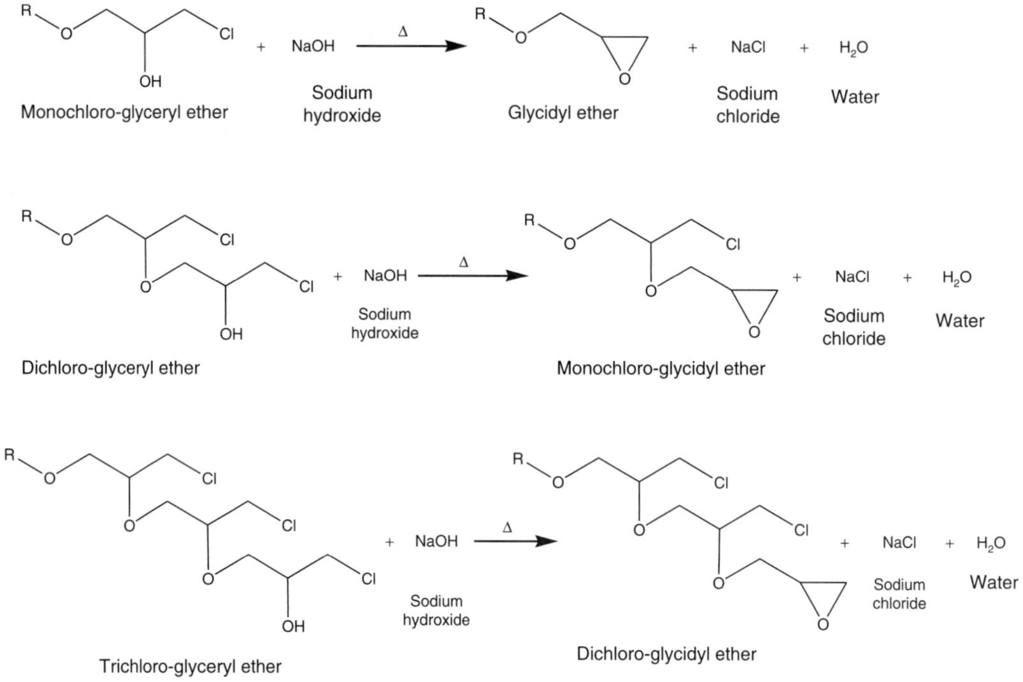

FIGURE 8.3 Epoxidation of chloro-glyceryl ethers to glycidyl ethers.

8.3.1 CHEMISTRY

The chloro-glyceryl ethers are treated with an alkali metal hydroxide solution, commonly sodium hydroxide, to epoxidize the hydroxyl group on the ether with the neighboring chlorine atom as seen in Figure 8.3. This reaction produces an undesired by-product of sodium chloride and water. In addition to epoxidizing the chloro-glyceryl ethers, another benefit of this reaction is the neutralization and destruction of the catalyst, which will prevent any further reactions from occurring.[3]

Care must be taken with this reaction to prevent the hydrolysis of the epoxide. The basic solution in the presence of water at high temperatures and extended reaction times can hydrolyze the epoxide into hydroxy glyceryl ether, effectively replacing the chlorine atoms on the chloro-glyceryl ethers, Figure 8.3, with hydroxyl groups. These hydroxyl groups are unable to be converted into sulfonates by Streckerization.

8.3.2 PROCESS

The preceding epoxidation reaction can be performed in a batch, semibatch, or continuous process. Similar to that of the glyceryl ether process, a batch process is preferred for several reasons. One reason is that if the epoxidation does not achieve the desired conversion, it is a simple matter to add additional alkali metal hydroxide solution and continue the reaction. Sodium or potassium hydroxide is the preferred epoxidation reagent due to economic considerations.

This epoxidation reaction proceeds very slowly. To increase the rate of reaction, an excess quantity of the alkali metal hydroxide is used, 25–50% molar excess or greater, and temperatures between 160 and 200°F are recommended.[7] Higher temperatures have been used to speed up the reaction further, provided proper precautions are taken to handle the vapor pressure generated during the reaction. The other rate-limiting factor for this reaction is that the chloro-glyceryl ethers are immiscible with the sodium hydroxide aqueous solution, thus requiring thorough mixing for the two layers to contact one another.

Production of Glyceryl Ether Sulfonates   163

The immiscible phases are a detriment to the speed of the reaction, but are very useful during separation. After the reaction, the agitation can be turned off and layers allowed to settle. Typically, glycidyl ethers will settle at the top, whereas the water, sodium chloride, and excess sodium hydroxide will settle at the bottom. The density of the glycidyl ethers does depend on the structure of the starting materials and the degree of oligomerization and thus may settle to the bottom. The aqueous solution can then be drained from the glycidyl ethers, and if desired, the ethers can be further washed with water to remove any remaining sodium chloride or unreacted sodium hydroxide. This aqueous solution will need to be neutralized and disposed of properly. If not for this step in the production pathway, the sodium chloride removed during settling would have remained in the final sulfonated product.

If a batch reactor is used in the production of the glycidyl ethers, then the settling operation can be performed in the same vessel. It is tempting to consider doing the entire reaction up to this point in the same vessel to reduce the capital cost. However, the alkali metal hydroxide used in this epoxidation reaction will introduce moisture into the vessel which will neutralize the catalyst used in the previous glyceryl ether reaction, reference Section 8.2.2. Therefore, it is preferable that the glyceryl ethers be placed in a separate vessel for the epoxidation reaction preventing any moisture from interfering with the initial glyceryl ether production.

8.4 STEP 3: PRODUCTION OF GLYCERYL ETHER SULFONATE

The chemistry and process descriptions described in the following sections are, primarily, based on Refs 8 and 9. The specific reaction conditions should be evaluated depending on the specific glycidyl ether used. As an example, the use of an ethoxylated alcohol versus a fatty alcohol during the production of the glyceryl ether has a large effect on the processing conditions. An alkoxylated glycidyl ether can be sulfonated in the same manner as the glycidyl ether, but with a few processing advantages such as lower reaction temperatures, lower pressures, and higher active concentrations without adversely affecting the viscosity.[10]

8.4.1 CHEMISTRY

The glycidyl ethers mentioned earlier are sulfonated by a process known as Streckerization. The recognized Streckerization reaction uses an alkali metal sulfite, which will sulfonate any of the remaining chlorine groups on the oligomers. However, the sulfite does not react with the epoxide group. The solution to this dilemma is a slightly modified form of Streckerization where an alkali metal bisulfite (versus sulfite) is used to sulfonate the epoxide group (Figure 8.4).[7] The preferred sulfonating reagents are a mixture of sodium sulfite and sodium bisulfite, although other alkali metals can be used but will alter the properties of the final surfactant due to the different cation present.

A mixture of sodium sulfite and sodium bisulfite is initially prepared. This can be done by either mixing an aqueous solution of the two reagents or reacting sodium bisulfite with a limiting amount of sodium hydroxide to yield the desired amount of sodium sulfite. Either way, the preferred pH of the solution is between 8 and 10 to provide good reaction completeness.[5] This pH range is suggested throughout the remaining process.

The aqueous sulfite/bisulfite solution is reacted with the glycidyl ethers to produce the final sulfonated product. As seen in Figure 8.4, the epoxide group as well as any chlorine atoms on the higher oligomers are sulfonated. For each mole of sodium sulfite that is reacted, an equimolar amount of sodium chloride is produced. Looking at the chemistry, it is easy to surmise that the chloro-glyceryl ethers from step 1 (Section 8.2) could have been used in this reaction with only sodium sulfite, but an excess of the undesired salt by-product would have been formed.

The final sulfonated product is typically characterized by three main analytical tests. The first is a cationic SO_3 titration, which will measure the total amount of active compound present. The

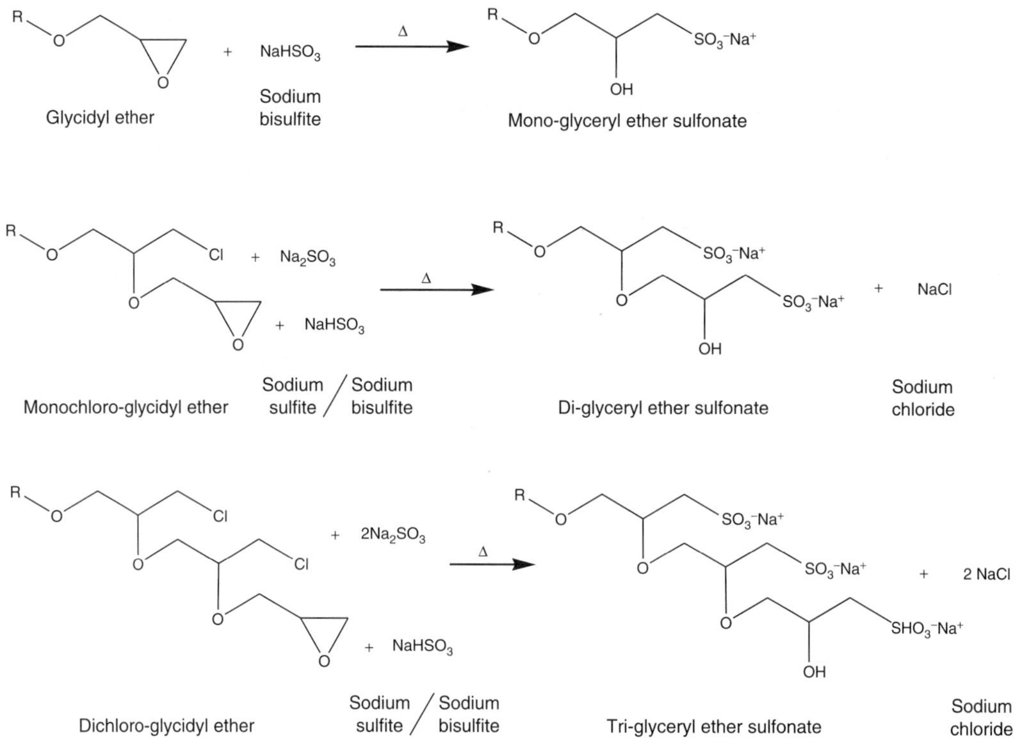

FIGURE 8.4

Chemistry for the production of glyceryl ether sulfonates from glycidyl ethers.

second is unsulfonated material determined by an extraction method, which will give the residual amount of unreacted ether, residual alcohols, and hydrolyzed ether. This method does not identify the type of unsulfonated material, only the total quantity present. The third is a moisture analysis to determine the amount of water in the final product.[5] The reaction should be tailored to the specific glycidyl ether being used to maximize and minimize the active compound and unsulfonated material, respectively. The amount of water present in the final solution is primarily an economic concern—on one hand the water adds to shipping costs, but on the other, it lowers the viscosity and makes the product easier to handle and store.

8.4.2 PROCESS

The process for this modified Streckerization is quoted in the patents as typically being a batch setup, although a continuous process can also be used. First, the desired amount of sodium bisulfite and sodium sulfite are added to the reactor followed by the desired amount of glycidyl ethers. This order of addition is not critical. In the reactions discussed earlier, an excess of one reactant is needed to get the desired conversion. However, in this step of the process, only a slight excess of the stoichiometric amount of sodium sulfite and bisulfite are necessary for good conversion.[7]

It is important to note here that once again an immiscible liquid system is encountered and good contact between the two phases is essential. A small amount of the alkyl glyceryl ether sulfonate product can be used as a seed to emulsify the two phases and provide more intimate contacting to increase the rate of reaction.[9]

An induction period is noticed during which the reaction does not occur at any appreciable rate. As the temperature is increased, the Streckerization reaction begins to accelerate in a very exothermic manner releasing large amounts of heat, which in turn accelerates the reaction. Temperatures around 300°F are quoted as being the point at which the reaction exotherm begins.[7] The

Production of Glyceryl Ether Sulfonates   **165**

actual temperature will depend on the quantity of sulfonate seed used as well as the structure of the glycidyl ethers. The temperature should be optimized to gain the maximum conversion and minimize any discoloration, which can occur at elevated temperatures.[5] Color of the final product can be improved by standard bleaching processes such as the use of hydrogen peroxide. Owing to the high temperature, the water present in the system from the aqueous sulfite/bisulfite solution will boil and generate a vapor pressure equal to that of water at that temperature. Therefore, this process must be performed in a suitable pressure vessel. This pressure adds benefit to the reaction to increase the miscibility of the two phases.

Moisture content is another key variable to be controlled during the production of glyceryl ether sulfonates. The minimum amount of water in the system should be enough to keep the sodium sulfite, with its limited solubility, completely dissolved in the system.[5] This is another advantage for using the modified Streckerization reaction since bisulfite is more soluble than sulfite, and as less sulfite is needed, the amount of water required is also reduced. The sulfonate product is also highly viscous with poor heat transfer. Therefore, an overall moisture concentration of 50% is recommended to lower the viscosity of the solution and allow for better temperature control.[11]

An aspect of the sulfonation process, which has little or no mention in literature, is the appropriate metallurgy for the vessels needed during this step in the production. The only reference found during the literature search is from David D. Whyte[7], who mentions that during example preparations, the final paste product had a strong sulfur dioxide odor and poor color indicating the corrosion of the equipment used in making the sample. However, this reference does not mention the metallurgy of the vessel being used. The reactants had a pH in the range of 5–8 as did the final product. Therefore, it is likely that the corrosion is due to the chloride chemistry versus the pH of the material.

8.5 ALTERNATIVE CHEMICAL PROCESSES

The earlier sections deal with the main process for making glyceryl ether sulfonates used in industry; however, multiple processes exist in the literature for making the surfactant as well as ether intermediates. An example is the sulfation of glyceryl ether alcohols, or monoalkyl glycerin ethers. These ethers can be prepared through hydrolysis of either chloro-glyceryl ethers or glycidyl ethers.[12] An alternative process is to perform a similar reaction that formed the chloro-glyceryl ethers (Figure 8.1) but use glycidol instead as shown in Figure 8.5. It should be noted that additional oligomerization then occurs on the hydroxyl group attached to the gamma carbon instead of the beta carbon.[13]

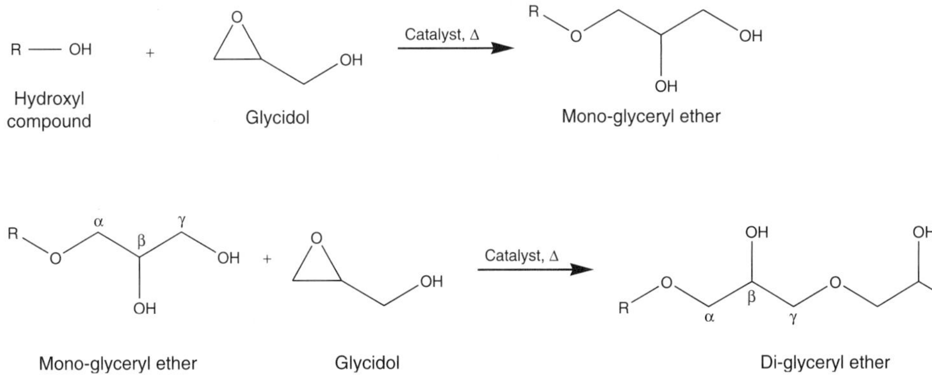

FIGURE 8.5 Formation of glyceryl ethers from glycidol.

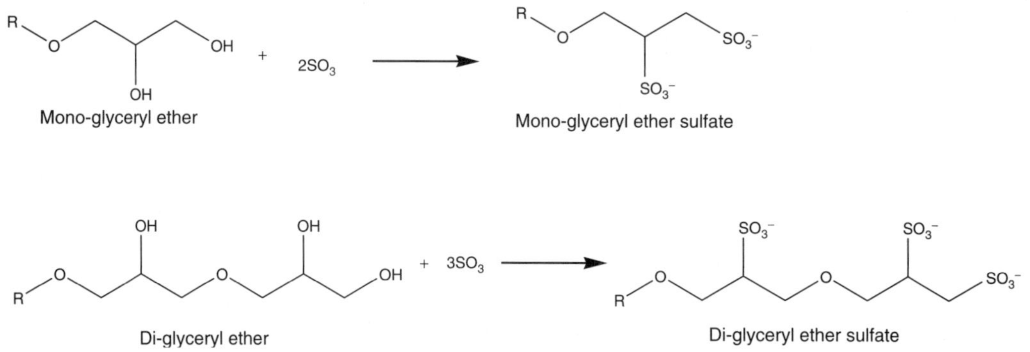

166   Handbook of Detergents/Part F: Production

FIGURE 8.6 Sulfation of glyceryl ethers.

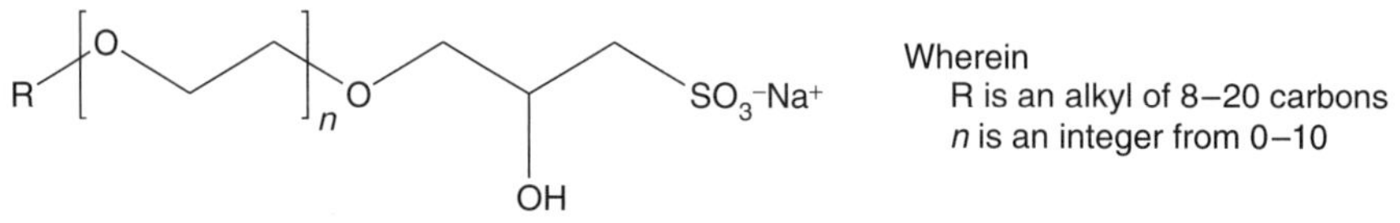

FIGURE 8.7 Generic structure of glyceryl ethers used in personal cleansing industry.

Wherein
R is an alkyl of 8–20 carbons
n is an integer from 0–10

These ethers can now be sulfated using any known sulfation agent such as chlorosulfonic acid, sulfur trioxide, pyrosulfate, or oleum to yield the final glyceryl ether sulfates (Figure 8.6) and then neutralized with an alkali metal hydroxide such as sodium hydroxide.[14] The main difference between this molecule and the one described earlier is that all of the hydroxyls will be sulfated whereas the earlier process retains one of the hydroxyl groups.

8.6 BENEFITS AND USES OF GLYCERYL ETHER SULFONATES

The majority of the uses of glyceryl ether sulfonates are in the personal cleansing industry as a cosurfactant. Specifically, they can be used in soap bars, liquid soaps, shampoos, and laundry detergents. The structure most quoted in patents for personal cleansing products is shown in Figure 8.7. Although the monomer form is the primary surfactant, these patents do acknowledge the dimer and trimer as being part of the overall surfactant composition. In fact, the monomer form is mentioned as having limited solubility whereas the higher polymeric forms, specifically the dimer, act as a solubilizing agent for the monomer.[5]

Probably the most impressive and published aspect of glyceryl ether sulfonates is their mildness. An *in vitro* collagen-swelling assay is typically used to determine the mildness of various personal cleansing products. This test has been able to clinically prove the mildness of glyceryl ether sulfonates as demonstrated by Table 8.1 with respect to erythema, tightness, and dryness.[15] The amount of mildness is also a function of the structure of the surfactant as demonstrated in Figure 8.8, which shows how mildness of the surfactant is affected by the average chain length of the alcohol used in its preparation. The data also show that the lower carbon chains of alkyl glyceryl ethers have higher lather potential than the heavier C18 chain length homolog.[16]

This surfactant is also known to be very tolerant in high electrolyte environments, such as hard water containing calcium and magnesium, by solubilizing the electrolytes and preventing them from precipitating out the other cosurfactants in the formulation. Hard water inhibits most surfactants, specifically fatty acid soaps, from foaming or dispersing the soap resulting in

TABLE 8.1
Results of Collagen Selling Test for Several Forms of Glyceryl Ether Sulfonates versus Other Synthetic Surfactants

Sample (1% Solution)	Collagen Swelling (μL Water/mg Film)
Sodium cocoyl isethionate	6.85 ± 0.10
Sodium cocomonoglyceride sulfate	6.90 ± 0.14
Sodium alcohol (C12–C13) glyceryl sulfonate	6.53 ± 0.26
Sodium alcohol (C14–C15) glyceryl sulfonate	4.81 ± 0.12
Sodium alcohol (C12–C13) ethoxy glyceryl sulfonate	5.91 ± 0.30
Sodium alcohol (C14–C15) ethoxy glyceryl sulfonate	3.85 ± 0.22

Source: Redrawn from Subramanyam, R. et al., U.S. Patent 5,310,508, 1994.

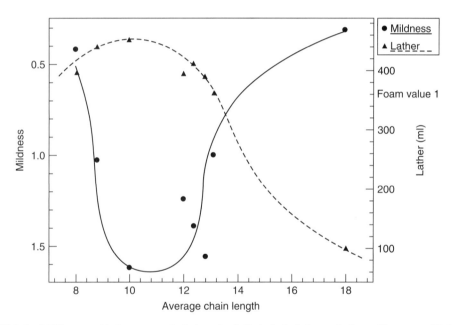

FIGURE 8.8 Mildness and lather versus chain length of alkyl alcohol glyceryl ethers. (Courtesy of Schwartz, J.R. et al., WO Patent 91/09931, 1991.)

a soap curd formation in bathtubs.[15] This benefit has been used not only in personal cleansing applications but also in enhanced oil recovery (Figure 8.9, parts a, b, and c from Refs 17, 9, and 18, respectively) processes.[18] This feature is also useful in the production of polymers through emulsion polymerization by not coprecipitating in the high electrolyte environment and thus decreasing impurities.[19]

Glyceryl ether sulfonates have not only shown to be highly effective wetting agents, but also increase the wet–dry tensile ratio of the resulting paper within the paper manufacturing industry.[19] Additionally, glyceryl ether sulfonates have been shown to be very good dye distributors in paper and textile industries with the percentage of dye leveling comparable or slightly better than the Dowfax™ 2A1, an industrial standard surfactant for this process.[19] The use of glyceryl ether sulfonates during froth flotation process for deinking paper results in greater fiber yield than commercial standards such as Lionsurf 727 and DI-600.[19]

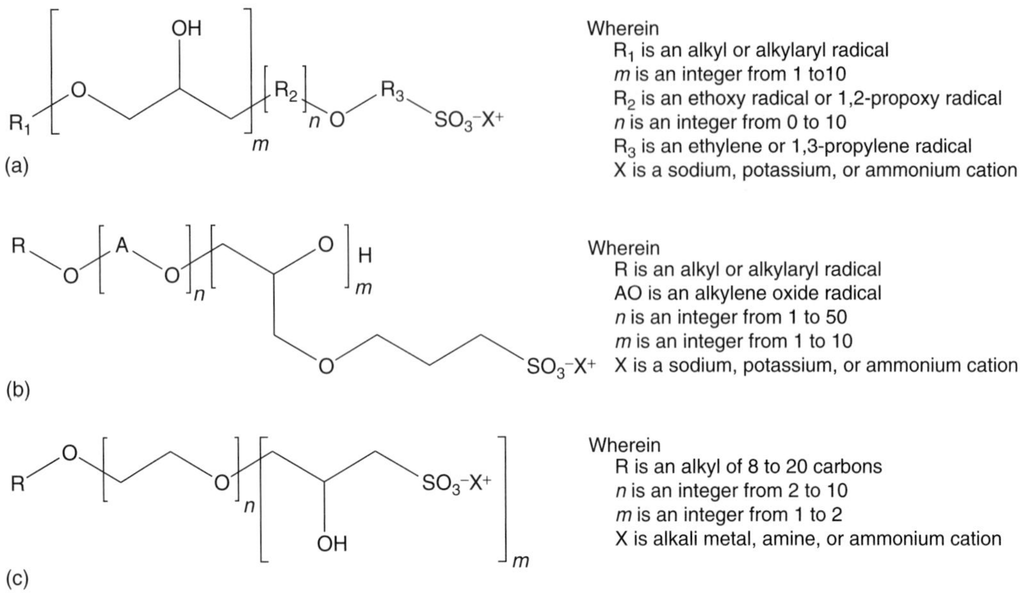

Forms diagram (a), (b), (c) with accompanying text.

(a)

Wherein
R₁ is an alkyl or alkylaryl radical
m is an integer from 1 to10
R₂ is an ethoxy radical or 1,2-propoxy radical
n is an integer from 0 to 10
R₃ is an ethylene or 1,3-propylene radical
X is a sodium, potassium, or ammonium cation

(b)

Wherein
R is an alkyl or alkylaryl radical
AO is an alkylene oxide radical
n is an integer from 1 to 50
m is an integer from 1 to 10
X is a sodium, potassium, or ammonium cation

(c)

Wherein
R is an alkyl of 8 to 20 carbons
n is an integer from 2 to 10
m is an integer from 1 to 2
X is alkali metal, amine, or ammonium cation

FIGURE 8.9 Forms of glyceryl ether sulfonates used in oil recovery. (Redrawn from (a) McCoy, D.R. U.S. Patent 4,466,891, 1984; (b) Naylor, C.G. U.S. Patent 4,436,672, 1984; and (c) Dewitt, W.J. et al., Canadian Patent 1077700, 1977.)

8.7 CONCLUSION

Glyceryl ether sulfonate is a somewhat generic term for a sulfonated surfactant, which has a glycerin ester bond associated with it. The chemistry is well established in patent literature and a wide range of homologs can be produced. Depending on the application, the structure of these surfactants can be tailored to provide a multitude of desired benefits for a vast array of applications.

The production uses a different method than the more common sulfonation systems such as chlorosulfonic acid, sulfur trioxide, pyrosulfate, or oleum. The Streckerization reaction utilizes a mixture of sulfite and bisulfite salts to sulfonate the ether intermediates into the final surfactant. The most commonly used process for the production of these surfactants is a multistep batch system, where each reaction step is performed in separate reactor. This prevents contamination of one step to the next as well as the introduction of certain species, which are a benefit to one step but a detriment to another—specifically water. The batch system's agitation allows for good mixing between two immiscible liquids, which is a common theme throughout the production. Although batch systems are the most common, the process can be adapted for continuous production.

REFERENCES

1. Kirstahler, A. US Patent 2,010,726, 1935.
2. Hueter, R. et al. US Patent 2,094,489, 1937.
3. Marple, K.E. et al. US Patent 2,327,053, 1943.
4. Marple, K.E. et al. US Patent 2,260,753, 1941.
5. Whyte, D.D. et al. US Patent 3,024,273, 1962.
6. Miyajima, T. et al. US Patent 6,437,196, 2002.
7. Whyte, D.D. et al. US Patent 2,989,547, 1961.
8. Cook, T.E. et al. US Patent 4,536,318, 1985.
9. Naylor, C.G. US Patent 4,436,672, 1984.
10. Subramanyam, R. et al. US Patent 5,672,740, 1997.
11. Whyte, D.D. Alkyl glyceryl ether sulfonates, in *Anionic Surfactants*, Surfactant Science Series Volume 7, Linfield, W.M., Ed., Marcel Dekker, New York, Chapter 14, 1976.

Production of Glyceryl Ether Sulfonates

12. Takaishi, N. et al. US Patent 4,465,869, 1984.
13. Berkowitz, S. US Patent 4,298,764, 1981.
14. Berkowitz, S. US Patent 4,217,296, 1980.
15. Subramanyam, R. et al. US Patent 5,310,508, 1994.
16. Schwartz, J.R. et al. WO Patent 91/09931, 1991.
17. McCoy, D.R. US Patent 4,466,891, 1984.
18. Dewitt, W.J. et al. Canadian Patent 1077700, 1977.
19. Rasheed, K. et al. US Patent 6,133,474, 2000.
20. Salvatore, J.S. US Patent 5,621,139, 1997.
21. Subramanyam, R. et al. US Patent 5,436,366, 1995.
22. Subramanyam, R. et al. US Patent 5,516,461, 1996.
23. Urata, K. et al. A convenient synthesis of long-chain 1-O-alkyl glyceryl ethers, *J. Am. Oil Chem. Soc.*, 65, 1299, 1988.
24. Urata, K. et al. The alkyl glycidyl ether as synthetic building blocks, *J. Am. Oil Chem. Soc.*, 71, 1027, 1994.
25. Najem, L. et al. Single step etherification of fatty alcohols by an epihalohydrin, *Synthetic Communications*, 24, 2031, 1994.
26. Lee, B.M. et al. US Patent 6,392,064, 2002.
27. Berkowitz, S. US Patent 4,269,786, 1981.
28. Jordan, N.W. et al. WO Patent 91/13958, 1991.
29. Schwartz, J.R. et al. WO Patent 91/13137, 1991.
30. Mills, V. et al. US Patent 2,988,511, 1961.
31. Cassidy, W.A. et al. WO Patent 91/09924, 1991.
32. Farris, R.D. et al. WO Patent 91/09923, 1991.
33. Richard, M.A. et al. US Patent 5,945,389, 1999.
34. Small, L.E. et al. US Patent 4,673,525, 1987.
35. Subramanyam, R. et al. US Patent 5,620,951, 1997.
36. Van Gaertner, R., et al. US Patent 3,228,979, 1966.
37. Behrens, J.R. et al. EP Patent 0,904,043, 1999.
38. Wells, R.L. US Patent 6,096,697, 2000.
39. Walker, R.D. et al. US Patent 2,970,963, 1961.
40. Parran, J.J. et al. US Patent 2,979,465, 1961.
41. Motley, C.B. et al. US Patent 5,783,200, 1998.

9 Manufacture of Syndet Toilet Bars

Paolo Tovaglieri

CONTENTS

9.1 Introduction 171
9.2 Syndet versus Soaps 172
    9.2.1 Appearance 172
        9.2.1.1 Composition Characteristics 172
        9.2.1.2 Behavior and Physical Characteristics 172
    9.2.2 Performance 173
9.3 Formulas 173
    9.3.1 Surfactants 173
        9.3.1.1 Fillers 176
        9.3.1.2 Additives 176
9.4 Process 176
    9.4.1 Manufacture of Syndet Bars 176
    9.4.2 Prerefining 177
    9.4.3 Mixing 177
    9.4.4 Refining 177
    9.4.5 Extruding 179
    9.4.6 Cutting 181
    9.4.7 Conditioning 181
References 181

9.1 INTRODUCTION

Since the advent of synthetic detergents, there has been active research to overcome a few drawbacks of traditional soap by replacing it with synthetic bars.

In fact, even if the economical values of raw materials play an important role, the basic reason for this replacement is due to the negative effect of the alkalinity of the traditional soaps on the skin.

The epidermis is covered with a fat coating, which mainly consists of unsaturated fatty acids, mixture of fatty acids, waxes and glycerides, and unsaponifiable matter.

This coating plays a relevant role in the protection of the skin. Moreover, the sebum, produced by the secretion of the sebaceous glands, plays a role in the protection and lubrication of the skin.

This acid protection barrier of the epidermis can be altered by an alkaline reaction such as an intensive or prolonged soaping. Three different approaches for basic composition of common toilet cakes that are mostly used are outlined.

Toilet soaps are obtained from either saponification of fat and natural oil mixture or neutralization of the similarly derived fatty acids.

Syndet bars are nonsoap-based materials and their active ingredients can be of widely different types. They can be derived from both natural oils and fats through chemical modification of their fatty acids, or the petrochemical industry.

Combo bars are products having active ingredients by "combination" of a synthetic and soap. The soap base should be at least 10%. Although combo bars are often considered cleansing bars based on soaps to which some lime-soap dispersants are added, their look and performance characteristics are generally closer to a syndet rather than a toilet soap. This is because the synthetic active ingredient is more effective than soap.

Free alkalinity from soap can be present. The addition of soap to a syndet base can be made for two main reasons.

1. To make the bar milder with a softer detergency effect of the active ingredients system
2. To use the soap as a plastifying/binding filler to obtain the bar

Nevertheless, there is no exact distinction between syndet and combo bars, both from the formula and performance points of view, that is, in the market area, the combo bars are more assimilated to syndets than to soaps.

9.2 SYNDET VERSUS SOAPS

Soap and synthetic toilet bars are used for the same purpose but they are relatively different in formulation types, behavior of the components, and performance of the products. To clarify the main characteristics that tend to basically differentiate the two products, a general idea is outlined in the following sections.

9.2.1 APPEARANCE

Syndets are more chalky or matte than soaps.

9.2.1.1 Composition Characteristics

Human skin has a slight acid reaction. Soaps are alkaline and they tend to defat the skin. Syndets are free from alkali and pH can be adjusted to neutral or slightly acidic. A true combo bar has, of course, an alkaline pH since it contains soap. Water-soluble inorganic salts are not usually recommended as fillers because of the possible efflorescence problem and heaviness of the cake.

The active matter in a synthetic bar is definitely lower than in soap bars. To complete the formula, a syndet needs filling materials that act as plasticizers or binders. The natural soap has an advantage over the active synthetic because it needs no fillers—being the surfactant itself, the largest part of the components with proper plastifying and binding characteristics. Soap is often included in a syndet formula for filling, detergency, and mildness.

9.2.1.2 Behavior and Physical Characteristics

Density of syndet is a bit higher than that of soap's. Wear rate of syndet is lesser than that of soap's for the same detergency effect and foam quantity.

It is quite rare for a detergent–filler combination to have the same temperature–plasticity relationship as an ordinary soap. In case of soap, there is only a small change in its plasticity, ranging between 35 and 40°C, whereas with syndet composition, large plasticity changes are encountered. For this reason, a careful control of extrusion and general process is required.

Manufacture of Syndet Toilet Bars

Crumbling of a synthetic extrusion bar is possible if there is a defect in the plasticizer and binder. Syndet hydrophilic groups are larger and are distributed over a longer section of the molecule, and their hydrophobic groups tend to be shorter, on an average, than the distribution found in soaps. This implies an increased solubility and decreased intermolecular adhesion in syndets when compared to soaps.

Water can readily permeate the mass of the already porous product to give a slushy tablet. Hydrophobic or water-repellent substances are required. Soaps do not tend to be porous, and the water penetration is very slow owing to the formulation of the well-known middle soap that has a jell-like structure.

9.2.2 PERFORMANCE

Synthetic lather perform well in soft water and hard water because the detergent salts with Ca and Mg ions are soluble in the resultant solution. Consequently, insoluble lime-soap precipitates are not formed and there is no flocculation, scum, or tub bath ring due to precipitation of Ca or Mg salts.

A more prolonged rinsing is required when a syndet is used rather than soap. Synthetic bars offer good resistance to cracking tendency. This phenomenon is not worthy of attention for syndets, although it is negatively important for soap.

Medicated tablets for people with extremely sensitive skin can be made through synthetic tablets having excellent cosmetic and dermatological properties.

Soft and slippery feel by washing with soap is partially due to some deposition of Ca and Mg salts, but mainly due to a swelling and softening effect of the alkali on the uppermost skin cells of the horny surface layer.

The soft layer of the hydrated part, that is, the slush that tends to form on toilet bars left in a moist environment, is generally called *slough*.

This is considered a negative property of syndets, which slough more than soaps due to both the higher solubility of their active ingredients and the higher permeability of the cake to the water. The latter seems to play the most important role in sloughing.

The slough rate trend should result as

$$Syndets > combo \ bars > soaps$$

Generally, those additives tending to increase the viscosity of the slough itself can reduce the sloughing of the cake (Figures 9.1 and 9.2).

9.3 FORMULAS

Most of the formula components in a soap cake are represented by fatty acid salts that act by themselves, as plasticizers and binders, directly and in the presence of a relatively small quantity of water.

On the contrary, the quantity of synthetic active ingredients is definitely lower when compared with soap. Moreover, its effect as a plasticizer and binder is generally lesser than a soap active. For this reason, fillers take up special importance in terms of their required high percentage and specific functions.

9.3.1 SURFACTANTS

A list of the main surfactants already used in syndet formulations is reported in Jungermann (1982) and Hollestein-Spitz (1982).

174 Handbook of Detergents/Part F: Production

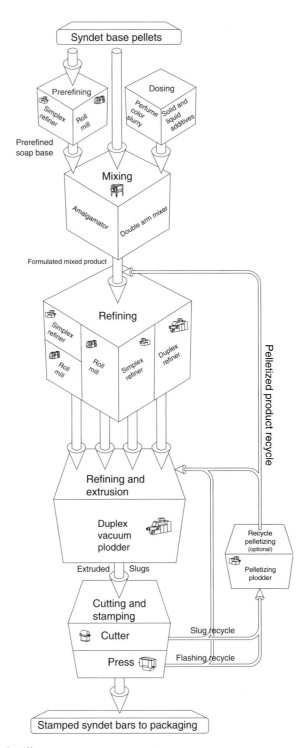

FIGURE 9.1 Synthetic toilet soap.

Manufacture of Syndet Toilet Bars

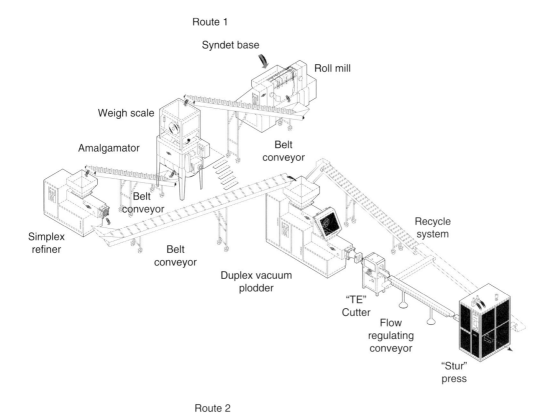
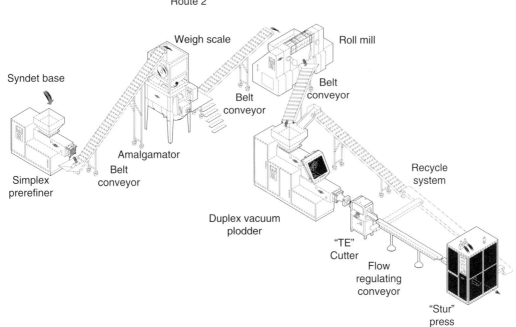

FIGURE 9.2 Manufacture of syndet toilet bars dual-function prerefining finishing line.

Following are the most commonly used surfactants:

RCOONa	Soap
ROSO$_3$Na	Alkyl sulfate
RCOOCH$_2$CH$_2$SO$_3$Na	Acyl isethionate (lauryl/mirystil isethionate = IGEPON AC 78)
RCON(CH$_3$)CH$_2$CH$_2$SO$_3$Na	Acyl *n*-methyl taurate (oleoyl *n*-methyl taurate = IGEPON T)
RCOOCH$_2$CHOHCH$_2$OSO$_2$Na	Cocomonoglyceridesulfate (COONa)
RCOOCH$_2$CH	Disodium monoalkyl sulfosuccinate (SO$_3$Na)

The functional group at the end of the alkyl chain tends to establish a major difference in the molecule's nature.

9.3.1.1 Fillers

Syndet bars contain a high percentage of fillers, which can act as inert materials, that is, pure fillers or more often as plasticizers or binders.

Such fillers can be fatty alcohols, fatty acids, soaps, fatty esters, destrine, starch, fatty amides, urea, glyceryl-sorbitol monodistearates, talcum, mannitol, puffed borax, paraffin, sodium sulfate, sodium carboxymethylcellulose, polypropylene glycol waxes, etc.

The quantity and quality of fillers affect the behavior of the mass at the manufacturing and extrusion stages and the characteristics such as appearance and performance of the final product.

9.3.1.2 Additives

These materials are always added in a relatively small percentage in the product composition. Their scope is to give, increase, or decrease some specific properties to the look or performance of the bar.

Following is a list of the most commonly used additives having specific functions:

Colors	To give a pleasant look
Perfume	To give a pleasant scent
Titanium dioxide	Whitener
Lanolin derivates	Emollient
Glycerine	Moisturizer
PEG	Moisturizer
Lecithin	Plasticizer
Ca/Zn stearate	Slippener
Polyox	Slippener
EDTA	Chelating
CMEA	Foam booster
TCC, TCCA, and TBS	Germicidal, deodorant
Al triformate	Sloughing reducer
Mg/Na chlorides and Mg sulfate	Solubility reducers
Lactic acid	pH adjuster
Tartaric acid	pH adjuster

9.4 PROCESS

9.4.1 MANUFACTURE OF SYNDET BARS

The production of syndet bars is to some extent more difficult than the manufacture of normal soaps. The difficulty is always in the area of refining and stamping. Properly formulated syndet bars can be produced with standard soap-finishing line machinery.

Manufacture of Syndet Toilet Bars

TABLE 9.1
Syndet Bar (Typical Formulas)

Composition	Percentage					
	Dry Skin	Oily Skin	Delicate Skin	Sensitive Skin	Intimate Hygiene	Baby
Synthetic base	95–96	96–97	95–96	95–96	95–96	95–96
Water	1–2	1–2	1–2	1–2	1–2	1–2
Perfume	0.3	0.25	0.27		0.45	0.18
Lecithin	x					
Vitamin F	x	x	x	x		x
PLC hydro soluble	x					
Hydropro	x					
Sulfur		x				
Salicylic acid		x				
Mink oil			x	x	x	x
Phenethyl alcohol				x		
H.A.D.				x	x	x
Irgasan				x	x	x
Dye	x	x	x		x	x

A brief description of the manufacturing steps and the most important differences between syndet and normal soap production is outlined in the following sections.

9.4.2 PREREFINING

Most refining problems occur when old, hardened, low-moisture content syndet base is used. If the initial moisture content is too low and makeup water is added in the amalgamator up to a predetermined optimum level, the first refining stage has difficulty in making the product work. It is recommended to plastify the syndet base and the added water through one refining step before mixing.

This will allow the use of only one refining machine after the amalgamator and will assure a much improved and easier refining action (Figure 9.3).

9.4.3 MIXING

The solid prerefined (if any) syndet base can be mixed with the minor liquid and other solid additives in a batch-type soap amalgamator to allow a very intimate mixing. The optimum mixing system (open-sygma or double-sygma type) depends on the product formulation and the batch size used.

High-grade formulated soap can cause difficulties in discharging the product from the standard open-sygma mixer and in conveying it to the refining stage; therefore, the use of double-sygma mixer is recommended (Figure 9.4).

9.4.4 REFINING

Refining the syndet base and all the additives into a homogeneous uniform mixture is the most critical and difficult processing step in syndet bar manufacturing.

Syndet refining, not unlike soap refining, can be performed with

- Roll mills
- Refining plodders
- Combination of refiners and roll mills

FIGURE 9.3 Simplex refiner.

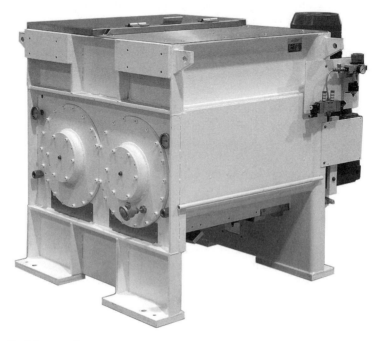

FIGURE 9.4 Double arm mixer.

Manufacture of Syndet Toilet Bars

Without going too much into equipment description details, it is enough to note that there are three-, four-, and five-roll mills, and single- and twin-worm simplex and duplex refiners.

To choose the best-suited refining system for a particular syndet bar production, the formula must be well established in advance and the moisture content of the syndet base and final syndet bar has to be kept within well-defined close ranges.

Finally, testing with a three-roll mill and simplex single-worm and simplex twin-worm refiners is essential. Only testing for adequate color dispersion, surface roughness, plasticity, and temperature assure the proper selection of refining equipment.

The operating variables are

- Water temperature for the roll mills and the refiners
- Refining screen size for the refiners

The temperature of water is very critical and depends largely on the water content of the product and, to a lesser degree, on line capacity, and the type and size of the refining machine used. In the refining plodders, refining screens of mesh number 16, 20, 28, or 50 can be used. The proper size depends on the product formulation, that is, the quantity and type of additives (Figure 9.5).

9.4.5 Extruding

All modern finishing lines have a duplex vacuum plodder for the final refining, compacting, and air-free extrusion of the product. These units consist of two plodders mounted in tandem and connected by a vacuum chamber. Twin-worm duplex vacuum plodders are the best for syndet bar lines. It is very important to note that the preliminary stage of the duplex vacuum plodder is the refining stage, whereas the final stage is for extrusion only. After passing the preliminary stage, which is the last refining stage, the product must be completely refined.

The extrusion stage must always be free from any screens and pressure plates to allow unrestricted, uniform extrusion. The extrusion head temperature ensures a smooth surface finish to the extruded slug (billet). The optimal temperature, anywhere between 60 and 90°C, depends on the line output and product formula (Figure 9.6).

FIGURE 9.5 Three roll-mill.

FIGURE 9.6 Duplex vacuum plodder.

FIGURE 9.7 Electronic cutter.

9.4.6 CUTTING

The continuously extruded slug, which leaves the duplex vacuum plodder, is cut into individual slugs of predetermined length. Single-blade electronic soap cutters are recommended for this operation (Figure 9.7).

9.4.7 CONDITIONING

The purpose of conditioning is to harden the slug surface before stamping. Most syndet bar formulations are sticky and, at times, soft for trouble-free stamping even with the use of refrigerated (chilled) stamping dyes.

Normally, conditioning tunnels are not used, but for certain products they should be considered.

REFERENCES

Hollestein-Spitz, *Journal of the American Oil Chemists' Society*, 59, October 1982.
Jungermann, *Cosmetics & Toiletries*, 97, July 1982.

10 Phosphate Ester Surfactants

David J. Tracy and Robert L. Reierson

CONTENTS

10.1 Introduction 183
10.2 Phosphation Reagents 184
    10.2.1 Phosphoric Anhydride 184
    10.2.2 Polyphosphoric Acid 185
    10.2.3 Phosphorus Oxychloride 187
10.3 Monoalkyl–Dialkyl Phosphate Ester Mixtures 187
10.4 Monoalkyl Phosphate Esters 189
    10.4.1 Polyphosphoric Acid Processes 189
    10.4.2 Hybrid Processes 190
    10.4.3 Phosphorus Oxychloride Processes 193
10.5 Dialkyl Phosphate Esters 194
10.6 Phosphobetaines 196
10.7 Conclusion 197
References 198

10.1 INTRODUCTION

The field of organophosphate ester chemistry is of ever-increasing interest and importance in the chemical industry. This is especially evident in the surfactant industry. As a result of environmental concerns and the changing needs of an aging population, opportunities exist for surfactants functioning as effective cleansing agents and possessing properties of low irritation, low toxicity, chemical stability, and biodegradability.

Phosphate esters are specialty surfactants fitting into a series of niche markets [1–4]. They can be used as

- Lubricants for textiles and leather processing
- Corrosion and rust inhibitors
- Antistatic agents
- Polymerization emulsifiers
- Wetting agents
- Dry cleaning detergents
- Hydrotropes
- Oil-field gellants
- Mild cleansing agents
- Medicaments
- Liquid ion exchangers

Recently, interest in phosphate esters has intensified because of their compatibility with skin and their inherently low-irritation characteristics [5–9]. These characteristics allow use in personal care–type applications such as body wash, shampoo, and oral care formulations. Such applications offer the potential for large-volume usage, which would move phosphate esters out of pure specialty-type uses to commodity-type products. With the increase in the price of oil, diesters are gaining importance as oil gellants in enhanced recovery.

Phosphate ester synthesis and properties have been reviewed earlier [1,10–13]. The current treatise will consequently concentrate on manufacturing techniques with special emphasis on recent advances in preparing high monoalkyl phosphate (MAP) and high dialkyl phosphate (DAP) esters.

This chapter discusses the methods for preparation of phosphate esters in their acidic form. The esters are frequently marketed as neutralized or partially neutralized salts. The counterions are alkali metals or organic amines, for example, triethanolamine. The salts possess greater hydrolytic stability than the free acids and are readily prepared from the free acids by neutralization.

10.2 PHOSPHATION REAGENTS

The principal phosphation reagents used to manufacture phosphate esters are phosphoric anhydride (P_4O_{10}), phosphorus oxychloride [$P(O)Cl_3$], and polyphosphoric acid. It is important to understand the physical and chemical characteristics of each raw material. The structure of the phosphation agent affects its reactivity and selectivity, resulting in variation in the monoester, diester, and triester product distribution.

10.2.1 Phosphoric Anhydride

Phosphoric anhydride is manufactured by direct air oxidation of phosphorus in the presence of excess oxygen [14]. The hot vapors are typically condensed into a fine, white, talc-like powder in a fluidized bed with a conical bottom, where the powder collects for periodic transfer directly to the shipping drums or stainless steel tote bins holding up to 1600 kg of the reagent. Electron diffraction [15] studies establish the structure as a tetrahedron. In the tetrahedron, illustrated in the following structures, each of the four phosphorus atoms is surrounded tetrahedrally by oxygen atoms, three oxygen atoms in each tetrahedron being common to three tetrahedra. Each phosphorus atom also contains an apical oxygen atom with a very short P–O bond of 1.39 A.

Phosphoric anhydride is a thermally stable, white, free-flowing powder. Phosphoric anhydride is very hydroscopic, acting as a powerful desiccating agent. Because of its hygroscopicity, phosphoric anhydride is prone to clumping through absorption of moisture from the atmosphere, thereby undergoing surface conversion to polyphosphoric acid. Care must be exercised to exclude moisture while charging phosphoric anhydride to the reaction mixture. Good agitation is also necessary for dispersion and uniform mixing of the powder. Adequate heat transfer coupled with efficient cooling

Phosphate Ester Surfactants

capacity aids in controlling the high heat of the reaction. If charged too rapidly to the reaction mixture, charring and discoloration will result. Phosphoric anhydride reacts preferentially with primary alcohols. Secondary alcohols are prone to undergo an undesirable side reaction, namely dehydration.

Typical specifications [16] for phosphoric anhydride, commercial name, P_2O_5, are

Phosphorus pentoxide (P_2O_5) (weight %)	99.5% Minimum
Iron (ppm)	2 Maximum
Heavy metals (ppm) (or Pb)	20 Maximum
Lower oxides (%) (as P_2O_3)	0.05 Maximum

Phosphoric anhydride is supplied commercially by Rhodia Inc. in the United States and Thermphos and Clariant in Europe.

10.2.2 POLYPHOSPHORIC ACID

Polyphosphoric acid is produced by a variety of methods—most commonly from wet phosphoric acid by evaporative concentration, or thermally by combustion of elemental phosphorus [17], especially in the presence of controlled amounts of moisture [18]. It can also be made by dissolution of phosphoric anhydride in syrupy phosphoric acid or reaction of phosphorus oxychloride and orthophosphoric acid with hydrogen chloride elimination. Polyphosphoric acid consists of a mixture of linear, oligomeric chains of alternating phosphorus and oxygen atoms [19].

$$H \!-\!\!\left[O\!-\!\!\overset{\displaystyle OH}{\underset{\displaystyle O}{\overset{|}{\underset{\|}{P}}}}\right]_n\!\!-\!OH$$

The strength of polyphosphoric acid is commonly expressed as a percentage of phosphoric acid or P_2O_5. As the value of n and the corresponding percentage of phosphoric acid increase, the material becomes progressively more viscous. The oily liquid first becomes rubbery and finally a glassy solid [14].

H_3PO_4 (%)	P_2O_5 (%)	Physical Form
100–113	72–82	Oily
113–124	82–90	Viscous, gummy
>124	90	Glassy, crystalline

Most frequently, polyphosphoric acid of about 115% (83–84% P_2O_5) is used. Compositions of concentrated polyphosphoric acids with varying levels of P_2O_5 are summarized in Table 10.1 [20].

The linear polymeric structure of polyphosphoric acid makes it less reactive than the higher-energy tetrahedral phosphoric anhydride. It more readily dissolves, usually requiring heat to promote reaction. The major suppliers of polyphosphoric acid are Innophos Inc., Israel Chemicals Ltd (ICL), Themphos, and Clariant.

TABLE 10.1

Percentage of Individual Polyphosphoric Acids in a Concentrated Polyphosphoric Acid Mixture

Total P_2O_5 (%)	Number of P Atoms in the Low-Molecular Polyacids														High Poly	Tri Meta	Tetra Meta
	1	2	3	4	5	6	7	8	9	10	11	12	13	14			
67.4	100																
68.7	99.7	0.33															
70.4	96.2	3.85															
71.7	91.0	8.86	Traces														
73.5	77.1	22.1	0.79														
73.9	73.6	25.1	1.34														
75.7	53.9	40.7	4.86	0.46													
77.5	33.5	50.6	11.5	2.86	0.74	Traces											
79.1	22.1	46.3	20.3	7.82	2.26	1.02	0.34										
80.5	13.8	38.2	23.0	13.0	6.86	3.38	1.67	1.03	0.22								
81.0	12.2	34.0	22.7	14.6	8.42	4.36	2.27	1.41	0.56	Traces							
81.2	10.9	32.9	22.3	15.0	9.36	5.41	2.85	1.75	0.97	0.36	0.05						
82.4	7.32	23.0	19.3	15.9	12.3	8.21	5.73	3.89	2.52	1.36	0.91	0.14			Traces		
84.0	3.92	11.8	12.7	12.0	10.5	8.97	7.99	6.62	5.63	4.54	3.72	3.03	2.46	1.68	6.63		
85.0	2.28	6.36	7.32	8.01	8.17	7.67	7.22	6.93	6.42	5.89	5.27	4.69	3.99	3.83	16.9		
85.3	1.87	4.73	6.33	6.58	6.66	6.71	6.36	6.11	5.88	5.46	5.07	4.90	4.64	4.38	25.6		
86.1	1.46	2.81	3.74	4.43	4.52	4.77	4.79	4.93	4.67	4.54	4.67	4.63	4.38	4.17	43.5	Traces	
87.1	0.83	1.81	2.17	2.53	3.09	3.39	3.46	3.33	3.55	3.47	3.45	3.52	3.26	3.24	61.1	Traces	
87.9	0.50	0.82	1.56	1.76	1.72	2.03	2.13	2.26	2.07	2.26	2.06	2.20	1.99	2.30	76.4	0.42	0.11
89.4	1.88	1.52	0.7	0.61	0.62	0.68	0.54	0.71	0.36	1.03	0.98	1.16	1.23	1.37	86.8	1.17	0.41

Source: Huhti, A.-L. and Gartaganis, P.A., *Can. J. Chem.*, 34, 785, 1956. With permission.

Phosphate Ester Surfactants

10.2.3 PHOSPHORUS OXYCHLORIDE

Phosphorus oxychloride is tetrahedral in structure [21]. It is prepared by the reaction of phosphorus pentachloride with phosphoric anhydride [22] or by oxidation of phosphorus trichloride with air or oxygen [16, p. 535].

Phosphorus oxychloride is a highly reactive, low-viscosity liquid reagent, which evolves hydrogen chloride on reaction, requiring environmental pollution control. The reagent frequently leads to generation of alkyl chloride as a by-product. It is very useful for laboratory preparations of mono- and diesters. However, it is the reagent of choice for manufacture of trialkylesters. The major suppliers of phosphorus oxychloride are Rhodia Inc., Lanxess, Israel Chemicals Ltd (ICL), and Thermphos.

10.3 MONOALKYL–DIALKYL PHOSPHATE ESTER MIXTURES

Traditionally, phosphate esters are prepared by the reaction of phosphoric anhydride with an alcohol or alcohol ethoxylate [23–26]. The reaction, when stoichiometric proportions of 6 moles of alcohol per mole of P_4O_{10} (four equivalents of phosphorus) are used, theoretically produces a 50:50 molar ratio of MAP and DAP. These mixtures are frequently referred to as sesquiphosphates.

$$P_4O_{10} + 6\ ROH \rightarrow 2(RO)_2P(O)OH + 2ROP(O)(OH)_2$$

The overall reaction is very complex because of the numerous intermediates generated, but can be very easily rationalized based on the tetrahedral structure of phosphoric anhydride as shown in the general, simplified scheme in Figure 10.1, where R would be the alkyl and R' a hydrogen. The initial alcohol molecule reacts with one of the phosphorus atoms of the tetrahedron, resulting in cleavage of an adjacent phosphorus–oxygen bond. The tetrahedron is opened to a bicyclic intermediate. Because of ring strain, these species are never detected under normal conditions. Successive reactions with alcohol molecules lead to monocyclic and then to acyclic intermediates. Each intermediate has a characteristic solubility and reactivity, and species in addition to those shown are likely involved. The conversions continue at progressively lower rates until all the anhydride (P–O–P) bonds are consumed. The branched sites, on which there are three P–O–P anhydride bonds, are more reactive than the internal linear or cyclic centers, with two anhydride bonds, which in turn are more reactive than the simple pyrophosphate. The net result is a 50:50 molar ratio of MAP to DAP.

The preferred ratio of alcohol (alcohol ethoxylate) to phosphoric anhydride (P_4O_{10}) is 5.0:6.2 [25], based on composition and cycle time considerations. In practice, the weight ratio of MAP to DAP ester can be manipulated between 65:35 and 35:65 by adjusting the alcohol or water proportions. This effect can be envisioned by substituting water for a portion of the ROH in the sequence. The product distribution also depends on when the water is added, with or after the alcohol [4,27]. The MAP content is increased by even traces of water in either the phosphoric anhydride or alcohol (alcohol ethoxylate).

This process will not produce a high (>90%) monoester. However, the monoester content can be raised to 65%. Use of water leads to phosphoric acid formation as a side reaction. Thus a more accurate representation is the following equation:

$$(6\text{-}n)ROH + nH_2O + P_4O_{10} \rightarrow xH_3PO_4 + y(RO)P(O)(OH)_2 + z(RO)_2P(O)OH$$

where $x + y + z = 4$.

Reaction of the MAP product with residual alcohol under forcing conditions to remove the water can lead to a diester.

$$ROP(O)(OH)_2 + ROH \rightarrow (RO)_2P(O)OH + H_2O$$

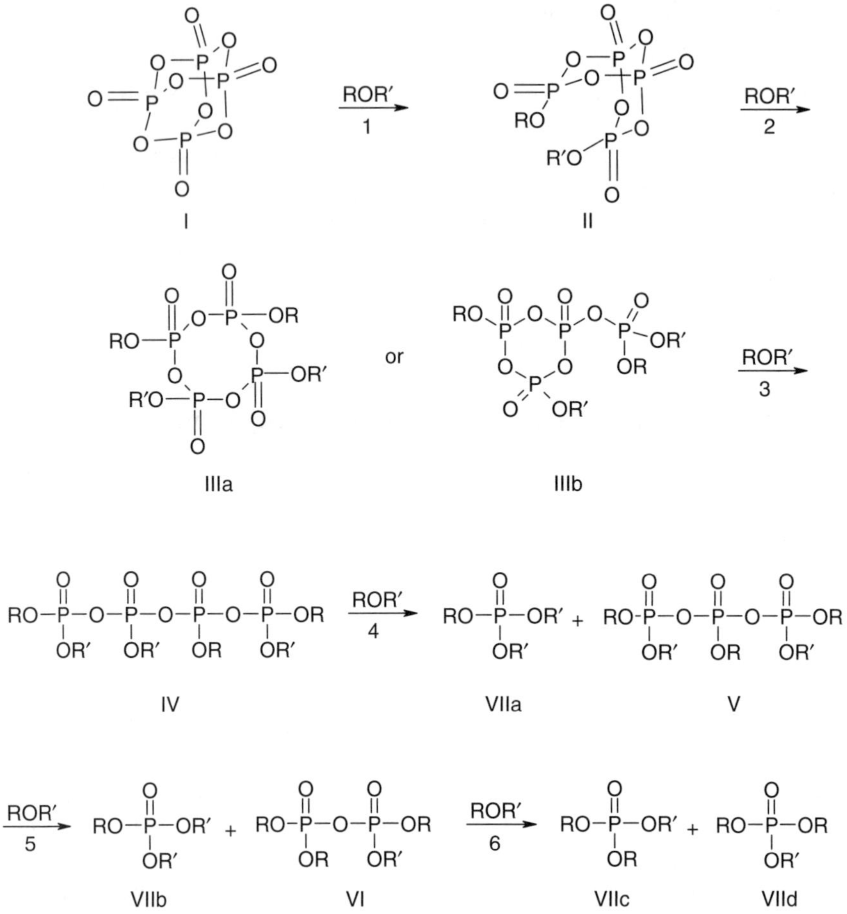

FIGURE 10.1 Simplified phosphoric anhydride reaction sequence.

This reaction, however, is not very efficient [28]. Consequently, commercial production of high dialkyl esters by this route is not feasible.

The principal phosphate ester components, unreacted alcohol, free phosphoric acid, MAP ester, and DAP, as well as the ester ratios are frequently reported in the literature in terms of weight. Because of the molecular weight differences between monoalkyl and dialkyl esters, the weight ratios are thus skewed toward the diesters. Another difficulty in this approach, evident in blends of alcohols or their ethoxylates, is that although each group forms its characteristic monoalkyl ester, each DAP is formed from a more random combination of any two groups, depending on its concentration and reactivity. The concept thus loses its precision, and the total weight of the DAPs can then only be approximated by taking the average molecular weights of the species present.

In contrast, the total molar phosphate content can be determined by titration and the molar ratios between the individual ortho- and pyrophosphate species can be measured by ^{31}P nuclear magnetic resonance (NMR) spectroscopy to allow the molar compositions to be calculated accurately and free from interferences. These molar ratios are a function of the reaction parameters and do not vary significantly as the alcohol or ethoxylate composition might change over a reasonable range in a homologous series, whereas the weight ratios would, even for a comparable molar composition. The difficulty in comparing compositions based on different starting materials could, therefore, be reduced by using molar ratios. However, since the weight ratios are prevalent in the literature references, they will be used as necessary in this review also.

Phosphate Ester Surfactants 189

In a typical process [24], the alcohol (4–9 mol of alcohol ethoxylate) is placed in a reactor and 1 mol P_4O_{10} is added with vigorous agitation with cooling at a controlled rate to keep the temperature below 50°C. The reaction is highly exothermic. The P_4O_{10} reacts as it dissolves with ~90% conversion during the addition period. The reaction mixture is heated to 100°C and held ~5 h to complete the reaction.

To minimize color, an inhibitor such as hypophosphorus acid, sodium hypophosphite [25,29], bis(hydroxymethyl) phosphinic acid [30], or sodium borohydride [31] can be employed. The products after phosphation can be bleached by hydrogen peroxide [29,32,33].

Use of hypophosphorus acid requires a scrubber or burner system to handle the toxic, pyrophoric phosphine, which is generated by its thermal decomposition [34].

Phosphation of alcohol ethoxylates leads to dioxane formation. Dioxane forms by the strong acid-catalyzed intramolecular cyclization of two polyether units with concomitant chain scission. Dioxane levels increase with the reaction temperature and poor dispersion or addition of the strong Lewis acid, phosphoric anhydride, rapidly, results in its persistence longer and at a higher level in the mixture. The final dioxane level also is commonly higher for products containing longer polyether chains. The dioxane, if present in the product at an undesirable level, can be reduced by use of a nitrogen sparge, vacuum distillation, or by use of a thin-film evaporator [35]. Neutralization of residual phosphoric acid before sparging or vacuum distillation is reported to facilitate reduction of dioxane levels to below 5 ppm [36]. Environmental control requires the dioxane to be scrubbed or passed through a burner.

The manufacture of phosphate esters through phosphoric anhydride has been adapted to a continuous process using a heated sump reactor [37].

Mixtures of MAP and DAP esters prepared by the reaction of phosphorus oxychloride and lauryl/myristic alcohol [38] have also been described. However, the methodology is more expensive and not practiced commercially.

10.4 MONOALKYL PHOSPHATE ESTERS

MAP ester compositions have received significant attention because of their combination of excellent detergency and abundant, lubricious foam production with low irritation. These properties are particularly desirable in cosmetic and personal care applications. The dialkyl ester contributes to a reduction in water solubility, foam generation, and detergency properties. Hence, considerable effort has been devoted to the development of economical and simpler methods to produce high monoalkyl ester compositions.

10.4.1 POLYPHOSPHORIC ACID PROCESSES

High monoalkyl esters are traditionally made through polyphosphoric acid reaction with an alcohol or alcohol ethoxylate [39–42]. Commercial polyphosphoric acid ranges between 105 and 117%. The monoalkyl to dialkyl ratio increases as the reaction temperature and the strength of the polyphosphoric acid decrease, with the limit at pyrophosphoric acid [28]. Direct esterification with orthophosphoric acid is not viable because of poor selectivity [43]. Polyphosphoric acid (115%) is most commonly employed; its chain length distribution would be approximated by the 84% P_2O_5 entry in Table 10.1. The alcohol reacts at an anhydride site on the chain, cleaving the polyphosphoric acid chain at that point. Large amounts of orthophosphoric acid are inherently produced because each chain produces, on average, one molecule of phosphoric acid. The amount of phosphoric acid is equal to $1/n$ (see phosphoric acid structure) where n equals the average polymer chain length [44]. Reaction of simple alcohols with an equimolar amount of 117% polyphosphoric acid is reported to produce 21.0–23.8 mol% phosphoric acid in good agreement with the 23.2 mol% predicted (see Table 10.1) for 117% polyphosphoric acid [44]. In commercial practice, to produce esters with low unreacted alcohol (alcohol ethoxylate), an excess of polyphosphoric acid is required.

For some applications, relatively high residual levels of alcohol ethoxylate surfactant are desirable; hence, excess ethoxylate is employed. Numerous products have evolved containing varying levels of residual phosphoric acid and alcohol (alcohol ethoxylate) to meet niche application requirements. Typically [40,42], 0.3–1.5 mol of alcohol (alcohol ethoxylate) reacted with 1 mol of P_2O_5 (0.15–0.75 mole ROH per equivalent P) in 115% polyphosphoric acid at 50°C. The reaction is heated to 105–110°C and held for 1–2 h. The product is 70–90 wt% monoester, 9–28% diester, and 0.1–12% nonionic alcohol ethoxylate [42].

Personal care applications require minimal phosphoric acid and residual alcohol levels, in addition to high MAP content. These requirements are necessary to minimize skin-irritation properties and to facilitate ease of formulation [9]. The presence of phosphoric acid, for example, produces salt on neutralization, which may contribute to solubility or stability (phase separation) problems on formulation. Consequently, considerable effort has been devoted to purification of polyphosphoric acid–derived materials.

Two methods are described in Henkel [45,46] patents. In one case, unreacted alcohol is stripped from the reaction mixture after partial neutralization of the phosphate esters. In the second case, unreacted alcohol is removed by steam distillation, and the phosphoric acid is removed by partial neutralization followed by phase separation. Unreacted alcohol causes undesirable odor when formulated, particularly in personal care applications. Alcohol has also been removed by a thin-film steam distillation. Residual alcohol level was shown to be reduced from 6 mol% to <0.5 mol% [47].

Kao [48] has reported an elaborate, extractive process to produce 96 wt% MAP containing only 2.3% phosphoric acid and <1% each of DAP and unreacted alcohol. The process uses a fivefold molar excess of polyphosphoric acid (105%) to alcohol, thereby minimizing unreacted alcohol in the product. Because of the high viscosity of the intermediate product, the reaction is carried out in hydrocarbon (hexane) solvent. On completion of the reaction, additional hexane is added and the mixture is extracted with isopropanol. The hexane and aqueous layers are then separated. The hexane layer containing the MAP ester is extracted for the second time with aqueous isopropanol, then any residual water and isopropanol are removed by azeotropic distillation with continuous hexane addition. The hexane fraction is cooled, allowing the monoalkyl ester to crystallize. It is filtered and the residual hexane is distilled for recycle. The water–isopropanol extracts containing phosphoric acid are then stripped to recover the phosphoric acid for recycle. This batch process can be adapted to a continuous operation. Several refinements of this process have been published [49–51].

10.4.2 HYBRID PROCESSES

Industrially simpler processes are required for economical manufacture of high MAP compositions in the desired performance ranges. The need for elimination of expensive purification and the attendant reagent/solvent recycle steps prompted investigation of direct processes with modified or hybrid phosphation reagents. Although high-purity DAPs may be necessary for premium performance applications, comparison of foaming and solubility properties to composition leads one to the following conclusions. Important surfactant properties such as foaming decrease markedly below a 70:30 weight ratio of MAP to DAP, but exhibit only a modest improvement above 90:10. A very good performance is observed in the range 80:20–90:10 [9]; hence, this would seem to be a more realistic price-performance target range.

Initially, multiple-step processes evolved, which could be conducted sequentially in a single reactor. These processes alternate the addition of phosphoric anhydride, water, alcohol, and hydrogen peroxide in a complex sequence to produce high MAP esters. To fill the need for a solvent to maintain reasonable viscosity throughout the process but avoid the necessity of its removal, the phosphate ester mixture itself could be used as a solvent [32,33].

Subsequently, as a result of extensive studies on the effect of water on the composition of the final product, a two-step process was developed, in which optimum results were obtained if the reactant ratios fell within defined limits for each step [52].

Phosphate Ester Surfactants

In the first step, the alcohol, phosphation reagent, and water (or its equivalent in phosphoric or polyphosphoric acid, calculated as a composite of phosphoric anhydride and water) are combined in accord with the following equation:

$$[B + C]/A = 1.2 - 2.8$$

where

A = moles of phosphation agent, expressed as P_2O_5
B = moles of water, including that found in the phosphation agent
C = moles of alcohol

Since the "P_2O_5" phosphation reagent requires 3 mol of alcohol plus water as reactants to react completely (6 mol per mol of P_4O_{10}), it is seen to be purposely in excess with the intent of preferentially forming the monoalkyl pyrophosphate intermediate in the first step. After conversion of the more reactive phosphation reagent and intermediates to pyrophosphates, the remaining amount of alcohol, D, is added in accord with the following equation in the second step:

$$[B + C + D]/A = 2.9 - 3.1$$

This completes the conversion of the monoalkyl pyrophosphate intermediate to monoalkyl orthophosphates as illustrated in the following sequence and description:

ROH (1.0 mol) + H$_3$PO$_4$(115%) (0.63 mol P$_2$O$_5$) (1.0 mol water) $\xrightarrow[1\,h]{50°C}$ Mixture

Mixture + P$_2$O$_5$ (0.37 mol) $\xrightarrow[8\,h]{80°C}$
$$RO-\overset{\overset{\displaystyle O}{\|}}{\underset{\underset{\displaystyle OH}{|}}{P}}-O-\overset{\overset{\displaystyle O}{\|}}{\underset{\underset{\displaystyle OH}{|}}{P}}-OH$$

$$RO\overset{\overset{\displaystyle O\ \ O}{\| \ \ \|}}{\underset{\underset{\displaystyle OH\ OH}{| \ \ |}}{POP}}OH + ROH\ (1.0\ mol) \xrightarrow[12\,h]{80°C} RO-\overset{\overset{\displaystyle O}{\|}}{\underset{\underset{\displaystyle OH}{|}}{P}}-OH\ (2.0\ mol)$$

To 189.0 g of lauryl alcohol (1.0 mole) is added 107.7 g of 115% orthoequivalent (83.3% P_2O_5) polyphosphoric acid (1.0 mole of water, 0.632 mole of P_2O_5) and reacted at 50°C for 1 hour. To this reactive solution, 52.2 g of phosphorus pentoxide (0.368 mole) is added gradually, and the reaction carried out at 80°C for 8 hours. To this solution, 189.0 g of lauryl alcohol (1.0 mole) is added and the reaction is further heated to 80°C and held for 12 hours. The reaction was analyzed by titration and found to comprise 82.1 molar % monolauryl phosphate, 8.1 molar % dilauryl phosphate and 9.9 molar % orthophosphoric acid. [52]

The weight ratio of MAP to DAP is above 80:20; thus, good performance properties are observed without the need of further purification. The reaction mixture is very viscous, especially in the early stages, requiring special mixing apparatus to maintain adequate fluidity.

Recently, even simpler processes have been reported. Phosphation reagent strengths, substantially greater than the 117% commercially available polyphosphoric acid, have been developed. A composite is prepared by blending phosphoric anhydride with phosphoric acid under moderate temperatures, compared to those required to produce a clear, high-viscosity product with the traditional distribution of oligomeric chains. Phosphoric acid free of water is preferred (105–116%) to avoid the necessity of compensating for water in the composition. Lower strengths of 75 or 85% are also useful, but the use of only water with the phosphoric anhydride, although still feasible, is not recommended. The intention is to convert only the most reactive, strained polycyclic species derived from P_4O_{10} into monocyclic species or relatively short-chain intermediates [53] without proceeding to the formation of the long-chain components characteristic of the traditional polyphosphoric acid of comparable strength. This hybrid reagent is not simply a physical suspension of phosphoric anhydride in polyphosphoric acid, because it dissolves more readily and can be combined more rapidly with alcohols without the highly exothermic reaction or discoloration characteristic of phosphoric anhydride [54]. It can be pumped at slightly elevated temperatures and is stable for reasonable periods as long as anhydrous conditions and moderate temperatures are maintained.

Phosphate esters are analyzed most effectively by phosphorus ^{31}P NMR spectroscopy. The monoester, diester, phosphoric acid, pyrophosphates, and evidence of higher phosphates can easily be identified and quantified [27]. Analysis by ^{31}P NMR spectroscopy [53] of a phosphation reagent composition prepared from a 2:1 molar ratio of H_3PO_4/P_4O_{10} (122.5% polyphosphoric acid equivalent or 88.7% P_4O_{10}) shows that 87 mol% of the phosphorus atoms exist as internal or chain groups with only a trace of residual phosphoric acid. Furthermore, ~11 mol% exist as chain end groups and 2 mol% as branched groups. Signals characteristic of higher polycyclic structures such as phosphoric anhydride itself, at -60 ppm, are absent [53].

Although this composite reagent consists of a range of intermediates, the average composition can be rationalized as being, at least initially, phosphate-substituted tri- and tetrametaphosphoric acids, such as IIIa and IIIb of Figure 10.1, in which R$'$ would still be hydrogen, but R would be a $-P(O)(OH)_2$ group (i.e., ROR$'$ is H_3PO_4) also shown in the following structure:

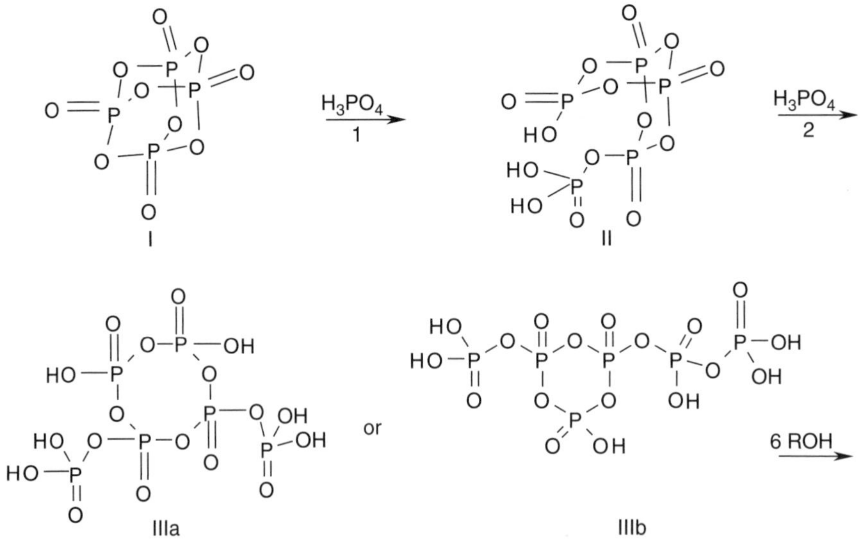

The hybrid reagent can be added to the reactant alcohol, the alcohol can be added to the reagent, or the reagent can be selectively generated (*in situ*) in the reactant alcohol, provided temperatures are maintained at which polyphosphoric acid and phosphoric anhydride do not react significantly

Phosphate Ester Surfactants

with the alcohol, but predominantly with each other [55,56]. After the combination of the phosphation reagent components is complete, the temperature is raised to the reaction temperature and the conversion to products is accomplished. Phosphation hybrid reagent strengths of about 119–123% produce products with a mono- to dialkyl weight ratio above 80:20, with residual phosphoric acid and alcohol levels below 6 wt%.

The combination and reaction of the phosphoric acid with phosphoric anhydride actually moderates the violent reactivity of the anhydride, converting it to less aggressive phosphation intermediates, described earlier. Additional advantages include the elimination of the large excesses required for 115% polyphosphoric acid processes, the viscosity limitations characteristic of earlier processes are resolved without the use of a solvent, and less product degradation is observed. The reactivity moderation is especially manifested by the lower levels of 1,4-dioxane produced, for instance, in the phosphation of a nonylphenol—9 mol ethoxylate with the hybrid reagent compared to phosphoric anhydride itself [57]. This patent also illustrates that hybrid phosphation reagent strengths in excess of 131% (in practice, above 134%) can be prepared and used *in situ*, depending on the ethoxylate selected and the desired final ester composition.

The strength of the hybrid phosphation reagent is a primary variable in establishing the product distribution in the final phosphate mixture. At the low end, it approaches a typical polyphosphoric acid–based composition; at the high end (P_4O_{10} is equivalent to 138.08% "polyphosphoric acid"), it approaches a sesquiphosphate composition characteristic of a phosphoric anhydride–based process. Thus, another feature of this hybrid phosphation reagent is that the phosphate ester compositions between the traditional extremes are now easily available through this simple process to allow optimization of the monoalkyl to dialkyl ratio composition to the performance needs of the application. This adds an additional dimension to new product development, with MAP:DAP molar ratios of ~90:10 to 52:48 available for hydrophobes from C_6 to C_{30}, degrees of ethoxylation (or propoxylation) from 0 to 50 (including behenyl alcohol and tristyrylphenol ethoxylates) combined with a variety of traditional or special [47] bases for neutralization.

A description of this simple *in situ* process follows [58]. To 7924 g dodecanol (42.5 mol), sequentially add 2265 g 115% polyphosphoric acid (26.5 mol) and 1129 g phosphoric anhydride (3.98 mol, 15.9 mol phosphorus) with rapid agitation and good cooling, allowing maximum temperatures of 42 and 59°C, respectively, during the addition periods. The phosphation reagent strength is 122.7%, expressed as polyphosphoric acid (88.9% P_4O_{10}), and the alcohol to phosphorus ratio is 1.0. The mixture is heated to 85°C and maintained for 11 h. On cooling to 65°C and bleaching by hydrogen peroxide, a clear colorless product is obtained. Analysis of the product shows it to be 4.7 wt% phosphoric acid, 80.2 wt% MAP, and 14.7 wt% DAP, for a mono/di weight ratio of 84.5:14.5, with 0.4% residual nonionic (dodecanol).

This process has been further refined and commercialized to consistently produce phosphate esters with molar monoalkyl:dialkyl ester ratios of about 90:10, and both free acid and alcohol contents are <6 wt% such that further purification steps to remove them are not necessary.

Additional versions of the earlier process [59–61] as well as processes based on phosphation of alcohol/water mixtures with phosphoric anhydride [62,63] have been published. The first group includes added intermediate or postreaction purification steps and steam stripping or other methods to remove residual alcohol. The products resulting from the water/phosphoric anhydride routes are generally not characterized by the combination of a high mono- to dialkyl ratio and low-residual phosphoric acid and alcohol levels associated with the phosphoric acid/phosphoric anhydride hybrid phosphation reagent processes.

10.4.3 PHOSPHORUS OXYCHLORIDE PROCESSES

Phosphorus oxychloride is frequently used to prepare laboratory samples of MAPs, but rarely on a commercial scale because of the required large excess of expensive reagent, high levels of corrosiveness, and the need to scavenge the hydrogen chloride. The synthesis is described by Imokawa [6] and summarized by the following equation:

$$ROH + POCl_3 \rightarrow RO\text{--}P(O)Cl_2 + HCl$$

$$\xrightarrow{2H_2O} ROP(O)(OH)_2 + 2HCl$$

The reaction usually involves use of inert solvents and requires purification, for example, multiple extractions [64,65]. Yields are also generally low. Furthermore, the reaction frequently generates alkyl chlorides as by-products.

A recently developed process by Kao [66] teaches the use of a metal salt of an acid having a pKa value of −5 to 13. The metal salt traps the hydrogen chloride produced. Yields >97% with over 99% pure MAP are claimed.

10.5   DIALKYL PHOSPHATE ESTERS

Certain applications require products possessing a higher DAP content than attainable from the sesquiphosphate mixtures derived from phosphoric anhydride. Multiple reaction sites and a lack of selectivity characteristic of the phosphation reagents described earlier make them unsuitable for the preparation of high dialkyl content esters.

If high purity is essential, DAPs can be prepared by the oxidation of the corresponding dialkyl phosphite [67], which is in turn prepared from phosphorus trichloride. To avoid phosphorus halides, a phosphation reagent can be prepared from the reaction of lower trialkyl phosphates with phosphoric anhydride, yielding a range of intermediates, depending on the conditions and molar ratio [68].

$$2(RO)_3PO + P_4O_{10} \rightarrow 1(RO)_6P_6O_{12}$$

$$4(RO)_3PO + P_4O_{10} \rightarrow 2(RO)_6P_4O_7$$

$$8(RO)_3PO + P_4O_{10} \rightarrow 6(RO)_4P_2O_3$$

This sequence, proceeding through metapoly-, tripoly-, and pyrophosphate intermediates, seems to follow pathways similar to that shown in Figure 10.1 for the reaction of phosphoric acid with phosphoric anhydride except that the alkyl groups are transferred instead of the hydrogen atoms. In Figure 10.1, for the ROR′, R would be, for instance, $-CH_2CH_3$ and R′ would be $-P(O)(OCH_2CH_3)_2$. In the following structure, $R = -CH_2CH_3$.

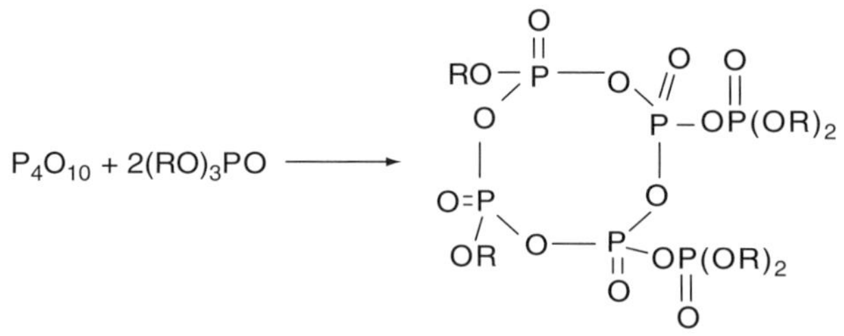

$$P_4O_{10} + 2(RO)_3PO \longrightarrow$$

Reaction of the preceding intermediate mixture with alcohols leads to mixed ester products characterized by high DAP, some trialkyl phosphate, and low MAP content with only traces of phosphoric acid.

Phosphate Ester Surfactants

Additional support for this mechanism is found by analogy to the mechanism reported for reaction of phosphoric anhydride with diethyl ether [69] to form Langheld ester [70].

$$P_4O_{10} + R_2O \longrightarrow$$

$$\xrightarrow{R_2O}$$

$$\xrightarrow{R_2O} \quad (RO)_2POPOPOP(OR)_2$$

where $R = -CH_2CH_3$.

Mixed DAP esters based on this chemistry are important as gellants for enhanced oil recovery [71–73]. Phosphoric anhydride is reacted with triethyl phosphate in a hydrocarbon solvent and the slurry heated to 100°C for 1 h. A blend of octanol and decanol is then added in a second step to the intermediate mixture and the reaction is continued for 2 h at 130°C [73]. The resulting product is blended with diesel oil and cross-linked to a viscous pumpable gel with aluminum isopropoxide or trivalent metal. The product, by reaction of phosphoric anhydride directly with a blend of ethanol, octanol, and decanol, does not perform because of a much higher MAP content [71]. A dialkyl content (mixed ethyl alkyl monobasic ester) of 75–85 mol% is preferred and advantages are reported for use of a hexanol–decanol blend in the finishing step [71].

The availability of relatively pure MAPs allows conversion to a mixture of unsymmetrical DAPs. For example, reaction of monolauryl phosphate with a fivefold excess of dimethyl sulfate in an aqueous potassium hydroxide is reported to yield lauryl methyl phosphate in 84% yield [74]. Epoxides are also popular alkylating agents. Reaction of a salt of monolauryl phosphate with ethylene oxide [75,76], epichlorohydrin [77], glycidol, or glycidol methacrylate [78] is reported to yield unsymmetrical alkyl phosphoglycol derivatives. The 3-chloro-2-hydroxypropyl ester adduct from epichlorohydrin can be further derivatized. Use of the monosodium or monopotassium salt is important in controlling selectivity to the unsymmetrical DAP.

196 Handbook of Detergents/Part F: Production

DAPs can also be prepared through phosphorus oxychloride [79], including lauryl phospho-glycol, which acts as a phospholipid mimic [80].

10.6 PHOSPHOBETAINES

With the increasing demand for skin care and repair products, the personal care market is evolving toward even milder, essentially nonirritating cleansing formulations. To design such skin-compatible compounds, it is only natural to attempt to imitate the functional characteristics of the biochemical surfactants, phospholipids. The phosphocholine moiety is the most common hydrophilic group, usually coupled to a diglyceride hydrophobe unit as in lecithin, a phospholipid commercially extracted from egg and soybean products.

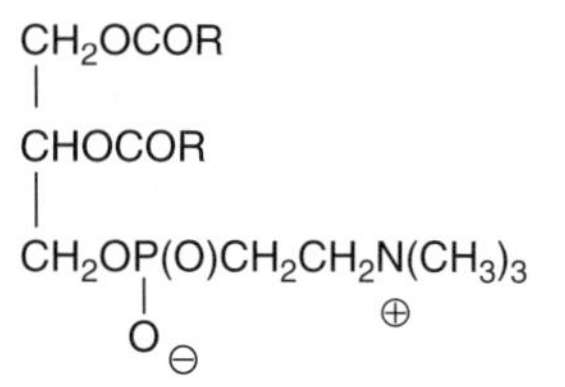

$$
\begin{array}{l}
CH_2OCOR \\
| \\
CHOCOR \\
| \\
CH_2OP(O)CH_2CH_2\overset{\oplus}{N}(CH_3)_3 \\
\quad | \\
\quad O_{\ominus}
\end{array}
$$

Phosphobetaines are synthetic analogues of phospholipids. They, in general, possess outstanding foaming and cleansing properties and are compatible with human tissue, exhibiting low skin and eye irritation and low oral toxicity [81].

As in earlier classes of specialty phosphates, pure research grade materials can be prepared by a phosphorus oxychloride route [80]. The more commercially feasible syntheses employ epoxides, particularly epichlorohydrin, as a coupling agent between the quaternary ammonium group and the alkyl phosphate. The order of reaction is unimportant, either the amine or the MAP can be reacted first with the epichlorohydrin followed by the other [77,78,82].

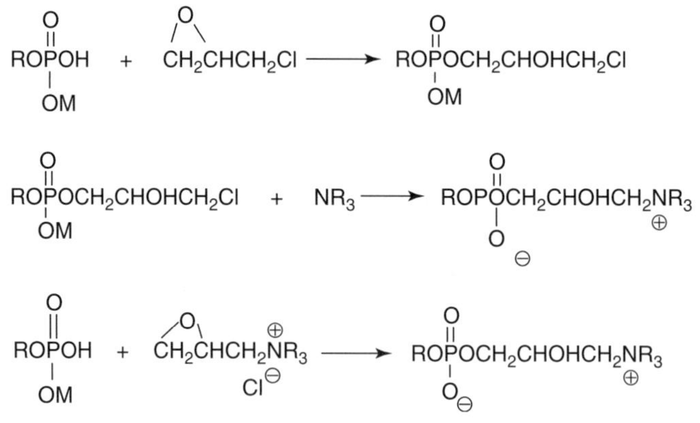

$$
\underset{OM}{\overset{O}{\underset{\|}{ROPOH}}} + CH_2CHCH_2Cl \longrightarrow \underset{OM}{\overset{O}{\underset{\|}{ROPOCH_2CHOHCH_2Cl}}}
$$

$$
\underset{OM}{\overset{O}{\underset{\|}{ROPOCH_2CHOHCH_2Cl}}} + NR_3 \longrightarrow \underset{O_{\ominus}}{\overset{O}{\underset{\|}{ROPOCH_2CHOHCH_2\overset{\oplus}{N}R_3}}}
$$

$$
\underset{OM}{\overset{O}{\underset{\|}{ROPOH}}} + \underset{Cl^{\ominus}}{CH_2CHCH_2\overset{\oplus}{N}R_3} \longrightarrow \underset{O_{\ominus}}{\overset{O}{\underset{\|}{ROPOCH_2CHOHCH_2\overset{\oplus}{N}R_3}}}
$$

Alternatively, the use of starting materials with the hydrophobic chain on the amine allows it to become the coupling group [83–86]. Typically, sodium dihydrogen phosphate is condensed in an aqueous environment with epichlorohydrin. The reaction is carried out at 30–50% solids at a pH of 4–5 at 80–85°C. The sodium 3-chloro-2-hydroxypropyl phosphate intermediate (see sequence in the following structure) is further reacted with a tertiary amine or amidoamine.

Phosphate Ester Surfactants

The quaternization reaction is carried out in an aqueous solution over a 3–4 h period at 90–95°C yielding the phosphobetaine in yields >97%. A second approach involves reaction of a tertiary aminoalcohol with phosphoric anhydride [87–89].

These latter species, based on fatty amines or fatty amidoamines, also have a reversed structure in which the nitrogen has become the coupling group to the hydrophobic chain and the phosphorus is at the end of the hydrophile. Currently, no significant advantage has been demonstrated favoring either molecular array.

10.7 CONCLUSION

Phosphate ester research in recent years has focused on high MAPs and DAPs. Recent advances in process technology are described.

For 40–50 years, phosphate esters occupied the role of relatively small-volume anionic surfactants. They have been unable to effectively compete economically and possessed lesser surface activity than sulfates. Consequently, they have primarily been used in industrial, household, industrial, and institutional (HI&I), and agricultural applications. The recent advance in processes for high MAPs and the development of phospholipid-type molecules have provided products with increased surface activity, mildness, and superior skin compatibility. Consequently, interest is beginning to develop in the personal care industry.

With the current capability to control the reaction parameters to produce products over a wide range of compositions (mono/diesters), the broad range of properties available through phosphate esters is accessible for optimization of the product performance to needs of each specific application to distinguish and differentiate the ultimate consumer products. As the average age of the population in the developed countries increases, the desire and demand for personal care and cosmetic products that combine good detergency and skin feel with exceptional mildness and moisturization that are effective, nonirritating, and stable at skin pH of about 5.5 also increase. Phosphate esters, particularly monoalkyl enriched esters, meet these demands and are, therefore, poised for substantial growth, especially in the case of sensitive skin, as in infant, child, and oral care applications [90,91], and in adults who desire to maintain the health and condition of their skin against natural deterioration with age.

REFERENCES

1. Burnette, T. W., *Nonionic Surfactants*, Vol. 1 (ed. M. J. Schick), Marcel Dekker, New York, 1996, p. 384.
2. Mayhew, R. L. and F. Krupin, *Soap Chem. Spec.*, *38*(4), 55, 1962.
3. Mayhew, R. L. and F. Krupin, *Soap Chem. Spec.*, *38*(5), 80, 1962.
4. Hochwalt, C. A., J. H. Lum, J. E. Malowan, and C. P. Dyer, *Ind. Eng. Chem.*, *34*, 20, 1942.
5. Imokawa, G. and H. Tsutsumi, *J. Am. Chem. Soc.*, *55*, 839–843, 1978.
6. Imokawa, G., *J. Am. Oil Chem. Soc.*, *56*, 604, 1979.
7. Imokawa, G., *J. Soc. Cosmet. Chem.*, *31*, 45, 1980.
8. Burns, T. and T. Schamper, *Soap Cosmet. Chem. Spec.*, 48, September 1992.
9. Imokawa, G., H. Tsutsumi, T. Kurosaki, M. Hayaski, and J. Kakuse, U.S. Patent 4,139,485 to Kao Soap Co., 1979.
10. Jungermann, E. and H. C. Silberman, *Anionic Surfactants, Part II*, Vol. 7 (ed. W. M. Linfield), Marcel Dekker, New York, 1976, p. 497.
11. Wasow, G. W., *Anionic Surfactants*, Vol. 56 (ed. H. W. Stache), Marcel Dekker, New York, 1995, p. 551.
12. Papp, F. D. and W. E. McEwen, *Chem. Rev.*, *58*, 321, 1958.
13. Tracy, D. J. and R. L. Reierson, *J. Surfactants Deterg.*, *5*, 169, 2002.
14. Boenig, I. A., M. M. Crutchfield, and C. W. Heitsch, *Encyclopedia of Chemical Technology*, 3rd Ed., Vol. 17, Wiley, New York, 1982, p. 518.
15. Hampson, G. C. and A. J. Stosick, *J. Am. Chem. Soc.*, *60*, 1814, 1938.
16. Technical Data Rhodia Inc., Phosphorus Products, North America CN 7500, Cranbury, NJ, April 2, 1996.
17. Bettermann, G., *Ullmann's Encyclopedia of Industrial Chemistry*, Vol. A19, 5th Ed. (ed. B. Elvers, S. Hawkins, and G. Schulz), VCH, Heidelberg, 1991, p. 478.
18. Hudson, R.B., U.S. Patent 4,309,394 to Monsanto Company, 1982.
19. Thilo, V. E. and R. Sauer, *J. Prakt. Chem.*, *4*, 324, 1957.
20. Huhti, A.-L. and P. A. Gartaganis, *Can. J. Chem.*, *34*, 785, 1956.
21. Pauling, L., *The Nature of the Chemical Bond*, 2nd Ed., Cornell University Press, Ithaca, NY, 1940, pp. 84, 224.
22. Booth, H. S., C. G. Seegmiller, and C. A. Seabright, *Inorganic Synthesis*, Vol. II, McGraw-Hill, New York, 1946, pp. 151–155.
23. Bertsch, H., U.S. Patent 1,900,973 to H. Th. Bohme AG, 1933.
24. Nunn Jr., L. G. and S. H. Hesse, U.S. Patent 3,004,056 to GAF Corp., 1961.
25. Nunn Jr., L. G., U.S. Patent 3,004,057 to GAF Corp., 1961.
26. Lazarus, A. K., U.S. Patent 3,487,130 to FMC Corp., 1969.
27. Kurosaki, T., J. Wakatsuki, T. Imamura, A. Matsunaga, H. Furugaki, and Y. Sassa, *Commun. J. Com. Esp. Deterg.*, *19*, 191, 1988.
28. Nelson, A. and A. Toy, *Inorg. Chem.*, *2*(4), 775, 1963.
29. Britain Patent 740,955 to Albright and Wilson Ltd., {C.A. *50*, 10759f (1955)}, 1955.
30. Schenck, L. M. and L. G. Nunn Jr., U.S. Patent 3,629,377 to GAF Corp., 1971.
31. Japan Patent 56,079,695 to Kao Soap Co., {C. A. *95*, 221706g (1981)}, 1981.
32. Davis, G. J., U.S. Patent 4,115,483 to Stauffer, 1978.
33. Via, F. A. and S. Y. Liu, U.S. Patent 4,126,650 to Stauffer, 1978.
34. Van Wazer, J. R., *Phosphorus and Its Compounds, Vol. 1: Chemistry*, Interscience Publishers, New York, 1964, p. 363.
35. Sasa, Y., T. Fujita, and S. Myamoto, Japanese Kokai 63 246357A2 to Kao Corp., 1988.
36. Katz, M. and C. Talley, U.S. Patent 4,375,437 to GAF Corp., 1983.
37. Nehmsmann, L. J. and L. M. Schenck, U.S. Patent 3,776,985 to GAF Corp., 1973.
38. DeWitt, G., U.S. Patent 2,005,619 to E. I. du Pont de Nemours and Co., 1935.
39. Woodstock, W. H., U.S. Patent 2,586,897 to Victor Chemical Works, 1952.
40. Mansfield, R. C., U.S. Patent 3,235,627 to Rohm and Haas Co., 1966.
41. Dupre, J. and R. C., Mansfield, U.S. Patent 3,312,624 to Rohm and Haas Co., 1967.
42. Eiseman Jr., F. S. and L. M. Schenck, U.S. Patent 3,331,896 to GAF Corp., 1967.
43. Kurosaki, T., H. Furugaki, M. Takeda, A. Mamba, and J. Wakatsuki, *Oil Chem.*, *39* (4), 259, 1990.
44. Clarke, F. and J. Lyons, *J. Am. Chem. Soc.*, *88*, 4401, 1966.

Phosphate Ester Surfactants

45. Uphues, G., J. Ploog, and K. Bischof, U.S. Patent 4,874,883 to Henkel, 1989.
46. Uphues, G. and J. Ploog, U.S. Patent 4,866,193 to Henkel, 1989.
47. Reierson, R., P. Herve, S. Soman, and R. Eng, PCT Patent WO 02/098549A2 to Rhodia, Inc., 2002.
48. Kurosaki, T., J. Wakatsuki, H. Furugaki, and K. Kojima, U.S. Patent 4,670,575 to Kao Corp., 1987.
49. Aimono, K., T. Fujita, T. Funeno, and Y. Sasa, Jpn. Kokai 03 188,089 to Kao Corp., 1991.
50. Tsuyutani, S. K. Aimono, T. Funeno, and T. Fujita, Jpn. Kokai, 05 148,276 to Kao Corp., 1993.
51. Takemura, K. and K. Aimono, Jpn. Kokai, 05 271,255 to Kao Corp., 1993.
52. Kurosaki, T. and A. Manba, U.S. Patent 4,350,645 to Kao Soap Co., 1982.
53. Reierson, R. L., U.S. Patent 5,554,781 to Rhone-Poulenc, 1996.
54. Reierson, R. L., U.S. Patent 6,136,221 to Rhodia Inc., 2000.
55. Reierson, R. L., U.S. Patent 5,550,274 to Rhône-Poulenc, 1996
56. Reierson, R. L., U.S. Patent 5,554,781 to Rhône-Poulenc, 1996.
57. Reierson, R. L., U.S. Patent 5,463,101 to Rhone Poulenc, 1995.
58. Reierson, R., European Patent 0675076A2 to Rhone Poulenc, 1995.
59. Matsunaga, A., A. Fujiu, S. Tsuyutami, T. Nozaki, and M. Ueda, U.S. Patent 6,407,277 to Kao Corp., 2002.
60. Matsunaga, A., A. Fujui, and S. Tsuyani, U.S. Patent 6,034,261 to Kao Corp., 2000.
61. Tsuyutani, S., K. Shibata, and K. Aimono, U.S. Patent 5,883,280 to Kao Corp., 1999.
62. Schroeder, W. and W. Ruback, *Tenside Surf. Deterg.*, *31*, 413, 1994.
63. Sanyo Kasei Kogyo, K. K., Method of Synthesizing Phosphate Esters, Patent Application No. 38-49701, August 12, 1966.
64. Cramer, F. and M. Winter, *Chem. Ber.*, *92*, 2761, 1959.
65. Modro, A. M. and T. A. Modro, *OPPI Briefs*, *24*(1), 57, 1992.
66. Shara, T., S. Yano, and K. Kita, U.S. Patent 5,565,601 to Kao Corp., 1996.
67. Frohlen, H., H. Block, H. Moretto, and P. Schmidt, European Patent 0562365A2, 1993.
68. Woodstock, W. H., U.S. Patent 2,402,703 to Victor Chemical Works, 1946.
69. Burkhardt, G., M. Klein, and M. Calvin, *J. Am. Chem. Soc.*, *87*(3), 59, 1965.
70. Langheld, K., *Bericht*, *43*, 1857, 1910.
71. Huddleston, D. A., U.S. Patent 4,877,894 to Nalco Chemical Corp., 1989.
72. McCabe, M., L. Norman, and J. Stanford, European Patent Appl. 0551021 A1, 1993.
73. Gross, J., U.S. Patent 5,190,675 to Dowell Schlumberger Inc., 1993.
74. Yamaki, K., H. Takada, and Y. Fujikura, Japanese Kokai 04 134088 A2 to Kao Corp., 1992.
75. Ishikawa, Y., K. Yamaki, and A. Kondo, Japanese Kokai 04 222,801 to Kao Corp., 1992.
76. Imai, K., J. Kametami, T. Imamura, K. Yamaki, and T. Kurosaki, Japanese Kokai 03 275,800 A2 to Kao Corp., 1991.
77. Wakatsuki, J., T. Katoh, and T. Kurosaki, U.S. Patent 4,740,609 to Kao Corp., 1988.
78. Wakatsuki, J., T. Kato, T. Kurosaki, and T. Imamura, U.S. Patent 4,736,051 to Kao Corp., 1988.
79. Haupke, K. and F. Wolf, *J. Prakt. Chem.*, *33*(4), 206, 1966.
80. de Jongh, H. and B. de Kruijff, *Biochem. Biophys. Acta*, *1029*(1), 105, 1990.
81. Lindemann, M. K. O., U.S. Patent 4,382,036 to Johnson & Johnson Baby Products Company, 1983.
82. Wakatsuki, J., T. Katoh, and T. Kurosaki, U.S. Patent 4,774,350 to Kao Corp., 1988.
83. Lindemann, M. K. O., R. L. Mayhew, A. O'Lenick Jr., and R. Verdichio, U.S. Patent 4,215,064 to Johnson & Johnson Company and Mona Industries Inc., 1981.
84. O'Lenick Jr., A. and R. L. Mayhew, U.S. Patent 4,283,542 to Mona Industries Inc., 1981.
85. Lukenbach, E. R. and R. R. Tenore, U.S. Patent 4,617,414 to Johnson & Johnson Baby Products Company, 1986.
86. Zimmerer, R. E., U.S. Patent 3,507,937 to Procter & Gamble Company, 1970.
87. Tsubone, K., N. Uchida, H. Niware, and K. Hondo, *J. Am. Oil Chem. Soc.*, *66*(6), 829, 1989.
88. Tsubone, K. and N. Uchida, *J. Am. Oil Chem. Soc.*, *67*(6), 394, 1990.
89. Tsubone, K. and N. Uchida, *J. Am. Oil Chem. Soc.*, *67*(3), 149, 1990.
90. Warburton, S., R. Reierson, T. Domke, and A. Gabbianelli, *Designed Phosphate Esters Compositions for High Performance Applications*, Sixth World Surfactants Congress (CESIO), Berlin, June 20–23, 2004.
91. Reierson, R. and T. Domke, U.S. Patent Application Serial No. 10/783721, 2004.

11 Production of Methyl Ester Sulfonates

Norman C. Foster, Brian W. MacArthur,
W. Brad Sheats, Michael C. Shea, and Sanjay N. Trivedi

CONTENTS

11.1 Background 201
11.2 Acid versus Neutral Bleaching 202
11.3 Methyl Ester Sulfonates Upstream: The Feedstock 203
11.4 Commercial Methyl Ester Sulfonates Production 204
    11.4.1 Plant Capacity 205
    11.4.2 Methyl Ester Sulfonates' Sulfonation Raw Material
          Characteristics 205
11.5 Methyl Ester Sulfonate Chemical Reactions and By-Products 206
11.6 General Methyl Ester Sulfonates' Process Conditions 207
11.7 Feed and Product Specifications 208
    11.7.1 Methyl Esters 208
          11.7.1.1 Methyl Ester Sulfonates 208
11.8 Methyl Ester Sulfonates Process Descriptions 208
    11.8.1 Basic Sulfonation Process 208
    11.8.2 Methyl Ester Sulfonic Acid Digester 209
    11.8.3 Methyl Ester Stabilization and Bleaching 210
    11.8.4 Neutralizer System 211
    11.8.5 Methyl Ester Sulfonates' Turbo Tube Dryer System 212
    11.8.6 Dried Product Cooling System 213
    11.8.7 Dryer Auxiliary Systems 214
    11.8.8 Methanol Recovery System 216
    11.8.9 By-Products and Safety 216
11.9 Conclusions 218
References 218

11.1 BACKGROUND

A great deal of interest is currently focused on methyl ester sulfonates (MES) from palm and coconut derivatives with the increase in crude oil prices and the resultant increase in the prices of petrochemicals. MES offers an environment friendly and viable alternative to the old workhorse surfactant linear alkylbenzene sulfonic acid/sulfonate (LAS/LABS), which is derived from linear alkylbenzene (LAB).

Lion Corporation (Lion),[1] Stepan Company (Stepan), and Chemithon Corporation (Chemithon)[2] have patented technologies for manufacturing MES. MES is produced commercially using technology based on "acid bleaching," which is described in this chapter—in Japan by Lion and in the United States by Stepan and Huish Detergents Inc. (Huish). New plants are coming on line in China, Europe, and Southeast Asia. Huish, which has a plant capacity of 80,000 tons per annum (TPA),[3] uses the Chemithon and Lurgi technologies for MES and methyl esters (ME), respectively. Huish produces commercial quantities of MES in a free-flowing powder form from palm oil, LION uses C14–C16 to produce a powder form of MES and STEPAN produces liquid MES from C12–C14 oils. These MES products are being formulated for both liquid and powder finished detergents.

The appeal of MES is based on its origin from a renewable oleo-based raw material, its excellent biodegradability, improved calcium hardness tolerance, and good detergency. The challenges for MES in detergent use include low-foam characteristics. Progress has already been made with liquid formulations as current enzyme-based laundry formulations use pH in the range of 7.5 to 8.5, so are not a problem for MES. Further, MES enzyme stability is superior to that of LAS. The availability of MES in a dry free-flowing powder or flaked form in recent years has overcome most of the manufacturing issues as the product can be directly added to the detergent formulation in a post-addition step. The issue of low foam can be addressed by the inclusion of a lauric chain length or by the addition of foam boosters such as lauryl sulfate, ethoxylated lauryl sulfate,[4] or α-olefin sulfonate.

The price of LAB is linked to the price of oil, and the long-term trend has moved upward with some significant short-term spikes. MES is an oleo-based feedstock and has until recently followed a different long-term trend. Recent activity related to the development of biodiesel production has caused prices of palm oil to vary, although many expect long-term planned production increases to stabilize the price spread. Oleo-based raw materials are projected to be favorably priced in comparison with petrochemical raw materials and therefore the cost advantage of MES relative to LAS is likely to be significant and increasing in the future.

Owing to the foregoing considerations, there has recently been an increased focus by global detergent producers in formulating MES-based products.[5,6,7] Formulation successes include the production of stable MES-based liquid laundry detergents and MES-based detergent bars[8,9] in North America and Latin America, respectively. European detergents will find a synergy when using MES for their moderate-temperature wash cycles. In the emerging markets of India and China, one of the challenges would be formulating the products with enhanced foaming properties.

The process for manufacturing MES requires bleaching the product to achieve acceptable colors. The Huish plant uses Chemithon acid bleaching technology with hydrogen peroxide as the bleaching agent. The process uses an excess of methanol to minimize disalt formation and to improve the predying characteristics for producing MES in a dry form. Lion, Stepan, and Chemithon have patents for their acid bleaching MES processes. A study of C16–C18 MES[10] by Lion showed a superior performance of the surfactant as compared to LAS or alcohol sulfate (AS) under low-temperature wash conditions and water hardness levels of approximately 100 parts per million (ppm) ($CaCO_3$). The use of C14 improves the cold-water performance. The C16 MES produced by Huish is used in premium detergent powders (Safeway) and in popular detergent packs (Wal-Mart, COSTCO).[11] Lion uses a C1416 MES in a mixed active surfactant system in their compact detergent formulations.

11.2 ACID VERSUS NEUTRAL BLEACHING

Acid bleaching processes have demonstrated superior product quality, especially for palm stearin–based products where acid bleaching yields lower color products (<20 Klett) with disalt levels in the 4% range (100% active basis). Additionally, the acid bleaching process is a rapid reaction, which allows a continuous process with a total residence time of ~2 h. However, neutral bleaching is very slow and requires storage of large volumes of material containing both methanol and peroxide for periods of up to 24 h. For a commercial-scale plant, there is significant risk in storing such a mixture in large tanks with the associated free space containing a potentially flammable vapor. All commercially demonstrated MES processes to date incorporate acid bleaching. Unresolved issues

Production of Methyl Ester Sulfonates

with neutral bleach systems include methanol in the product, relatively high color after bleaching, and high disalt levels.

In addition to its proven acid bleach systems, Lion has a patented[12] neutral bleach process that introduces an inorganic chemical, sodium sulfate, to the sulfonation system, which then reduces color body formation in the digestion step. A slight excess of stoichiometric methanol is added to the sulfonic acid and digested at elevated temperatures to break the adduct down to nondisalt forming products. The product is then neutralized and bleached under pressure at 120°C for several hours to make a neutral paste.

A similar process is offered by Desmet Ballestra that uses the Henkel neutral bleach MES process. High-temperature digestion is followed by the addition of a slight excess of stoichiometric methanol to the sulfonic acid, then neutralization in a more dilute form and bleaching with hydrogen peroxide for up to 24 h. Drying is accomplished with a wiped film evaporator.

11.3 METHYL ESTER SULFONATES UPSTREAM: THE FEEDSTOCK

The viability of MES is dependent on the availability of the ME at a reasonable price. The Asia Pacific region, where Malaysia and Indonesia lead the world's palm oil production, offers a stable source for the palm oils or derivatives for producing ME. The interest in biodiesel from palm oil and the need for biodiesel to have a low cold filter plugging point (CFPP) will provide a ready supply of C16 ME that can be subsequently converted into a surfactant.

Detergent-grade ME can be made in several ways. The simplest option is to transesterify the oil or fat (palm, palm olein, or palm stearin) with methanol and use a fractionated C16 or C16–C18 after removal of the unsaturated components. An innovative process fractionates the C16 stream, minimizing the need to hydrogenate the resultant product before making the MES. The isolated C18 ME can be used as biodiesel to blend with diesel fuel. ME can also be made by transesterifying a palm oil stream after the extraction of minor components (vitamin E, β-carotene, etc.), separating the glycerine and fractionating the stream to isolate the C16 ME. The transesterification step is a low-pressure, low-temperature process that has been used by the oleochemical industry. An alternative source of ME can come from the esterification of fatty acids, palm fatty acid distillate, and acid oils derived from the processing of palm oil. However, the esterification process requires a catalyst as well as higher pressure and temperature, resulting in a more expensive processing step.

Refined, bleached, and deodorized palm stearin (RBDPS) is used as a raw material for ME manufacture. ME technology producers include Lurgi Gmbh, Crown Iron Works Company, Archer Daniels Midland Company, Desmet Ballestra, and a number of other companies.

The C16 feed is preferred for ME sulfonation because of a high degree of saturation that occurs naturally, which contributes to its low requirement for hydrogenation to reduce the iodine value. However, the high-level degree of saturation of C16 ME makes it undesirable for use as biodiesel because the highly saturated C16 ME has a high freezing point. Low freezing point, CFPP, is an essential requirement for biodiesel. As more and more RBDPS, which contains a large fraction of saturated C16, is used to manufacture biodiesel, more C16 ME must be fractioned from the biodiesel to meet CFPP specifications. Thus, the growing demand for biodiesel will make the ≤C16 feed more readily available for use in detergent applications. Furthermore, the solubility and detergency of the C16 MES is excellent for many formulations of liquids and powders.

Generally, the lower the molecular weight of the ME, the easier it is to achieve low color and low disalt content in the sulfonated product. Dark color is undesirable since consumers equate light colors with product purity. Therefore, light color is mandatory if MES is to become a replacement for traditional surfactants in consumer products. Low conversion of ME to MES is an economic issue since it effectively increases the cost of MES. The principal conversion problem has been the breaking of the ester bond in MES to form methanol and α-sulfo sodium carboxylate (disalt). Disalt, although technically a surfactant, has poor surfactant properties as compared with MES.

The "benchmark" feeds indicated in Table 11.2 have been extensively tested in both laboratory[13–16] and commercial[17–19] settings and are known to produce MES products of low color and disalt content suitable for use in liquid and powder products. The C16 is preferred over C16–C18 for a sulfonation feedstock because it makes a better color MES and has proven product quality and excellent detergency characteristics; it will become more widely available as a feedstock, the economics of producing detergent-grade ME favors C16 (requires less hydrogenation).

11.4 COMMERCIAL METHYL ESTER SULFONATES PRODUCTION

The MES production method described in this chapter is the commercially proven acid bleach type in service at Chemithon and Lion facilities, and to a lesser extent at Stepan. Figure 11.1 illustrates the unit operations in the Chemithon MES process.

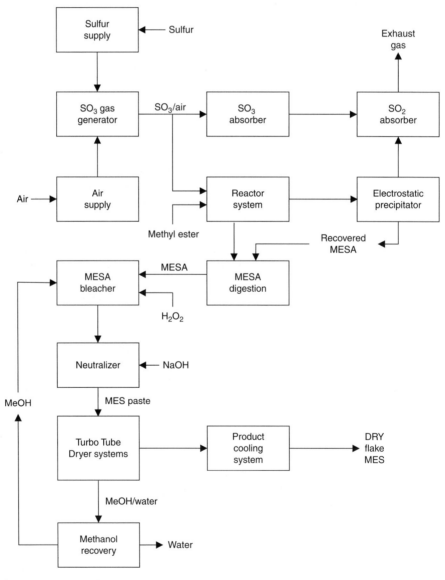

FIGURE 11.1 Overview of ME sulfonation plant.

Production of Methyl Ester Sulfonates

TABLE 11.1
Benchmark Feed Characteristics

ME	Benchmark C16 ME		Benchmark C16–C18 ME	
	Typical	Specification	Typical	Specification
Molecular weight	270		280.7	
Iodine value (cg I/g ME)	0.03	<0.05	0.10	<0.15
Free fatty acid (wt%)	0.03	<0.06	0.05	<0.1
Unsulfonatable (wt%)	0.3	<0.5	0.3	<0.5
Moisture (wt%)	0.01	<0.03	0.05	<0.06
Other organics[a] (wt%)	0.05	<0.1	0.05	<0.1
Nickel (ppm wt)	<0.2	<0.5	<0.2	<0.5
Color (Klett)	<20		<20	
Carbon chain length (wt%)				
<C12	0		0	
C12	0.2		0	
C14	1.7		4.0 (maximum)	
C16	94.6		59.0 (minimum)	
C18	3.5		40.0 (maximum)	
>C18	0		<1	

[a] Other organics include glycerides, glycerine, and methanol.

11.4.1 PLANT CAPACITY

A 5 t/h nominal production capacity (5000 kg/h) based on 100% active content forms the basis for the following tables. The 5 t design fits current industry requirements for capacity and economics. The actual capacity of the MES plant is highly dependent on both the molecular weight and the quality of the ME raw material. The ultimate capacity of an MES plant is determined by the amount of sulfur that can be burned and converted into SO_3. The plant design is based on a fixed number of moles per hour of ME of a specified molecular weight.

A convenient way to express the capacity of a sulfur burning plant is to say that it will produce X kilogram moles of SO_3 per hour. When reacting ME with a fixed number of moles of SO_3, the value in kilogram per hour of MES produced will change as the molecular weight of ME changes. Additionally, MES is unique among sulfonates in that its production requires a reaction of ME with a large excess of SO_3. The exact ratio depends on the quality of the ME feed and can vary from 1.15 to 1.25 mol of SO_3 required per mole of ME feed. Less-pure feeds will require higher mole ratios (MRs) of SO_3 to ME and result in reduced plant capacity.

This discussion is based on the quality of the C16 ME feed, as shown in Table 11.1. ME feed of this quality will require approximately 1.15–1.2 mol of SO_3 per mole of ME.

11.4.2 METHYL ESTER SULFONATES' SULFONATION RAW MATERIAL CHARACTERISTICS

Sulfur—Recovery grade is recommended, Bright Frasch is acceptable with the following specifications:

Ash	≤0.05 wt%
Carbon	≤0.05 wt%
Water	≤0.01 wt%

Caustic Soda

Concentration	≥50 wt%
Iron (Fe)	≤10 ppm (wt)

Sulfate (Na_2SO_4)	≤0.1 wt%
Chloride (NaCl)	≤1.5 wt%
Appearance	Clear liquid, colorless to slight yellowish green
Process Water, Treated Zeolite	
Hardness	≤100 ppm $CaCO_3$
Chloride content	≤10 ppm (wt)
Iron (Fe)	≤10 ppm (wt)
Methanol (MeOH)	
Purity	≥98.5 wt%
Chloride content	≤0.10 wt%
Hydrogen Peroxide (H_2O_2)	
Concentration	50 wt%
Nitrogen (N_2)	
Purity	≥99 wt%

11.5 METHYL ESTER SULFONATE CHEMICAL REACTIONS AND BY-PRODUCTS

The absorption of sulfur trioxide by ME in the falling-film reactor shown in the reaction in Reaction 11.1 is rapid to form the intermediate II, which is commonly termed an *adduct* or *anhydride*. Intermediate II is in equilibrium with a form that activates the α-carbon for sulfonation in the reaction in Reaction 11.2 to form intermediate III. Intermediate III must undergo a "rearrangement" as shown in Reaction 11.3 to release sulfur trioxide during the digestion step after the falling-film reactor to form the desired methyl ester sulfonic acid (MESA) (IV). The released sulfur trioxide will then convert the remaining intermediate II to intermediate III. If the intermediate III is not converted to MESA (IV) before neutralization, hydrolysis of the ester occurs, forming disalt (V) as shown in Reaction 11.4.

Reaction in Reaction 11.3 is completed by the reaction of sulfur trioxide with intermediate II as shown in Reaction 11.2. Once intermediate II is consumed, reaction in Reaction 11.3 slows down appreciably. For typical MRs of sulfur trioxide to ME, the amount of intermediate III varies from 10 to 20%. This can be minimized by long and hot digestion of the sulfonic acid, which creates dark colors, or by addition of an alcohol as per the reaction in Reaction 11.5 to react with the remaining intermediate III before neutralization to form the desired MESA (IV).

Neutralization of MESA (IV) to form MES (VI) is shown in Reaction 11.6. However, if the pH of neutralization is not controlled, the MES (VI) product can be hydrolyzed to form disalt (V) as shown in Reaction 11.7. This reaction produces both disalt (V) and methanol. Thus, minimizing the yield of disalt (V) requires completion of the reaction of the intermediate III to MESA (IV) before neutralization as well as precise control of the bleaching and neutralization conditions to prevent large conversion of MES (VI) to disalt (V) and methanol.

An MES production plant (see Figure 11.1) includes both sulfonation and MES systems. Sulfonation systems are fairly typical and include a sulfur supply, air supply, SO_3 gas generator, sulfonation, effluent gas treatment (includes electrostatic precipitator, SO_2 absorber, and effluent gas filter), and computer control systems and a motor control panel. An SO_3 absorber system or other means of absorbing SO_3 gas during plant startup may be included. MES systems include a MESA digestion, MESA bleaching, MESA neutralizer, Turbo Tube® dryer*, steam ejector vacuum, product cooling, peroxide treatment, sodium sulfite, and methanol recovery systems, in addition to a computer control system and motor control panel. The sulfonation system described in this chapter uses a Chemithon annular falling-film reactor[20–22] designed for a nominal 7% (volume) SO_3 in air, although other types of reactors could be used.

---

* Turbo Tube® Dryer is a Registered Trademark of the Chemithon Corporation.

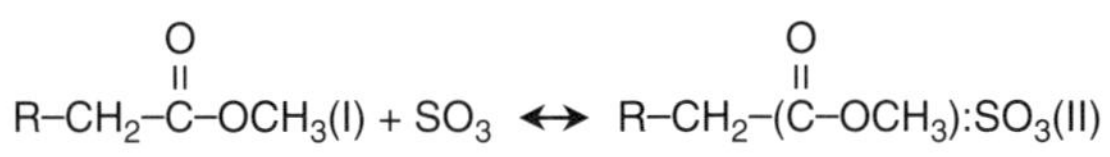

Production of Methyl Ester Sulfonates

$$R-CH_2-\overset{\overset{\displaystyle O}{\|}}{C}-OCH_3(I) + SO_3 \leftrightarrow R-CH_2-(\overset{\overset{\displaystyle O}{\|}}{C}-OCH_3):SO_3(II)$$

REACTION 11.1

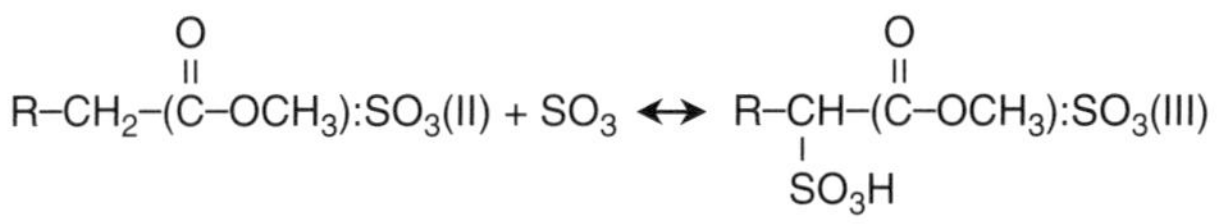

$$R-CH_2-(\overset{\overset{\displaystyle O}{\|}}{C}-OCH_3):SO_3(II) + SO_3 \leftrightarrow R-\underset{\underset{\displaystyle SO_3H}{|}}{CH}-(\overset{\overset{\displaystyle O}{\|}}{C}-OCH_3):SO_3(III)$$

REACTION 11.2

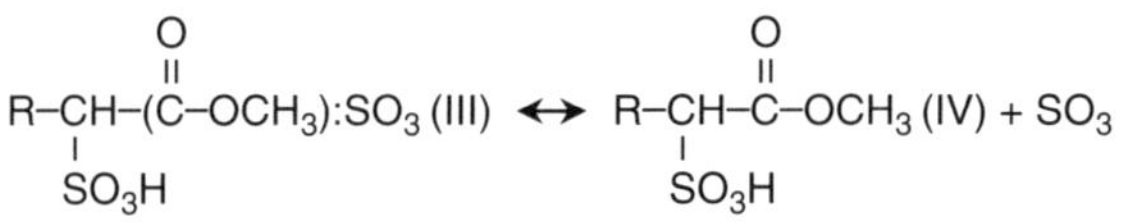

$$R-\underset{\underset{\displaystyle SO_3H}{|}}{CH}-(\overset{\overset{\displaystyle O}{\|}}{C}-OCH_3):SO_3 \text{ (III)} \leftrightarrow R-\underset{\underset{\displaystyle SO_3H}{|}}{CH}-\overset{\overset{\displaystyle O}{\|}}{C}-OCH_3 \text{ (IV)} + SO_3$$

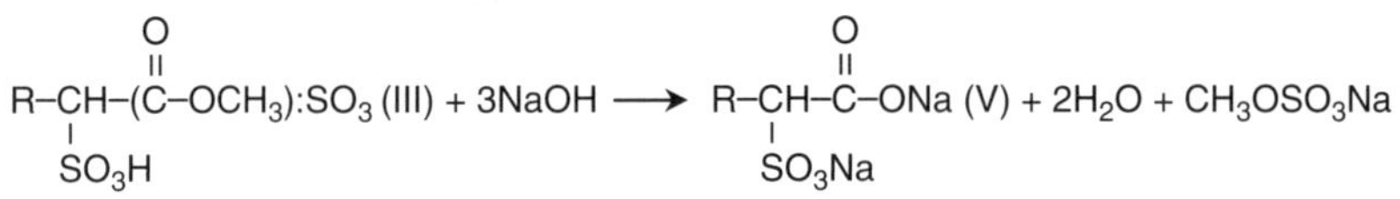

REACTION 11.3

$$R-\underset{\underset{\displaystyle SO_3H}{|}}{CH}-(\overset{\overset{\displaystyle O}{\|}}{C}-OCH_3):SO_3 \text{ (III)} + 3NaOH \longrightarrow R-\underset{\underset{\displaystyle SO_3Na}{|}}{CH}-\overset{\overset{\displaystyle O}{\|}}{C}-ONa \text{ (V)} + 2H_2O + CH_3OSO_3Na$$

REACTION 11.4

$$R-\underset{\underset{\displaystyle SO_3H}{|}}{CH}-(\overset{\overset{\displaystyle O}{\|}}{C}-OCH_3):SO_3 \text{ (III)} + CH_3OH \longrightarrow R-\underset{\underset{\displaystyle SO_3H}{|}}{CH}-\overset{\overset{\displaystyle O}{\|}}{C}-OCH_3 \text{ (IV)} + CH_3OSO_3H$$

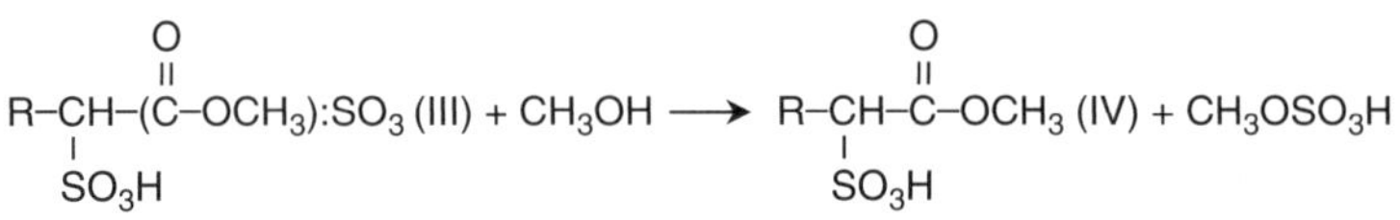

REACTION 11.5

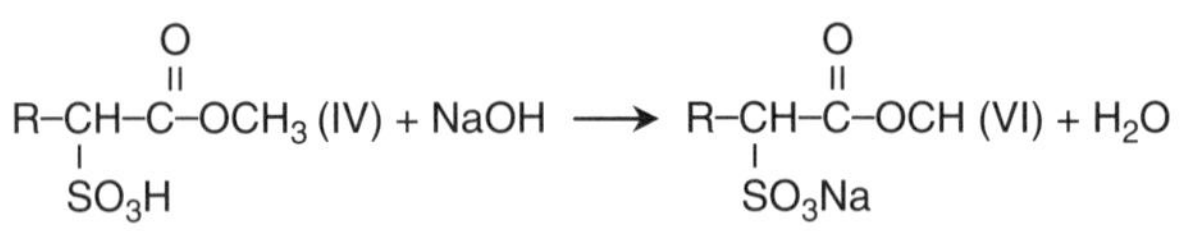

$$R-\underset{\underset{\displaystyle SO_3H}{|}}{CH}-\overset{\overset{\displaystyle O}{\|}}{C}-OCH_3 \text{ (IV)} + NaOH \longrightarrow R-\underset{\underset{\displaystyle SO_3Na}{|}}{CH}-\overset{\overset{\displaystyle O}{\|}}{C}-OCH \text{ (VI)} + H_2O$$

REACTION 11.6

$$R-\underset{\underset{\displaystyle SO_3Na}{|}}{CH}-\overset{\overset{\displaystyle O}{\|}}{C}-OCH_3 \text{ (VI)} + NaOH \longrightarrow R-\underset{\underset{\displaystyle SO_3Na}{|}}{CH}-\overset{\overset{\displaystyle O}{\|}}{C}-ONa \text{ (V)} + CH_3OH$$

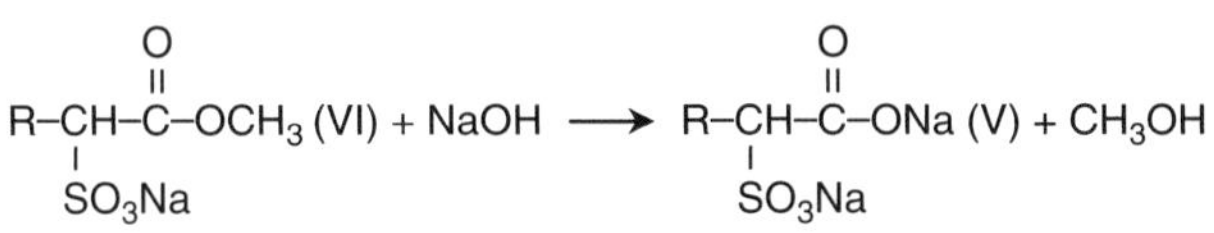

REACTION 11.7

11.6 GENERAL METHYL ESTER SULFONATES' PROCESS CONDITIONS

The reactor inlet SO_3 gas concentration is 7% (volume) and the reactor inlet gas temperature is ~42°C. The ME feedstock is supplied to the reactor at a temperature ranging from 40 to 56°C, well above the freezing point of ME feedstocks. The mass flow of reactants is controlled to maintain a fixed MR of SO_3 to ME, typically ranging from 1.15 to 1.25. The choice of MR is dependent on the expected selectivity of the particular ME to side reactions and by-product formation. These include oxidation of the alkyl chain by SO_3, sulfonation of some of the resulting olefin sites, formation of methyl sulfuric acid, and hydrolysis of the ester to form disalt.

The MESA is transferred to an acid digester system where it rapidly reaches digestion temperature. After the MESA is digested, the methanol (30–35 wt%, digested MESA basis) and 50% hydrogen peroxide are ratio added with the MESA into the MES bleacher. A significant amount of exothermic reaction occurs in the bleacher. The acid bleaching step requires 1–1.5 h—more bleaching time can further reduce color as long as all of the available peroxide is not consumed. The excess methanol effectively limits the production of disalt and significantly reduces the viscosity of the mixture, which improves mixing and heat transfer through the bleaching process and subsequent neutralization process.

Bleached MESA is forwarded to the neutralizer where a controlled proportion of 50% sodium hydroxide is admixed with the bleached MESA and a large recycle stream of neutralized paste. Neutralized MES paste continuously discharges to a dryer where the excess water and methanol are removed. The dryer functions as a distinct processing system that processes concentrated pastes at rates to match the sulfonation system output. The MES inlet temperature for the drying process is ~145°C and operates under vacuum conditions of 120–200 torr. No significant increase in disalt occurs at the elevated process temperatures of the dryer and the product color is stable. Primary goals for any MES production facility are to achieve the lowest colors and the highest ratio of active matter to disalt (for the "benchmark" C16—21.5 parts active:1 part disalt).

11.7 FEED AND PRODUCT SPECIFICATIONS

11.7.1 METHYL ESTERS

The benchmark feed characteristics of ME are given in Table 11.1.

11.7.1.1 Methyl Ester Sulfonates

Table 11.2 summarizes the benchmark product characteristics of MES.

11.8 METHYL ESTER SULFONATES PROCESS DESCRIPTIONS

11.8.1 BASIC SULFONATION PROCESS

A detailed description of typical sulfonation systems[23,24] is omitted from the production descriptions contained in this chapter. However, all MES processes share the common use of a falling-film sulfonation system for the initial reaction of ME with air/SO_3. The processes differ in the treatment of MESA after the sulfonation reactor. The sulfonation systems shown in Figure 11.1 are generic for

TABLE 11.2
Benchmark Product Characteristics

MES	Benchmark C16 Dry MES Product	Benchmark C16–C18 Dry MES Product
Molecular weight	372	382.7
Total active (wt%) (MES + disalt)	91.0	88.5
Disalt (wt%)	4.7	5.8
Methanol (wt%)	0.1	0.1
Moisture (wt%)	2.0	2.6
Free ME (wt%)	2.3	2.8
Final color (Klett) (5 wt%)	29–40	40

Production of Methyl Ester Sulfonates

all MES processes. As most commercial sulfonation plants are based on sulfur burning, the following description is written for such type of equipment.

The sulfur supply system is designed to supply a steady, measured, and known flow of sulfur to the SO_3 gas plant. The purpose of the air supply system is to supply a constant flow of dry ($-70°C$ dew point), relatively low-pressure (1–1.2 bar) air to the SO_3 gas plant. In the SO_3 gas generator, the sulfur is burned in the dry air to form SO_2 gas, which is subsequently cooled and converted into SO_3 gas. This SO_3 gas is cooled, filtered, diluted, and sent to either the SO_3 absorber system or the reactor system/sulfonator. The entire SO_3 generation process is designed to supply the sulfonator with a known flow of SO_3 gas at relatively low pressure.

In the sulfonator, the organic raw material is precisely metered into the falling-film reactor where it reacts with the SO_3 gas to form sulfonic acid. The sulfonation reaction is exothermic and water-cooled heat exchange surfaces are provided to remove the heat of reaction and control the temperature of the sulfonic acid. The sulfonic acid and depleted air are separated, the air is sent to the effluent gas treatment system where contaminants are removed before discharge into the atmosphere. The sulfonic acid is digested and further processed as described in Sections 11.6 and 11.8.3.

The control of contact of the organic and SO_3 gas is the most critical part of the process. It is important that it should be controlled on both an overall (macro) and a point-by-point (micro) scale. Sulfonation equipment suppliers have devoted decades and tens of thousands of effort hours to the perfection of falling-film reactor design with the goal of controlling the ratio of organic to SO_3 gas and the temperature of the reaction.

Capacities for commercial sulfur burning air/SO_3 sulfonation units range from 250 to 20,000 kg/h of 100% active detergent, although capacity of MES plants are rarely lesser than 3 million tons/year for economic reasons. As MES product quality has low sensitivity to SO_3 gas concentration, an MES sulfonation plant is usually designed to sulfonate with ~6–7% (volume) SO_3, close to the maximum for conventional sulfonation gas plant designs.

11.8.2 Methyl Ester Sulfonic Acid Digester

Production of MESA is a complex process because sulfonation proceeds in two steps. The first step is a rapid reaction of the ME with two molecules of SO_3 to form an adduct, which is accomplished in a standard sulfonator. The second process step is the rearrangement of the adduct to form MESA in the MESA digester (see Figure 11.2). The adduct also releases the second molecule of SO_3 that

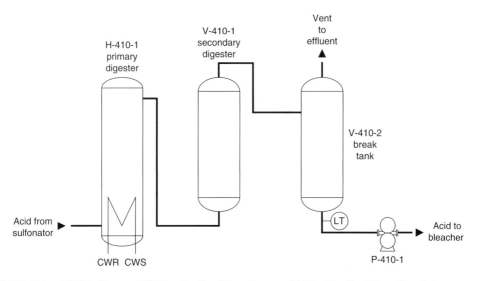

FIGURE 11.2 MESA digester (CWR—Cooling Water Return; CWS—Cooling Water Supply).

is then free to further react with another ME molecule. The second reaction is slow and requires extended digestion time and elevated temperature for completion.

The acid darkens as a consequence of the MESA digestion process to a degree that is characteristic of the specific ME raw material. A proprietary bleaching step is required to lighten the final product color to acceptable formulation levels. The unbleached acid is continuously metered to the MESA bleaching system, where reaction with hydrogen peroxide in the presence of methanol takes place at precisely controlled conditions.

11.8.3 Methyl Ester Stabilization and Bleaching

The digested MESA is metered to the MESA bleaching system by mass flow. The MESA bleacher (see Figure 11.3) reacts with the MESA formed during sulfonation and digestion steps with methanol and hydrogen peroxide to stabilize the ME and reduce the color. The primary function of the bleacher is to reduce the color of the final product with hydrogen peroxide (H_2O_2). The mass flow of MESA determines the mass flow rates for methanol and hydrogen peroxide injection. A large excess of methanol is admixed into the MESA to stabilize the ME against hydrolysis. Methanol also inhibits the formation of undesirable α-sulfo carboxylic acid, or disalt (see Figure 11.4), and modifies the MESA viscosity.

The patented acid bleaching process is safe and effective. The process uses a two-phase bleacher digester in which methanol is boiled, then condensed, and refluxed, maintaining a high vapor transport rate to constantly remove oxygen and any organic flammables from the vessel. Dry nitrogen (N_2) is dosed into the vapor phase as an inert diluent. The nitrogen flow is governed by an oxygen sensor to maintain the vent gas line in a nonflammable composition. The methanol boiling rate is used to control process heat removal. The bleacher operating pressure determines the boiling temperature. A large cooling coil is provided for positive temperature control.

The benefit of the high bleaching temperatures is shown in Figure 11.5. The color of the MESA leaving the bleaching reaction is substantially less at the higher-temperature range. As seen, the yield of disalt does increase slightly as the bleaching temperature is increased, but at these optimal conditions, the impact is very slight.

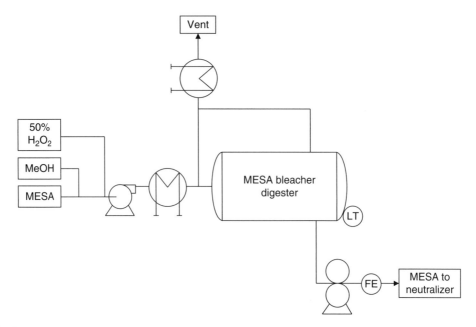

FIGURE 11.3 MESA bleacher (LT—Level Transmitter; FE—Flow Meter).

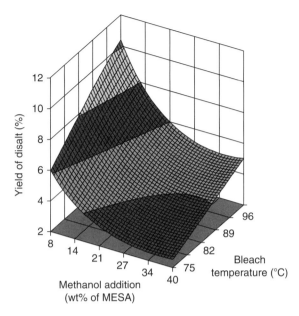

FIGURE 11.4 Yield of disalt from bleaching.

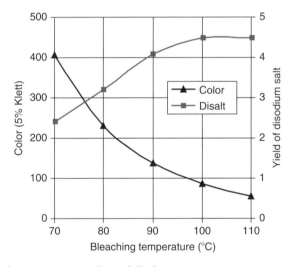

FIGURE 11.5 Effect of temperature on color and disalt.

11.8.4 Neutralizer System

The process for neutralization of MESA in the MES neutralizer system (see Figure 11.6) is unique. The composition of neutral paste includes methanol and some minor compounds that tend to form precipitates, which requires engineering features that are unique to MES neutralizer systems. The MESA neutralizer maximizes MES stability and minimizes disalt formation.

The mass flow of MESA into the neutralizer provides the primary process control signal, and the flow of caustic soda is controlled in a fixed ratio to the inlet MESA. Process heat is removed by the heat exchanger in the neutralization loop. A cooling water recirculation system is provided and is equipped to control the temperature of the cooling water supplied to the heat exchanger.

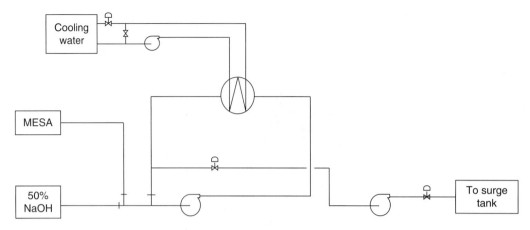

FIGURE 11.6 MES neutralizer system.

The presence of alcohol in MESA during neutralization reduces the viscosity of the high-active neutral product from a very thick paste to a thin liquid. This reduction in viscosity greatly improves mixing, which eliminates points of high local pH that will cause disalt formation. In addition, the presence of alcohol inhibits hydrolysis; therefore, the level of disalt is further reduced.

11.8.5 METHYL ESTER SULFONATES' TURBO TUBE DRYER SYSTEM

The term *drying* as applied to this process means removing both water and methanol from the remainder of the composition. Neutralized MES paste is metered from the surge tank to the MES Turbo Tube® Dryer (TTD) System (U.S. patents[25,26] and worldwide counterpart patents) at a constant flow rate and under a controlled back pressure (see Figure 11.7). The paste stream is heated to a controlled temperature selected to provide sufficient stored energy in the liquid to flash vaporize both water and methanol as the composition enters the drying tubes. In the dryer, the paste passes through a common chamber above an array of drying tubes where it is forced through a series of small orifices arranged at the inlet of each drying tube. The sudden reduction of pressure allows the water and methanol to vaporize and the expansion causes sufficient pressure drop on each path to efficiently distribute the paste in substantially equal flows to all of the drying tubes. A flow of injection steam is fed through the drying tubes along with the MES composition to further promote even flow distribution and to help maintain the dryer tubes in a clean and effective state.

The initial flash vaporization is controlled by the temperature and pressure of the paste at the inlet of the TTD. It is preferable to vaporize sufficient water and methanol to instantly establish a two-phase flow regime inside the drying tube. The released vapor is the continuous phase and solids are dispersed as particulate and propelled down the drying tube by the expanding vapor. The drying tubes discharge the MES composition and vapor phase into a separation tank, which is held under vacuum. Since the flash evaporation of water and methanol has a cooling effect on the residual paste, additional heat is applied to the drying tubes by a steam jacket that surrounds the tubes with pressure-regulated saturated steam. This maintains the dryer tube walls at a set temperature, preventing any condensation of vapor and adding more heat to the composition to continue the vaporization of water and methanol from the MES composition.

The dry MES composition (adduct) discharging from the drying tubes is in a molten state and is collected in the throat of the discharge plodder at the bottom of the separation tank. The plodder is a variable-speed, double-screw design that has a steam-jacketed liquid discharge end; therefore, the MES can be kept molten and fluid. Since the separation tank is maintained under vacuum, the plodder provides means to seal the product outlet and to move the molten MES paste from the

Production of Methyl Ester Sulfonates 213

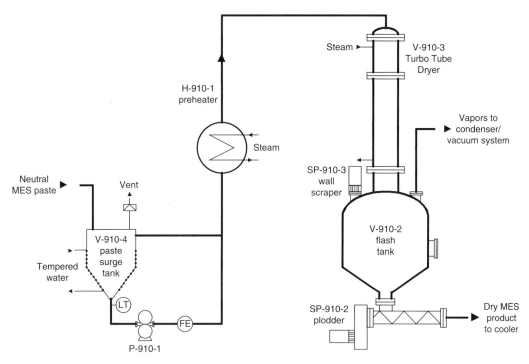

FIGURE 11.7 MES Turbo Tube Dryer System.

vacuum to discharge at atmospheric (or slight positive) pressure. The dried MES plodded from the TTD typically contains 2–4% water and less than 0.1 wt% residual methanol. Exact values for these parameters depend on the operating conditions.

The vapor phase is separated and continuously drawn off the separation tank and passes through a condenser and vacuum system such as steam ejectors or vacuum pumps. Condensate is then collected and supplied to a methanol recovery system described in Section 11.8.8.

11.8.6 Dried Product Cooling System

The hot, plastic MES product extruded from the TTD plodder must be cooled to solidify the composition. Cooling the molten MES product can be accomplished continuously with a double-chilled belt cooler or chilled drum flaker (see Figure 11.8) that forms a 2- or 0.5-mm-thick flakes, respectively.

The MES product is discharged from the plodder at 100–120°C and is distributed onto a cooler where it is chilled below its freezing point temperature (for C16 MES, ~32°C) forming a flaked product. The drum flaker is preferred for the cooling step because this machine is able to continuously produce a 0.5-mm-thick flake, which can be postadded and incorporated into powder detergent formulations with little or no further processing. The thicker flakes from a double-chilled belt cooler require milling with screening steps for separation of oversized and fine flakes before postaddition, and therefore significant additional processing is required.

A positive displacement pump is used to push the molten MES through a distribution header, depositing the material along the drum flaker above the applicator roll, as seen in Figure 11.8. The applicator forms a uniform thin film of the molten MES on the cooling drum. Coolant (typically chilled water) is applied to the inner surface of the cooling drum by means of a spray tube that is inserted in the centerline of the cooling drum. The film of MES adheres to the outer surface of the cooling drum and solidifies as it rides around to the knife, which is placed with some tension

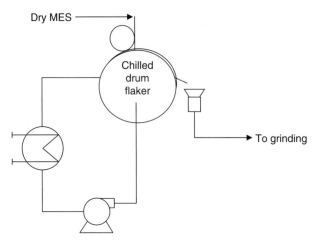

FIGURE 11.8 Chilled drum flaker.

very close to the surface of the cooling drum. The speed of the cooling drum is adjusted so that the MES film is completely solidified and the knife assists the separation by lifting off the film that breaks up into fairly uniform flakes. Ultimately, final flake thickness is determined by the applicator roll drive speed and clearance to the cooling drum surface, which can be precisely set for a 0.5 mm clearance. The drum flaker is very precisely machined and manufactured to provide critical strength and structural rigidity so that such small clearances can be established even for full-scale machines that may be up to 4.5 m in length and for such applications where 100°C (or more) temperature differences exist.

11.8.7 Dryer Auxiliary Systems

Some auxiliary equipment is required to create a vacuum and separate the methanol and water removed from the product in the vacuum and condensate collection system (see Figure 11.9). These auxiliary systems include a steam ejector vacuum system, peroxide treatment system, and tempered water system.

The TTD operates in a vacuum with ~100 torr for drying MES paste. As the vacuum increases, the operating temperature of the TTD decreases. The recommended method of generating this level of vacuum is to use a steam ejector vacuum system as shown in Figure 11.9 (may be single- or two-stage, depending on specific conditions). Condensate containing water, methanol, a small percentage of hydrogen peroxide, and unreacted feed flows into a collection tank. The methanol/water is then pumped to the methanol recovery system (see Figure 11.12).

Sodium sulfite is required in the process to react with the small amount of hydrogen peroxide in the condensate. It is also used as a safety measure by adding it to the top of the methanol distillation tower to prevent organic peroxide formation.

Plants equipped with continuous sulfur dioxide absorption towers can produce sodium sulfite solution that can be used for this process step. A sodium sulfite supply system (see Figure 11.10) is provided to store the sodium sulfite solution. An optional sulfite dissolver tank may be provided for dissolving dry sodium sulfite with water in case additional sodium sulfite may be required. Sodium sulfite solution is injected into the methanol/water condensate solution before injecting into the condensate surge tank inlet (see Figure 11.11) so that it thoroughly mixes into the solution. This ensures that the solution in the tank is free of peroxides and safe for temporary storage of the condensate solution.

Production of Methyl Ester Sulfonates

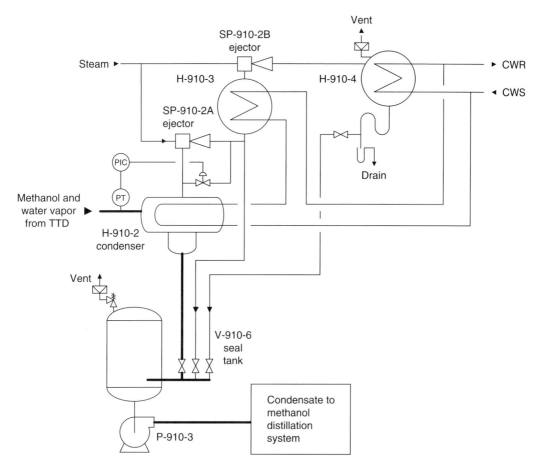

FIGURE 11.9 Vacuum system (steam ejector type) and condensate collection.

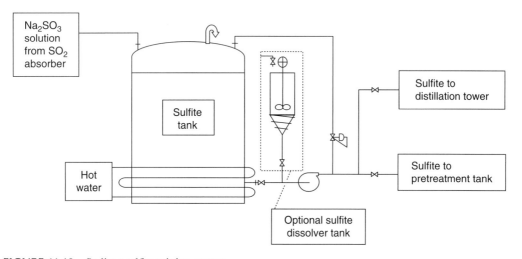

FIGURE 11.10 Sodium sulfite mixing system.

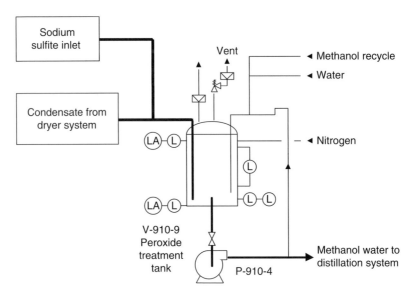

FIGURE 11.11 Condensate surge tank.

11.8.8 Methanol Recovery System

The methanol and water solution is transferred from the condensate collection tank to the condensate surge tank as illustrated in Figure 11.11. The condensate surge tank is designed for temporary storage of the methanol/water solution before its supply to the distillation tower.

The methanol/water solution is then pumped over to the distillation system where methanol is separated for recycling to the process (see Figure 11.12). The solution first enters a preheater where it is heated to the boiling point and then is fed to the middle of the distillation tower. The distillation column is equipped with multiple beds of structured packing. Overhead vapors pass through a reflux condenser, which returns a portion of the overhead (methanol stream) to the tower and discharges a portion as recovered methanol. The tower bottoms stream (wastewater) is discharged to the water treatment system. The distillation tower system may be located remotely from the MES plant and has its own control systems local to the system. The distillation system may be taken offline periodically for cleaning procedures.

11.8.9 By-Products and Safety

The safety issues in MES processing involve the presence of flammables and oxygen in the system together. Extensive hazard and operability (HAZOP) studies of MES processes have been conducted to meet rigorous safety and environmental protection standards. The Huish MES plant has been underwritten by factory mutual underwriters and already has more than 5 years of trouble-free operation; the older Lion and Stepan plants have operated safely for longer periods.

Methanol is present in the process after sulfonation and has been measured at several tenths of 1% in the digested acid before bleaching or before any alcohol addition. Methanol in the system can form hydrogen methyl sulfate by reaction with free sulfur trioxide or the adduct of MESA and sulfur trioxide. The hydrogen methyl sulfate can then react with methanol to form dimethyl sulfate (DMS) as shown in Reaction 11.1. Although ppm levels of DMS have been detected in the bleached sulfonic acid, testing has verified that none is detectable in neutral MES because DMS is destroyed by the addition of NaOH during the neutralization step.

$$CH_3SO_4H + CH_3OH \rightarrow CH_3SO_4CH_3 + H_2O \qquad (11.1)$$

Production of Methyl Ester Sulfonates

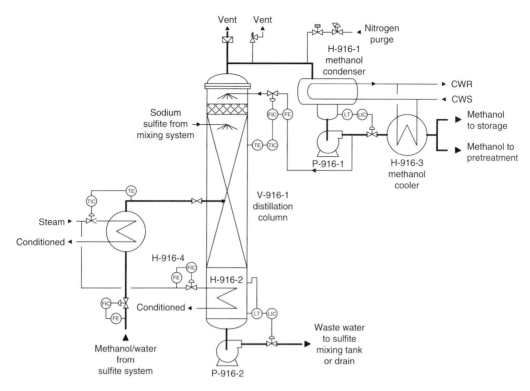

FIGURE 11.12 Methanol recovery system.

The DMS can further react with methanol to form hydrogen methyl sulfate and dimethyl ether (DME) as shown in Reaction 11.2. DME has been detected in the digested acid before bleaching or methanol addition. This chemistry can occur with any alcohol present in the system. DME is removed from the system by a N_2 purge of the bleacher digester. The DME/N_2 stream is sent to the plant boiler where it is burned.

$$CH_3SO_4CH_3 + CH_3OH \rightarrow CH_3SO_4H + CH_3OCH_3 \tag{11.2}$$

Hydrogen peroxide in the bleaching step can decompose to form free oxygen. Some decomposition will occur during acidic bleaching, as in the Chemithon process, or under neutral bleaching conditions. When the volatile gases are eventually vented from the system, the oxygen will be present with flammable materials such as alcohols or ethers. A flammable atmosphere is avoided by the use of nitrogen purging to control the concentration of oxygen in the vessel vapor spaces and vent systems. Oxygen sensors constantly monitor oxygen concentration. Additionally, sodium sulfite is added into the dryer condensate system and into the methanol distillation system to ensure destruction of any organic peroxides that may be formed.

Sodium sulfite is used to destroy traces of peroxide in the methanol/water mixture before distillation and recovery of methanol. Because the presence of sulfite in MES may complicate formulation of finished consumer products, the possibility that sulfite from the recycled methanol may be present in the MES product has been investigated. Commercial MES made using the Chemithon MES process was analyzed for sodium sulfite. No detectable sodium sulfite was found. To verify that sulfite is not present, the commercial MES sample was also analyzed for residual hydrogen peroxide. Approximately 0.1% peroxide was detected, indicating that no sodium sulfite was present.

11.9 CONCLUSIONS

MES has strong appeal because ME is a relatively inexpensive, natural, and renewable feedstock. MES provide superior surfactant properties at low cost and therefore a strong economic incentive to substitute MES for traditional surfactants in many applications. Production capacity of MES will increase in the coming years. More formulation work based on the concentrated and ultraconcentrated forms of MES can be expected utilizing different types of ME feedstock.

The process of converting ME into MES needs to produce a finished product with acceptable colors and with minimal disalt content. Production of high-quality MES requires more sophisticated equipment than LABS or fatty alcohol sulfate. Even so, MES systems have been safely and efficiently operated and controlled for comparatively high production level. These plants also have relatively higher production rates than those for "typical" products. Although more training and attention must be allocated for operation and maintenance of MES plants versus typical sulfonation plants due to the greater number of unit operations and generally higher level of complexity, it can be minimized and managed with effective operation procedures and policies and comprehensive preventive maintenance programs.

Although the process challenges for making good-quality MES have been addressed, there is still scope for technology upgrades to minimize the consumption of bleach and solvents used in the process. There is also a need to produce the MES as a granule or bead to facilitate its use in a postaddition process. This will provide the detergent formulators with the flexibility for producing mixed active detergent surfactants and optimizing their formulations based on the market dynamics of the raw materials.

REFERENCES

1. Ogoshi, T., Kusumi, Y., Bleaching Method for Sulfonic Acid, U.S. Patent No. 3,997,575, 1976.
2. Hovda, K.D., Sulfonation of Fatty Acid Esters, U.S. Patent No. 5,587,500, 1996.
3. Watkins, C., All Eyes Are on Texas, *INFORM*, 12, 1152–1159, 2001.
4. Yamada, K., Matsutani, S., Separation and Identification of Colored Substances in Sulfonated Fatty Acid Methyl Ester, *84th AOCS*, Anaheim, CA, April, 1993.
5. Satsuki et al., Performance and physicochemical properties of alpha-sulfo fatty acid methyl esters, *JAOCS*, 69, 672–677, 1992.
6. Schambil, Schwuger, Physico-chemical properties of alpha-sulpho fatty acid methyl esters and alpha-sulpho fatty acid di salts, *Tenside Surfact. Det.*, 27(6), 380–385, 1990.
7. Drozd, J.C., Use of Sulfonated Methyl Esters in Household Cleaning Products, *Proceedings of World Conference on Oleochemicals.*, pp. 256–268, March, 1991.
8. Stirton, A.J. et al., Synergistic Tallow-Based Detergent Compositions, U.S. Patent No. 3,632,517, 1972.
9. Droz, J.C. et al., Liquid laundry detergents based on soap and alpha-sulfo methyl esters, *JAOCS*, 68, 59–62, 1991.
10. Okumura et al., Mechanism of Sulfonation of Fatty Acid Esters with Sulfur Trioxide and Properties of Alpha-Sulfo Fatty Acid Esters, *Proceedings of the 7th CID in Moscow*, 1976.
11. Foster, N.C., Methyl ester sulfonates in commercial detergents, *SÖFW J.*, 130–134, 2004.
12. Horie, K., New Process of Methyl Ester Sulfonate and Its Application, *Proceedings of the 6th CESIO in Berlin,* 2004.
13. Stein, W., Baumann, H., Alpha-sulfonated fatty acids and ester: Manufacturing process, properties and applications, *JAOCS*, 52, 323–329, 1975.
14. Hovda, K.D., Methyl Ester Sulfonation: Process Optimization, *Proceedings of PORIM International Palm Oil Congress*, September, 1993.
15. Hovda, K.D., Effect of Methyl Ester Feedstock on Sulfonate Quality, *Proceedings of 1994 International Seminar on Surfactants and Detergents*, Xian, November 10–13, 1994.
16. Hovda, K.D., The Challenge of Methyl Ester Sulfonation, *Household '96*, Sao Paulo, June, 1996.
17. MacArthur, B.W., Brooks, B., Sheats W.B., Foster, N.C., Meeting the Challenge of Methyl Ester Sulfonation, *Proceeding of World Conference on Palm and Coconut Oils for the 21st Century*, Bali, 1998.
18. Sheats, W.B., MacArthur, B.W., Methyl Ester Sulfonate Products, *CESIO*, 2000.

19. Foster, N.C. Manufacture of Methyl Ester Sulfonates and Other Derivatives, in *SODEOPEC: Soaps, Detergents, Oleochemicals, and Personal Care Products*, L. Spitz, Editor, AOCS Press, Urbana, IL, 2004.
20. Brooks, B. et al., U.S. Patent No. 3,257,175, 1966.
21. Brooks, B. et al., U.S. Patent No. 3,427,342, 1969.
22. Brooks, B. et al., U.S. Patent No. 3,350,428, 1967.
23. Foster, N.C., Sulfur burning SO_3 gaseous sulfonation technology, in *Treatise of Fats, Fatty Acids and Oleochemicals*, O.P. Narula Editor, Chapter 6, Vol. 1, Industrial Consultants (India), New Delhi, 1995.
24. Foster, N.C., Sulfonation and sulfation processes, in *Soaps and Detergents: A Theoretical and Practical Review*, AOCS Press, Urbana, IL, 1996.
25. Duvall, L.R., Brooks, B., Jessup, W., U.S. Patent No. 5,723,433, 1995.
26. Brooks, B., Jessup, W., MacArthur, B.W., U.S. Patent No. 6,058,623, 2000.

12 Amphoteric Surfactants: Synthesis and Production

David J. Floyd and Mathew Jurczyk (edited by Uri Zoller)

CONTENTS

12.1 Introduction 221
12.2 Background 221
    12.2.1 Description 221
    12.2.2 Physical and Functional Properties 222
12.3 Selective Characteristics 222
12.4 Amphoteric Surfactants and Their Synthesis 223
    12.4.1 Acyl/Dialkyl Ethylenediamines and Derivatives 223
        12.4.1.1 Amphoacetates 223
        12.4.1.2 Amino Propionates 227
        12.4.1.3 Carboxyamphoterics 228
        12.4.1.4 Acyl/Dialkyl Ethylenediamine 228
    12.4.2 N-Alkylamino Acids 228
    12.4.3 Miscellaneous Amphoterics 231
        12.4.3.1 Betaines 231
        12.4.3.2 Amine Oxides 235
12.5 Conclusion 236
References 236

12.1 INTRODUCTION

Amphoteric surfactants comprise a broad range of compounds, which display nonionic, cationic, or even anionic tendencies depending on pH or in-use conditions. Betaines, imadazoline-derived amphoacetates, alkylamino propionates, and glycinates are generally included in this category. Amine oxides, which may exhibit nonionic or cationic characteristics depending on pH conditions, are also included in this category.

Amphoterics were among these many specialty surfactants commercialized after World War II [1]. Since then, these surfactants have been mainly used in personal care products because of their unsurpassed mildness and low eye-sting properties. Today, most mild skin cleansers and shampoos contain an amphoteric surfactant as a major component.

12.2 BACKGROUND

12.2.1 DESCRIPTION

Amphoteric surfactants are surfactants where the charge changes as a function of the pH value of the formulation in which they are used. They are generally regarded as mild surfactants, but this may not always be true. Amphoteric surfactants build complexes in combination with anionic surfactants and these complexes are milder than each of the individual surfactants [2].

Substances are classified as amphoteric only if the charge on the hydrophobe changes as a function of the pH. Surfactants that carry a cationic charge on the hydrophobe changes as a function of the pH; those that carry a cationic charge in strongly acidic media carry an anionic charge in strongly basic media, and form zwitterionic species at intermediate pH values, are amphoteric as shown in the following:

$$[RNH_2CH_2CH_2COOH]^+ X^-$$

Low pH; cationic hydrophobe

$$[RN^+H_2CH_2CH_2COO^-]$$

Intermediate pH; zwitterionic hydrophobe

$$[RNHCH_2CH_2COO]^- B^+$$

High pH; anionic hydrophobe

In these structures, X^- represents an unidentified anion, for example, Cl^- and B^+ an identified cation, for example, K^+. Amphoterics must be examined individually and at specific pH values, otherwise they may be misidentified [3].

The amphoteric surfactants are relatively expensive products compared to anionic surfactants. Thus, it is not surprising that they are primarily being utilized at low concentrations in cosmetic formulations. A review of 438 shampoos of the U.S. market reveals that appreciable quantities, for example, >5% of alkylamido betaines and imidazolinium surfactants were found in only 8.7% and 13.5%, respectively, of the investigated shampoos. Alkyl betaines were found in a limited number of cases, whereas sulfo betaines were not found in this study [4,5]. Since these studies were made, the market has focused on mild products. The change in the U.S. market from bar to liquid soaps has increased the use of alkylamidopropyl betaines.

12.2.2 PHYSICAL AND FUNCTIONAL PROPERTIES

Although the variety and range of amphoteric surfactant types is quite large, there are certain properties these materials have in common. These include lime soap dispersancy and hard water tolerance; hydrolytic stability over broad acidic and alkali pH ranges; water solubility and wetting characteristics; compatibility with cationic, anionic, and nonionic surfactants; foam and viscosity enhancement; emulsification capability; good detergency in hard water; and mild toxicology profile and irritation reduction in combination with anionic surfactants. Thus, stability and performance at acidic and alkaline pH extremes are a signature characteristic of amphoteric surfactants [6]. Commercial amphoterics such as dihydroxyethyl alkyl glycinate are considered excellent thickeners for strongly alkaline formulation cleaners and amine oxides enjoy similar properties [7].

12.3 SELECTIVE CHARACTERISTICS

1. *Interface behavior.* The concentration of a water-soluble surfactant at an interface (such as with air) is higher than its concentration in the bulk (aqueous) solution. This accumulation is responsible for a variety of surface phenomena such as wetting characteristics, lowered interfacial tension, and sometimes a change in surface charge.

Amphoteric Surfactants: Synthesis and Production 223

2. *Micelle formation.* The second characteristic of surfactants is their tendency to aggregate and form micelles. In aqueous systems, the size of the surfactant aggregates increases as the concentration increases.

3. *The critical micelle concentration (CMC).* A surfactant forms micelles once its bulk concentration exceeds the CMC. The CMC plays a key role in surfactants chemistry; its value drops with increasing hydrophobicity and accordingly controls the concentration of monomeric surfactants in aqueous systems.

4. *Isoelectric point.* True amphoteric surfactants are characterized by their ability to vary their net charge, according to pH conditions. Compared to betaines, which have a permanent positive charge on the quaternized nitrogen atom, true amphoterics have both a carboxyl- and an amine group that can be protonated. Amphoterics derived from imidazoline can exist in anionic (alkaline conditions), cationic (acidic conditions), or zwitterionic form (around the isoelectric point at pH 5.3).

5. *Surfactancy.* Amphoteric surfactants have the ability to reduce the surface tension of water, as do all other surfactants. The reduction of surface tension depends on the pH of the solution. At pH 6, close to the isoelectric point of most amphoteric surfactants, the net charge of the molecule is zero, which allows for closer packing of surfactant molecules at the interface and produces a great reduction in surface tension.

6. *Foaming.* Amphoterics are excellent foamers and foam stabilizers under a variety of conditions. Foamability is essentially insensitive to water hardness. All of these characteristics have their implications on the designs of the production processes of the various amphoterics in the detergent market.

12.4 AMPHOTERIC SURFACTANTS AND THEIR SYNTHESIS

Amphoteric surfactants can be subdivided into three major classes as follows:

1. Acyl/dialkyl ethylenediamines and derivatives
2. N-alkylamino acids
3. Miscellaneous products
   a. Betaines
   b. Amine oxides

The classical types of amphoteric surfactants are presented in Figure 12.1.

The most important shampoo hair conditioners are the alkylamido alkylamines. As a rule, they are complex mixtures derived from the reaction of alkyl-substituted imidazolines with chloroacetic acid or ethyl acrylate [8]. Similar to the acylated protein derivatives, these amphoteric surfactants exhibit detergency, are compatible with anionic detergents, and reportedly form complex salts with anionics. These complexes reportedly do not sting the eyes and are employed in baby shampoos. These amphoterics in combination with anionic detergents leave the hair conditioned after rinsing.

12.4.1 ACYL/DIALKYL ETHYLENEDIAMINES AND DERIVATIVES

Recent chemical studies have demonstrated that imidazolines are formed during the synthesis of these substances from a long-chain carboxylic acid (or a derivative) with aminoethylethanolamine (AEEA) (NH_2–CH_2–CH_2–NH–CH_2–CH_2–OH) (Figure 12.2).

12.4.1.1 Amphoacetates

Amphoteric surfactants based on carboxymethylation of fatty imidazolines or fatty imidoamines generated by their hydrolysis are well established as extremely mild surfactants [9].

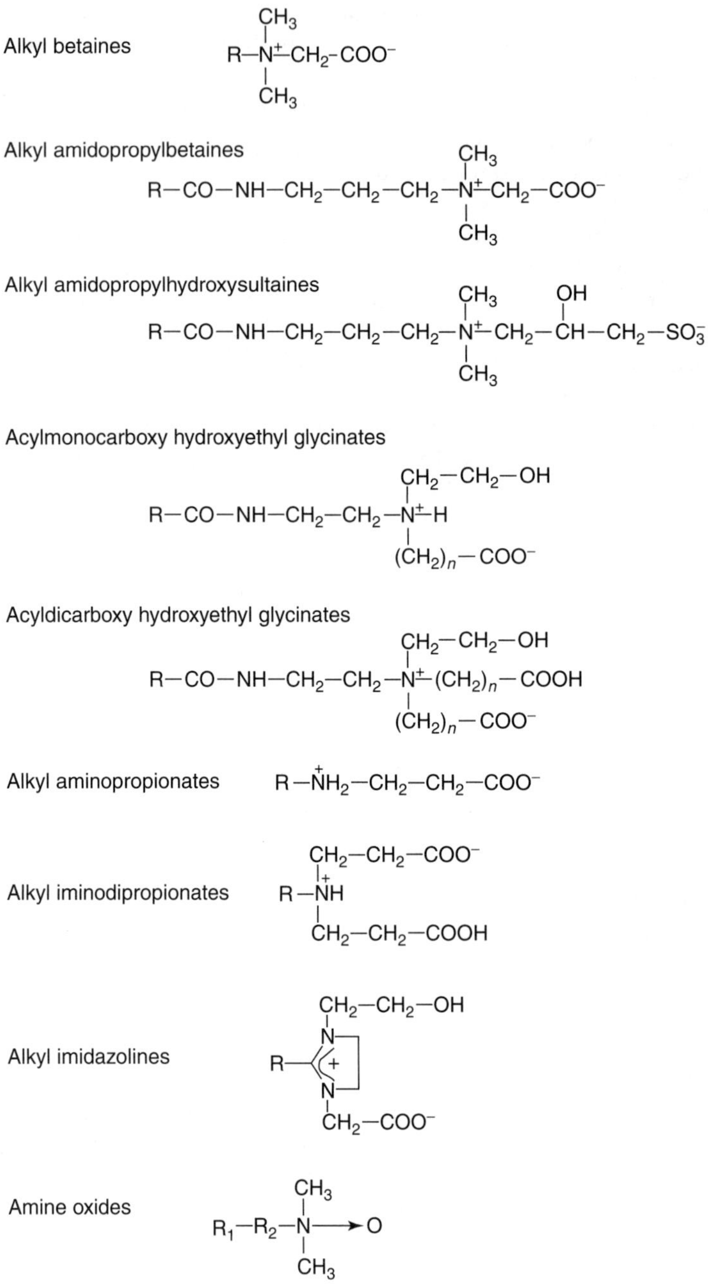

Handbook of Detergents/Part F: Production

Alkyl betaines

$$R-\overset{\overset{\displaystyle CH_3}{|}}{\underset{\underset{\displaystyle CH_3}{|}}{N^{\pm}}}-CH_2-COO^-$$

Alkyl amidopropylbetaines

$$R-CO-NH-CH_2-CH_2-CH_2-\overset{\overset{\displaystyle CH_3}{|}}{\underset{\underset{\displaystyle CH_3}{|}}{N^{\pm}}}CH_2-COO^-$$

Alkyl amidopropylhydroxysultaines

$$R-CO-NH-CH_2-CH_2-CH_2-\overset{\overset{\displaystyle CH_3}{|}}{\underset{\underset{\displaystyle CH_3}{|}}{N^{\pm}}}CH_2-\overset{\overset{\displaystyle OH}{|}}{CH}-CH_2-SO_3^-$$

Acylmonocarboxy hydroxyethyl glycinates

$$R-CO-NH-CH_2-CH_2-\overset{\overset{\displaystyle CH_2-CH_2-OH}{|}}{\underset{\underset{\displaystyle (CH_2)_n-COO^-}{|}}{N^{\pm}}}H$$

Acyldicarboxy hydroxyethyl glycinates

$$R-CO-NH-CH_2-CH_2-\overset{\overset{\displaystyle CH_2-CH_2-OH}{|}}{\underset{\underset{\displaystyle (CH_2)_n-COO^-}{|}}{N^{\pm}}}(CH_2)_n-COOH$$

Alkyl aminopropionates

$$R-\overset{+}{N}H_2-CH_2-CH_2-COO^-$$

Alkyl iminodipropionates

$$R-\overset{\overset{\displaystyle CH_2-CH_2-COO^-}{|+}}{\underset{\underset{\displaystyle CH_2-CH_2-COOH}{|}}{N}}H$$

Alkyl imidazolines

$$R-\underset{N}{\overset{N}{\diagdown}}\overset{+}{\diagup}\quad\begin{array}{c}CH_2-CH_2-OH\\ \\CH_2-COO^-\end{array}$$

Amine oxides

$$R_1-R_2-\overset{\overset{\displaystyle CH_3}{|}}{\underset{\underset{\displaystyle CH_3}{|}}{N}}\longrightarrow O$$

FIGURE 12.1 Amphoteric surfactants used in cosmetics.

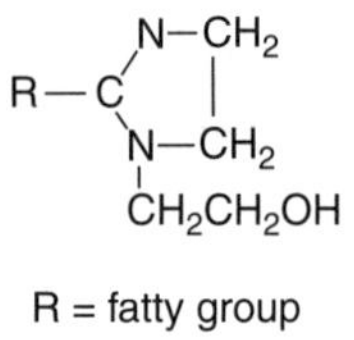

$$R-C\underset{\underset{\underset{\displaystyle CH_2CH_2OH}{|}}{N-CH_2}}{\overset{N-CH_2}{\diagup}}$$

R = fatty group

FIGURE 12.2 Structure for fatty imidazoline. (From Manheimer, H.S., U.S. Patent 2,2773,068, 1956.)

Amphoteric Surfactants: Synthesis and Production   **225**

Although most of the properties and applications of these classes of surfactants are reported widely in the literature, the actual chemistry and composition of commercial products is not well understood.

Synthesis of imidazolinium amphoteric surfactants was reported in a patent awarded to Manheimer in 1950 [10], in which 1-(2-hydroxyethyl)-2-alkyl-2-imidazoline (hereafter called imidazoline) was reacted with sodium monochloroacetate (SMCA). The Miranol company was the first to engage in large-scale production of these products in 1947, and it was around that time when Miranol's products came to be recognized for their mildness and ability to reduce the irritation of anionic surfactants.

The initial reports for these products did not recognize that the imidazoline ring opens during the carboxymethylation step [11]. Recently, the actual structures of these products have been identified through analytical work, based on degradation of commercial products followed by derivatization and various chromatographic techniques [12,13].

The synthesis of amphoacetates and amphodiacetates as described in the early Manheimer patents consisted of two distinct steps: synthesis of the hydroxyethyl-imidazoline and carboxymethylation of the imidazoline with SMCA. In the first step, the fatty acid or the corresponding ester was condensed with AEEA at an elevated temperature and reduced pressure (Figures 12.3a through 12.3c). The reaction proceeded stepwise through the amide state (Figure 12.3, b1 and b2), followed by ring closure. The main component detected after this reaction was imidazoline (Figure 12.3c) with traces of noncyclic amidoamines.

Some by-products may be formed (Figures 12.3d through 12.3f) depending on the reaction conditions. The diamides exhibiting high melting points and low water solubility (Figures 12.3d and 12.3e) need to be minimized to get a final product of high clarity and good storage stability. The symmetrical diamide (Figure 12.3e) is derived from ethylenediamine that may be present in AEEA and can be eliminated by using high-quality raw materials. In contrast, the asymmetrical amide (Figure 12.3d) is formed by the reaction of the primary raw materials and may be minimized by using suitable reaction conditions and maintaining process control. Formation of the amidoesters (Figure 12.3f) is observed only during the intermediate stages of synthesis and is not typically present in the finished imidazoline.

During the second step of amphoacetate or amphodiacetate preparation, the imidazoline (Figure 12.3c) is reacted with SMCA in an aqueous medium. This reaction is carried out under alkaline conditions at moderate temperature. According to the literature, the cocoamphoacetate is produced by the reaction of equimolar quantities of SMCA and imidazoline. Cocoamphodiacetate can also be produced by using the reactants in a 2:1 ratio (Table 12.1). It is worth noting that in water and under alkaline conditions, imidazoline undergoes hydrolysis and generates a mixture of amidoamines (Figure 12.3; b1 and b2). Although the formation of two different amidoamines is possible, it was established that the linear amidoamine b2 (Figure 12.3) is formed under highly alkaline conditions (pH >10). This tertiary amidoamine (b2) may be formed in variable amounts depending on reaction conditions, but it is typically a minor component. The hydrolysis of the imidazoline ring takes place to produce high yields of the linear amidoamine b2 [14]. Noticeably, under typical carboxymethylation conditions, SMCA may hydrolyze to form sodium glycolate (see Figure 12.3i), which may react further with SMCA to form the corresponding diglycolate. During the carboxymethylation step at least three main reactions occur concurrently: the opening of the imidazoline ring leading, predominantly, to the linear amidoamine (Figure 12.3b) and the carboxymethylation of the linear amidoamine by reaction with SMCA that, predominantly, produces the compound 3h; the hydrolysis of SMCA to form sodium glycolate (3i); and a possible further reaction of glycolate with this reagent to produce diglycolate, which is typically a minor component. At present, there is no evidence for the direct reaction of imidazoline with SMCA followed by ring opening to produce the final product. It was further established that there was no noticeable change in the amount of the monocarboxymethylated species even when the SMCA to amidoamine ratio was substantially

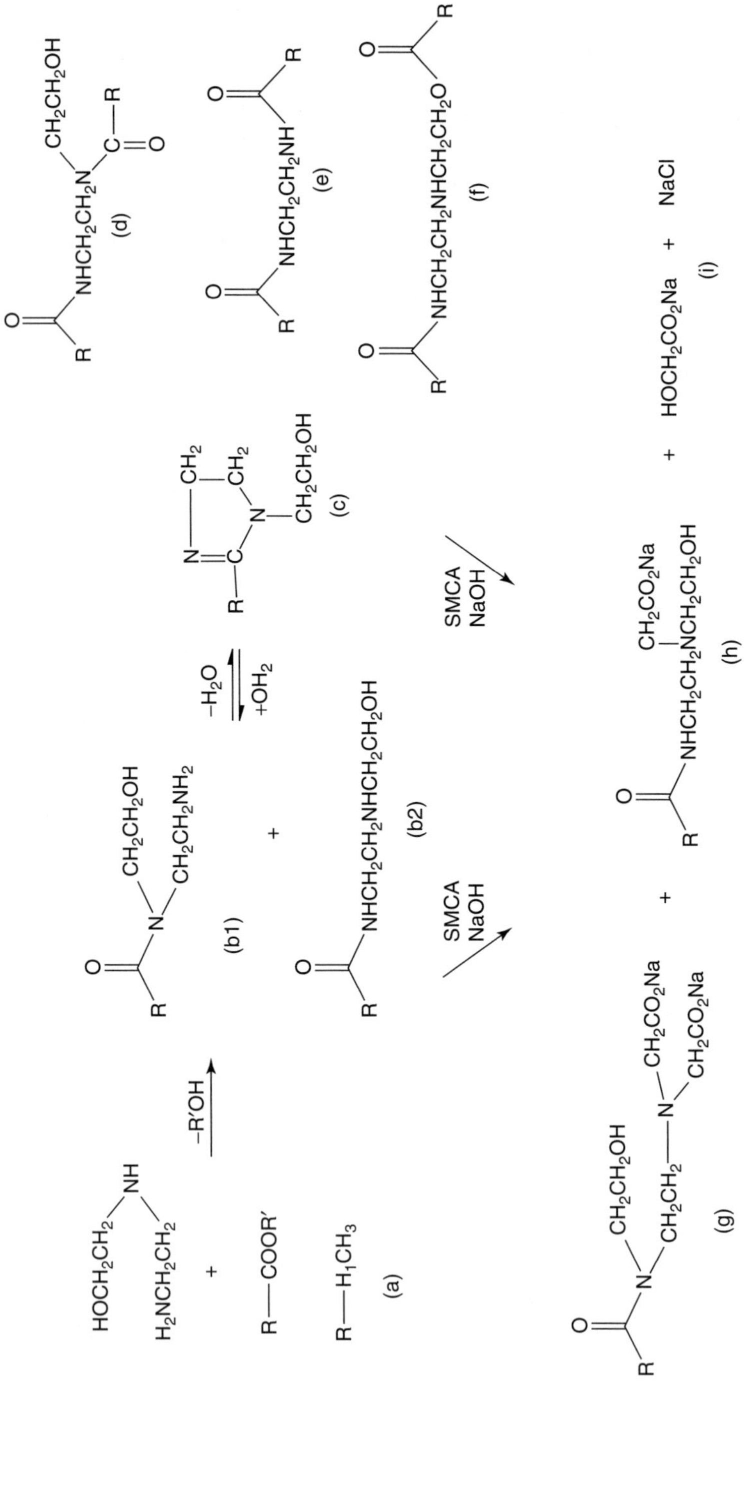

FIGURE 12.3 Major reaction pathways in the formation of imidazoline and its carboxymethylation.

Amphoteric Surfactants: Synthesis and Production

TABLE 12.1
U.S. Patent Citations, 1991–1997

	Betaines	Amphoterics (Others)	Amine Oxides
Consumer or I&I			
Detergent, undifferentiated liquids	15	4	7
Hard-surface cleaners	12	14	5
Light-duty liquids, including dish wash	3	2	3
Fabric and laundry cleaners	5	5	7
Shampoos	3	7	1
Skin cleansers and gels	6	6	0
Hair treatment products and dyes	2	9	0
Oral care and food	4	9	3
Pharmaceutical and clinical	6	10	2
Industrial and agricultural			
Agricultural products	1		
Metallurgy	4	6	0
Photography	3	6	0
Inks	3	0	3
Magnetic recording devices	2	1	0
Corrosion inhibitors	1	1	0
Mining and well treatment	0	6	0
Paper	0	6	0
Flame retardants	0	3	0
Electronics	0	5	0
Latex and rubber	0	8	0
Leather and textile process	0	4	0
Cellulose processing	0	0	13
Waste treatment	0	3	0
Biocides	1	2	2
Fiberglass	0	1	0

increased (from 3:1 to 5:1). Thus, it appears that the reactivity of the monocarboxylated linear amidoamine (3h) is diminished due to the electronic effect and increased steric hindrance around the reaction site, preventing the quaternization from taking place. These results indicate that the dicarboxylated species can be derived only from a structure such as b1 (Figure 12.3) and cannot be formed by further carboxymethylation of the monocarboxymethylated species 3h under these reaction conditions. In brief, most of the cocoamphoacetates are derived from near-equimolar quantities of imidazoline and SMCA containing unreacted amidoamine (3b2) in addition to the active species (3h) and by-products. This is due to concurrent hydrolysis of SMCA during the course of the carboxymethylation reaction.

Cocoamphodiacetates, made by using a larger excess of SMCA, typically contained cocoamphoacetate as the main surfactant component with a diminished amount of the amidoamine (3b2) and an increased content of glycolate. Thus, the increased amount of SMCA is used predominantly for driving the carboxymethylation reaction to completion with little effect on the structure of the major surfactant species. The typical composition of commercial cocoamphoacetates and cocoamphodiacetates are given in Table 12.2.

12.4.1.2 Amino Propionates

Amino propionates are preferably produced by the addition of methyl acrylate on primary fatty amines [15–17]. Depending on the amount of acrylate added, mono- and diadducts are obtained in

TABLE 12.2
Typical Compositions of Commercial Cocoamphoacetates and Cocoamphodiacetates

	Cocoamphoacetate	Cocoamphodiacetate
Solids (%)	45	50
Sodium chloride (%)	7.5	13
Coco fatty acid (%)	<0.7	<0.7
Glycolates (%)	<4	<8
Mono/di ratio	>9	>9
SMCA (ppm)	Variable	100–2000
Color (Gardner)	<5	<5
Viscosity (cP)	Variable	2–100 K
Actives as % solids	Variable	62

varying ratios. The monoadducts are obtained, primarily, by the heating of carbonate-free alkylamine with 1.1 mol of methyl acrylate at 100°C in an autoclave for several hours. This is followed by removal of the unreacted acrylate by vacuum distillation, and hydrolysis of the ester adduct with either alkali or acid.

12.4.1.3 Carboxyamphoterics

Carboxyamphoteric surfactants based on fatty alkyl imidazolines ("imidazolinium" surfactants) make up a large part of the amphoteric surfactants. The very divergent interpretation of their chemical structure is partly attributable to little-developed analytical procedures in the past, but is also a consequence of special processing methods by different manufacturers. Materials of this surfactant class are based on the imidazolines obtained by the condensation of fatty acids, or their esters, with AEEA. In Figures 12.4 and 12.5, a summary of the synthesis of amphoteric surfactants based on imidazolines [4,8] is presented.

12.4.1.4 Acyl/Dialkyl Ethylenediamine

The most important products of this group are the coco derivatives—cocoamphoglycinate, cocoamphocarboxyglycinate, cocoamphopropionate, and cocoamphocarboxypropionate resulting from complete hydrolyzation of the imidazoline structure during the production process. In the first production step, AEEA is reacted with fatty acid to produce an amide. During this stage, the imidazoline ring is formed. The next step is the reaction with sodium chloroacetate. This leads to the production of a glycinate or, in case of reaction of 2 mol of sodium chloroacetate, the form of carboxyglycinate [18]. As in the case of the betaines, salt (NaCl) is a by-product of the reaction. When acrylic acid is used in the second step, salt-free products, which are the propionates, can be obtained.

12.4.2 N-ALKYLAMINO ACIDS

This group of amphoteric surfactants is derived from various amino acids, and its members do not possess the hydroxyethyl grouping. Alkylation of the primary amino groups of an amino acid leads to secondary and tertiary amines that, as a rule, are more basic than the original primary amine. In addition, some of the alkyl substituents may carry a second amino group that provides an additional basic center.

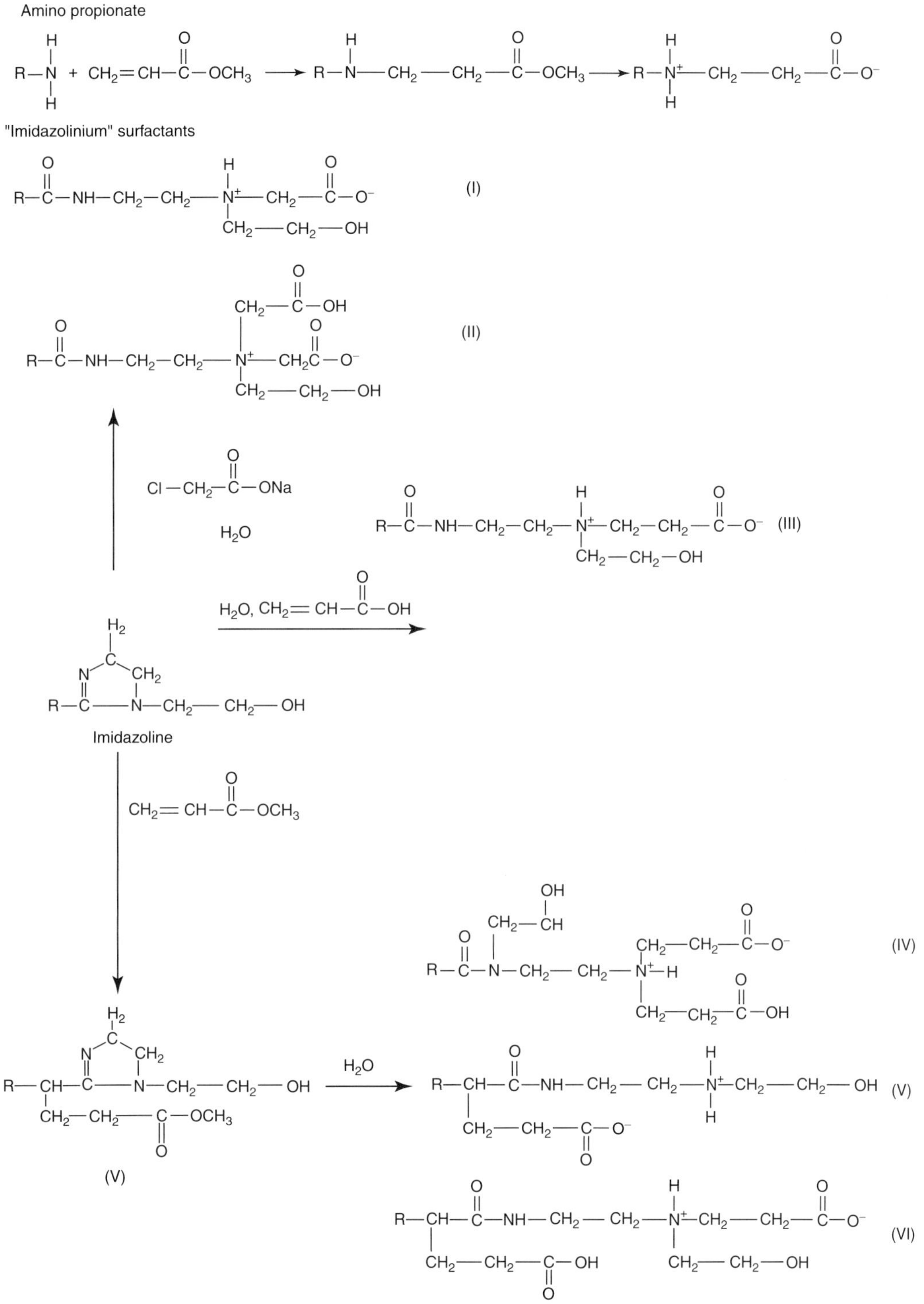

Amphoteric Surfactants: Synthesis and Production

229

FIGURE 12.4 Carboxy amphoterics (alkylamino carboxylic acids). (From Rieger, M.M., *Cosmet. Toil.* 99, 61–67, 1984.)

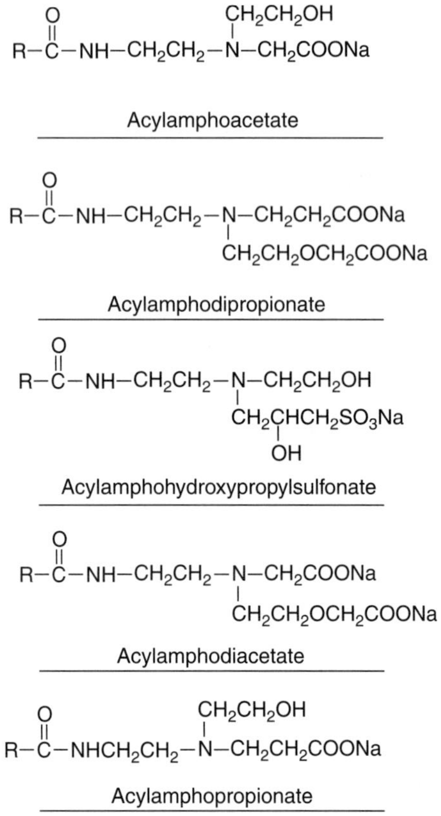

$$R-\overset{\overset{\displaystyle O}{\|}}{C}-NH-CH_2CH_2-\overset{\overset{\displaystyle CH_2CH_2OH}{|}}{N}-CH_2COONa$$

Acylamphoacetate

$$R-\overset{\overset{\displaystyle O}{\|}}{C}-NH-CH_2CH_2-\overset{\overset{\displaystyle N}{|}}{\underset{\displaystyle CH_2CH_2OCH_2COONa}{}}-CH_2CH_2COONa$$

Acylamphodipropionate

$$R-\overset{\overset{\displaystyle O}{\|}}{C}-NH-CH_2CH_2-\overset{\displaystyle N}{\underset{\displaystyle \underset{\displaystyle OH}{|}}{\overset{\displaystyle |}{CH_2CHCH_2SO_3Na}}}-CH_2CH_2OH$$

Acylamphohydroxypropylsulfonate

$$R-\overset{\overset{\displaystyle O}{\|}}{C}-NH-CH_2CH_2-\overset{\displaystyle N}{\underset{\displaystyle CH_2CH_2OCH_2COONa}{|}}-CH_2COONa$$

Acylamphodiacetate

$$R-\overset{\overset{\displaystyle O}{\|}}{C}-NHCH_2CH_2-\overset{\overset{\displaystyle CH_2CH_2OH}{|}}{N}-CH_2CH_2COONa$$

Acylamphopropionate

FIGURE 12.5 Further illustrative structures of imidazolines. (From Rieger, M.M., *Cosmet. Toil.* 99, 61–67, 1984.)

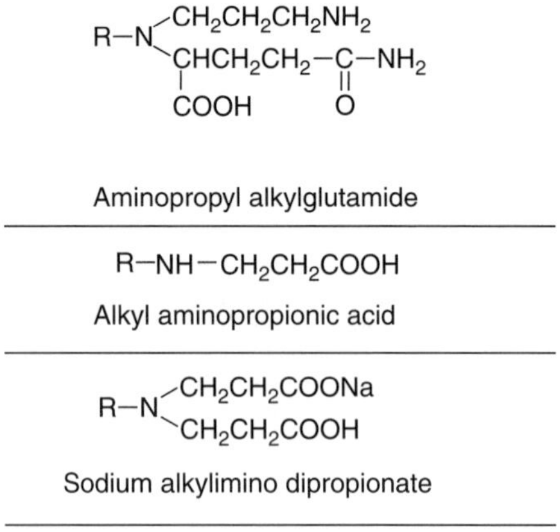

$$R-N\overset{\displaystyle CH_2CH_2CH_2NH_2}{\underset{\displaystyle \underset{\displaystyle COOH}{|}}{CHCH_2CH_2-\overset{\overset{\displaystyle }{\|}}{\underset{\displaystyle O}{C}}-NH_2}}$$

Aminopropyl alkylglutamide

$$R-NH-CH_2CH_2COOH$$

Alkyl aminopropionic acid

$$R-N\overset{\displaystyle CH_2CH_2COONa}{\underset{\displaystyle CH_2CH_2COOH}{}}$$

Sodium alkylimino dipropionate

FIGURE 12.6 Structures of N-alkylamino acids.

Most of the N-alkylamino acids are alkyl derivatives of β-alkaline or of β-*N*(2-carboxyethyl)-alanine. Only a few of them exhibit structures analogous to those of the natural α-amino acids. The compounds in this group are available as free carboxylic acids or more commonly as the corresponding sodium salts. Although these amino acids and their salts are available as solids, they are also distributed in the solution form. Their structures are shown in Figure 12.6.

Amphoteric Surfactants: Synthesis and Production   **231**

12.4.3   MISCELLANEOUS AMPHOTERICS

12.4.3.1   Betaines

Amphoteric surfactants are characterized by a molecular structure containing two different functional groups, with anionic and cationic characters, respectively [19]. Most amphoteric surfactants are able to behave like cationic surfactants in acidic medium, and like anionic surfactants in alkaline medium. However, betaines are different in that they cannot be forced to assume anionic active behavior through an increase in the pH value [20,21]. Figure 12.7 shows structures of the most widely used amphoteric surfactants, as produced, in dependence of the pH value.

The number of chemically conceivable amphoteric surfactant structures is limited in practice to the derivatives of easily accessible and, therefore, economical raw materials. Relative to the amphoteric surfactants with the combination of cationic nitrogen atom and carboxyl groups, the so-called sulfo betaines with the sulfonate group, and also the phosphato betaines with the phosphate group as anionic component, play only a subordinate role.

Starting materials for the preparation of these surfactants are alkylatable N-compounds (long-chain alkylamines) and alkylation reagents such as sodium chloroacetate, acrylic acid, and sodium chloro hydroxypropane sulfonate.

12.4.3.1.1   Alkyl Betaines

Alkyl betaines may be considered homologues of betaine, for example, trimethyl ammonium acetate. They are prepared by condensation of an alkyl dimethylamine with sodium chloroacetate (Figure 12.8). Commercial betaines are usually 30% active products containing ~6% sodium chloride, which is a by-product of the reaction. Depending on the pH, the alkyl betaines can be cationic or anionic surfactants.

12.4.3.1.2   Alkylamido Betaines

Alkylamido betaines are the most common of the betaine types. They have become the most important type of secondary surfactants, especially cocamidopropyl betaine (CAPB) [22], which can also

FIGURE 12.7   Structures of amphoteric surfactants.

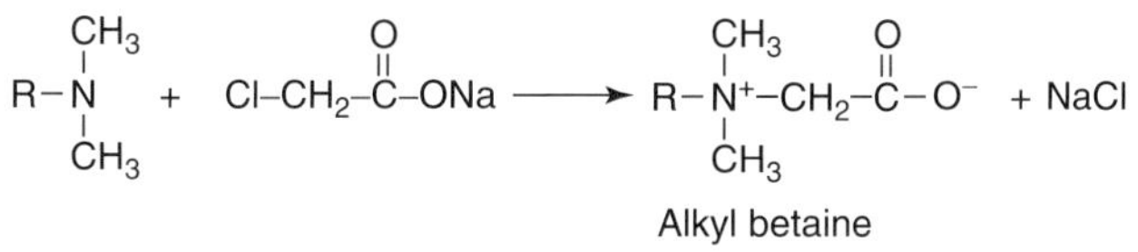

FIGURE 12.8   Alkyl betaine reaction.

be referred to as coco fatty acid amidopropyl betaine [23,24]. Since its introduction in the 1960s, CAPB has become an essential secondary surfactant.

12.4.3.1.2.1 Amide Formation and Purification

The methods for producing fatty acid amidopropyl betaines basically follow a scheme that has shown validity for some time. Forming the fatty acid amide is the first step (Figure 12.9); carboxymethylating the amide is the second step (Figure 12.10).

For CAPB, the first step is a reaction of 3-aminopropyl-dimethylamine (DMAPA) with either fatty acids, fatty acid methyl esters, or directly with natural fats (fatty acid glycerin esters). The predominant source oils used determine the fatty acid composition of the betaine, which corresponds to that of the oil. Table 12.3 outlines the typical fatty acid composition of CAPB.

To use DMAPA very efficiently and avoid large amounts of liquid waste, one should recycle the amine, which is partly distilled from the reaction medium together with the methanol or water formed, during the transformation of the fatty acid methyl esters or the fatty acids.

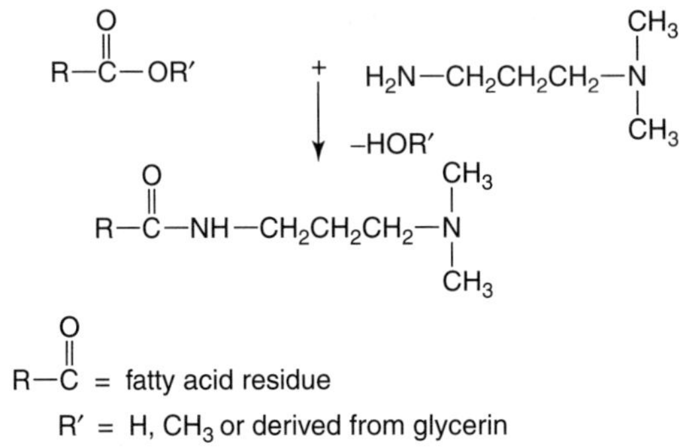

FIGURE 12.9 Betaine production, amidation.

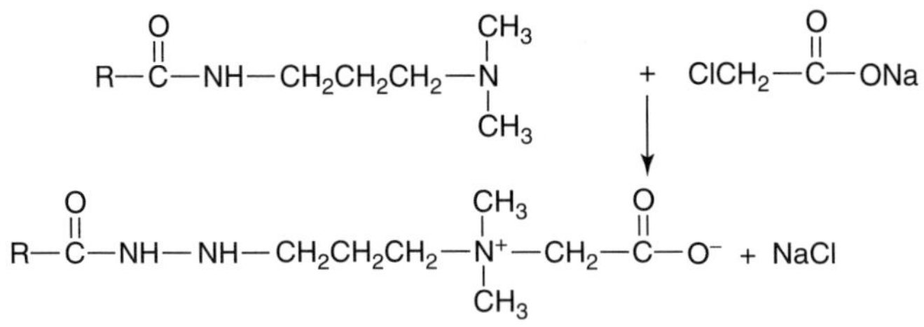

FIGURE 12.10 Betaine production, carboxymethylation.

TABLE 12.3

Fatty Acid Composition of CAPB

Fatty Acid (Chain Length)	Weight % (Approximate Values)
8	7
10	6
12	49
14	19
16	9
18	10

Amphoteric Surfactants: Synthesis and Production

When the CAPB synthesis starts from a fat source, glycerin forms as a by-product. As glycerin has no negative effect on most applications and may even be desired, a glycerin content of 2–3% is normal in betaine solutions. If desired, some glycerin can be removed from the amidoamine by phase separation, leaving a glycerin content of ~1% in the final betaine solution.

The excess DMAPA used in the amidation reaction can be distilled out of the amidoamine, leaving a residual of ~0.05%. All amidoamines, whether derived from fatty acid esters or fatty acids, contain small amounts of nonamidated fatty acids that are carried over into the betaine.

Betaines with specialized fatty acid compositions can also be made, including lauric acid amidopropyl betaine, caprylic/capric acid, amidopropyl betaine, and betaines from topped fatty acids that contain none or only very few short-chain components, typically the C_8 or C_{10} fractions.

12.4.3.1.2.2 Carboxymethylation

The amidoamine is carboxymethylated in an aqueous medium with chloroacetic acid or its sodium salt. The reaction leads directly to the betaine, normally obtained as a 30% aqueous solution. Sodium chloride, formed as a by-product, is present in most betaine solutions at ~5%. Apart from the reaction with the tertiary nitrogen atom, the chloroacetic acid can also be hydrolyzed in a side reaction to form glycolic acid. The glycolic acid content can reach levels of more than 1% in the betaine solution; however, the level is typically considerably lower (~0.1%).

During the past few years, research on the production of betaines has primarily concentrated on improving the carboxymethylation process. These efforts have focused on manufacturing purer products and minimizing by-products [25]. Other efforts have focused on increasing concentration levels or reducing the water content in betaine solutions; all these efforts primarily involve fine-tuning the carboxymethylation step.

12.4.3.1.2.3 Minimizing Contaminants

As the tertiary amines are consumed during carboxymethylation, the pH value falls considerably. This slows the reaction rate proportionally because free amines are also removed from the reaction medium by protonation as well as carboxymethylation. By maintaining an alkaline pH throughout the reaction, carboxymethylation can be carried out faster and more completely, with less amidoamine in the final product [26]. This technique can give amidoamine contents of <0.3% in the betaine. Some types of CAPB have higher amidoamine contents, up to 3%, for special applications.

Chloroacetic acid and the dichloroacetic acid present in the reagent used in small amounts are both unwanted by-products in betaines (as well as in other zwitterionic or amphoteric surfactants such as amphoglycinates) because of their toxicity. Chloroacetic acid is almost completely depleted during the carboxymethylation reaction, but it is almost inert under the typical reaction conditions. By submitting betaines to additional posttreatment steps, the residual chloroacetic acids can be reduced. This can be achieved either by reacting at alkaline pH or by additional treatment with ammonia or amino acids that reduces the amount of monochloroacetic acid [27].

Using sulfonating reagents is another possible way to minimize the chloroacetic acids. Finally, both mono- and dichloroacetic acid can be hydrolytically decomposed simply by exposure to high temperatures (>120°C). Levels of <5 ppm chloroacetic acid and <10 ppm dichloroacetic acid can be thus achieved. There are, however, very few products in the market that fulfill CAPB purity requirements of amidoamine content <0.3%, monochloroacetic acid content <5 ppm, and dichloroacetic acid content <10 ppm [22].

Numerous attempts have been made to obtain flowable betaine solutions with increased active matter; in some cases, by adding other surfactants [28–32]. Other ways to achieve highly concentrated betaines include adding solvents or using additional salts not normally contained in betaine solutions, such as sodium citrate, trimethylglycine (natural betaine: methanaminium, 1-carboxy-*N*, *N*, *N*-trimethyl-, inner salt [33], or nitrilotriacetate).

A relatively simple way to obtain more highly concentrated betaine solutions is to adjust the free fatty acid content. By adding small amounts of fatty acid to the betaine with the amidoamine solution, one can produce betaine contents of 34–36% and detergent-active substances, including fatty acids, of 36–38%.

By spraydrying aqueous betaine solutions, it is possible to obtain highly concentrated betaine products, typically consisting of 80–85% fatty acid amidopropyl betaine, 13–15% sodium chloride, and 0.3–3.0% water. Spraydrying typically uses betaines with low C_8/C_{10} alkyl components (Venzmer, J., Personal communication, 1996).

The details of betaine production demonstrate that marketed solutions will contain by-products. Determining the types and amounts of such incidental components helps us characterize the betaine's quality. There is considerable effort within the industry to develop materials having reduced impurities that would otherwise cause undesirable physiological side effects. According to this procedure, a fatty acid amide dialkylamine was quaternized with carboxylic acids, or other salts, in an aqueous solution until levels of organically bound chlorine are reduced to ≤ 10 ppm [34].

12.4.3.1.3 Imidazolinium Betaines

Imidazolinium betaines were initially obtained from the reaction of fatty alkyl imidazolines with sodium chloroacetate [10,11]. Their preparation in this manner, however, is not practiced. They can be prepared by the reaction of imidazoline with acrylic acid (Figure 12.11). The corresponding carboxyl ethyl betaine may be obtained, in ~65% yield, by the electrophilic addition of acrylic acid on tertiary amine [35] (Figure 12.12).

The addition of acrylic acid to high-percentage imidazolines improves the existence of an imidazolinium betaine in the model substance 1-hydroxyethyl-2-heptyl-imidazolinium-3-ethyl carboxylate through purification by vacuum distillation [36]. The preparation is carried out by heating the imidazoline derivative for several hours with a 5% excess of acrylic acid in the absence of water. More than 65% of the imidazolinium structure is retained depending on the length of the fatty alkyl chain. Open-chain reaction products are also formed, by application of excess acrylic acid, depending on the reaction temperature.

12.4.3.1.4 Sulfo Betaines

The classic preparation method of sulfo betaines utilizes the conversion of tertiary amines with propane sultone. Following the discovery that propane sultone is carcinogenic, this reaction is of limited

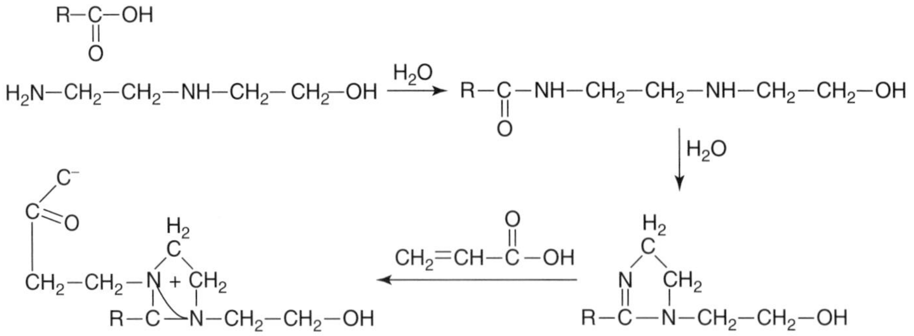

FIGURE 12.11 Imidazolinium betaine formation.

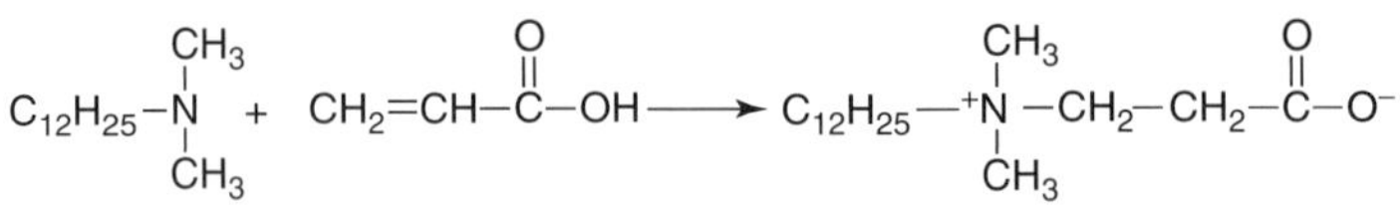

FIGURE 12.12 Carboxyethyl betaine reaction.

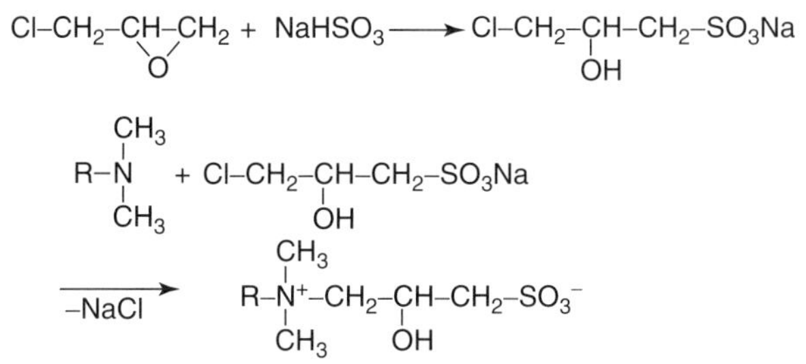

Amphoteric Surfactants: Synthesis and Production

FIGURE 12.13 Hydroxysulfo betaine reaction.

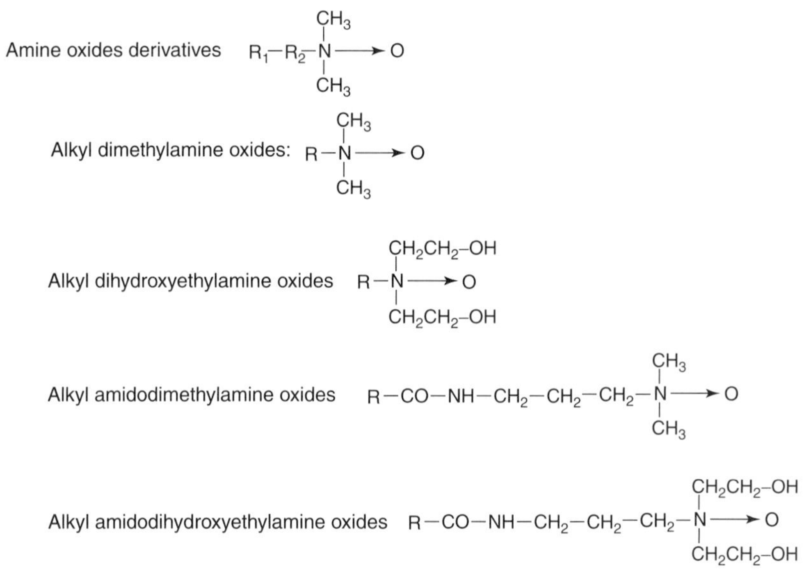

FIGURE 12.14 Types of amine oxides.

importance today. Hydroxysulfo betaines (lv), (lvi), and (lvii) are obtained from tertiary amines and chloro hydroxypropane sulfonic acid [37–39]. The latter compound is obtained from epichlorohydrine (which is also toxic) by the reaction with sodium hydrogen sulfite: The preparation is carried out by heating an aqueous solution of equivalent amounts of tertiary amine and chloro hydroxypropane sulfonate to 80–130°C in an autoclave if necessary. The reactions involved are shown in Figure 12.13.

12.4.3.2 Amine Oxides

Unlike betaines, amine oxides are never anionic. However, they do show cationic or nonionic behavior depending on the pH, and therefore behave quite similarly [40]. Amine oxides are produced by the oxidation of tertiary alkylamines; for example, dimethyl or ethoxylated amines with hydrogen peroxide. They are colorless liquids and free from salt (NaCl). As such, they are excellent foamers. The various types of amine oxides are shown in Figure 12.14.

As with betaines and other amphoterics, efforts continue in the development of amine oxides with reduced levels of residual impurities. A process for preparing long monoalkyl-chain amine oxides with low nitrite and nitrosamine levels [41] and another process [42] involving the oxidation of C_{12}–C_{13} dimethylamine with hydrogen peroxide in the presence of malic acid and diethylenetriaminepentaacetate (DTPA) have been described. The resulting amine oxide contains (the undesired) nitrite levels below 1 ppm [42].

12.5 CONCLUSION

Amphoteric surfactants are a class of chemicals whose functionality and utility to the industry is showing a definite growth, finding increased amounts used in personal products because they allow the manufacturer to use a great number of versatile feedstocks/raw materials to produce formulations that are both functional and safe [43]. The wide range of chemistries that are encompassed by the class of surfactants afford product-development chemists the opportunity to tailor the performance of their products for specific applications. This is evident by the broad range of industries interested in these technologies.

REFERENCES

1. Price, S. N. C. 50 years of surfactants. *Cosmet. Toil.* 110: 49–66, 1995.
2. Zeidler, U., G. Reese. *In-vitro*-Test zur Hautvertraglichkeit von Tensiden. *Arztliche Kosmetol.* 13: 39–45, 1983.
3. Rieger, M. Surfactant interactions with skin. *Cosmet. Toil.* 110: 31–50, 1995.
4. Fiedler, H. P. *Seifen-le-Fette-Wachse.* 110: 512–515, 1984.
5. Fox, C. Shampoo components—1985. *Cosmet. Toil.* 100: 31–46, 1985.
6. Amphoterics, betaines, amine oxides, nonionics. *McIntyre* 6, Undated.
7. Sauer, J. D. Amine oxides: storage and handling. In: J. M. Richmond, ed. *Cationic Surfactants Organic Chemistry.* New York: Marcel Dekker, 1990, pp. 284–286.
8. Rieger, M. M. The structure of amphoterics derived from imidazoline. *Cosmet. Toil.* 99: 61–67, 1984.
9. Final report on the safety assessments of cocamphoacetate, cocamphopropionate, cocoamphodiacetate, and cocamphodipropiojnate. *J. Am. Coll. Toxicol.* 9: 2, 1990.
10. Manheimer, H. S., J. J. McCabe. U.S. Patent 2,528,378, 1950.
11. Manheimer, H. S. U.S. Patent 2,2773,068, 1956.
12. Lomax, E. Analysis of amphoteric surfactants in recent developments in the analysis of surfactants, In: M. R. Porter, ed. *Handbook of Surfactants.* London: Elsevier Applied Science, 1991.
13. Takano, S., K. J. Tsuji. *Am. Oil Chem. Soc.* 60: 1798–1807, 1983.
14. Zougsh, L., Z. Zhuangyu. *Tenside Surf. Deterg.* 31: 128–131, 1994.
15. General Mills, Inc. U.S. Patent 2,468,012, 1945.
16. General Mills, Inc. U.S. Patent 2,810,752, 1957.
17. General Mills, Inc. U.S. Patent 2,814,643, 1957.
18. Ellis, P. R., P. J. Derian, R. Vokov. Amphoteric surfactants—the next generation. *Euro Cosmet.* 2: 14–16, 1994.
19. Bluestein, B. R., C. L. Hilton. *Amphoteric Surfactants.* Vol. 12, New York: Marcel Dekker, 1982.
20. Moore, C. D. Ampholytic surface active agents. *J. Soc. Cosmet. Chem.* 11: 13–25, 1960.
21. Ploong, U. *Seifen-le-Fette-Wachse.* 108: 373, 1982.
22. Grüning, B., H. Küseborn, L. Leidreiter. Cocamidopropyl betaine. *Cosmet. Toil.* 112: 67–76, 1997.
23. Hoffman, K. Jahrbuch fur den Parktiker. *Verlag F Chem. Ind. H. Ziolkowsky,* 5, 1973.
24. Hüttinger, R. *Goldschmidt Informiert.* 32: 39, 1975.
25. Goldschmidt, Th. U.S. Patent 5,470,992, 1995.
26. Bade, V. DE Patent 29.26.479, 1992.
27. Uphues, G., U. Ploog, K. Bishof, K. Kenar, P. Saldek. DE Patent 39.39.264, 1992.
28. Bade, V. DE Patent 36.13.944.
29. Bade, V. DE Patent 37.26.322.
30. Bade, V. DE Patent 38.26.654.
31. Messenger, T., D. E. Mather, B. M. Phillips. U.S. Patent 4,243,549, 1981.
32. Uphues, G., P. Neumann, A. Behler. DE Patent 42.32.157.
33. Hamann, J., H. J. Kyhle, K. Start, W. Wehner. EU Patent 656 346.
34. Goldschmidt, Th. U.S. Patent 5,345,906, 1994.
35. LeBerre, A., A. Delacroix, Synthesis of 1-(2-sulfoethyl)-pyridinium betaine. *Bull. Soc. Chim. Fr.* 7/8: 2404, 1973.
36. Hitz, H., W. Schnfer. *Seifen-le-Fette-Wachse.* 109: 20, 1983.

Amphoteric Surfactants: Synthesis and Production

37. Shell, British Patent 1541427, 1979.
38. Fernley, G. W. *J. Am. Oil Chem. Soc.* 55: 98–103, 1978.
39. Shell, DOS Patent 2431031, 1974.
40. Roerig, H., R. Stephan. Amine oxides and their applications. *La Rivista Italiana Delle Sostanze Grasse.* 68: 317–321, 1991.
41. The Procter & Gamble Company. U.S. Patent 5,583,258, 1996.
42. The Procter & Gamble Company. U.S. Patent 5,498,373, 1996.
43. *Technical and Product Development Data.* Dayton, NJ: Miranol, 1987, p. 13.

13 Production of Alkanolamides, Alkylpolyglucosides, Alkylsulfosuccinates, and Alkylglucamides

Bernhard Gutsche and Ansgar Behler

CONTENTS

13.1 Fatty Alkanolamides 239
13.2 Alkylpolyglucosides 241
13.3 Fatty Acid Glucamides 242
13.4 Sulfosuccinates 244
13.5 Summary 245
References 246

13.1 FATTY ALKANOLAMIDES

The alkanolamides are condensates of fatty acids and alkanolamines; they are nonionic surfactants. Based on different alkanolamines such as monoethanolamine (MEA), diethanolamines (DEA), and monoisopropylamine (MIPA), and derivatives from oils and fats, they are "hybrids" from petrochemistry and from renewable oleochemistry.

Fatty alkanolamides have many different applications, such as detergents, foam boosters, viscosity builders, and stabilizers. These are mostly in liquid formulations in applications such as dishwashing and shampoos. The fatty alkanolamides have also been evaluated as alternatives to nonylphenol ethoxylates [1].

The total amount of fatty alkanolamides produced in 1989 was 107,000 t declining to ~60,000 metric tons in 2001. The prediction for the world market for 2006 was around 55,000 t, declining especially in personal care applications (about −5 to −7%) based on calculations of Bizzari [2].

The main reason for the declining quantities in the total is new formulations in the personal care market, in which, especially in leave-on products, alkanolamides from MEA are limited to 10% (for details, see Refs 3 and 4).

Different pathways based on different raw material sources are used in technical processes. The original route as described in patents by W. Krichevsky in 1937 is based on the direct condensation of fatty acids with alkanolamines in excess. For the synthesis of diethanol amides, the reaction scheme is shown as follows:

$$RCOH + HN(CH_2CH_2OH)_2 \rightleftharpoons RCN(CH_2CH_2OH)_2 + H_2O$$

The fatty acid is condensated with the alkanolamines without a catalyst. To get high yields of the product, separation of water is essential. To avoid the formation of by-products or even the decomposition of the reactants, the reaction temperature is limited to 430 K. An improved process based on fatty acid methyl esters and stoichiometric amounts of amines was proposed by E. Meade in 1949.

$$\text{RCOCH}_3 + \text{HN(CH}_2\text{CH}_2\text{OH)}_2 \underset{\text{catalyst}}{\rightleftarrows} \text{RCN(CH}_2\text{CH}_2\text{OH)}_2 + \text{CH}_3\text{OH}$$

Using fractionated methyl ester and alkaline catalyst, a "transamidation" with no excess of one component is possible—the equilibrium is shifted to the alkanolamide by continuously distilling off the by-product methanol, due to its high vapour pressure this is much easier than to separate the water formed in the Kritchevsky process.

The reaction temperature of the Meade process is ~360–390 K; a faster reaction can be achieved using a vacuum ramp during conversion. Because of the better quality of these amides, they are often called as *super-amides*.

Farris [5] reported different qualities for both pathways; the product yield from fatty acid is only ~55%; the main by-products are free DEA (22%) and amine soap and amino-amine (10% each). About 90% yield was reached in the pathway based on methyl ester; the main by-products are free DEA (5%) and amide ester (4%). By using triglycerides directly, still 80% yield can be reached.

These data should be taken today just as an indication for process economics; better reaction and separation technology increased the quality and economics of fatty alkanolamide production (Figure 13.1).

An example for a new process improvement in the past years was patented by Oftring et al. (BASF) [6]. They developed an improved extraction process for the transamidation with triglycer-

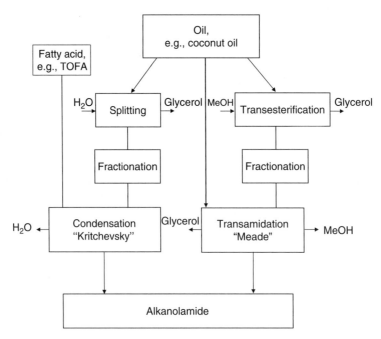

FIGURE 13.1 Alkanolamides: pathways from different raw material to final products (TOFA—Tall oil fatty acid).

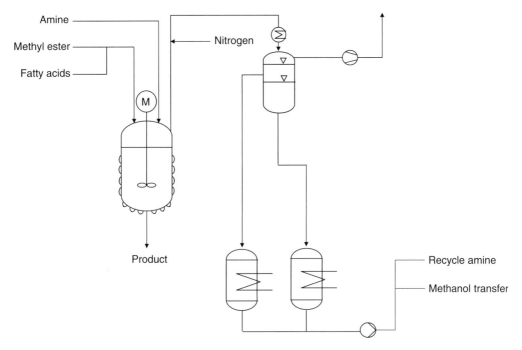

FIGURE 13.2 Alkanolamide production plant. (From Adami I., Ballestra S.p.A. company brochure and additional information, September 2005.)

ides. The glycerol formed during the reaction has to be removed from the organic phase to get high yields in fatty alkanol amides.

The fatty acid route (using 2 mol of amine per mol of fatty acid) is still in use because of the excellent solubility in water and organic solvents; the product is used in low concentrations in aqueous media as a textile detergent, shampoo ingredient, rust inhibitors, and fuel oil additive.

The superamides are very good thickeners for liquid detergents and shampoos containing sodium lauryl sulfate [7] or ether sulfate [8]. Alkanolamides are very often produced using standardized multipurpose reactor configuration. An example is shown in Figure 13.2 [9].

13.2 ALKYLPOLYGLUCOSIDES

The development of detergents based on the combination of fats and oils as the hydrophobic part and carbohydrates as the hydrophilic part gained importance with the discussion to reduce the petrochemical (fossil) resources in detergent production [10]. Important criteria for the use of these derivatives are price, quality, and availability of the raw material, and the processing costs [11].

One successful solution to overcome the problem of nonselective derivatization of carbohydrates was the glucosidation found by Fischer [12]. The technical process is an adaptation of Fischer's synthesis using long-chain fatty alcohols and a controlled degree of polymerization. So alkyl polyglycosides (APG) with different application properties can be produced.

Short-chain APGs (alcohol chain length C8/C10) are used in bathroom/toilet and window cleaners. Medium-chain (C12/C14) APGs have their main application field in detergents for dishwashing, laundry, and personal care [11]. For a life-cycle inventory of APGs, see in Ref. 13. For the technical synthesis of APGs, two pathways have been studied in the pilot plant stage, the direct synthesis and the two-step transacetalization process. The butanolysis followed by a transacetalization with fatty alcohol can be used for many carbohydrates from starch to glucose or glucose monohydrate. The

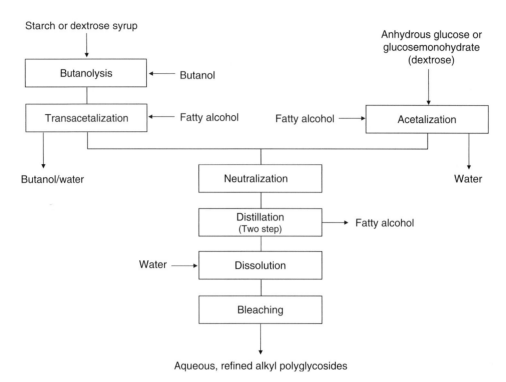

FIGURE 13.3 Pathways for APG production. (From Eskuchen, R. and Nitsche, M., in *Alkyl Polyglycosides*, Hill, K. H., von Rybinski, W., and Stoll, G. (Eds.), VCH, Weinheim, 1996. With permission.)

direct glycosidation starts with glucose or its monohydrate. The simplified flow diagram for both the pathways is shown in Figure 13.3.

Figure 13.3 indicates the more complex process, which starts from starch or dextrose syrup compared to the direct acetalization using anhydrous glucose or glucose monohydrate. The key problem for the carbohydrate source is the increasing price from starch to glucose monohydrate, so an optimization of total manufacturing costs is necessary between plant and processing costs on one hand and carbohydrate costs on the other. The relevance of APGs might even become more important, expecting an increase in prices for mineral oil–based raw materials.

The flow sheet of the technical process is shown in Figure 13.4. For a comprehensive description of critical steps of this technology, see Refs. 14 and 15; the main points are control of the solid–liquid reaction between the glucose and the fatty alcohol with acidic catalysis avoiding a fast polymerization of glucose, the good water removal, the economic separation and recycle of the excess fatty alcohol, and the treatment of the highly viscous APG melt to get the desired product quality with low polydextrose content.

13.3 FATTY ACID GLUCAMIDES

The synthesis to produce fatty acid glucamides involves the reaction of glucose with methylamine, under reductive conditions, to form the corresponding *N*-methylglucamine. In the subsequent reaction step, this intermediate is converted with fatty acid methyl ester to the corresponding fatty acid amide. When compared with the APG, fatty acid glucamides are composed of only a single carbohydrate molecule attached to the fatty acid chain (Figure 13.5).

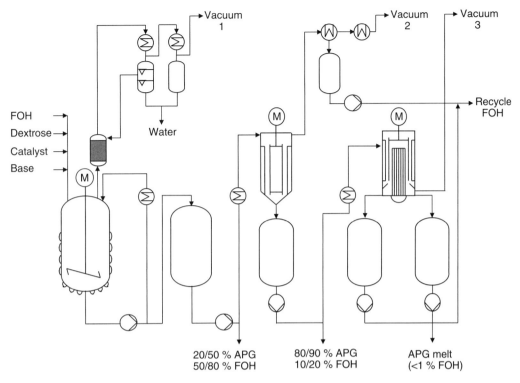

FIGURE 13.4 APG production plant (FOH—Fatty alcohol) (From Eskuchen, R. and Nitsche, M., in *Alkyl Polyglycosides*, Hill, K. H., von Rybinski, W., and Stoll, G. (Eds.), VCH, Weinheim, 1996. With permission.)

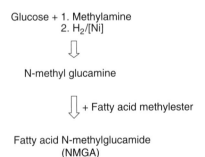

FIGURE 13.5 Chemical pathway for fatty glucamide production. (From Eskuchen, R. and Nitsche, M., in *Alkyl Polyglycosides*, Hill, K. H., von Rybinski, W., and Stoll, G. (Eds.), VCH, Weinheim, 1996. With permission.)

So fatty acid glucamides are less soluble and tend to crystallize more easily from aqueous solutions. Figure 13.6 shows the manufacturing scheme for the production of fatty acid glucamides.

To avoid significant amounts of unreacted *N*-methylglucamine, which could be considered as potential precursors for nitrosamines, Procter & Gamble developed an optional reaction with acetic anhydride in the finished product. Free secondary amines can be acetylated in this step, and the resulting acetates can remain in the final product [16,17].

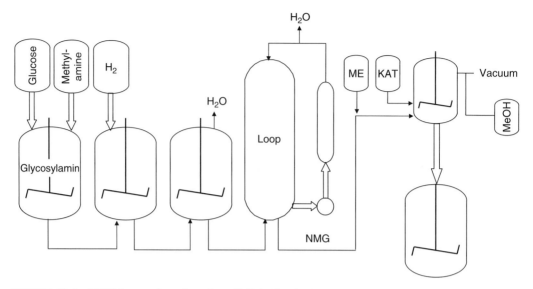

FIGURE 13.6 NMGA manufacturing plant (KAT—Catalyst). (From Ruback, W., Schmidt, S., and van Bekkum (Eds.), *Carbohydrates as Organic Raw Materials III*, VCH, Weinheim, 1996.)

13.4 SULFOSUCCINATES

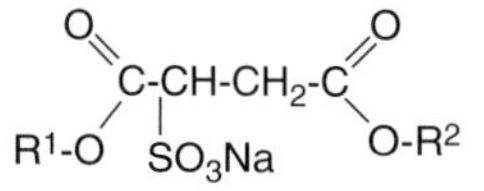

Sulfosuccinates (sulfosuccinic acid esters) are anionic surfactants that are accessible on the basis of maleic acid anhydride. One can distinguish mono- and dialkyl esters of the sulfosuccinic acid (Figure 13.7).

FIGURE 13.7 Dialkyl ester of sulfosuccinic acid.

Both mono- and diesters are obtained in a two-step process. In the first reaction step, maleic acid anhydride is esterified with compounds containing hydroxyl groups to mono- or diester. Both liquid and solid maleic acid anhydrides are used in manufacturing in such a way that it is added to alcoholic compound.

Although diesters are mainly produced with alcohols, many different raw materials with hydroxyl groups are used in the case of monoesters. Fatty alcohols, fatty acid alkanolamides, and its oxethylates are most commonly used [18]. Usual esterification catalysts such as *p*-toluenesulfonic acid are suitable as catalysts for diester production.

In the second reaction step, the maleic acid ester is sulfated with an aqueous sodium sulfite solution to obtain the corresponding sulfosuccinate. Since the ester bond is sensible to hydrolysis, pH value of the aqueous sulfosuccinate must be carefully adjusted between 5 and 7.5 (Figure 13.8).

In the case of the sulfosuccinic acid monoester, two regioisomeric sulfosuccinates are possible (Figure 13.9).

It was detected by H nuclear magnetic resonance (NMR) analysis that the β position is preferred during sulfation. The ratio of β/α is approximately 4:1 [19]. A process flow sheet is shown in Figure 13.10.

Sulfosuccinates are used in many different fields of application. For comprehensive overviews, see Refs 18,20–22. Sulfosuccinic acid dialkyl esters are weakly foaming surfactants with good wetting power. In particular, products on the basis of octanol or 2-ethyl hexanol are distinguished by their outstanding wetting properties. Even at a low concentration, they can cause a considerable reduction in the surface tension of aqueous solutions [21]. Sulfosuccinic acid dialkyl esters, on the basis of alcohols with fewer than nine carbon atoms, are water soluble. Branched alkyl groups increase the solubility [23].

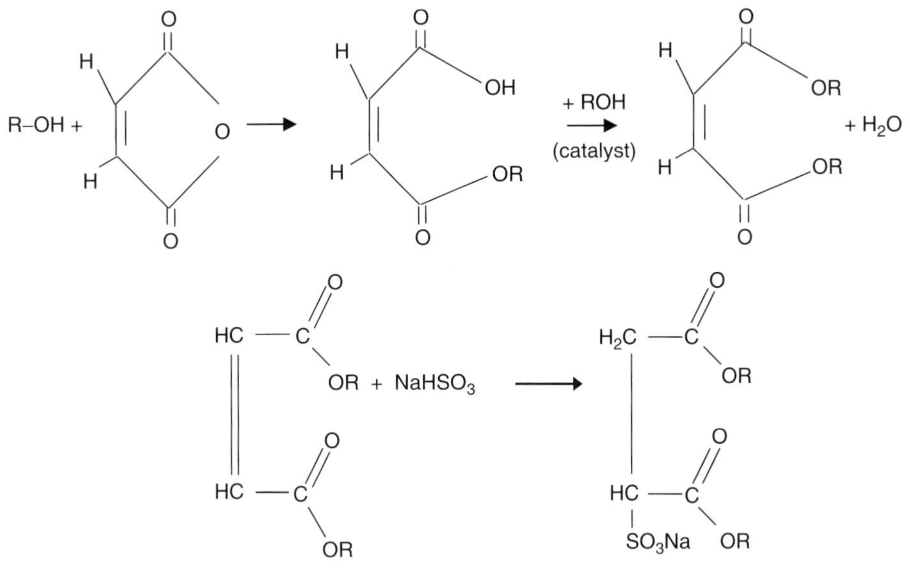

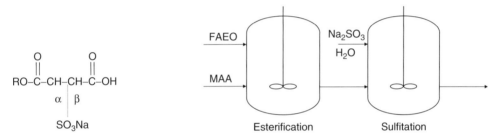

FIGURE 13.8 Sulfosuccinates—Reaction scheme.

FIGURE 13.9 Regio isomers of sulfosuccinates.

FIGURE 13.10 Manufacturing process scheme for sulfosuccinates (FAEO—Fatty alcohol ethoxylate; MAA—Maleic acid anhydride).

With regard to household products, the application of sulfosuccinic acid dialkyl esters is restricted to specific glass cleaners, for example, spectacle lenses or windscreens as well as carpet shampoos.

In contrast to sulfosuccinic acid dialkyl esters, sulfosuccinic acid monoalkyl esters are good foaming surfactants. Especially, products on the basis of oxethylated fatty alcohols, for example, lauryl/myristyl polyoxyethylene (3) alcohol, exhibit an outstanding skin compatibility [24]. Because of their mildness to skin, large quantities of sulfosuccinic acid monoalkyl esters are used in personal care products such as shower gels, shampoos, and skin-cleaning agents [25,26]. Because of the hydrolysis-sensitive ester bond, their application is limited to a pH range of 6–8. In the industrial sector, sulfosuccinic acid monoalkyl esters are, for example, used as emulsifiers for emulsion polymerization. Both the mono- and the dialkyl esters are readily biodegradable and show low toxicity [18].

13.5. SUMMARY

This chapter describes the production of mild surfactants based on fatty acids, fatty acid methyl esters, and fatty alcohols from the oleochemistry in combination with petrochemical compounds such as alkanolamines, maleic acid anhydride, or glucose/glucose derivatives.

246 Handbook of Detergents/Part F: Production

Besides, the basics of chemistry such as the reaction scheme process information including basic plant design are discussed. The reference list provides also an information about the relevant technologies.

REFERENCES

1. Tox Ecology (Ed.), *Alternatives to Nonylphenol Ethoxylates*, Final Report 28, For Environment Canada, May 2002.
2. Bizzari, S. N., Fatty alkanolamides, CEH data summary, *Chemical Economics Handbook*, SRI International, Menlo Park, CA, 2002.
3. Langian, R. S., Final report on the safety assessment of cacamide MEA, *Int. J. Toxicol.* 18(Suppl. 2), 9, 1999.
4. Johnson, W., Andersen, F. A., Amended final report on the safety assessment of cocamide DEA, *J. Am. Coll. Toxicol.* 15, 527, 1996.
5. Farris, R. D., Methyl esters in the fatty acid industry, *J. Am. Oil Chem. Soc.* 56, 770–773, 1979.
6. Oftring, A., Oetter, G., Baur, R., Borzyk, O., Burkhart, B., Ott, C., and aus dem Kahmen, M., US Patent 6.034.257, 2000.
7. Rosen, M. J. (Ed.), Characteristic Features of Surfactants. In: *Surfactants and Interfacial Phenomena (2294)*, Wiley, New York, 2004, p. 24.
8. Cognis Care Chemicals, Comperlan COD data sheet for coconut fatty acid diethanol-amide, Rev. 5.03.2004, www.cognis.com.
9. Adami, I., Production, processing and uses of methyl esters and derivatives, Ballestra S.p.A. Company Brochure and Additional Information, Milano, Italy, September 2005.
10. Ames, G. R., Long-chain derivatives of monosaccharides and oligo saccharides, *Chem. Rev.* 60, 541, 1960.
11. Hill, K. H. and Rhode, O., Sugar based surfactant of consumer products and technical applications, *Fett/Lipid.* 101, 25, 1999.
12. Fischer, E., Über die glycoside der alkohole, *Ber. Dtsch. Chem. Ges.* 26, 2400, 1893.
13. Hirsinger, F. and Hill, K. H., Life-cycle inventory of alkyl polyglycosides in detergent alcohols, in: *Alkyl Polyglucosides,* Hill, K. H., ven Rybinski, W., and Stoll, G. (Eds.), VCH, Weinheim, 1996.
14. Eskuchen, R. and Nitsche, M., Technology and production of alkyl polyglucosides, in: *Alkyl Polyglycosides,* Hill, K. H., von Rybinski, W., and Stoll, G. (Eds.), VCH, Weinheim, 1996.
15. Ruback, W., Schmidt, S., and van Bekkum (Eds.), *Carbohydrates as Organic Raw Materials III*, VCH, Weinheim, 1996.
16. Laughlin, R. G., Fu, Y.-C., Wireko, F. C., Scheibel J. J., and Munyon, R. L., The physical science of *N*-dodanoyl-*N*-methylglucamine, in: *Novel Surfactants, Preparations, Applications and Biodegradability*, Homberg, K. (Ed.), Marcel Dekker, New York, 1998, pp. 1–30.
17. Scheibel, J. J., Connor, D. S., Shumate, R. E., and St. Laurent, J. C. T. R. B., EP-B 0558515, US-A 5984262, 1990.
18. Domsch, A. and Irrgang, B., Anionic sufactants, in: *Surfactant Science Series*, Vol. 56, Stache, H. W. (Ed.), Marcel Dekker, New York, 1995, p. 501.
19. Henkel KGaA, unpublished results.
20. Milne, J. A., R., Trends in the application for sulfosuccinate surfactants, *Soc. Chem. (Ind. Appl. Surf.).* 77, 76, 1990.
21. O'Lenick, A. J. and Smith, W. C., Sulfosuccinates chemistry and applications, *Soap Cosmet. Chem. Spec.* 64, 36, 1988.
22. Schoenberg, T., Sulfosuccinate surfactants. Mildness and low cost have spurred growth in body, hair and skin cleansers, *Cosmet. Toil.* 105, 105, 1989.
23. Falbe, J. (Ed.), *Surfactants in Consumer Products*, Springer, Berlin, 1987, p. 83.
24. Kaestner, W., Anionic surfactants. Local tolerance (animal tests): Mucos membranes and skin. Marcel Dekker, New York, Surfactant Science Series (1980), Vol. 10 (Anionic Surfactants: Biochem. Toricol. Dermatol.) 139–307.
25. Gawelek, G., *Tenside*, Akademie Verlag, Berlin, 1975, p. 232.

14 Production of Hydrotropes

Robert L. Burns (edited by Uri Zoller)

CONTENTS

14.1 Introduction 247
14.2 Synthesis of Sulfonic Acid Salts 247
References 251

14.1 INTRODUCTION

Alkylbenzene sulfonates and alkylnaphthalene sulfonates act as hydrotropes to modify solubilities, viscosities, and other properties of surfactants and surfactant formulations. Sodium toluenesulfonate (STS) and sodium xylene sulfonate (SXS) are the best-known hydrotropes, however, alkylnaphthalene sulfonates also possess hydrotropic properties. Hydrotropicity is the amount of different hydrotropes required to achieve clear solutions. The amount of different hydrotropes, for example, SXS, STS, sodium cumene sulfonate (SCS), and others, in light-duty liquid (LDL) detergent formulations is given in Figure 14.1. There seems to be no trend; different hydrotropes are (more or less) similarly efficient in different formulations.

Hydrotropes are also used for modifying the viscosity of surfactant formulations. Some heavy-duty liquid (HDL) formulations are clear without hydrotrope(s), but have a high viscosity. Adding ~10% of SXS, STS, SCS, and other hydrotropes (as salts) reduces the viscosity substantially (Figure 14.2).

The toluene, xylene, cumene, and alkylnaphthalene sulfonates, specifically considered in this chapter, are generally prepared by sulfonation of the corresponding aromatic hydrocarbon followed by neutralization with the appropriate base. Gilbert's monograph on sulfonation reactions [1] forms the primary basis for the discussion on their synthesis, that is, the sulfonation of the aromatic portion of the starting raw materials—alkylbenzenes and alkyl naphthalenes.

14.2 SYNTHESIS OF SULFONIC ACID SALTS

The sulfonation reaction is thought to be an S_E2 (electrophilic substitution) reaction in which monomeric sulfur trioxide (SO_3) is the reacting species. Two moles of monomeric SO_3 successively first react to attack the aromatic ring and then to protonate the incipient sulfonate group on the aromatic ring. The SO_3 also functions as a base for removal of the proton, thus forming a pyrosulfonic acid. The pyrosulfonic acid intermediate sulfonates a second mole of aromatic compound, leading to the desired product, an aromatic sulfonic acid, and to some sulfone formation. The proposed equations for toluene sulfonation are presented in Figure 14.3.

Portions of this chapter were presented as a poster session at the AOCS National Meeting in Seattle, WA in May 1997 and at Soaps, Detergents, and Oleochemicals: An AOCS International Conference in Fort Lauderdale, FL in October 1997.

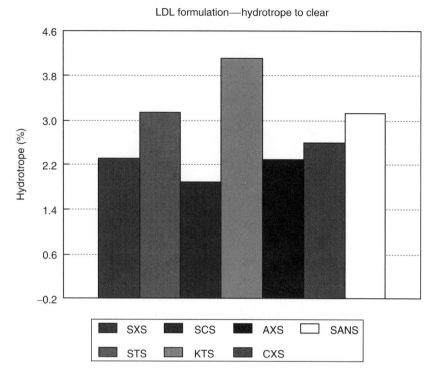

FIGURE 14.1 LDL formulation—hydrotrope to clear.

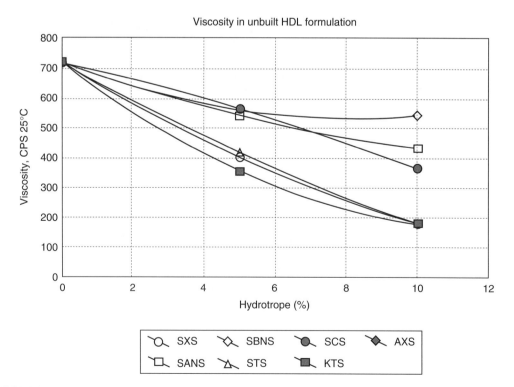

FIGURE 14.2 Viscosity in unbuilt HDL formulation.

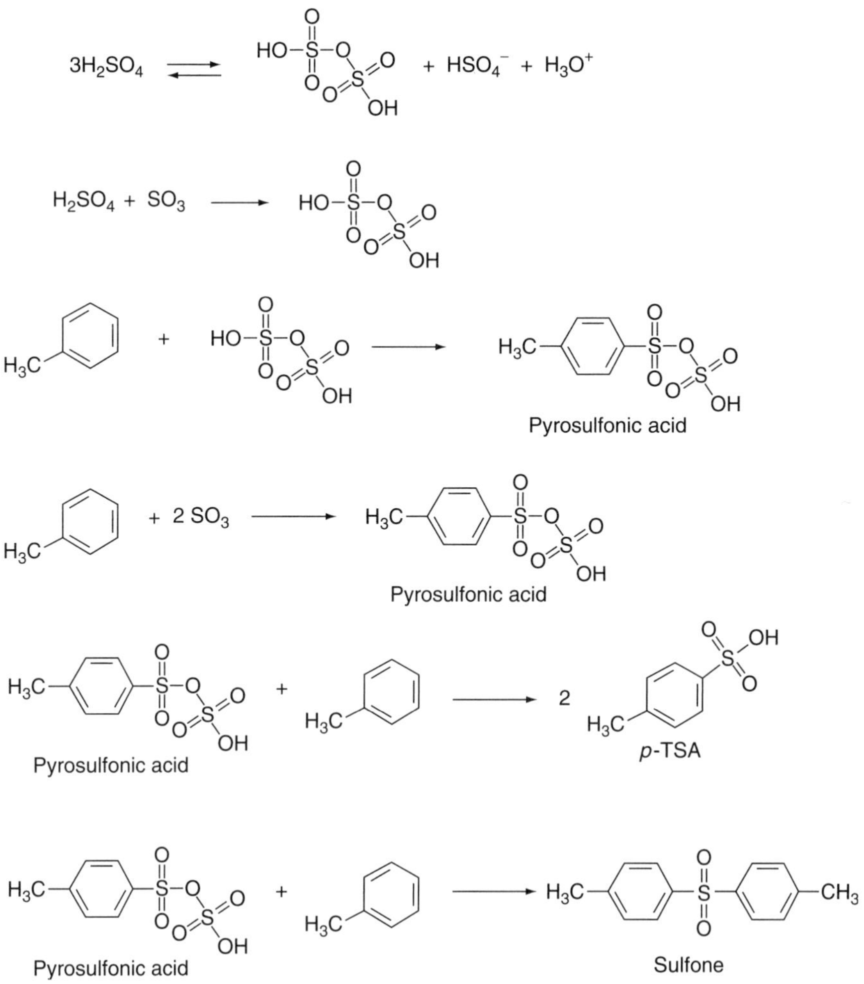

Production of Hydrotropes

FIGURE 14.3 Toluene sulfonation equations.

Gilbert [1, p.1] states that all the common sulfonating agents can be thought of as compounds of SO_3. For example, sulfuric acid can be thought of as $SO_3 \cdot H_2O$ and chlorosulfonic acid as $SO_3 \cdot HCl$.

Sulfuric acid can be used to sulfonate aromatic compounds directly. This process produces small quantities of the sulfone by-product, however, the rate of sulfonic acid formation decreases with the increasing water concentration, which occurs as a result of the consumption of the sulfuric acid and the formation of water as the reaction proceeds. The reaction stops when H_2SO_4 reaches a concentration at which the rates of sulfonation and desulfonation are equal, that is, equilibrium has been attained. This concentration is dependent on the specific compounds undergoing reaction and is sometimes referred to as the pi (π) factor.

The pi factor was originally conceived as the concentration of SO_3 at equilibrium [2], however, because sulfuric acid can be considered to be the monohydrate of SO_3, π can be usefully expressed mathematically as follows:

$$\pi = [H_2SO_4]/H_2SO_4] + [H_2O]$$

where the concentrations in the square brackets are the concentrations in the total reaction mass. It can be determined easily for any system in equilibrium by simple measurement of the acid and water concentrations.

One method that allows for a nearly complete utilization of the reactants involves the removal of water, as it is formed during the sulfonation reaction, using excess hydrocarbon and an azeotropic distillation procedure [1, pp.72, 73]. The water/hydrocarbon azeotrope is distilled out, phase separated, and the hydrocarbon is returned to the reaction vessel. In the laboratory, this is conveniently done using a Dean–Stark trap. A similar method returns hot hydrocarbon vapors to the reaction vessel. These methods are only effective for hydrocarbons which boil in a reasonable range, ~100–140°C. Hydrocarbons which boil at higher temperatures can be sulfonated at reduced pressure to allow the hydrocarbon to boil.

A second approach involves the use of excess acid to ensure that the acid concentration will remain above the π factor until all of the hydrocarbon substrate has reacted. Since many sulfonic acids are relatively insoluble in sulfuric acid of intermediate strength, between 50 and 80%, suitable dilution will cause a phase separation with the dilute sulfuric acid at the bottom and the sulfonic acid on the top, allowing convenient removal of the former [2, pp. 362, 363]. For other hydrocarbons, excess acid is neutralized and the resulting inorganic salt is separated from the sulfonate product. This approach results in nearly quantitative yields on the hydrocarbon basis.

A third procedure carries out the sulfonation reactions at high temperatures and pressures. Because the π-factor effect is minimized at high temperatures and pressures, good yields can be obtained under these conditions, although colored by-products are often formed.

Yet another procedure uses a chemical method for removing water as it is formed during sulfonation, increasing sulfonic acid yields, by adding thionyl chloride ($SOCl_2$) to produce hydrochloric acid (HCl) and sulfur dioxide (SO_2). This approach is suitable for laboratory use, whereas the other procedures outlined earlier have industrial and commercial applications in actual production processes of the hydrotropes based on short-chain alkylbenzene sulfonates.

Sulfonation of aromatic hydrocarbons can be accomplished by using SO_3 directly in the reaction, usually in a continuous sulfonater. This reaction is instantaneous and does not require the removal of water from the reaction. However, it gives a higher yield of sulfones and other unwanted by-products. Chemical inhibitors can reduce sulfone production in this reaction. Interestingly, the sodium salt of the sulfonic acid, the actual hydrotrope, inhibits sulfone formation [3]. Inorganic sulfites [4] as well as acetic acid [5] can also be used.

Another method for sulfonation involves the use of oleum, a solution of SO_3 in sulfuric acid. The process is similar to the use of sulfuric acid, but this process generates less water than all of the sulfuric acid processes and so should, in principle, be faster. However, sulfone generation is generally higher than with sulfuric acid.

Chlorosulfonic acid, $ClSO_3H$, is quite convenient for laboratory use, but generates a mole of HCl for each mole of sulfonic acid formed. Reaction of an aromatic hydrocarbon with 2 moles of chlorosulfonic acid will result in the formation of the sulfonyl chloride [1, p. 84].

The positional isomer distribution in toluenesulfonic acid is influenced by the sulfonation temperature [6,7]. For example, at 75°C with sulfuric acid, the distribution is ~75% para, 6% meta, and 19% ortho.

Aromatic sulfonic acids can be converted to their respective salts by neutralization with an appropriate base. For example, reaction of potassium hydroxide (KOH) with a sulfonic acid produces a potassium sulfonate; sodium hydroxide (NaOH) and the sulfonic acid produce a sodium sulfonate; and ammonium hydroxide (NH_4OH) and the aromatic sulfonic acid produce ammonium sulfonate. A wide range of aromatic sulfonic acid salts can be produced from various aliphatic and aromatic amines and metal cations. The neutralization is conveniently done in water, since the sulfonates, and even more so the short alkyl chain hydrotropes, are generally water soluble to the extent of 30–50%.

Postprocessing can involve a solvent extraction to remove sulfones and treatment with carbon or chemical bleach to reduce color. If solvent extraction is done, traces of solvent must be removed by distillation, usually as the azeotrope. Treatment with lime may be necessary if a low-sulfate product is desired.

REFERENCES

1. Gilbert, E.E., *Sulfonation and Related Reactions*, Interscience, New York, 1965, p. 62.
2. Gilbert, E.E. and Groggins, P.H., in *Unit Processes in Organic Synthesis*, P.H. Groggins, Editor, McGraw-Hill, New York, 1958, p. 338.
3. Sobczak, E., Badzynski, M., Chajnacki, S. and Blanowicz, K., Optimization of sulfonation of benzene to benzenesulfonic acid by means of gaseous sulfur trioxide, *Prace Wydzialu Nauk Technicznych, Bydgoskie Towarzystwo Naukowe, Seria A*, 13, 109–112, 1979.
4. Robin and Schulte, US Patent 3,789,067.
5. Rueggeberg, W.H.C., Sauls, T.W. and Norwood, S.L., On the mechanism of sulfonation of the aromatic nucleus and sulfone formation, *J. Org. Chem.*, 20, 455–465, 1955.
6. Suter, C.M., *The Organic Chemistry of Sulfur*, Wiley, New York, 1944, p. 203.
7. McNeil, D., Sulfonation of Toluene and Production of Cresols from Toluene, in *Toluene the Xylenes and their Industrial Derivatives*, E.G. Hancock, Editor, Elsevier, New York, 1982, p. 211.
8. Matson, T.P. and Berretz, M., *Res. Dev. Dep.*, Conoco, Inc., Ponca City, OK, USA. *Soap, Cosmetics, Chemical Specialties*, 55(12), 41–42B, 1979.
9. Johnson, G. and Oxynes, B.L. Modeling of a Thin Film Sulfur Trioxide Sulfonization Reactor. *Ind. Eng. Chem. Process Des. Develop*, 13(1), 6–14, 1974.

15 Production of Ethylene Oxide/Propylene Oxide Block Copolymers

Elio Santacesaria, Martino Di Serio, and Riccardo Tesser

CONTENTS

15.1 Introduction 253
15.2 Classification and Applications 254
15.3 Reaction Mechanisms and Kinetics 256
    15.3.1 Catalyst Formation 257
    15.3.2 Initiation 257
    15.3.3 Propagations 258
    15.3.4 Proton Transfer 259
15.4 Processes 262
    15.4.1 Synthesis 262
        15.4.1.1 Pluronic Surfactant 263
        15.4.1.2 Tetronic Surfactant 263
        15.4.1.3 All-Heteric Surfactant 263
    15.4.2 Analysis 264
    15.4.3 Reactors 265
    15.4.4 Safety and Alkylene Oxide Handling 267
References 269

15.1 INTRODUCTION

Polyoxyalkylene block copolymers represent an important class of nonionic surfactants with different applications in the field of detergency. Even if, in principle, these compounds can be synthesized by the polymerization of several cyclic ethers such as, for example, ethylene oxide (EO), propylene oxide (PO), tetrahydrofuran, or 1,2-butylene oxide, in this chapter, our attention is focused exclusively on the derivatives of EO and PO. The initiators of the polymerization vary considerably and are mainly distinguished on the basis of their functionality. In most cases, for products with applications for detergency, tetrafunctional initiators can be adopted.

In this chapter, different aspects of EO/PO block copolymers production are reviewed, starting with Section 15.2 describing general product classification and related field of application. This is followed by Section 15.3 describing the relevant reaction mechanisms and kinetics that represent the principal background of the corresponding industrial production and processes. Section 15.4.1 summarizes briefly the synthesis and typical recipes adopted. Section 15.4.2 reports the main analytical techniques that are used for both process control and product characterization. In Section 15.4.3, the industrial production is dealt by examining the existing processes and their characteristics. Section 15.4.4 concludes this chapter by describing some relevant aspects of industrial safety and handling of alkyleneoxides.

15.2 CLASSIFICATION AND APPLICATIONS

Polyalkylene oxide block copolymers can be classified in different ways [1]. An arbitrary classification can be established, for example, by considering (as a first criterion) the number of reactive hydrogen atoms that are present in the molecules of the initiator. Following this approach, different classes of nonionic copolymer surfactants can be identified by grouping these compounds as follows: (1) class I consists of derived products, for example, from 1-butanol or other monofunctional alcohols; (2) class II consists of the derivatives of ethylene glycol, propylene glycol, or other bifunctional initiators; (3) class III consists of the trifunctional glycerol-derived surfactants and so on. Accordingly, an initiator can have up to eight reactive hydrogen atoms (of the hydroxyl groups) so that the corresponding eight classes may be used for an exhaustive classification purpose.

The compounds present in each of these basic categories can be further distinguished on the basis of the polymer structure that is obtained as a consequence of different reaction sequential steps. Depending on the addition strategies of alkylene oxides (AO), different polymers can be obtained— an all-heteric (totally random) copolymer in which the sequence of alkylene oxide units is statistically organized; a block copolymer in which relatively long alternating sequences of alkylene oxide units can be identified; and a combination of the two (heteric and block) extreme situations. The great influence exerted by the addition procedure of alkylene oxides on the surfactant structure also affects the properties—different addition sequences result in completely different behavior and properties of the obtained copolymers.

More specifically, the subclasses can be classified as follows:

All-block subgroup. This first subgroup consists of the nonionic surfactants that are formed by the addition of one alkylene oxide to an initiator followed by another reaction step in which a different alkylene oxide is added. Further reaction steps can take place with different or already used oxyalkylenes. The resulting copolymers have a structure consisting of blocks of oxyalkylene units in the chain. The length of these blocks can be adjusted by controlling the operative conditions of each reactive step.

All-heteric subgroup. This subgroup consists of the copolymeric nonionic surfactants that are formed when a mixture of two or more alkylene oxides is added to an initiator. The first reaction step can be followed by other reactive events in which the oxyalkylenic mixture can be changed both in the number of components and their relative amounts. The resulting product is a copolymer that has a random distribution of oxyalkylenic units in the whole chain.

Block–heteric or heteric–block subgroup. This subgroup of nonionic surfactants includes those polyols that are formed by the addition, in an initial reaction step, of a pure alkylene oxide to an initiator and then a mixture of two or more alkylene oxides. If these two reaction steps are reversed, the resulting product is of the heteric–block type—the heteric oxyalkylene chain is formed first with a random distribution of oxide units and then follows the addition of a pure alkylene oxide giving place to a block-type portion of the chain.

The classification described is related to the wide class of compounds formed by the addition of alkylene oxide units to an initiator and the final use of the derived products can be very different, since the resulting properties are quite different. The compounds more widely used in the field of detergency are those obtained by starting with an initiator having at the most four reactive hydrogen atoms. Table 15.1 shows an example of EO/PO copolymers classification together with the uses, trademark names, and main producers. In this table, the acronyms used for the subclass identification (RPE, REP, EPE, and PEP: this nomenclature is explained in detail in Ref. 1) correspond to the sequence of the oxyalkylene addition reactions; RPE means a nonionic surfactant formed by the

Production of Ethylene Oxide/Propylene Oxide Block Copolymers

TABLE 15.1

Classification of EO/PO Block Copolymers, Uses, Commercial Names, and Main Producers

Class	Initiator	Subclass	Uses	Trademark	Producer
I	Aliphatic monohydric alcohol	Heteric RPE	Emusifiers	Tergitol XJ, XD, and XH	Dow
				Pluriol A.PE	BASF
II	Propylene glycol	Block EPE	Defoamers, wetters, solubilizers, detergents, foamers	Pluronic	BASF
				Symperonic PE	ICI
				Dowfax	Dow
				Tergitol L	Dow
	Ethylene glycol	Block PEP	Defoamers, wetters, detergents	Pluronic R	BASF
				Symperonic PE R	ICI
III	Glycerol	RPE	Demulsifers, defoamers	Polyglycol 112–2	Dow
	Trymethylolpropane	Block RPE	Demulsifer, defoamers	Pluracol TPE	BASF
IV	Ethylene diamine	Block RPE	Defoamers, low foam wetters, emulsifiers	Tetronic	BASF
		Block REP	Metal cleaning	Tetronic R	BASF

addition of PO to an initiator followed by the addition of a block of EO units; REP is a product similar to the former but with a reverse order of oxyalkylenic reactants addition. Both RPE- and REP-type compounds contain two blocks of alkylene oxides in the chain. Triblock products are those of the EPE type, prepared by an initial addition of PO to propylene glycol and a second step of EO reaction. Another triblock surfactant, can be identified as PEP, has the two lateral blocks constituted by PO repeating units, whereas the central block is made of EO units. Moreover, block copolymers are not uniform and alkylene chain lengths are unevenly distributed for both ethoxylated and propoxylated chains.

As seen in Table 15.1, the most important EO/PO block nonionic surfactants are those derived from propylene glycol and ethylenediamine as initiators, commercially known as Pluronics and Tetronics. Their general formulae can be written as follows:

$$H-[OCH_2CH_2]_x-[OCHCH_2]_y-[CH_2CH_2O]_x-H$$
$$\underset{CH_3}{|}$$

In these structures, x and y represent the average numbers of oxyethylenic and oxyopropylenic units, respectively, in the polymeric blocks and a more or less wide distribution exists around these mean values. Concerning the average molecular weights, these products are characterized by values of up to 10,000 Da although the majority fall in the 3000–5000 Da range.

About 80 manufacturers produce nonionic surfactants in an appreciable amount around the world and about 20 of these producers are international companies [2].

15.3 REACTION MECHANISMS AND KINETICS

Different classes of catalysts, operating with different mechanisms, are able to promote the reactions of ethoxylation and propoxylation [3]:

$$ROH + nAO \rightarrow RO(AO)_{n-1}AOH$$

(AO = EO or PO).

It is possible to induce, for example, polymerization in the presence of an acid catalyst of both the Bronsted [4] and Lewis type [5]. It is also possible to favor the reaction by using transition metal complexes [6,7]. Despite the large number of papers and patents published in the literature dealing with the various mentioned catalysts, the syntheses of polyethylene glycols, polypropylene glycols, and EO–PO copolymers for the detergent industry are still performed in the presence of alkaline catalysts such as KOH and NaOH.

EO/PO copolymers are synthesized at 120–180°C, 2–6 bar in semibatch reactors where pure alkylene oxide or a mixture of them is fed over time. Lower temperatures are used when PO is fed to limit the intervention of secondary reactions such as the hydroxyl dehydration [8] and allyl alcohol formation [9].

These secondary reactions together with the presence of impurity delimited the molecular weight of EO/PO copolymer using basic catalysts (20,000 and 5,000 Da for multifunctional and monofunctional starters, respectively) [10]. However, these limits are much higher than the ones of the molecular weight of commercial products used as detergents and surfactants [1]. In practice, the molecular weights for industrial processes are limited by the lowering of reaction rate due to the dilution of catalyst as the polymerization proceeds [10].

The reaction of hydroxyl dehydration, which eliminates the chain terminal hydroxyl is a drawback in the case of production of polyols for polyurethane industry, therefore several catalysts apart from potassium and sodium hydroxide have been proposed. These catalysts, including hydroxide of rubidium, cesium, barium and strontium, or double metal cyanide, are much more expensive and toxic than classical catalysts [11]. Recently, a new class of catalyst has been

Production of Ethylene Oxide/Propylene Oxide Block Copolymers

proposed by BASF based on the use of an aluminum phosphate catalyst (*bis*(di-isobutylaluminum)-methylphosphonate) [11].

The mechanism of ethoxylation and propoxylation of a monofunctional starter containing hydroxyl groups, catalyzed by a metal hydroxide, can be summarized as follows [12].

First of all, the metal hydroxide is added to the reactor. A concentrated aqueous solution is generally adopted.

15.3.1 Catalyst Formation

The catalyst is formed *in situ* by the reaction of KOH with the starter, forming an ion pair:

$$ROH + MOH \rightarrow RO^- M^+ + H_2O \uparrow$$

After the introduction of the catalyst, the reactor is purged with N_2 stream at high temperature to remove oxygen (for safety reasons) and water. The water is an impurity, which is to be eliminated because it is a starter for the formation of polyglycols. In the subsequent step, the alkylene oxide is fed to the reactor.

15.3.2 Initiation

The alkoxide anion reacts with alkylene oxide:

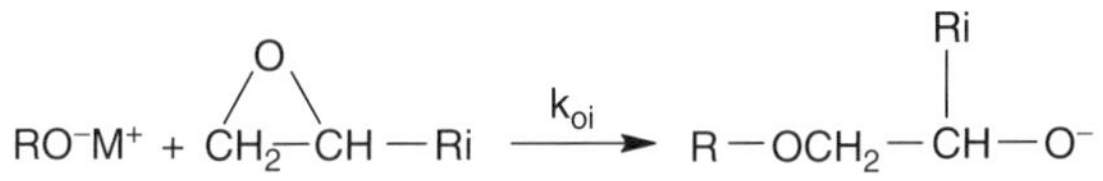

$$RO^-M^+ + CH_2\!-\!CH\!-\!Ri \xrightarrow{\ k_{oi}\ } R\!-\!OCH_2\!-\!\underset{Ri}{CH}\!-\!O^-$$

with $i = 1$ or 2, $R_1 = H$, $R_2 = CH_3$.

The following reaction can be classified as SN2 nucleophilic substitution:

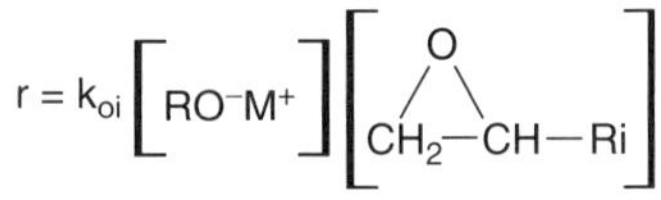

$$r = k_{oi}\left[RO^-M^+\right]\left[CH_2\!-\!CH\!-\!Ri\right]$$

The values of the kinetic constants depend on the nucleophilicity of the alkoxide anion of the starter [3] and are different for ethoxylation (k_{01}) and propoxylation (k_{02}) [12] (in Table 15.2, the values of the initiation kinetic constants for different starters are shown).

TABLE 15.2

Mean Initiation Kinetic and Proton Transfer Equilibrium Constants for Ethoxylation (k_{01}, Ke_{01}) and Propoxylation (k_{02}, Ke_{02}) Reactions of Different Starters

	k_{01}		k_{02}			
Starters	ln A (cm³/mol/min)	E (kcal/mol)	ln A (cm³/mol/min)	E (kcal/mol)	Ke_{01}	Ke_{02}
1-Octanol [10]	24.6 ± 0.3	13.0 ± 1.8	26.8 ± 0.1	15.6 ± 0.5	2.0 ± 0.3	3.5 ± 0.4
2-Octanol [10]	27.9 ± 0.2	16.8 ± 1.0	29.2 ± 0.2	19.0 ± 1.9	2.2 ± 0.1	2.5 ± 0.3
1-Dodecanol [13]	24.4 ± 0.3	13.3 ± 0.2	—	—	4.7 ± 0.1	—
Ethylene glycol [11]	37.5 ± 2.6	21.7 ± 1.9	30.8 ± 1.9	18.6 ± 1.5	1.0	0.7 ± 0.1

15.3.3 PROPAGATIONS

The anions formed may be involved in successive propagation steps such as

$$RO-(AO)_{k-2}-CH_2-\underset{\underset{R_i}{|}}{CH}-O^-M^+ + \underset{\underset{}{}}{CH_2\overset{O}{\overset{\diagup\diagdown}{-}}CH}-R_j \xrightarrow{k_{ij}} RO-(AO)_{k-2}-CH_2-\underset{\underset{R_i}{|}}{CH}-O-CH_2-\underset{\underset{R_j}{|}}{CH}-O^-M^+$$

with k = 2, n; i = 1 or 2; j = 1 or 2; R_1 = H, R_2 = CH_3.

These reactions are also SN2 nucleophilic substitutions:

$$r = k_{ij} \left[CH_2\overset{O}{\overset{\diagup\diagdown}{-}}CH-R_j \right] \left[RO-(AO)_{k-2}-CH_2-\underset{\underset{R_i}{|}}{CH}-O^-M^+ \right]$$

In propagation reactions, we can recognize two different types of reactive systems corresponding to four kinetic constants that can be considered independent from the starter and chain length [3,12,13]:

k_{11} Ethoxylation of a primary hydroxyl in the growing chain
k_{21} Ethoxylation of a secondary hydroxyl in the growing chain
k_{12} Propoxylation of a primary hydroxyl in the growing chain
k_{22} Propoxylation of a secondary hydroxyl in the growing chain

Some experimental values of k_{11} and k_{22} constants are shown in Figure 15.1. The constant k_{12} and k_{21} are related to the reaction of addition from one to another alkylene oxide, or in the case of reaction of a mixture of epoxides. The value of k_{12} can be assumed equal to the initiation constant of propoxylation of 1-octanol, whereas the value of k_{21} can be assumed equal to the initiation constant

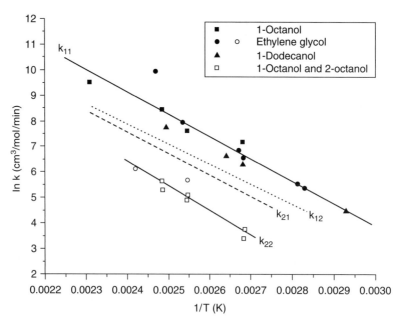

FIGURE 15.1 Arrhenius plot for the ethoxylation and propoxylation kinetic constants of primary and secondary hydroxyls.

of ethoxylation of 2-octanol [13]; the calculated values are also shown in Figure 15.1 for a useful comparison.

We can observe from Figure 15.1 that the rate constant of ethoxylation is greater than the rate constant of propoxylation, and alkoxylation rate constant of a primary hydroxyl is greater than that of a secondary one.

15.3.4 Proton Transfer

The oxyalkylene anions are also involved in the following proton-transfer reactions:

$$RO^-M^+ + RO-(AO)_{k-1}-CH_2-\overset{Ri}{\underset{|}{CH}}-OH \underset{}{\overset{k_{eoi}}{\rightleftarrows}} ROH + RO-(AO)_{k-1}-CH_2-\overset{Ri}{\underset{|}{CH}}-O^-M^+$$

The proton-transfer equilibrium constants depend on the relative acidity of both the substrate and the growing chain and also on the relative stability of the corresponding ionic couples, and their values strongly influence the oligomer distribution [3,14–17]. Some experimental proton-transfer equilibrium constants are given in Table 15.2.

The different reaction rates of alkylene oxides strongly influence the distribution and size of alkylene oxide block and when we plan a new synthesis, these kinetic aspects should be taken into account.

By using the already described kinetic model and related kinetic constants together with the correct values of epoxide solubilities, it is possible to describe the evolution with the time of epoxide consumption and oligomer distribution, as can be seen, for example, in Figures 15.2 and 15.3 [12].

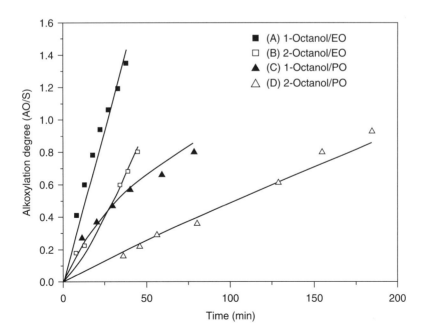

FIGURE 15.2 Alkoxylation degree as a function of time for ethoxylation (A, B) and propoxylation (C, D) kinetic runs performed on both 1-octanol and 2-octanol. All the runs were performed at 120°C, 2 atm, and 2% mol of KOH as catalyst. Symbols and continuous lines are experimental data and model prediction, respectively.

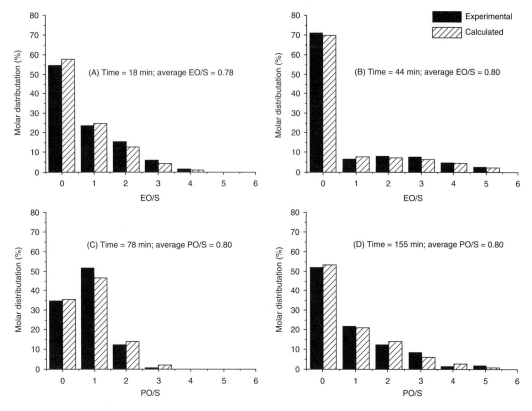

FIGURE 15.3 Oligomer distributions for the same kinetic runs mentioned in Figure 15.2. In all the cases, the value of AO/S is ~0.80 mol AO/mol substrate. The distributions are referred to the times indicated in the figure and a comparison between experimental and model is provided.

The broad oligomer distribution observed in Figure 15.3 for the ethoxylation of a secondary alcohol is the consequence of the value of initiation constant that is lower than propagation ($k_{01} < k_{11}$). On the contrary, the oligomer distribution in the case of ethoxylation of a primary alcohol is narrower, because in this case, the initiation and propagation rate constants are quite similar ($k_{01} = k_{11}$). The same situation is operative in the propoxylation of a secondary alcohol ($k_{02} = k_{22}$). In the propoxylation of a primary alcohol, peaked oligomer distributions are obtained as a consequence of the higher value of initiation with respect to the propagation rate constant ($k_{02} > k_{22}$).

Another example of the influence of the difference in kinetic constant ($k_{21} < k_{11}$) is the necessity for high degree of ethoxylation (mole of EO/mole of substrate hydroxyl) to obtain EO/PO block copolymer with high concentration of primary hydroxyls when the EO is fed after PO reaction, as shown in Figure 15.4 [18].

In the earlier kinetic scheme, we did not consider the asymmetric characteristic of the PO molecule. Propylene oxide could generate a primary or secondary alcohol when ring opening occurs as a consequence of a nucleophilic attack on the methylene (a) or methyne group (b), respectively:

Production of Ethylene Oxide/Propylene Oxide Block Copolymers

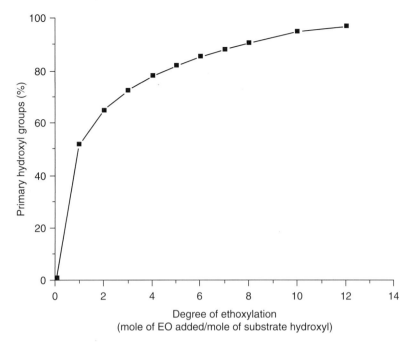

FIGURE 15.4 Percentage of primary hydroxyl groups as a function of ethoxylation degree in the case of block polymer in which the first step is the PO reaction and EO is fed subsequently.

We have simplified the kinetic scheme considering that the contribution of the reaction pathway (2) is negligible [12,19].

Moreover, it must be pointed out that in the case of a multifunctional starter, the kinetic scheme is much more complicated. As an example, we can see that the following scheme is operative in the propoxylation of ethylene glycol [13]:

$$EG \xrightarrow{PO} A_1 \begin{array}{c} \xrightarrow{PO} A_2 \xrightarrow{PO} A_3 \xrightarrow{PO} A_4 \xrightarrow{PO} \\ \searrow^{PO} \searrow^{PO} \searrow^{PO} \\ \xrightarrow{PO} S_2 \xrightarrow{PO} S_3 \xrightarrow{PO} S_4 \xrightarrow{PO} \end{array}$$

$$EG = HO-CH_2-CH_2-OH \qquad A_i = HO-CH_2-CH_2-O\left[CH_2-\underset{CH_3}{\overset{|}{CH}}-O\right]_{i-1}CH_2-\underset{CH_3}{\overset{|}{CH}}-OH$$

$$S_i = HO-\underset{CH_3}{\overset{|}{CH}}-CH_2-O\left[\underset{CH_3}{\overset{|}{CH}}-CH_2-O\right]_j CH_2-CH_2-O\left[CH_2-\underset{CH_3}{\overset{|}{CH}}-O\right]_{i-j-2}CH_2-\underset{CH_3}{\overset{|}{CH}}-OH$$

Notwithstanding the complexity of the kinetic scheme, it is possible to develop a kinetic model that is useful in describing the behavior of these systems [13]. This behavior is similar to the one occurring for the polyethoxylated chain reacting with PO during the preparation of PEP block copolymers.

Concerning EO/PO random copolymers, defining the composition in all the possible situations is necessary to know the reactivity of both primary and secondary hydroxyls with EO and PO, as already mentioned.

In the synthesis of PO/EO copolymer using an amine as starter, the metal hydroxide is added to the reactor after the complete formation of a monoadduct [19]:

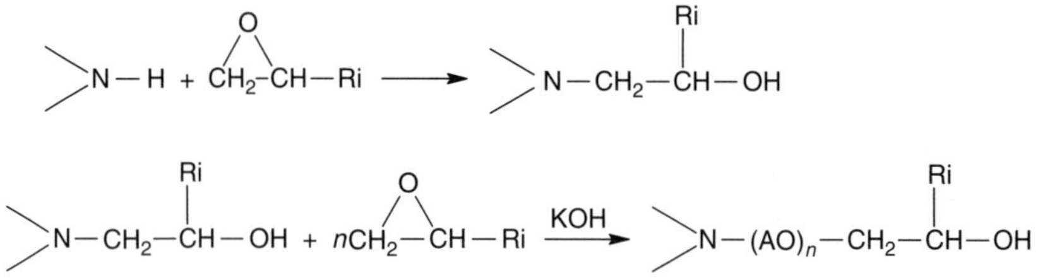

In this case, the acidity of the proton of the used amine is very important for the rate of the monoadduct formation, because the formation of hydrogen bond gives place to an increase of electrophilicity of methylene group of alkylene oxide, favoring the reaction [20]:

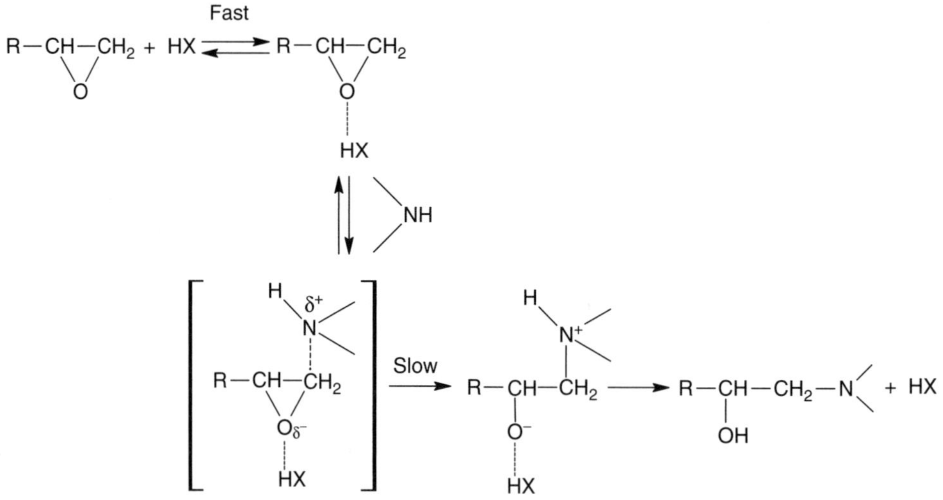

After the first step and catalyst feeding, the used solvent (normally water) has to be carefully eliminated by stripping to avoid by-product formation.

15.4 PROCESSES

Many aspects must be considered in the production processes of EO/PO block copolymers concerning raw materials, their storage and subsequent purification, reactor configurations, separation technologies, chemical and physical analysis, and quality and safety standards.

In this section, we have focused our attention on four aspects that are the most important from the industrial perspective—the synthesis and typical industrial recipes adopted; principal analytical methods for process control and product characterization; main reactor types used, their characteristics, and involved technologies; and aspects connected to the safety and hazards in alkyleneoxide processing.

15.4.1 SYNTHESIS

In this section, two typical laboratory and one pilot plant syntheses are described as examples for the production of an EPE Pluronic and a Tetronic surfactant and an all-heteric random copolymer, respectively.

Production of Ethylene Oxide/Propylene Oxide Block Copolymers 263

The different examined systems, published in the patent literature, can easily be scaled up to industrial size since all are semibatch systems.

15.4.1.1 Pluronic Surfactant

The synthesis equipment consists of a round-bottom flask with a total volume of 1 L, equipped with a mechanical stirrer, thermometer, reflux condenser, and a line for the addition of alkylene oxides in successive steps. The flask was initially charged with 57 g of propylene glycol and 7.5 g of anhydrous sodium hydroxide as a catalyst. The flask was purged with nitrogen to remove air and heated to a temperature of 120°C under stirring. When the sodium hydroxide was completely dissolved, the feeding of PO was started at a rate such that it ultimately reacted completely. The product was cooled under a nitrogen atmosphere and the NaOH catalyst was neutralized with sulfuric acid. The final product, after filtration, was a polypropylene-glycol insoluble in water with an average molecular weight of 1620 g/mol as determined by hydroxyl number [22–24]. The product obtained in the reaction step desoribed earlier was submitted to ethoxylation reaction in the same apparatus. An amount of 500 g of this product was charged in the flask together with 5 g of anhydrous sodium hydroxide and heated at 120°C. After the complete dissolution of the catalyst, EO addition was started. A total amount of 105 g of EO was added that corresponds to 17.4% by weight compared to the final EPE block copolymer produced [21].

15.4.1.2 Tetronic Surfactant

In this example, ethylenediamine was used as a tetrafunctional initiator for the addition reactions of the PO and EO in the first and second steps, respectively, performed with a procedure similar to the Pluronic surfactant [21]. Owing to the basic characteristics of nitrogen-containing compounds, the first addition step in which PO is reacted was carried out in the presence of water (~10% by weight). Both the condensation of PO and subsequent reaction with EO was carried out with sodium hydroxide as a catalyst. In Table 15.3, the carbon soil removal value of some of the obtainable products is given as a function of the composition of copolymers [19].

15.4.1.3 All-Heteric Surfactant

Butyl alcohol was used as the initiator in this reaction and 9.072 kg of it was introduced in the reactor with 0.408 kg of sodium hydroxide as a catalyst, successively dissolved. The amount of

---

TABLE 15.3
Carbon Soil Removal Value (0.1% Concentration at 90°F) of EO/PO Copolymers Obtained Using Ethylenediamine as Starter

Molecular Weight of Oxypropylene Chains	Wight Percent of Oxyethylene Chains	Carbon Soil Removal Value (0.1% Concentration at 90°F)
2975	25	284
2975	52	335
2975	75	185
2975	85	153
5965	25	293
5965	50	255
5965	75	143
5965	85	113

Source: Lester, G. L., and Gross, I. Nitrogen containing polyoxyalkylene detergent compositions. U.S. Patent 2,979,528.

EO fed to the autoclave was 21.546 kg together with the same quantity of PO over a time of 1.4 h and humidity content of ~0.2% by weight. The total pressure was 1.7 atm and temperature was kept constant at 107°C. The product obtained was then neutralized with aqueous sulfuric acid and volatile residue was removed by vacuum distillation at 180°C at 2 mm Hg of pressure. The average molecular weight of the obtained product is 467 g/mol, as evaluated by means of the hydroxyl end group method [22–25].

15.4.2 ANALYSIS

Various experimental techniques are regularly used for the characterization of the nonionic block polyoxyalkylene surfactants with the general aim to obtain information regarding the chemical composition of the complex mixtures formed by the implementation of synthesis steps. More specifically, these techniques are dev oted to the determination of polymer-chain structure, molecular weight distribution, EO/PO ratio, EO and PO block length distribution, amount of residual reactants, any by-product presence, concentration, and type of terminal groups. All these characteristics contribute to the overall bulk properties of the surfactant mixture and the topic of their determination has been extensively studied in the literature. Particularly, in Refs 26 and 27, a comprehensive review of the analytical methods applied to EO/PO block copolymers is presented. However, in this chapter, a general overview of the main analytical techniques for the characterization of EO/PO copolymers is presented, especially regarding analysis used for both process and quality control as a useful tool for industrial production. Major details are described in *Handbook of Detergents/Part C: Analysis.*

A particularly important property of the EO/PO block copolymers is the average molecular weight obtained in the reaction steps that greatly affects the final properties of the end-use product. The most widely used determination of average molecular weight and degree of oxyethylation of a surfactant mixture is the classical method determination of hydroxyl number. This method consists of an esterification reaction, of a known weighted sample, with phthalic or acetic anhydride that convert all the hydroxyl groups present into ester and carboxylic acid end groups according to the following reaction:

$$(RCO)_2O + R'OH \rightarrow R'OOCR + RCOOH$$

The excess of added anhydride is hydrolyzed with water, whereas the acid groups formed can be titrated with a standard solution of potassium hydroxide in comparison with a blank titration.

Different standardized methods have been developed for the hydroxyl value determination and are easily available in the literatures [23–24] with information regarding interferences, solutions preparation procedure, and other experimental details.

Another useful experimental technique for the characterization of EO/PO nonionic surfactant is chromatography, which allows information to be obtained for both the overall EO/PO ratio and oligomer distributions. Even if gas chromatography (GC) cannot be directly applied to surfactant mixtures due to the very low volatility of the compounds present, an application of this technique is in the chemical cleavage of block copolymers through reaction with HBr or HI. The resulting mixture of bromopropane and bromoethane, or alternatively iodopropane and iodoethane, can be easily resolved and quantitatively analyzed on a GC column allowing the determination of EO/PO ratio [28,29]. However, high-performance liquid chromatography is the ideal tool for the evaluation of the nonionic surfactant product distribution [30], especially in the case of a single oligomer-family-like homopolymers. The copolymers give rise to more complex chromatograms, and the difficulties in the interpretation of the elution sequence and attribution of the response factors limit the application of this technique mainly to homopolymers. Another drawback that represents a limitation in the application of this analytical method is the availability of standard pure substances for calibration purposes.

Production of Ethylene Oxide/Propylene Oxide Block Copolymers

As a general characterization technique for nonionic surfactants, the nuclear magnetic resonance (NMR) can be considered, due to the highly recognizable chemical shifts of hydrogen and carbon atoms in various organic molecules. In the case of a mixture such as a block EO/PO surfactant, the NMR spectra presents a high degree of complexity and information that can be extracted from it has a statistical significance and does not reflect the structure of a single and well-defined molecule. NMR results for an EO/PO surfactant must be interpreted in a broader context that also involves other spectroscopic and physicochemical techniques such as chromatography. A particular NMR-based analysis is represented by the determination of the EO/PO ratio in a polyoxyalkylene surfactant, described as a routine analysis [31] even if this method has been applied to EO/PO adducts of fatty alcohols and alkylphenols. The protonic NMR can also be used for the determination of the relative abundance of primary and secondary hydroxyl groups in copolymers [32]. The amount of information collected on EO/PO copolymer surfactant structures can be increased if ^{13}C-NMR is used instead of protonic NMR. With this technique, EO/PO distribution along the copolymer chain can be determined together with information on primary/secondary hydroxyl ratios. Moreover, ^{13}C-NMR can be used for the differentiation of block structures between PEP and EPE block sequences [33].

15.4.3 Reactors

To satisfy the high-quality standard requirements for EO/PO block copolymers and wide range of surfactant products that can be obtained, various processes have been industrially developed. The production processes due to the high toxicity and chemical instability of the starting materials must also meet, as a common characteristic, very rigid safety standard of manufacturing. The majority of nonionic surfactants are synthesized in semibatch-operated reactors, whereas continuous reactors are rarely adopted in the industrial practice. Reactor conditions are a pressure of up to 2–6 bars, temperature in the range 120–180°C, and maximum reactor volume of 60–70 m^3. Basically, three types of reactors are used for the synthesis and they can be distinguished with respect to the gas–liquid contacting principle and related devices [34].

The more widely used type of reactor is the mechanically agitated tank, a general scheme of which is given in Figure 15.5a. This reactor is equipped with internal cooling coils and external jacket or half-tubes for thermal control. The feed of alkylene oxide is sparged into the liquid phase just below the stirrer. The amount of raw materials charged and feeding rate of EO/PO must control the liquid-phase volume expansion, which occurs during the reaction. In a typical synthesis, the volume of the reaction mixture can expand by a factor of 3 or more and this value must be compatible with the available reactor volume. Moreover, the volume increase involves a decrease in the catalyst concentration and consequent slow down of the reaction rate. A possible variant to this reactor configuration is shown in Figure 15.5b, which illustrates the scheme of a reactor having an external heat exchanger that allows a better temperature control, especially for high production volumes. This kind of reactor, in both its possible arrangements, has a limited efficiency of mass transfer between gaseous and liquid phases resulting in a relatively low performance in yield and productivity.

A second category of industrially used reactor is the jet-loop reactor. A typical configuration for the jet-loop reactor is presented in Figure 15.6 and is characterized by the presence of an injector nozzle inserted in a circulation loop. Through this injector, the gaseous and liquid phases are intimately contacted by dispersing the first into the second. To achieve good performances in terms of mass-transfer rate, an efficient pumping device has to be installed for introducing into the system the amount of energy enough for optimum phase contacting. The injector nozzle is fully immersed in the liquid phase in such a way that a great turbulence of the liquid is realized. The adoption of the technology of jet-loop reactor allows a certain degree of scalability, when high reaction rates are required, which is facilitated by the possibility of fitting two circulating loops and injector nozzles into the reactor body.

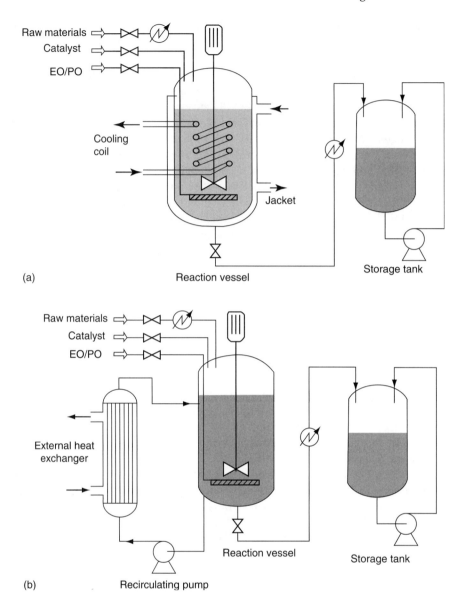

FIGURE 15.5 Schemes of a mechanically agitated (a) jacketed reactor with internal cooling coil and (b) reactor with external heat exchanger and circulation loop.

A further possibility of reactor configuration for the synthesis of EO/PO block polymers is represented by the spray reactor, as schematically illustrated in Figure 15.7. In this case, the liquid reaction mixture is finely dispersed into a gaseous phase constituted of EO or PO. Also in this reactor configuration, an external circulation loop is present comprising a circulating pump and thermal exchange device for an accurate temperature control. The reactor can be optimally operated in a way that the liquid droplets can reach saturation in EO or PO during their passage in the gaseous atmosphere (mass-transfer zone), arriving in the liquid phase (reaction zone) with the maximum load of gaseous reactant [35]. The spray reactor ensures a good mass-transfer rate and an easy scalability (multiple spray nozzles can be mounted in the top of the reactor) so that the reactor can be used in high-capacity plant in which high throughput is required [35].

A simplified drawing of the alkoxylation plant of Scientific Design Company, as plant example, is shown in Figure 15.8 [36]. The flow diagram shows EO, PO, and starter feeds; vessels

Production of Ethylene Oxide/Propylene Oxide Block Copolymers

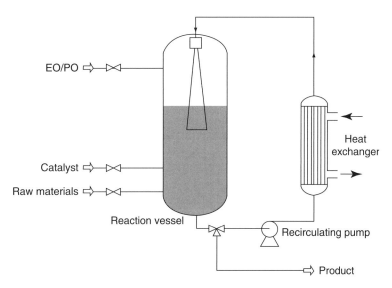

FIGURE 15.6 Scheme of a jet-loop reactor with injector nozzle and external heat exchanger placed on the circulating line.

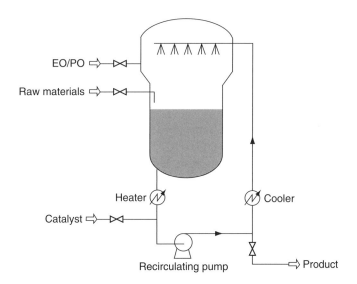

FIGURE 15.7 Scheme of a spray reactor with circulating line.

for preparation of the catalyst and neutralizing agent; vacuum system; and dual loops for higher growth ratio products.

15.4.4 Safety and Alkylene Oxide Handling

As mentioned earlier, the most commonly employed alkylene oxides for producing block nonionic surfactants are the very reactive three-membered cyclic ethers such as EO and PO. Particularly, EO is highly flammable, explosive in some condition, and also toxic and an irritant for skin and eyes. For these reasons, the reactor employed in the synthesis, described earlier, must be equipped with sophisticated safety devices, process monitoring systems, and distributed control systems (DCS) to keep the reaction in safety condition and avoid gas-phase decomposition and the so-called runaway

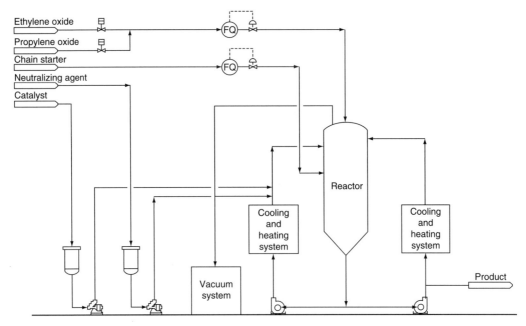

FIGURE 15.8 Scheme of an alkoxylation plant. (Courtesy of Scientific Design Company.)

reactions. It must be pointed out that, coupled with reliable hardware safety devices, other valuable tool that must be available for an efficient process control (particularly from safety point of view) is represented by physicochemical models of the reacting system and reactor. These models contain information regarding various aspects such as the kinetics of the reaction, mass, and heat transport properties of the system, and physicochemical properties such as EO solubility and mixture density. If properly used, the tool of mathematical modeling can furnish information about the optimal reactor operation in safety conditions [35].

The mentioned safety considerations, particularly stringent for the reaction section of the plant, must be extended also to other sections in which storage, transportation, and handling of EO are involved. In the following, the main sources of hazard connected to the use of EO in a production plant are summarized [37]:

- *Fast decomposition reactions.* EO, in certain conditions, can decompose explosively, also in the absence of air or oxygen, into carbon monoxide and methane [37]. In the presence of rust, the decomposition produces methane, carbon dioxide, and water even if other decomposition products such as ethane, ethylene, and hydrogen have been observed [38,39]. This phenomenon can be prevented by dilution with a suitable inert gas. Not only nitrogen is usually chosen, but also methane and other diluents have been used. The dilution amount depends on temperature, pressure, and expected ignition source and duration [40]. The most thorough discussion of the EO decomposition process is presented in Ref. 39. The inert gas must be used also for blanketing of vessels and lines, and details about this procedure are given in Refs 38 and 40.
- *Flammability and explosivity.* EO is highly flammable and explosive and mixtures of EO/air exhibit a very large flammability range. The minimum value cited for the lower flammable limit of EO/air mixtures is 2.6% by volume [37], whereas the upper flammable limit is typically stated as 100% by volume, since pure EO can decompose in the absence of air and oxygen. The autoignition temperature of EO in air at a total pressure of 1 bar is 429°C [37].

Production of Ethylene Oxide/Propylene Oxide Block Copolymers

- *Polymerization reactions.* EO can give rise to very exothermic polycondensation reactions in the presence of concentrated alkaline substances or metal oxides [41,42]. These reactions if not properly prevented or controlled can evolve as runaway reactions or, more generally, cause reactor and line fouling due to polymer formation.
- If a small amount of catalyst is introduced in the reactor due to a human error, the reaction rate is quite low and EO and PO accumulate in the liquid phase due to their relative high solubility in the reaction mixture [43]. The same situation occurs when the reactor temperature is much lower than the one indicated by the temperature-control device, when it is out of order. In this last case, the EO and PO accumulation is due to the fact that solubility strongly increases by decreasing the temperature. In both cases, initially reaction occurs very slowly in the former and latter cases for the low catalyst concentration and low temperature, respectively. As the polymerization reaction is highly exothermic, the liquid phase in both cases gradually increases its temperature and this increases exponentially the reaction rate and pressure in the reactor, giving rise to a runaway behavior. When the system reaches these conditions, a temperature control is not more feasible and safety apparatus (sensors, relief valves, and DCS) must intervene for discharging the reactor and draining out the excess of EO or PO to another vessel to reduce the pressure. The most frequently occurring causes for the establishment of runaway reactions have been identified in the wrong choice of the operating conditions [43] or failure in the cooling system connected to the reactor [34]. In each of the mentioned cases, the availability of reliable simulation tools, incorporating kinetic and solubility data, is of great importance for the reactor sizing and design and plant operations in safety conditions.

REFERENCES

1. Schmolka, I. R. 1967. Polyalkylene oxide block copolymers. In: Schick, M. J. ed. *Nonionic Surfactants.* London: Edward Arnold Publishing, New York: Marcel Dekker, pp. 300–371.
2. Kosswig, K. 1987. Surfactants. In: *Ullmann's Encyclopedia of Industrial Chemistry.* Berlin: Wiley-VCH, Vol. A25, pp. 796–797.
3. Santacesaria, E., Iengo, P., Di Serio, M. 1999. Catalytic and kinetic effects in ethoxylation processes. In: Karsa, D. R. ed. *Design and Selection of Performance Surfactants.* Sheffield, Scotland: Sheffield Academic Press.
4. Parker, R. E., Isaacs, N. S. 1959. Mechanisms of epoxide reactions. *Chem. Rev.* 59:737–799.
5. Penckzec, S., Kubisa, P., Matyiaszewsky, K. 1985. Cationic ring opening polymerization. 2-Synthetic application. In: *Advances in Polymer Science*, Berlin: Springer, Vols. 68/69.
6. Pruit, M. E., Baggett, J. M. 1955. Catalysts for the polymerization of olefin oxides. U.S. Patent 2,706,181.
7. Edwards, C. L. 1987. Preparation of nonionic surfactants. European Patent 228,121.
8. Dege, G. J., Harris, R. L., MacKenzie, J. S. 1959. Terminal unsaturation in propylene glycol. *J. Am. Chem. Soc.* 81:3374–3379.
9. Price, C. C., Carmelite, D. D. J. 1966. Reactions of epoxides in dimethyl sulfoxide catalyzed by potassium tert-butoxide. *J. Am. Chem. Soc.* 88:4039–4044.
10. Whitmarsh, R. H. 1996. Synthesis and modification of POA block copolymers. In: Nace, V. M. ed. *Nonionic Surfactant, Polyoxylakylene Block Copolymers.* New York: Marcel Dekker, Vol. 60, pp. 1–30.
11. Dexheimer, E. M. 2003. Production of polyetherols using aluminium phosphonate catalysts. U.S. 2003/0139568 A1.
12. Di Serio, M., Vairo, G., Iengo, P., Felippone, F., Santacesaria, E. 1996. Kinetics of ethoxylation and propoxylation of 1- and 2-octanol catalyzed by KOH. *Ind. Eng. Chem. Res.* 35:3848–3853.
13. Di Serio, M., Tesser, R., Dimiccoli, A., Santacesaria, E. 2002. Kinetics of ethoxylation and propoxylation of ethylene glycol catalyzed by KOH. *Ind. Eng. Chem. Res.* 41:5196–5206.
14. Weibull, B., Nicander, B. 1954. The distribution of compounds in the reaction between ethylene oxide and water, ethanol, ethylene glycol, or ethylene glycol monoethyl ether. *Acta Chem. Scand.* 8:847–858.

15. Santacesaria, E., Di Serio, M., Garaffa, R., Addino, G. 1992. Kinetics and mechanism of fatty alcohol polyethoxylation 1. The reaction catalyzed by potassium hydroxide, *Ind. Eng. Chem. Res.* 31:2413–2418.
16. Santacesaria, E., Di Serio, M., Garaffa, R., Addino, G. 1992. Kinetics and mechanism of fatty alcohol polyethoxylation 2. Narrow-range polyethoxylation obtained with barium catalysts, *Ind. Eng. Chem. Res.* 31:2419–2421.
17. Nissen, D., Straehle, W., Marx, M. 1984. Process for the preparation of cellular and non-cellular polyurethane elastomers. U.S. Patent 4,440,705.
18. Schilling, F. C., Tonelli, A. E. 1986. Carbon-13 NMR determination of poly(propylene oxide) microstructure. *Macromolecules* 19:1337–1343.
19. Lester, G. L., Gross, I. 1961. Nitrogen containing polyoxyalkylene detergent compositions. U.S. Patent 2,979,528.
20. Holubka, J. W., Bach, R. D., Andrès, J. L. 1992. Theoretical study of reaction of ethylene oxide and ammonia. A model study of the epoxy adhesive curing mechanism. *Macromolecules* 35:1189–1192.
21. Lunstedt, L. G., Ile, G. 1954. Polyoxyalkylene compounds. U.S. Patent 2,674,619.
22. Siggia, S., Starke Jr., A. C., Garis Jr., J., Stahl, C. R. 1958. Determination of oxyalkylene groups in glycols and glycol and polyglycol ethers and esters. *Anal. Chem.* 30:115–116.
23. International Organisation for Standardization, ISO 4327. 1979. E. Geneva., Switzerland
24. International Organisation for Standardization, ISO 4326. 1980. E. Geneva., Switzerland
25. Roberts, F. H., Va, W., Fife, H. R. 1947. Mixtures of polyoxyalkylene monohydroxy compounds and methods of making such mixtures. U.S. Patent 2,425,755.
26. Cross, J. ed. 1987. *Nonionic Surfactants: Chemical Analysis.* New York: Marcel Dekker, Vol. 19.
27. Kalinoski, H. T. 1996. Chemical analysis of polyoxyalkylene block copolymers. In: Nace, V. M. ed. *Nonionic Surfactants: Polyoxylakylene Block Copolymers.* New York: Marcel Dekker, Vol. 60, pp. 31–66.
28. Mathias, A., Mellor, N. 1966. Analysis of alkylene oxide polymers by NMR spectrometry and by gas-liquid chromatography. *Anal. Chem.* 38:472–477.
29. Stead, J. B., Hindley, A. H. 1969. Analysis of oxyethylene/oxypropylene copolymers by chemical fission and gas chromatography. *J. Chromatogr.* 42:470–475.
30. Garti, N., Kaufman, V. R., Avraham, A. 1987. Analysis of nonionic surfactants by high-performance liquid chromatography. In: Cross, J, ed. *Nonionic Surfactant: Chemical Analysis.* New York: Marcel Dekker, Vol. 19, pp. 225–283.
31. Gronski, W., Hellmann, G., Wilsch-Irrgang, A. 1991. Carbon 13 NMR characterization of ethylene oxide/propylene oxide adducts. *Makromol. Chem.* 192:591–601.
32. LeBas, C. L., Turley, P. A. 1984. Primary hydroxyl content in polyols: evaluation of two NMR methods. *J. Cell. Plast.* 20:194–199.
33. Heatley, F., Luo, Y-Z., Ding, J-F., Mobbs, R. H., Booth, C. 1988. A carbon-13 NMR study of the triad sequence structure of block and statistical copolymers of ethylene oxide and propylene oxide. *Macromolecules* 21:2713–2721.
34. Koernig, W. (2000). Production of alkoxylated surfactants. In: *Proceedings 5th World Surfactants Congress CESIO.* Firenze, Italy, May 29 to June 2.
35. Dimiccoli, A., Di Serio, M., Santacesaria, E. 2000. Mass transfer and kinetics in spray tower loop absorbers and reactors. *Ind. Eng. Chem. Res.* 39(11):4082–4093.
36. Courtesy of Scientific Design Company.
37. Buckles, C., Chipman, P., Cubillas, M., Lakin, M., Slezak, D., Townsend, D., Vogel, K., Wagner, M. Ethylene Oxide User's Guide Second Edition, available on www.ethyleneoxide.com (May 2004).
38. Rebsdat, S., Mayer, D. 1987. Ethylene oxide. In: *Ullmann's Encyclopedia of Industrial Chemistry.* Berlin: Wiley-VCH. Vol. A10, pp. 117–135.
39. Britton, L. G. 1990. Thermal stability and deflagration of ethylene oxide. *Plant/Oper. Prog.* 9:275–286.
40. Dever, J. P., George, K. F., Hoffman, W. C., Soo, H. 1994. Ethylene oxide. In: *Kirk-Othmer Encyclopedia of Chemical Technology.* New York: Wiley, 4th ed., Vol. 9, pp. 915–959.
41. Schonfeld, N. 1969. *Surface Active Ethylene Oxide Adducts.* New York: Pergamon Press.
42. Stull, D. R. 1977. Fundamental of fire and explosion. *AICHE Monogr. Ser.,* 73(10):67–68.
43. Santacesaria, E., Di Serio, M., Tesser, R. 1995. Role of ethylene oxide solubility in the ethoxylation processes. *Catal. Today* 24:23–28.

16 Production of Oxyethylated Fatty Acid Methyl Esters

Jan Szymanowski

CONTENTS

16.1 Introduction 271
16.2 Catalysts for Oxyethylation of Fatty Acid Methyl Esters 273
16.3 Synthesis of Oxyethylated Fatty Acid Methyl Esters 275
16.4 Some Properties of Oxyethylated Fatty Acid Methyl Esters 281
References 282

16.1 INTRODUCTION

Fatty acids obtained from natural renewable resources are important raw materials for nonionic surfactants although a major portion of surface-active agents consumed today in industrialized countries are derived from petrochemical sources. Moreover, a trend to rely more and more on surfactants from naturally renewable resources has developed and is gaining momentum.

Natural fats are converted into fatty acids or their methyl esters in typical processes of hydrolysis or methanolysis, respectively. The latter now seems more attractive due to a growing interest in ecological diesel fuel.

$$\begin{array}{ccc}
CH_2OOCR & & CH_2OH \\
| & H^+ & | \\
CHOOCR & \xrightarrow{H_2O} & CHOH \quad + \quad 3RCOOH \\
| & T, p & | \\
CH_2OOCR & & CH_2OH
\end{array} \qquad (16.1)$$

$$\begin{array}{ccc}
CH_2{-}OOCR & & CH_2OH \\
| & K_2CO_3 & | \\
CH{-}OOCR \quad + \quad 3CH_3OH & \rightleftharpoons & CHOH \quad + \quad 3RCOOCH_3 \\
| & & | \\
CH_2{-}OOCR & & CH_2OH
\end{array} \qquad (16.2)$$

Hydrogenolysis of fatty acids and their methyl esters gives alcohols used for the manufacture of nonionic surfactants.

$$RCOOCH_3 + 2H_2 \Rightarrow RCH_2OH + CH_3OH \qquad (16.3)$$

$$RXH \;+\; nH_2C \underset{\displaystyle O}{\overset{\displaystyle \diagdown\;\diagup}{\text{---}\; CH_2}} \longrightarrow RX(CH_2CH_2O)_nH \qquad (16.4)$$

X in Equation 16.4 can have different meanings depending on the oxyethylated compound, for example, X = O in alcohols.

This three-step process could be simplified if surfactants were directly obtained from fatty acids or their methyl esters. Direct synthesis from fats and oils would be the best choice. However, such a process is hardly realistic due to the instability of glycerin at the high temperature of the process.

The direct oxyethylation of fatty acids [1–4] gives a complex mixture with enormous amounts of polyoxyethylene glycols (PEGs) and their diesters. The first reaction step in which ethylene glycol monoester is formed occurs without any catalyst. The contribution of the catalytic reaction is negligible. The equilibrium of the proton transfer reaction is totally shifted toward a nonactivated ester with an equilibrium constant of approximately 0.001 [4]. This means that further steps of oxyethylation cannot occur before the total conversion of the fatty acid. The succeeding steps of oxyethylation occur in a typical way in the presence of an alkaline catalyst according to the typical S_N2 mechanism giving desired surfactants. However, the transesterification is also catalyzed by an alkaline catalyst giving PEGs and their diesters. The former having two active terminal hydroxyl groups react further, consuming significant amounts of ethylene oxide (EO). Thus, the process is not attractive due to economic aspects and poor quality of the resulting products.

Up to early 1990s, it was assumed that oxyethylation can occur only when the hydrophobic reagent had a labile hydrogen [1–5]. Thus, fatty acid methyl esters (FAME) were not considered as a raw material for the direct synthesis of nonionic surfactants with a polyoxyethylene chain. However, esters of fatty acids and PEG monomethyl ethers were known and their properties were described [6]. They were synthesized in a two-step process. Methanol was oxyethylated to PEG monomethyl ether that was then converted into the final product by transesterification with FAME or by esterification with fatty acids, carried out in the presence of an alkaline B or an acid catalyst, respectively. Esters of typical nonionics were synthesized in similar ways and their properties were described [7–9].

$$CH_3OH + nH_2C \overset{\diagdown}{\underset{O}{\diagup}} CH_2 \longrightarrow CH_3O(CH_2CH_2O)_nH \tag{16.5}$$

$$CH_3O(CH_2CH_2O)_nH + RCOOCH_3 \xrightarrow{B} RCO(OCH_2CH_2)_nOCH_3 + CH_3OH \tag{16.6}$$

$$CH_3O(CH_2CH_2O)_nH + RCOOH \xrightarrow{H^+} RCO(OCH_2CH_2)_nOCH_3 + H_2O \tag{16.7}$$

Fats could also be used in the transesterification step.

$$\begin{matrix} CH_2OOCR \\ | \\ CHOOCR \\ | \\ CH_2OOCR \end{matrix} + 3R^1O(CH_2CH_2O)_nH \underset{\longleftarrow}{\overset{K_2CO_3}{\rightleftharpoons}} \begin{matrix} CH_2-OH \\ | \\ CH-OH \\ | \\ CH_2-OH \end{matrix} + 3RCO(CH_2CH_2O)_nOR^1 \tag{16.8}$$

The pioneer works of Hoechst [10], Clariant [11–14], Vista Chemical Company [15–19], and Lion Corporation [20] have demonstrated that the direct reaction between FAME and EO can be carried out when a new type of catalyst is used. More detailed studies were described by Hama et al. [21–23].

$$RCOOCH_3 + nH_2C \overset{\diagdown}{\underset{O}{\diagup}} CH_2 \longrightarrow RCO(OCH_2CH_2)_nOCH_3 \tag{16.9}$$

Production of Oxyethylated Fatty Acid Methyl Esters   **273**

The first oxyethylene unit is inserted between the carbonyl carbon and the methoxy group [23]. The following oxyethylene groups are also inserted between the carbonyl carbon and the oxygen atom of the $(OCH_2CH_2)_nOCH_3$ group. Thus, the process shows important similarities to the polymerization of ethylene carried out in the presence of Ziegler–Natta catalysts. It is surprising that as many as a few decades were needed to transfer the knowledge from polymers to surfactants.

16.2 CATALYSTS FOR OXYETHYLATION OF FATTY ACID METHYL ESTERS

The reaction of FAME with EO needs a weakening or even a breaking of the bond between the carbonyl carbon and the oxygen of the methoxy group [24]. It is well known that the hydrolysis of esters composed of carboxylic acids and primary alcohols (e.g., fatty acids and methanol) occurs according to the $B_{AC}2$ or $A_{AC}2$ mechanisms in the presence of an alkaline or an acidic catalyst, respectively.

$$(16.10)$$

$$(16.11)$$

The characteristic feature of these reactions is the stability of the intermediate complexes that can even be considered as intermediate products. The stability of the cationic intermediate product is so high that an exchange between the oxygen atom and a solvent molecule occurs. Does it mean that such an exchange with a molecule of EO is also possible? This is a problem to be solved in the near future. Each step of the process carried out in the presence of an acidic catalyst is reversible. It means that the original ester or an ester-type product can be regenerated. The intermediate product formed in the presence of an alkaline catalyst is also stable, but each reaction step is irreversible.

All this means that an acidic catalysis seems prospective for a direct reaction of FAME with EO. However, for obvious reasons, typical mineral and Lewis acids cannot be used.

During the past 20 years, several new catalysts were prepared for oxyethylation of alcohols [25]. Some of them can also be used for oxyethylation of FAME. Alkoxylates of metal IIA group modified with mineral acids (mainly sulfuric and phosphoric acids) are simple but effective catalysts.

The catalyst can be formed directly in the oxyethylation reactor from an inorganic component, and alcohol is used as the hydrophobic starter. However, in this case, a long initiation period is needed to start the reaction. More conveniently and more safely, which is very important in large commercial plants, the catalyst is obtained separately according to the following reaction schemes:

$$MX + 2R'OH \rightleftharpoons (R'O)_2 M + 2HX \tag{16.12}$$

where $X = O$, OH, or H_2; and $R' = C_1 – C_3$.

$$(R'O)_2 M + 2ROH \rightarrow (RO)_2 M + 2R'OH \uparrow \tag{16.13}$$

$$(R'COO)_2 M + 2ROH \rightarrow (RO)_2 M + 2R'COOH \uparrow \tag{16.14}$$

$$2(CH_3COO)_2 Ca + 2ROH + H_2SO_4 \rightarrow (ROCaO)_2 SO_2 + 4CH_3COOH \uparrow \tag{16.15}$$

Exemplary catalyst can be prepared by using isopropanol dried with nitrogen. In the anhydrous isopropanol, calcium acetate is dissolved at a temperature of 105°C. The obtained catalyst is modified by using sulfuric acid (molar ratio of calcium acetate to sulfuric acid is equal to 2). Volatile components (water, isopropanol, and acetic acid) are then distilled under decreased pressure.

Aluminum alkoxides, also modified with mineral acids, can be used as catalysts [15–19,25]. Effective catalysts are obtained from hydrotalcite

$$Mg_a Al(OH)_b (CO_3)_2 dH_2O \tag{16.16}$$

by applying the conditions $1 < a < 5$, $b > c$, $(b + 2c) = 2a + 3$, and $0 < d < 10$ or from other layered minerals of the general formula:

$$M(II)_{1-x} M(III)_x (OH)_2 A_a B_b z H_2O \tag{16.17}$$

where M(II) can be Mg, Zn, Ca, Fe, Co, Cu, Cd, Ni, or Mn; M(III) stands for Al, Fe, Cr, Bi, or Ce; A denotes an anion of organic carboxylic acid; and B is an inorganic anion such as CO_3^{2-}, HCO_3^-, SO_4^{2-}, NO_3^-, NO_2^-, PO_4^{3-}, OH^-, or halide (mainly Cl^-); and $0.1 \leq x \leq 0.5$, $0 < a \leq 0.5$, $0 \leq b \leq 0.5$, $0 < a + b \leq 0.5$, or $0 \leq z \leq 10$.

Hydrotalcite obtains its catalytic activity after calcination at 400–600°C [25–29]. Very high catalytic activity is displayed by hydrophobized hydrotalcites in which carbonate anions are replaced by various carboxylic acids [27,28]. Such catalysts can be obtained by treating a dispersion of hydrotalcite in a low-molecular alcohol (e.g., isopropanol) or even in water with a carboxylic acid, especially on dissolving it in isopropanol. Boiling such a mixture under reflux for 1–2 h removes CO_2 and introduces hydrophobic mono- and dicarboxylic acids containing even as many as above 20 carbon atoms in a molecule. Hydrotalcites hydrophobized with carboxylic acids show higher catalytic activities than those obtained by calcination. More narrow-range distributed oxyethylates are also obtained.

When using natural hydrotalcite, calcination gives catalysts having too many Al sites that can, however, be deactivated by partial poisoning with alkali. Increased poisoning gives catalysts with lower activities, but more narrow distributed products are formed.

A more convenient method for synthesis of metal oxide composites is the impregnation of the support (usually magnesium oxide) with various salts (nitrates and carbonates of Al(III), Ga(III), In(III), Ti(III), Co(III), Sc(III), La(III), and Mn(II)) [25,30]. An alternative method is the precipitation of metal hydroxide mixtures from solutions of appropriate salts, usually nitrates of magnesium and aluminum, but also In(III), Tl(III), Co(III), Sc(III), La(III), and Mn(II) instead of Al(III) [20] and Zn(II), Sn(II), Ti(IV), Sb(III), Zr(IV), and Ba(II) [31] with the atomic ratio of Mg(II) to the second metal component being equal to 1:0.05. Calcination gives the final catalyst. The composite MgO–Al$_2$O$_3$ catalysts can contain the third component from the VIA, VIIA, or VIII group, especially Cr, Mn, or Fe [32]. The atomic ratio of Al to Al + Mg and the third component to the total sum of the three components should equal to approximately 0.5 and 0.4, respectively.

The catalytic activity of metal oxide composites depends both on the acidity and basicity of the surface that can be characterized by the level of ammonia sorption and pK_a. The catalyst should adsorb 100–200 μmol/g NH$_3$ and its pK_a should be in the range 15–17.5 [31]. Catalysts having low basicity and acidity of the surface have too low activities, but a lot of undesired by-products (dioxane and PEGs) are formed when their acidities are too high. The activity of Al–Mg composite oxide catalyst increases with an increase in the calcination temperature and reaches a maximum of approximately 700°C [22]. Further increase in the calcination temperature results in reduced activities, attributed to the sintering of the catalyst surface. The catalyst activity increases with increasing aluminum content, but less narrow distributed products are formed. Filtration of the catalyst is not always possible. Therefore, addition of water (180 g of water/1063 g of oxyethylation product) and the use of activated clay or diatomaceous earth as a filter aid are proposed [33].

16.3 SYNTHESIS OF OXYETHYLATED FATTY ACID METHYL ESTERS

Oxyethylation of FAME can be carried out in typical equipments used for oxyethylation (Figure 16.1) [34]. The process consists of a few traditional steps: dosing of a hydrophobe and catalyst, drying of the hydrophobe starter at a temperature of ~150°C with simultaneous purging of nitrogen, further heating of the reaction mixture to 175–185°C, successive dosing of EO, cooling of the postreaction mixture, and venting to remove EO with nitrogen (Figure 16.2). The mole ratio of EO to the hydrophobic starter can be changed to a large degree to obtain products of desired hydrophobicities. There are no specific differences in oxyethylation between FAME and alcohols. The amounts of catalyst used (0.5 wt/wt%) are also similar. The most important difference is in the reaction temperature. Oxyethylation of alcohols in the presence of typical basic catalysts, for example, NaOH can be carried out at as low a temperature as 135°C. FAME are less reactive than alcohols in oxyethylation, and the process must be carried out in the temperature range 175–185°C. Moreover,

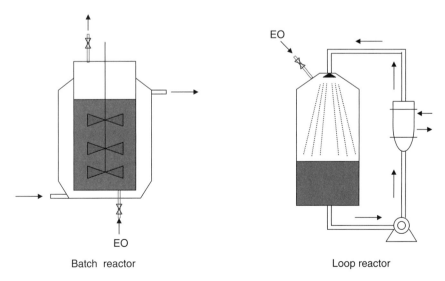

FIGURE 16.1 Semibatch reactors used for oxyethylation. (Dimiccoli, A., Di Serio, M., and Santacesaria, E. *Proceedings of the 5th World Surfactants Congress*, Firenze, Federchimica Assobase—P.I.T.I.O, 1: 99–110, 2000.)

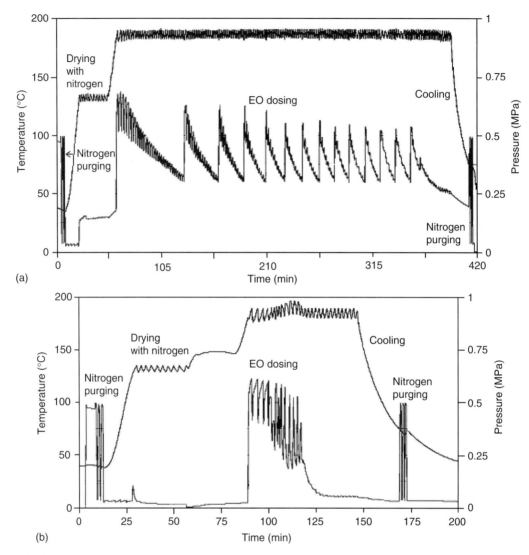

FIGURE 16.2 Oxyethylation of (a) dodecanol and (b) methyl dodecanoate in the presence of calcium-based catalyst at 185°C.

the distribution of the polyoxyethylene chain in oxyethylated FAME depends on temperature and decreases with an increase in temperature up to 185°C [35–37]; narrow distributed polyoxyethylene products are then obtained. Under similar conditions, oxyethylation of alcohols occurs faster and shorter reaction times are needed compared to FAME.

Lower reactivity of FAME and increased temperature may cause some hazardous problems with the initiation of the process connected with the dissolution of the calcium-based catalyst [37–39]. The reaction does not proceed rapidly after the first portions of EO are dosed, and some initiation period is observed (Figure 16.2). This period decreases when the temperature is increased up to 175–185°C. Induction period depends on the hydrophobe and is shorter for alcohols, but longer for oxyethylation of rapeseed FAME compared to methyl dodecanoate (MD).

Initiation period is also observed in Figure 16.2 in which successive dosing of EO is presented. The first drop from 700 to 300 kPa needed ~50 min for oxyethylation of MD. The following periods are significantly shorter, decreasing to ~10 min.

Production of Oxyethylated Fatty Acid Methyl Esters

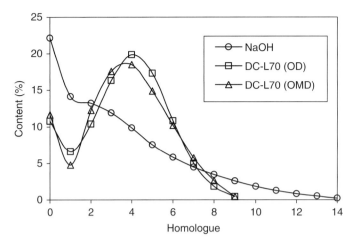

FIGURE 16.3 Comparison of homologue distribution in OD and OMD for NaOH and calcium-based catalyst (DC-L70) (average degree of oxyethylation is equal to 4).

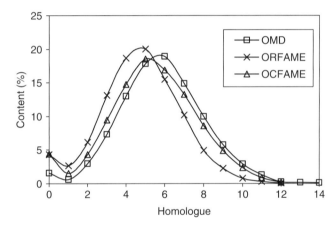

FIGURE 16.4 Comparison of homologue distribution in OMD, oxyethylated coco fatty acid methyl esters (OCFAME), and oxyethylated rapeseed fatty acid methyl esters (ORFAME) (average degree of oxyethylation is approximately 6).

Figures 16.3 and 16.4 show the distribution of polyoxyethylene homologues having a different number of oxyethylene groups. Distribution depends on the catalyst and the hydrophobic starter. The traditional catalyst (NaOH) gives very broad distributed oxyethylated dodecanol (OD). The main product is unreacted alcohol (~22% for average degree of oxyethylation equal to 4). Calcium-based catalyst gives narrow distributed products. The distributions of OD and oxyethylated methyl dodecanoate (OMD) are similar. Some important differences are only observed in the contents of both the starter and the first homologues for low degrees of oxyethylation. The content of unreacted MD is low, it depends on average degree of oxyethylation (n) and is in the range 10–12% and 2–5% for $n = 4$ and 6, respectively.

Oxyethylation is a moderately fast gas–liquid reaction in which mass transfer and reaction are consecutive, often rate-determining processes. A kinetic model [34,40–42] was proposed for oxyethylation of various starting materials (alcohols, alkylphenols, and fatty acids), carried out in the presence of a typical alkaline catalyst. Oxyethylation of nonylphenol carried out in an autoclave under common conditions and vigorous mixing (up to 3000 rpm) is controlled by diffusion.

The same applies to the oxyethylation of FAME. Thus, the reaction depends on the rate of EO transport to the liquid phase. The delivered EO is immediately consumed in successive parallel reactions giving products with various numbers of oxyethylene groups. The rate constants of the first, second, third, and nth reaction steps are denoted by k_0, k_1, k_2, and k_{n-1}, respectively. However, the contribution of the reaction depends on the process temperature and degree of oxyethylation. An increase in temperature enhances the role of diffusion, whereas an increase in oxyethylation degree has an opposite effect caused by an increased concentration of EO dissolved in the liquid phase. All this means that the computed constants cannot be considered as typical kinetic constants but only as relative estimates of consecutive reaction steps.

$$\text{RCOOCH}_3 + \text{H}_2\text{C} \overset{\text{O}}{-} \text{CH}_2 \xrightarrow{k_0} \text{RCO(OCH}_2\text{CH}_2)\text{OCH}_3 \tag{16.18}$$

$$\text{RCO(OCH}_2\text{CH}_2)\text{OCH}_3 + \text{H}_2\text{C} \overset{\text{O}}{-} \text{CH}_2 \xrightarrow{k_1} \text{RCO(OCH}_2\text{CH}_2)_2\text{OCH}_3 \tag{16.19}$$

$$\text{RCO(OCH}_2\text{CH}_2)_2\text{OCH}_3 + \text{H}_2\text{C} \overset{\text{O}}{-} \text{CH}_2 \xrightarrow{k_2} \text{RCO(OCH}_2\text{CH}_2)_3\text{OCH}_3 \tag{16.20}$$

$$\text{RCO(OCH}_2\text{CH}_2)_{n-1}\text{OCH}_3 + \text{H}_2\text{C} \overset{\text{O}}{-} \text{CH}_2 \xrightarrow{k_{n-1}} \text{RCO(OCH}_2\text{CH}_2)_n\text{OCH}_3 \tag{16.21}$$

The rates of the preceding consecutive reaction steps (Equations 16.18 through 16.21) depend on the reactant concentrations (the second-order reactions) according to the following relationships:

$$\frac{-dC_{\text{MD}}}{dt} = k_0 C_{\text{MD}} C_{\text{EO}} \tag{16.22}$$

$$\frac{dC_{\text{MDEO}_i}}{dt} = k_0 C_{\text{MD}} C_{\text{EO}} - k_1 C_{\text{MDEO}_i} C_{\text{EO}} \tag{16.23}$$

$$\frac{dC_{\text{MDEO}_i}}{dt} = k_{i-1} C_{\text{MDEO}_{i-1}} C_{\text{EO}} - k_i C_{\text{MDEO}_i} C_{\text{EO}} \tag{16.24}$$

where MD denotes methyl dodecanoate and MDEO_i stands for oxyethylated methyl dodecanoate having "i" oxyethylene groups.

Modeling of a semibatch reactor (Figure 16.1) enables to determine the reaction rate pseudoconstants. For lack of physical data, a number of assumptions have to be made. The volume of the liquid phase is the function of composition, temperature, pressure, and mass of EO reacted with raw material. At a constant temperature ($185 \pm 5°C$), the volume of the liquid phase increases due to an increased solubility of EO. However, the rate of change is relatively low compared to the reaction rate. The universal functional activity coefficient (UNIFAC) method [43] was used to calculate the activity coefficients. The method was adopted for the heterogeneous liquid–liquid–vapor system as the limited solubility of liquid components was observed. The

TABLE 16.1
Evaluated Ratios of Successive Pseudo-Rate Constants k_i/k_0 (Average Oxyethylation Degree [n])

Parameters Used for Evaluation	k_1/k_0	k_2/k_0	k_3/k_0	k_4/k_0	k_5/k_0	k_6/k_0	k_7/k_0	Reference
P, C, $n = 4$	3.1	2.5	2.3	2.1	2.0	—	—	49
P, $n = 0.6$	3.7	2.3	2.2	1.9	1.8	1.75	1.75	46
C, $n = 0.6$	1.0	5.8	13.3	23.3	31.7	36	40	38

physical and chemical properties of MD and its oxyethylated homologues were also estimated using the UNIFAC method. Satisfactory agreement between experimental and estimated values was observed for MD. Fitting the experimental profiles of the reaction pressure versus time and the composition of the obtained products with those obtained from integration by minimizing the appropriate square form gives the evaluated reaction rate pseudoconstants. The model fits the experimental results obtained for various degrees of oxyethylation well, that is, for a very low degree when only a single dose of EO is considered [36,44,45], but also for the actual degrees of oxyethylation when the dosing of EO is automatically repeated up to the desired average oxyethylation degree [46,47].

The introduction of the first and second oxyethylene groups is the slowest and quickest, respectively. The second step is rapid and the pseudo-rate constant is approximately three times as high as the rate constant for the first step. The rate constants for the successive steps are lower than the second-step rate constant, but they are always higher than the first-step rate constant (approximately twice as high).

It is obvious that the determined values of rate constants can actually be different from those given in Table 16.1, but the character of the observed changes is correct.

Similar sets of rate constants were obtained when pressure profile was used for estimation. However, when the pseudo-rate constants were computed only from the composition of oxyethylated rapeseed FAME, a different order of changes was reported. It shows the importance of pressure profile simulation that must be considered simultaneously with product composition.

The reaction pathway carried out in the presence of calcium-based catalysts is not well explained. Cox and Weerasooriya [18] proposed the following reaction scheme for FAME oxyethylation, carried out in the presence of calcium alkoxylate:

$$(R'O)_2M + H_2C\overset{O}{\overset{\diagup\diagdown}{-}}CH_2 \longrightarrow R'OCH_2CH_2OM-OR' \tag{16.25}$$

$$RCOOCH_3 + R'OCH_2CH_2OM-OR' \rightarrow RCOOCH_2CH_2OR' + CH_3OM-OR' \tag{16.26}$$

$$CH_3O-M-OR' + n\,H_2C\overset{O}{\overset{\diagup\diagdown}{-}}CH_2 \longrightarrow CH_3O(CH_2CH_2O)_x M(OCH_2CH_2)_y OR' \tag{16.27}$$

where M denotes Ca (but can also be Al).

The proposed scheme assumes oxyethylation of calcium alkoxylate and its oxyethylated homologues, followed by transesterification of oxyethylated methyl alkoxylates with FAME.

Oxyethylation of the catalyst may proceed in each alkoxyl group, giving products with a different total number of oxyethylene groups and their various distributions in both the chains.

The modeling of such a process scheme carried out for a semibatch reactor shows that the content of oxyethylated FAME increases with time only when free FAME are present in the reaction mixture. No accumulation of the successive homologues is simulated and the content of each succeeding homologue is lower than that of the homologues having fewer oxyethylene groups. Thus, the discussed scheme cannot simulate the distribution of homologues in the oxyethylated FAME.

Another explanation can be the high stability of the cationic intermediate product and direct reaction with EO.

$$R-\overset{\overset{\displaystyle O}{\|}}{C}-OR' \;\underset{-H^+}{\overset{+H^+}{\rightleftharpoons}}\; R-\overset{\overset{\displaystyle OH}{|}}{\overset{+}{C}}-OR' \;\underset{EO}{\overset{+EO}{\rightleftharpoons}}\; \left[\begin{array}{c} \overset{\displaystyle OH}{|} \\ R-\overset{|}{C}-OR' \\ \overset{|}{\underset{H_2C-CH_2}{\overset{O}{\diagup\!\diagdown}}} \end{array} \right] \;\overset{-H^+}{\rightleftharpoons}\; R-\overset{\overset{\displaystyle O}{\|}}{C}-OCH_2CH_2OR' \tag{16.28}$$

Each step of the process carried out in the presence of an acidic catalyst is reversible. This means that the original ester or an ester-type product can be regenerated.

The calcium-based catalyst probably contains the following components:

$$(RO)_2Ca \quad \underset{b}{ROCaOSO_2OH} \quad \underset{c}{ROCaOSO_2OCaOR} \tag{16.29}$$

$$\underset{a}{}$$

Components a and c may catalyze oxyethylation of FAME according to the reaction scheme proposed by Cox and Weerasooriya [18]. Component b may act as an acidic catalyst. However, component c may also be considered as an acidic catalyst due to the following equilibrium and formation of component d:

$$\underset{c}{ROCaO-\overset{\overset{\displaystyle O}{\|}}{\underset{\underset{O}{\|}}{S}}-OCaOR} \;\rightleftharpoons\; \underset{d}{ROCaO-\overset{\overset{\displaystyle O}{\|}}{\underset{\underset{|\underline{O}|}{|^-}}{\overset{+}{S}}}-OCaOR} \tag{16.30}$$

A bulky cation (Equation 16.28) may be coordinated to the carbonyl oxygen atom instead of a proton. Owing to steric hindrance, such a cation cannot be transferred to the oxygen atom of the alkoxyl group. As a result, the C–OR' bond is weakened, R'OH is not formed, and the oxyethylene group enters between the carbonyl carbon atom and the oxygen atom of the alkoxyl group. The problem needs further thorough studies.

Oxyethylation carried out in the presence of metal oxide composites occurs probably with dissociative chemisorption of the ester [22]. The dissociative chemisorption of $RCOOCH_3$ gives an acyl cation and a methoxy anion bonded to the negative and positive sites of the surface. EO is polarized by interaction with Al acid sites and $^-OCH_2CH_2OCH_3$ is formed. Each succeeding propagation step occurs slower because each oxygen atom of the polyoxyethylene chain can interact with Al acidic sites, increasing the stability of the adsorbed (polyoxyethylene) methoxylate and decreasing the number of active sites.

Recombination of ions gives successive homologues of oxyethylated FAME.

$$
\underset{\substack{\text{O} \\ \| \\ \text{RC}-\text{OCH}_3}}{} \rightleftharpoons \quad \underset{\substack{\text{O} \\ \| \\ \text{RC}^+ \quad {}^-\text{OCH}_3 \\ \overline{} \\ -\text{O}-\text{Al}-\text{O}-\text{Mg}-}}{}
$$

$$
\downarrow \qquad\qquad (16.31)
$$

$$
\underset{\substack{\text{O} \\ \| \\ \text{RC}(\text{OCH}_2\text{CH}_2)_n\text{OCH}_3}}{} \rightleftharpoons \quad \underset{\substack{\text{O} \\ \| \\ \text{RC}^+ \quad {}^-\text{OCH}_2\text{CH}_2(\text{OCH}_2\text{CH}_2)_{n-1}\text{CH}_3 \\ \overline{} \\ -\text{O}-\text{Al}-\text{O}-\text{Mg}-}}{}
$$

There is little information available on the formation of by-products in the direct oxyethylation of FAME. Hama et al. [21,22] did not register any by-products, probably the high-performance liquid chromatography (HPLC) used was not sensitive enough. According to Equation 16.31, recombination of chemisorbed ions gives successive homologues of polyoxyethylated FAME. However, in the presence of moisture, fatty acids and methanol can be formed. Thus, PEG mono- and diesters, PEGs, and PEG monomethyl ethers can be formed as by-products in a reaction with EO.

The type of main by-products and their content depends on the catalyst type, its activity, and reaction conditions. PEG monomethyl ethers $(\text{CH}_3\text{O}(\text{CH}_2\text{CH}_2\text{O})_n\text{OH})$ and PEG diesters $(\text{RCOO}(\text{CH}_2\text{CH}_2\text{O})_n\text{OCR})$ were identified in oxyethylated FAME obtained in the presence of calcium-based catalyst [48]. Liquid chromatography at a critical point of adsorption enabled the separation of PEG or PEG monomethyl ethers [48]. The presence of PEG monomethyl ethers was additionally proven using the matrix-assisted laser desorption ionization time-of-flight mass spectrometry (MALDI-TOF-MS). However, typical PEGs were not identified using this technique. Size exclusion chromatography proved the presence of PEG diesters [48]. The contents of these products depended on the catalyst used and on the reaction conditions.

16.4 SOME PROPERTIES OF OXYETHYLATED FATTY ACID METHYL ESTERS

Oxyethylated FAME exhibits the surface-active and usage properties similar to typical nonionic surfactants, including oxyethylated alcohols [25,49–64]. It is obvious that there are some differences, connected with the somewhat different structure, that is, the presence of a terminal methyl group. As a result, oxyethylated FAME show lower foaming ability, lower clouding points, and different tendency for gelation.

The crucial parameter is, however, the hydrolytic stability of oxyethylated FAME. In the studies performed at room temperature, the samples at pH 3, 5, 7, and 9 were all >98% active after 34 days [13]. At 40°C over a 34-day period at pH 5 and 7, the sample showed only ~2.5% and 4% hydrolysis, respectively. The samples at extreme pH conditions (3 and 9) showed the highest degree of hydrolysis with approximately 10% of the sample being hydrolyzed after 34 days. But 10% aqueous solutions of oxyethylated (60%) methyl tetradecanoate showed that, after more than 11 weeks of storage at 40°C, the degree of hydrolysis was 4% and 13.5% at pH 7 and 9, respectively [18]. These results suggest that the products containing oxyethylated FAME should be formulated at a pH of 9 or below. Hama et al. [21] also confirmed the hydrolytic stability of OMD solutions (1%) in a pH range from 3 to 8 stored at 60°C for 24 h. Rapid hydrolysis was obtained at pH >12.

Oxyethylated FAME show higher biodegradability and lower aquatic toxicity compared to oxyethylated alcohols [65]. Thus, they have a high potential to become an important group of surfactants as new prospective surfactants. However, the development of a new surfactant technology is a slow process. Methyl ester surfactants, narrow-range oxyethylated alcohols, alkyl polyglucosides, and *N*-methylglucamide have been known for many years, but they emerged in the detergent industry only recently. All these surfactants including oxyethylated FAME are derived from oleochemical raw materials and are considered to be "environmentally compatible surfactants."

REFERENCES

1. Schick, M. J. 1967. *Nonionic Surfactants, Surfactant Science Series*, Vol. 1. ed. New York: Marcel Dekker.
2. Schick, M. J. 1987. *Nonionic Surfactants: Physical Chemistry, Surfactant Science Series*, Vol. 23. ed. New York: Marcel Dekker.
3. Santacesaria, E., Iengo, P., Di Serio, M. 1999. *Design and Selection of Performance Surfactants, Annual Surfactants Review*, Vol. 2. Karsa, D. R., ed. Shefield: Shefield Academic Press, pp. 168–215.
4. Di Serio, M., Di Martino, S., Santacesaria, E. 1994. Kinetic of fatty acids polyethoxylation. *Ind. Eng. Chem. Res.* 33: 509–514.
5. Van Os, N. M. 1997. *Nonionic Surfactants: Organic Chemistry, Surfactant Science Series*, Vol. 72. ed. New York: Marcel Dekker.
6. Mukerjee, P., Mysele, K. J. (1971). *Critical Micelle Concentration of Aqueous Surfactant Systems*, Nat. Stand. Ref. Ser., Nat. Bur. Stand., Coden, AL.
7. Miesiac, I., Szymanowski, J. 1988. Micellization properties of ethoxylated carboxylic acid esters. *Tenside Surfact. Deterg.* 25: 174–179.
8. Szymanowski, J., Miesiac, I., Jerzykiewicz, W. 1980. Synthesis and properties of esterification products of some oxyethylated alcohols and alkylphenols with fatty acids, *Fett. Wiss. Technol.* 82: 244–249.
9. Szymanowski, J., Miesiac, I. 1985. Polyoxyethylene glycol monoalkyl ethers and esters of carboxylic acids. *Tenside Surfact. Deterg.* 22: 230–236.
10. Wehle, D., Kremer, G., Wimmer, I. 1995. Verfahren zur Herstellung von Alkoxylaten unter Verwendung von Esterverbindungen als Katalysator. German Patent 4,341,576.
11. Hermann, K., Behler, A., Friedrich, K., Raths, H. C. 1990. Verwendung von calcinierten Hydrotalciten als Katalysatoren für die Ethoxylierung bzw. Propoxylierung von Fettsäureestern. German Patent 3,914,131.
12. Behler, A., Folge, A. 1997. German Patent 19,611,999.
13. Littau, Ch., Miller, D. 1998. Methyl ester ethoxylates. *SOFW-J.* 124: 690–697.
14. Behler, A., Raths, M. L., Guckenbiehl, B. 1966. Neue Zutwicklungen auf dem Gebiet der nichtionischen Tenside. *Tenside Surfact. Deterg.* 33: 64–68.
15. Leach, B. E., Lin, J., Aeschbacher, C. L., Robertson, D. T., Sandoval, T. S., Weerosooriya, U. 1993. Process for alkoxylation of esters and products produced there from. U.S. Patent 5,220,046.
16. Weerasooriya, U., Aeschbacher, C. L., Leach, B. E., Lin, J., Robertson, D. T. 1995. Process for alkoxylation of esters and products produced there from. Reaction of esters with an alkoxide calcium catalysts. U.S. Patent 5,386,045.
17. Cox, M. F., Weerasooriya, U. 1998. Impact of molecular structure on the performance of methyl ester ethoxylates. *J. Surfact. Deterg.* 1: 11–22.
18. Cox, M. F., Weerasooriya, U. 1997. Methyl ester ethoxylates. *J. Am. Oil Chem. Soc.* 74: 847–859.
19. Leach, B., Shannon, M., Wharry, D. 1988. Alkoxylation process using calcium-based catalysts: alkoxylated alcohols, calcium compound and organoaluminum compound. U.S. Patent 4,775,653.
20. Nakamura, H., Hama, I., Fujimori, Y., Nakamoto, Y. 1994. Method and manufacturing of fatty acid esters of polyoxyalkylene alkyl ethers. U.S. Patent 5,374,750.
21. Hama, I., Okamoto, T., Nakamura, H. 1995. Preparation and properties of ethoxylated fatty methyl ester nonionics. *J. Am. Oil Chem. Soc.* 72: 781–784.
22. Hama, I., Samasoto, H., Okamoto, T. 1997. Influence of catalyst structure on direct ethoxylation of fatty methyl esters over Al–Mg composite oxide catalyst. *J. Am. Oil Chem. Soc.* 74: 817–822.
23. Hama, I., Okamoto, T., Hidai, E., Yamada, Y. 1997. Direct ethoxylation of fatty methyl ester over Al–Mg composite oxide catalyst. *J. Am. Oil Chem. Soc.* 74: 19–24.
24. Szymanowski, J. 2001. Ethoxylation of fatty acid methyl esters. *Riv. Ital. Sostanze Grasse* 5: 279–284.
25. Bialowas, E., Szymanowski, J. 2004. Catalysts for oxyethylation of alcohols and fatty acid methyl esters. *Ind. Eng. Chem. Res.* (in press).

26. Hermann, K., Behler, A., Endres, H., Friedrich, K. 1989. Use of calcined hydrotalcites as ethoxylation or propoxylation catalysts. German Patent 3,843,713.
27. Raths, H. C., Breuer, W., Friedrich, K., Hermann, K. 1994. Use of hydrophobized hydrotalcites as catalysts for ethoxylation or propoxylation. U.S. Patent 5,292,910.
28. Wolf, G., Burkhart, B., Lauth, G., Trapp, H., Oftring, A. 1998. Preparation of alkoxylation products in the presence of mixed hydroxides modified with additives. U.S. Patent 5,741,947.
29. Breuer, W., Raths, H. C. 1994. Hydrophobized double layer hydroxide compounds. U.S. Patent 5,326,891.
30. Nakamura, H., Nakamoto, Y., Fujimori, Y. 1991. Alkoxylation catalyst. U.S. Patent 5,012,012.
31. Imanaka, T., Tanaka, T., Kono, J., Nagumo, H., Tamaura, H. 1997. Alkoxylation catalyst, process for preparation of the catalyst and process for preparing alkoxylate with the use of the catalyst. U.S. Patent 5,686,379.
32. Okamoto, T., Uemura, S., Hama, I. 2003. Alkoxylation catalyst and method for producing the same, and method for producing alkylene oxide adduct using the catalyst. U.S. Patent 6,504,061.
33. Hama, I., Okamoto, T., Sasamoto, H., Nakamura, H. 1998. Method of producing an alkylene oxide adduct of a compound having one or more active hydrogens. U.S. Patent 5,750,796.
34. Hreczuch, W. 1999. Comparison of the kinetics and composition of ethoxylated methyl dodecanoate and ethoxylated dodecanol with narrow and broad distribution of homologues. *J. Surfact. Deterg.* 2: 287–292.
35. Hreczuch, W. 2002. Temperature-related reaction kinetics and product composition of ethoxylated fatty acid methyl esters. *J. Chem. Technol. Biotechnol.* 77: 511–516.
36. Alejski, K., Bialowas, E., Hreczuch, W., Trathnigg, B., Szymanowski, J. 2003. Oxyethylation of fatty acid methyl esters. Molar ratio and temperature effects. Pressure drop modeling. *Ind. Eng. Chem. Res.* 42: 2924–2933.
37. Hreczuch, W., Bekierz, G., Pyżalski, K., Waćkowski, J., Tomik, Z., Szymanowski, J., Rolnik, K., Roguska, D., Siwek, Z., Wietrzyńska-Lalak, Z., Dziwiński, E., Stempińska, T., Domarecki, W., Zdunek, A. 2000. Pat PCT Polish Patent Appl. No. 338,530.
38. Hreczuch, W., Bekierz, G., Pyżalski, K., Wackowski, J., Tomik, Z., Szymanowski, J., Rolnik, K., Roguska, D., Siwek Z., Wietrzyńska-Lalak, Z., Trathnigg, B., Dziwiński, E., Stempińska, T., Domarecki, W., Zdunek, A. 2001. An alkoxylation catalyst and a method to manufacture the alkoxylenation catalyst, Global Patent Appl. No. PCT/PL 01/00020.
39. Di Serio, M., Tesser, R., Felippone, F., Santacesaria, E. 1995. Ethylene oxide solubility and ethoxylation kinetics in the synthesis of nonionic surfactants. *Ind. Eng. Chem. Res.* 34: 4092–4099.
40. Di Serio, M., Vairo, G., Iengo, P., Felippone, F., Santacesaria, E. 1996. Kinetics of ethoxylation and propoxylation of 1- and 2-octanol catalyzed by KOH. *Ind. Eng. Chem. Res.* 35: 3848–3853.
41. Santacesaria, E., Di Serio, M., Gelosa, D., Lisi, L. 1990. Kinetics of nonylphenol polyethoxylation catalyzed by potassium hydroxide. *Ind. Eng. Chem. Res.* 29: 719–725.
42. Dimiccoli, A., Di Serio, M., Santacesaria, E. 2000. Key factors in ethoxylation and propoxylation technology. *Proceedings of the 5th World Surfactants Congress.* Firenze, Federchimica Assobase—P.I.T.I.O. 1: 99–110.
43. Fredelslund, A., Gmehling, J., Rasmussen, P. 1977. *Vapor–Liquid Equilibrium Using UNIFAC.* Amsterdam: Elsevier.
44. Alejski, K., Bialowas, E., Hreczuch, W., Szymanowski, J. 2003. Modeling of fatty acid methyl esters oxyethylation. *Riv. Ital. Sostanze Grasse* 80: 317–322.
45. Alejski, K., Bialowas, E., Hreczuch, W., Szymanowski, J. 2003. Kinetic model of oxyethylation of fatty acid methyl esters. *Polish J. Chem. Tech.* 4: 50, 51.
46. Alejski, K., Bialowas, E., Hreczuch, W., Trathnigg, B., Szymanowski, J. 2003. Oxyethylation of fatty acid methyl esters. Kinetics and modeling. *Comunicaciones 33 Jornadas Anuales del CED*, Barcelona, *Proc. Jorn. Com. Esp. Deterg.* 33: 185–191.
47. Alejski, K., Bartkowiak, P., Szymanowski, J. 2004. Modeling of fatty acid methyl ester oxyethylation in a semi-batch reactor with successive dosing of ethylene oxide. *Tenside Surfact. Deterg.* 41: 130–134.
48. Trathnigg, B., Hreczuch, W. 2000. Characterization of fatty ester ethoxylates by coupled chromatographic techniques. *Proceedings of the 5th World Surfactant Congress.* Firenze, 1: 472–480.
49. Lin, S. Y., Tsai, R. Y., Lin, L. W., Chen, S. I. 1996. Adsorption kinetics of $C_{12}E_8$ at the air–water interface: adsorption onto a clean interface. *Langmuir* 12: 6530–6536.
50. Chang, H. C., Hsu, C. T., Lin, S. Y. 1998. Adsorption kinetics of $C_{10}E_8$ at the air–water interface. *Langmuir* 14: 2476–2484.

284   Handbook of Detergents/Part F: Production

51. Hreczuch, W., Szymanowski, J., Bekierz, G., Pyżalski, K. 1999. Comparison of the synthesis and composition of directly oxyethylated fatty acid methyl esters and oxyethylated alcohols with narrow and broad range distribution of homologues. *Proceedings XXIX Journadas Anuales del CED*. Barcelona, 351–358.

52. Siwek, Z., Hreczuch, W., Szymanowski, J. 1999. Comparative gas chromatography analysis of ethoxylated fatty acid methyl esters with packed and capillary columns. In: *Proceedings of the 4th World Conference and Exhibition on Detergents: Strategies for the 21st Century*, Montreux, 1998, Cahn, A., ed. Champaign, IL: AOCS Press, Vol. 215, pp. 273–383.

53. Szymanowski, J., Makowska, D., Hreczuch, W. 1999. Surface activity of ethoxylated methyl dodecanoate. In: *Proceedings of the 4th World Conference and Exhibition on Detergents: Strategies for the 21st Century*, Montreoux, 1998, Cahn, A., ed. Champaign, IL: AOCS Press, Vol. 7, pp. 255–260.

54. Makowska, D., Hreczuch, W., Szymanowski, J. 1999. Comparison of surface activity of oxyethylated methyl esters of fatty acid and oxyethylated alcohols. *Proceedings XXIX Jornadas Annales del CED*. Barcelona, 367–376.

55. Makowska, D., Hreczuch, W., Zimoch, J., Bogacki, M. B., Szymanowski, J. 2001. Surface activity of the mixtures of oxyethylated methyl dodecanoate and sodium dodecylbenzenesulfonate. *J. Surfact. Deterg.* 4: 121–126.

56. Makowska, D., Materna, K., Hreczuch, W., Szymanowski, J. 2000. Synthesis and properties of oxyethylated methyl dodecanoates. *Proceedings of the 5th World Surfactants Congress*. Firenze, 1: 392–401.

57. Hreczuch, W., Trathnigg, B., Dziwiński, E., Pyżalski, K. 2000. Direct ethoxylation of a longer chain aliphatic ester. *Proceedings of the 5th World Surfactants Congress*. Firenze, 1: 357–366.

58. Behler, A., Syldath, A. 2000. Fatty acid methyl ester ethoxylates—a new class of nonionic surfactants. *Proceedings of the 5th World Surfactants Congress*. Firenze, 1: 382–391.

59. Hreczuch, W., Szymanowski, J. 2001. Synthesis of ethoxylated fatty acid methyl esters. Discussion of reaction pathway. *Proceedings XXXI Jornadas Anuales del CID*. Barcelona, 167–177.

60. Makowska, D., Sobczynska, A., Zimoch, J., Szymanowski, J. 2001. Interfacial activity of ethoxyethylated fatty acid methyl esters. Mixtures with sodium dodecylbenzene sulphonate. *Proceedings XXXI Jornadas Anuales del CID*. Barcelona, 325–336.

61. Hreczuch, W. 2001. Ethoxylated rape seed oil acid methyl esters as new ingredients for detergent formulations. *Tenside Surfact. Det.* 38: 72–79.

62. Zimoch, J., Hreczuch, W., Trathnigg, B., Meissner, J., Bialowas, E., Szymanowski, J. 2002. Detergency and dynamic surface tension of oxyethylated fatty acid methyl esters. *Tenside Surfact. Det.* 39: 8–16.

63. Bialowas, E., Hreczuch, W., Trathnigg, B., Szymanowski, J. 2003. Static and dynamic surface tension of oxyethylated rape seed fatty acid methyl esters. In: *Reinventing the Industry: Opportunities and Challenges*. Cahn A., ed., Champaign, IL: AOCS Press, pp. 166–172.

64. Bialowas, E., Szymanowski, J. 2003. Static and dynamic surface tension of aqueous solutions containing oxyethylated rape seed methyl esters synthesized at various temperatures. In: *Comunicaciones 33 Jornadas Annales del CED*, Barcelona 2003, *Proc. Jorn. Com. Esp. Deterg.* 33: 193–201.

65. Szwach, I., Hreczuch, W., Fochtman, P. 2002. Biodegradability, ecotoxity and prognostic evaluation of environmental fate of oxyethylated dodecanol and oxyethylated rapeseed acid methyl esters as non-ionic surfactants. In: *Proceedings of the 4th World Conference on Detergents: Strategies for the 21st Century*, Cahn, A., ed. Champaign, IL: AOCS Press, pp. 163–166.

17 Production of Silicone Surfactants and Antifoam Compounds in Detergents

Anthony J. O'Lenick, Jr. and Kevin A. O'Lenick

CONTENTS

17.1 Background 286
17.2 Chemistry for Synthesis of Raw Materials 286
    17.2.1 Rochow Process 286
17.3 Silicone Derivatives 287
    17.3.1 Nomenclature 287
        17.3.1.1 Construction 288
17.4 Hydrophobic Silicone Compounds 288
    17.4.1 Silicone Fluids..288
        17.4.1.1 Softening Benefits of Silicone Fluids.........................289
        17.4.1.2 Process and Technology for Silicone
                Fluids (Equilibration)...289
    17.4.2 Silanols..290
        17.4.2.1 Process and Technology 290
    17.4.3 Alkyl Fluoro Silicone Fluids...290
17.5 Silicone Antifoam Compounds...291
    17.5.1 Background 291
        17.5.1.1 Mechanism of Antifoam 291
    17.5.2 Antifoam for Detergents 292
    17.5.3 Process for Making Antifoam...292
17.6 Silicone Surfactants 293
    17.6.1 Background 293
    17.6.2 Dimethicone Copolyol 296
        17.6.2.1 Process and Technology 297
17.7 Relative Economics..299
17.8 Future Trends 299
    17.8.1 Antifoam Compounds..299
17.9 Conclusion..299
References 299

17.1 BACKGROUND

Silicone compounds have been known since 1860, but were of little commercial interest until the 1940s.[1] This is partially due to the lack of cost-effective production processes. The development of the Rochow process[2] in the mid-1940s made silicone compounds commercially viable. Over the years, silicone compounds have received growing acceptance in many applications, but have been used only in very specific applications in detergent formulations. This is in large part due to the higher cost of the silicone surfactant relative to the fatty or petroleum-based detergents that have been the workhorse in the industry. However, silicone compounds have enjoyed usage in detergent products where the attributes of the silicone-based compound provide some unique property in formulation. Namely, either silicone compounds augment traditional surfactants, providing a boost in detergency or another property such as softness, soil repellency, fiber lubrication, or provide very effective anti-foam attributes to the detergent formulation. Despite their high price per pound, silicone compounds are generally used at low concentrations, making their use cost-effective in many applications.

The term *silicone* is actually a misnomer. It was incorrectly thought that the early silicone polymers were silicon-based ketones, hence the contraction silicone. Despite this error, the term is still widely used and accepted. The vast majority of the volume of silicone compounds used in all market segments are silicone fluids. They are the oldest and perhaps the best understood of the silicone compounds in use today.

In addition to silicone fluids, there has been a recent explosion in the availability of chemically modified silicone compounds that provide conditioning, softening, irritation mitigation, barrier properties, emulsification, and antifoam properties.[3] Many of these desirable effects are not achievable using traditional hydrocarbon-based compounds or silicone fluids. Alternatively, many of these new multifunctional compounds are effective at concentrations far below those needed to achieve the same effect using traditional compounds.

17.2 CHEMISTRY FOR SYNTHESIS OF RAW MATERIALS

17.2.1 ROCHOW PROCESS

One of the major synthetic pathways used to make silicone compounds is the Rochow process. This process is named after Eugene G. Rochow, the father of silicone chemistry. The process is very capital-intensive and practiced on a commercial scale by only six companies worldwide.

The ratio of each specific chlorosilane produced using this process is dependent on the stoichiometry, catalyst, promoter, and exact conditions utilized. Most major manufacturers try to optimize the yield of dichloro-dimethyl silane, the raw material for the synthesis of linear polymers. The synthesis of chlorosilanes[1] is as follows:

$$Si + CH_3Cl \longrightarrow \begin{array}{ll} (CH_3)_2-Si-Cl_2 & 70-90\% \\ (CH_3)_3-Si-Cl & 3-15\% \\ CH_3-Si-(Cl)_3 & 3\% \\ Si-(CH_3)_4 & 0.1\% \\ CH_3HSiCl_2 & 1-3\% \\ CH_3HSiCl & 0.5\% \\ \text{Others} & \end{array}$$

The products of the Rochow process are hydrolyzed by reaction of the chlorosilane mixture with water to produce a mixture called hydrolyzate:

$$\text{Rochow process products} + \text{water} \xrightarrow[-HCl]{} \text{hydrolyzate}$$

Production of Silicone Surfactants and Antifoam Compounds 287

17.3 SILICONE DERIVATIVES

Hydrolyzate is a major raw material from which other silicone compounds are prepared. Silicone derivatives can then be made using hydrolyzate or its various fractions. The unit operations used to fractionate, polymerize, and otherwise refine hydrolyzate result in many commercially important compounds including silicone fluids, gums, cyclic compounds, silanol compounds, organofunctional silicone compounds, and silicone surfactants.

17.3.1 NOMENCLATURE

Silicone compounds are generally referred to a nomenclature system, which gives a unique name to each type of substitution pattern around the silicon atom. If the methyl group (Me) is replaced with another substituent, the group on which the replacement occurs is given a * (star). The following structure is typical of this nomenclature:

```
          Me
          |
   Me---Si----O--        M unit
          |
          Me


          Me
          |
   Me---Si----O--        M* unit
          |
          R


          Me
          |
 --O---Si----O--         D unit
          |
          Me


          Me
          |
 --O---Si----O--         D* unit
          |
          R


          O--
          |
 --O---Si----O--         T unit
          |
          Me


          O--
          |
 --O---Si----O--         T* unit
          |
          R


          O--
          |
 --O---Si----O--         Q unit
          |
          O--
```

17.3.1.1 Construction

Each of the units in the preceding structure are linked together into a polymer chain using chain extension polymerization. The resulting polymer themselves are also identified by a shorthand nomenclature. If the substitution is within the silicone polymer, that is, based on D*, a comb polymer is achieved.

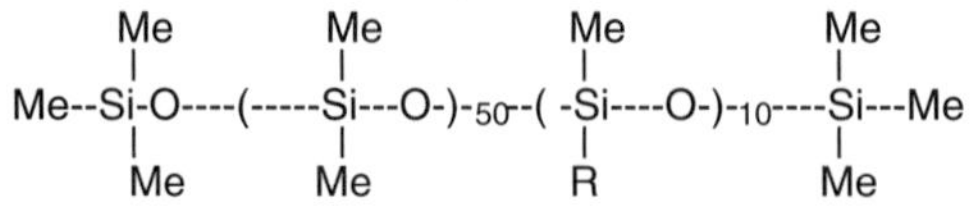

$$
\begin{array}{ccccccc}
& Me & & Me & & Me & & Me \\
& | & & | & & | & & | \\
Me{-}Si{-}O{-}{-}{-}{-}({-}{-}{-}{-}{-}Si{-}{-}{-}O{-}){-}_{50}{-}({-}Si{-}{-}{-}{-}O{-}){-}_{10}{-}{-}{-}{-}Si{-}{-}{-}Me \\
& | & & | & & | & & | \\
& Me & & Me & & R & & Me
\end{array}
$$

The preceding polymer is referred to as $M\ D_{50}\ D^*_{10}\ M$. Me is methyl. If, however, the substitution is achieved using a M* compound, a terminal polymer results.

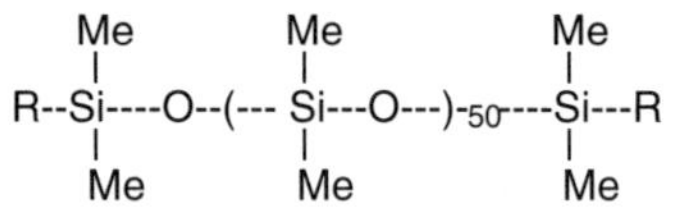

$$
\begin{array}{ccccc}
& Me & & Me & & Me \\
& | & & | & & | \\
R{-}Si{-}{-}{-}{-}O{-}{-}({-}{-}{-}Si{-}{-}{-}O{-}{-}{-}){-}_{50}{-}{-}{-}Si{-}{-}{-}R \\
& | & & | & & | \\
& Me & & Me & & Me
\end{array}
$$

This polymer is referred to as $M^*\ D_{50}\ M^*$. Me is methyl.

Compounds can be synthesized that have substitution in both the terminal and the comb positions. Such compounds are called a mixed polymer.

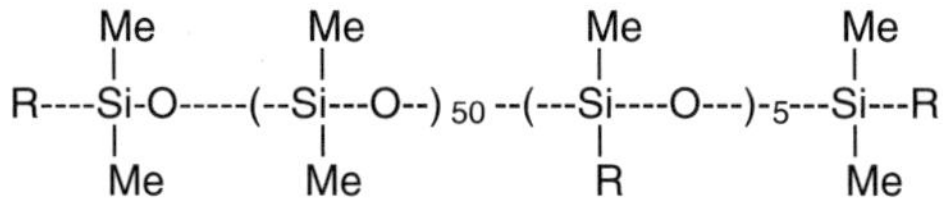

$$
\begin{array}{ccccccc}
& Me & & Me & & Me & & Me \\
& | & & | & & | & & | \\
R{-}{-}{-}{-}Si{-}O{-}{-}{-}{-}{-}({-}{-}Si{-}{-}{-}O{-}{-})\ _{50}\ {-}{-}({-}{-}{-}Si{-}{-}{-}{-}O{-}{-}{-}){-}_{5}{-}{-}{-}Si{-}{-}{-}R \\
& | & & | & & | & & | \\
& Me & & Me & & R & & Me
\end{array}
$$

This polymer is referred to as $M^*\ D_{50}\ D^*_5\ M^*$. Me is methyl.

17.4 HYDROPHOBIC SILICONE COMPOUNDS

17.4.1 SILICONE FLUIDS

Silicone fluids have only methyl (Me) substitution and are composed only of M and D units. Therefore, they are M—D—M polymers. The structure of a silicone fluid is as follows:

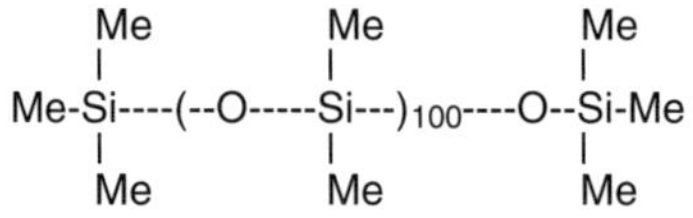

$$
\begin{array}{ccccc}
& Me & & Me & & Me \\
& | & & | & & | \\
Me{-}Si{-}{-}{-}{-}({-}{-}O{-}{-}{-}{-}{-}Si{-}{-}{-})_{100}{-}{-}{-}{-}O{-}{-}Si{-}Me \\
& | & & | & & | \\
& Me & & Me & & Me
\end{array}
$$

where Me is methyl.

The shorthand for this compound would be $MD_{100}M$. Knowing the number of D units in a fluid will allow for the calculation of the molecular weight of a specific molecule. This in turn will have a direct effect on viscosity. Consequently, silicone fluids are sold by viscosity, which is an indication of, and is much easier to determine than molecular weight.

Viscosity	Approximate Molecular Weight
50	3780
100	5970
350	13,650
1000	28,000
10,000	62,700
60,000	116,500

Production of Silicone Surfactants and Antifoam Compounds

One of the significant properties of silicones that make it of interest in detergents is the ability to lower surface tension at the interface. Fatty surfactants lower the surface tension from 72 dynes/ cm^2 to around 30 dynes/cm^2. Silicone compounds lower the surface tension to 20 dynes/cm^2. This reduction leads to improved fiber-to-fiber lubrication and softening. The difficulty is to incorporate silicone into detergents used in aqueous systems. One approach is to use emulsion technology and another is to use coacervate technology. A coacervate is a spherical aggregation of silicone molecules making up a colloidal inclusion, which is held together by hydrophobic forces. More plainly stated it is usually a little ball of silicone that is formed by the repulsion of water by something like an oil. Any insoluble silicone can be made useful using coacervate technology.

17.4.1.1 Softening Benefits of Silicone Fluids

There has been a growing market acceptance of liquid laundry formulations that combine detergent and fabric softening functions into single products. Commonly called softergents, these formulations are finding increased market acceptance. However, the development of cost-effective softergents is not without challenge to the formulator. Since most softergent formulations incorporate cationic materials such as monoalkyltrimethyl ammonium halide, which serves as a softening agent, and traditional anionic detergents, problems may occur. Quaternary ammonium compounds often form complexes with anionic surfactants and precipitate out of solution. The result can be unsightly deposits on the surface of fabrics or entrapped within fibers. In the washing solution itself, the formation of these complexes depletes the concentration of cleaning surfactants and softening agents available to treat the fabric and results in ineffective product performance.

To minimize anionic–cationic interaction in softergent systems, the ratio of components may be changed. But here again, the net effect of this change is either less than optimal levels of anionic surfactants, resulting in poor cleaning, or less effective softening, due to less optimal cationic content.

This problem has been overcome by the addition of silicone softening agents such as 2000-viscosity silicone fluid, which can improve overall softergent performance by providing the desired softening effect without interacting with anionic components of the formulation.

Amine functional silicones can also provide softening benefits comparable to those obtained with cationic rinse cycle softeners, but these silicones are expensive and can cause yellowing in fabric after repeated washing. Silicone fluids in softergent formulations do not cause yellowing.

U.S. Patent 4,767,548 to Dow Corning discloses the use of a silicone fluid in a dryer sheet. U.S. Patent 5,524,269 to Lever Brothers discloses the use of a silicone fluid on dryer sheets and that the incorporation of a dimethicone copolyol, which improves the uniformity of the sheet and deposition on the fabric.[4]

17.4.1.2 Process and Technology for Silicone Fluids (Equilibration)

Silicone fluids can be made by either anionic (acid catalyst) or cationic (base catalyst) polymerization of cyclic silicone compounds and MM. Free radical initiators are not useful in the reaction, because of the nature of the silicone bond. The reactions are shown as follows:

$$100 \ Me\text{-}(Si\text{-}O)_4 + Me\text{-}Si\text{--}O\text{-}Si\text{--}Me \xrightarrow{2\% \ H_2SO_4} Me\text{-}Si\text{---}(\text{-}O\text{--}Si\text{-})_{400}\text{--}O\text{---}Si\text{---}Me$$

Cyclic — mm — Silicone fluid

where Me is methyl.

The equilibration can be carried out at ambient temperatures with good agitation. The acid is neutralized and the residual MM stripped off.

17.4.2 SILANOLS

Silicone compounds which have a Si—OH bond are referred to as silanols. These compounds are also referred to as dimethiconol in the personal care market. Silanols (Si—OH) are very different from carbanols (CH$_2$—OH) in that the former easily undergo condensation reactions to form higher-molecular-weight silanols and water. Carbanols do not undergo this reaction readily, since the resulting compounds will be ethers. Silanols conform to the following structure:

$$
\begin{array}{ccccc}
& Me & & Me & & Me \\
& | & & | & & | \\
H\text{-}O\text{-}Si\text{----}(&\text{--}O\text{-----}Si\text{---})_a&\text{----}O\text{--}Si\text{-}O\text{-}H \\
& | & & | & & | \\
& Me & & Me & & Me
\end{array}
$$

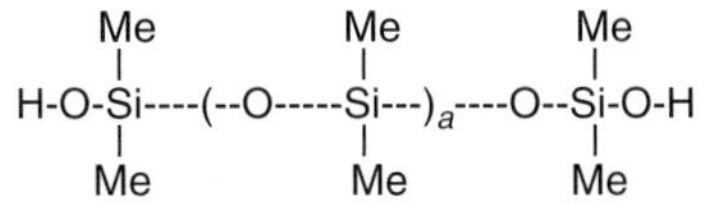

Silanols are reactive silicone compounds which when reacted with silica gives very effective antifoam compounds. Silanol compounds have also been used in softergent applications, in many cases giving better, more durable softening than silicone fluids.

17.4.2.1 Process and Technology

Silanols can be made by either anionic (acid catalyst) or cationic (base catalyst) polymerization of cyclic silicone compounds. Free radical initiators are not useful in the reaction, because of the nature of the silicone bond. The reaction can be shown as follows:

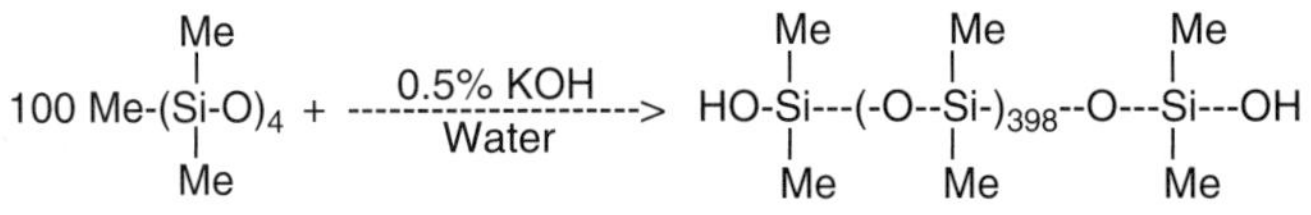

$$
\begin{array}{l}
Me \\
| \\
100\ Me\text{-}(Si\text{-}O)_4\ +\ \xrightarrow[\text{Water}]{0.5\%\ KOH}\ \ HO\text{-}Si\text{---}(\text{-}O\text{--}Si\text{-})_{398}\text{--}O\text{---}Si\text{---}OH \\
| \\
Me
\end{array}
$$

Silanols can be made into emulsions that can be added to softeners. They provide an excellent hand without yellowing.

17.4.3 ALKYL FLUORO SILICONE FLUIDS

Recently, a new series of fluoro-containing silicone fluids have been introduced. In addition to being nonyellowing softeners, these materials have been found to provide outstanding fiber to metal lubrication and good spreadability on surfaces. These compounds conform to the following structure:

$$
\begin{array}{ccccccc}
& Me & & Me & & Me & & Me \\
& | & & | & & | & & | \\
Me\text{---}Si\text{-}(&\text{--}O\text{---}Si\text{--})_x&\text{-----}(\text{-}O\text{---}Si\text{--})_y&O\text{------}Si\text{--}Me \\
& | & & | & & | & & | \\
& Me & & (CH_2)_{17} & & (CH_2)_2 & & Me \\
& & & | & & | \\
& & & Me & & (CF_2)_n\text{-}CF_3
\end{array}
$$

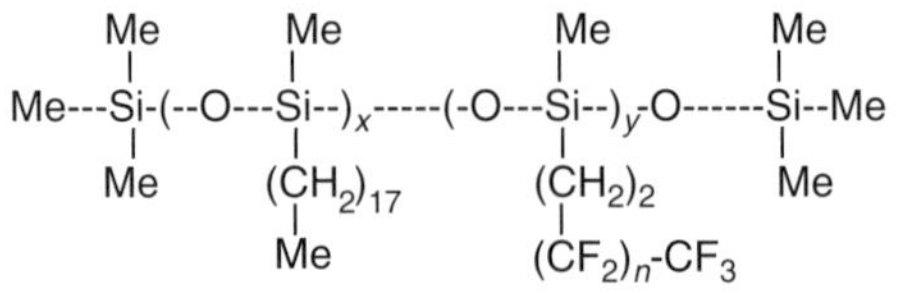

These products dramatically reduce fiber to metal friction and make outstanding ironing aids. Their superior fiber for metal lubrication helps prevent piling of fabrics and therefore can extend garment life.[5]

17.5 SILICONE ANTIFOAM COMPOUNDS

17.5.1 BACKGROUND

By virtue of their structure, surfactants perform several functions in aqueous solution. Often, however, there are processes in which one uses a surfactant for a desired property and does not want the other properties inherent in the surfactant. For example, one may want detergency without foam. Modification of the surfactant molecule offers minimal relief. Consequently, antifoam compounds are added to many processes.

Antifoam agents are divided into three classes: (a) those compounds used in industrial applications; (b) those compounds used in applications sanctioned under 21 C.F.R. 173.105, 173.340, or 173.300; and (c) those compounds which have been modified to meet specific performance requirements.

Since most silicone compounds are water insoluble, they simply float on water as oily liquids. This attribute of water insolubility of many silicone compounds makes them useful as antifoam compounds. The term *antifoam* is generally used to denote a compound with the ability to prevent foam formation. In contrast, the term *defoamer* is generally used to denote a material which will knock down existing foam. Finally, a deaerating agent will break up entrained foam. Although some types of compounds are better when used in defoaming applications and some compounds are better when used in antifoaming applications, most compounds have properties which make them useful in both applications.

Foam in machine dish and laundry applications needs to be carefully controlled. Too much foam can overflow the machine. There are several inherent properties of antifoam compounds desired for detergent applications. The antifoam must have the ability to

1. Remain effective over a wide range of temperature
2. Be stable over a wide range of pH
3. Be nonspotting, that is, to be rinsed off completely with water
4. Be safe for contact with food

Antifoam compounds for use in detergents can be found in many types of formulated products including dishwashing detergents and rinse aids, home laundry formulations, home laundry softeners, window cleaners, alkaline metal degreasers, hard-surface cleaners, acid cleaning of metals, bottle-cleaning formulations, and many others.

A major application for antifoam compounds is in aqueous formulations such as detergents and softeners. A low concentration of antifoam compound is present in the formulation to allow more efficient filling of the bottles. In many operations, foam will cause the bottle to overflow, shutting down the bottling operation.

Antifoam compounds are often used during the wash out and clean-up operations in detergent plants. In addition, many detergent plants use antifoam compounds in their effluent to control foam.[6]

17.5.1.1 Mechanism of Antifoam

There are two mechanisms by which antifoam compounds work. The first is by destroying interfacial films, and the other is by impairing foam stability. The former is more commonly used and more effective in most applications. A layer of antifoam, by virtue of its insolubility, ends up in between the bubble and where it contacts the water. This dislodges the bubble and breaks it.

Silicone fluids *per se* have both antifoam and defoaming attributes, they can be modified by reaction with silica to make significantly more efficient antifoam compounds. Silicone-based antifoam compounds for use in detergents are composed of two major components: silicone fluid and hydrophobic silica. The fluid polymer acts as a carrier to deliver the silica particles to the foam air–water interface, where film rupture then occurs.

Very efficient antifoam compounds can be prepared by the reaction of silanol compounds with silica. The so-called *in situ* hydrophobized silica makes very efficient antifoam compound.

Silicone fluid is clear, colorless; insoluble in aqueous media; has a very low surface tension of 22 dyn/cm; spreads spontaneously in most aqueous solutions; chemically inert; and difficult to emulsify. The hydrophobic particulate silica is extremely effective at breaking bubbles by disrupting the contact point between the foam and water.

The performance of silicone-based antifoam compounds is independent of water hardness. They are effective at very low addition levels in all types of surfactant systems normally present in detergent formulations and across a wide range of use conditions. Furthermore, silicones cause no yellowing on fabric. Thus, silicone-based antifoam compounds have a number of benefits over soap-based foam control systems.

Hundred percent–active silicone-based antifoam compounds are normally referred to as silicone antifoam compounds. If the silicone antifoam is in water, it is referred to as antifoam emulsions. Mixtures of silicone antifoam compounds with nonaqueous dispersion or delivery systems also exist, to aid their dispersion in aqueous media.

The term *emulsion* applied to aqueous silicone antifoam is a misnomer. The compositions are actually thickened dispersions. Addition of water will cause them to separate into two layers. If water is added, the dispersion needs to be rethickened with polyacrylate or a similar thickener.

17.5.2 Antifoam for Detergents

Owing to the turbulent washing action of the domestic front-loading automatic washing machines and low water levels used during wash and rinse cycles, foam generation during washing can be considered. It is essential, however, that too much foam during the wash be avoided to ensure good cleaning power. Thus, the inclusion of foam control agents in heavy-duty detergents has led to the "low suds" powder and liquid brands sold for home laundry.

For formulators of laundry detergents, there is a constant challenge to meet the changing needs of the consumer market. High cleaning performance and ingredient compatibility are two important aspects of the desired product profile, but high foam generation is a frequently encountered problem.

The unique properties of silicone-based antifoam compounds have led to their wide use by the detergent industry. They are active at low use levels in all types of surfactant systems present in detergent formulations. Many custom-formulated antifoam compounds have been developed for use in detergents.

U.S. Patent 5,589,449 to Dow Corning discloses a particulate foam control agent on a zeolite carrier. This composition is designed to make the formulation of antifoam into a powdered detergent easier.

17.5.3 Process for Making Antifoam

Silicone antifoam compounds are made by hydrophobizing silica in silicone fluid. The process is conducted at a temperature of ~200°C in the presence of a base catalyst, typically KOH. The powdered silica is dispersed into silicone fluid and mixed well. The KOH is added and dispersion is heated. Once the reaction temperature is reached, water gets distilled. Over a period of 4–8 h, the dispersion goes from opaque to translucent.

Depending on the application, the amount of silica ranges from 10 to 20%, with 15% being typical. The resulting translucent material is commonly called antifoam compound and can be used at 100%, but typically is made into a thickened emulsion at 10 and 30%.

Production of Silicone Surfactants and Antifoam Compounds   **293**

The performance of the compound and resulting emulsions is dependent on the amount of silica added and viscosity of the silicone fluid used. In cases where a defoaming is desired as a function of temperature of wash, a silicone glycol can be added to help in dispersability, a silicone glycol when heated will become insoluble and consequently aid in defoaming. On cooling, it will become soluble again aiding in rinsing.

In addition to antifoam compounds, various modified formulator-friendly antifoams are available. These include encapsulated antifoam compounds. A method of encapsulating antifoam is by placing it into powders that can be made granular. Silicone compound is a translucent liquid, which can be hard to place into powder detergents. Placing the compound on a material that can be ground to a specified particle size makes formulation easier.

17.6   SILICONE SURFACTANTS

17.6.1   BACKGROUND

To make silicone fluids useful in aqueous systems, there are various emulsions available. The use of an emulsion makes it easier to handle, and use in formulations, but there are issues related to emulsion stability, which must be addressed. Specifically, the addition of surfactants by the formulator to a purchased emulsion may shift the Hydrophile–Lipophile Balance (HLB) and split the emulsion.

HLB SYSTEM

HLB, the so-called Hydrophile–Lipophile Balance, is the ratio of oil-soluble and water-soluble portions of a molecule. The system was originally developed for ethoxylated products. Listed in Table 17.1 are some approximations for the HLB value for surfactants as a function of their solubility in water. Values are assigned based upon that table to form a one-dimensional scale, ranging from 0 to 20.

We are using the generic term *hydrocarbon* to designate the oil-soluble portion of the molecule. This generic term includes the more specific terms *fatty*, *lipid*, and *alkyl*.

There are two basic types of emulsions envisioned by the current HLB system. They are oil in water (O/W) and water in oil (W/O). The phase listed first is the discontinuous phase. That is, it is the phase that is emulsified into the other. Upon mixing of the two phases with a surfactant present, the emulsifier forms a third phase as a film at the interface between the two phases being mixed together. It is also predicted that the phase in which the emulsifier is most soluble will become the continuous phase. The continuous phase need not be the predominant quantity of material present. There are emulsions where the discontinuous phase makes up a greater weight percent than the continuous phase. A simple test is if the emulsion is readily diluted with water, water is the continuous phase.

TABLE 17.1

Solubility in Water	HLB Value	Description
Insoluble	4–5	Water-in-oil emulsifier
Poorly dispersible (milky appearance)	6–9	Wetting agent
Translucent to clear	10–12	Detergent
Very soluble	13–18	Oil-in-water emulsifier

CALCULATION OF HLB

The HLB system, in its most basic form, allows for the calculation of HLB using the following formulation:

$$HLB = \frac{\text{Percentage of Hydrophile by weight of molecule}}{5}$$

Example: Oleyl alcohol 5 E.O.

$$\frac{\text{Molecular weight Hydrophile (5) (44)} = 220}{\text{Total molecular weight of molecule}} = 45.0\%$$

$$HLB = \frac{45\%}{5} = 9.0.$$

APPLICATION OF HLB

One can predict the approximate HLB needed to emulsify a given material and make more intelligent estimates of which surfactant or combinations of surfactants are appropriate to a given application. When blends are used, the HLB can be estimated by using a weighted average of the surfactants used in the blend.

HLB NEEDED TO EMULSIFY[2]

Acetophenone	14	Lanolin	12
Acid, lauric	16	Lauryl amine	12
Acid, oleic	17	Mineral spirits	10
Beeswax	9	Nonylphenol	14
Benzene	15	Orthodichlorobenzene	13
Butyl stearate	11	Pine oil	16
Carbon tetrachloride	16	Toluene	15
Castor oil	14	Xylene	14
Chlorobenzene	13	Kerosene	14
Cottonseed oil	6	Cyclohexane	15
Petrolatum	7	Chloronated paraffin	8

For those materials that are not listed above, it is recommended that the oil be tested using specific blends of know *n* emulsifiers. This allows the formulator to calculate the HLB needed to emulsify the nonlisted oil.

The appearance of the emulsion is dependant on the particle size of the discontinuous phase. Particle size is listed in nanometers.

Particle Size	
Size	Appearance
>1	White
0.1–1.0	Blue white
0.05–0.1	Translucent
<0.05	Transparent

In addition, emulsions have a limited freeze-thaw stability. Finally, there is an equilibrium between the silicone, emulsifier, and substrate being treated. Often, the emulsifier also has detergent properties and majority of the silicone ends up in the wash water.

Although silicone compounds have been known for years and fatty surface-active agents have been known even longer, it was not until recently that an attempt has been made to use silicone as

Production of Silicone Surfactants and Antifoam Compounds

the hydrophobe in preparing surface-active agents, free-radical polymers, and other compounds which are the result of applying classical detergent chemistry to suitable silicone compounds. This approach overcomes many of the traditional difficulties and formulation problems encountered when working with silicone fluids due to their lack of solubility in many solvents. Silicone fluids not only lack solubility in water but are also insoluble in many solvents including mineral oil. This limited solubility, coupled with their tendency to defoam, limits the suitability of silicone oils to various applications. Synthetic modification of the silicone molecule results in new multifunctional products, useful in a wide range of detergent systems.

There are a wide variety of fatty surfactants, which differ in both structure and functional properties, available to the formulator. This allows for greater formulation latitude and creation of products, which are optimized for detergent applications. The use of silicone compounds requires the synthetic modification of the molecule to make it useful in application areas where a water soluble or dispersible material is needed. Too often in the past, the formulator has had to accept many of the drawbacks of the use of silicone oils in formulations, or leave them out altogether. The ability to make friendly silicone formulator has led to the synthesis of many new silicone-based surfactants. Many of the newer detergent products already in the market contain these materials, and more will be in the future.

To make a surface-active molecule, one needs to have both a water-soluble and an oil-soluble portion in the molecule. The traditional oil-soluble portion of the molecule is fatty. The silicone surfactants substitute or add on silicone-based hydrophobicity. This results in materials that are more easily formulated into detergent systems and have the improved substantivity, lower irritation, improved softening, and other attributes of silicone and properties one expects from the fatty surfactant. In molecules where silicone is predominate, the functional attributes of silicone will predominate. If the molecule has both a silicone and fatty hydrophobe present, it will function with attributes of both of the materials. This allows for the formulation of a wide variety of products, which have oil, water, silicone, or variable solubility.

It has been suggested that detergent systems which contain surfactants having fat-, silicone-, and water-soluble portions give better detergency over a wider range of soils than those based on fatty surfactants only. A three-dimensional HLB system has recently been proposed.

As one looks over the plethora of fatty surfactants available in the market today, one is overwhelmed with the possibilities. One may ask, why are there so many classes of surfactants available? The answer clearly is that the different classes of surfactants function in different applications. For example, fatty quats are generally used for softening and conditioning fatty alcohol sulfates for detergency. It would be very difficult for a formulator to develop products, which have all the properties using only one class of fatty compounds.

Silicone surfactants were limited primarily to dimethicone copolyol compounds before 1989. These materials are analogous to fatty alcohol alkoxylates. Although useful in some applications, there were many applications where they were simply ineffective.

It is therefore not surprising then that an entire series of surfactants, which are based on silicone as a hydrophobe containing surfactant functional groups, similar to those seen in traditional surfactants would be developed. In some instances, silicone is incorporated into a surface-active agent, with a polyoxyalkylene or hydrocarbon portion of the molecule.

Since the word silicone has been used synonymously with silicone fluid, several misconceptions have arisen about using silicone surfactants in many detergent applications. The following properties of silicone fluids have been attributed to all silicone compounds and consequently are misconceptions.

1. Silicone compounds defoam minimizing their use in formulations.
2. All silicone compounds are difficult to formulate.
3. All silicone compounds are water-insoluble.
4. All silicone compounds are mineral oil–insoluble.
5. All silicone compounds are greasy.
6. Silicone compounds are only oil phases.

7. All silicone compounds are liquids.
8. All silicone compounds polymerize.
9. Silicone compounds are not analogous to carbon chemistry, making comparison meaningless.
10. Silicone compounds are of limited use in formulations.

By selecting the correct silicone polymers, the limitations discussed can be overcome and desirable attributes of silicone enhanced.

17.6.2 DIMETHICONE COPOLYOL

Dimethicone copolyol is the oldest and perhaps the best-understood silicone surfactant. They conform to the following structure:

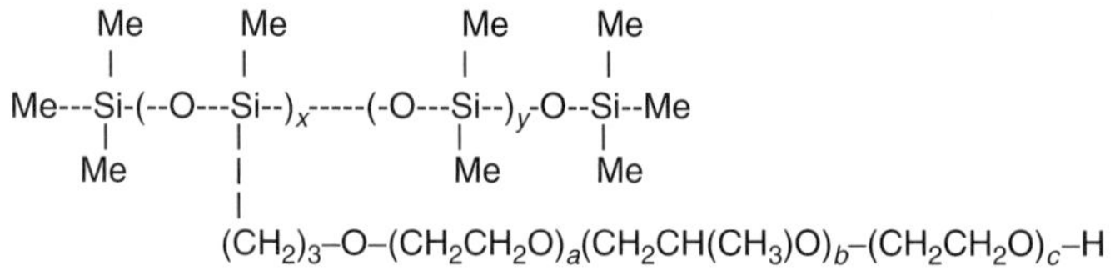

$$\begin{array}{cccc} Me & Me & Me & Me \\ | & | & | & | \\ Me\text{---}Si\text{-(--O---}Si\text{--)}_x\text{-----(-O---}Si\text{--)}_y\text{-O--}Si\text{--}Me \\ | & | & | & | \\ Me & | & Me & Me \\ & | & & \\ & (CH_2)_3\text{--O--}(CH_2CH_2O)_a(CH_2CH(CH_3)O)_b\text{--}(CH_2CH_2O)_c\text{--H} \end{array}$$

There are several key features of dimethicone copolyols, which allow for the synthesis of compounds with varied properties. They include (i) the linkage to silicone through three methlyene groups (CH_2), (ii) the presence of polyoxyalkylene oxide in the molecule (the a, b, and c units), (iii) the presence of "D" units in the molecule (the x and y values), and (iv) the total molecular weight of the molecule.

The three methylene groups are a result of the fact that allyl alcohol derivatives were used to make the compound and render the molecule stable to hydrolysis at the Si–C bond. Fewer carbon atoms between the Si and O molecules result in a dimethicone copolyol with decreased hydrolytic stability.

The polyoxyalkylene groups allow for the preparation of dimethicone copolyol compounds, which have varying water solubility. In addition, the introduction of propylene oxide into the molecule results in increasing the oil solubility. It is also quite significant where the propylene oxide is introduced into the polyoxyalkylene oxide chain. If the propylene oxide is introduced at the end of the chain, better hydrolytic stability in alkali is achieved. If however it is introduced first, the resulting dimethicone copolyol has improved liquidity.

The number of "D" units introduced into the molecule, relative to the number of "D*" units containing the water-soluble polyoxyalkylene oxide group determines many of the key functional attributes of the dimethicone copolyol. These include (i) whether the product will make silicone in water emulsion or water in silicone emulsion, (ii) the hand of the product on fabric, (iii) the gloss on hard surfaces, (iv) the durability to washing, (v) the solubilization properties with cationic softeners, and (vi) bleach and alkali stability.

The molecular weight of the dimethicone copolyol is also critical to surfactant properties. If one looks at a series of products having the same equivalent molecular weight but different total molecular weights, a trend develops. The compounds with low molecular weight (up to 500 molecular weight units or MWU) are the best wetting agents. Those of intermediate molecular weight are emulsifiers and compatabilizers (~5000 MWU). Those having still higher molecular weights (50,000 MWU) are good softeners and conditioners.

Dimethicone copolyols when dissolved in water orientate themselves at the air/water interface. The minimum energy state is with the silicone portion of the molecule orientated out of the water. Consequently, the minimum expected surface tension should approach that of silicone fluid, which is ~20 dyn/cm. This is in fact what has been observed. The dimethicone copolyol compounds,

Production of Silicone Surfactants and Antifoam Compounds

which result in the lowest surface tensions in aqueous solution, are low-molecular-weight dimethicone copolyols. Molecular packing at the surface is thought to be the reason.

Dimethicone copolyols exhibit an inverse cloud point phenomenon as an aqueous solution is heated. This same phenomenon is observed with ethoxylated fatty alcohol. The hydrogen bonding of the water with the polyoxyethylene portion of the molecule causes the cloud point. The inverse cloud point of the molecule is related to the length of the ethylene oxide chain and not the number of D units or molecular weight. The term *inverse cloud point* refers to the temperature at which a clear solution develops turbidity on heating. Cloud point is a phenomenon, relating to turbidity, which develops on cooling.

Since dimethicone copolyol compounds have their minimum solubility above their inverse cloud point, they can be used as temperature-sensitive antifoam compounds. The dimethicone copolyol will function as an emulsifier below its high cloud point, but defoam when it becomes insoluble above it. Since the dimethicone copolyol becomes soluble again on cooling, detergent formulations can be made which have minimal foam when used at high temperature, but rinse well at low temperatures.

17.6.2.1 Process and Technology

17.6.2.1.1 Hydrosilylation

Dimethicone copolyols are made by a process referred to as hydrosilylation. It is carried out according to the following structure where Me is methyl:

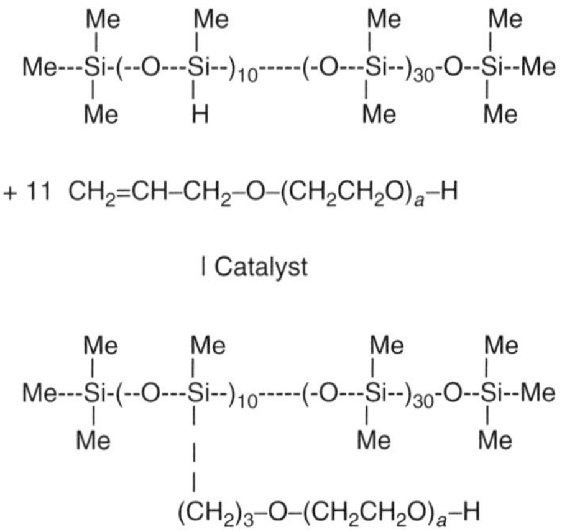

$$
\begin{array}{cccc}
Me & Me & Me & Me \\
| & | & | & | \\
Me{-}{-}{-}Si{-}({-}{-}O{-}{-}{-}Si{-}{-})_{10}{-}{-}{-}{-}{-}({-}O{-}{-}{-}Si{-}{-})_{30}{-}O{-}{-}Si{-}{-}Me \\
| & | & | & | \\
Me & H & Me & Me
\end{array}
$$

$$+\ 11\ \ CH_2{=}CH{-}CH_2{-}O{-}(CH_2CH_2O)_a{-}H$$

$$|\ Catalyst$$

$$
\begin{array}{cccc}
Me & Me & Me & Me \\
| & | & | & | \\
Me{-}{-}{-}Si{-}({-}{-}O{-}{-}{-}Si{-}{-})_{10}{-}{-}{-}{-}{-}({-}O{-}{-}{-}Si{-}{-})_{30}{-}O{-}{-}Si{-}{-}Me \\
| & | & | & | \\
Me & & Me & Me \\
& | & & \\
& (CH_2)_3{-}O{-}(CH_2CH_2O)_a{-}H & &
\end{array}
$$

The reaction involves a silanic hydrogen group (Si–H) and vinyl-containing compound, preferably a molecule undergoing alpha unsaturation ($CH_2{=}CH{-}CH_2$). In a typical process, the compound containing the Si–H group is added to a solvent, commonly isopropanol or toluene. The resulting solution is then heated to 85°C and a hydrosilylation catalyst, typically chloroplatinic acid (H_2PtCl_6), is added. The reaction is exothermic. The temperature increases to 95°C. The reaction mass is stirred for ~2 h. During this time, silanic hydrogen concentration drops to zero. The reaction is cooled to 65°C and sodium bicarbonate is added to neutralize the solution. The solution is filtered through a 4-micron pad and any residual solvent is distilled off.[7]

17.6.2.1.2 Applications

Dimethicone copolyol has been used to help make stable emulsions and dispersions of silicone fluid in detergent applications. In addition, the fact that the dimethicone copolyol is nonionic makes it compatible with both cationic and anionic systems. Consequently, one would expect to encounter the use of these materials to couple silicone fluid into fabric softeners and laundry detergent systems.

In fact, U.S. Patent 4,818,421 to Colgate Palmolive discloses that a silicone-fluid-free cationic fabric softener contains a dimethicone copolyol. The dimethicone copolyol surprisingly improves the fabric-softening capabilities of the fatty quat.

The synthesis of compounds, which are analogs of fatty bases surfactants, has been completed for many compound types.

COMPARISON OF HYDROCARBON COMPOUNDS WITH SILICONE-MODIFIED COMPOUNDS

Anionic Compounds

Hydrocarbon Products	Silicone Products
Phosphate esters	Silicone phosphate esters[1,2]
Sulfates	Silicone sulfates[3]
Carboxylates	Silicone carboxylates[4,5]
Sulfosuccinates	Silicone sulfosuccinates[6,7]

Cationic Compounds

Hydrocarbon Products	Silicone Products
Alkyl quats	Silicone qlkyl quats[8]
Amido quats	Silicone amido quats[9]
Imidazoline quats	Silicone imidazoline quats[10]

Amphoteric Compounds

Hydrocarbon Products	Silicone Products
Amino proprionates	Silicone amphoterics[11]
Betaines	Silicone betaines[12]
Phosphobetaines	Silicone phosphobetaines[13]

Nonionic Compounds

Hydrocarbon Products	Silicone Products
Alcohol alkoxylates	Dimethicone copolyol
Alkanolamids	Silicone alkanolamids[14]
Esters	Silicone esters[15–17]
Taurine derivatives	Silicone taurine[18]
Isethionates	Silicone isethionates[19]
Alkyl glycosides	Silicone glycosides[20]

References

1. U.S. Patent 5,149,765, September 1992 to O'Lenick.
2. U.S. Patent 4,724,248, February 1988 to Dexter et al.
3. U.S. Patent 4,960,845, October 1990 to O'Lenick.
4. U.S. Patent 3,560,544, February 1971 to Haluska.
5. U.S. Patent 5,296,625, March 1994 to O'Lenick.
6. U.S. Patent 4,717,498, January 1988 to Maxon.
7. U.S. Patent 4,777,277, November 1998 to Colas.
8. U.S. Patent 5,098,979, March 1992 to O'Lenick.
9. U.S. Patent 5,153,294, October 1992 to O'Lenick.
10. U.S. Patent 5,196,499, February 1993 to O'Lenick.
11. U.S. Patent 5,073,619, December 1991 to O'Lenick.
12. U.S. Patent 4,654,161, March 1987 to Kollmeier.
13. U.S. Patent 5,237,035, August 1993 to O'Lenick.
14. U.S. Patent 5,070,171, December 1991 to O'Lenick.

Production of Silicone Surfactants and Antifoam Compounds   **299**

15. U.S. Patent 5,070,168, December 1994 to O'Lenick.
16. U.S. Patent 4,724,258, issued February 1988 to Dexter.
17. U.S. Patent 6,338,042, December 2002 to O'Lenick.
18. U.S. Patent 5,280,099, January 1994 to O'Lenick.
19. U.S. Patent 5,300,666, April 1994 to O'Lenick.
20. U.S. Patent 5,120,812, June 1992 to O'Lenick.

17.7 RELATIVE ECONOMICS

Silicone-derived products are between four and ten times more expensive than their fatty counterparts. Therefore, the use of these materials in formulations needs to be driven either by (i) properties that cannot be achieved using traditional materials or (ii) the fact that they can be used at very low concentrations in formulations. The former is the case when in addition to detergency properties, emulsification properties can be improved or softened, gloss or other properties can be achieved. Because of the economics, it is highly unlikely that the concentration of a silicone surfactant in a detergent system will ever be over a few percentage by weight. In the case of antifoam compounds, the concentration ranges from 50 ppm to 0.05% by weight.

17.8 FUTURE TRENDS

17.8.1 ANTIFOAM COMPOUNDS

To meet the changing needs of the consumer as well as the increasing environmental concerns over energy usage and effects of waste detergents, there have been considerable changes in machine design and detergent product form and formulation. Almost every component of the detergent formulation has been changed or modified to maximize cleaning performance.

For example, surfactant systems have been changed to provide optimum performance. The introduction of new builder systems—enzymes, bleach activators, and new polymers for reduced soil redeposition—has also maximized soil removal from mixed fabric bundles during lower temperature washes in less wash water. In addition, new product forms such as concentrated liquids and powders with in-wash dosing containers have resulted in better usage and dispersion of detergent at low wash temperatures.

Some of these changes have resulted in the need for more effective foam control systems, whereas the soaps traditionally used have been largely replaced by other options.

The unique properties of silicones make them ideal for formulating high-performance, versatile antifoam compounds. New modifications of silicone surfactants will continue to be developed. They will be optimized for particular functionality in detergent formulations.

17.9 CONCLUSION

The development of improved, cost-effective silicone products will be driven by the development of new and higher efficiency processes. The cost/benefit ratio for the use of silicones will drive their use. Silicones will be chosen when they can enable a product that cannot be obtained using other less-expensive technologies. There are two parts to this—first is cost, which is process driven and second is effectiveness in the formulation, which is formulation driven. Only by addressing these two will develop new breakthrough products.

REFERENCES

1. O'Lenick, A., *Silicones for Personal Care*, Allured Publishing, Carol Stream, IL, 2003, pp. 13, 23.
2. Rochow, E., *Chemistry of Silicones*, Wiley, New York, 1946, p. 125.

3. Voorhoeve, H.J., *Organosilanes: Precursors to Silicones*, 1967, p. 28.
4. Ushakova, V., Van Roy, B., *External Validation of Silicone Technology for Fabric Care*, Form 27-11145-01, 2003, p. 4.
5. U.S. Patent No. 5,473,038 to Siltech Inc.
6. Siltech Inc. Bulletin, *Silicone Fluids, Emulsions Antifoam and Specialties*, 1989.
7. Marciniec, B., *Comprehensive Handbook of Hydrosilylation*, Pergamon Press, Oxford, 1992, p. 9.

18 Production of Fluorinated Surfactants by Electrochemical Fluorination

Hans-Joachim Lehmler

CONTENTS

18.1 Introduction 301
    18.1.1 Properties of Fluorinated Surfactants 301
    18.1.2 Nomenclature of Fluorinated Surfactants 303
18.2 Perfluoroalkanesulfonyl Surfactants 303
    18.2.1 Perfluorooctanesulfonyl Fluoride and Its Homologues 305
    18.2.2 Perfluoroalkanesulfonic Acids and Their Metal Salts 307
    18.2.3 Synthesis of Perfluorooctanesulfonamide and Its Homologues 308
        18.2.3.1 Reaction of Perfluoroalkanesulfonyl Halides with
                Ammonia and Amines 308
        18.2.3.2 Alkylation of Perfluoroalkanesulfonamides 310
    18.2.4 Perfluoroalkanesulfonamide-Derived Surfactants 310
        18.2.4.1 Anionic Surfactants 310
        18.2.4.2 Nonionic Surfactants 311
        18.2.4.3 Cationic and Zwitterionic Surfactants 312
18.3 Perfluorocarboxylate Surfactants 313
    18.3.1 Synthesis of Perfluorocarboxylate Fluorides by Electrochemical Fluorination 314
    18.3.2 Perfluorocarboxylate-Derived Surfactants 314
        18.3.2.1 Anionic Surfactants 314
        18.3.2.2 Nonionic Surfactants 315
        18.3.2.3 Perfluorocarboxylate-Derived Cationic and
                Zwitterionic Surfactants 316
18.4 Fluorinated Surfactants and Their Environmental Impact 318
Acknowledgments 318
References 318

18.1 INTRODUCTION

18.1.1 PROPERTIES OF FLUORINATED SURFACTANTS

Fluorinated surfactants have unique properties that make them highly suitable for many industrial processes,[1–3] use in consumer products,[1–3] and biomedical[2,4,5] and research applications such as novel material synthesis applications.[6] Several unique characteristics of the fluorine atom contribute to the properties of surfactants with a highly fluorinated hydrophobic tail (or fluorinated surfactants for short). These unique characteristics include the following: fluorine is the most electronegative element; the fluorine–fluorine bond is weak, whereas the fluorine–carbon bond is among the strongest known covalent bonds; the van der Waals radius of fluorine is small; and fluorine has three nonbonding electron pairs.[2]

The stability of the fluorine–carbon bond and the shielding effect of the nonbonding electron pairs of the fluorine atoms render perfluorinated alkyl chains highly stable toward chemical, biological, and thermal degradation—a useful feature of fluorinated surfactants. Perfluoroalkanesulfonic acids (R_FSO_3H) and perfluorocarboxylates (PFCAs) (R_FCO_2H) are thermally the most stable fluorinated surfactants. For example, perfluorooctanesulfonic acid (PFOS) and the corresponding potassium salt are stable at temperature as high as 400°C.[7,8] Perfluoroalkanesulfonic acids and PFCAs (as well as their metal salts) also have remarkable stability toward oxidants, acids, and alkali and can be used under conditions where hydrocarbon surfactants are degraded.

Another important property of fluorinated surfactants is their high surface activity. For example, the minimum surface tension of perhydrocarbon surfactants is 25–35 mN/m, whereas much lower surface tensions of 15–25 mN/m can be achieved with analogous perfluorocarbon surfactants.[9] Similarly, the critical micelle concentration (CMC) of fluorinated surfactants is always below the CMC of analogous perhydrocarbon surfactants. As a general rule, the CMC of fluorinated surfactants is equal to that of an analogous hydrocarbon surfactant of 50% greater tail length ("1.5 rule").[10] From an application point of view, the lower CMC of perfluorocarbon surfactants implies that less-fluorinated surfactant is needed to achieve a substantial reduction in surface tension, a fact that offsets the higher costs of fluorinated surfactants. Within a homologous series of fluorinated surfactants, the minimum achievable surface tension decreases with the increasing length of the perfluorinated tail up to a chain length of C_7F_{15}. No further reduction in surface pressure can be observed for longer perfluoroalkyl chains, thus making C_7F_{15} and C_8F_{17} the economically most relevant fluorinated tails.[2,8]

Fluorinated surfactants are both hydrophobic and lipophobic.[11] For example, potassium perfluorooctanesulfonate—an industrially important surfactant[12]—forms a third phase with octanol and water, and it is impossible to determine its octanol–water partition coefficient.[13,14] Similar to fluorocarbon–hydrocarbon bulk solvent mixtures,[15] mixed binary systems containing a perfluorocarbon surfactant and a structurally related hydrocarbon surfactant are known to behave nonideally, that is, exhibit phase separation in insoluble monolayers at the air–water interface[16–19] or form two types of micelles simultaneously in solution—one type is fluorocarbon-rich and the other is hydrocarbon-rich.[11,20] This nonideal behavior of fluorocarbon–hydrocarbon surfactant mixtures is used in firefighting foams and powders—an important technical application of fluorinated surfactants.[2,14]

A myriad of structurally different fluorinated surfactants have been synthesized industrially (and in the laboratory) to utilize the unique properties of a highly fluorinated hydrophobic tail. On the basis of their chemical structure, fluorinated surfactants can be divided into perfluorinated and partially fluorinated surfactants.[2] Perfluorinated surfactants do not contain hydrogen atoms or heteroatoms such as oxygen in the hydrophobic tail. For example, salts of perfluoroalkanesulfonic acids and PFCAs are truly perfluorinated surfactants. Partially fluorinated surfactants contain methylene (CH_2) or benzene (C_6H_4) groups. This distinction is important because the presence of hydrogen atoms in the tail not only alters the physicochemical properties of a surfactant, but also decreases its stability toward chemical, biological, and thermal degradation.[2] Fluorinated surfactants obtained by anionic oligomerization of fluorinated oxiranes (e.g., 2,2,3-trifluoro-3-trifluoromethyl-oxirane) do not fit into either category and should be considered as a third class of fluorinated polyether surfactants.

This chapter is an introduction to industrial- and laboratory-scale syntheses of surfactants derived from fluorinated intermediates obtained by electrochemical fluorination. The chemistry involved in the preparation of these compounds has been established for many decades and, because of its confidential nature, is mostly described in the patent literature. Recent interest in the

Production of Fluorinated Surfactants

production of these compounds results from the detection of fluorinated surfactants in the environment, thus warranting an in-depth review of the production of fluorinated surfactants derived by electrochemical fluorination. Other processes, especially radical telomerization of fluorinated ethenes (e.g., 1,1,2,2-tetrafluoroethene), have also been used to synthesize fluorinated surfactants. The synthetic strategies reported for the production of fluorotelomer-derived surfactants are similar to the ones obtained by electrochemical fluorination and are not the focus of this chapter. Refs 2, 8, and 21 give an in-depth description of industrial processes based on radical telomerization and other miscellaneous processes.

18.1.2 NOMENCLATURE OF FLUORINATED SURFACTANTS

Before discussing the preparation of fluorinated surfactants, it is helpful to discuss the nomenclature of fluorinated surfactants. The International Union of Pure and Applied Chemistry (IUPAC) names of fluorinated surfactants are long and impractical to use because the IUPAC name of a fluorinated carbon chain typically includes the position and number of fluorine atoms. For example, the position and number of the fluorine substituents of a $-(CF_2)_5CF_3$ chain is correctly referred to as "1,1, 2,2,3,3,4,4,5,5,6,6,6-tridecafluoro" or simplified as "tridecafluoro." To simplify the nomenclature, perfluorinated alkyl chains are frequently referred to as "perfluoro," for example, "perfluorohexyl" or "perfluorohexane." Many important fluorinated surfactants have a single hydrophobic chain and a head group, for example, $-SO_3^-$ and COO^-, in position 1 of the alkyl chain. The position of the head group is normally omitted from the name of the respective surfactant. With partially fluorinated compounds, it is frequently easier to indicate the position of hydrogen atoms, for example, by providing their position followed by H and not the fluorine atoms. For example, $1H,1H$-perfluorooctanol has the molecular formula $CF_3(CF_2)_6CH_2OH$. The IUPAC name of this alcohol is 2,2,3,3,4, 4,5,5,6,6,7,7,7-pentadecafluoro-octan-1-ol. This simplified terminology for fluorinated compounds is used in this chapter for easy reading. Refs 22 and 23 give more details on the nomenclature of fluorinated compounds.

18.2 PERFLUOROALKANESULFONYL SURFACTANTS

Surfactants derived from perfluorooctanesulfonyl fluorides and homologous perfluoroalkanes sulfonyl fluorides have been produced for over 40 years and represent an important class of fluorinated surfactants.[24] The simplified industrial process outlined in Figure 18.1 illustrates the large variety of fluorinated perfluorooctanesulfonyl surfactants that can be prepared from this important intermediate.[2,14,25,26] The worldwide major producer of perfluorooctanesulfonyl fluoride was the 3M Company (United States).[14] On the basis of 3M Company's estimates, ~3665 t of perfluorooctanesulfonyl fluoride were manufactured worldwide in 2000. In the same year, 1820 t of perfluorooctanesulfonyl fluoride were produced or imported into the United States. A large portion of perfluorooctanesulfonyl fluoride was used to manufacture products for surface treatments and paper protection. Only a small quantity of perfluorooctanesulfonyl fluoride was converted into perfluoroalkanesulfonic acid surfactants. It is estimated that 91 t of PFOS and its salts were sold as finished products in 2000; 151 t of firefighting foams containing PFOS chemicals were produced in the same year. Owing to environmental concerns, the 3M Company voluntarily began to phase out the production of perfluorooctanesulfonyl fluoride by the end of the year 2000 and all production in the United States ceased in 2002.[24]

As shown in Figure 18.2, the majority of the perfluorooctanesulfonyl fluoride was converted on-site by the 3M Company through the sulfonamide into N-methyl or N-ethyl perfluorooctanesulfonamidoethanol.[26] These two compounds (and most of the other intermediates shown in Figure 18.2) were primarily used as intermediates for the preparation of fluoropolymers in protective treatments for carpets, upholstery, apparel, and leather products. Several of the other compounds derived from perfluorooctanesulfonyl fluoride were used for the production of surfactants. For example, the

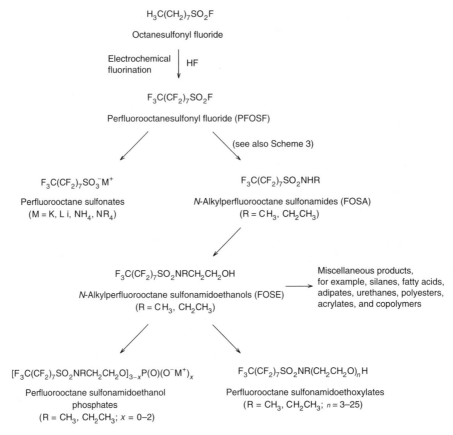

FIGURE 18.1 Simplified production pathway of fluorinated surfactants derived from perfluorooctanesulfonyl fluoride. (Adapted from Kissa, E., *Fluorinated Surfactants and Repellents,* Marcel Dekker, New York, 2001; Environment Directorate, OECD, Hazard assessment of perfluorooctane sulfonate (PFOS) and its salts. Report ENV/JM/RD(2002)17/FINAL, Organization for Economic Cooperation and Development, Paris, 2002; Schultz, M.M. et al., *Environ. Eng. Sci.*, 20, 487, 2003.)

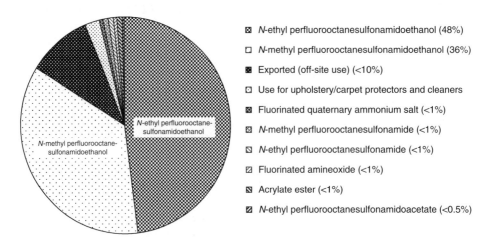

FIGURE 18.2 Chemical end use of perfluorooctanesulfonyl fluoride produced by the 3M Company as a consumed intermediate for the production of other chemicals. (Based on 3M Company, Voluntary use and exposure information profile perfluorooctanesulfonyl fluoride (PFOS). U.S. EPA Administrative Record AR226-0576, U.S. EPA, Washington, DC, 2000.)

fluorinated quaternary ammonium salt $F_{17}C_8SO_2NH(CH_2)_3NMe_3^+Cl^-$ was converted into surfactants for the mining and oil well industry; the fluorinated aminoxide $F_{17}C_8SO_2NH(CH_2)_3N(O)Me_2$ was used in the production of firefighting foams; and N-ethyl perfluorooctanesulfonamidoacetate $F_{17}C_8SO_2N(C_2H_5)CH_2COO^-K^+$ was employed to make a household cleaner. Only a small quantity of the fluoride was used off-site.

18.2.1 PERFLUOROOCTANESULFONYL FLUORIDE AND ITS HOMOLOGUES

As mentioned earlier, perfluorooctanesulfonyl fluoride and its homologues are prepared by electrochemical fluorination of organic compounds.[2,14] During the process, which was invented by Simons,[27,28] the corresponding hydrocarbon sulfonyl fluoride or chloride is fluorinated in anhydrous HF by applying a direct electrical current with a voltage typically between 5 and 7 V.[7,29–32] The electrochemical fluorination process replaces the carbon–hydrogen bonds of the hydrocarbon sulfonyl fluoride with carbon–fluorine bonds following the general formula:

$$C_nH_{2n+1}SO_2F + (2n+1)F^- \xrightarrow{HF} C_nF_{2n+1}SO_2F + (2n+1)H^+ + (4n+2)e^-$$

Figure 18.3 shows a diagram of an electrochemical fluorination plant.[33] The central component of this plant, the electrolytic cell, is shown in Figure 18.4. This cell consists of a nickel box surrounded by a steel cooling jacket. The polythene base of the cell is fitted with a product drain port and its polythene head is fitted with a starting material inlet port and an exit port, which is connected to a nickel reflux condenser. Nickel stubs through the cell head allow sealed electrical connections to the electrode package, the conductivity cell, and an auxiliary reference electrode circuit. The electrode package consists of alternating nickel anodes and cathodes separated by polythene spacers. In a typical fluorination experiment, the preconditioned cell containing anhydrous hydrogen fluoride is

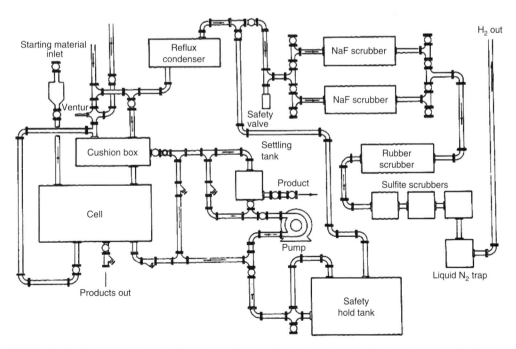

FIGURE 18.3 Schematic outline of an electrochemical fluorination plant. (Reproduced from Drakesmith, F.G. and Hughes, D.A., *J. Appl. Electrochem.*, 9, 685, 1979. With permission of Springer Science and Business Media.)

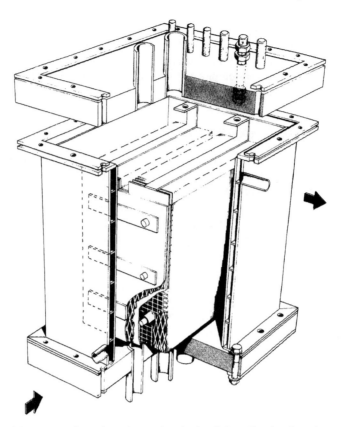

FIGURE 18.4 Partial cutaway view of an electrochemical cell for a fluorination plant. (Reproduced from Drakesmith, F.G., and Hughes, D.A., *J. Appl. Electrochem.*, 9, 685, 1979. With permission of Springer Science and Business Media.)

loaded with octanoyl chloride dissolved in hydrogen fluoride and electrolyzed at 0°C using an anode potential of ~4.3 V.[33] This cell temperature represents a compromise between the convenience of handling liquid hydrogen fluoride, which has a boiling point of 19°C, and increasing conductivity of anhydrous hydrogen fluoride at higher temperatures.[2]

Short-chain sulfonyl fluorides can be obtained in excellent yields and good purity by electrochemical fluorination; however, the yield of this fluorination decreases steadily with increasing chain length (87%, 79%, 25%, and 12% for perfluoromethane-, perfluoroethane-, perfluorooctane-, and perfluorodecane-sulfonyl fluoride, respectively). The electrochemical fluorination process used by the 3M Company till 2002[24] yielded between 30 and 45% of the linear perfluorooctanesulfonyl fluoride in addition to higher and lower straight-chain homologues (7%), branched perfluoroalkyl-sulfonyl fluorides (18–20%); linear, branched, and cyclic perfluoroalkanes; and ethers (20–25%) as well as tars and other by-products including hydrogen (10–15%).[26]

As a result of the side reactions, industrial perfluorooctanesulfonyl fluorides are complex mixtures containing only approximately 70% of the *n*-perfluorooctanesulfonyl compound. These mixtures are typically not purified before use for the production of fluorinated surfactants. Because of the similar boiling points, it is not feasible to separate linear perfluorooctanesulfonyl fluoride from its branched analogues by fractional distillation; however, it is possible to obtain pure linear perfluorooctanesulfonyl fluorides by fractional crystallization at −20°C.[34]

A detailed mechanistic study of the carbon-chain isomerization during the electrochemical fluorination of two short-chain 1-sulfonyl fluorides, 2-methylpropane- and 2-methylethane-1-sulfonyl fluoride has been reported.[35] This study gives insight into the complexity of this process and the

Production of Fluorinated Surfactants

various side products formed during electrochemical fluorination of hydrocarbon sulfonyl fluorides. For example, the electrochemical fluorination of 2-methylpropane-1-sulfonyl fluoride results in the formation of perfluorobutane-1-sulfonyl fluoride, perfluoro-2-methylpropane-1-sulfonyl fluoride, perfluorobutane-2-sulfonyl fluoride, and perfluoro-2-methylpropane-2-sulfonyl fluoride in 61.7, 8.1, 1.5, and 0.1 mol% (based on the total mixture). In addition, small quantities of PFCA fluorides, perfluoroalkanes, SO_2F_2, SOF_2, and partially fluorinated products are formed. A similarly complex product pattern was observed for 2-methylethane-1-sulfonyl fluoride.[35]

The by-products formed during the electrochemical fluorination of octanesulfonyl fluoride—an industrially relevant process—has not been characterized in detail; however, the electrochemical fluorination of the corresponding chloride has been studied.[7] Similar to the short-chain sulfonyl fluorides, a number of breakdown products are formed during the fluorination of this long-chain chloride. These include short-chain perfluoroalkanesulfonyl fluorides containing 1–7 carbon atoms, perfluorooctane and other short-chain perfluoroalkanes, SO_2F_2, SF_6, PFCAs containing 1–7 carbon atoms, as well as longer-chain products. Typically, these by-products (such as perfluorooctane) are formed by the cleavage of the carbon–sulfur bond and then by the cleavage of carbon–carbon bonds; however, the cleavage of the carbon–sulfur bond is not required for the breakdown of the carbon chain. The formation of PFCAs suggests the electrochemical oxidation of the perfluorocarbon chain to PFCA fluorides. These fluorides are subsequently hydrolyzed by moisture in the air after removal of the reaction mixture from the electrolysis cell.[7]

The alkanesulfonic acids themselves are not suitable starting materials for electrochemical fluorination because the yields of perfluoroalkanesulfonic acids obtained from the free acids are much lower compared to the alkanesulfonyl fluorides and chlorides.[7,36] Another major disadvantage of this approach is the formation of water, which may result in the formation of explosive oxygen difluoride (OF_2).

The yield of the electrochemical fluorination can reportedly be improved by the addition of butadiene sulfones.[30] This additive is highly soluble in anhydrous HF and increases the conductivity of the electrolyte solution. Butadiene sulfone itself is fluorinated to perfluorobutanesulfonyl fluoride and, therefore, needs to be added continuously to the reaction. The use of this additive increases the yield of perfluorooctanesulfonyl fluoride by ~70%.[30] It may act by trapping radicals formed at the anode, but the mechanism by which it increases the yield of the electrochemical fluorination of long-chain sulfonic acid derivatives is unknown.

18.2.2 PERFLUOROALKANESULFONIC ACIDS AND THEIR METAL SALTS

Perfluoroalkanesulfonic acid salts (e.g., potassium, lithium, ammonium, and diethanol ammonium salts) are useful surfactants[12] and are prepared by the hydrolysis of the fluoride as shown in Scheme 18.1. Long-chain perfluoroalkanesulfonyl fluorides are only slowly hydrolyzed by water. For example, perfluorooctanesulfonyl fluoride is minimally hydrolyzed by water at 180°C, even

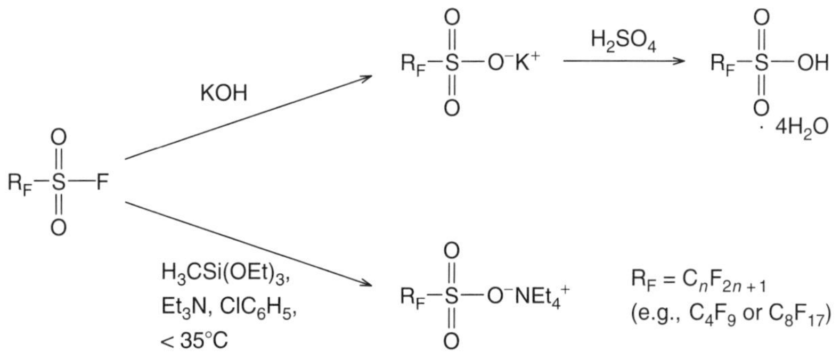

SCHEME 18.1

after several days.[7] Short-chain sulfonyl fluorides are more readily hydrolyzed, which is thought to be a result of their higher solubility in water. Potassium perfluorooctanesulfonate can be synthesized by the hydrolysis of the fluoride with aqueous potassium hydroxide.[7,12,37] The potassium salt is only slightly water-soluble (~2% at 25°C) and precipitates from the aqueous solution. Other bases such as sodium hydroxide[38] and barium hydroxide[7,29,31] in water have also been used to synthesize salts of perfluoroalkanesulfonic acids. Calcium hydroxide[39,40] or calcium oxide[41] are also suitable bases for the hydrolysis of perfluoroalkanesulfonyl fluorides and can be used in continuous processes where calcium fluoride precipitates, thus resulting in fluoride-free perfluoroalkanesulfonate potassium or ammonium salts. Similarly, sulfonyl chlorides can be hydrolyzed in good yields to the corresponding potassium salt using KOH in ethanol water at 50°C.[42] If desired, the free acid (such as PFOS) can be released from the salt by treatment with concentrated sulfuric acid (Scheme 18.1).[7,29] The free acid can then be separated from the sulfuric acid by careful fractional distillation.[7]

A challenge for industrial syntheses of perfluoroalkanesulfonate tetraalkylammonium salts is the impurities present in technical perfluorooctanesulfonyl fluoride. These impurities cause side reactions that result in tar formation, thus lowering the production yield.[43] One patent claims the preparation of the desired perfluoroalkanesulfonate tetraalkylammonium salts in excellent yields and good purities by reaction of the sulfonyl fluoride with alkoxysilanes (e.g., triethoxymethylsilane) and trialkylamines (e.g., triethylamine) in a suitable organic solvent (e.g., chlorobenzene, ether, or toluene) as outlined in Scheme 18.1.[43,44] Under these conditions, the ammonium salts crystallize from the reaction mixture, whereas the fluorosilane by-products remain dissolved in the organic solvent.

18.2.3 SYNTHESIS OF PERFLUOROOCTANESULFONAMIDE AND ITS HOMOLOGUES

As outlined in Figures 18.1 and 18.2, perfluoroalkanesulfonyl fluorides, especially perfluorooctanesulfonyl fluoride, are industrial intermediates that are converted into the corresponding sulfonamides.[2,14,26] These sulfonamide intermediates are ultimately converted into a whole host of fluorinated surfactants, but have also been used in large quantities for the preparation of surface treatments or paper protectants.[14,24,26] The following sections discuss important aspects of the synthesis of perfluoroalkanesulfonamides followed by an overview over the synthesis of selected perfluoroalkanesulfonyl surfactants (Section 18.2.4).

18.2.3.1 Reaction of Perfluoroalkanesulfonyl Halides with Ammonia and Amines

Perfluoroalkanesulfonamides have been claimed as intermediates for the synthesis of fluorinated surfactants. Their synthesis by reaction of the corresponding perfluoroalkanesulfonyl fluoride (or chloride) with liquid ammonia is prototypical for the synthesis of alkylated perfluoroalkanesulfonamide surfactants.[45-48] This reaction initially forms a complex ammonium salt as shown in Scheme 18.2. The desired amide is obtained by dissolving the crude product in dioxane and

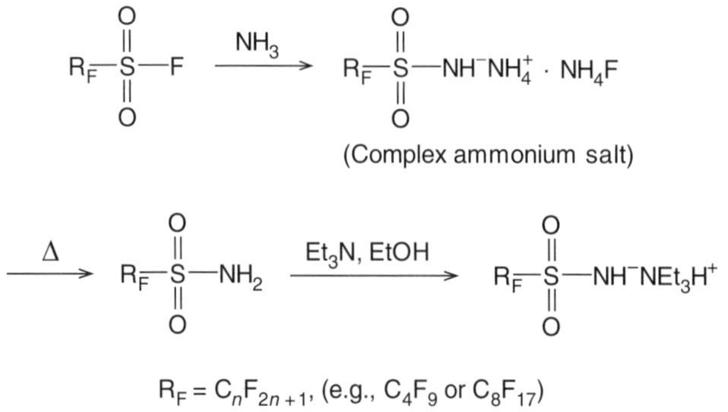

$$R_F{-}\overset{\overset{\displaystyle O}{\|}}{\underset{\underset{\displaystyle O}{\|}}{S}}{-}F \xrightarrow{\ NH_3\ } R_F{-}\overset{\overset{\displaystyle O}{\|}}{\underset{\underset{\displaystyle O}{\|}}{S}}{-}NH^-NH_4^+ \cdot NH_4F$$

(Complex ammonium salt)

$$\xrightarrow{\ \Delta\ } R_F{-}\overset{\overset{\displaystyle O}{\|}}{\underset{\underset{\displaystyle O}{\|}}{S}}{-}NH_2 \xrightarrow{\ Et_3N,\ EtOH\ } R_F{-}\overset{\overset{\displaystyle O}{\|}}{\underset{\underset{\displaystyle O}{\|}}{S}}{-}NH^-NEt_3H^+$$

$$R_F = C_nF_{2n+1}, \text{ (e.g., } C_4F_9 \text{ or } C_8F_{17})$$

SCHEME 18.2

Production of Fluorinated Surfactants   **309**

exchanging $R_FSO_2NH^-$ with chloride by passing anhydrous hydrochloric acid through the solution. The resulting ammonium chloride and fluoride are filtered off to give the sulfonamide.[45] A more straightforward approach utilizes the fact that the complex fluoride salts are thermally unstable and extracts the sulfonamide with (boiling) diethyl ether.[46–48]

N-mono- and N,N-difunctionalized perfluoroalkanesulfonamides (such as the industrially important N-methyl[49] and N-ethyl amides[50]) are synthesized analogously by the reaction of the sulfonyl fluoride (or chloride) with the corresponding amine in the presence of a base.[47,49–54] This reaction can be accomplished using a broad range of solvents and reaction conditions. For example, water or aqueous dimethylformamide (DMF) with sodium or potassium hydroxide,[55,56] an organic solvent (e.g., tetrahydrofuran [THF], diethyl ether, or dichloromethane) with an organic (e.g., sodium ethylate, triethylamine, or pyridine) or inorganic base (e.g., potassium carbonate or hydroxide),[57–62] or direct reaction of excess amine with the neat fluoride[47,63] has been employed successfully. The optimal temperature range for these reactions reported in the patent literature is 10–60°C.[61,62] Lower temperatures are, however, used in laboratory syntheses to reduce side reactions leading to coloration.

Similar to the unsubstituted sulfonamides (Scheme 18.2), the synthesis of N-mono-functionalized perfluoroalkanesulfonamides from the respective fluoride and a primary amine initially results in the formation of an ammonium salt. The alkylated perfluoroalkanesulfonamide can be separated from the salt after acidification with concentrated hydrochloric or sulfuric acid,[56,63] addition of an aqueous solution of an alkali salt of a volatile organic acid (e.g., sodium acetate),[61] or treatment with an aqueous solution of an alkali hydroxide (e.g., sodium hydroxide).[61] Selected examples of syntheses of perfluoroalkanesulfonamides from mono-, di-, and polyamines are shown in Scheme 18.3.[61,62,64,65]

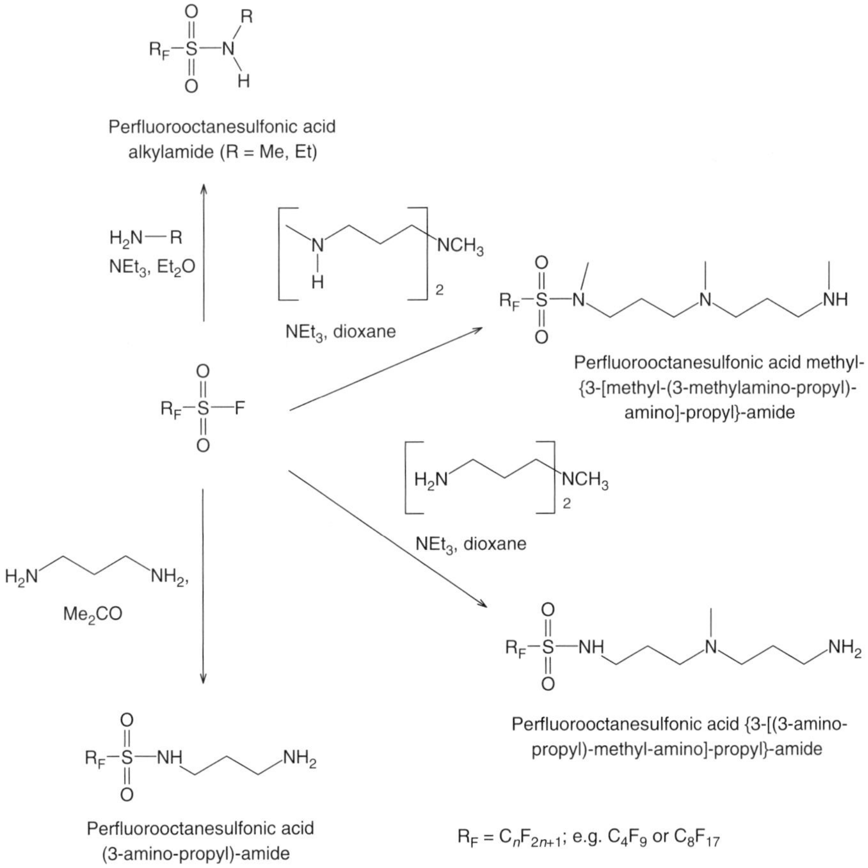

SCHEME 18.3

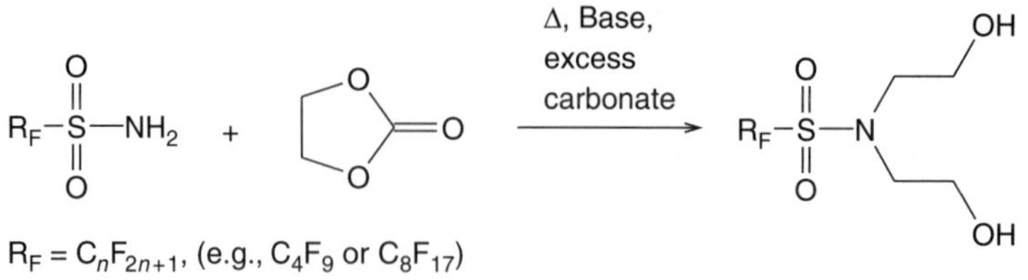

SCHEME 18.4

Perfluorooctanesulfonamidoethanol derivatives, which are important intermediates for the synthesis of surfactants, cannot be synthesized by the reaction of ethanolamine derivatives with perfluoroalkanesulfonyl fluorides because of the formation of several products that cannot be separated.[30] An important side reaction is the intermediate formation of sulfonyl esters by reaction of the fluoride with the OH group.[36,66] The sulfonyl esters are excellent alkylating agents that participate in various side reactions, for example, the alkylation of ethanolamine derivative under formation of a sulfonate anion.[44]

18.2.3.2 Alkylation of Perfluoroalkanesulfonamides

On the basis of a literature report, the alkylation of perfluoroalkanesulfonamides $R_FSO_2NH_2$ does not proceed neatly;[67] however, there are several claims of direct alkylations of perfluoroalkanesulfonamides in the patent literature. Perfluoroalkanesulfonamides react with excess 1,3-dioxolan-2-one (ethylene carbonate), oxirane, and related compounds in the presence of a base to yield the bis(hydroxyethyl)sulfonamides as shown in Scheme 18.4.[30,66] Similarly, the reaction of a perfluoroalkanesulfonamide with an equimolar amount of chloroethanol[68] or 1,3-dioxolan-2-one[30] has been reported to result in perfluoroalkanesulfonamidoethanols.

The N-alkylation of N-substituted (e.g., *N*-methyl and *N*-ethyl) perfluoroalkanesulfonamides with halogenated alcohols (e.g., chloroethanol) is an important reaction in organic chemistry[69] and can be achieved with K_2CO_3,[67,70–73] triethylamine,[74] NaH,[75] or with sodium methoxylate[54,68] or ethoxylate.[54] Several patents claim that perfluoroalkanesulfonamidoethanols can be obtained by the reaction of a sulfonamide with ethylene carbonate, oxirane, and related compounds at elevated temperatures in the presence of a base, for example, potassium hydroxide, pyridine, triethylamine, or potassium carbonate.[30,66,76] Industrially, the most important process is the alkylation of *N*-methyl and *N*-ethyl perfluoroalkanesulfonamide with ethylene carbonate (see Figure 18.2).[26]

18.2.4 PERFLUOROALKANESULFONAMIDE-DERIVED SURFACTANTS

18.2.4.1 Anionic Surfactants

The simplest class of anionic perfluoroalkanesulfonamides surfactants with the general structure $R_FSO_2NH^-M^+$ can be obtained by the reaction of suitable intermediates such as perfluoroalkanesulfonamides or N-alkylated perfluoroalkanesulfonamides with a base (e.g., sodium hydroxide, triethylamine, or propylamine) in a solvent (e.g., water or water-miscible organic solvents) at temperatures between 20 and 70°C (Scheme 18.2).[77] Although these salts can be obtained from both unalkylated and monoalkylated sulfonamides, salts derived from unsubstituted sulfonamides are of particular interest and have been claimed to lower the surface tension of water to 20 dyn/cm at concentrations as low as 0.2 g/L.

Perfluoroalkanesulfonamides (see Scheme 18.3 for typical examples) can be converted into anionic surfactants by alkylation with chloroacetic-[65,78] or chloroethanesulfonic acid[78,79] derivatives. The glycine derivatives obtained by this process are not only industrial surfactants (i.e., *N*-ethyl perfluorooctanesulfonamidoacetate[80]), but also important environmental contaminants.[23] In addition

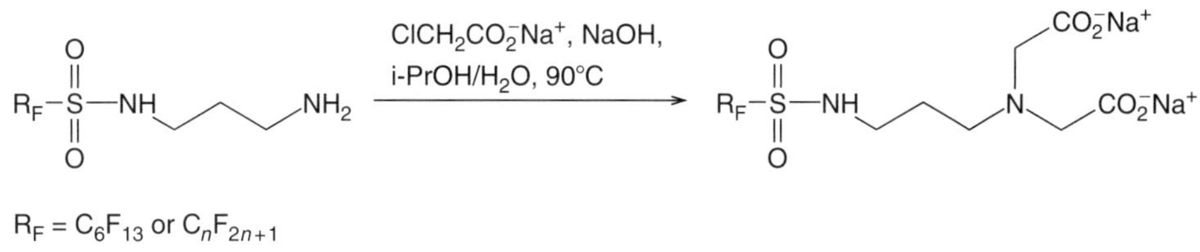

Production of Fluorinated Surfactants   311

SCHEME 18.5

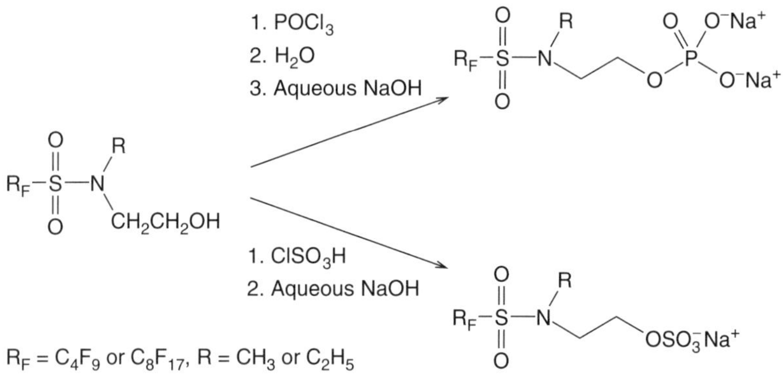

SCHEME 18.6

to the free acids, the respective metal salts (e.g., sodium salts) or alkyl esters can be employed in these alkylation reactions. The synthesis of (carboxymethyl-[3-(perfluorohexanesulfonylamino)-propyl]-amino)-acetic acid disodium salt from the respective amide and sodium chloroacetate is shown as an example in Scheme 18.5.[65] This approach has also been claimed for the synthesis of tricarboxylic acid–type, amphoteric surfactants.[64] Another important example is the alkylation of N-ethyl perfluorooctanesulfonamide with chloroacetic acid ethyl ester followed by base-catalyzed hydrolysis of the ester.[75,78]

Perfluoroalkanesulfonamidoethanols are another type of intermediates that can be converted into anionic surfactants. For example, highly surface-active phosphates with the general structure $R_FSO_2NRCH_2CH_2OP(=O)(O^-M^+)_2$ can be obtained by the reaction of an N-alkylated perfluoroalkanesulfonamidoethanol with $POCl_3$. The resulting dichloride is hydrolyzed with water and neutralized with an organic or inorganic base as shown in Scheme 18.6.[81,82] Another synthetic approach utilizes H_3PO_4 and P_4O_{10} instead of $POCl_3$ as the phosphate precursor.[76] Similarly, perfluoroalkanesulfonamidoethanols can be converted into sulfonic acids by reaction with $ClSO_3H$ as illustrated in Scheme 18.6 for the synthesis of 2-[ethyl-(perfluorooctanesulfonyl)-amino]-ethanesulfonic acid.

18.2.4.2   Nonionic Surfactants

N-alkylated (e.g., N-methyl and N-ethyl) perfluoroalkanesulfonamides can be converted into nonionic surfactants with a polyoxyethylene head group by reaction of the sulfonamide with excess oxirane in the presence of a base. This synthesis approach is similar to the one outlined in Section 18.2.3.2 for the synthesis of perfluoroalkanesulfonamidoethanols. The reaction of N-methylperfluorobutanesulfonamide with oxirane yields the desired polyoxyethylene surfactants in the presence of potassium hydroxide at 100–120°C (Scheme 18.7).[66] A different approach to similar polyoxyethylene surfactants claims the reaction of perfluoroalkanesulfonamidoethanols with oxirane in the presence of sodium hydroxide as a catalyst (Scheme 18.7).[83] Both approaches can also be extended to other oxirane derivatives,[66,83] or use perfluoroalkanesulfonamides such as the ones shown in Scheme 18.3 as starting materials.[84] Polyoxyethylene groups can also be attached to

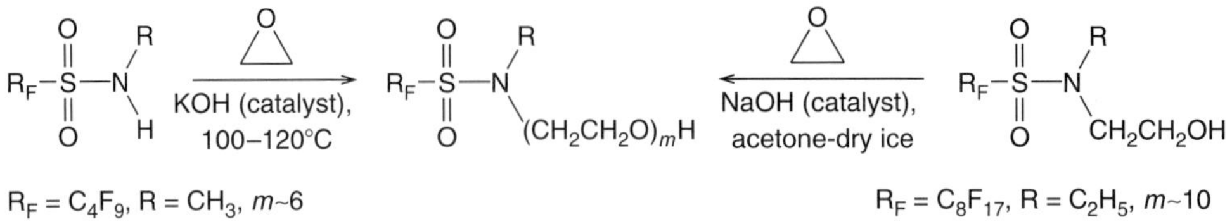

SCHEME 18.7

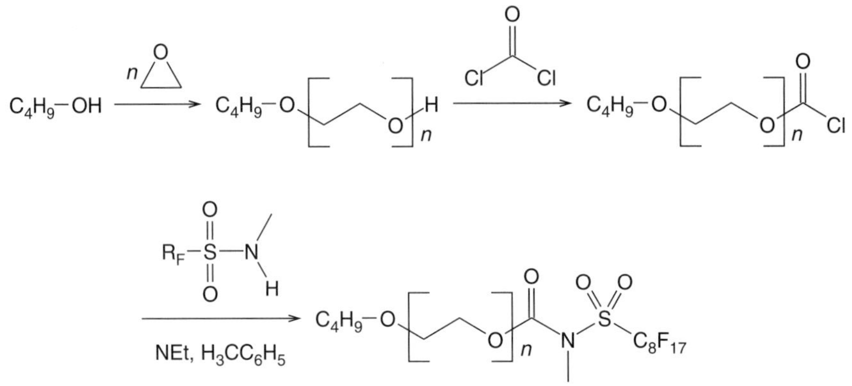

SCHEME 18.8

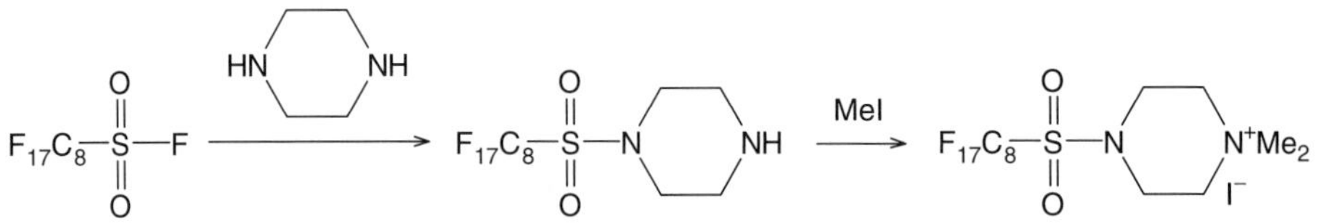

SCHEME 18.9

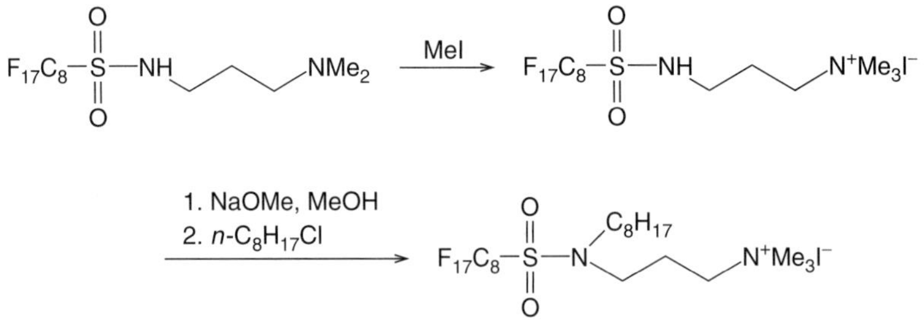

SCHEME 18.10

a perfluoroalkanesulfonamide through a carbamate linkage (Scheme 18.8).[8,85] This approach offers the advantage of mild reaction conditions and does not require high pressure.

18.2.4.3 Cationic and Zwitterionic Surfactants

Perfluorinated, cationic and zwitterionic surfactants are synthesized by the alkylation of perfluoroalkanesulfonamides containing a secondary[53,62] or tertiary amine[62,78,86] function. Two examples illustrating the versatility of this approach are shown in Schemes 18.9 and 18.10. Perfluorooctanesulfonyl

Production of Fluorinated Surfactants

fluoride is reacted with piperazine and the resulting amide is alkylated with methyl iodide to yield a dimethylammonium iodide (Scheme 18.9).[53] This and similar compounds are known to lower the surface tension of diesel fuel.

N-alkylated perfluorooctanesulfonamides such as PFOS (3-dimethylamino-propyl)-amide can be functionalized to yield cationic surfactants (Scheme 18.10). For example, the methylation of this amide yields fluorinated quaternary ammonium surfactant that were manufactured by the 3M Company (Figure 18.2).[26] These ammonium surfactants can be further alkylated to yield cationic surfactants with both a fluorophobic and a lipophobic chain.[86] Similarly, zwitterionic surfactants are synthesized by using chloroacetic acid, 3-propanesulfonic acid, or related compounds instead of methyl iodide as an alkylating agent.

18.3 PERFLUOROCARBOXYLATE SURFACTANTS

Analogous to perfluoroalkanesulfonyl fluorides, PFCA fluorides are useful starting materials for the synthesis of PFCAs, fluorinated esters, amides, or other intermediates. These PFCA intermediates can subsequently be converted into numerous fluorinated surfactants. The industrial production pathways for these important intermediates are outlined in Figure 18.5.[2,14,25] As outlined in Section 18.3.1, the key step is the preparation of PFCA fluorides by electrochemical fluorination from the corresponding alkanoyl chlorides or fluorides.[33,87–90] Compared to perfluorooctanesulfonyl surfactants (see Section 18.2), the production volume of PFCA fluorides by electrochemical fluorination was much smaller. For example, the U.S. production of ammonium perfluorooctanoate

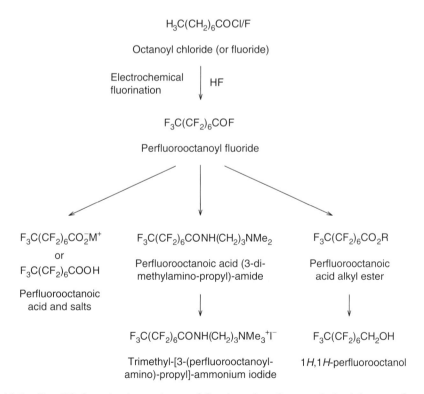

FIGURE 18.5 Simplified production pathway of fluorinated surfactants derived from perfluorooctanoyl fluoride. (Adapted from Kissa, E., *Fluorinated Surfactants and Repellents*, Marcel Dekker, New York, 2001; Environment Directorate, OECD, Hazard assessment of perfluorooctane sulfonate (PFOS) and its salts. Report ENV/JM/RD(2002)17/FINAL, Organization for Economic Cooperation and Development, Paris, 2002; Schultz, M.M. et al., *Environ. Eng. Sci.*, 20, 487, 2003.)

(APFO) by electrochemical fluorination was ~227 t/year.[91] Owing to environmental concerns, the 3M Company voluntarily began to phase out the production of PFCA fluorides at the end of the year 2000 and, with the exception of a small amount of perfluorooctanoic acid (PFOA) for internal use,[1] production of PFCAs by electrochemical fluorination ceased in the United States in 2002.[24]

18.3.1 SYNTHESIS OF PERFLUOROCARBOXYLATE FLUORIDES BY ELECTROCHEMICAL FLUORINATION

The electrochemical fluorination of the corresponding hydrocarbon carboxylic acid fluorides or chlorides follows the general reaction scheme:

$$C_nH_{2n+1}COF + (2n + 1)F^- \xrightarrow{\text{HF}} C_nF_{2n+1}COF + (2n + 1)H^+ + (4n + 2)e^-$$

Similar to perfluoroalkanesulfonic acids, the yield of the fluorination reaction decreases with the increasing chain length of the starting material. For example, the electrochemical fluorination of acetyl fluoride yields 76% of perfluoroacetyl fluoride, whereas the fluorination of octanoyl fluoride yields only 10% of perfluorooctanoyl fluoride.[2] The yield is also lower when the readily available acid chlorides, and not the fluorides, are employed as starting materials.[89] A major side reaction of the fluorination of acid chlorides such as octanoyl chloride is the formation of cyclic products.[33,87] Other by-products and impurities present in PFCA fluorides obtained by electrochemical fluorination include short-chain PFCA fluorides, perfluoroalkanes, COF_2, and HF. Perfluorooctanoyl fluoride, the most important PFCA, was industrially synthesized from the corresponding chloride despite the lower yield and formation of cyclic by-products.[92]

PFCAs can also be synthesized by electrochemical fluorination of the corresponding alkanoic acid; however, the yield of this reaction is only 10–20%. Slightly higher yields can be obtained by fluorination of the corresponding anhydrides. For example, acetic acid and acetic acid anhydride form perfluoroacetyl fluoride in a 17% and 32% yield, respectively.[90] A major disadvantage of this approach is the formation of water that may form explosive oxygen difluoride. Simple alkanoic acids were, therefore, never useful starting materials for the industrial synthesis of PFCA fluorides.

18.3.2 PERFLUOROCARBOXYLATE-DERIVED SURFACTANTS

18.3.2.1 Anionic Surfactants

PFCA salts (e.g., the sodium and ammonium salts) are an industrially important class of anionic PFCA surfactants.[2,92] Electrochemical fluorination has been employed to manufacture APFO, an emulsifier used in the production of fluoropolymers[2] and other PFCAs.[2,92,93] Hydrolysis of the acid fluorides obtained by electrochemical fluorination results either in the free carboxylic acids or the respective salts. A major problem associated with the hydrolysis of PFCA fluorides is the presence of HF from the electrochemical fluorination and the hydrolysis of the PFCA fluorides with water.[40] The technical process of isolating fluoride-free PFCA derivatives has, therefore, been optimized to reduce the fluoride content of the final product to trace amounts. An example of such a process is shown in Figure 18.6.[40]

In this example, hydrolysis of perfluorobutanoyl fluoride results in an acidic solution containing the free acid and HF. Addition of potassium hydroxide results in the formation of perfluorobutanoic acid potassium salt and potassium fluoride. Concentration of the aqueous solution of both salts results in precipitation of perfluorobutanoic acid potassium salt, whereas potassium fluoride, because of its good aqueous solubility (920 kg/L at 18°C), remains in solution. The precipitate is dried and treated with sulfuric acid (98%) and silicon dioxide at 100°C in a stream of nitrogen to remove silicon tetrafluoride, thus removing the remaining free fluoride. Distillation yields the desired perfluorobutanoic acid in 99.98% purity with <1 ppm of free fluoride. The filtrate containing mostly potassium salt is treated with calcium hydroxide, resulting in the precipitation of

Production of Fluorinated Surfactants

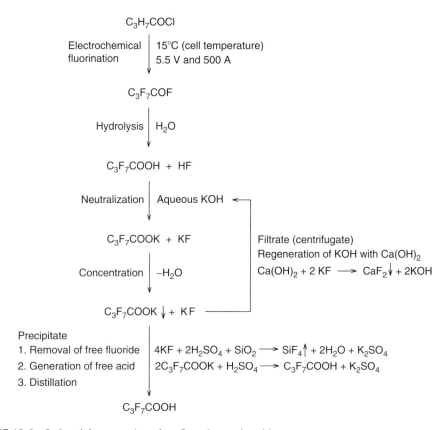

FIGURE 18.6 Industrial preparation of perfluorobutanoic acid.

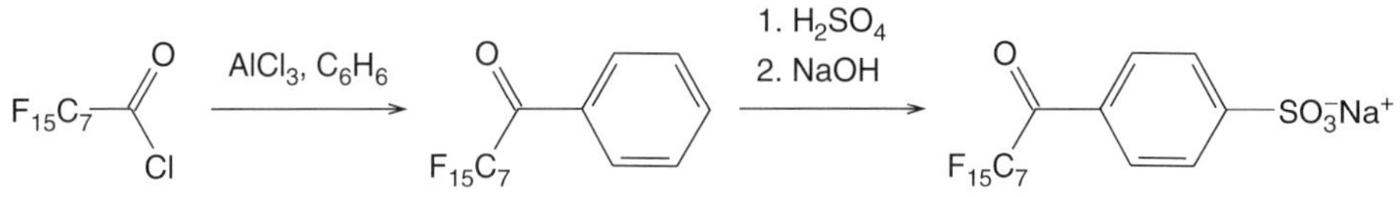

SCHEME 18.11

water-insoluble calcium fluoride and the regeneration of potassium hydroxide, which is reused in the neutralization step.

Anionic surfactants with the general structure $R_F CONH-X-COONa$ (e.g., $X = -(CH_2)_5-$) have been reported in the patent literature.[94] These surfactants are synthesized from the corresponding isopropyl ester and 6-amino-hexanoic acid sodium or ammonium salt ($H_2N(CH_2)_5COOM$). Another class of anionic surfactants derived from PFCA derivatives are perfluoroacylbenzenesulfonates (Scheme 18.11).[95] This group of surfactants is synthesized by Friedel–Crafts acylation of benzene with a PFCA halide (e.g., perfluorooctanoyl chloride) in the presence of at least one equivalent of a Lewis acid (e.g., anhydrous aluminum chloride). This reaction proceeds smoothly and in good yields at subambient or ambient temperatures. The perfluoroacylbenzene is sulfonated with oleum or sulfur trioxide and neutralized with a base (e.g., sodium hydroxide).

18.3.2.2 Nonionic Surfactants

PFCA derivatives can be reacted with ethoxylated amines to yield surface-active amides.[21,96,97] For example, the reaction of a diethoxylated amine with PFOA methyl ester yields the desired amide in

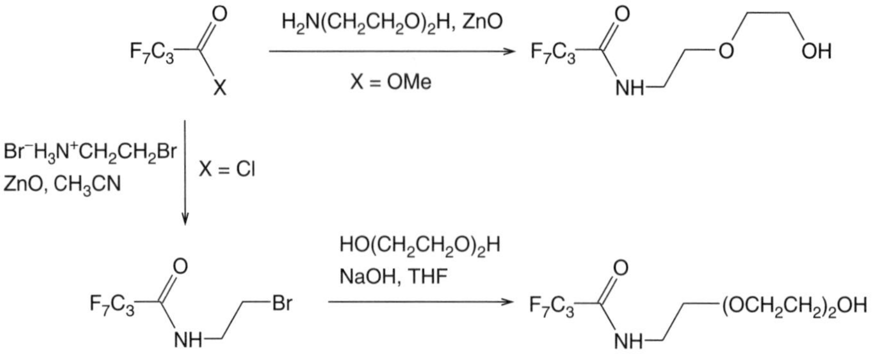

SCHEME 18.12

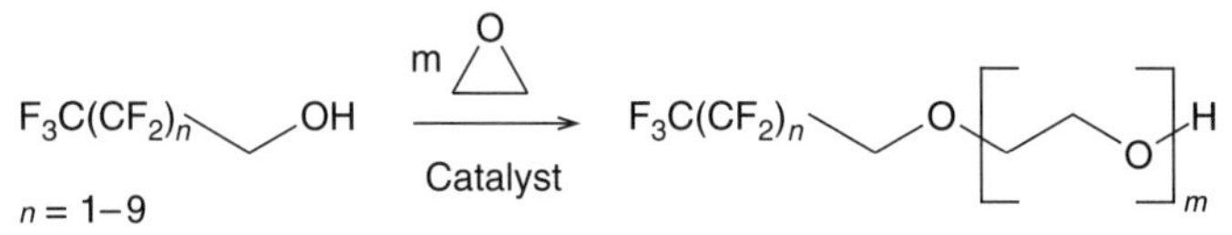

SCHEME 18.13

the presence of a catalytic amount of zinc oxide (Scheme 18.12).[96] Alternatively, a functionalized PFCA can be modified to yield the desired nonionic surfactant, for example, by the ethoxylation of an amine function as described in Section 18.2.4.2 (Scheme 18.7) or, as shown in Scheme 18.12, by the alkylation of excess diethylene glycol with a bromine-substituted PFOA amide.[96]

As shown in Figure 18.5, $1H,1H$-perfluoroalkanols are an important group of intermediates that can be obtained from PFCA fluorides obtained by electrochemical fluorination (or, alternatively, by a telomerization process[2]). These alcohols are readily converted into polydisperse nonionic polyoxyethylene surfactants by reaction with oxirane (or analogues) in the presence of a catalyst (Scheme 18.13).[98–100] Borontrifluoride etherate is the classical catalyst for this oxyethylation reaction,[98] but there are claims for the utilization of other catalysts such as aluminum alkoxides,[99] sodium hydroxide,[101] and acidic catalysts.[100]

Monodisperse fluorinated surfactants with a defined number of oxoethylene groups have been synthesized by condensing a $1H,1H$-perfluoroalkanol and a polyoxyethylene with $(Me_2N)_3P/KPF_6$.[102,103] Hexamethyl phosphoric triamide (HMPA), a potential human carcinogen, is formed during this reaction, thus limiting the usefulness of this approach. Another laboratory-scale approach synthesizes fluorinated polyoxyethylene surfactants by coupling the $1H,1H$-perfluoroalkanol with an (oxyethylene) glycol tosyl methyl ether of the general structure $4\text{-}MeC_6H_4SO_2\text{-}O(CH_2CH_2O)_mOMe$.[104] As shown in the example in Scheme 18.14, the major challenge of this approach is the synthesis of an (oxyethylene) glycol tosyl methyl ether with a defined number of oxyethylene groups m. In this approach, tetrakis(oxyethylene) glycol needs to be protected with a benzyl group, alkylated with 1-chloro-2-methoxy-ethane, deprotected by hydrogenation and activated by tosylation before coupling with the $1H,1H$-perfluoroalkanol.

18.3.2.3 Perfluorocarboxylate-Derived Cationic and Zwitterionic Surfactants

Cationic PFCA surfactants are obtained through PFCA amides, which are readily obtained from the corresponding PFCA fluorides or chlorides[96,105,106] and esters.[96] The synthesis of these surfactants is typically simple and requires only a few synthetic steps; however, the amide function of these perfluorinated amides is relatively labile because of the strong electron-withdrawing effect of the CF_2-group adjacent to the amide group, thus posing a limitation for the usefulness of these surfactants.[97]

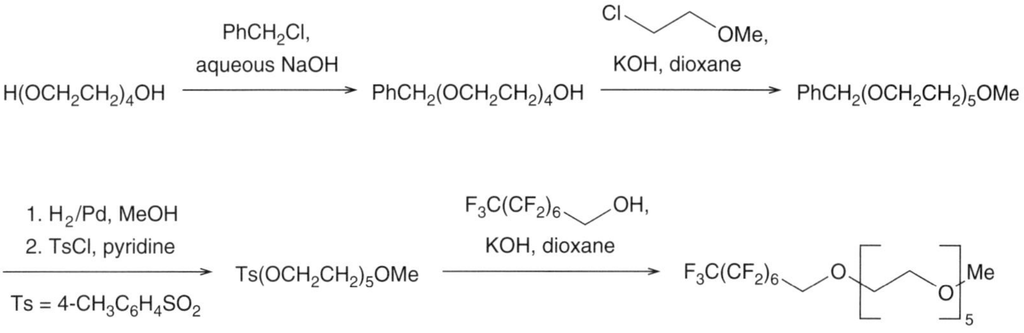

Production of Fluorinated Surfactants

SCHEME 18.14

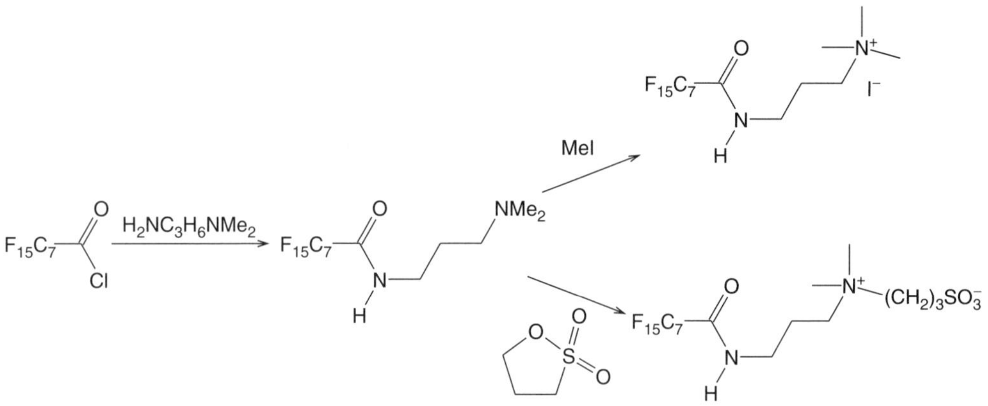

SCHEME 18.15

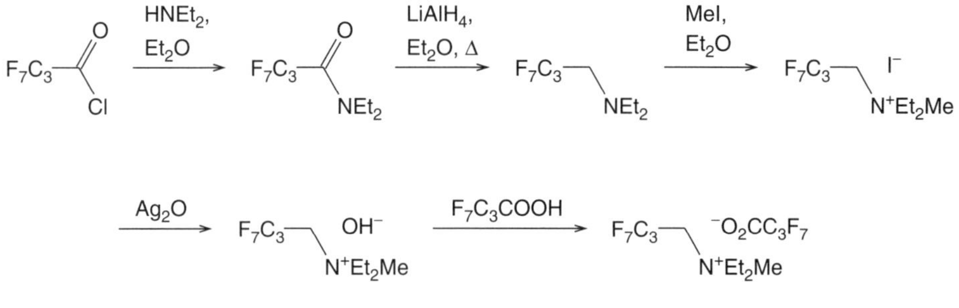

SCHEME 18.16

Similar to cationic PFOS surfactants (Section 18.2.4.3), an important approach reacts a PFCA derivative with an amine of the general structure $H_2N(CH_2)_nNR^1R^2$ (with $n \geq 2$ and R^1 and R^2 being an alkyl group or hydrogen) (Scheme 18.15).[107] The resulting amine-functionalized PFCA amide is alkylated, such as with methyl iodide, to yield a highly surface-active and typically water-soluble quaternary ammonium salt. Zwitterionic carboxy- or sulfobetaine surfactants are obtained when the amine function is alkylated with chloroacetic acid[108] or 3-chloropropanesulfonic acid,[107] respectively. As shown in Scheme 18.15, propane sultone can be used to synthesize sulfobetaines; however, this starting material is a suspected human carcinogen and, therefore, not an acceptable starting material for the synthesis of surfactants.

Another approach synthesizes fluorinated quaternary ammonium surfactants in a three-step synthesis from a PFCA derivative (e.g., the chloride) and a secondary amine.[105,106] An example of such a synthesis is shown in Scheme 18.16.[106] In this example, the N,N-dialkylated amide is synthesized from perfluorobutanoyl chloride and diethylamine. The amide is subsequently reduced with lithium aluminum hydride and the resulting tertiary amine is alkylated with methyl iodide to yield a quaternary ammonium iodide. This iodide can be converted into the hydroxide or acid salts by conventional methods, for example, treatment with silver(I) oxide followed by the addition of perfluorobutanoic acid. Quaternization of the intermediate fluorinated tertiary amines can also be alkylated with ω-chloroalkanoic acid esters or sodium salts, thus resulting in fluorinated zwitterionic surfactants.[105]

18.4 FLUORINATED SURFACTANTS AND THEIR ENVIRONMENTAL IMPACT

Fluorinated surfactants and related compounds have been used commercially for over four decades. As mentioned earlier, their fluorinated tails are highly stable toward chemical, biological, and thermal degradation—a characteristic of persistent environmental contaminants. It is, therefore, not surprising that fluorinated compounds, especially perfluorooctanesulfonic and PFCA derivatives, can be found in environmental matrices such as surface water, wildlife, and humans,[14,23,109] thus raising concerns about the environmental and human health effects of fluorinated environmental contaminants such as PFOA or PFOS. The toxicity of these and related compounds is only poorly investigated at this time and many questions still remain.[14,109–112] The environmental and public health concerns arising from the widespread environmental contamination had (and have) implications for the production and use of fluorinated surfactants. For example, the 3M Company ceased its production of perfluorooctanesulfonyl fluoride and related compounds in 2002.[14] Instead, the 3M Company is commercializing replacement products based on perfluorobutanesulfonate—a compound with very low toxicity and a very low potential to bioaccumulate. Other replacements for PFOS and related compounds include non-PFOS fluorosurfactants as well as silicone and hydrocarbon-based surfactants.[1]

ACKNOWLEDGMENTS

The author gratefully acknowledges support from the National Institute of Environmental Health Sciences (ES12475) and the National Science Foundation (DMR-0210517).

REFERENCES

1. Environment Directorate, OECD, Results of survey on production and use of PFOS, PFAS and PFOA, related substances and products/mixtures containing these substances. Report ENV/JM/MONO(2005)1, Organization for Economic Cooperation and Development, Paris, 2005.
2. Kissa, E., *Fluorinated Surfactants and Repellents*. Vol. 97, Marcel Dekker, New York, 2001.
3. 3M Company, Fluorochemical use, distribution and release overview. U.S. EPA Administrative Record AR226-0550, U.S. EPA, Washington, DC, 1999.
4. Lehmler, H.-J. et al., Liquid ventilation—a new way to deliver drugs to diseased lungs? *Chemtech*, 29, 7, 1999.
5. Riess, J.G. and Krafft, M.P., Fluorocarbons and fluorosurfactants for *in vivo* oxygen transport (blood substitutes), imaging, and drug delivery, *MRS Bull.*, 24, 42, 1999.
6. Tan, B. et al., Controlling nanopore size and shape by fluorosurfactant templating of silica, *Chem. Mater.*, 17, 916, 2005.
7. Gramstad, T. and Haszeldine, R.N., Perfluoroalkyl derivatives of sulphur. Part VI. Perfluoroalkanesulphonic acids $CF_3(CF_2)_nSO_3H$ ($n = 1–7$), *J. Chem. Soc.*, 2640, 1957.
8. Klein, H.G. et al., Fluorine-containing surfactants based on electrochemical fluoridation and possibilities for introducing them in surface treatment techniques, *Metalloberfläche*, 29, 559, 1975.
9. Kunleda, H. and Shinoda, K., Krafft points, critical micelle concentrations, surface tension, and solubilizing power of aqueous solutions of fluorinated surfactants, *J. Phys. Chem.*, 80, 2468, 1976.
10. Shinoda, K. et al., Physicochemical properties of aqueous solutions of fluorinated surfactants, *J. Phys. Chem.*, 76, 909, 1972.
11. Mukerjee, P. and Yang, A.Y.S., Nonideality of mixing of micelles of fluorocarbon and hydrocarbon surfactants and evidence of partial miscibility from differential conductance data, *J. Phys. Chem.*, 80, 1388, 1976.

Production of Fluorinated Surfactants

12. 3M Company, Voluntary use and exposure information profile perfluorooctane sulfonic acid and various salt forms. U.S. EPA Administrative Record AR226-0928, U.S. EPA, Washington, DC, 2000.
13. 3M Company, Octanol/water coefficient: Perfluorooctanesulfonate. U.S. EPA Administrative Record AR226-0050, U.S. EPA, Washington, DC, 2000.
14. Environment Directorate, OECD, Hazard assessment of perfluorooctane sulfonate (PFOS) and its salts. Report ENV/JM/RD(2002)17/FINAL, Organization for Economic Cooperation and Development, Paris, 2002.
15. Patrick, C.R., Physicochemical properties of highly fluorinated organic compounds, *Chem. Brit.*, 7, 154, 1971.
16. Iimura, K.-I. et al., Micro-phase separation in binary mixed Langmuir monolayers of *n*-alkyl fatty acids and a perfluoropolyether derivative, *Langmuir*, 18, 10183, 2002.
17. Imae, T. et al., Phase separation in hybrid Langmuir–Blodgett films of perfluorinated and hydrogenated amphiphiles. Examination by atomic force microscopy, *Langmuir*, 16, 612, 2000.
18. Ge, S. et al., Aggregation structure and surface properties of immobilized organosilane monolayers prepared by the upward drawing method, *J. Vac. Sci. Technol. A*, 12, 2530, 1994.
19. Meyer, E. et al., Friction force microscopy of mixed Langmuir–Blodgett films, *Thin Solid Films*, 220, 132, 1992.
20. Mukerjee, P., Fluorocarbon–hydrocarbon interactions in micelles and other lipid assemblies, at interfaces, and in solutions, *Colloid. Surface. A*, 84, 1, 1994.
21. Schuierer, E., Fluorine containing surfactants, synthesis and properties, *Tenside Surfact. Det.*, 13, 1, 1976.
22. Banks, R.E. *Fluorocarbons and Their Derivatives*. MacDonald, London, 1970.
23. Lehmler, H.-J., Synthesis of environmentally relevant fluorinated surfactants—a review, *Chemosphere*, 58, 1471, 2005.
24. 3M Company, Phase-out plan for POSF-based products. U.S. EPA Administrative Record AR226-0588, U.S. EPA, Washington, DC, 2000.
25. Schultz, M.M. et al., Fluorinated alkyl surfactants, *Environ. Eng. Sci.*, 20, 487, 2003.
26. 3M Company, Voluntary use and exposure information profile perfluorooctanesulfonyl fluoride (PFOS). U.S. EPA Administrative Record AR226-0576, U.S. EPA, Washington, DC, 2000.
27. Simons, J.H., U.S. 2519983, 1950 (CAN 45: 380).
28. Simons, J.H. et al., Electrochemical process for the production of fluorocarbons, *J. Electrochem. Soc.*, 95, 47, 1949.
29. Ignat'ev, N. et al., Comparative electrochemical fluorination of ethanesulfonyl chloride and fluoride, *Acta Chem. Scand.*, 53, 1110, 1999.
30. Niederpruem, H. et al., Hydroxyalkylation of perfluoroalkanesulfonamides, *Liebigs Ann. Chem.*, 11–19, 1973.
31. Burdon, J. et al., Fluorinated sulphonic acids. Part I. Perfluoro-methane-, -octane-, and -decane-sulphonic acids and their simple derivatives, *J. Chem. Soc.*, 2574, 1957.
32. Hollitzer, E. and Sartori, P., The electrochemical perfluorination (ECPF) of propanesulfonyl fluorides. Part I. Preparation and ECPF of 1-propanesulfonyl fluoride and 1,3-propanedisulfonyl difluoride, *J. Fluorine Chem.*, 35, 329, 1987.
33. Drakesmith, F.G. and Hughes, D.A., The electrochemical fluorination of octanoyl chloride, *J. Appl. Electrochem.*, 9, 685, 1979.
34. Meussdoerffer, J.N. et al., DE 2238152, 1974 (CAN 80: 120258).
35. Ignat'ev, N.V. et al., Carbon-chain isomerization during the electrochemical fluorination in anhydrous hydrogen fluoride—a mechanistic study, *J. Fluorine Chem.*, 124, 21, 2003.
36. Gramstad, T. and Haszeldine, R.N., Perfluoroalkyl derivatives of sulfur. VII. Alkyl trifluoromethanesulfonates as alkylating agents, trifluoromethanesulfonic anhydride as a promoter for esterification, and some reactions of trifluoromethanesulfonic acid, *J. Chem. Soc.*, 4069, 1957.
37. Hebert, G.N. et al., Method for the determination of sub-ppm concentrations of perfluoroalkylsulfonate anions in water, *J. Environ. Monit.*, 4, 90, 2002.
38. Zhang, L. et al., Absolute rate constants of alkene addition reactions of a fluorinated radical in water, *J. Am. Chem. Soc.*, 124, 6362, 2002.
39. Van Dyke Tiers, G., The chemistry of fluorocarbon sulfonic acids. I. Preparation of anhydrides and sulfonyl halides, *J. Org. Chem.*, 28, 1244, 1963.
40. Aramaki, M. et al., DE 3829409, 1989 (CAN 111: 99369).
41. Wechsberg, M. and Niederpruem, H., DE 2319078, 1974 (CAN 82: 57463).
42. Feiring, A.E. et al., Synthesis of partially fluorinated monomers and polymers for ion-exchange resins, *J. Fluorine Chem.*, 93, 93, 1999.

43. Mitschke, K.H. et al., DE 2658560, 1978 (CAN 89: 163065).
44. Beyl, V. et al., New reaction of perfluoroalkylsulfonyl fluorides, *Liebigs Ann. Chem.*, 731, 58, 1970.
45. Meussdoerffer, J.N. and Niederpruem, H., Bis(perfluoralkanesulfonyl)imides $(R_FSO_2)_2NH$, *Chem. Ztg.*, 96, 582, 1972.
46. Podol'skii, A.V. et al., Preparation of perfluoroalkanesulfonamide ammonium salts, *Russ. J. Org. Chem.*, 17, 1242, 1990.
47. Roesky, H.W. et al., Sulfur–nitrogen compounds. 29. *N*-trifluoromethylsulfonyl compounds, *Z. Naturforsch.*, 25b, 252, 1970.
48. Bussas, R. and Kresze, G., Reactions of *N*-sulfinyl compounds. XV. Synthesis of 2-alkenesulfinic acid amides—structure–reactivity relations for *N*-sulfinyl compounds as enophiles, *Liebigs Ann. Chem.*, 545, 1982.
49. 3M Company, Voluntary use and exposure information profile N-methyl perfluorooctanesulfonamide. U.S. EPA Administrative Record AR226-0581, U.S. EPA, Washington, DC, 2000.
50. 3M Company, Voluntary use and exposure information profile N-ethyl perfluorooctanesulfonamide. U.S. EPA Administrative Record AR226-0580, U.S. EPA, Washington, DC, 2000.
51. Chen, Q.-Y. and Qiu, Z.-M., Studies on fluoroalkylation and fluoroalkoxylation. Part 10. Electron-transfer-induced reactions of perfluoroalkyl iodides and the dialkyl malonate anion and β-fragmentation of the halotetrafluoroethyl radical, *J. Fluorine Chem.*, 31, 301, 1986.
52. Huang, T.-J. et al., 1,1,2,2-Tetrafluoro-2-(polyfluoroalkoxy)ethanesulfonyl fluorides, *Inorg. Chem.*, 26, 2604, 1987.
53. Katritzky, A.R. et al., Design and synthesis of novel fluorinated surfactants for hydrocarbon subphases, *Langmuir*, 4, 732, 1988.
54. Lyapkalo, I.M. et al., Study of unusually high rotational barriers about S–N bonds in nonafluorobutane-1-sulfonamides: The electronic nature of the torsional effect, *Helv. Chim. Acta*, 85, 4206, 2002.
55. Benefice-Malouet, S. et al., Catalytic preparation of perfluoroalkanesulfonamides by addition of ethylamine to perfluoroalkanesulfonyl chloride, *J. Fluorine Chem.*, 31, 319, 1986.
56. DeChristopher, P.J. et al., Simple deaminations. V. Preparation and some properties of *N*-alkyl-*N*, *N*-disulfonimides, *J. Org. Chem.*, 39, 3525, 1974.
57. Takeuchi, Y. et al., Enantioselective fluorination of organic molecules. I. Synthetic studies of the agents for electrophilic, enantioselective fluorination of carbanions, *Chem. Pharm. Bull.*, 45, 1085, 1997.
58. Kas'yan, A.O. et al., Amines and sulfonamides containing an adamantane fragment. Theoretical and experimental study, *Russ. J. Org. Chem.*, 33, 985, 1997.
59. Trepka, R.D. et al., Acidities and partition coefficients of fluoromethanesulfonamides, *J. Org. Chem.*, 39, 1094, 1974.
60. Cho, B.T. and Chun, Y.S., Catalytic enantioselective reactions. Part 12. Enantioselective addition of diethylzinc to aldehydes catalyzed by zinc complexes modified with chiral β-sulfonamido alcohols, *Synthetic Commun.*, 29, 521, 1999.
61. Reitz, G. and Boehmke, G., DE 2601375, 1977 (CAN 87: 133886).
62. Reitz, G. et al., DE 2457754, 1976 (CAN 85: 77689).
63. Zhu, S.-Z. et al., Synthesis of dialkyl *N*-alkyl-*N*-perfluoroalkanesulfonyl phosphoramidates, *Phosphorus Sulfur*, 89, 77, 1994.
64. Kamei, M. et al., JP 58201752, 1983 (CAN 100: 174444).
65. Hashimoto, Y. and Kamei, M., JP 59046252, 1984 (CAN 101: 9130).
66. Mitschke, K.H. et al., DE 2832346, 1980 (CAN 93: 25904).
67. Hendrickson, J.B. et al., New "Gabriel synthesis of amines", *Tetrahedron*, 31, 2517, 1975.
68. Ahlbrecht, A.H. and Brown, H.A., US 2803656, 1957 (CAN 52: 11247).
69. Hendrickson, J.B. et al., Uses of the triflyl [trifluoromethanesulfonyl] group in organic synthesis. A review, *Org. Prep. Proced. Int.*, 9, 173, 1977.
70. Harris, J.M. et al., Hydrogen abstraction from silylamines: An investigation of the 1,2-migration of the trimethylsilyl group in aminyl radicals, *J. Chem. Soc., Perkin Trans. 2*, 2119, 1993.
71. Dohle, W. et al., Fe(III)-catalyzed cross-coupling between functionalized arylmagnesium compounds and alkenyl halides, *Synlett*, 1901, 2001.
72. Ostwald, R. et al., Catalytic asymmetric addition of polyfunctional dialkylzincs to β-stannylated and β-silylated unsaturated aldehydes, *J. Org. Chem.*, 59, 4143, 1994.
73. Lutz, C. et al., Enantioselective synthesis of 1,2-, 1,3- and 1,4-aminoalcohols by the addition of dialkylzincs to 1,2-, 1,3- and 1,4-aminoaldehydes, *Tetrahedron*, 54, 6385, 1998.
74. Scozzafava, A. et al., Arylsulfonyl-*N*, *N*-diethyl-dithiocarbamates: A novel class of antitumor agents, *Bioorgan. Med. Chem. Lett.*, 10, 1887, 2000.

Production of Fluorinated Surfactants

75. Kawase, M. et al., Use of the triflamide group for Friedel-Crafts acylation of N-(β-phenethyl)amino acids to 3-benzazepine derivatives, *Heterocycles*, 45, 1121, 1997.
76. Harrison, S.S. and Hunt, K.B., EP 683267, 1995 (CAN 124: 149102).
77. Mitschke, K.H. and Niederpruem, H., DE 2921142, 1980 (CAN 94: 105293).
78. Kimura, C. et al., Preparation and surface active properties of perfluorooctanesulfonamide derivatives, *Yukagaku*, 31, 448, 1982.
79. Hashimoto, Y., JP 57209259, 1982 (CAN 98: 162829).
80. 3M Company, Voluntary use and exposure information profile perfluorooctane sulfonamido ethyl acetate. U.S. EPA Administrative Record AR226-0578, U.S. EPA, Washington, DC, 2000.
81. Heine, R.F., FR 1317427, 1963 (CAN 59: 54279).
82. Mitschke, K.H. and Niederpruem, H., DE 2713498, 1978 (CAN 90: 38530).
83. Ahlbrecht, A.H. and Morin, D.E., US 2915554, 1959 (CAN 54: 28271).
84. Reitz, G. and Boehmke, G., DE 2639473, 1978 (CAN 88: 190086).
85. Meussdoerffer, J.N. et al., DE 2238740, 1974 (CAN 85: 22318).
86. Kamei, M. et al., JP 54041817, 1979 (CAN 91: 107666).
87. Prokop, H.W. et al., Analysis of the products from the electrochemical fluorination of octanoyl chloride, *J. Fluorine Chem.*, 43, 277, 1989.
88. Prokop, H.W. et al., Process improvements in the electrochemical fluorination of octanoyl chloride, *J. Fluorine Chem.*, 43, 257, 1989.
89. Brice, T.J. et al., U.S. 2713593, 1955 (CAN 50: 28114).
90. Scholberg, H.M. and Bryce, H.G., U.S. 2717871, 1955 (CAN 49: 82967).
91. Fluoropolymer Manufacturers Group, Fluoropolymer manufacturers group presentation slides. U.S. EPA Administrative Record AR226-1094, U.S. EPA, Washington, DC, 2002.
92. 3M Company, Voluntary use and exposure information profile perfluorooctanoic acid and salts. U.S. EPA Administrative Record AR226-0595, U.S. EPA, Washington, DC, 2000.
93. Prevedouros, K. et al., Sources, fate and transport of perfluorocarboxylates, *Environ. Sci. Technol.*, 40, 32, 2005.
94. Hayashi, T. and Osai, Y., JP 54044610, 1979 (CAN 91: 76024).
95. Dreher, B. et al., DD 239788, 1986 (CAN 106: 215901).
96. Afzal, J. et al., Syntheses of perfluoroalkyl N-polyethoxylated amides, *J. Fluorine Chem.*, 34, 385, 1987.
97. Selve, C. et al., Monodisperse perfluoro-polyethoxylated amphiphilic compounds with two-chain polar head—preparation and properties, *Tetrahedron*, 47, 411, 1991.
98. Greiner, A. and Gerhardt, W., DD 111522, 1975 (CAN 84: 123793).
99. Yang, K. et al., U.S. 4490561, 1984 (CAN 102: 115550).
100. Naik, A.R., DE 2261681, 1973 (CAN 79: 93723).
101. Tatematsu, R. et al., Preparation and properties of some fluoroalcohol-ethylene oxide adducts, *Yukagaku*, 26, 367, 1977.
102. Leempoel, P. et al., EP 51527, 1982 (CAN 97: 127041).
103. Selve, C. et al., Synthesis of homogeneous poly(oxyethylene) perfluoroalkyl surfactants. A new method, *Tetrahedron*, 39, 1313, 1983.
104. Achilefu, S. et al., Monodisperse perfluoroalkyl oxyethylene nonionic surfactants with methoxy capping: Synthesis and phase behavior of water/surfactant binary systems, *Langmuir*, 10, 2131, 1994.
105. Nivet, J.B. et al., Synthesis and bioacceptability of fluorinated surfactants derived from F-alkylated tertiary amines, *Eur. J. Med. Chem.*, 27, 891, 1992.
106. Husted, D.R., U.S. 2727923, 1955 (CAN 50: 74151).
107. Kimura, C. et al., Preparation and surface active properties of perfluorooctanamidopropylenedimethyl-amine derivatives, *Yukagaku*, 31, 464, 1982.
108. Hayashi, T. and Otoshi, Y., JP 54062990, 1979 (CAN 91: 125254).
109. Kumar, K.S., Fluorinated organic chemicals: A review, *Res. J. Chem. Env.*, 9, 50, 2005.
110. Kennedy, G.L., Jr. et al., The toxicology of perfluorooctanoate, *Crit. Rev. Toxicol.*, 34, 351, 2004.
111. Kudo, N. and Kawashima, Y., Toxicity and toxicokinetics of perfluorooctanoic acid in humans and animals, *J. Toxicol. Sci.*, 28, 49, 2003.
112. Lau, C. et al., The developmental toxicity of perfluoroalkyl acids and their derivatives, *Toxicol. Appl. Pharmacol.*, 198, 231, 2004.

19 Detergent Processing

A. E. Bayly, D. J. Smith, Nigel S. Roberts,
David W. York, and S. Capeci

CONTENTS

19.1 Introduction 324
    19.1.1 Surfactant 326
    19.1.2 Product Forms 326
        19.1.2.1 Powders 327
        19.1.2.2 Compact Powders 327
        19.1.2.3 Tablets 328
        19.1.2.4 Liquid Detergents 328
        19.1.2.5 Liquid Pouches 329
        19.1.2.6 Syndet Bars 329
        19.1.2.7 Other Product Forms 331
19.2 Detergent Powder Processing 331
    19.2.1 Processes for Making Detergents 331
        19.2.1.1 Spray Drying 331
        19.2.1.2 Agglomeration in Detergent Processing 337
        19.2.1.3 The Finishing of Dry Laundry Detergents 345
19.3 Heavy-Duty Liquid Detergents 350
    19.3.1 Introduction 350
        19.3.1.1 Transformation Requirements 350
    19.3.2 Basic Process Options 351
        19.3.2.1 Batch 351
        19.3.2.2 Continuous Operations 352
        19.3.2.3 Choosing the Optimal Process to Meet Market Demands 353
        19.3.2.4 Late Product Differentiation 353
    19.3.3 Other Design Considerations 354
        19.3.3.1 Micro 354
        19.3.3.2 Recycle 354
    19.3.4 Conclusion 354
19.4 Unit Dose Detergent Process Technologies 354
    19.4.1 Introduction 354
        19.4.1.1 Scope of This Section 354
        19.4.1.2 Review of the Evolution of Unit Dose Detergent Products 354
    19.4.2 Hard Compressed Tablets 355
    19.4.3 Soft Compressed Tablets 359

19.4.4	Water-Soluble Pouches	360
	19.4.4.1 Single Compartment	360
	19.4.4.2 Dual and Multicompartment	362
19.4.5	Summary and Trends	363
Reference		363

19.1 INTRODUCTION

The manufacture of laundry detergents has changed significantly over the past 20 years, but the appearance of the products has not although a number of different product forms have been introduced. Even with the growth of the liquid form, the main form is still that of a powder and remains the largest growth area globally, especially in the developing regions.

Powders have a number of advantages in that they contain a relatively large proportion of inorganic materials, such as builders (water hardness removers), alkalinity sources (e.g., sodium carbonate), and bleaches (sodium percarbonate), which tend to be cheaper than petro- and oleo-derived materials. In addition, due to their low moisture content, it is much easier to include a range of more complex or metastable ingredients such as bleach and enzymes and keep them stable during the trade. This is important since the products must remain stable for about 12 months and beyond after making. Products can experience wide ranges of temperature during transportation, particularly when exposed to sunshine, as well as in warehouses and retail outlets. Temperatures can vary from -10 to $+50°C$. However, this then requires careful control of moisture ingress since the products are normally made with very low relative humidity and they tend to attract water vapor from outside the package.

Another advantage of the powder form is the formulation consistency. When segregation is well managed there is little danger of separation of ingredients, whereas liquids are more prone to phase separation, and although a lot of progress has been made in this area, phase separation is still poorly understood.

There is also, especially in the developing countries, a large market for syndet bars. These are similar to soap bars, but are much bigger with the necessary ingredients for cleaning clothes. They are mainly used in hand washing, where they can be easily applied in a concentrated form, and also for rubbing soiled garments to assist in cleaning.

Recently, a number of more convenient forms have been introduced with varying success, mainly driven by local consumer preferences. These include highly compacted powders (such as Kao's Attack in Japan), uniform pelletized products (such as Henkel's MegaPerls in Germany), and unit dose tablets and liquid pouches (such as Procter & Gamble's (P&G) Ariel Liquitabs in the United Kingdom). Each of these forms presents unique challenges during their manufacturing. Although in each case, the same principles drive the process innovation and manufacturing operation. These include

Capital minimization. Since this consumer area is of considerable size, the manufacturing plants tend to have large throughputs, necessitating substantial capital investment. This not only impacts on product cost but also has business risk for new ventures and the time to market in a fast moving, highly competitive market. Thus, there is a drive for high production rates and low capital investment, typically pushing the boundaries of experience in areas such as spray drying, agglomeration, tabletting, and packaging.

In addition to the need for high production rates, there is also the need for high capacity utilization and process reliability. Given the high number of unit operations in a detergent plant, each of which can stop the process, it is important to understand and eliminate causes of downtime. Today, the best plants operate with a capital utilization of over 95%.

Formulation flexibility and agility. Typically, one plant will make a range of different products, either under different brand names or under a range of custom-made "own-label"

Detergent Processing 325

products. This is made more complicated in that, typically a lot of the ingredients are sourced from a number of different suppliers and some of these raw materials can vary in terms of activity and quality. Thus, a process must be flexible enough to cope with such variations, while still producing the consistent quality a consumer demands. In addition, the competitiveness of the market place means that competitors are continually developing new formulations and new ingredients to gain an edge in the market place. This again results in challenges regarding the equipment design and plant layout to allow fast and low cost changes to new formulations.

A recent trend has been in the area of customization, whereby the consumers prefer to have a number of different variants to meet their desire for variations such as different perfume types. Achieving this without a large expansion of capital investment is a significant challenge to large-scale producers.

Finally, inventory costs are such that plants no longer have the luxury of producing one formulation for a number of days before switching to another. Typically, plants are required to make a number of different products within one shift, which puts pressure on being able to rapidly change from one product to the other without losing out on quality.

Packaging. With few exceptions, the product is packed on-site into cartons, bottles, or sachets as appropriate. Since numerous sizes are sold on the market, a number of packaging lines are needed. Typically, a packaging line will not run at the same speed as the production line, and therefore, a buffer is needed between the two operations. Even if this were not the case, hard linking of the two is not recommended as packing lines tend to have much more breakdowns than the production line. Coping with this buffer stock and feeding the material to the right packing lines is a considerable logistics and control challenge.

Initiative management. Another area is that of managing change from one product range to a new one to stay ahead of competition. A modern, large production facility is very complex and bringing in new ingredients can not only add to the capital cost but also impact on the logistical operation of the plant. Mathematical models are now used to predict any bottlenecks and resolve them ahead of any building work as well as minimize the impact of construction work on the day-to-day operations of a plant.

Another balance has to be made between buying a turnkey unit from one of the manufacturers and custom design. Typically, the larger producers have such an extensive experience of operating their plants plus a knowledge of the new ingredients that custom design is the preferred option. It also aids in the maintenance of trade secret information.

Quality control. Consumers expect consistent quality, and it is much harder to gain new consumers than maintain existing ones. Thus, maintaining quality is very important. Although the pressure to reduce inventories means that it is costly to hold the finished product while it is being checked for quality before releasing to be shipped. Consequently, a lot of research has been done, and is still being developed, on rapid product characterization tools to minimize the hold up time of a product once it is made. In addition, online monitoring and feed-forward control systems greatly minimize the chances of producing off-specification material. A similar attention to raw material quality is just as important in this area.

Operating costs. Highly automated systems can run with low manning requirements but typically need highly qualified people to maintain and monitor the processes, and require a high capital investment. Highly manual systems are much more flexible and require lower investment, but higher levels of manning. Getting the balance right to minimize cost, while maintaining flexibility and quality is very complex even for a simple recipe, and therefore, each situation needs to be analyzed individually, taking into account the varying local factors including import duties, scale of operation, and labor costs.

Energy costs have always been a factor, and the growing cost of energy is having a big impact on the process design and plant operation in the continued drive toward reducing added cost and providing the consumer with a better value.

Safety. Finally, all these must be achieved without compromising on safety, for the worker, equipment, and environment. Hygiene is particularly important for the health of the worker, especially to protect against fine, alkaline dusts, and enzyme particles in the atmosphere, whether from particulate dust or aerosols in liquid-making products. Spray drying combustible material has led to fires in the spray drying towers, resulting in costly repairs and downtimes. Environmental emissions can be kept to a minimum by good process operations and material capture and internal recycle.

The introduction of liquids has resulted in a new concern—that of bacterial contamination. Although detergents are designed to be nondigestible and therefore consumer safety is not a problem, contamination can lead to off odors and off-putting appearance of liquid products leading to consumer dissatisfaction, and this has to be avoided.

19.1.1 SURFACTANT

Although each product form encompasses a number of different raw materials, the one material that is common to all forms is linear alkyl benzene sulfonate (LAS). This has been the main workhorse for over 50 years, providing the basic surfactancy demanded for cleaning. It has the advantages of being available in large volumes, is relatively cheap, has a good safety and biodegradability profile, and is solid at room temperature. It is also, typically the only raw material that is chemically altered before adding to the rest of the ingredients, and is summarized in the following text. The rest of the ingredients are aqueous solutions such as sodium silicate, powders such as zeolite of ~5–100 µm, or special particles such as bleach materials with a particle size of ~500 µm to match that of the blown powder.

Typically, plants would buy in the alkyl benzene and sulfonate on-site using either oleum or liquid SO_3 before neutralizing with caustic solution. The reactions are highly exothermic and can cause the coloration of the product if temperatures are not controlled. Thus, a dominate bath approach was used to control the quality since consumers do not associate a brown-coloured product with good cleaning power!

Concerns regarding the safety of transporting aggressive chemicals such as oleum, as well as maintaining a good process control has led to a number of different process options being used.

First, the large producers have tended toward producing SO_3 on-site by burning sulfur and using falling film reactors for sulfonation. This is a high capital investment and a complex process to run and maintain, but in terms of product cost it more than pays back.

For smaller producers, a different approach is taken. Fortunately, the acid form of LAS is quite stable and has a low viscosity at an active level of ~98%, and is less aggressive compared to oleum. This material is shipped in drums or bulk to detergent plants for use in product making.

It has one other advantage in that it can be neutralized with sodium carbonate without overheating as the reaction rate is relatively slow. This "dry neutralization" reaction is used to good effect in making high-density surfactant agglomerates for high-density products with relatively low capital since the surfactant contains very low amounts of water, and therefore, there is no need to include drying equipment to remove water from the system.

19.1.2 PRODUCT FORMS

The main product forms, with their pros and cons and the specific challenges in manufacturing are covered separately in the following sections.

19.1.2.1 Powders

As mentioned previously, powders are the largest volume produced globally and are still seen as the form with the largest growth potential. The main process is to form a concentrated, aqueous slurry of most of the raw materials and spray dry it in a countercurrent tower. This has the advantage of producing a relatively free-flowing powder with most of the ingredients contained inside, and therefore is less prone to the effects of segregation. The high surface area provides rapid dissolution rates and the process is highly flexible with respect to formulation. Typical bulk density of the product is ~500 g/L with median particle size of ~400 μm. Both these can be varied slightly by changing process conditions.

Certain ingredients such as perfumes, enzymes, and bleach cannot be processed in this way, and therefore, they are added to the spray dried product (blown powder) afterward.

Typically, the products are white but some producers add dyes to form colored blown powders. This has the disadvantage of requiring cleanouts between products if the same unit has to produce a white product as well.

These units typically require high capital investment, and therefore need a large market to provide satisfactory economics.

Products are sold in a variety of packaging forms, from single-use sachets to >10 kg cartons. Sachets have an advantage of providing a good moisture barrier protection when compared to cartons, although a lot of cartons are made with an internal polymer layer to provide some resistance to moisture ingress.

19.1.2.2 Compact Powders

The move to lower volume products really started in Japan (Figure 19.1), where more compact boxes were more desirable by consumers due to the lower storage space requirement.

This then moved to numerous other developed markets. The desired bulk density was beyond that achievable by normal spray drying, and thus various process innovations were developed. These included taking the blown powder and compacting it, using roller compactors, followed by grinding and sieving to achieve the right density. Although being relatively simple, the process was not very efficient due to the recycle level, and thus, other processes such as agglomeration were developed. The extension of this concept was reached in Germany where Henkel (Figure 19.2) produced a novel-looking product by extruding the ingredient mix into noodles, which were then broken up and rounded.

FIGURE 19.1 Kao's Attack® product.

FIGURE 19.2 Henkel's Megaperls®.

This latter process is simple in concept but technically challenging since extrusion demands certain rheological properties and the low surface area/volume is not conducive to fast dissolution. Thus, numerous process and formulation innovations have been required to achieve the desired product quality, the magnitude of the challenge being evident from the large number of patents in this area.

19.1.2.3 Tablets

The laundry tablet form is the ultimate in compaction. This also makes dosing simpler for the consumer and is less messy. However, there is a balance to be made in that high degrees of compaction are typically needed to ensure a tablet that is strong enough to resist fracture on handling, yet, this level of compaction restricts the rate at which a tablet can dissolve fast enough in the relatively low-shear environment of the washing process. Typically, compaction is kept to the minimum required to form a strong enough tablet, and various coatings and ingredients are added, respectively to improve the fracture resistance during transportation and to enhance dissolution by causing the tablet to swell and crack once it hits the water.

Controlling the physical properties of the product to be compactable at low pressure, as well as being able to do this at high enough production rates to be financially viable places a number of tough technical challenges on the process engineer and the manufacturing plant.

Despite the growing acceptance of this form in the automatic dishwashing market, it is of limited acceptance in the laundry market.

19.1.2.4 Liquid Detergents

Although being widely used for light-duty cleaning for a long time, the so-called heavy-duty liquids (HDL, Figure 19.3) have only been of significant impact for ~20 years.

This was not helped by the early design of horizontal drum washing machines, where the product leaked out of the dispenser drawer and into the machine sump, thereby significantly reducing the amount of material in the wash drum for cleaning. Various producers resolved this with machine dosing devices until the machine manufactures caught up, and it is a good example of the interrelationship between the producers and the machine manufacturers.

Liquids have numerous advantages over powders: they are quick to dissolve and do not have a tendency to cause product gels in dispenser draws or machines—a major consumer concern. The lack of inorganic components reduces the residues on fabrics for an improved fabric feel, and this product form has grown especially in North America.

The big challenge is the ingredient compatibility, especially between enzymes and bleach stability in an aqueous matrix, as well as phase stability. In the latter case, it is possible for various

FIGURE 19.3 P&G's Tide® liquid detergent.

FIGURE 19.4　P&G's Ariel® liquitabs.

ingredients to separate out on aging and thus provide poor performance for the consumer. These product forms are much less flexible compared to powders and therefore demand a much greater degree of interaction between the product designers and the process engineers for success.

They are manufactured either in batches or continuously and there are strong proponents for both processes. However, one advantage is that capital investment for small-scale operations can be much lower than for the spray dried granules, and thus, small producers can enter the market more easily with this form.

19.1.2.5　Liquid Pouches

The initial popularity of tablets led to the logical question as to whether liquid detergents could be made available in a unit dose form. The resolution came with the development of a polymer pouch using polyvinyl alcohol (PVOH) as the container (Figure 19.4).

PVOH has some unique properties in that it is very water soluble, but it can cope with small amounts of water in the entrapped liquid; it is highly elastic, and therefore can be stretched, filled, and then be made to spring back to form a taut pillow shape; and its chemistry can be modified substantially to make it compatible with the detergent product. Nevertheless, achieving high production rates with low leakage levels has been a considerable challenge. Typical problems include quality control of the incoming film to avoid pinholes and defects for preventing leakage on stretching and being able to seal the pouches at high production speeds that are required to keep the capital investment low. Recycling off-standard products and returns from the trade is also more challenging and makes the whole operation more complex. Despite the difficulties, there are a large number of consumers who prefer this simple, no mess means of dosing product to the washing machine.

19.1.2.6　Syndet Bars

These bars started out basically as a block of solidified soap in a convenient to make and handle form, often shaved for placing into washing machines to aid dissolution. However, with the development of synthetic surfactants and detergency aids such as polyphosphates, the developed markets moved toward synthetic detergents, which are sold either in large bar form, of ~500 g, or can also be cut into sections for customers with limited cash (Figures 19.5 and 19.6). Here, the solid bar is anything but rapidly soluble, which is benefitial for hand wash and long usage life as the bar is used to scrub clothes as well as generate high concentrations for difficult soil removal.

A laundry detergent bar, used directly on stains for hand washing of fabrics, is one of the oldest versions of laundry detergent and is still very prevalent in developing countries. The earliest versions of laundry bars were made of soap. Since the late 1960s, synthetic detergent bars started replacing the traditional soap bars. In Philippines, India, and other Far East countries, synthetic detergent bars are the dominant form. In Latin America, Africa, and China, soap bars are still the major form of bars used. Recently, a trend toward soap-syndet combo bars is gaining ground in some markets. However, conversion to granules and powders is taking place at a steady pace in all the regions of the world.

FIGURE 19.5 P&G's Mr. Clean® syndet bars.

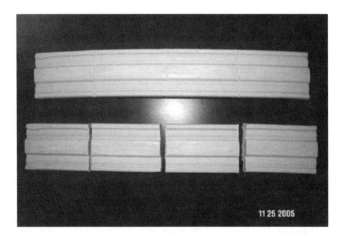

FIGURE 19.6 Typical unwrapped syndet bars.

The basic bar-making process involves neutralization of LAS sulfonic acid (in the case of LAS, containing bars) using sodium carbonate, and mixing it with all other ingredients in a particular sequence followed by extrusion of the mixed mass through an extrusion die. The bars coming out of the extruders are cut to the desired size and stamped with the logo and cooled in trays or on a belt in a cooling tunnel before packing. During processing, the actives are exposed to relatively high moisture, high temperatures, and high-shear stress. An alternative process is to compression mould the mixed mass. This process is more compatible with many actives such as bleach and enzyme due to the relatively low-temperature processing. These processing conditions, and the need to maintain chemical and physical stability of the bar through repeated wetting/drying cycles (as the bar is used over multiple days) represent key challenges to the formulator.

Synthetic detergent bars deliver much of the same chemistry to the wash as heavy-duty granules. The primary cleaning power comes from anionic surfactants along with phosphate builders such as sodium tripolyphosphate (STPP) and tetra sodium pyro phosphate (TSPP). In India, Latin America, and Africa, LAS is used as the main surfactant, whereas in the Philippines, a combination of alcohol sulfates (preferably coconut based) and LAS is used as the surfactant system. In premium bars, STPP is the preferred builder as it offers, in addition to excellent hardness binding capacity,

Detergent Processing

very good processability and control of in-use physical properties such as bar wear rate and sogginess or mushiness. In low-priced bars, a small amount of STPP is sometimes used for processability reasons. Performance-enhancing actives, such as chelants, polymers, fluorescent whritening agent (FWA), and recently, enzymes and bleach are often included as well, as are sodium carbonate, sodium sulfate, inorganic fillers such as calcite, talc, and clay. Carbonate and sometimes sodium silicate provide the necessary alkalinity, whereas the inorganic fillers give "body" to the bar and also act as process aids.

Surfactant level varies from 10–15% in low-priced bars to 20–30% in premium bars. Where a combination of surfactants is used, as in the Philippines, typically a mixture of alcohol sulfate and LAS is used in a ratio of 50:50–85:15. Bars containing only AS as the surfactant are very brittle and need a high amount of humectants such as glycerine or addition of hydrotropes to reduce the brittleness. STPP/TSPP level ranges from 0–3% in low-priced bars to 15–30% in premium bars. The moisture level in the bars varies from 3 to 12%, carbonates typically from 10 to 25%, and fillers such as calcite, talc, and clay fill the rest of the formulation. Because the presence of free water in the bars can lead to mushiness during production or use, a variety of desiccants and adsorbent materials have also been added to bars to control the moisture. Examples include phosphorous pentoxide, sulfuric acid, boric acid, and calcium oxide as well as a variety of clays.

19.1.2.7 Other Product Forms

Various novel forms such as gels, foams, and sheets have been tried in the past, each presenting unique challenges. However, none of these has resonated with consumers to date, and therefore, these remain niche products of low volume and will not be considered further in this chapter.

19.2 DETERGENT POWDER PROCESSING

Detergent powder processing, as mentioned earlier, is composed of a number of basic operations such as spray drying, agglomeration, and finished product making and handling. Each operation is considered in the following in text.

19.2.1 PROCESSES FOR MAKING DETERGENTS

19.2.1.1 Spray Drying

19.2.1.1.1 Introduction
Spray drying is the most important process used in the manufacture of detergent granules. It is the process route by which the main component of the vast majority of granular products is produced and the spray dried powder properties dominate the physical characteristics of the product. In the developed world, it survived the rise of the compact, agglomerate-based products, in the 1990s, the consumer preferring the lower-density product offered by spray drying. In the developing world, granular, spray dried products are the detergents of choice, and increasing prosperity in these regions has driven increased production volumes. Spray drying processes are capital intensive and typically quite large as shown in Figure 19.7.

The detergent spray drying process itself is well established. It was introduced about 60 years ago. Over the years, the process has been optimized considerably. In particular, major improvements have been made to production reliability, whereby plant utilization has increased from ~45 to over 90% in some locations due to the application of reliability engineering tools to minimize downtimes in production. The production rate of individual units has also increased quite dramatically as limits are understood and overcome by interventions such as airflow modifications and multilevel spraying. Rates of over 80 t/h are now achieved in single spray drying towers today, although smaller tower rates can be as low as 1 t/h. These two factors, along with radically improved transportation networks, have led to some significant consolidation of production units in the developed regions of the globe.

FIGURE 19.7 A typical manufacturing unit for spray dried detergent powders.

Despite its maturity, the spray drying process still has many challenges ahead. Formulations continue to change at an ever-increasing pace and frequently push the boundaries of the known operating envelope. Additionally, as the market becomes more segmented, the number of formulas increase. This, together with the drive to just-in-time production schedules (to minimize inventory) drives down production run length. Thus, start-up and shutdown times on these large-scale production units need to be significantly reduced to maintain plant utilization as well as good product quality.

19.2.1.1.2 Blown Powder Formulation
The spray dried granules are often known as blown powder. The components of blown powder are those of the detergent formulation that are robust to the operating temperatures within the tower. The core components are as follows:

Anionic surfactant—most often LAS
Builder—zeolite or STPP
Inorganic salts—sodium sulfate, sodium carbonate, sodium silicate
Polymers—polyacrylate, carboxy methyl cellulose

The properties of the slurry and the subsequent blown powder are dominated by the interaction of the LAS, which forms a liquid crystalline phase with the other ingredients within the slurry. Consequently, small quantities of some minor ingredients such as polymers, hydrotropes, and cosurfactants can have a significant effect on the slurry properties such as its rheology, and therefore, the amount of water required in the crutcher mix.

The structure of the blown powder is also dictated by the crystallization of the inorganic phases. In this respect, silicate and polyacrylate both improve granule crispiness and toughness.

19.2.1.1.3 Process Description
The spray drying process enables efficient, countercurrent, contact of an atomized detergent slurry with hot air, producing a detergent granule. The process itself can be split into five sequential operations as follows (Figure 19.8):

Detergent Processing

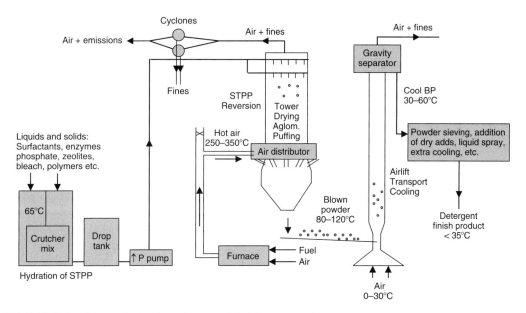

FIGURE 19.8 Schematic drawing of a spray dried detergent unit.

Slurry making
Pumping
Atomization
Drying
Cooling and classification

These operations are considered in the following text. Raw material storage, handling and dosing are common to many industries are not considered in this chapter.

19.2.1.1.3.1 Slurry Making
The objective of this process is to make a homogeneous slurry of consistent composition and aeration, with minimum water content, that is, the lowest possible drying load. The first challenge of incorporating powder and liquid raw materials is achieved by a continuous or batch mixer, known in the industry as a crutcher. Both of these types of mixer have been optimized by experience, and advocates for both methods will argue their benefits, for example, in minimizing aeration or in their ability to handle lower water content slurries. In reality, there are rarely two mixers that are the same and it is unclear whether either has a fundamental advantage. Over recent years, as production scales have increased and control technology improved, the systems for dosing these mixers have become highly automated and more accurate. In addition, the increase in production rates in the drying tower has meant considerable work to reduce the dosing and mixing times, and therefore avoid large capital investment and plant shutdowns during upgradation. From the crutcher, the slurry is transferred to a holding tank, sometimes known as a drop tank; here, further mixing occurs and the slurry ages allowing various phase formations and crystallization processes to take place. Filtration is carried out to achieve further homogenization and ensure that the spray nozzles do not block the final stage of slurry making. Typically, magnetic filters are used followed by disintegrators, which incorporate a filter with a smaller orifice size than the nozzles. Systems for adjusting slurry temperature and aeration, by injecting stream or air into the slurry, can be incorporated at any point in the process.

19.2.1.1.3.2 Sodium Tripolyphosphate Products
STTP was, for a long time a significant part of detergent formulations. Not only is it good at controlling water hardness, it also has good sequestering properties and provides buffering capacity and a crisp, robust granule. It is still used in many countries and can present processing challenges.

The reason for this is that it hydrates to the hexahydrate during the slurry making and continues to hydrate after the making process. Such hydration reduces the free water in the mix and thus can have a big impact on viscosity. The material has two different crystalline phases—I and II. The former hydrates at a much faster rate than the second. The hydration rates in both cases are sensitive to temperature also—lower temperatures giving more hydration.

Too much hydration in the slurry can lead to extremely high viscosities, even beyond the processing capability (instances are known whereby the slurry has set solid due to overhydration). In addition, the dehydration of the hydrate during spray drying can lead to degradation of the STPP to poorer performing molecules. This is known as reversion.

In contrast, too little hydration can lead to the production of a poorly hydrated blown powder. This then tends to pick up moisture and hydrate on storage and can lead to overheating and caking. Therefore, good control of STPP hydration is crucial for steady operation and for producing a good-quality product. Thus, it is important to have a consistent quality of the raw material.

19.2.1.1.3.3 Pumping

As a result of the low moisture content of the slurry and the short mixing times, the slurry can have inorganic lumps, which would block the atomization nozzles. The typical way of resolving this is to pump the mix through a filter and provide a disintegrator to break up any such lumps before passing to the main slurry pump. To achieve the high pressures (up to 100 bar) required for atomization, one or more piston pumps are typically used. In most cases these require a booster pump, typically a positive displacement pump, to keep them efficiently filled with viscous aerated slurry.

19.2.1.1.3.4 Atomization

The objective of the atomization process is to create drops small enough to dry in the spray drying tower. This is done with a number of high-pressure nozzles known as hollow-cone pressure swirl nozzles. These nozzles are distributed at one or more levels within the spray drying tower and have to be sufficiently distant from the wall to avoid buildup caused by wet drops sticking before they have dried sufficiently. For this reason, and for reasons of residence time, smaller towers typically run with smaller nozzles.

Two problems are often encountered with spray nozzles: nozzle wear and nozzle blockages.

Nozzle wear. The abrasive slurry causes the nozzle tips to become enlarged and rounded with time, even with the hard tungsten carbide or yttrium carbide nozzles that are typically used. This wear causes bigger drops, wall buildup, and poor product properties. This is particularly the case with zeolite-based slurries where nozzles must be carefully monitored and changed every few weeks.

Nozzle blockages. Occasionally nozzles get blocked, therefore several spare nozzles are usually kept available during operation. With a suitable filter size, good operating, and clean-out practices, these occurrences can be minimized. However, most towers have inspection windows that allow operators to monitor atomization during operation. Alternatively, constant monitoring of the spray dried particle size can indicate upstream problems before their impact can get out of control.

Typically, a nozzle will produce a wide range of fine droplets, which do not have time for the surface tension to pull them into a sphere before they are dried. In addition, the different sizes tend to agglomerate in the turbulence, and thus, a spray dried granule is more like an irregular-shaped bunch of grapes. Problems can occur if the nozzles are too close to one another in the tower, leading to overagglomeration and hence too many coarse particles. Figure 19.9 shows a typical sample of granules.

On further magnification, the knobbly shape and internal porous structure can be seen as in Figure 19.10—a scanning electron microscope (SEM) picture.

Detergent Processing 335

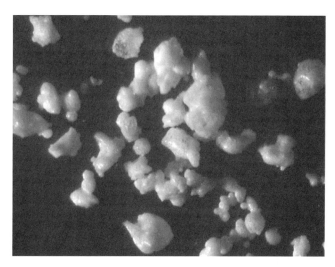

FIGURE 19.9 Typical spray dried detergent particles.

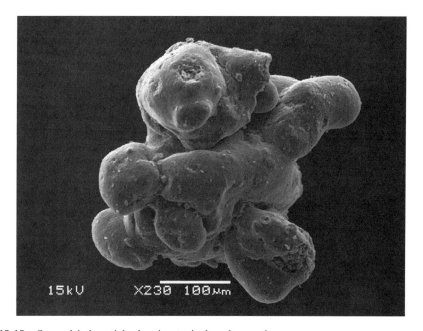

FIGURE 19.10 Spray dried particle showing typical agglomeration.

19.2.1.1.3.5 Drying

Countercurrent spray drying towers are used to dry the droplets and operate with an inlet temperature of ~300°C. The tower design has not changed much over the years; a typical tower outline is shown in the schematic Figure 19.8. The cylindrical section of the diameter is typically in the range 3–10 m. The geometry of the tower around the hot air inlets is dictated by the need to minimize any contact of the hot air with the built up product, and thereby reducing the amount of browning that occurs. This is an important area since higher air temperatures improved thermal efficiency and lead to increased production rates. However, if the distribution is poor, overheating of the powder occurs, leading to potential exothermic runaway and subsequent fires in the tower.

The hot air is generated by an oil- or gas-fired furnace and is uniformly distributed around the tower circumference using a plenum ring. Care must be taken for oil furnaces since poor combustion can lead to black specks entering the tower and causing discoloration of the blown powder. Failure to get a uniform distribution of air or temperature can lead to poor product quality, whereby some granules can be underdried and sticky and some can be overdried and brown. The inlet air can enter radially or with a swirl, that is, some tangential component of velocity. The swirl tends to stabilize the airflow and decrease the exhaust temperature a little, improving thermal efficiency. The exhaust air is ducted out of the top of the tower and any fine product carried over is removed using technologies such as cyclones, bag filters, and electrostatic precipitators.

19.2.1.1.3.6 Cooling and Classification

The powder temperature exiting the tower base is typically over 70°C. This needs to be reduced to allow temperature-sensitive additives to be mixed in. Typically, this is done in an airlift, which transports the powder to the top of the building. A gravity separator is then typically used to disengage the particles from the air. This type of system also performs an initial classification of the powder as large lumps are not transported up the airlift and fine particles do not disengage from the gravity separator. (These fine particles are subsequently removed by a bag filter.) Further classification is often required to remove coarser particles. This is typically done using mechanical screens.

19.2.1.1.4 Operation
19.2.1.1.4.1 Overview: Control and Changeovers

The detergent drying process is a large-scale process, and in modern installations digital control systems are used to control the plant. However, the control of product properties such as density and blown powder moisture tends not to be fully automated, partly due to the multivariable nature of the product properties and lack of robust measurement systems. Therefore, there tends to be a human operator responsible for the starting-up, center lining, and shutting down of the process.

Multiple formulations tend to be run on one tower and with increasing formula numbers, run times for a particular formulation are decreasing. Therefore, the challenge for the tower operators is to quickly bring the system to centerline and acceptable product properties, thereby minimizing the amount of material that has to be recycled back to the slurry.

19.2.1.1.5 Powder Properties

In the majority of granular detergent formulations, the physical properties of the products are determined by the blown powder properties. The key properties are as follows:

Density
Particle size distribution
Cake strength—flow properties
Solubility

Density is typically 250–550 kg/m^3 and determined by many variables such as the formulation, slurry air content, and process conditions. Density is an important quality item since most consumers dose by volume, and therefore, the bulk density is the prime variable to control. Typically, droplets tend to form a skin on the outside as they are dried, leading to the internal water diffusion out of the drop being very slow. In certain cases, this water can turn to steam and expand the droplet causing puffing, and hence, reduced density. The extent to which this can occur will depend on the water content of the slurry, air temperature, and chemical composition. Generally, lower density is caused by a higher slurry water content, higher air temperatures, and a high concentration of film-forming materials. Incorporation of air is also used at high pressures to increase this expansion. However, care must be taken not to overdo this effect since too rapid an expansion can lead to the fracture of the granule and result in an excess of fine material.

An additional factor in bulk density is the ease with which the particles pack. One practice is to spray perfumes and de-dusting agents onto the powder. If these are not absorbed into the granules

Detergent Processing 337

the liquid can stay on the outside of the particles, leading to poor flow properties and a reduction in bulk density, which is difficult to control.

Particle size distribution. A typical volume-based median particle size is ~400–500 μm. The spread of the distribution is typically wide with some significant fine and coarse powder resulting from the agglomeration process, which takes place within the tower as wet drops collide. Acceptable ends of the spectrum typically range from <5% below 100 μm to <2% above 800 μm. The former is for dustiness of the powder and subsequent dispersion when wet. The latter is to control appearance and dissolution of the powder over time during the washing process.

Cake strength. The cake strength as measured by a uniaxial shear test is often used as a measure of flowability and granule stickiness. These are important properties for both postprocessing and consumer acceptance. The cake strength tends to be an intrinsic function of the formulation and blown powder moisture. Apart from this, the process conditions tend to have only a minor effect.

Solubility. The consumer prefers a fully soluble granule, therefore insoluble residues left by a granule are undesirable. The key factor for solubility tends to be formulation and component interactions, although process conditions can play a role as well. Some of the better-known interactions that can cause insolubles involve sodium silicate—a common ingredient. Poor mixing control with acidic ingredients can result in the production of insoluble silica, and interactions can also occur with zeolite formulations, which will also cause large insolubles. In both cases, careful control of mixing and pH is needed to avoid these problems.

19.2.1.1.5.1 Makeup

One of the common factors that can impact good operation is the attachment of wet slurry to the tower walls, where they dry and form a base for more droplets to stick. This makeup can lead to large lumps inside the tower, reducing thermal efficiency. In the worst case, the makeup near the hot air inlet temperature can result in the overheating of the dried slurry and generate brown and black speckles, which ruin the appearance of the product, leading to shutdown and cleanout. If this is not done, the powder can self-ignite. Makeup is most often caused by poor nozzle positioning and poor airflow control. It is therefore important to ensure these are resolved during the commissioning of a spray drying tower.

19.2.1.1.5.2 Future Developments

As mentioned earlier, the current spray drying process is very well established but is still undergoing constant change and optimization in the drive to reduce costs and increase formulation flexibility and product quality. This is most relevant to the large production units, which make a number of different formulations. As new tools become available, they are applied to these units. Thus, reliability engineering has increased plant utilization, CFD modeling has improved tower airflow control, and process control advances has reduced start-up and shutdown times. Further improvement in modeling capability will allow much better control of product quality and allow the operating windows to be widened in the drive for more cost-effective density and moisture control.

19.2.1.2 Agglomeration in Detergent Processing

19.2.1.2.1 Introduction

Agglomeration is a very widely used process in detergent making. Agglomerators range from high-shear, high-speed mixers to low-shear fluid beds. Processes can range from small-scale batch making to large-scale continuous production. In other words, agglomeration is simply the "sticking together" of smaller particles into some sort of combined entity. This is often not difficult—the real challenge lies in making agglomerates of the desired properties. Agglomeration has been widely used in the last 20 years or so as an alternative to spray drying. Spray drying is well suited for making lower-density particles of more thermally stable materials at high rates. It is most cost effective at high production rates due to the high amount of capital required. Agglomeration is more suited to making higher-density granules at lower volumes and potentially contain more thermally sensitive ingredients. Many detergents are made by combinations of spray dried and agglomerated materials.

19.2.1.2.2 Agglomeration

Agglomeration consists of sticking powder particles with a (liquid) binder. It is possible to agglomerate soft solids where a separate binder is not added, but for this discussion we will restrict the scope to a situation where we have discrete powders and binders. This covers the vast majority of detergent applications. One trite but easily overlooked point is that the use of process aids needs to be minimized for cost reasons, hence, the choice of powders or binders is constrained. Detergent agglomerates include granules where the binder is the active component and levels need to be maximized as well as granules where the powder(s) is the active component and the binder level needs to be minimized.

19.2.1.2.3 Significant Factors in Detergent Agglomeration

Academic research in the last decade has become much more relevant in explaining typical detergent processes. Agglomeration of glass ballotini with dilute glycerol solutions did not provide much help in understanding agglomeration of high or very high viscosity binders often coupled with some chemical/physical change. Agglomeration regime maps in particular are helpful in understanding how the *deformability* of the powder/binder mix and the *degree of saturation* control how the agglomeration proceeds.

This is well covered in a review article by Iveson et al.[1] The agglomeration regime map in Figure 19.11 shows the agglomeration regime as a function of the deformability of the powder: binder system and the degree of saturation.

Deformability (how easily a granule deforms in the mixer). The more deformable the powder: binder mix is in the mixer, easier it is for particles to coalesce and agglomerate on impact due to the viscous dissipation of kinetic energy. The advantage of thinking in terms of

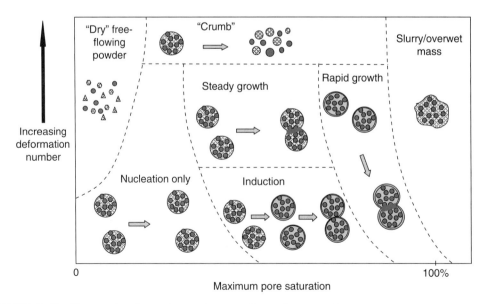

FIGURE 19.11 Phase regime for agglomeration according to Iveson et al.

Detergent Processing
339

deformability is that it combines both liquid, powder and process mixing unit properties. Deformability includes a term for the velocity of the collision of two particles and this is controlled by the mixer operation. Focusing on just one component, such as the binder, can be misleading. Nonionic surfactants, which are often low viscosity liquids at ambient conditions, can be successfully agglomerated with a very fine powder such as silica. They cannot be agglomerated with a coarser powder such as sodium carbonate. This is because the very fine powder will make a mix stiffer than the same weight of a coarse powder.

Pore saturation. The more saturated the system is with binder, easier it is for binder to be forced onto the surface during a collision and assist in the coalescence of the particles. If a porous powder is used then the binder can be absorbed *into* the powder and not be on the surface to affect the agglomeration.

One simple way to interpret the map—for a given powder and mixer—is to think of the "*y*"-axis as showing *decreasing* binder viscosity and the "*x*"-axis showing the *amount of binder* added.

Any agglomeration process is a journey around the map. During a batch agglomeration process, we move from left to right as the binder is progressively added. An overwet mass can often be recovered by adding more powder to reduce the effective pore saturation and move the regime from right to left.

$$\text{Maximum pore saturation} \quad S_{max} = w\rho_s(1-\varepsilon_{min})/\rho_l\varepsilon_{min}$$

$$\text{Deformation number} \quad \text{De} = \rho_g U_c^2/2Y_g$$

where
w = mass ratio of liquid to solid
ρ_g = density of granules
ρ_s = density of solid particles
ρ_l = density of liquid
Y_g = dynamic yield stress of granule
ε_{min} = characteristic porosity of the granule
U_c = collision velocity in mixer

19.2.1.2.4 Description of the Agglomeration Regimes

Free-flowing powder. The mix is very easy to deform (i.e., low viscosity binders) and there is not enough binder to form any agglomerates. The mix just looks like the powder.

Crumb. Crumbs are very weak agglomerates that are easily formed and broken. The binder does not have the strength to hold the agglomerate together. As binder levels increase or the mass in further worked we will move directly into the slurry regime.

Slurry or overwet regime. We now have a wet dough.

Nucleation only. We have small agglomerates or cores dispersed in the powder. These small agglomerates have been formed by the initial dispersion of the binder. However, there is not enough binder or the binder is so hard that no further agglomeration will happen.

Induction growth. This region is where the agglomerates are quite hard and have a low tendency to coalesce on impact. Growth is mostly by accretion of powder onto the surface of bigger particles.

Steady growth. The mix is deformable enough and there is enough binder that granules will tend to coalesce and grow.

19.2.1.2.5 Generic Description of a Detergent Agglomeration Process

In any process, some or all of the powders need to be introduced into one or more mixer(s). Some or all of the binder is also introduced into the mixer. The binder needs to be dispersed and contacted with the powders. The higher the viscosity of the binder the harder it is to do this. High-shear choppers need to be used to disperse high viscosity binders. The powder needs to be mixed with the binder and energy inputted to drive the agglomeration process. The granules may need postmaking treatment such as drying and cooling followed by size classification and reblending of the undesired material. This recycle can easily affect the process.

19.2.1.2.6   Factors to Consider in Agglomeration

1. *Initial viscosity of the binder.* This controls dispersion into the powder. Low viscosity binders are easier to disperse. High viscosity binders need mechanical action to break up and disperse them.
2. *Viscosity of the binder in the powder mass.* This may be different to the initial viscosity if the binder undergoes a chemical or physical change such as solidification. It is harder to mix powders into a stiff binder. The higher the binder viscosity the more binder will be needed.
3. *Ability of the binder to "wet" the powder.* Less binder is needed if it does not wet the powder(s) easily. This is easily checked by placing a drop of binder onto the powder(s) and seeing if it disperses into the powder(s). Poor wetting ability is often shown by the binder, which remains as a droplet on the powder surface.
4. *Particle size and particle size distribution of the powders.* Small particles have more ability to coat and stiffen the binder, hence requiring more binder for agglomeration. The most robust agglomerates are made by powders with a wide particle size distribution due to the tighter packing of powder particles occurring in the agglomerate.
5. *"Deformability" of the binder/powder system.* Mixes that are soft and easy to deform agglomerate more rapidly than less deformable mixes. The deformability of an agglomerating mass is a function of the combination of powder properties, especially size, binder properties, and mixer operation (especially speed).
6. *Type and condition of mixer used (high- versus low shear).* The mixer disperses the binder and provides the energy for agglomeration. Understanding the mixer design is important. In particular, "pinch points" where the agglomerating mass is compacted and subjected to high shear can drive the whole agglomeration process.
7. *Chemical/physical reactions during agglomeration.* In a holt-melt agglomeration process, the hot-melt binder typically is of low viscosity initially but immediately starts to harden on contact with the cooler powder. Hence, the deformability of the agglomerating mass is constantly changing.
8. *Amount of recycle required.* Agglomeration processes driven by coalescence of particles will always give a distribution of particle sizes due to the probabilistic nature of the collisions. Hence many agglomeration processes require size classification to give an acceptable final product. The oversize and fines have to be recycled into the system. The recycle of coarse particles back to the mixer can promote further agglomeration and lead to process instability. The recycle of high levels of fines back to the mixer can reduce agglomeration. However, it can be beneficial for stability to recycle some finished agglomerates into the process to act as agglomeration "cores."
9. *Use of dusting.* Fine powder, such as zeolite, can be introduced into a mixer to control the agglomeration process. The powder temporarily coats the surface of the agglomerating particles reducing their tendency to coalesce. Continued mixing reduces this as the dusting layer gets worked into the agglomerates leaving the surface sticky again.
10. *Ambient conditions.* These often seem to be overlooked, but can be very significant. Summer conditions are warmer and more humid than winter conditions. Raw materials will be warmer, and the increased process temperatures will generally increase ease of agglomeration. Increased humidity will affect processes that use aqueous solutions as binders.

Detergent Processing

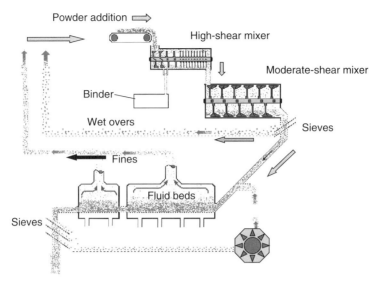

FIGURE 19.12 Schematic of a typical two-stage agglomeration process. (Adapted from Iveson, S., J. Litster, K. Hapgood, B. Ennis, *Powder Technol.*, 117, 3–39, 2001.)

Absolute—when compared to relative—humidity is often more helpful when comparing conditions.

11. *Make-up in the mixer.* Operation of a mixer for extended periods will cause make-up on the sides and tools. If this is extensive, it will tend to increase agglomeration by increasing the mechanical energy introduced into the powder. The frictional heat generated by the mixer will make the granules more deformable. Makeup on tools can disrupt the powder flow pattern. For example, an angled paddle in a paddle mixer will not be able to impart axial motion to the powder if it has excessive make-up. Tool fatigue can occur if tools come into contact with the wall make-up. Finally, makeup will tend to get pulled from walls and become oversized. Although, in some processes the friction from some wall makeup is necessary to help impart the necessary shear to get agglomeration to occur.

19.2.1.2.7 Dense Laundry Process

Figure 19.12 shows a generic description of a two-stage agglomeration process with recycle streams. Generally, the "wet" overs act as agglomeration seeds. The greater the proportion of large, soft particles recycled back to the agglomerator, the smaller the operating window.

19.2.1.2.8 Process Control Strategies

Process control in agglomeration is intended to maintain a consistent granule powder: binder ratio as well as particle size. Raw materials do not have constant properties and all raw material specifications include acceptable ranges whether for purity, particle size or some other quality parameter. Hence, agglomeration conditions alter from batch to batch. Temperature can also vary. Fresh powders can be delivered at high temperatures if the supplier is nearby. Powder feeders do not give completely consistent flow rates but rather oscillate around a set point. In continuous processes, this can be important. The mean residence time in a high-speed continuous mixer can easily be only a few seconds. In this situation, a short "blip" in a powder feeder can completely alter what happens in that mixer. Ambient conditions also change. Finally, so do the operators! One source of variability that is rarely mentioned is the fact that, whatever may be the procedure, different operators operate equipment differently.

Process control alters the degree of agglomeration by altering the deformability of the material in the agglomerator or changing the saturation of the mass. If the degree of agglomeration is too great then the deformability or the saturation needs to be decreased. Options for both batch and continuous processes are as follows:

Reduce the binder:powder ratio.
Make the binder less deformable, for example, cool surfactant pastes.
Make the powder finer, for example, increase the degree of grinding if applicable.
Alter the speed of the mixer.
Altering the binder:powder ratio gives an immediate response but changes the composition of the resulting agglomerate. In systems where there is a drying step, one possibility is to introduce water into the mixer as an additional binder as well as the desired binder. Then the water level can be used to control the degree of agglomeration without altering the final agglomerate composition.

Some processes include an inline-grinding step. In these situations, the degree of grinding or the proportion of the material that is ground can be altered. The finer the particles, the less deformable the agglomerating mass will be and, hence, reduced agglomeration.

Some processes use a fine dusting agent to control agglomeration. This is used mostly in processes where there are multiple mixers.

19.2.1.2.9   Mixer Operation

Mixer operation can provide some process control but not as much as to alter the material. In batch mixers, mixing time and binder addition time can be altered. If an agglomerating mass is more deformable, then the energy input from a mixer can be reduced, for example, by reducing the speed. In horizontal mixers where make-up becomes significant, the speed can often be reduced to reduce agglomeration.

In processes where there are multiple mixers, the addition position of the binder can be varied. For example, one company describes a process that combines a high-speed mixer and a fluid bed agglomerator. The binder—in this case consisting of acidic form of LAS (HLAS)—is introduced into both the mixers, and the proportion of HLAS introduced into each mixer can be altered to control the process. In general, moving binder addition location from the initial to a subsequent mixer will reduce the degree of agglomeration.

19.2.1.2.10   Mixer Operation in Fluidized Bed Agglomeration

In fluid bed agglomerators, spray nozzle height can often be used as a process control tool. This is because the height affects the intensity of the spray flux as well as the time taken for the binder droplets to impact on powder particles. The higher the nozzle is above the fluidized bed, greater the footprint, and hence, lower the spray flux.

If the flux is too low, agglomeration is slow because the amount of binder is insufficient to cause coalescence. In such cases, coating of the particle can occur as the liquid spreads over the particle surface. If the flux is too high, the powder surfaces become overloaded with the binder causing localized excessive agglomeration, often leading to "bogging" of the fluidized bed. The flux is a function of the spraying rate and the nozzle position.

In fluid bed processes where aqueous solutions are sprayed onto a fluidized bed, the spray flux and the position of the nozzles are important. If the addition rate is too low, the binder becomes dried and nonsticky before particles have a chance to coalesce. Hence, agglomeration is reduced. If the spray nozzles are too high above the surface of the fluidized bed, binder droplets can become dried before they hit a powder particle. Again agglomeration is reduced.

Detergent Processing

19.2.1.2.11 Common Detergent Agglomerate Ingredients

Powders

Sodium carbonate/bicarbonate/carbonate-containing minerals	Mostly used to neutralize sulfonic acid in "dry neutralization" processes (see Section 19.2.1.2.13 for more details)
Zeolites—especially Types 4A, and X	Fine powder with particles <4 μm. Very useful as an absorbent powder
Phosphate—STPP	Used in some agglomerates. A very good moisture sink, hence, can be used to provide *in situ* drying
Tetra acetyl ethylene diamine (TAED) powder	Commonly used bleach activator. Strong dust explosive. Agglomerated with polymer solutions
Silica	An extremely "absorbent" powder due to its very small size. Can be used to agglomerate liquids or soft solids such as nonionic surfactants. Useful as a dusting aid
Spray dried blown powder	Agglomeration of blown powder can be used to increase bulk density
Sulfate	Used as a filler. High-density, nonabsorbent powder. Will hydrate at <32°C. Used as a "filler" in some agglomeration processes but does not have much influence on the processes
Salt	Used as filler. Does not have much effect on processes

Liquids

HLAS	Most common and available surfactant globally. HLAS is the acid form of the LAS (see Section 19.2.1.2.13 for more details)
Surfactant pastes, for example, alkyl sulfates and alkyl ethoxylated sulfates	Often very high viscosity, high concentrated solutions
Polymer solutions, for example, polyvinyl pyrolidone, PVOH, polycarboxylates (e.g., Sokalan CP5)	Good binders. Hardening mechanism is by drying
Molten waxy materials, for example, poly ethoxy glycols (PEGs)	Used as hot-melt binders. Hardening mechanism by solidification
Nonionic surfactants	Most of them are very hard to process as liquids at room temperature. No hardening mechanism Agglomerates either need to be very porous and absorbent with nonionic sprayed on or a very fine powder such as silica needs to be used
Silicate solutions	A good polymeric binder. Cheaper than many polymers
Water	Cheap—can be used where the powder contains some soluble binder or can be used as a control tool where other binders are used

19.2.1.2.12 High- and Low-Density Agglomerates and Agglomeration

The bulk density of an agglomerate is a function of its shape, particle size distribution, surface stickiness, and internal porosity. An agglomerate that has a very irregular shape will pack less well than a more rounded agglomerate. Hence, its bulk density will be lower. The internal porosity of an agglomerate is a function of the raw materials and the compaction and working the agglomerate is subjected to during processing.

Fluid bed agglomeration is widely used to give lower-density agglomerates. The low-shear environment of the fluidized bed gives a more open agglomerate structure. In a high-shear mixer, the granules are compacted and this internal porosity is "squeezed out." In addition, granules become rounded resulting in better packing and an increase in density. It is possible to measure a significant decrease in pore size as surfactant agglomerates go through a mixer such as a ploughshare. Low-shear mixers such as fluid beds or drum mixers are used to give lower-density agglomerates. In these cases, the binder needs to be "predispersed" by, for example, a spray nozzle.

In an interesting process, a surfactant is foamed before agglomeration. It is important that the air bubbles in the surfactant are small (preferably <5 m) so that they are retained during processing. One preferred option is to foam a hot-melt surfactant such as an alkyl polyglucoside.

If an agglomerate is very sticky then it will not pack well and result in a low-bulk density. Dusting will result in an increase in density as well as a decrease in stickiness.

The ability of an agglomerate to retain porosity in a mixer is controlled by its deformability. Hence, there is an inverse relationship between the agglomerate deformability and density.

There is no universally agreed definition of what is a high-shear mixer and what is a low-shear mixer. In patent literature, there is universal agreement that high-speed horizontal mixers such as the Lodige CB series are high-shear mixers. Fluid beds and mix drums are low-shear agglomerators. Many mixers have high-speed choppers or dispersion bars to disperse liquids, which can be responsible for inputting much of the total energy. There is no clear agreement for describing such mixers.

19.2.1.2.13 "Dry Neutralization": An In-Depth Discussion

The most common detergent agglomeration process involves the neutralization of linear alkylbenzene sulfonic acid (Figure 19.13) with sodium carbonate or bicarbonate or a combination of the two. LAS is the global "workhorse" surfactant due to its cost and effectiveness. In many regions, it is the only surfactant that is available in quantity. Dry neutralization was the first surfactant agglomeration process. Ironically, HLAS-based processes are probably the most active area of development due to its suitability for use in developing countries and cost advantages in developed countries.

It is to be noted that HLAS—the acid form of the surfactant—is stable. It is a dark, viscous liquid but is easily pumpable at temperatures >35°C. It reacts with sodium carbonate/bicarbonate to form the sodium salt and water and carbon dioxide. Sodium LAS is a solid at low water contents, and hence, HLAS can be neutralized and agglomerated with powdered sodium carbonate or carbonate-containing minerals to form a free flowing agglomerate. The lack of water means that no drying step in needed although some cooling step, such as a fluid bed drier, is often required. The equipment required can be simple and cheap, and hence, cost effectiveness is high.

Agglomeration of HLAS with sodium carbonate is actually a complex process due to the changes in the binder properties during the agglomeration. This is best understood by looking at the changes that the binder undergoes. It starts off as a relatively low viscosity liquid, which is (relatively) easily dispersed by a chopper or a dispersion bar or by the powder motion. As soon as the HLAS droplets come into contact with an alkaline surface such as sodium carbonate, the HLAS begins to neutralize to solid LAS. Hence, the deformability of the agglomerating mass decreases with time. As the agglomeration proceeds, there is less and less available carbonate surface to neutralize the HLAS liquid, and hence, the agglomerating mass gets stickier and stickier. The color of the agglomerates gets darker as well due to the unreacted HLAS.

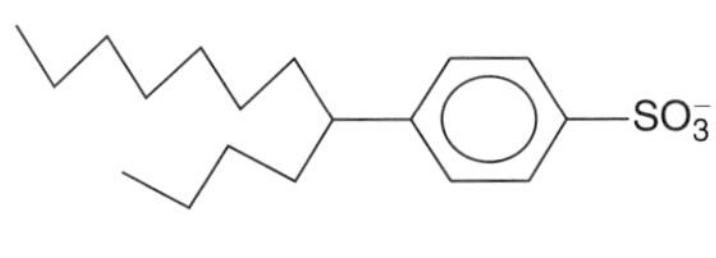

FIGURE 19.13

Detergent Processing   **345**

Practical points about LAS agglomeration by HLAS neutralization:

Owing to the inefficiency of the solid:liquid reaction, a significant excess of carbonate is required to get thorough neutralization. A HLAS:carbonate ratio of 1:2.5 to 1:3.5 is typical. The exact ratio in an agglomerate is a function of the particle size of the carbonate and the available surface area. The ratio of 1:2.5 would apply to ground carbonate. Other powders can be mixed with the carbonate to reduce the amount of carbonate required but can give stickier agglomerates needing additional dusting.

The use of anhydrous carbonate gives higher active agglomerates. This is because the (hydrophobic) HLAS does not wet the hydrated carbonate as well as the anhydrous carbonate. This applies to zeolite as well. Typically, an LAS:carbonate agglomerate could reduce activity from 20 to ~18% due to hydrated powders.

The use of low levels of water in the HLAS (2–5%) speeds up the rate of neutralization dramatically. Levels of water >10% cause the HLAS to gel. A level of 2–3% gives an optimum of increased neutralization without affecting physical properties negatively.

Poorly neutralized agglomerates are stickier and can negatively affect any perfumes that are sprayed on.

The neutralization reaction can take hours to complete. However, the vast bulk of the neutralization takes place in <1 min.

Rapid addition of HLAS can lower the achievable activity as the HLAS has not had time to react.

HLAS-based agglomerates are made in a very wide range of agglomerators.

19.2.1.3   The Finishing of Dry Laundry Detergents

Finishing is the final step in the detergent manufacturing process, whereby all the constituent parts of the detergent product are combined together. This results in the final product that is packed and sold to the consumer.

Finishing generally involves the blending together of some or all of the following dry detergent components into a uniform, homogeneous mix:

Spray dried detergent granules
Surfactant agglomerates
Bleaches and bleach activators
Enzymes
Buffers and fillers
Speciality particles

In addition, various liquid sprays may be applied to the product such as

Perfumes
Nonionic surfactants
Colorants

For smaller operations, finishing can be accomplished in a batch mixer. For larger operations, the use of continuous mixing equipment is more commonly employed.

19.2.1.3.1   Finishing Transformations
Within the mixing unit, the following transformations can commonly occur:

Mixing
Absorption
Heat transfer

Segregation
Agglomeration
Attrition
Solidification

The extent to which any of these transformations occurs is linked to the final product being made, the type of mixer, it's mode of operation (e.g., batch versus continuous) and it's operating conditions (including mixing time).

19.2.1.3.2 Batch Mixing

For relatively small operations, producing <3–5 T/h of final product, batch mixing would normally be selected. In this case, each mix is a discrete batch and care needs to be taken to ensure consistent raw material additions and mixing conditions are applied to each batch (Figure 19.14).

In batch mixing, low-level additives are often preweighed before being added to the mixer. This increases the weighing accuracy and allows multiple small additions to be combined together before dosing into the mixer. High-level additives can also be weighed separately or directly into the mixer using the mixer load cells. For materials that are regularly used (e.g., spray dried granule and sodium carbonate), a feed hopper may be installed above the mixer. When required, material is fed from the hopper into the mixer by a slide gate or rotary valve.

Batch mixing allows a high level of manual operation and low level of automation. It is, therefore, advantageous when starting a new product line or when capital expenditure needs to be controlled. However, accurate records and checklists need to be maintained to ensure all additions are correctly weighed out and added to each batch. With higher investment, a fully automated system can be designed with ingredients added by hoppers using intermediate scales or the mixer load cell. In both cases, care needs to be taken to ensure adequate distribution of low-level ingredients.

Various batch mixers can be used for the final blending of powder detergents. Those commonly used include ribbon blenders, paddle mixers, V-blenders, drum mixers, and ploughshare mixers. Ribbon blenders are of low cost and useful for simple dry mixing operations, which do not require significant liquid additions. In cases where higher levels of liquid sprays are required, a higher level of powder fluidization or free powder surface is required to ensure good liquid distribution. This is the characteristic of the paddle mixer or drum mixer.

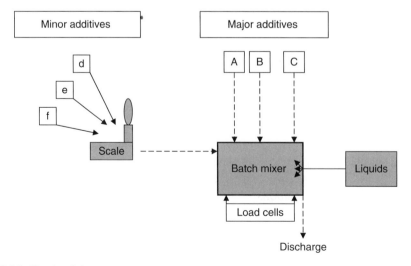

FIGURE 19.14 Batch mixing.

Detergent Processing

In cases where liquid sprays are added, it is critical that the liquid impinges on powder and not on the walls of the mixer, baffles, or mixing arms. Overspray onto these areas can result in make-up, generation of oversized particles, poor distribution of the liquid ingredient, and poor-quality finished products, as well as frequent shutdowns for cleaning. Liquid spraying should commence only after all the powders have been added to the mixer. The nozzle selection and orientation of the liquid sprays also need to be carefully designed. For common low viscosity liquids (e.g., nonionic, perfume, and colorant solutions), a single fluid (pressure) nozzle should provide sufficient atomization to evenly disperse the liquid into the powder. For higher viscosity materials, two fluid, air atomized nozzles may be used.

The time taken for mixing depends on the design of the mixer. In cases with high fluidization, mixing efficiency is high and mix times can be very short. In such mixers, extended mixing can result in granule breakup and densification, and therefore overmixing needs to be avoided. When assessing the total cycle time of a batch mixer, the following need to be included:

Time for addition of all dry powder raw materials (including time for preweighing of minor ingredients)
Mixing time
Time for all liquid spray-ons
Aging time (if required)
Discharging

After mixing is completed, the batch mixer is completely discharged and is ready for the next batch. Whether the mixer is still running during discharge will be a function of the flow characteristics of the final product and the mixer design. The finished, fully mixed product is collected in bins or other temporary storage containers before quality assurance checks and packing.

19.2.1.3.3 Continuous Mixing

For larger operations, continuous mixing is used. In this case, a continuous feed of all the powder and liquid ingredients is fed to the mixing unit (Figure 19.15).

Upstream processes in the larger detergent manufacturing plants are usually continuous (e.g., spray drying and agglomeration). Their production rates are, however, not always consistent, with some fluctuations in rate occurring over time. To even out these variations, it is common for some buffering to be applied before mixing, often through the use of a surge hopper. Thereafter, these materials and all the other dry powder feeds can be controlled using metered feed systems.

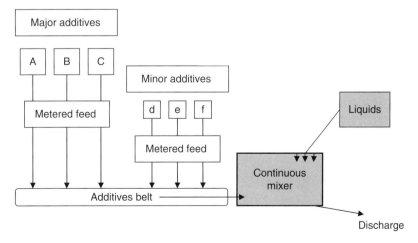

FIGURE 19.15 Continuous process for addition and mixing for additives.

The major ingredient (often the spray dried granule) is often taken as the master additive, with all other addition rates slaved to this material.

The feed systems used vary depending on the nature of the material being fed. For high-level additions of cohesive powders such as agglomerates and spray dried granules, a belt feeder is commonly used. For free flowing, robust ingredients such as carbonate, sulfate, and perborate, a screw feeder can readily be used. Other feeder designs are available and are used for handling specialized materials. In all cases, the feeders either use integrated weighed cells to directly measure addition level or those that have been previously calibrated for the material being fed.

Given that a large number of different dry additives are used in laundry detergents, the number of feed systems in a production plant can be high. To simplify the feeding of all these materials into the mix drum, an additives belt is often installed, which then allows all ingredients to be added on to one belt in a proportional "sandwich" before discharging into the mixer. Usually the bulk ingredients are added first, with the smaller additives layered on top. For operations with multiple additives, a combination of interconnected belts can be used.

A common design of continuous mixer is a rotating cylindrical drum. By adjusting the feed rate, angular rotation speed of the drum, and its inclination to the horizontal, the fill level and residence time can be controlled. As the angular rotation speed is increased, the flow regime in the drum changes accordingly and more turbulent flow is obtained. For good mixing, turbulent flow is required, however, if the drum speed is too high the powder will start to centrifuge and no further mixing occurs. Insertion of baffles within the drum can increase the level of mixing and provide greater turbulence. However, baffles do increase the powder retention in the drum at brand changeover and levels of general make-up.

19.2.1.3.4 Liquid Spray-ons

The drum needs to be sized according to its duty. If high levels of liquid or multiple liquids are to be sprayed on, then a longer drum is required. This not only allows more space to site the spray-on arms, but also provides some time for the powder to *age* (with absorption or solidification of liquids) before discharge. Without *aging*, the risk of powder caking downstream of the mixer can increase.

Liquid sprays need to be sited to avoid adjacent sprays from overlapping. If overlap occurs, it may result in overwetting of the powder surface leading to caking and poor product characteristics. It is also important to ensure that the spray is directed onto the powder and does not impinge on the drum wall. Where spray is on the drum wall, excess make-up can occur in the drum and balling can occur in the final product.

In mixers where multiple sprays are used, the sprays need to be positioned down the drum in a controlled order, with suitable spacing between materials. An *aging* zone should also be included before discharge (Figure 19.16).

Nozzle selection for the liquid sprays is dependent on the nature of the liquid to be sprayed (e.g., its viscosity), the level of liquid required in the final product, how it interacts with the final product,

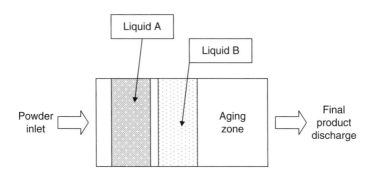

FIGURE 19.16 Liquid spray zones.

Detergent Processing

and the overall production rate. For low viscosity liquids, (e.g., perfumes, nonionic, and colorant solutions), pressure atomization is normally sufficient. For higher viscosity liquids, it may be necessary to use two fluid, air atomized nozzles to ensure adequate atomization.

Full cone, flat spray, and hollow cone nozzles with a wide range of nozzle angles (e.g., $0-110°$) are available. The selection of the nozzle for the given duty will depend on whether full, partial, or selective coverage of the final product by the liquid is required. For instance, where a speckled appearance is required from a colorant spray-on, a narrow angled hollow cone nozzle might be selected. This would only color a proportion of the particles. However, using the same colorant to fully cover the powder, a wide-angled full cone nozzle might be selected. As the nozzle angle increases, or distance between the nozzle and powder bed increases, the risk of spray impinging on the drum wall increases.

19.2.1.3.5   Finished Product Quality

There are a number of aspects that are important during the subsequent handling of the finished product. First, flow properties are important since the product is often stored in bulk containers until the packing lines are available. Second, the stability of the finished product must be tested to ensure that the product quality does not deteriorate on aging. Typical deterioration includes the decomposition of bleach and enzymes, primarily occurring once the bleach decomposes. One of the most stable forms of bleach is sodium perborate monohydrate. This is very stable to temperature, dissolves well, and is also relatively stable to high humidities. It suffers from a cost disadvantage to sodium percarbonate, as well as having boron, which is viewed by environmentalists with concern. Finished products are metastable and can degrade on aging. This is strongly influenced by the amount of water the powder picks up, and therefore, a number of different protection methods are used. Simplest is the poly bag, which is used a lot in developing countries. Here, the moisture barrier properties of the bag provide a reasonable storage life for the product. Cardboard boxes are very common in developed markets. These have inferior moisture barrier properties, and therefore often the board is laminated with polyethylene as a layer between the board to enhance product stability. In addition, the moisture level of the blown powder is usually low to provide some moisture sink for the product again.

Typical problems with respect to aging include caking of the product as well as chemical degradation of the enzymes and bleach, as well as reduction in odor quality as the perfume raw materials react with the bleach or other chemicals in the product.

19.2.1.3.6   Dust Control/Explosion Protection

Raw material feed systems. For materials which are potential sensitizers, (e.g., enzymes), it is advisable to fully enclose the raw material unloading station and feeder system in a booth with dedicated dust control. This ensures that any release of dust is contained. Operators entering the enclosure should do so with appropriate personal protective equipment.

For materials that are potentially dust explosive, nitrogen blanketing can be used for protection against ignition sources. Once diluted within the final product mix, this risk is usually removed.

Mix drum system. To minimize the risk of powder release and protect operators from dust exposure, dust control should be provided on the mixer. This is normally provided at the powder inlet and outlet. Adding the extraction within the mixer is not recommended. This has a tendency to remove excess levels of powders, especially if located near fluidized areas or close to liquid spray-ons.

The spray-on and associated atomization of some liquids (e.g., perfumes) can create an explosive atmosphere. The risk of explosion is low, but to protect the unit and operators the mixer should be equipped with explosion relief panels, water deluge, or some form of commercial dust explosion suppression.

350   Handbook of Detergents/Part F: Production

19.3 HEAVY-DUTY LIQUID DETERGENTS

19.3.1 INTRODUCTION

HDLs are generally clear, isotropic, homogenous, and thermodynamically stable. These products have only one well-defined, desired state. There are some exceptions to this. Some liquid products and detergents have moved to structured or heterogeneous formulations. These span a wide range of manifestations such as products with dispersed microcrystals to products with suspended particles. For these products, the quality of the product can vary as a result of the process history. Typically, these products are made on the same processes as conventional detergents. Therefore, the rest of this discussion focuses on the production of isotropic products and only touches on the further complications with these other products as appropriate.

With these types of products, much of the burden of the final product quality comes from the choice and control of the formulation. Small changes in key ingredients (e.g., solvents) or impurities (e.g., salts) can make a big difference in the phase behavior of the product leading to poor physical properties, appearance, and stability. Thus, the process engineer and the product designer must work closely together to establish the formulation and insure quality control.

The challenge for the process is to complete the necessary transformations and drive to the thermodynamic equilibrium. The transformations involved are generally not highly complex. Many production units involve only simple hardware and operations. Process complexity comes from demands to make a wide range of products with high efficiency, a large degree of flexibility, and ultimately at high volumes. Although these provide great challenges, they are the elements that drive consumer satisfaction and have fueled the growth of this sector in many markets.

This section provides a broad overview of HDL processing. The discussion begins with the transformational requirements of the product and basic process options followed by market demands and how these affect the choice and design of the process.

19.3.1.1 Transformation Requirements

Production of typical liquid detergents begins with chemical transformations. Many of the more complex chemical reactions are handled during the production of the raw materials, for example, sulfation/sulfonation of anionic surfactants. Chemical transformations in the final production are generally limited to some bulk neutralization and pH adjustment. For example, some materials are supplied in an intermediate, unneutralized form (e.g., HLAS). The biggest concern with these reactions is process control; pH specifications of the finished detergent are near neutral and quite tight to insure product safety and provide an environment conducive to chemical stability of enzymes and other ingredients. The process engineer must also deal with these same chemical stability concerns in process, avoiding in process swings in pH that degrade other materials in the formulation (e.g., alkyl sulfates are susceptible to hydrolysis at extreme pHs). Depending on the formulation, a mixture of cations may be used for neutralization (sodium, calcium, magnesium, etc.) and the process needs to drive the product to the desired, equilibrium salts.

The remainder of the key transformations are related to physical chemistry and state of the product. Here, surfactant-phase transitions can be a particularly important consideration. Anionic surfactants may be supplied in (or converted *in situ* to) a concentrated form in a lamellar, liquid crystalline phase. The final product is typically in a micellar phase. The transition may involve a middle phase that can have a very high, complex rheology and slow phase migration kinetics. Similarly, intermediate compositions can have much higher ingredient concentrations than the finished detergent. For example, water activity and ionic strength can be an order of magnitude different in these intermediates. This can drive unusual phase states that can be difficult to manage. However, these transitional states can provide an environment that enhances the kinetics of difficult transformations such as dissolution of fatty acids (e.g., high temperature and high solvent concentration).

Completion of all these transformations requires just a few basic operations such as metering, mixing, heat transfer, and control. The mixing requirement is generally not very demanding—low to moderate levels only. In fact, heterogeneous products often require limited shear or work input. However, the process control requirements can be quite demanding. Order of addition is important to manage the intermediate states, and pH/stability requirements sometimes drive very tight limits on formula recipe. All these operations can be achieved in either batch or continuous systems. The following section describes each operation and the factors influencing its choice.

19.3.2 Basic Process Options

19.3.2.1 Batch

The simplest and most often utilized process is the batch process. These operations vary in size and complexity. The most basic operation is a simple agitated tank. Additional mixing can be provided by a recirculation loop. The loop can also include a heat exchanger to control the temperature of the batch. The recipe is controlled by several options. For smaller, less sophisticated operations, materials can be weighed and added manually. Larger, more sophisticated operations utilize online weighing/metering. Online systems include mounting the mix vessel on load cells, having separate weigh vessels or totalizing flow meters on the inlet streams. All these options affect the capital- and operating cost of the process. Figure 19.17 shows a typical batch process operation.

The biggest advantages of batch operations are their simplicity and flexibility. These processes can be adapted to the market size and labor costs in the region; they can be very basic, manual operations, or sophisticated, highly automated operations. They can manage a wide range of formulae, orders of addition, and operating conditions at each step. These parameters can be varied almost indefinitely without hardware changes (although there may be an impact to batch cycle time or capacity). Contract or toll processors sometimes use batch operations to make different types of products (e.g., laundry detergent and dish soap) with proper cleanouts.

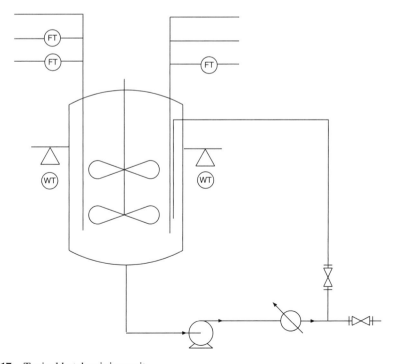

FIGURE 19.17 Typical batch mixing unit.

The biggest disadvantage of batch operations is their fixed scale. Once a batch tank size is chosen, it is difficult to run smaller campaigns. Each weighing/metering approach has a minimum amount that can be added accurately and there is a minimum level that can be properly mixed. There is also some hold up in the batch that cannot be pumped out at the end. This is fixed based on the size of the mixer as well as the hold up volume in the transfer lines to the bottle-filling lines, and therefore, making smaller batches in the same mixer increases the scrap or contamination as a percent of the production.

Similarly, higher volume production is limited by batch cycle times. The batch cycle time is a function of the kinetics of the transformations and transport operations (e.g., how quickly the ingredients and heat can be added and removed from the batch). The engineer can generally manage the residence time it takes to complete a transformation about as well in a batch as any other process. Thus, the unique limitations of batch processing are the transport steps. These tend to be pretty inefficient as all the materials must be moved twice, once in and once out, and both steps use operating time on the mixer.

Along these lines, batch processes offer engineering challenges because they have cyclical demands. For example, a material feed may occur in 1% of the batch cycle time. Thus, the feed rate is 100× what it would be if it were added to a continuous process making the same overall production rate. Similarly, there may be a short step in the batch cycle for cooling. The utility systems must be designed to deliver the entire cooling duty of the batch during that short step.

Finally, the mixing in a batch process can be somewhat limited and variable. Agitators impart a wide range of shear in a large vessel. This is not well suited to handling transformations with critical mixing requirements and is not well suited to high or complex rheologies. Detergent compositions are also subject to aeration during mixing in a batch. Aeration makes it difficult to accurately meter the product, which leads to issues in the packing operation. Disengagement of the air can also lead to separation of a dispersed phase in a heterogeneous (structured) product.

19.3.2.2 Continuous Operations

The alternative to batch operations is a continuous liquids process. These are more sophisticated and more geared to larger-scale operations. They tend to be well suited for complex rheologies and imparting a controlled mixing environment (intense or gentle). A typical process sketch is shown in Figure 19.18.

Depending on the product state and requirements at each point along the process, a number of unit operations can be employed. For mixing, these range from simple inline static mixers (as shown in Figure 19.18) to any of a variety of high-shear mixers. Similarly, heat transfer can be accomplished by a variety of heat exchanger designs including shell and tube and plate and frame-type designs. Here, the choice depends on the rheology of the mixture at that point in the process as well as ease of cleaning.

The biggest advantage of the continuous process is its scale: its ability to be optimized and produce large volumes at relatively low capital. This also implies its biggest disadvantage: lack of flexibility to a wide variety of formulae at low volumes. Generally, wide changes in addition levels or order of addition requires hardware modifications. The size of the process and the product hold up

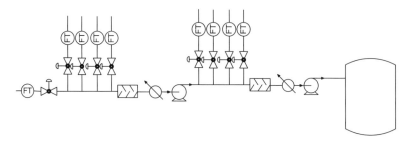

FIGURE 19.18 Schematic of multiple addition points and inline static mixers.

Detergent Processing

make transitions from one product to another a challenge. The hold up either becomes scrap or off-quality material that must be dealt with. The product quality requirement and economics of scrapping places a practical limit on the minimum campaign or the number of formulae it can handle.

19.3.2.3 Choosing the Optimal Process to Meet Market Demands

In developing markets, either low volume markets or markets where HDL share is low, the manufacturer typically wants to enter the market at a low investment and test the market. At low to modest volumes, basic batch and continuous processes can be relatively inexpensive. However, bulk-handling systems and sophisticated controls can drive up the total capital by several fold. Continuous processes demand more of these by their very nature. Batch systems can be designed to be completely manual and, thus, are often viewed as the lowest capital cost way into low volume production. The cost of labor has to be considered while operating such a process.

To grow a new market or drive a share in existing markets, detergent manufacturers are tending to provide more customized product, tailored to their target consumer. In the past, this meant a few brands with broadly different benefits. Over time, it evolved to the same group of brands with a range of color and scent-based line extensions. In today's world, there is an increasing move to offer a wider range of performance variation to meet the unique needs of consumers better. This approach cascades into a number of demands on the manufacturing operation. First and simply, it means that the process must be able to handle a wider variety of formulations and raw materials.

Second, it means the minimum production campaign must decrease. As the number of variants grows, each new variant is inherently a smaller fraction of the total production. This means that the production campaigns have to get smaller, or the time it takes to cycle through the whole portfolio increases. Larger, less frequent runs translate into higher inventory in the supply chain. This is an important consideration as the cost of carrying inventory can be high. Also, the supply chain will be reacting to the increased number of variants by trying to reduce inventory per variant to keep the total inventory levels constant. In the end, it is quite plausible for a manufacturer to hit their limiting case: where the minimum production campaign makes more product than the target inventory per variant and there will be high financial pressure to make process changes to reduce the size of each campaign.

Another way detergent manufacturers are winning in these situations is by constantly upgrading their lineup of products. This means a more frequent stream of formula changes in manufacturing, which will also increase as the number of variants grows. In all likelihood, the complexity of the upgrades will also increase with more variants because of the wider range in formulations.

On the whole as a market grows, the process must deal with both increasing production and an increasing variety of products. Therefore, the manufacturer is faced with the choice of optimizing on scale (typically with a continuous process) or flexibility (typically with a batch process). Frequently, the process choice is simply the best value process based on the volume and market status at the time of the investment.

19.3.2.4 Late Product Differentiation

Some manufacturers have looked for ways to achieve both scale advantages and support a large portfolio of variants. This can be done with a concept known as late product differentiation. Here, the majority of the ingredients are consolidated in a few base formulations and the bases are then topped off with key differentiating ingredients, dyes, and perfumes. This concept can be used with either batch or continuous processes depending on the application.

Most often, the base-making process is a large-scale continuous operation. The process is lined out on the base: pH, temperature, mixing, etc. are all carefully driven to the target. At the end of the process, the differentiating ingredients are added to make the desired finished product. This final part of the process is designed to be small, easy to line out, and therefore, forgiving to changeovers. With this approach, short runs can be made of each required variant to support a large product lineup and allow a fairly high volume run of the base.

19.3.3 OTHER DESIGN CONSIDERATIONS

19.3.3.1 Micro

Most liquid detergents are sold in an aqueous solution, and therefore, are subject to the potential for microbial growth. Formulators generally create an environment that is hostile by properly balancing the actives in the formula or adding an antimicrobial agent. Even so, these systems demand good, sanitary designs. The basis for these designs are to avoid stagnant areas (to avoid an area for micro growth), provide good cleanability, and sanitization capability (to eliminate any growth if it occurs). There are standard design techniques and equipment for accomplishing this. The food industry is an excellent source for such information.

19.3.3.2 Recycle

There are a number of sources of off-quality material in any operation, such as process system changeovers, packing system changeovers, process upsets, warehouse, and trade returns. Ideally, all these sources would be eliminated. However, some of these are out of the control of the process engineer and even the manufacturer, therefore complete elimination is often impractical. If material costs are low, the product can be targeted for an alternative use or disposed of. For premium products, the preferred option is to recycle the product to recover its value.

Incorporating recycle into the detergent matrix is not difficult from a transformation point of view. The primary concern is the impact of the recycle on the quality of the new finished product. Of key concern for liquid detergents is managing the impact on physical characteristics such as stability, odor, color, and viscosity. At low levels of recycle, these issues are usually negligible. If the recycle stream gets large, it requires control or management to insure there is no impact to finished product (FP) quality.

19.3.4 CONCLUSION

The transformations involved in liquid detergent production are generally less technically complex than other detergent forms. Success of the form and the individual producers/marketers has been driven by the pace and volume of new products. Thus, the challenge is in developing a very efficient and agile manufacturing system.

19.4 UNIT DOSE DETERGENT PROCESS TECHNOLOGIES

19.4.1 INTRODUCTION

19.4.1.1 Scope of This Section

This section is concerned with detergent-making processes for the manufacture of products in a unit dose form. The sale of household detergents as a unitized dosage has become a significant sector of the market, with all the principal detergent manufacturers offering some products in this form. The most common product type is the tablet form and in recent years this has been joined by products based on a water-soluble film. A broad review of the typical processes used to produce both these types of unit dose is provided in the following sections.

19.4.1.2 Review of the Evolution of Unit Dose Detergent Products

As a preface, it is worth taking a moment to review the evolution of the unit dose form for detergents. It is generally true, and never more so than for unit dose, that the product and process are inexorably linked. Particularly, the physical properties of the desired formulation plays a significant role in determining the suitability of the material for conversion into a unit dose form. It is, therefore, nearly always the case that the product design must evolve through a number of iterations to achieve a suitable balance of product, process, and affordability. The unit dose process development

Detergent Processing 355

engineer therefore has a unique role to play in delivering the desired consumer experience at an economically viable price.

The desire to sell products as a unit dose originated purely to provide the consumer with enhanced convenience. Measuring by the consumer is eliminated and the potential annoyance associated with dust and spills. In new unit dose forms, additional advantages are being exploited such as the ability to separate incompatible actives. Fabric cleaning detergent tablets were the first unit dose form to appear in the market and the first examples date back to as early as the 1960s when products such as Salvo tablets were sold in the United States. These early detergent tablets suffered from poor rates of dissolution, which contributed to the demise of the form. It was only in the late 1980s before detergent tablets appeared again, and this time, they created a successful segment of the market. By the turn of the century, significant volumes of detergents were being sold in tablet form, a particularly successful segment being for automatic dishwashing tablets. At this time, the unit dose success was limited to Europe, whereas in the United States numerous detergent tablets had failed to create a significant market segment. A recent unit dose development is that of water-soluble pouches, which are typically produced using polyvinyl alcohol (PVOH) film. In Europe, the first launch of this form was for liquid fabric cleaning detergents and later in the United States for automatic dishwashing. The most recent developments using PVOH for automatic dishwashing products have more than one PVOH compartment with different ingredients.

19.4.2 Hard Compressed Tablets

The manufacture of compressed dosage forms (Figure 19.19) for pharmaceutical products has been known for over 200 years and it is this process technology, which has been employed to make by far the vast majority of detergent products in unit dose form. The fundamental aspects of compaction are common to all compacted materials and the tabletting machines used by different industries are also basically similar. The standard steps in uniaxial die compaction are as follows:

1. Flow of the desired amount of powder into the die
2. Application of force
3. Compaction, that is, reduction of voids and increase in bulk density and bond formation
4. Removal of force
5. Ejection of the tablet

Figure 19.20 shows a typical strain curve for the compaction cycle as a function of stress. In principle, it would be possible to predict quantitatively the mechanisms from the physical and chemical properties of the powders and the applied forces. In reality, this is not yet feasible.

The initial powder density in the die will depend on the particle size distribution and particle shape but for many powders it is likely to be about 50% of the intrinsic density. After the application

FIGURE 19.19

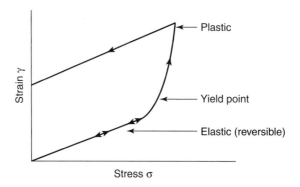

FIGURE 19.20 Typical strain curve as function of stress.

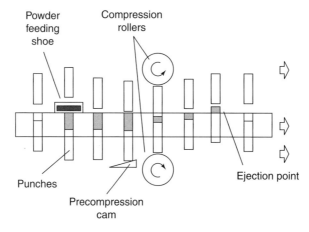

FIGURE 19.21 Cycle of a rotary tablet.

and removal of the compressive force, the average density would have increased to ~85% of the maximum. However, for most laundry tablets it is a lot less due to the need for very rapid dissolution.

During compression within a die, there are a multitude of physical and mechanical processes occurring. These include powder flow, percolation, friction, lubrication, fracture, elastic, viscous, and plastic deformation. As a result of these transformations, a powder is converted into a tablet in a matter of some few hundred milliseconds.

The stages of a typical production rotary tablet press are shown diagrammatically in Figure 19.21.

In developing a detergent tablet, the key concerns for the process engineer is to achieve the optimum balance of strength to dissolution. Tablet strength is important for packing operations in the plant and subsequent distribution to the trade and dissolution is critical for the performance of the product and to avoid highly undesirable product residues at the end of the wash. In addition, there are also a number of less obvious factors to be considered such as

> *Powder flow ability.* To achieve a consistent weight, it is important that the feed powder has consistent flow properties. Older presses often use purely gravity-based feed systems, whereas more modern presses may have assisted feed systems with features such as rotary paddles in the feeding shoe. The assessment and factors influencing powder flow are well documented in the literature, but for tabletting the final test must be on the production scale press.

Compression properties. The compressibility of the desired actives is unlikely to be suitable without some modifications. The compressibility of each ingredient can be significantly modified by adjusting particle size distribution and porosity. Typically narrow, fine particle size distribution (PSDs) require high forces and produce a tablet with low porosity and, therefore, poor solubility. Granular PSDs, produced by processes such as agglomeration or spray drying are much more suitable for tabletting. During the production of pharmaceutical tablets, the raw materials are always pretreated by a batch granulation (agglomeration) process for this reason. In addition, additives may be used to improve both solubility (disintegration) and tablet strength. Disintegrants work by either swelling on contact with water to break apart the tablet or they are themselves very soluble. Binders are generally "sticky" deformable materials, which will act to stick together the less deformable active ingredients. This facilitates the production of either harder tablets at the same compression force or improved disintegration rates for the same hardness produced at a lower compression force.

Punch sticking. Typical detergent formulations are often quite cohesive compared to many pharmaceutical formulations. The result is that the product sticks to the upper punch of the press. The problem is restricted to the upper punch since this punch leaves the product with a vertical movement while the tablet leaves the lower punch with a shearing force at ejection. The result is that the powder very rarely remains on the lower punch. Punch sticking is a major problem in production for both product quality and production reliability causing frequent shutdowns for cleaning. Adjusting the formulation is one approach to solving this problem but one generally has limited degrees of freedom in this respect. The problem is generally solved by the process design. Four common options are described in the following:

Punch coatings need to be tested by trial and error. Options range from polished chrome plate to Teflon.

Continuous twist system in both punches in the precompression and compression phases.

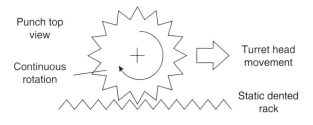

Cleaning brushes to remove makeup from the punches. It is very important to place close attention to the means for removing the brushed off powder to avoid accumulation of powder within the press using this approach.

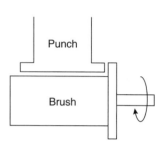

Liquid spray to bring antiadhesive silicon oil to the surface of the punches (top and bottom). The oil is sprayed during a few seconds several times during the day. Magnesium stearate powder is also used for the same purpose.

The process/quality control strategy for the manufacture of detergent tablets is largely based on the approach developed for pharmaceutical tablets. The tablet weight is controlled by the fill cam position. In modern presses, this will be a motorized adjustment and is linked by a feedback control loop based on the compression force. Typically, this can achieve a weight variation of less than $+/- 4\%$ (three relative standard deviations), since for a given material the compression force is linked to slight changes in the powder density. In recent years, some press vendors are also offering systems that periodically sample and automatically weigh tablets.

One enhancement of the tablet press is the capability to make multilayer tablets. This option was developed some decades ago for pharmaceutical tablets and it is now often seen in automatic dishwashing tablets. The multiple layers are achieved by simply replicating the fill/compression stations on the press. Hence, the sequence for a two-layer tablet would be fill layer 1/compress/fill layer 2/compress/eject. To achieve a good bonding between the layers, the final compression is always at the highest force.

Since the hard compressed technology has been much developed for pharmaceuticals, production systems based on this approach are likely to run at high process reliabilities (>80%).

Detergent Processing

19.4.3 SOFT COMPRESSED TABLETS

For some applications, such as laundry tablets in front-loading washing machines, the requirements for tablet dissolution are extremely demanding due to the low wash temperature, low agitation, and short wash times. In this application, some detergent manufacturers produce very weak tablets that are able to disintegrate very rapidly. The problem with this approach is that the tablet is too weak and sustains significant damage in the packing operations and shipment to the trade. One solution to this problem is to coat the tablet to boost the strength while still maintaining very quick disintegration. The application of the coating is typically performed while the tablets are passed along some form of perforated conveyor belt using a spray or liquid "curtain" method. Standard pharmaceutical drum-coating methods are not suitable since the tablets are too big and too weak.

To give some perspective as to the magnitude of the difference between "hard" and "soft" compressed tablets, typical hard detergent tablets (such as those produced for automatic dish washing) are compressed at pressures over 40,000 kN/m^2, whereas "soft" tablets use pressure as low as 2,000 kN/m^2.

An additional problem that is often much more significant with soft compression tablets is that of material sticking to the upper punch. This is more common with the production of weak tablets since the bonding forces between the particles themselves and those between particles and the punch surface are of a similar magnitude, and hence, some powder is likely to remain on the surface of the upper punch when it pulls up before ejection. Potential solutions to this problem include those mentioned earlier for hard-pressed tablets, one of the best options being the twisting punch. Unfortunately, a twisting punch will not work if the tablet is not circular! An alternative approach is to use single-sided tablet presses that have an upper compression plate instead of a punch. This plate can rotate thereby minimizing sticking. This type of tablet press is also commonly used for the production of stock cubes for cooking where the ingredients are very sticky. The principals of a typical single-sided press are shown in Figure 19.22.

All the considerations described earlier for hard-compressed tablets apply to soft tablets but it is evident due to the nature of the "weak" tablet that some concerns may be more problematic. In particular, the generation of dust in the environment of the press can be a big production problem. If the removal of the dust is not addressed, over time, it will lead to a reduction in process reliability. Although it is not impossible to achieve high process reliabilities, it is certainly more challenging and requires a high degree of attention to the details of all aspects of the press design and upstream and downstream integration with the powder feeding and packing operations.

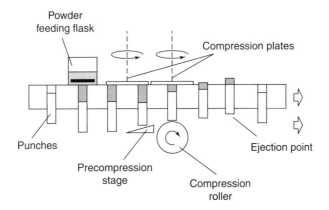

FIGURE 19.22 Schematic of tablet making process (direction right to left).

19.4.4 WATER-SOLUBLE POUCHES

19.4.4.1 Single Compartment

Compressed tablets have the obvious disadvantage of only being able to deliver as much liquid as may be absorbed within the powder. This is typically <10% w/w. Driven by the same advantages of the solid unit dose form, a number of detergent manufacturers have developed unitizing processes that can handle liquids. These products are all based on using polyvinyl alcohol (PVOH)-based films. PVOH is a good choice since it has good mechanical properties for pouch-forming processes, has reasonably rapid solubility in water, and is available in commercial quantities at a price that is consistent with detergent products (Figure 19.23).

19.4.4.1.1 Vertical Forming Processes

The first versions of liquid-filled PVOH pouches used a common existing process technology called vertical form fill seal (VFFS). This technology (Figure 19.24) was well established and employed on a wide range of products from tea bags to large bags of detergent powders. The disadvantage of

FIGURE 19.23

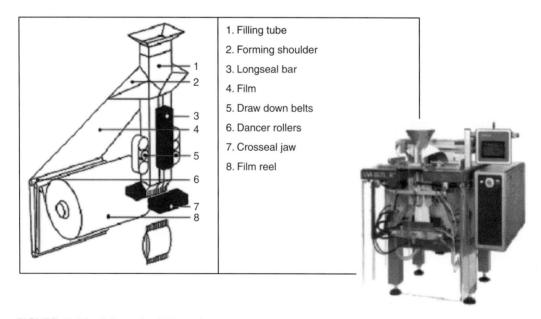

1. Filling tube
2. Forming shoulder
3. Longseal bar
4. Film
5. Draw down belts
6. Dancer rollers
7. Crosseal jaw
8. Film reel

FIGURE 19.24 Schematic of film path.

Detergent Processing · 361

this process is that it is quite capital intensive for the large outputs required for detergent unit dose products and consequently the profitability was generally poor. This is due to the fact that only one pouch is made at a time. In addition, as can be seen from Figure 19.24, this process is essentially a continuous folding of the film. Since it is impractical to fill the so-formed pouch 100%, and since the film is not under tension, the resultant product is quite flexible or "floppy." It was found that the majority of consumers did not like this floppy appearance.

19.4.4.1.2 Horizontal Forming Processes

As a natural evolution from the VFFS process, manufacturers of PVOH pouches have moved production lines to more efficient processes based on equipment, making more than one pouch at a time. These are often referred to as multilane converters. This technology is a development that is based on packaging technology in which deformable film materials are formed into multiple moulds to produce packages. This type of process is most often run in an horizontal mode although there are examples where the same principles are applied to a rotary cylindrical design. An example which will be familiar to the reader is that of prepacked sandwiches, here the semi-rigid part containing the sandwich is produced using the horizontal principle. PVOH pouches may be produced on this type of equipment using the following steps:

PVOH film, from a roll, is accurately positioned above a set of moulds having the desired size of the final product.

A combination of vacuum and heat is used to draw the film into the mould.

The product (liquid or powder) is dosed into the mould.

A second roll of PVOH film is applied to the top surface of the mould.

A suitable combination of heat and pressure is applied to seal the second film to the top surface of the first film.

The set of pouches are cut along the seal lines to form individual pouches.

The pouches are finally ejected from the moulds.

In addition to improving the capital efficiency compared to the VFFS process, the horizontal forming process involves creating some tension in the film, which will at least partially relax postprocessing. This results in a product with a neat, more rigid appearance, which is preferred by most consumers. Although the process may appear simple from the preceding description, this is still a relatively new venture for the detergent manufacturers and the equipment vendors, there being many details of the process that require careful attention to produce a consistent output at high process reliabilities.

PVOH film quality. If the product is to contain a liquid, it is vital that a high-quality source of film is used. Prior uses of PVOH film were less sensitive to small defects in the film, and therefore, it will probably be necessary to work closely with your film supplier to ensure that your quality requirements are clearly understood.

Weight control. Compared to the tabletting process, this process is less flexible with regard to weight control since the "mould" is fixed. This is typically not an issue for liquid products since the density is very consistent, however for powder, fluctuations in the powder density will result in changes in the powder weight in each pouch. For the process engineer, this must be addressed by upstream control of the powder-making process and in trying to maintain very consistent conditions on the pouch-making line. However, it is very unlikely that the high degree of weight control achievable on a tablet press can be reproduced by this type of process.

Leaker detection. If the product is a liquid, then leaking pouches caused by poor sealing or film quality must be detected as soon as possible. For the consumer, it is clearly imperative to avoid a shipping product where some pouches are leaking and in production, leakers can cause major downtime problems. For example, if liquid gets into the vacuum system this can result in significant lost time to cleanout and if the vacuum pump is not protected by a "catch pot" it can even result in the destruction of an expensive vacuum pump.

The production of water-soluble pouches has many similarities with high-speed packing lines such as those used to produce disposable nappies (diapers), and the process engineer working in this field would do well to visit such types of production facilities to compare their approaches to quality control and process optimization. Here, we are talking about aspects ranging from high-speed vision systems for product inspection to high-accuracy servo drive motor systems to ensure reproducible and highly accurate process manipulations.

For comparative output to a tabletting line, the capital cost is higher and the process reliability is likely to be lower. This is not an inherent limitation of this technology but is rather a reflection of the relative infancy compared to tabletting.

19.4.4.2 Dual and Multicompartment

As the market for detergents in PVOH pouches became somewhat established, leading manufacturers developed more complex products as a means to differentiate their product from others in this very competitive market. Two prime examples are P&G's Cascade ActionPacs®/Fairy Active Bursts (Figure 19.25) and Reckitt Benckiser's Finish Quantum® (Figure 19.26).

19.4.4.2.1 Horizontal Forming Processes
The horizontal forming process for PVOH pouches can be adapted to produce pouches with more than one section or compartment. For obvious reasons, precise details are proprietary to each manufacturer; however, it is evident from the product itself as to the basic principles employed.

19.4.4.2.2 Moulding Processes
A relatively recent development for unit dose products based on PVOH is a process based on injection moulding. This is generally a very common and well-known process for products ranging from plastic automotive parts to children's toys.

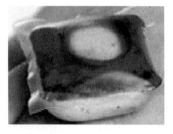

FIGURE 19.25 Version of dishwashing unit dose.

FIGURE 19.26 Another version of dishwashing unit dose.

Detergent Processing

The advantage of injection moulding is that quite complex shapes can be easily produced, provided that the injection material can be made to flow into the mould. For the application to unit dose detergents, there are two principle challenges to be addressed by the process development engineer who desires to employ this process for PVOH unit dose detergents. First, formulating a PVOH mix that is conducive to injection moulding. PVOH alone will not be suitable and therefore additives that do not adversely effect the desired final product properties of the PVOH such as its ability to dissolve rapidly must be found. Second, the complexities of dosing multiple ingredients into the injection-moulded PVOH part at viable production line speeds without contamination from one to the other. The existence of such a product in the market demonstrates that these challenges can be solved.

19.4.4.2.3 Recycle Considerations
In the manufacture of traditional nonunit dose detergents, the recycle or "reblend" of an off-specification product was an important necessity for both environmental and economic considerations but it was not an area of the process that required complex process innovation. A box of powder or a bottle of liquid may be simply opened and added back to normal production at a low level. Even with the introduction of tablets, this did not create significant difficulties since tablets are easily converted back to powder in a grinder ready for addition back to the production of fresh powder.

Conversely, with the introduction of water-soluble unit dose pouches, the recycle question immediately became much more complex. It is an inherent element of these processes (except the VFFS) that the production of pouches results in a side edge trim being the outer edges of the film, which is held during the process to ensure that the web of film is under positive control. This may only be a few percent of the total film used but nevertheless, it must either be scrapped or returned to the film producer for reblending. A more complex challenge is the difficulty in dealing with off-specification pouches since there is no immediately obvious means to recover the product separately from the film. The pouches could be cut open in some type of knife mill but it is not easy to cleanly separate the film from the active product. The challenge becomes even more difficult for the new dual compartment pouches containing powder and liquid.

19.4.5 SUMMARY AND TRENDS

The last decade has seen an explosion of unit dose product forms in the household detergent market. It is still most popular in Europe where some segments have passed the 50% mark for unit dose such as automatic dish washing in Germany where unit dose accounts for more than 60% of the market. Fast-moving consumer goods are a very competitive market, and in a very established area such as detergents, unit dose has become the prime vehicle for manufacturers to seek to establish competitive advantage and to be seen at the leading edge of innovation. This trend, driven by marketing, has created a very exciting environment for the process development engineer working in this field—the technical challenges of unit dose being in addition to those related to the production of the base powder or liquid.

Since the launch of the first simple tablets, the unit dose form has become more complex almost every year thereafter. Some of them include dual layer tablets, tablets with inserts, and, recently, the multicompartment PVOH pouches. The need to remain competitive will continue to drive more complex forms of unit dose products for the foreseeable future.

REFERENCE

1. Iveson, S., J. Litster, K. Hapgood, and B. Ennis, Nucleation, growth and breakage phenomena in agitated wet granulation processes- a review. *Powder Technology*, 117(2001), 3–39.

20 Production of Quaternary Surfactants

Ansgar Behler

CONTENTS

20.1 Introduction 365
20.2 Cationics Based on (Tri)alkylamines 367
20.3 Cationics Based on Alkanolamine Esters (Esterquats) 370
20.4 Quaternization 371
References 372

20.1 INTRODUCTION

Quaternary surfactants are simply characterized in that they bear a cationic charge in the molecule [1–5,6]. Typically, one or more long-chain hydrophobic hydrocarbon groups with a C-chain length between C6 and C22 are linked to a cationic group. Several cationic groups are known (Figure 20.1). But products based on phosphonium and sulfonium that can be used as phase-transfer catalysts [7] are less important. Products containing an ammonium group are the only significant ones within the group of cationic surfactants.

In western Europe, the consumption of surfactants in 2005 amounted for 2.9 million tons. This amount is distributed to the different surfactant classes as follows [8]:

Anionics	40%
Nonionics (ethoxylates)	44%
Cationics	9%
Amphoterics	2%
Other nonionics (e.g., alkanolamides)	5%

Thus, cationic surfactants, short cationics, represent the third largest group of surfactants. Generally, cationic surfactants are prepared by reacting a tertiary amine with an alkylation agent (Figure 20.2).

Many different cationics with an ammonium group in the molecule are known. In general, they differ in the alkyl or aryl chains linked to the ammonium group and the counterion. Several alkylation or quaternization agents are known (Figure 20.3).

The most important ones are dimethyl sulfate and methyl chloride.

Among the starting tertiary amines also many different substances are known. Figure 20.4 summarizes some of them.

The most important ones are alkylamines and esters of alkanolamine, which are discussed in detail in the following text.

365

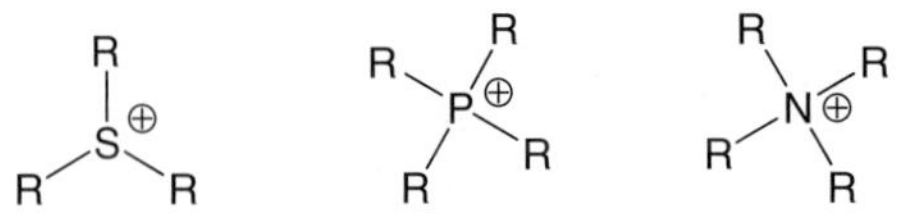

FIGURE 20.1 Cationic surfactants based on sulfur, phosphorous, and nitrogen.

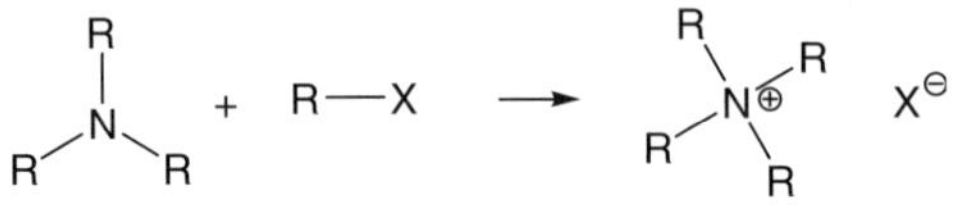

FIGURE 20.2 Preparation of cationic surfactants.

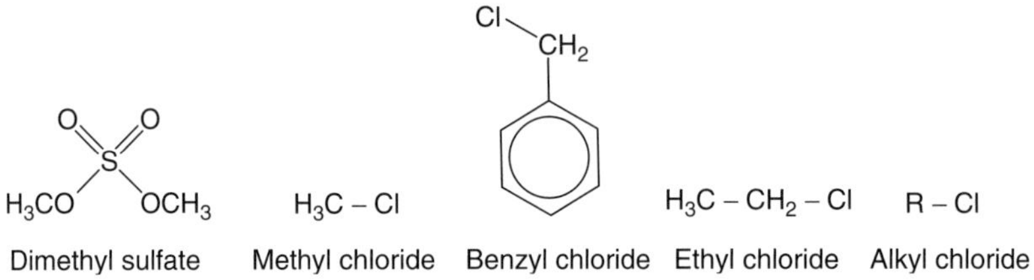

Dimethyl sulfate   Methyl chloride   Benzyl chloride   Ethyl chloride   Alkyl chloride

FIGURE 20.3 Common alkylation agents.

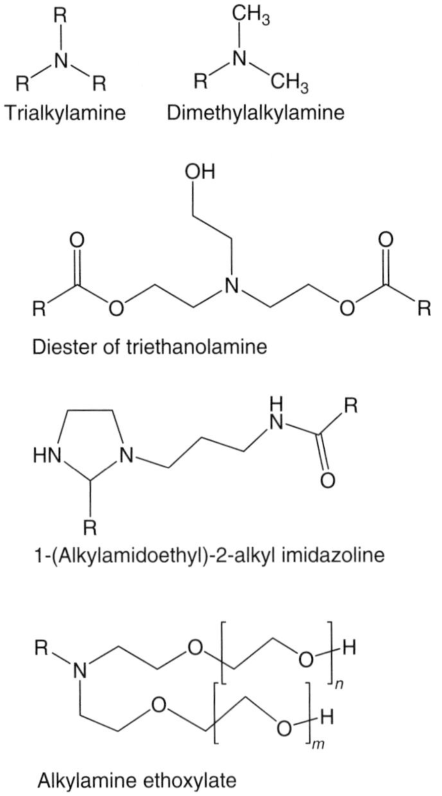

Trialkylamine   Dimethylalkylamine

Diester of triethanolamine

1-(Alkylamidoethyl)-2-alkyl imidazoline

Alkylamine ethoxylate

FIGURE 20.4 Common tertiary amines as starting materials for cationic surfactants.

Production of Quaternary Surfactants

20.2 CATIONICS BASED ON (TRI)ALKYLAMINES

Fatty amines with alkyl chains from C8 to C22 are used as raw materials for this type of cationics. Alkylamines with two long alkyl chains and one methyl group in the molecule are of significant interest. The first commercially available cationic surfactant distearyl dimethyl ammonium chloride (DSDMAC), which was introduced in 1949 as a fabric softener (Figure 20.5), belongs to this group.

Several manufacturing processes are described for fatty amines [9]. The most significant one regarding volume starts with fatty acid (Figure 20.6).

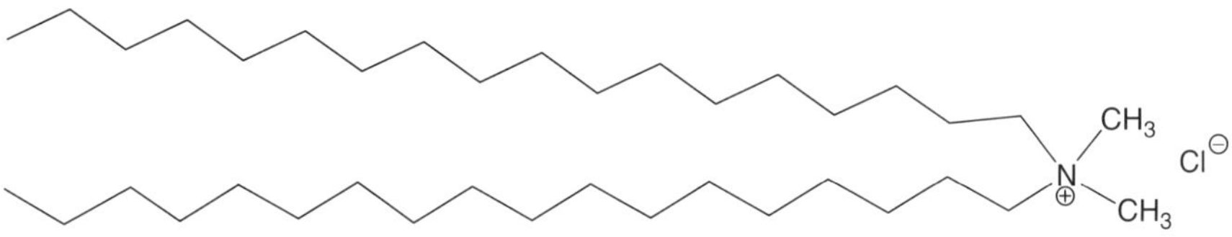

FIGURE 20.5 Structure of DSDMAC.

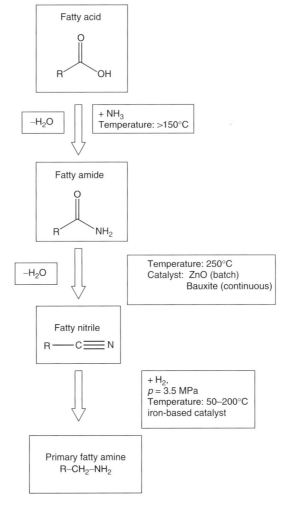

FIGURE 20.6 Manufacturing process for primary fatty amines.

In the first step, fatty acid reacts with ammonia to form ammonium soap, which then dehydrates to the corresponding fatty amide at temperatures above 150°C. At 250°C, this fatty amide dehydrates again, in the presence of suitable catalysts (e.g., ZnO for a continuous process and bauxite for a batch process), to form the fatty nitrile. Hydrogenation of the nitrile at high pressures and temperatures between 50 and 200°C, again by using suitable catalysts, for example, based on nickel or noble metals, provides the primary fatty amine. Both continuous and batch processes carried out by using different type of catalysts are described [10–12]. To suppress the formation of secondary or tertiary amines as by-products, ammonia is added during the hydrogenation step.

Secondary fatty amines can be prepared directly from primary amines by the removal of ammonia at reduced pressure, passing the hydrogen through the reaction mixture, and using suitable catalysts (nickel or copper based) at temperatures between 150 and 220°C [13] (Figure 20.7).

Dimethylalkylamines are also made by reductive alkylation (methylation) of primary amines with formaldehyde/hydrogen and nickel catalysts [14,15] (Figure 20.8).

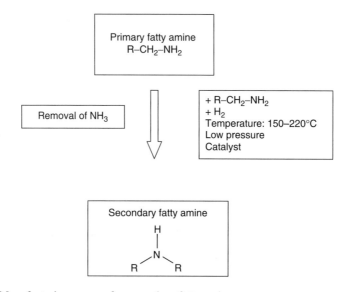

FIGURE 20.7 Manufacturing process for secondary fatty amines.

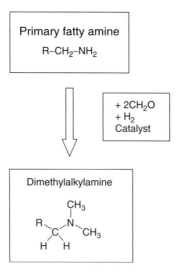

FIGURE 20.8 Manufacturing process for dimethylalkylamine.

Production of Quaternary Surfactants

In a similar process, dialkylmethylamines are prepared by reductive methylation of secondary amines (Figure 20.9).

Distearylmethyl amine, prepared by this procedure, is the precursor for DSDMAC mentioned earlier.

Dimethylalkyamines are also available by two different routes starting with fatty alcohol. In the first route, the fatty alcohol is converted directly to dimethylamine by using a copper catalyst [16,17] (Figure 20.10).

In the second route, the fatty alcohol is first converted to the corresponding alkyl chloride by reaction with phosphorous trichloride or other chlorination agents. This alkyl chloride then reacts with dimethylamine to form the desired dimethylalkylamine [18] (Figure 20.11).

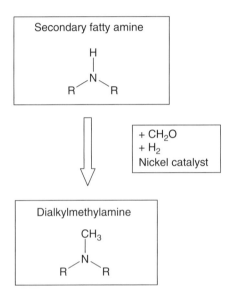

FIGURE 20.9 Manufacturing process for dialkylmethylamine.

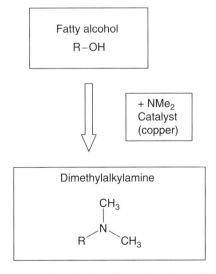

FIGURE 20.10 Manufacturing process for dimethylalkylamine through fatty alcohol.

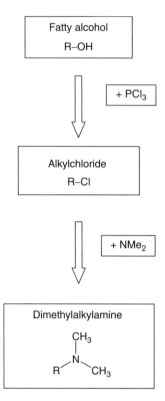

FIGURE 20.11 Manufacturing process for dimethylalkylamine through alkyl chloride.

20.3 CATIONICS BASED ON ALKANOLAMINE ESTERS (ESTERQUATS)

Esterification of alkanolamines with fatty acids leads to the formation of esteramines, which can be quaternized to the so-called esterquats [19,20] (Figure 20.12).

In contrast to alkyl-based cationics, esterquats contain ester groups in the molecule. Owing to this hydrolytically unstable function in the molecule, esterquats are much better biodegradable than alkyl-based cationics. In the European detergent industry, this fact and its lower toxicity to water organisms [21] led to a nearly complete replacement of DSDMAC by esterquats in household rinse cycle softeners since 1991.

Esteramines can be prepared by using different alkanolamines. The most important ones are triethanolamine and methyldiethanolamine (Figure 20.13).

Esterification of these alkanolamines with fatty acid is carried out at temperatures up to 200°C. Released water is removed by stripping with an inert gas or under vacuum. Often an acid catalyst is used (e.g., hypophosphorous acid) [22] (Figure 20.14).

When 2 mol of fatty acid, usually with alkyl chains from C12 to C22, is reacted with 1 mol of triethanolamine, a statistical diester is formed, which has approximately the following composition:

Triethanolamine	5%
Monoester	22%
Diester	43%
Triester	30%

Methyl esters, triglycerides, and acid chlorides can be used to prepare the esteramines as an alternative to fatty acid as well.

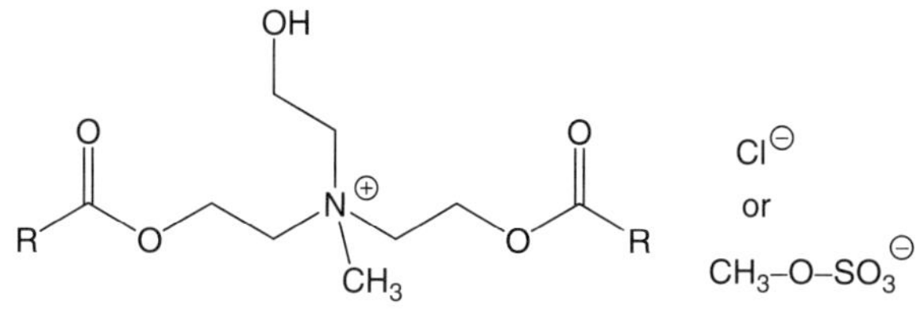

Production of Quaternary Surfactants 371

FIGURE 20.12 Structure of an esterquat based on triethanolamine.

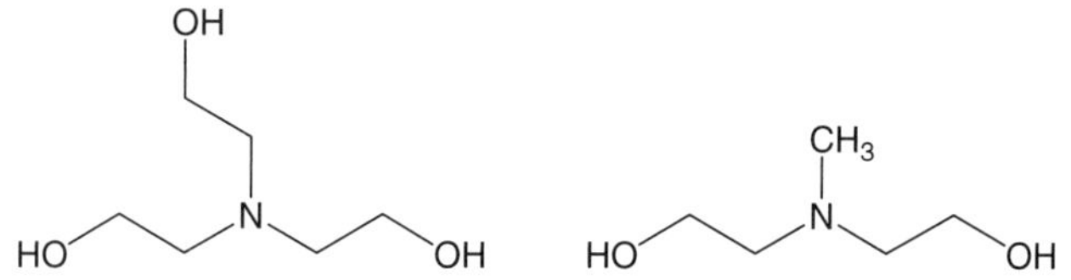

FIGURE 20.13 Structure of triethanolamine and methyldiethanolamine.

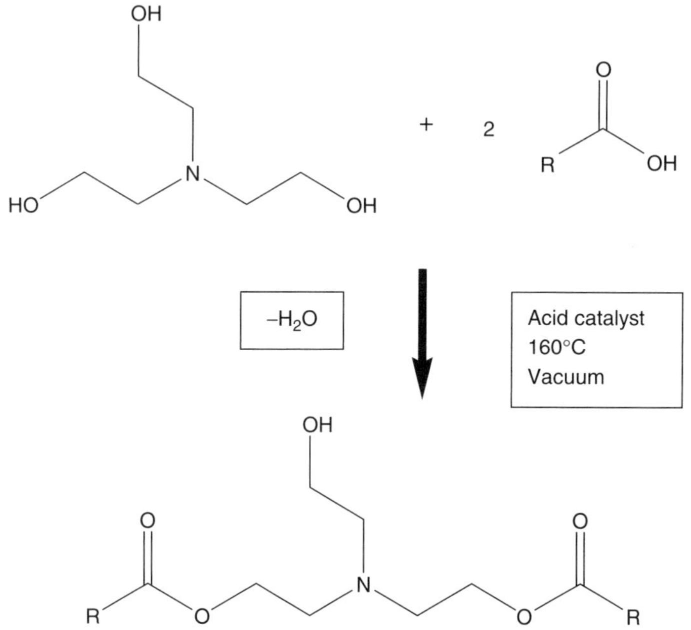

FIGURE 20.14 Preparation of triethanolamine diester.

20.4 QUATERNIZATION

The most significant process to prepare cationic surfactants involves the reaction of a tertiary amine with an alkylation agent (see Figures 20.2 and 20.3) [23]. Several other different types of tertiary amines, discussed earlier, can also be used (Figure 20.4).

Generally, the tertiary fatty amine with a typical alkyl chain length ranging from C12 to C22 is charged into the reactor with a solvent, for example, water, isopropanol, ethanol, ethylene glycol, propylene glycol, or long-chain alcohols. Owing to possible corrosion problems, especially if a

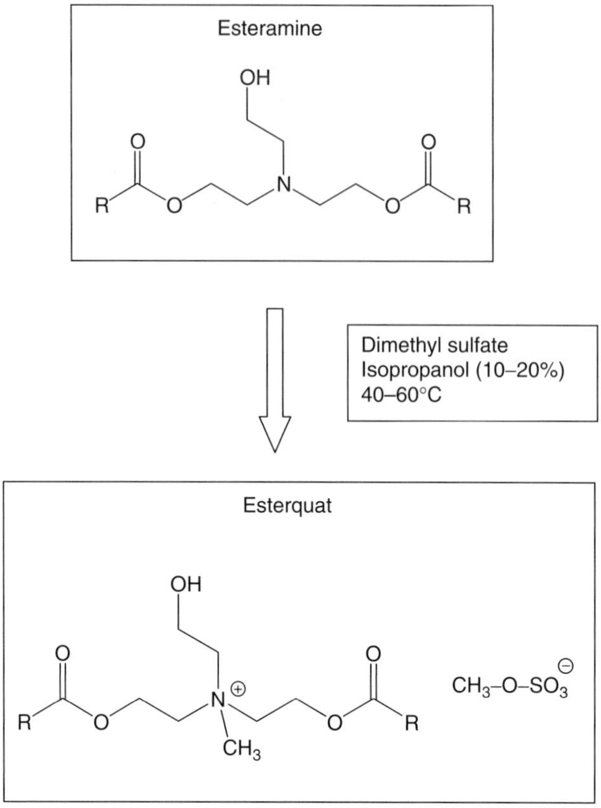

FIGURE 20.15 Manufacturing process for esterquats.

chloride is used as an alkylation agent, the reactor must be glass lined or stainless steel. At temperatures between 40 and 100°C, the alkylation agent is added in several portions till the desired degree of quaternization is achieved. In the case of methyl chloride, the reaction is carried out under pressure (3–8 bars).

Figure 20.15 illustrates the typical manufacturing process for an esterquat [24]. In this case, the starting tertiary amine is a diester of triethanolamine and C16/18 fatty acid. To this esteramine, isopropanol is added as a solvent to lower the viscosity of the reaction mixture. The quaternization is carried out by slowly adding dimethyl sulfate in several portions at temperatures between 40 and 60°C. A slightly understoichiometric amount of dimethyl sulfate is used to ensure that all of the toxic dimethyl sulfate can react with esteramine, and thus avoiding traces of dimethyl sulfate in the end product.

REFERENCES

1. Jungermann, E. (ed.), *Cationic Surfactants*, Surfactant Science Series, vol. 4, Marcel Dekker, New York, 1970.
2. Richmond, J.M. (ed.), *Cationic Surfactants—Organic Chemistry*, Surfactant Science Series, vol. 34, Marcel Dekker, New York, 1990.
3. Rubingh, D.N., Holland, P.M. (eds), *Cationic Surfactants—Physical Chemistry*, Surfactant Science Series, vol. 37, Marcel Dekker, New York, 1991.
4. Franklin, F., et al., Cationic and amine-based surfactants, in *Oleochemical Manufacture and Applications*, Gunstone, F.D., Hamilton, R.J. (eds), chap. 2, CRC Press, Boca Raton, FL, 2001.

5. Behler, A., et al., Industrial surfactant syntheses, in *Reactions and Synthesis in Surfactant Systems*, Texter, J. (ed.), Surfactant Science Series, vol. 100, Marcel Dekker, New York, 2001, p. 30.
6. Dery, M., Quarternary ammonium compounds, in *Kirk Othmer Encyclopedia of Chemical Technology*, 4th ed., vol. 20, Wiley, New York, 1997.
7. Dehmlow, E.V., Dehmlow, S.S., *Phase Transfer Catalysis*, Verlag Chemie, Weinheim, 1980.
8. Chemical Sectors: Detergents: CESIO, www.cefic.org.
9. Visek, K., Amines (Fatty), in *Kirk Othmer Encyclopedia of Chemical Technology*, 4th ed., vol. 2, Wiley, New York, 1997.
10. Waddleton, N., GB Patent, 1,321,981 (4.7.1973).
11. Henkel&Cie, GmbH, GB Patent, 1,153,919 (4.6.1969).
12. Specken, G.A., US Patent, 3,574,754 (13.4.1971).
13. Glankler, G.W., Nitrogen derivatives, *JAOCS* 56, 802A–805A, 1979.
14. Billenstein, S., Blaschke, G., Industrial production of fatty amines and their derivatives, *J. Am. Oil. Chem. Soc.* 61, 353, 1984.
15. Lion Akzo Company Ltd., US Patent, 4,845,298 (4.7.1989).
16. Ethyl Corp., US Patent, 4,994,622, 1991.
17. KAO Corp., US Patent, 5,696,294, 1997.
18. Millmaster Onyx Corp., US Patent 3,548,001, 1970.
19. Krüger, G., Boltersdorf, D., Overkempe, K., Esterquats, in *Novel Surfactants*, Holmberg, K. (ed.), Surfactant Science Series, vol. 74, chap. 4, Marcel Dekker, New York, 1998.
20. Puchta, R., Krings, P., Sandkühler, P., A new generation of softeners, *Tenside Surf. Deterg.* 30, 186, 1993.
21. Versteeg, D.J., An environmental risk assessment for DTDMAC in the Netherlands, *Chempshere* 24, 641, 1992.
22. Henkel KgaA, WO 91/01295 (17.7.1989).
23. Franklin, F., et al., Cationic and amine-based surfactants, in *Oleochemical Manufacture and Applications*, Gunstone, F.D., Hamilton, R.J. (eds), CRC Press, Boca Raton, FL, 2001, pp. 40–43.
24. Krüger, G., Boltersdorf, D., Overkempe, K., Esterquats, in *Novel Surfactants*, Holmberg, K. (ed.), Surfactant Science Series, vol. 74, Marcel Dekker, New York, 1998, pp. 118–121.

21 Production of Detergent Builders: Phosphates, Carbonates, and Polycarboxylates

Olina G. Raney

CONTENTS

21.1 Phosphates 375
    21.1.1 Sodium Tripolyphosphate 375
        21.1.1.1 Production...376
    21.1.2 Pyrophosphates 377
21.2 Sodium Carbonate 378
    21.2.1 Production..378
        21.2.1.1 Natural Soda Ash..378
        21.2.1.2 Synthetic Soda Ash..380
21.3 Polycarboxylates...381
    21.3.1 Emulsion Polymerization..381
        21.3.1.1 Raw Materials..382
        21.3.1.2 Production Processes 382
Acknowledgments...384
References..384

21.1 PHOSPHATES

Two forms of sodium phosphates have been used as detergent builders: sodium tripolyphosphate (STPP), which is the more commonly used form, and tetrasodium pyrophosphate (TSPP).

21.1.1 SODIUM TRIPOLYPHOSPHATE

STPP, one of the builders that is commonly used in detergents, primarily functions to soften the water by sequestering the hardness of mineral ions in the washing solution and preventing their deposition on the fabric and washing machine. It is also effective in removing particulate matter. In addition, STPP helps prevent redeposition of soil on the washed fabrics and provides a buffer for the wash solution at a desired pH. The phosphate composition in the detergent is typically formulated for effectiveness in both hard and soft water regions.

Phosphates, although performing effectively as builders, have come under increased scrutiny due to their adverse effects on the environment. Since phosphates are nutrients, their presence in the discharge water can result in eutrophication—a process that causes growth of plant organisms in the water, leading to reduced amounts of oxygen available to aquatic life. This has led to efforts to reduce the level of phosphates in detergents so as to limit such harmful effects to the environment.

21.1.1.1 Production

STTP, $Na_5P_3O_{10}$, also known as pentasodium triphosphate or sodium triphosphate, is produced from phosphates obtained by mining phosphate rock—a material containing phosphate minerals in amounts sufficient for commercial production. The phosphate is generally obtained from the mineral apatite. Both surface and underground mining processes are used to mine phosphate rock. Both mechanical and chemical processing methods are required to produce the phosphate [1].

Phosphorus is synthesized from calcined phosphate rock, silica, and coke in an electric furnace. Phosphorus is usually burned in air and the resulting oxide is dissolved in water to produce phosphoric acid. A mixture of 2 mol of disodium phosphate and 1 mol of monosodium phosphate is typically used to produce STTP as shown in the following reaction:

$$2Na_2HPO_4 + NaH_2PO_4 \rightarrow Na_5P_3O_{10} + 2H_2O$$

The temperature range for the reaction is 300–550°C.

The starting material is usually a monophosphate solution stoichiometrically adjusted by neutralization of phosphoric acid with caustic soda or soda ash. The solution is dehydrated using single- or two-stage processes to produce STTP [1,2].

21.1.1.1.1 Two-Stage Process

The two-stage process, shown in Figure 21.1, involves dehydration of the solution in the first stage to produce an anhydrous monophosphate mixture during which diphosphate is formed [2]. The second stage results in the condensation to sodium tripolyphosphate. Spray dryers are typically used for the dehydration of monophosphate. Rotary kilns are commonly used for the conversion into tripolyphosphate.

21.1.1.1.2 Hoechst-Knapsack Process

Rotary kilns or spray dryers are used in the single-stage process. The Hoechst-Knapsack process shown in Figure 21.2 involves spraying of the monophosphate solution at 1–2 MPa into a stainless-steel spray tower [2–4]. The solution moves downward with the burner gases and is quickly dehydrated and

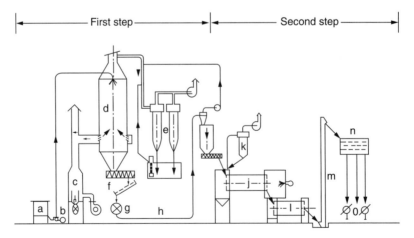

FIGURE 21.1 Two-stage process for production of pentasodium triphosphate. (a) Receiver, (b) heavy-duty pump, (c) combustion chamber, (d) spray tower, (e) west dust removal from flue gas, (f) sie, (g) mill, (h) pneumatic conveyor, (i) silo, (j) rotary kiln, (k) flue gas purification, (l) condensation, (m) bucket chain, (n) sie, (o) bagging scale. (Reprinted from Shrodter, K., Bettermann, G., Staffel, T., Klein, T., Hofmann, T., *Ullman's Encyclopedia of Industrial Chemistry*, p. 127, Wiley-Vch, Weinheim, Germany, 2003. With permission.)

Production of Detergent Builders

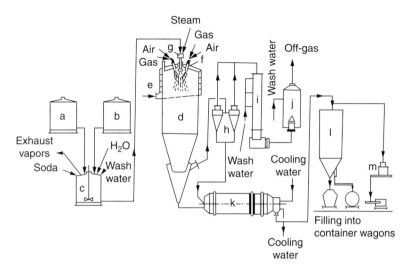

FIGURE 21.2 Production of pentasodium triphosphate by Hoechst-Knapsack process. (a) Caustic soda; (b) phosphoric acid; (c) neutralization; (d) spray tower; (e) cooling jacket; (f) burner; (g) spray nozzle; (h) cyclone; (i) cooling pipe; (j) wash tower; (k) rotary cooling drum; (l) product silo; (m) bag filling and weighing machine. (Reprinted from Shrodter, K., Bettermann, G., Staffel, T., Klein, T., Hofmann, T., *Ullman's Encyclopedia of Industrial Chemistry*, p. 127, Wiley-Vch, Weinheim, Germany, 2003. With permission.)

converted to tripolyphosphate. The majority of the tripolyphosphate is collected in the tower cone, whereas the fine particles are separated from the off-gas by the cyclones. These gases are purified using a liquid wash. The final product is a free-flowing powder product with a tripolyphosphate content of up to 98%.

21.1.1.1.3 FMC Processes
The FMC process involves spraying the monophosphate solution upward from the bottom of the tower [5]. The particles travel to the flame zone at the top of the tower and the hot gas stream carries them downward.

Another FMC process involving a countercurrent rotary kiln has been described in the literature [6]. The solution is carried countercurrent to the gas stream, which is produced in a combustion chamber outside the rotary kiln. This process has better heat utilization when compared to the cocurrent process where flow of solution and gas stream are in same direction. However, a negative effect is the occurrence of caking at the charging point.

Various modifications to the commercial process have been suggested in the patent literature. These are primarily focused on the production of STTP with particular properties such as low bulk densities [7], increased abrasion resistance [8], and improved solubility properties [9,10]. Crystalline or coarse tripolyphosphate particles can also be obtained by spraying seed crystals with the desired characteristics during the production process [11]. The use of rotary kilns to obtain a granular product in both stages of the two-stage process has been discussed previously [12]. In this case, the intermediate product is ground and water is added before calcination.

Production methods involving low-temperature modifications have also been described [13,14]. Also, the use of a fluidized-bed reactor for the production of tripolyphosphate from the dried monophosphate has been proposed [15].

21.1.2 Pyrophosphates

TSPP ($Na_4P_2O_7$), also known as tetrasodium diphosphate, has found application as a detergent builder and as a water softener, although in significantly smaller volumes than STTP, and its use has declined over the past 10 years. A drawback of this builder is the formation of insoluble salts on reaction with water hardness ions, which leads to the undesirable effects of incrustation on fabrics and washing machine [16].

The production process for TSPP involves the reaction of phosphoric acid with soda ash that results in a disodium hydrogen phosphate (Na_2HPO_4), which is then calcined at temperatures in the range 300–900°C to produce TSPP [2]. The chemical reaction involved is as follows:

$$2Na_2HPO_4 \rightarrow Na_4P_2O_7 + H_2O$$

The two-stage process involves dehydration of a concentrated monophosphate solution to produce anhydrous disodium hydrogen phosphate, which is then converted to sodium pyrophosphate in a rotary kiln. The process used for the production of STTP can typically be used for sodium pyrophosphate.

Another pyrophosphate, tetrapotassium pyrophosphate ($K_4P_2O_7$), also known as tetrapotassium diphosphate, is used as a builder in liquid detergent formulations at levels in the range 20–25% despite its higher cost [2]. These levels are due to its significantly higher water solubility properties when compared to the sodium salt as well as the stability of the pyrophosphate anion. Tetrapotassium pyrophosphate functions as a dispersant and binds multivalent cations, keeping them in solution.

The production process for tetrapotassium pyrophosphate is similar to that of the sodium salt described earlier, produced by dehydration of dipotassium hydrogen phosphate (K_2HPO_4) at temperatures in the range 350–460°C.

21.2 SODIUM CARBONATE

Sodium carbonate or soda ash is incorporated as an ingredient in detergents to function as an inexpensive builder as well as provide the alkalinity desired in the washing solution. The soda ash removes calcium from hard water in the form of precipitated calcium carbonate and magnesium carbonate. These precipitates can cause incrustations on fabrics and washing machines. However, these incrustations can be minimized by the use of polycarboxylates as dispersants in the detergent formulation [16]. Soda ash can also be used in combination with other builders such as zeolites and silicates in the detergent formulation.

The use of sodium carbonate is considered to be environmentally safe. There are concerns, however, regarding synthetic soda ash production, which can result in calcium chloride as well as ammonium and heavy metal compounds in the wastewater. The high alkalinity of sodium carbonate is also a consumer safety concern [16].

21.2.1 PRODUCTION

Soda ash or commonly known as sodium carbonate (Na_2CO_3) is obtained from naturally occurring deposits as well as from chemical production processes. While natural deposits of soda ash have been known as early as about 3500 BC, the beginning of the Industrial Revolution in the late eighteenth century led to increased demand for soda ash resulting in the development of various processes such as the Leblanc, Solvay, and other processes [17].

21.2.1.1 Natural Soda Ash

Soda ash can be obtained from natural deposits of sodium carbonate, primarily from the mineral trona. Production of soda ash by this method has become more important since World War II. Natural soda ash plants are located in the western regions of the United States in Nevada, California, and Wyoming, with largest known natural deposits of trona in the world occurring in Green

Production of Detergent Builders 379

River, Wyoming. In 1991, 30% of total world production of soda ash was attributed to natural soda ash.

Natural soda ash can be processed from trona by refining the ore, which contains approximately 90% pure sodium sesquicarbonate ($Na_2CO_3 \cdot NaHCO_3 \cdot 2H_2O$) to produce soda ash with greater than 99% purity. Two methods used in the refining process are the sesquicarbonate process and the monohydrate process with the monohydrate process being the more commonly used method. Approximately 1.0 t of processed soda ash is produced by 1.8 t of trona ore.

The monohydrate process developed to produce soda ash from trona is illustrated in Figure 21.3. The process first involves crushing the trona ore followed by calcination in gas-fired calciners at 150–300°C to remove water and carbon dioxide as shown in the following reaction:

$$Na_2CO_3 \cdot NaHCO_3 \cdot 2H_2O \text{ (trona)} + \text{heat} \rightarrow 3Na_2CO_3 + CO_2 + 5H_2O$$

The product, now containing about 85% soda ash and 15% insoluble matter, is dissolved in hot water, and the hot solution is fed to crystallizers where sodium carbonate monohydrate crystals ($Na_2CO_3 \cdot H_2O$) are precipitated at temperatures between 40 and 100°C. The crystals are then sent to hydroclones and excess water is removed in centrifuges. The product is then dried in steam dryers at 150°C to produce anhydrous dense soda ash having bulk density about 1.0 g/mL.

Soda ash crystals with lighter bulk densities of about 0.89/mL are produced by the sesquicarbonate process. In this process, the trona ore is crushed, dissolved in hot mother liquor, and the solution is clarified, filtered, and sent to cooling crystallizers where crystals of sodium sesquicarbonate ($Na_2CO_3 \cdot NaHCO_3 \cdot 2H_2O$) are precipitated. The crystals are hydrocloned, centrifuged, and

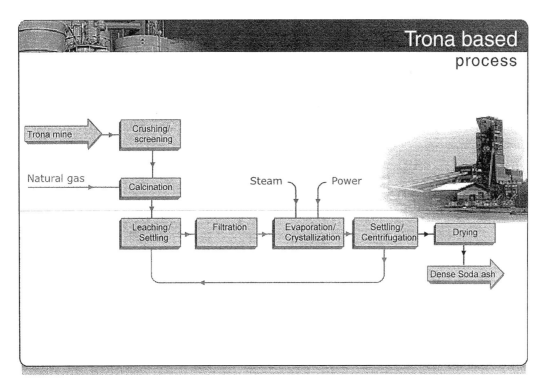

FIGURE 21.3 Flow diagram for trona-based process for production of soda ash. (Courtesy of Solvay Chemicals.)

calcined at temperatures in the range 110–175°C to produce light to intermediate grades of soda ash. Calcination of the sodium sesquicarbonate at 350°C can be used to produce dense soda ash [17].

21.2.1.2 Synthetic Soda Ash

Soda ash can be produced synthetically by one of several processes that use commonly available raw materials such as limestone, salt, and coke. Some of the these processes include the Leblanc process, Solvay process, electrolytic process, caustic carbonation process, Caprolactum pyrolysis, ammonium chloride process, new Asahi (NA) process, Akzo process, Ormiston Mining process, and Huls process. The Solvay process is the most commonly used method for the manufacture of synthetic soda ash. This method is discussed in the following text. For details of all the processes see Ref. 17.

Majority of soda ash is produced by synthetic processes with most using the Solvay process [18]. Also known as the ammonium-soda process, the Solvay process was developed by Alfred and Ernest Solvay in 1861 as an inexpensive process to commercially produce soda ash. Soda ash is produced in this process from salt, limestone, and coke, using ammonia as a catalyst. The successful commercialization of this process is attributed to the use of carbonating towers patented by Solvay.

The flow diagram for the Solvay process is shown in Figure 21.4. Limestone calcined with coke is used to produce carbon dioxide (Equation 21.1) and calcium oxide for the recovery of ammonia (Equations 21.1, 21.3, 21.9). The brine solution is then saturated with ammonia and carbon dioxide gas to produce ammonium bicarbonate, which then reacts with the salt to form sodium bicarbonate and ammonium chloride. The sodium bicarbonate, which precipitates, is filtered and calcined at 175–225°C to produce light soda ash, having bulk density in the range 0.51–0.62 g/mL. Dense soda ash, with bulk density of 0.76–1.06 g/mL can be produced by hydrating light soda ash.

The gases produced in the process are recycled into the liquid phase. Ammonia is recovered from the solution containing ammonium chloride by treating with milk of lime. By-products of the

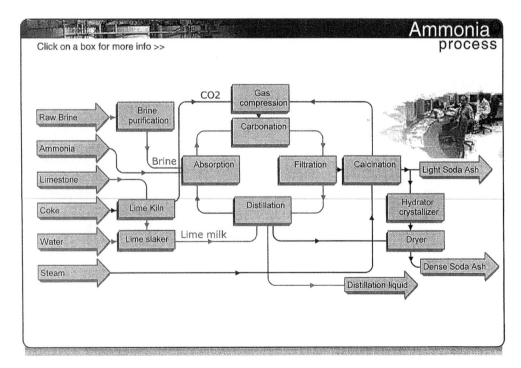

FIGURE 21.4 Flow diagram for Solvay Ammonia-based process for production of soda ash. (Courtesy of Solvay Chemicals.)

Production of Detergent Builders

process are sodium chloride and calcium chloride, the disposal of which has been associated with environmental problems. The Solvay process is energy, labor, and capital intensive. Owing to these issues, production of synthetic soda ash by this method has been decreasing worldwide [17].

The reactions involved in the Solvay process are as follows:

$$CaCO_3 \rightarrow CaO + CO_2 \tag{21.1}$$

$$C + O_2 \rightarrow CO_2 \tag{21.2}$$

$$CaO + H_2O \rightarrow Ca(OH)_2 \tag{21.3}$$

$$NH_3 + H_2O \rightarrow NH_4OH \tag{21.4}$$

$$2NH_4OH + CO_2 \rightarrow (NH_4)_2CO_3 + H_2O \tag{21.5}$$

$$(NH_4)_2CO_3 + CO_2 + H_2O \rightarrow 2NH_4HCO_3 \tag{21.6}$$

$$NH_4HCO_3 + NaCl \rightarrow NH_4Cl + NaHCO_3 \tag{21.7}$$

$$2NaHCO_3 \rightarrow Na_2CO_3 + CO_2 + H_2O \tag{21.8}$$

$$2NH_4Cl + Ca(OH)_2 \rightarrow 2NH_3 + CaCl_2 + H_2O \tag{21.9}$$

21.3 POLYCARBOXYLATES

Polycarboxylates, also known as polyacrylates, are polymers having a carbon–carbon backbone with attached carboxyl groups. In household consumer products, the anionic polycarboxylates most commonly used are copolymers of acrylic acid and maleic anhydride, copolymers of acrylic acid and methacrylic acid, or polymers of acrylic acid. Polycarboxylates find use as detergent builders and as partial replacements for polyphosphates. However, they are very expensive for use in high concentrations as builders in detergent formulations and also have poor biodegradability [16].

The polycarboxylate builders inhibit the growth of inorganic crystals, which would be formed in their absence during the washing process when phosphate-free or low-phosphate detergents are used. Another function of the polycarboxylates is their ability to disperse soil as well as prevent redeposition of the soil and minimize incrustations of insoluble salts on the fabric. Polycarboxylates are also used as dispersants and to provide shine in formulations for automatic dishwashing machines.

Polycarboxylates are produced primarily by emulsion polymerization. Water, monomers, emulsifers, initiators, and if needed, modifiers, are materials used in the process [19,20]. Initiators for the reaction are radical formers such as peroxides, other percompounds, or azo starters. Polymerization can also be initiated photochemically, by gamma rays or by electron beams [21].

21.3.1 EMULSION POLYMERIZATION

Polycarboxylates with high molecular weights are obtained by the process of emulsion polymerization, which also results in high polymerization rates. The heat of polymerization is easily removed through the aqueous phase since the viscosity is typically low. This production method typically results in polymers with up to 60 wt% of solids and molecular mass in the range 10^5–10^6. These are

382
Handbook of Detergents/Part F: Production

advantageous over the solution polymerization method. Also, this method does not use flammable solvents, therefore, there is no safety hazard [19–21].

21.3.1.1 Raw Materials

The raw materials required for the production of polycarboxylates using the emulsion polymerization process are water, monomers, emulsifiers, initiators, and if required, modifiers.

Water. Distilled or completely dimineralized water is preferred since electrolytes can affect emulsion stability.

Monomers. The properties required of the polycarboxylates determine the monomer composition used. Examples of monomers used are the acrylates of the general formula $CH_2=CH–COOR$, where R is an allyl group.

Emulsifiers. These serve several purposes in emulsion polymerization. Emulsifiers are needed for the preparation of the monomer emulsion and are also used later for stabilizing the final dispersion. The choice of the emulsifier, amount added, and the method of addition determines the particle size of the dispersion, which is typically in the range 50–1000 mm.

For the emulsification process, anionic and nonionic emulsifiers are commonly used, whereas cationic or amphoteric compounds are used less often. Ionic and nonionic emulsifiers are frequently combined to improve dispersion stability.

Anionic emulsifiers. Typically used anionic emulsifiers are sodium, potassium, or ammonium salts of fatty acids and sulfonic acids; alkali salts of alkyl sulfates, ethoxylated and sulfated or sulfonated fatty alcohols; alkyl phenols; and sulfodicarboxylate esters. These emulsifiers are typically used at levels of 0.2–5 wt% based on monomers.

Nonionic emulsifiers. Commonly used nonionic emulsifiers are ethoxylated fatty alcohols and alkyl phenols.

Cationic emulsifiers. Ammonium, phosphonium, and sulfonium compounds with at least one long hydrocarbon chain can be used.

Colloids. Starch, cellulose derivatives, or polyvinyl alcohol can also be used for the stabilization of dispersions.

Initiators. Initiators are typically water-soluble peroxo compounds such as alkali persulfates, ammonium persulfate, or hydrogen peroxide. Polymerization temperatures for these initiators is usually between 50 and 85°C. Amounts of initiator used are in the range 0.2–0.5 wt% based on the monomer. The initiator concentration can improve the stability of the dispersion by providing additional stabilization, or it can adversely affect the dispersion stability due to the electrolyte content. These can also lower the molecular mass of the product.

Polymerization can occur below 50°C if redox systems are used for initiation [22,23]. Hydrogen peroxide and alkali persulfate can be used as oxidizing agents. Reducing agents include iron(II) sulfate, sodium thiosulfate, or sodium bisulfite. Catalysts such as iron(II) salts can also be used in small amounts. The ratio of oxidizing agent to reducing agent has to be optimized experimentally.

In cationic dispersions, alkali persulfates cannot be used since salts can form. Instead, water-soluble organic peroxides or hydrogen peroxide is typically used with polymerization pH in the range 2–7.

Modifiers. Modifiers can be added to reduce the molecular mass of the polymer. Halogen-containing compounds such as carbon tetrachloride, carbon tetrabromide, bromoform, or thiol such as butyl or dodecyl thiol are commonly used modifiers [21].

21.3.1.2 Production Processes

Polycarboxylates are produced in batch and semibatch processes as shown in Figure 21.5. All production processes are performed under nitrogen since polymerization is inhibited by oxygen.

Production of Detergent Builders

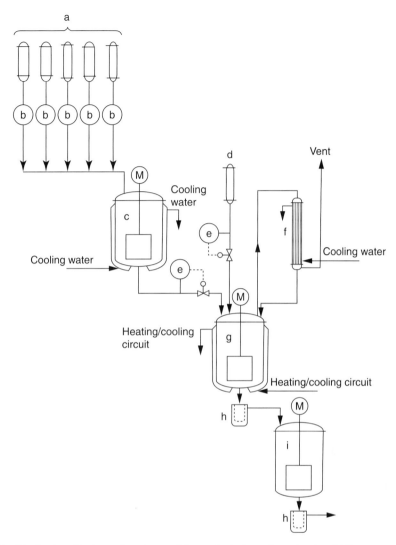

FIGURE 21.5 Production of polyacrylates by emulsion and solution polymerization. (a) Storage tanks for monomers, distilled water, emulsifiers; (b) flow meter; (c) mixing tank; (d) storage tank for initiator; (e) flow registration control; (f) cooler; (g) polymerization reactor; (h) filter; and (i) product storage tank. (Reprinted from Penzel, E., *Ullman's Encyclopedia of Industrial Chemistry*, p. 1, Wiley-Vch, Weinheim, Germany, 2003. With permission.)

21.3.1.2.1 Batch Process

The batch process involves the addition of the monomer emulsion to the reaction vessel, which is cooled below 15°C. This is followed by the addition of the redox initiator, usually as an aqueous solution. The metering of the initiator can sometimes control the polymerization rate. The polymerization reaction results in an initial temperature rise during the first half an hour to one hour and then falls because of the cooling and removal of the heat of reaction.

An advantage of the batch process is the larger space–time yield. Some of the disadvantages are inefficient utilization of the cooling capacity and the inability to reproduce the procedure and product properties.

The batch can be divided into two or more partial batches when cooling is insufficient so as to prevent the reaction from going out of control. A process in which monomers and initiators are added in an infinite number of small steps is known as continual feed (semibatch) process [24].

384   Handbook of Detergents/Part F: Production

This process allows easier and more accurate temperature control of the reaction. In actual practice, mixed batch and semibatch processes are used.

21.3.1.2.2   Semibatch Processes
21.3.1.2.2.1 Semibatch Process with Emulsion Feed
In this process, the reaction vessel contains some water initially which is heated to the desired reaction temperature. About 5–10% of the monomer emulsion and initiator solution is then added as the initial batch to the reactor.

In another procedure, some of the emulsifier and water are added, and the remaining emulsifiers are used to prepare the monomer emulsion. The batch is polymerized for 15–30 min and the number of particles and particle size are determined in this batch.

Polymer properties can be changed by varying the proportion of the emulsifier in the initial batch and feed. A dispersion consisting of fine particles can be used as "seed" instead of an initial batch. Then the monomer emulsion and initiator solution are added from separate tanks for 2–3 h [25].

21.3.1.2.2.2   Semibatch Process with Monomer Feed
In this process, water and emulsifiers are first added to the reactor. Then a continuous stream containing the monomer mixture, initiator solution, and other auxiliary material is fed into the reaction vessel. Specific properties of the product can be adjusted by small changes in the solution such as, temperature, pH, and delay in addition of certain monomers [26]. Another advantage of this system is that the properties of the resulting dispersions remain constant from batch to batch when polymerization is performed under specified conditions. However, this process is more complex in terms of temperature control and feed regulation.

21.3.1.2.3   Equipment
Raw materials are contained in storage tanks and fed through metering devices into the mixing tank, which contains a stirrer and has a cooling jacket. All vessels, stirrers, and all parts that can come into contact with the product are typically made of stainless steel. The inside wall of the reactor is required to be as smooth as possible to prevent deposition of scale. The reactor may have a capacity of 30 m^3 or higher. The stirrer design depends on the batch size and dispersion viscosity. The dispersion is discharged through a filter and pumped to a storage tank where the solids content and pH is adjusted. Stabilizers and microbiocides are also added. The final product is transported to the customer in polyethylene-lined metal drums, polyethylene vessels, or tank cars.

ACKNOWLEDGMENTS

The author wishes to acknowledge Dr. Noel Boulos and Dr. Kirk Raney for their assistance in preparation and review of this manuscript.

REFERENCES

1. Bartels, J.E., T.M. Gurr, *Phosphate rock in Industrial Minerals and Rocks*, 6th ed., D.D. Carr (ed.), 1944, pp. 751–764.
2. Shrodter, K., Bettermann, G., Staffel, T., Klein, T., Hofmann, T., *Ullman's Encyclopedia of Industrial Chemistry*, Sixth Completely Revised Edition, Vol. 26, p. 127, Wiley-Vch, Weinheim, Germany, 2003.
3. Hoechst, US 4 534 946, 1985.
4. Knapsack-Griesheim, DE 1018394, 1954.
5. FMC, US 3 661 514, 1970.
6. Fuchs, R.J., *Proc. Int. Congr. Phosphorus Comp.* 1977, 201.
7. Knapsack-Griesheim, DE 1 002 742, 1955.
8. EI. Reduction Comp., US 3 356 447, 1963.
9. FMC, US 3 054 656, 1958.

Production of Detergent Builders

10. Knapsack, FR 1 484 560, 1965.
11. FMC, US 4 656 019, 1987.
12. Monsanto, US 3 233 967, 1962.
13. Chem. Fabr. Budenheim, DE 965 126, 1950.
14. Comp. de Saint-Gobain, US 3 030 180, 1958.
15. Stauffer, US 3 210 154, 1962.
16. Rieck, H.-P., in *Powdered Detergents*, M.S. Showell (ed.), Marcel Dekker, New York, 1998, pp. 43–108.
17. Kostick, D.S., *Soda Ash in Industrial Minerals and Rocks*, 6th ed., D.D. Carr (ed.), 1994, pp. 929–958.
18. Thieme, C., *Ullmann's Encylcopedia of Chemical Industry*, Sixth Completely Revised Edition, Vol. 33, p. 187, Wiley-Vch, Weinheim, Germany, 2003.
19. Markert, G., *Angew. Makromol. Chem.* **123/124**, 1984, 285.
20. Alexander, A.E., Napper, D.H., *Prog. Polym. Sci.* **3**, 1971, 145.
21. Penzel, E., *Ullman's Encyclopedia of Industrial Chemistry*, Sixth Completely Revised Edition, Vol. 28, p. 1, Wiley-Vch, Weinheim, Germany, 2003.
22. Kern, W., *Makromol. Chem.* **1**, 1947/48, 199.
23. Fryling, C.F., Follett, A.E., *Polym. Sci.* **6**, 1951, 59.
24. Min, K.W., Ray, W.H., *J. Macromol. Sci.Rev. Macromol.Chem.* **C11**, 1974, 1977.
25. Snuparek J., Jr., *Acta Polym.* **32**, 1981, 368.
26. Suetterlin, N., *Macromol Chem.Phys.Suppl.* **10/11**, 1985, 403.

22 Production of Silicates and Zeolites for Detergent Industry

Harald P. Bauer

CONTENTS

22.1 Sodium Silicates...388
    22.1.1 Introduction 388
    22.1.2 Anhydrous Sodium Silicate Lumps..388
        22.1.2.1 Introduction..388
        22.1.2.2 Principles of the Soda Ash Fusion Process 389
        22.1.2.3 Raw Materials..390
        22.1.2.4 Open-Hearth Regenerative Furnace Process............................391
        22.1.2.5 Rotary Furnace Process..393
    22.1.3 Sodium Silicate Solutions..393
        22.1.3.1 Introduction..393
        22.1.3.2 Hydrothermal Dissolution of Anhydrous Sodium Silicate Lumps.......394
        22.1.3.3 Hydrothermal Production Process from Quartz Sand
                and Caustic Soda..395
    22.1.4 Hydrous Sodium Polysilicates..396
        22.1.4.1 Introduction..396
        22.1.4.2 Manufacturing of Hydrous Sodium Polysilicate Powders...................397
        22.1.4.3 Manufacturing of Hydrous Sodium Polysilicate Granules...................397
    22.1.5 Cogranules of Hydrous Sodium Polysilicate and Sodium Carbonate...................398
        22.1.5.1 Introduction..398
        22.1.5.2 Manufacturing of Sodium Polysilicate-Sodium
                Carbonate-Cogranules..399
    22.1.6 Anhydrous Sodium Metasilicate 399
        22.1.6.1 Introduction..399
        22.1.6.2 Production of Anhydrous Sodium Metasilicate by the Direct
                Fusion Route 400
        22.1.6.3 Production of Anhydrous Sodium Metasilicate
                by Solid-State Reaction..400
        22.1.6.4 Granulation of Sodium Metasilicate Solution 400
    22.1.7 Sodium Metasilicate Pentahydrate 401
        22.1.7.1 Introduction..401
        22.1.7.2 Crystallization from Sodium Metasilicate Solution.................401
    22.1.8 Crystalline Layered Sodium Disilicate 402
        22.1.8.1 Introduction..402

22.1.8.2   Manufacturing Process of Crystalline Layered Sodium Disilicates 403
22.1.9   Requirements of Resources 404
22.2   Zeolites..404
22.2.1   Zeolite NaA 404
22.2.1.1   Introduction...404
22.2.1.2   Raw Materials..405
22.2.1.3   Principles of the Zeolite a Production Process.....................406
22.2.1.4   Continuous Process Operation versus Batch-Type Process.................409
22.2.1.5   Clay Conversion Process 409
22.2.1.6   Zeolite NaA Production by Hydrogel Process.......................409
22.2.1.7   Requirements of Resources... 411
22.2.2   Zeolite P 411
22.2.2.1   Introduction 411
22.2.2.2   Production of Zeolite P by Hydrogel Process 411
22.2.3   Zeolite AX 412
22.2.3.1   Introduction... 412
22.2.3.2   Production of Zeolite AX by Hydrogel Process...................... 412
Acknowledgments.. 412
References 412

22.1   SODIUM SILICATES

22.1.1   INTRODUCTION

To date, silicates have been acknowledged ingredients in the detergent industry. This is because they can most prominently offer a source of alkalinity and pH buffering capacity [1,2]. Some of the types can further assist the detergent builders with the deflocculation of soil and the prevention of its redeposition. Other types are able to act directly as water softeners and meet the standards of detergent builders [3]. Another important property is their ability to encounter the corrosion of glass and a range of metals [1,4] in the washing and cleaning process. In the detergent manufacturing process, they are appreciated as valuable processing aids during agglomeration of detergent granules or the adjustment of the bulk density of detergent powders [5].

Commercially, sodium silicates are of the highest interest among alkali silicates. They are large volume chemicals available in various physical forms, grades, and compositions. Sodium silicates can be delivered not only as aqueous solutions but also as fluffy powders and in granular forms of high bulk density. The solid grades range from amorphous to crystalline forms and from water-free to hydrous types. Crystalline materials disclose their identity in the form of characteristic patterns in x-ray diffraction experiments, whereas amorphous silicates do not.

The common way to characterize the composition of sodium silicates is to use the ratio of silica to alkali by weight. The first refers to the SiO_2 content of the material, and the latter to its Na_2O content. This characteristic value is often called modulus. The ratio between silicon dioxide and sodium oxide can also be expressed on a molar basis. The conversion of modulus and molar ratio is possible by the respective formula weights. In case of sodium silicates, the modulus has to be multiplied by a conversion factor of 1.032 to obtain the molar ratio [1,5,6–13].

22.1.2   ANHYDROUS SODIUM SILICATE LUMPS

22.1.2.1   Introduction

Anhydrous glass lumps or cullets are of industrial importance because they are the first commercially available downstream silicate products. Although they lack typical builder properties, they are important intermediate raw materials for all manufacturers of sodium silicate solutions and solid hydrous sodium silicates.

They may be produced with weight ratios of SiO_2/Na_2O from 0.97 up to 4.1 [1,14], but the two ratios that are commercially most available are 2.00 and 3.30. Glasses with weight ratios of 2.0–2.1 are often known as alkaline glasses and those with weight ratios of 3.3–3.4 as neutral or siliceous glasses because of their higher content of SiO_2 [1,11,13,15]. Although they are in pure form colorless to white crystal-like pieces or lumps, the commercial alkaline glasses display pale yellowish to brown colors due to divalent iron traces. Neutral glasses are light bluish-green colored by small amounts of trivalent iron [5,13,16]. Anhydrous glasses are x-ray amorphous and do not exhibit a clear melting point. They begin to soften at temperatures above 550–670°C [5]. Softening temperatures depend on the composition, that is, the molar ratio and can be extracted from phase diagrams [1,5,7,12,17]. Alkaline glasses are slightly soluble in boiling water but neutral glass lumps are almost insoluble in both cold and boiling water [13,16]. The more siliceous sodium silicates are glasses that resemble typical noncrystalline solid solutions [6].

22.1.2.2 Principles of the Soda Ash Fusion Process

A variety of sodium silicate compounds is produced by the soda ash fusion process, which is also known as furnace process or kiln process (see Figure 22.1). The reaction of quartz sand and sodium carbonate (soda ash) is described by Equation 22.1:

$$Na_2CO_3 + n\,SiO_2 \xrightarrow{heat} Na_2O \cdot n\,SiO_2 + CO_2 \qquad (22.1)$$

The composition of such products may range from $Na_2O \cdot SiO_2$ to $Na_2O \cdot 4.2SiO_2$, and the weight ratio ranges most frequently from 2.0 to 3.5. High modulus sodium silicates can be produced most easily by the furnace process.

The temperature of the melt has to be high enough to provide first, a reasonable quartz dissolution rate in the molten bath and second, a manageable low melt viscosity. The latter also depends on the chosen silica to alkali ratio. Fusion temperatures of 1300–1400°C for alkaline glass and 1400–1500°C for neutral glass are common (see Table 22.1). Carbon dioxide is driven off by the reaction of alkali carbonate with silica. The main part of carbon dioxide is liberated at 700°C [1,5–7,11,12,14,15,18–21].

In an older fusion process, sodium sulfate and fine coal were used as raw materials instead of soda ash (see Table 22.1). It was common in Europe in the past, but has now been rendered obsolete [1,5,12,14,15]. Two key factors for this may be the liberation of large amounts of sulfur dioxide and products of a lower purity than those from the soda ash fusion process.

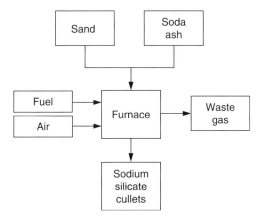

FIGURE 22.1 Process diagram of solid sodium silicate production by the soda ash fusion process.

TABLE 22.1
Detergent Sodium Silicates: An Overview About Products and Production Processes

Product	Raw Materials	Production Process	Comments
Anhydrous sodium silicate lumps	Quartz sand, soda ash; quartz sand, sodium sulfate, coal	Soda ash fusion process	Product: solid, amorphous, water insoluble, $SiO_2:Na_2O$ = ca. 2.0 (a) 3.3 (b) process temperature: (a) 1300–1400°C, (b) 1400–1500°C
Anhydrous sodium silicate lumps	Quartz sand, soda ash	Rotary furnace process	—
Sodium silicate solutions	Anhydrous sodium silicate lumps	Hydrothermal dissolution of anhydrous sodium silicate lumps	Product: liquid, amorphous, SiO_2:Na_2O = ca. 3.3; process temperature: 120–150°C
Sodium silicate solutions	Quartz sand, caustic soda	Hydrothermal dissolution of quartz sand	Product: liquid, amorphous, SiO_2:Na_2O = ca. 2.0–2.6; process temperature: ca. 200°C
Hydrous sodium polysilicate powders	Sodium silicate solution	Spray- and drum drying	Product: solid, amorphous, water soluble, $SiO_2:Na_2O$ = ca. 2.0–2.6; process temperature: ca. 200°C
Hydrous sodium polysilicate granules	Hydrous sodium polysilicate powder	Roll compaction	Product: solid, amorphous, water soluble, $SiO_2:Na_2O$ = ca. 2.0–2.6
Hydrous sodium polysilicate granules	Sodium silicate solutions	Turbodryer process; agglomeration	—
Cogranules of hydrous sodium polysilicate and sodium carbonate	Sodium silicate solution, sodium carbonate	Agglomeration	Product: solid, amorphous, water soluble, 55 wt% soda ash, 29 wt% sodium silicate (amorphous), 16 wt% water; process temperature: ca. 70–95°C
Anhydrous sodium metasilicate	Quartz sand, soda ash	Direct fusion process	Product: solid, crystalline, water soluble, $SiO_2:Na_2O$ = 1.0; process temperature: 900–1088°C
Anhydrous sodium metasilicate	Quartz sand, soda ash	Solid-state reaction	Product: solid, crystalline, water soluble, $SiO_2:Na_2O$ = 1.0; process temperature: ca. 850°C
Anhydrous sodium metasilicate	Sodium silicate solution	Agglomeration	Product: solid, crystalline, water soluble, $SiO_2:Na_2O$ = 1.0
Sodium metasilicate pentahydrate	Sodium silicate solution	Crystallization, agglomeration	Product: solid, crystalline, water soluble, $SiO_2:Na_2O$ = 1.0; process temperature: ca. 54–60°C
Crystalline layered sodium disilicate powder; sodium phyllosilicate	Quartz sand, caustic soda	Hydrothermal dissolution of quartz sand, spray drying, crystallization	Product: solid, crystalline, water soluble, $SiO_2:Na_2O$ = 2.0; process temperature: ca. 600–800°C
Crystalline layered sodium disilicate granules	Crystalline layered sodium disilicate powder	Roll compaction	Product: solid, crystalline, water soluble, $SiO_2:Na_2O$ = 2.0

22.1.2.3 Raw Materials

22.1.2.3.1 Sodium Carbonate

Soda ash is one of the two basic raw materials that are necessary for the production of sodium silicate via the furnace process. Over half of the world's soda ash output is consumed by the production of glass [22,23].

Production of Silicates and Zeolites 391

Sodium carbonate manufacturing in the United States starts nearly entirely from naturally occurring trona mineral deposits. Trona ($Na_2CO_3 \cdot NaHCO_3 \cdot 2H_2O$), for example, Wyoming trona ore, is extracted and refined mainly by the monohydrate process.

In Europe and in many other parts of the world, the ammonia-soda process (Solvay process; see Chapter 21) is used to produce synthetic soda ash according to Equation 22.2:

$$2NaCl + CaCO_3 \rightarrow Na_2CO_3 + CaCl_2 \qquad (22.2)$$

The production cycle starts with the extraction of sodium chloride. About 20% of the world's salt consumption goes into soda ash production [24]. The next step after rock salt mining is the production and purification of brine yielding a concentrated aqueous sodium chloride solution [8,25–27]. A parallel step is the production of carbon dioxide gas by calcination of limestone. The brine is treated with ammonia and carbon dioxide under precipitation of the less-soluble sodium hydrogencarbonate. Ammonia is recovered by mixing the mother liquor with calcium hydroxide and stripping off the ammonia with steam. Thermal decomposition of sodium hydrogencarbonate yields synthetic soda ash [8,20,22,23,28–38]. The output of soda ash produced by the ammonia-soda process amounts to about two-thirds of the world production [22,23].

22.1.2.3.2 Quartz Sand

The common source of silica is quartz sand or quartz flour that meets glass melting quality standards. The suitable particle size for glass sand is 60–2000 μm [39], preferably 100–500 μm [13,39,40]. Quartz powder, also known as quartz or silica flour has a grain size of 2–60 μm and is produced by dry grinding [39].

The purity of quartz sand required for the manufacture of soluble silicate has to be high, for example, the iron content should not exceed 300 ppm [7,11,41]. The primary contaminant is Fe_2O_3, followed by titanium oxide (rutile), zirconium oxide, and chromium oxide. Typical analysis data of high-quality grades are SiO_2 ca. 99.0–99.8%, Fe_2O_3 ca. 0.01–0.05%, and Al_2O_3 ca. 0.05–0.5% [40].

Sand extraction usually includes a purification-washing step to remove loosely adhering impurities such as clays and other impurities. To achieve quality standards, it may be necessary to use multistep processing including friction washing, flotation, chemical, and physical purification [11,40].

In Europe, a big deposit of almost pure sand that requires no processing is situated in the Maastricht-Aachen area of Holland and Germany. In the United Kingdom, the purer sands are found at Kings Lynn, Loch Aline, Oakamoor, and Redhill [1].

22.1.2.4 Open-Hearth Regenerative Furnace Process

For large-scale production of sodium silicate cullets, the mixtures of sand and soda ash are fused in open-hearth regenerative furnaces having a design that is conventional for the glass manufacturing industry.

This kind of furnaces goes back to Siemens Martin regenerative furnaces invented and designed for steelmaking [1,6,7,13,14,18,41,42].

Typical regenerative furnaces have tank-melting areas of 40–60 m^2. Typical dimensions include a length of 15 m, width of 5 m, and height of 3 m. Capacities are on the order of 60–150 t, which are of similar size to those used in the glass industry [1,10,17].

The output of a tank furnace with 50 t fused melt volume is about 60 t of silicate per day. This figure reflects optimization from 3.0 t/m^2 of hearth area per day in the mid-1970s to approximately 4.5 t/m^2 of hearth area per day in the mid-1980s [1,5,14].

Smaller furnaces may be operated batchwise, but continuous melting processes are used most frequently [41].

Sand and soda are drawn from hoppers, dosed via belt weighers and homogenized in screw conveyer mixers (see Figure 22.2). A normal shrinkage in the weight of the charge (ca. 10%) is due

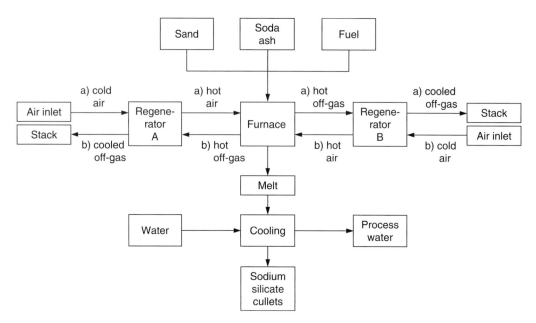

FIGURE 22.2 Regenerative furnace process for the production of solid sodium silicate.

to the loss of gases and volatilization of alkali oxides. It has to be compensated by the initial weight of the feed. The raw material mixture is adjusted in humidity by the addition of water to get a water content of ca. 5%. The solid is then fed continuously to the furnaces by means of cooled screw conveyors and is distributed to the whole surface of the bath to yield an optimum energy transfer [10,11,14].

The process is characterized by the energy saving principle of preheating the incoming air by hot combustion gases. Regenerative heat exchangers as well as recuperative heat recycling systems may be used.

Siemens-type continuous tanks are heated by combinations of heavy fuel oil and air or gas and air. Cold air is blown into a regenerative chamber set A by fans, forced air inlet valves, and dampers (see Figure 22.2, setting a). Thus, it is heated to ca. 1200°C. Then it is mixed with liquid fuel or gas and burned in the respective burners [14,18,43]. The water-cooled burners form flames, which burn across the bath. Temperatures at the roof above the center of the flame can reach up to 1700°C.

Burners and regenerator chambers operate in pairs. Cross-firing ports in the sidewalls of the furnace are connected to the regenerator chambers. The hot combustion gases are withdrawn on the side opposite to the burners into checker chamber set B and the bricks are heated there (see Figure 22.2, setting a). After leaving them, the waste gases are directed to flues that lead to consecutive waste heat boilers and eventually to the stacks.

When regenerative chambers B achieve a defined temperature, the burners are switched off and the airflow direction is reversed (see Figure 22.2, setting b). Fresh cold air is conducted to the hot regenerative chambers B. It is heated there, mixed with fuel, and fed to the burners. Hot waste gas is directed to regenerative chambers A to be cooled and eventually passed to the stacks.

The direction of gas streams is changed every 15–30 min.

In a recuperator made from metal, the hot flue gases are separated from one another and led by a countercurrent to cold inlet air [43].

The melted material flows gradually through the furnace and evolves carbon dioxide. The fused water glass is continuously drawn from the furnace by a siphon overflow at a temperature of 1050–1100°C [6,13,44].

Production of Silicates and Zeolites

The melt flows onto a moving conveyer into an endless chain of steel molds forming cast pieces of ca. 8 cm in diameter and having a thickness of 1–2 cm. The material solidifies to a semitransparent solid. It cools during transportation and often shatters into fragments (lumps, cullets) when it is sprayed with a stream of cold water. Chilled to ca. 450°C, the cullets may be transported to a hopper for further storage, distribution, or grinding steps or directly conducted to the water glass dissolution process [5,6,11,13,14].

Owing to their high alkali content, water glass melts are much more corrosive to furnace linings than the common bottle (soda lime) glasses. Therefore, selection and disposition of refractories are essential. Furnaces are lined with high-quality refractory bricks (e.g., made from silimanite or blocks of alumina–zircona–silica [AZS]). Silimanite is an aluminium silicate, which is normally precalcinated to ca. 1550°C to convert the silimanite to sinter-mullite. The latter has high thermal shock resistance and a very low coefficient of thermal expansion. The softening temperature under load is in the order of 1550–1650°C. Highly refractory fusion cast may also be made from AZS [1,10,11,13,17,41].

Improvements in the water glass production process have been driven by the development of better refractory materials, thus realizing enhanced furnace lives. Today, 3 years between rebuilds would be regarded as normal, with an intermediate repair after operation of 18 months. Along with this, the use of liquid fuel having a higher specific caloric value allows higher temperatures, giving higher melt performance. Further, objects of optimization are to decrease energy consumption by increasing the regenerative systems for a better energy recovery from the combustion gases, the optimization of the furnace design in positioning and number of burners at the hearth, and the production of steam from the waste gases [1,45].

22.1.2.5 Rotary Furnace Process

Although they are well known for the potassium silicate production, rotary furnaces can be used for the production of alkali silicate glasses in general (see Table 22.1). Typically, a continuously operated rotary furnace is installed at a slight slope of 2–7° to the horizontal. It rotates at 0.5–1.0 rpm. The mixture of raw materials is charged at the top. The heating may be by gas or liquid fuel from the lower end of the furnace. The silicate glass melt has a more or less high viscosity that depends on the silica to alkali ratio. It flows in a countercurrent toward the burning flame. At the lower end of the rotary kiln, the product is withdrawn from the outlet and may be cooled and processed further to downstream products such as sodium silicate solutions [1,10,18,44,46].

22.1.3 SODIUM SILICATE SOLUTIONS

22.1.3.1 Introduction

Sodium silicate solutions are produced to satisfy the demand for silicates, which are readily soluble in water at ambient temperatures. The main benefits are their alkalinity but they also function as a processing aid in detergent manufacturing processes. In spray drying processes, they can serve as a binder and stabilizer of the detergent beads. In agglomeration processes, they function as a granulation agent. Further properties of silicates are their bleach stabilizing effect by masking heavy metal ion traces and their anticorrosion activity by virtue of the formation of protective silicate coatings on metal surfaces.

The most common silicate solution grades correspond to silica to alkali ratios of 2.0 and 3.3. But the possibility to blend standard grades makes a whole spectrum of compositions commercially available with ratios ranging from 1.5 to 3.85.

The clear to cloudy, colorless aqueous solutions vary in specific gravity and viscosity. Viscosity of silicate solutions increases with the increasing solids content as well as with increasing ratio. In contrast, for any particular concentration and ratio there is a minimum in viscosity. Increases in viscosity can be rather abrupt. This is especially valid in the case of the more siliceous solutions.

The implication is that the maximum reasonable viscosity that can be handled readily limits the solids levels of the commercial solutions. Thus, an alkaline glass of a 2.00 ratio by weight can be handled in solutions that contain up to 54% solids, whereas the upper limit of a neutral water glass with ratio 3.30 is 39% active matter [1,5–7,11,12,16,47,48].

22.1.3.2 Hydrothermal Dissolution of Anhydrous Sodium Silicate Lumps

The reactivity of anhydrous sodium silicates to water determines the parameters of the production processes that lead to sodium silicate solutions. Solubility depends on the silica to alkali ratio and the particle size of the silicate lumps as well as on the conditions that are applied, such as the concentration of the system and its temperature and pressure.

Solid sodium silicates of ratios below 2 are hygroscopic and tend to cake in moist air upon storage. They dissolve fairly well in water. Lumps having a ratio of over 2.0, and in particular above 2.5–3.0, do not dissolve in contact with boiling water. These siliceous silicates have to be dissolved with water under pressure.

Surprisingly, silicate cullets are less readily soluble in large amounts of water than in small amounts of water. Therefore, it is common to heat an excess of solid glass with water and maintain a high glass-to-water ratio in the dissolver to achieve a satisfactory dissolution rate and high output. There is no saturation point in the accepted sense and the concentration attained is determined, for example, by the intended viscosity and the given reaction time [1,5,9,11–13,16].

The first step in the dissolution of anhydrous silicate is an ion exchange between the alkali ions in the glass and the hydrogen ions of the surrounding water. The glass surface gets covered by a protective layer of silanol groups, while the pH of aqueous solution rises. In the second step, the alkalinity drives the depolymerization of the silicate particles [5].

It is customary to dissolve lump glass using either rotary or stationary pressure dissolvers (see Table 22.1 and Figure 22.3). A stationary dissolver may be designed in the form of a large vessel of 15–25 m^3 volume, which is heated directly with superheated steam to temperatures of 150°C and 5 bar pressure.

During the process, the vessel is charged with cullets and sealed. Water is added and steam is injected until the desired working pressure is obtained. Concentrations and ratios of solutions are monitored with the use of charts, which calculate the ratio and solids content from the density and

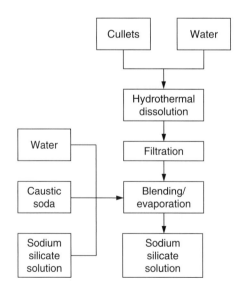

FIGURE 22.3 Hydrothermal dissolution of sodium silicate cullets.

Production of Silicates and Zeolites

sodium oxide content. The reaction is completed when the silicate solution reaches the desired solids level. The liquor is then blown over to a collecting tank.

The lumps may also be passed into a rotary dissolver of a volume of about 15 m³. The dissolution process then occurs at 120°C upon injection of steam of 4–7 bar pressure.

It is possible to utilize the remaining heat of the 400°C hot lumps that come directly from the fusion process. Therefore, they may be added to additional water in rotary or stationary pressure dissolvers and heated with steam of 4 bar pressure to 150°C. The added water may be recycled from the cullet-cooling step, and the superheated steam may be produced by utilization of hot exhaust gases from the melt furnace [1,6,7,9,11,14,18,44].

Crude water glass solutions from the dissolving process may be slightly turbid due to solid impurities of very fine unreacted sand particles and amorphous particles of aluminum, magnesium, and calcium silicates. The liquor is clarified at elevated temperatures in settling tanks. Higher-quality standards can be met by the addition of a filter aid such as diatomaceous earth and filtering through filter presses (see Figure 22.3). The products are brilliantly clear solutions [1,5–7,9,11,13].

Solutions of intermediate ratios may be adjusted by blending standard liquors of ratios 2 and 3, respectively (see Figure 22.3). Caustic soda can be added to produce more alkaline sodium silicate solutions and silica to obtain more siliceous types [1,5,6,13,18,21]. The concentration in the filtered silicate solution may be increased to the desired density by passing it through evaporators under reduced pressure. Water may be added to achieve products of a lower solids level [1,5,6,11].

22.1.3.3 Hydrothermal Production Process from Quartz Sand and Caustic Soda

Silicate solutions having a molar ratio of up to ca. 2.65 can be manufactured from quartz sand and concentrated caustic soda solution by a hydrothermal reaction according to Equation 22.3:

$$2NaOH + nSiO_2 \xrightarrow{heat} Na_2O \cdot nSiO_2 + H_2O \qquad (22.3)$$

This direct dissolution process, also known as pressure process, runs under forced conditions, that is, elevated temperature and pressure (see Table 22.1 and Figure 22.4). The shortcomings of this

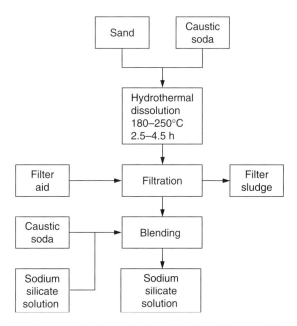

FIGURE 22.4 Hydrothermal process for the production of sodium silicate solutions.

process are a product spectrum that is limited to the alkaline end of silica to sodium oxide ratios and the higher raw material costs of sodium hydroxide versus sodium carbonate. These disadvantages are usually offset by the lower energy costs of the autoclave process. It is beneficial to include it into a process chain, for example, for the production of hydrous sodium silicates, layered sodium disilicates, or zeolites. The molar ratio of the obtained hydrothermal liquor depends on the applied temperature, the corresponding pressure, and the excess of silica. The ratio may be fine-tuned toward higher values by blending with more siliceous water glass solutions and to lower values by the addition of caustic soda solution. Solid levels of 48 wt% and a weight ratio of 2 are typical product properties [7,8,11,12,15,17–19].

One of the starting materials is quartz sand of low particle size or silica flour of similar purity as the types used for the fusion process. Typical quality requirements can be 99 wt% SiO_2 and particle sizes from 100 to 700 μm, for example, 90% particles below 500 μm. A low content of organic carbonaceous matter is of particular importance because these components are not decomposed to the same extent as in the fusion process. Cristobalite, although of minor importance, may be an alternative to attain higher molar ratio products to run the process at lower temperatures or realize reduced reaction times [7,12,15,21,49–51].

The other major raw material is caustic soda solution typically produced from sodium chloride by chlor-alkali electrolysis. A type that is frequently used for this purpose is of 50 wt% concentration [8,24–27,52–61].

The raw materials are charged to an agitated reaction vessel, for example, a nickel-clad cylindrical autoclave, equipped with a stirring device and heated by an exterior steam jacket or by direct injection of superheated steam (e.g., 16 bar, see Figure 22.4). The reaction is usually performed with an excess of sand. Thus, a molar ratio of SiO_2 to Na_2O of 2.15:1 may be necessary to obtain a water glass solution having a molar ratio of SiO_2 to Na_2O of 2.04:1. After the autoclave is sealed, the reaction mixture is heated to the scheduled temperature and maintained at this temperature for the reaction cycle. Reaction temperatures of 180–250°C (typically 200°C) at the related water pressure and reaction periods of the order of about 2.5–4.5 h may be required. At the end of the reaction, the steam is shut off, the reaction vessel is vented, and the reaction mixture is passed into a collecting tank [11,15,17,19,21,44,45,50,51,62–64].

The suspension is filtered to remove all sludge resulting from the excess of sand and its impurities [8,50,62]. A filter aid of the perlite type may be added [51].

22.1.4 HYDROUS SODIUM POLYSILICATES

22.1.4.1 Introduction

Hydrated water-soluble silicates are also known as hydrous polysilicates or dried silicate solutions. The most common soluble silicates are sodium silicates [19].

They satisfy the demand for solid silicates, which are much more readily and significantly faster soluble in water at ambient temperature compared to the equivalent anhydrous powdered glasses. The dissolution velocity depends on the particle size, silica to sodium oxide ratio, and water content. They are capable of providing alkalinity to boost the action of the main detergent builders, decrease soil redeposition, and inhibit corrosion of metal surfaces. In addition to these abilities, they are beneficial aids in detergent processing, primarily by helping to shape the texture of the detergent powders as well as adjust their bulk density [1,7,18].

Hydrous sodium polysilicates are available as powders with bulk densities from 80 to 650 g/L and compacted granules with bulk densities in the range 750–950 g/L. Particle sizes are between 80 and 700 μm [2,11,65]. Made from sodium silicate solutions, they may contain 16–22 wt% moisture [1,5,7,11]. Water uptake is generally limited by their tendency toward caking at higher moisture levels. They are x-ray amorphous and have no defined stoichiometry [19,65]. The composition is described by the weight ratio of silica to sodium oxide, which varies from 1.5 to 4.0 from type to type. Frequently used types have ratios of 2.00–2.65 [1,2,65,66].

22.1.4.2 Manufacturing of Hydrous Sodium Polysilicate Powders

Hydrous sodium polysilicate powders are manufactured by removing a part of the water content of the corresponding silicate solutions through drum or spray drying (see Table 22.1) [1,7,8,17–19,21,49].

Bulk densities of spray-dried powders may depend on the chosen process conditions. Products from spray towers that are equipped with spray nozzles show bead forms with lower bulk density [5].

During the process, the aqueous silicate solution is introduced into the upper portion of the gas-fired spray dryer and passes through a spray nozzle or a disk atomizer (see Figure 22.5). The speed of the spray wheel may be about 11,000 rpm. The finely and evenly dispersed liquid comes into contact with upwardly directed hot air. Typical spray tower temperatures are about 180°C [21] with inlet temperatures of about 260–300°C and outlet air temperatures of above 100°C. The resultant spray-dried droplets adopt the form of hollow microspheres. The silicate particles are collected at the spray dryer's bottom and are withdrawn by a screw conveyor. The amorphous sodium silicate may have a bulk density on the order of 250–500 g/L, an $SiO_2:Na_2O$ molar ratio of 2.04:1, and an ignition loss on the order of 19–20%. Its mean particle size can be on the order of 100–200 μm. The material may be subjected to further milling to modify the form and density of the powder [51,63].

Silicate agglomerates of relatively high bulk density and large particle size can be prepared by recycling the fine particles that are normally produced in spray drying operations back into the spray dryer. As these particles fall through the dryer, they encounter droplets of sodium silicate solution that have been atomized into the dryer. The particles may adhere to one another and agglomerate as they dry [67].

22.1.4.3 Manufacturing of Hydrous Sodium Polysilicate Granules

22.1.4.3.1 Compaction of Hydrous Sodium Polysilicate Powders

Hydrous sodium silicate granules can be prepared from the spray-dried powders by compaction operations to enlarge bulk density to about 500–900 g/L (see Table 22.1). This property is essential for many applications, for example, automatic dishwash detergents [65,66].

Fluffy spray-dried material can be compacted into articles having a larger particle size and higher bulk density by the application of mechanical pressure or force. The compaction can be accomplished by known means such as tabletting presses and briquetting machines, but preferably by passing the powder between rotating rolls.

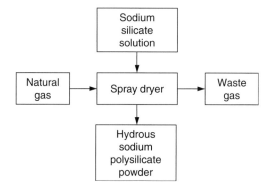

FIGURE 22.5 Spray drying process for the production of hydrous sodium polysilicate powders.

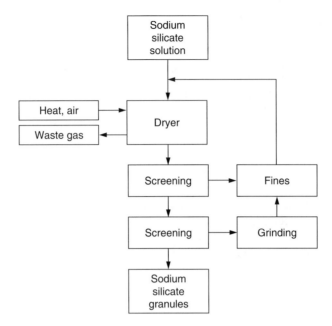

FIGURE 22.6 Turbodryer process for the production of sodium silicate granules.

Thus, spray-dried sodium silicate powder may be mixed with up to 10% of water glass solution before compaction. The sodium silicate powder may have a weight ratio of $SiO_2:Na_2O$ from about 1.6:1.0 to about 3.0:1.0 and a moisture content from about 17 to 20%. The water glass solution can have equal $SiO_2:Na_2O$ ratio and 30–50 wt% active matter.

The mixture is fed between compression rollers applying a pressure of 70–90 bar. Thereby, a compacted sheet or flakes of 0.8–12 mm thickness are formed. The flakes are crushed. Oversized granules that are larger than desired are recycled into the deagglomeration step. The fines reenter into the process at the roll-compactor feed. Bulk densities in the range of ~850 to 900 g/L can be realized [68–70].

22.1.4.3.2 The Turbodryer Process
The turbodryer process is a single-stage drying and granulation facility designed to avoid the two-stage process, which encompasses spray drying of sodium silicate solution and compacting the hydrous silicate powder (see Table 22.1 and Figure 22.6) [71,72]. Sodium silicate solution, for example, having a $SiO_2:Na_2O$ molar ratio of 2:1 and a solids content of 46 wt%, is sprayed into a horizontally mounted double-walled tubular drum. The latter contains a rotor, which is rotating at speeds from 200 to 1500 rpm. The drum is heated through the wall by a diathermic oil to a temperature in the range of about 150–200°C. Air is introduced into the lower region of the drum at a temperature on the order of 220°C. The blades of the rotor are arranged helically and oriented so as to centrifuge and convey the product being treated toward the outlet. The silicate product is drawn off of the drum. Fine and coarse material falling outside the desired particle size range can be recycled in the silicate solution without further processing such as milling.

22.1.5 Cogranules of Hydrous Sodium Polysilicate and Sodium Carbonate

22.1.5.1 Introduction

A particular answer to the demand for a low dust, low attrition, and low caking alternative to granular hydrous silicates having a favorable cost to performance ratio are the recently developed

Production of Silicates and Zeolites

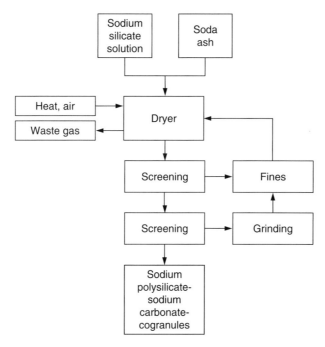

FIGURE 22.7 Production of cogranules of hydrous sodium polysilicate and sodium carbonate.

cogranules of hydrous sodium polysilicate and sodium carbonate. They are described to resemble pure silicates in respect to primary wash results and inhibition of fabric graying. Thus, they might be able to partly substitute zeolite A [66]. The composition of such cogranules is made up of 55% soda ash, 29% sodium disilicate, and 16% water, that is, the silicate to carbonate ratio is 1:2. The silicate constituents part has a $SiO_2:Na_2O$ molar ratio in the range 1.8–3.5 and shall reflect a specific distribution of silicate polyanion species. Density of the granular material is 900 g/L and the mean particle diameter is on the order of 700–800 μm [73,74].

22.1.5.2 Manufacturing of Sodium Polysilicate-Sodium Carbonate-Cogranules

The characteristic feature of the production process is that an aqueous solution of a mixture of sodium polysilicate and sodium carbonate is sprayed on a bed of rolling particles of identical composition as the liquor. The process is performed in a rotary device of granulation such as a granulation plate, turning drum, and preferably a mixer granulator (see Table 22.1 and Figure 22.7). The latter is able to apply strong shear forces to the mixture. The process temperature may be in the range 70–95°C. The combined drying–granulation process may be modified in that a pure sodium polysilicate solution is sprayed on the bed of rolling particles composed of fine sodium carbonate particles. The silicate is thus acting as a binder between the essentially anhydrous carbonate particles [73,74].

22.1.6 ANHYDROUS SODIUM METASILICATE

22.1.6.1 Introduction

In contrast to household detergents, industrial detergents contain much greater amounts of silicate. When dissolved in water, they give solutions of much higher pH than domestic formulations [1]. This is the domain of anhydrous sodium metasilicate. It is described by the chemical formula Na_2SiO_3, which corresponds to a one-to-one molar ratio of silicon oxide and sodium oxide ($Na_2O:SiO_2 = 1:1$). It has a crystalline structure and forms monoclinic, colorless crystals or a white powder of a specific gravity of 2.4 g/mL and a melting point of 1088–1089°C. It is soluble in cold and hot water and crystallizes from aqueous solutions as pentahydrate and nonahydrate, respectively. Commercially, it is shipped having a particle size of about 800 μm in diameter and a bulk density in the order of 1200 g/L [1,5,6,11,12,15,18,19,47,62,70,75].

22.1.6.2 Production of Anhydrous Sodium Metasilicate by the Direct Fusion Route

Anhydrous sodium metasilicate can be produced by the direct fusion of silica and alkali-bearing raw materials (see Table 22.1 and Figure 22.8). Quartz sand and soda ash are mixed in a one-to-one molar ratio of silicon oxide to sodium oxide. Both components are fused at temperatures in the range 900°C to above the melting point of metasilicate (1088°C). The product is then cooled, solidified, ground, and graded. Design and operation of the furnace needs special care to minimize the possible attack of refractories by the highly alkaline reaction mixture. Thus, the melt is usually contained within a barrier of unfused feed [1,5,10,76].

The refractories may corrode and there may be difficulties to complete the reaction. Therefore this process may be seen as less satisfactory to produce a high-quality product. Metasilicate produced by this route may bear a considerable proportion of insoluble material such as unreacted sand, natural impurities in sand, and refractory particles [1,7].

22.1.6.3 Production of Anhydrous Sodium Metasilicate by Solid-State Reaction

It has been frequently proposed to effect a solid-state reaction between the raw materials at 850–860°C in rotary or tunnel kilns or fluidized beds. The process temperature is above the melting point of soda ash but below that of metasilicate (see Table 22.1 and Figure 22.8) [10,11,13,77].

Process modifications may include producing granules from quartz flour, soda ash, and sodium silicate solution, the latter to act as a binder, by roll compaction, and subjecting them to calcination [78].

22.1.6.4 Granulation of Sodium Metasilicate Solution

Drying a solution of metasilicate in a special drum granulator drier is commonly used in industry (see Table 22.1) [15,79,80]. The aqueous sodium silicate solution is prepared from alkaline sodium silicate liquor and caustic soda (see Figure 22.9).

The product is derived from a purified silicate liquor and has a much lower level of impurities than products made by the direct fusion route. By virtue of the granulating process, the spray-coated product has essentially an onion-type-layered structure and a superior particle size distribution and is relatively dust free and more readily soluble in water [1].

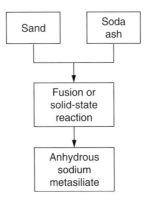

FIGURE 22.8 Direct fusion process for the production of anhydrous sodium metasilicate.

Production of Silicates and Zeolites

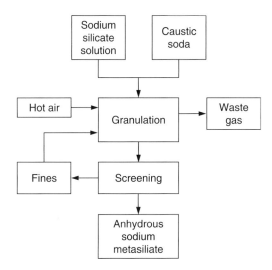

FIGURE 22.9 Granulation process for the production of anhydrous sodium metasilicate.

In the drum granulation process, a plurality of small solid particles of anhydrous metasilicate are fed into the end of a revolving drum, thus forming a moving bed.

Aqueous hot liquor at metasilicate composition is sprayed into the drum and collected by the granular feed. The moving mass of small particles is coated by a film of liquor, which is caused to crystallize as a layer of anhydrous sodium metasilicate.

Water that is liberated by the crystallization is dried out in a current of hot air. Thus, in the coating zone the granules grow and dry uniformly. At the end of the drum, they are screened and the undersized material are returned to the feed for reuse as nuclei. With this kind of granulation, over 70% of the product may be recycled to ensure uniform granule size [1,7,11,79,80].

An advantageous process modification may be to run the process with separated atomization and aging zones to maintain a defined content of humidity (2–6 wt%) in the product [81,82]. To facilitate a fluidized bed dryer may be another option [7,11].

22.1.7 Sodium Metasilicate Pentahydrate

22.1.7.1 Introduction

Sodium metasilicate exists in various hydrated forms. Metasilicate nonahydrate and metasilicate pentahydrate are the more well known thereof, but only the latter is of commercial importance. Sodium metasilicate pentahydrate is a white, free-flowing, granular or crystalline material, which is described by the chemical formula $Na_2SiO_3 \cdot 5H_2O$ with a melting point of 71.8–72.2°C and a specific gravity of 1.749 g/mL. Generally, granules of pentahydrate are used, which are delivered with a bulk density of 1000 g/L and a particle size of 1000 μm. They dissolve quickly in cold and hot water [1,5,7,11,12,15,19,47,70,75]. Sodium metasilicate octahydrate, $Na_2SiO_3 \cdot 8H_2O$, has a melting point of 48.35°C and a specific gravity of 1.672 [12]. Sodium metasilicate nonahydrate, $Na_2SiO_3 \cdot 9H_2O$, forms white, efflorescent rhombic crystals with a melting point of 48°C. It has virtually no commercial importance [5,6,12,75,76].

22.1.7.2 Crystallization from Sodium Metasilicate Solution

In a typical manufacturing process, sodium silicate solution of a weight ratio of 2 and caustic soda liquor (50 wt%) are mixed together to give a solution of metasilicate ratio (see Figure 22.10). Solid sodium hydroxide may be used instead of caustic soda liquor. This solution is concentrated to a

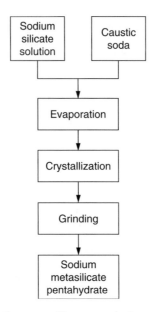

FIGURE 22.10 Crystallization of sodium metasilicate pentahydrate.

composition equivalent to that of the pentahydrate or nonahydrate by evaporation. The mixture is cooled and crystallized as a mass (see Table 22.1). Crystallization is accelerated by seeding the liquor. The solid crystalline mass is milled, graded, and packed [1,5,7,8,11,83].

Granular sodium metasilicate pentahydrate, which is dust free, free flowing, and of substantially uniform size can be prepared by agglomeration. Thus, molten sodium metasilicate pentahydrate is mixed to a larger amount of solid sodium metasilicate pentahydrate in a sigma-blade mixer. The liquor may also be injected into a moving bed of particles in a rotated drum or atomized to a fluidized bed, respectively. Good mechanical agitation and a temperature favorable to rapid crystallization, that is, 54–60°C, are critical. The molten material agglomerates the solid particles and then crystallizes, yielding agglomerated particles, which are then subjected to screening. Particles of the desired size are removed as product. Fines are recycled to the crystallization operation. Oversized ones are milled and then recycled to the crystallization stage [62,84,85].

22.1.8 CRYSTALLINE LAYERED SODIUM DISILICATE

22.1.8.1 Introduction

Crystalline layered sodium disilicates are the most recent development in the field of detergent silicates. In contrast to hydrous sodium polysilicates and metasilicates, they display a three-dimensional crystal lattice characterized by corrugated silicate layers, which are separated by sodium ions. The chemical formula is $Na_2Si_2O_5$. This reflects that the material is anhydrous and has a formal molar ratio of 2. Several crystal phases differ in the kind of corrugation of the silicate layers [3,21,66,86–88].

The sodium phyllosilicates are commercially available under the brand names ™SKS-6 and ™Purifeed. By virtue of the layered crystal structure, they exhibit ion exchange capability with a high selectivity for the water hardness ions and heavy metal ions, respectively. They are environmentally compliant because they are noneutrophic. These properties together with a high reserve alkalinity that buffers the pH of washing liquors in the favorable range make them prime candidates for modern multifunctional builders [3,21,86,87,89–91].

Alkalinity, heavy metal binding, and an anhydrous nature make them compatible with state-of-the-art detergent bleach systems [89,90].

The unique delayed solubility of layered sodium disilicates is the cause for an outstanding efficiency in the protection of fine glasses and china surfaces that makes them acknowledged ingredients in automatic dishwash detergents [4].

A high uptake of surfactants in connection with tolerance to humidity and the possibility to produce colored low-attrition coarse granules make them a preferred choice for colored speckles in the field of high-quality detergent tablets and granular powders [87,91,92].

Cogranules incorporating disintegration agents aim at detergent tablet applications, too [93].

Layered sodium silicate powders are shipped with a mean particle size of 120–130 μm at a bulk density of 400–600 g/L [3,86], whereas granules have a mean grain size on the order of 500–700 μm at a bulk density of 800–1000 g/L [3,94].

22.1.8.2 Manufacturing Process of Crystalline Layered Sodium Disilicates

The manufacturing process sequence to crystalline layered sodium disilicate may start with the production of sodium silicate solution of a $SiO_2:Na_2O$ ratio of 2 by the hydrothermal process (see Table 22.1 and Figure 22.11). Quartz sand is reacted at temperatures of 180–240°C with sodium hydroxide solution of 50 wt% strength. The reaction can be performed in a nickel-clad autoclave, which is equipped with a stirring device and heated by injection of steam. The reaction solution is filtrated after addition of perlite as the filter aid. The 45 wt% solids content of the final liquor was realized by dilution with water [21,49,51,95].

The sodium silicate liquor is treated in a spray drying process. The spray-dry tower is operated at inlet air temperatures of 200–300°C, exhaust gas temperatures on the order of 90–130°C, and residence times of the powder on the order of 10–25 s [3,49,51,95,96].

The crystallization of amorphous, water-containing sodium silicate to anhydrous crystalline layered sodium disilicate is crucial. It is effected in a further step. The formation of delta-crystal phase is an exothermic process at 685°C [89].

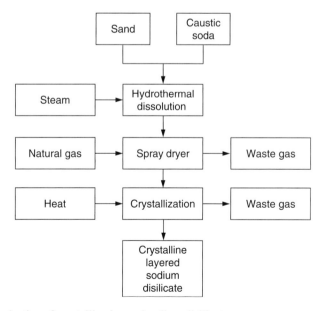

FIGURE 22.11 Production of crystalline layered sodium disilicates.

The spray-dried sodium silicate is heated in a rotary tubular kiln to temperatures on the order of 600–800°C. The kiln may be fitted with devices for moving solids and heated directly with flue gas in a countercurrent or externally via the wall. The latter approach allows the establishment of a plurality of different temperature zones in the interior ranging from 460 to 690°C.

The process encompasses an additional drying step, where the remaining water of the amorphous disilicate is liberated, and a subsequent annealing step, where the actual crystallization takes place. The crystalline sodium silicate emerging from the rotary kiln in the form of lumps is comminuted by means of mechanical mills and crushers to a powder grade of 100–150 μm particle size [3,21,49,51,66,95,96].

Granular grades are produced by compacting or agglomerating processes [3,49,66]. Thus, powder grade–layered sodium disilicate is mixed with 45 wt% water glass solution of molar ratio 2, which acts as a hardening aid. The mixture is processed on a roll compactor, followed by comminution of the flakes and screening to give dust-free granules (see Table 22.1) [97].

Agglomeration of powder size–layered sodium disilicate with acidic acrylic acid–maleic acid copolymer can be performed, for example, in high-shear ploughshare mixers. The pulverulent solid raw materials are agglomerated by the liquid components and afterdried in a fluidized bed at 120°C [4].

A process for the preparation of a colored granular sodium phyllosilicate compound consists of mixing coarse compressed alkali metal phyllosilicate granulates or agglomerates with a colorant and then spraying on an additive and water and, if desired, afterdrying it with a downstream fluidized bed [92].

22.1.9 REQUIREMENTS OF RESOURCES

Many industrial processes in Europe employed for the manufacturing of detergent ingredients have been assessed in life cycle inventories. A major target was to generate data for individual products that may be indicative of the latter's environmental impact. Consumption of energy and raw materials as well as emissions and the generation of waste were regarded as key indicators.

The production chain was traced back entirely to the extraction of raw materials to cover all production steps in a "cradle-to-factory-gate" approach. The collection of data coming from different production sites was condensed to average figures for every single detergent ingredient. Table 22.2 gives a compilation of selected data about the production of sodium silicate furnace lumps, furnace liquor, hydrothermal solution, spray-dried sodium silicate powder, sodium metasilicate pentahydrate, and crystalline layered sodium disilicate [6,20,94,98]. A direct comparison of the data sets may be misleading because of the different application fields and performance profiles of the underlying products.

22.2 ZEOLITES

22.2.1 ZEOLITE NaA

22.2.1.1 Introduction

Builders are of major importance for many detergent applications. In terms of production volume, sodium-A-type zeolite is one of the most important builders in household washing formulations.

The main function of a builder is to provide a rapid and efficient softening of the water of the wash liquor under a wide range of conditions [99]. For example calcium-binding capacity is 165 mg CaO per gram of zeolite. The binding capability of zeolite A for multivalent ions goes back to its particular crystal structure [100–103]. The porous structure is built up from cages connected by channels, and the constituent sodium ions are exchanged easily against divalent water hardness ions [104–108]. In the case of type-A zeolite, the cavities have a free diameter of about 1.14 nm and the free aperture of the cages is about 0.42 nm [18,66,104,108–115].

Production of Silicates and Zeolites

TABLE 22.2
Requirements of Resources for the Production of Sodium Silicates (Selected Data)

Data per 1 t of Product	Sodium Silicate Furnace Lumps[a]	Sodium Silicate Furnace Liquor[b]	Hydrothermal Sodium Silicate Solution[c]	Spray-Dried Sodium Silicate Powder[d]	Sodium Metasilicate Pentahydrate[e]	Layered Sodium Disilicate[f]
Soda ash	0.400 t	0.149 t				
Caustic soda (100 wt%)			0.209 t	0.362 t	0.380 t	
Sand	0.772 t	0.287 t	0.325 t	0.562 t	0.293 t	Total raw materials 1 t
Total energy consumed	11 GJ	4.6 GJ	5.4 GJ	18 GJ	10.6 GJ	24 GJ
Solid-waste generated	0.128 t	0.082 t	0.025 t	0.040 t	0.039 t	0.033 t
Carbon dioxide generated	1.1 t	0.43 t	0.29 t	0.89 t	0.57 t	1.3 t

[a] Anhydrous basis; weight ratio of 3.3; represents 260,000 t SiO_2 production capacity [20,98].

[b] 37 wt% active matter basis; weight ratio of 3.3; represents 330,000 t SiO_2 production capacity [6,98].

[c] 48 wt% active matter basis; weight ratio of 2.0; represents 91,000 t SiO_2 production capacity [98].

[d] 80 wt% active matter basis; weight ratio of 2.0; represents 15,000 t SiO_2 production capacity [98].

[e] 58 wt% active matter basis; weight ratio of 1.0; represents 38,000 t SiO_2 production capacity [98].

[f] 100 wt% active matter basis; molar ratio of 2.0 [94].

The stoichiometric formula $Na_{12}[(AlO_2)_{12}(SiO_2)_{12}] \cdot 27H_2O$ reveals a Si:Al-ratio of virtually 1 and about 20% by weight of interstitial water. The latter is continuously liberated upon heating the material [15,20,66,99–106,108–110,112–131].

"Low-silica" zeolites such as sodium type A, having a molar ratio of Si:Al near unity, contain the maximum number of cation exchange sites that balance the aluminum in the structure and thus have the highest possible cation exchange capacities [104,105,120]. Intermediate-silica zeolites, for example, of the faujasite type, have ratios of 2–5 and high silica zeolites, for example, ZSM types, have ratios of 10–50, respectively [104].

The water-softening kinetics, metered through the ion exchange rate is strongly influenced by particle size, too. Commercial type-A zeolite has a mean particle size of about 2–about 4 μm and a narrow particle size distribution [66,102,103,105,114,117,123,127,129,130]. A border grain diameter (i.e., limiting particle size) above 45 μm is designated as "grit." Such larger particles have to be avoided by careful adjustments of synthesis conditions [117,118,125,128–130,132–136].

The specific gravity of the cubic microcrystals is about 2 g/cm^3 and the bulk density of the powder is on the order of 170–450 g/L, preferably 350 g/L [102,103,106,112,121].

Further advantages of zeolite A are that they are noneutrophic, safe for humans and the environment, comparatively cheap, and easily incorporated into tower or nontower detergent manufacturing processes [99,105,115,123].

22.2.1.2 Raw Materials

The production of zeolite by the hydrogel route utilizes sodium silicate solutions, which have the two commercially most available silica to sodium oxide ratios of ca. 2.4 and ca. 3.5 [15,99,103,104,117,124,128–131,114,134–142].

Sodium silicate solution of a $SiO_2:Na_2O$ ratio of ca. 3.5 is predominantly produced from soda ash and quartz sand by the furnace process. The anhydrous sodium silicate cullets that are obtained by melting together these two basic raw materials are dissolved in water under hydrothermal conditions and the resulting water glass liquor is filtered and blended to the sodium silicate solution of desired purity and composition [1,5–21,44,45,47,98,99,143].

Soda ash is available from natural origin by the extraction of trona deposits and subsequent refining processes as well as synthetically from sodium chloride and calcium carbonate. The latter both are reacted together in the ammonia-soda process (Solvay process) to yield heavy soda ash [20,22,23,28–32,34,36,38].

Quartz sand of necessary purity is extracted from naturally occurring deposits to meet glass-melting standards [1,7,8,11,13,39,40,41,144–149].

Alkaline sodium silicate solution of a $SiO_2:Na_2O$ ratio of ca. 2 is largely manufactured by hydrothermal dissolution of quartz sand or flour in concentrated caustic soda solution [7,8,11–13,15,17–19,21,44,45,98,99,119,143].

The caustic soda, which is needed for both the hydrothermal production of sodium silicate solution and for the production of sodium aluminate is manufactured by the extraction of sodium chloride [24–27,142] and subsequent electrolysis of sodium chloride brine to yield caustic soda solution [20,52–61,120,142].

The other major raw material for zeolite production is the sodium aluminate solution. The production route to this important ingredient starts from bauxite minerals, for example, gibbsite [15,20,99,100,119,126,142,150–154]. The ore is extracted and purified via the Bayer process. The process encompasses the digestion of sodium aluminate by means of caustic soda and precipitation of purified aluminum trihydroxide [15,18,20,99,100,119,126,142,150–159].

Depending on the backward integration of the individual zeolite production process, the alumina source may be aluminum hydroxide as well as sodium aluminate [20,103,114,116,117,123, 124,129,130,132–136,139,141,152,158,160–165]. Usually, the synthesis starts with the dissolution of aluminum hydroxide in caustic soda solution at temperatures above 80°C and filtration (see Equation 22.4) [15,20,99,104,110,126,128,166].

$$NaOH + Al(OH)_3 \rightarrow NaAl(OH)_4 \ (aq) \tag{22.4}$$

Apart from its use in the synthesis of the raw materials, caustic soda is an important ingredient in the zeolite production process. During the crystallization step, it supplies the sodium ions and assists additionally in controlling the pH. It is commonly used as 50 wt% aqueous solution [15,99,116,126,139,166].

22.2.1.3 Principles of the Zeolite A Production Process

The zeolite producing industry saw the design and development of a multitude of processes for the manufacturing of zeolites. The individual embodiments aimed to meet the varying constraints of energy and raw material supply and product quality.

The predominant process, whether employed semicontinuously or batchwise is the alumino-silicate hydrogel route, sometimes referred to as the silicate route. Sodium silicate and sodium aluminate solutions are mixed together in the desired ratio in an alkaline aqueous solution (see Equation 22.5).

$$\text{Sodium aluminate (aq.)} + \text{sodium silicate (aq.)} \xrightarrow{\text{heat, NaOH}} \text{amorphous sodium aluminosilicate hydrogel} \tag{22.5}$$

Production of Silicates and Zeolites 407

Subsequent to the sol phase, the viscosity of the reaction mixture rises and a hydrogel of amorphous sodium aluminosilicate is formed. Precipitation is carried out in relatively dilute solution and the amorphous sodium aluminosilicate has typically a high water content [20,100,101,103,104,108–110,112,119,122,126,137].

The formation of the sodium aluminosilicate precursor is regarded to be critical. Its formation conditions and properties are considered responsible for the crystalline zeolites quality parameters. Two major targets of process optimization are to ensure the specified particle size, that is, to obtain a low-grit product, while keeping the time required for the process as short as possible.

Particle size mainly depends on the parameters chosen in gel precipitation, for example, the alkalinity of the precipitating solution ($Na_2O:H_2O$ ratio), its concentration and composition, temperature, mixing sequence, the rate of feeding of the different components, the mixing energy introduced to the reaction mixture by stirring or shearing, and the addition of seeds (see Table 22.3) [112,123,125,126].

Usually, the gel preparation is performed at temperatures between 50 and 80°C by a process that requires 0.5–1.5 h (see Table 22.3) [123,125,126].

Alkalinity is well known to assist mineralization. Therefore, a higher concentration of free sodium hydroxide is considered to be advantageous, resulting in an increase in the rate of formation of the zeolite's crystals as well as in the formation of small particles (see Table 22.3) [125,126,138].

TABLE 22.3
Parameters of the Sodium Zeolite A Production Process

	Precipitation of Amorphous Sodium Aluminosilicate Hydrogel	Aging of Amorphous Sodium Aluminosilicate Hydrogel	Crystallization of Sodium Zeolite A
Process time	Important; 0.5–1.5 h	5 min–3 h	2–10 h
Temperature	Affects particle size; 50–80°C	60–75°C	Affects process time; 90–95°C
Alkalinity	Affects particle size; free NaOH advantageous	—	Affects particle size, particle shape, process time; pH = 10–14
Concentration	Affects particle size, yield, process time; 90–98 mol-% H_2O	—	—
Composition	Affects particle size, yield, process time; 2–3.6 Na_2O: Al_2O_3:1.74–2.0 SiO_2:40–90 H_2O; excess of Al advantageous	—	—
Mixing sequence of raw materials	Affects particle size	—	—
Rate of feeding raw materials	Affects particle size	—	—
Mixing energy	Affects particle size, particle shape, by-product level; high shear-forces advantageous	—	Affects particle size and process time
Addition of seeds	Affects particle size	—	Affects process time
Recycling of crystallization liquor	—	—	Potentially advantageous
Tempering	—	—	Potentially advantageous

The way the reactants are mixed may affect the structure of the hydrogel. The latter undergoes dissolution during the crystallization process via a quasiequilibrium with the liquid phase. To assist this dissolution, it is advantageous to apply shear forces to the gel to keep the hydrogel particles as small as possible. This process is often described as "to stirr the system until homogenous" or "to break it into a nearly homogenous mix." In consequence, the rate of crystallization and the nature of the final product are determined in this stage. This includes a desired crystal size of below 10 μm and a rounded shape, as well as types and level of by-product impurities (see Table 22.3). The appropriate way to employ high-shear forces in this context is by the utilization of turbine mixers, high-speed mixers, or dispensers [108,113,127,128,130,132,134,136,137,139].

The reaction mixture of the synthesis of zeolite A can be varied within wide boundaries of the overall composition. It is necessary to find the best composition in terms of product quality, yield, and process duration. Usually, the water content is quite high and is on a molar basis between 90 and 98% (see Table 22.3) [140].

Optimized compositions of the reaction solution are described on a molar basis in the range 2.0–3.6 Na_2O:Al_2O_3:1.74–2.00 SiO_2:40–90 H_2O (see Table 22.3) [104,110,113,116,123–128,133,137,140]. Thus, compared to the product stoichiometry (SiO_2:Al_2O_3 = 2), an excess of aluminium is usually employed [125,126].

Besides shear force, the sequence and rate of adding the reactants are held responsible for the product's quality with respect to particle size and by-product content (see Table 22.3). Thus, a multitude of recipes is published, which cover features such as providing one reactant and feeding the other, feeding the other stepwise at different rates or at different concentration, mixing the raw materials simultaneously, or diluting the reaction mixture between addition steps. Generally, it appears to be important to maintain an excess of alumina. The latter is provided by sodium aluminate liquor [104,108,114,117,128,129,131,133–135,141].

The subsequent step to gel precipitation is its aging (see Table 22.3). Usually, the aging step takes 5 min to 3 h at 60–75°C. Elevated temperatures reduce the duration of the aging step [104,108,114,133,137].

Thereafter, the sodium aluminosilicate hydrogel is hydrothermally transformed into the highly-ordered crystalline sodium zeolite 4A (see Equation 22.6) [103,109,119,122,142].

$$\text{Amorphous sodium aluminosilicate hydrogel} \xrightarrow[-\text{NaOH}]{\text{heat, time}} Na_2O \cdot Al_2O_3 \cdot 2SiO_2 \cdot 4.5H_2O$$

$$\text{(crystalline sodium zeolite A)}$$

(22.6)

The required time depends on temperature, alkali concentration, addition of seeds, and agitation of the reaction mixture.

Temperatures for crystallization are usually slightly higher (70–100°C, preferably 90–95°C) than that for gel precipitation (see Table 22.3) [15,20,99,104,110,113,123,125–127,129,131,133–135, 137,139].

An increase in temperature leads to a decrease in crystallization time because the solubility of the aluminosilicate hydrogel is enhanced at higher temperatures. Thus, the aging is accelerated, incubation period shortened, and crystallization velocity increased. The latter increases steadily with time and drops after a relatively high crystallization degree has been obtained. The reaction is commonly monitored by means of x-ray powder diffraction and is stopped at this point [104,110,137].

Depending on the chosen settings of the other reaction parameters, the crystallization time is in the range 2–10 h, preferably less than 1 h (see Table 22.3) [15,99,104,108,110,123,125,131,133, 134,139].

Low flow velocity during crystallization, just as high as to avoid sedimentation, may result in undisturbed crystal growth [104,137]. This may be preferred over application of shear forces during crystallization (see Table 22.3) [118,130,132,134,136,141].

Production of Silicates and Zeolites 409

The alkalinity of the crystallization liquor can enhance crystallization by virtue of increased formation of the crystal nuclei. Further, it may lead to smaller crystals with a more rounded off shape (see Table 22.3). Both are desired properties for detergent builder applications. Thus, the preferred pH value is in the range 10–14 [104,128,137,140].

Further modifications of the crystallization process may encompass the addition of seeds, for example, by recycling crystallization liquor [113,126,137], and a further tempering step (see Table 22.3) [129,130,132,134–136].

The crystal-containing suspension may contain 120–200 kg of zeolite (dry matter basis) per m^3 [108,126].

22.2.1.4 Continuous Process Operation versus Batch-Type Process

The production of zeolites is frequently performed in batch reactors. Gel precipitation and crystallization are then carried out stepwise. A sodium aluminate solution is prepared first by mixing an alumina source with caustic soda. The silica-containing solution is then dosed to the alkaline alumina-containing solution [140].

The process may also be run semicontinuously. In this case, the hydrogel is prepared continuously and zeolite is crystallized batchwise in large reaction vessels [123].

A continuous production process requires the separation of gel precipitation and crystallization stages. The latter step lasts three times longer then gel precipitation and aging. High-shear forces are employed during gel precipitation and aging, whereas low flow velocity is advisable during crystallization [15,99,104]. Nevertheless, continuous crystallization tends to be difficult due to the metastability of the crystallization and dynamic processes, which result in the unforeseeable formation of impurities [108].

22.2.1.5 Clay Conversion Process

The hydrothermal conversion of modified clay mineral, for example, metakaolin, is another manufacturing route to zeolites. Although the raw material costs may be as much as 15% lower than that of the hydrogel route, the major shortcoming of the process is that iron impurities in the raw material may cause an undesirable coloring of the zeolite product [15,18,99–101,104,108,116,122,123,125,126].

22.2.1.6 Zeolite NaA Production by Hydrogel Process

Alumina trihydrate is dissolved in the first vessel (aluminate dissolver) by caustic soda solution (see Figure 22.12). The latter is often used in excess to prepare a stable sodium aluminate solution [15,99,119,123,128,132,139,140,142].

This alkaline solution is heated and mixed intimately with sodium silicate solution in the precipitation reactor at a temperature between 50 and 80°C (see Figure 22.12) [104,113,119,128,132, 140,142,167].

The aluminosilicate hydrogel–containing solution may then be pumped to a separate crystallization vessel of a size of up to 100 m^3 and above. The crystallization of zeolite A is usually carried out batchwise at temperatures of 90–95°C. It is accomplished after 40–60 min. Discontinuous processes may use the same reactor for precipitation and crystallization [15,99,104,108,110,116, 119,122,125–128,139,140,142].

Once the reaction is complete, the reaction mixture is passed over a suitable separator such as a vacuum belt filter, rotary filter, filter press, or decanter to separate crystals from the mother liquor and washed free of excess reactants (see Figure 22.12). The hot deionized wash water may be evaporator condensate, which is recycled from the mother liquor concentration process. Volumes and, therefore, evaporation costs can be minimized by using a countercurrent washing process. The effluent wash water has a pH value between 9 and 12. The mother liquor filtrate and wash water contain excess of sodium hydroxide and possibly an excess of alumina. The complete utilization of the mother liquor is essential in the manufacturing of zeolites. Therefore, it is concentrated in

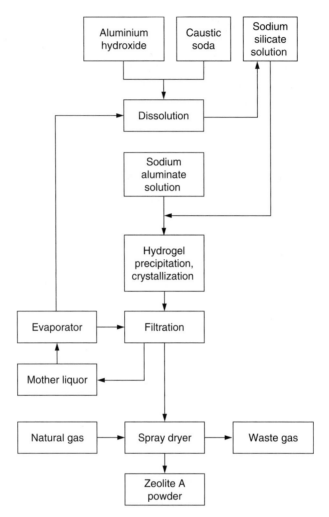

FIGURE 22.12 Zeolite NaA production by hydrogel process.

an evaporator as far as required. The concentrated stream may be reentered into the process and fortified with sodium aluminate or sodium hydroxide as well as being recycled to the aluminate-preparation unit (see Figure 22.12) [15,99,104,108,110,113,122,123,126,140,167].

A portion of the crystal-containing suspension may be recycled back to the reactor to serve as a seed to increase the rate of crystallization. Eventually, a filter cake with a product content of ca. 40–50% is obtained [15,20,99,104,108,110,113,116,122,123,126,128,140,167].

The wet filter cake is finally spray-dried to the commercial sodium A-type zeolite, which contains ~20 wt% of crystal water and has free-flowing powder characteristics (see Figure 22.12). It is stable for storage and delivery. The powder is best suited for detergent manufacturers that apply the dry-granulation technology for their detergent production [20,104,108,110, 113,119,121–123,126,140,143,167]. Zeolite A powder can be used in this context as a processing aid, for example, to adsorb liquid ingredients in the agglomeration step.

Some detergent manufacturers, predominantly those who make use of spray drying detergent production processes, prefer to obtain an aqueous dispersion of zeolite A (48–50%). This is stabilized against rapid sedimentation by the addition of 1.5% surface-active agent (3.0% on anhydrous solids). This way the detergent builder can be distributed as slurry of a density of 1.1–1.5 g/cm^3 [104,110,119,122,123,126,142]. For detergent manufacturers that have an integrated zeolite

Production of Silicates and Zeolites 411

production facility, it can be a cost benefit if they can mix the zeolite slurry with surfactants etc. and spray dry the mixture directly to a detergent powder. The zeolite slurry should be less attractive to nonintegrated producers due to higher transportation costs.

22.2.1.7 Requirements of Resources

The manufacturing processes of many detergent ingredients have been assessed in life cycle inventories for their energy and raw material consumption. Among the targets was to generate data about processes running in Europe for individual products and cover all production steps in a "cradle-to-factory-gate" approach. Average data were calculated from individual productions. The dataset on the production of zeolite NaA by the hydrogel route was provided from five major zeolite-producing companies. Calculated on an 80 wt% active matter basis, the production of a 1 t product consumed 0.269 t of caustic soda (100%), 0.509 t sodium silicate ($Na_2O \cdot 2SiO_2$ 100%), 0.485 t aluminum trihydroxide, and 21.2 GJ total energy, and generated a considerable amount of solid waste of 0.223 t and 1.1 t of carbon dioxide [20,109,142].

22.2.2 ZEOLITE P

22.2.2.1 Introduction

The search for sodium tripolyphosphate substitutes in the 1970s resulted in other candidate builder systems besides zeolite A. One of these is zeolite P, which is also known as zeolite MAP (it is a P-type zeolite having maximum aluminium content) and zeolite A24. It was not before the early 1990s that it became commercially available under the brand name ™Doucil 24 [66,143].

There are several crystal phases and each has some flexibility in the chemical composition [106,168,169]. The most frequently used formula is $Na_6[(AlO_2)_6SiO_2)_{10}] \cdot 12-15H_2O$ [106,109,126]. The silicon to aluminum ratio may vary between 1.0–2.5 [66,106,168,169] and 1.6–2.7 [106], respectively.

The molecular framework is different than that of zeolite A and its particles are smaller than those of the standard zeolitic builder. This implies that it exchanges calcium more rapidly and binds it more firmly than zeolite A does, especially at low temperatures. Its magnesium exchange is reported to be more rapid than that of 4A [66,109,115,168,170,171].

The smaller particle size of 1–1.5 µm and a morphology of pseudospherical aggregates result not only in an enhanced absorption capacity for surfactants, but also in a better dispersability in the washing liquor [66,109,170,171].

Another difference in zeolite A is the lower content of water and its lower availability. Both are the prerequisites for a better stability of peroxygen bleaches in detergent formulas. The active matter of commercial A24 zeolite is 90 wt% [66,109,143,170,171].

22.2.2.2 Production of Zeolite P by Hydrogel Process

The basic principles of P-type zeolite production are not unlike those of zeolite A, which are well known. It has been reported that zeolite P production is currently accomplished on converted zeolite A4 manufacturing plants.

The hydrogel synthesis is based on the conventional raw materials such as sodium aluminate, caustic soda, and commercial alkaline sodium silicate solution, for example, of a SiO_2/Na_2O ratio of 2. The preheated solutions are intimately mixed together to form a hydrogel. The compositions of such reaction mixtures may vary considerably within the range of ratios 2.0–6.5 Na_2O:2.0–8.0 SiO_2:Al_2O_3:94.0–187.9 H_2O. Crystallization times are significantly longer compared with those in the production of zeolite A. Zeolite P is known to be the final stable product in faujasite catalyst synthesis when the formation of zeolite X and Y is inhibited at increasing temperatures above 150°C [66,106,108,140,168,169,171,172].

Thus, a process setup for the preparation of zeolite P may be designed to provide a solution of sodium aluminate and caustic soda and heat the mix up to 90–95°C. Furthermore, a feeding of sodium silicate solution at the same temperature is accomplished with vigorous stirring. A bottom flat-bladed disc turbine stirring at 100 rpm is facilitated. The mixing schedule may work vice versa, too. The resulting silicate–alumina hydrogel is subjected to undergo aging and crystallization, for example, for 5 h at 95°C. Sufficient stirring has to be maintained to prevent solids from sedimentation. The crystalline sodium aluminosilicate thus formed is separated by the usual means, washed with water to a pH within the range 10–12.5, and spray dried [168].

An efficient recycling management of filtrates and mother liquor is also important in this case. Filtrates and mother liquor can be recirculated after optional concentration for the preparation of the initial reactant solutions, for example, the aluminum trihydrate dissolving step. It can be reentered into the process not only for the dilution of sodium silicate solution, but also directly into the hydrogel precipitation vessel [169].

The application of a seeding process may be advantageous. Thus, a slurry of P-type zeolite may be recirculated to the hydrogel precipitation vessel via the stream of recycled filtrate and mother liquor. As a result, a higher product yield can be realized [169,172].

22.2.3 ZEOLITE AX

22.2.3.1 Introduction

The search for enhanced builder properties has led to the development of zeolite AX. This is a cocrystallisate of 20% A-type zeolite and 80% zeolite LSX. It is a zeolite of the X-type having low silica content. Its chemical composition is reported as 10% Na_2O, 1% K_2O, 43% SiO_2, 32% Al_2O_3, and 23% H_2O [105].

The production of the new zeolite type is intended to overcome the shortcomings of zeolite LSX such as its considerable potassium content and a complicated production process. Features of the new builder are enhanced low-temperature builder action, that is, the almost complete exchange of calcium ions and a better magnesium exchange in comparison to A-type zeolite. Important physical properties of the material are particle sizes of 4.2 μm and a bulk density of 400 g/L [105,173].

22.2.3.2 Production of Zeolite AX by Hydrogel Process

Zeolite AX is prepared by the aluminosilicate hydrogel route. Therefore, a commercial sodium silicate solution (28.4% SiO_2 and 14.2% Na_2O) is fed to a mixture of sodium aluminate and potassium hydroxide at a temperature of 65°C. The amorphous alkali aluminosilicate hydrogel is maintained at the same temperature under agitation for 30 min and without agitation for 12 h. The product is filtered off, washed, and dried by the usual means [174,175].

Zeolite A is the largest volume zeolitic builder produced today. The new P-type zeolite and zeolite AX have different properties and aim at novel applications fields. Nevertheless, they have to prove their ability for world-scale commercialization.

ACKNOWLEDGMENTS

I am indebted to Clariant GmbH/Functional Chemicals Division for making the needed infrastructure available to me. My heartfelt thanks to my dear wife Susanne Lang-Bauer and children Eva-Maria Bauer and Claus-Jürgen Bauer for their patience, forbearance, and mental support while I was working on this chapter.

REFERENCES

1. Barby, D., Griffiths, T., Jacques, A. R., Pawson, D. 1977. Soluble Silicates and Their Derivatives. In: Thomson, R., Ed., *The Modern Inorganic Chemicals Industry*. London: London Chemical Society, pp. 320–352.

Production of Silicates and Zeolites

2. Morris, T. C. 1993. *Zeolite A and Hydrous Polysilicates for Today's Detergent Formulations.* AOCS Meeting, Anaheim, CA.
3. Dany, F. J., Gohla, W., Kandler, J., Rieck, H. P., Schimmel, G. 1990. Kristallines Schichtsilikat—ein neuer Builder. *SÖFW-Journal.* 20:805–808.
4. Mueller, T., Thewes, V., Bauer, H., Hardt, P., Holz, J., Berghahn, M., Mertens, R., Schimmel, G. 1998. Powder washing and cleaning components. EP0849355. Stockhausen Chem. Fab. GmbH, Clariant GmbH.
5. Engler, D. 1974. Soluble Silicates. *Seifen Öle Fette Wachse.* 100:165–170.
6. Lowenstein, F. A., Moran, M. K., Eds. 1975. *Faith, Keyes & Clark's Industrial Chemicals.* 4th ed. New York: Wiley-Interscience, pp. 755–761.
7. Falcone, J. S. 1982. Synthetic Inorganic Silicates. In: Grayson, M., Ed. *Kirk-Othmer Encyclopedia of Chemical Technology.* Vol. 20. 3rd ed. New York: Wiley, pp. 855–880.
8. Fawer, M. 1997. Life Cycle Inventories for the Production of Sodium Silicates. Report No. 241. St. Gallen: EMPA Eidgenössische Materialprüfungs—und Forschungsanstalt.
9. Hayes, J. 1974. Manufacture and Quality Control of Sodium Silicate. *Water Treatment and Examination.* 23(4):380–383.
10. Lagaly, G., Klose, D., Heinerth, E. 1982. Silicate. In: Bartholomé, E., Biekert, E., Hellmann, H., Ley, H., Weigert, W. M., Weise, E., Eds. *Ullmanns Encyklopädie der Technischen Chemie.* Vol. 21. 4th ed. Weinheim: Verlag Chemie GmbH, pp. 365–416.
11. Lagaly, G., Klose, D., Minihan, A., Lovell, A. 1993. Silicates. In: Elvers, B., Hawkins, S., Russey, W., Schulz, G., Eds. *Ullmann's Encyclopedia of Industrial Chemistry.* Vol. A23. 5th ed. Weinheim: VCH Verlagsgesellschaft mbH, pp. 661–719.
12. Merrill, R. C. 1947. Chemistry of the Soluble Silicates. *J. Chem. Educ.* 24:262–269.
13. Heinerth, E. 1959. Alkalisilicate In: Winnacker, K., Küchler, L., Eds, *Chemische Technologie. Band 2: Anorganische Technologie II.* 2nd ed. München: Carl Hanser Verlag, pp. 487–495.
14. Laufenberg, J., Novotny, R. 1983. Wasserglas—Herstellung und Anwendung. *Glastechnische Berichte.* 58:294–298.
15. Schweiker, G. C. 1978. Sodium Silicates and Sodium Aluminosilicates. *J. Am. Oil Chem. Soc.* 55:36–40.
16. O'Neil, M. J., Smith, A., Heckelmann, P. E., Obenchain, J. R., Gallipeau, J. A. R., D'Arecca, M. A., Eds. 2001. *The Merck Index.* 13th ed. Whitehouse Station, NJ: Merck & Co., Inc., p. 1547.
17. Heinerth, E. 1964. Technische Alkalisilicate. In: Foerst, W., Ed. *Ullmanns Encyklopädie der Technischen Chemie.* Vol. 15. 3rd ed. München-Berlin: Urban & Schwarzenberg, pp. 732–739.
18. Büchner, W., Schliebs, R., Winter, G., Büchel, K. H. 1984. *Industrielle Anorganische Chemie.* Weinheim: Verlag Chemie.
19. Christophliemk, P. 1985. Herstellung, Struktur und Chemie technisch wichtiger Alkalisilicate. *Glastechnische Berichte.* 58:308–314.
20. Grießhammer, R., Bunke, D., Gensch, C. O. 1996. Produktlinienanalyse Waschen und Waschmittel. Freiburg: Öko-Institut e.V., Geschäftsstelle, Postfach 6226, 79038 Freiburg i. Br.
21. Rieck, H. P. 1996. Natriumschichtsilicate und Schichtkieselsäuren. *Nachr. Chem. Tech. Lab.* 44:699–704.
22. Burridge, E. 2002. *Soda Ash.* European Chemical News. 77 (December 23):16.
23. Burridge, E. 2004. *Soda Ash.* European Chemical News. 81 (October 18–24):19.
24. Gibson, J. 2004. *Salt demand Growth on the Up.* European Chemical News. 81 (August 23):11.
25. Westphal, G., Kristen, G., Wegener, W., Ambatiello, P., Geyer, H., Epron, B., Bonal, C., Seebode, K., Kowalski, U. 1993. Sodium Chloride. In: Elvers, B., Hawkins, S., Russey, W., Schulz, G., Eds. *Ullmann's Encyclopedia of Industrial Chemistry.* Vol. A24. 5th ed. Weinheim: VCH Verlagsgesellschaft mbH, pp. 317–339.
26. Falbe, J., Regitz, M., Eds. 1998. *Römpp-Lexikon Chemie.* Vol. 4. 10th ed. Stuttgart, New York: Thieme, pp. 2824–2826.
27. Gülpen, E., Thieme, C., Url, H., Herde, W., Demel, W., Ambatiello, P., Lopau, O., Gallone, P., Seebode, K. 1979. Natriumchlorid. In: Bartholomé, E., Biekert, E., Hellmann, H., Ley, H., Weigert, W. M., Weise, E., Eds. *Ullmanns Encyklopädie der Technischen Chemie.* Vol. 17. 4th ed. Weinheim: Verlag Chemie GmbH, pp. 179–199.
28. Braune, G., Schneider, H. 1979. Natriumcarbonat und Natriumhydrogencarbonat. In: Bartholomé, E., Biekert, E., Hellmann, H., Ley, H., Weigert, W. M., Weise, E., Eds. *Ullmanns Encyklopädie der Technischen Chemie.* Vol. 17. 4th ed. Weinheim: Verlag Chemie GmbH, pp. 159–177.
29. Robertson, A. S. 1978. Sodium Carbonate. In: Grayson, M., Ed. *Kirk-Othmer Encyclopedia of Chemical Technology.* Vol. 1. 3rd ed. New York: Wiley pp. 866–883.

30. Falbe, J., Regitz, M., Eds. 1998. *Römpp-Lexikon Chemie*. Vol. 4. 10th ed. Stuttgart, New York: Thieme, pp. 2823–2824.
31. Schmidt, H. W., Rothe, H. J. 1960. Natriumcarbonat und Natriumhydroxyd. In: Foerst, W., Ed. *Ullmanns Encyklopädie der Technischen Chemie*. Vol. 12. 3th ed. München-Berlin: Urban & Schwarzenberg, pp. 643–662.
32. Sittig, M. 1968. *Inorganic Chemical and Metallurgical Process Encyclopedia*. London: Noyes Development Corporation, p. 653.
33. Pan, L. 1955. Method and apparatus for the production of sodium bicarbonate. 1955. US2866681. 09.05.1955. Chemical Construction Corporation.
34. Sittig, M. 1968. *Inorganic Chemical and Metallurgical Process Encyclopedia*. London: Noyes Development Corporation, p. 662.
35. Hansen, G., Mueller, H., Schmitt, O., Vogel, K. Method of calcining sodium bicarbonate. 1961. US3232701. 29.11.1961. Badische Anilin- & Soda-Fabrik A.G.
36. Sittig, M. 1968. *Inorganic Chemical and Metallurgical Process Encyclopedia*. London: Noyes Development Corporation, p. 665.
37. Devaux, A., Jean, M. Conversion of sodium chloride into sodium carbonate and ammonia chloride. US2843454. 26.07.1954. Societe Chimique de la Grande Paroisse.
38. Thieme, C. 1993. Sodium Carbonates. In: Elvers, B., Hawkins, S., Russey, W., Schulz, G., Eds. *Ullmann's Encyclopedia of Industrial Chemistry*. Vol. A24. 5th ed. Weinheim: VCH Verlagsgesellschaft mbH, pp. 299–316.
39. Flörke, O. W., Martin, B., Benda, L., Paschen, S., Bergna, H. E., Roberts, W. O., Welsh, W. A., Ettlinger, M., Kerner, D., Kleinschmit, P., Meyer, J., Gies, H., Schiffmann, D. 1993. Silica. In: Elvers, B., Hawkins, S., Russey, W., Schulz, G., Eds. *Ullmann's Encyclopedia of Industrial Chemistry*. Vol. A23. 5th ed. Weinheim: VCH Verlagsgesellschaft mbH, pp. 583–660.
40. Weiss, R., Paschen, S., Schober, P., Merz, G., Schlimper, H. U., Ferch, H., Kreher, A., Habersang, S. 1982. Siliciumdioxid. In: Bartholomé, E., Biekert, E., Hellmann, H., Ley, H., Weigert, W. M., Weise, E., Eds. *Ullmanns Encyklopädie der Technischen Chemie*. Vol. 21. 4th ed. Weinheim: Verlag Chemie GmbH, pp. 439–476.
41. De Jong, B. H. W. S. 1989. Glass. In: Elvers, B., Hawkins, S., Ravenscroft, M., Rounsaville, J. F., Schulz, G., Eds. *Ullmann's Encyclopedia of Industrial Chemistry*. Vol. A12. 5th ed. Weinheim: VCH Verlagsgesellschaft mbH, pp. 365–432.
42. King, R. J. 1978. Steel. In: Grayson, M. *Kirk-Othmer Encyclopedia of Chemical Technology*. Vol. 21. 3rd ed. New York: Wiley, pp. 552–556.
43. Gliemeroth, G., Müller, G. 1976. Glas und Glaseramik. In: Bartholomé, E., Biekert, E., Hellmann, H., Ley, H., Weigert, W. M., Weise, E., Eds. *Ullmanns Encyklopädie der Technischen Chemie*. Vol. 12. 4th ed. Weinheim: Verlag Chemie GmbH, pp. 317–366.
44. W Kuhr. 1998. Wasserglas—Herstellung und Anwendung. *Henkel Referate*. 34:7–13.
45. Novotny, R. 1985. Henkel Water Glass 1884 to 1984—Hundred Years of Production and Development. *Glastechnische Berichte*. 58(11):295–300.
46. Sugranes, J. F. 1975. Schmelzofen zur kontinuierlichen Herstellung von festen alkalischen Silikaten. DE2501850. Foret S. A.
47. Falcone, J. S., Spencer, R. W. 1975. Silicates expand role in waste treatment, bleaching, deinking. *Pulp&Paper*. 12:114–117.
48. Falbe, J., Regitz, M., Eds. 1999. *Römpp-Lexikon Chemie*. Vol. 6. 10th ed. Stuttgart, New York: Thieme, p. 4939.
49. Mühlenkamp, S. 1998. Glasklare Sache. *Process*. 11:158–159.
50. Novotny, R., Hoff, A., Schuertz, J. 1990. Process for the hydrothermal preparation of sodium silicate solutions. EP0380997. Henkel KGAA.
51. Kotzian, M., Schimmel, G., Tapper, A., Bauer, K. 1994. Process for the production of crystalline sodium disilicate in an externally heated rotary kiln having temperature zones. US5308596. Hoechst AG.
52. Bergner, D. 1994. Entwicklungsstand der Alkalichlorid-Elektrolyse Teil 1: Zellen, Membranen, Elektrolyte, Produkte. *Chemie Ingenieur Technik*. 66:783–791.
53. Bergner, D. 1994. Entwicklungsstand der Alkalichlorid-Elektrolyse Teil 2: Elektrochemische Größen, Wirtschaftliche Fragen. *Chemie Ingenieur Technik*. 66:1026–1033.
54. Bergner, D. 1997. 20 Jahre Entwicklung einer bipolaren Membranzelle für die Alkalichlorid-Elektrolyse vom Labor bis zur weltweiten Anwendung. *Chemie Ingenieur Technik*. 69:438–445.
55. Burridge, E. 2003. *Caustic Soda*. European Chemical News. 78 (June 02):18.

Production of Silicates and Zeolites

56. Gibson, J. 2004. *Caustic Soda Demand Breaks Ten-Year Record*. European Chemical News. 80 (May 24):10.
57. Harriman, S. 2002. *Getting the Balance Right*. European Chemical News. May 7, 76:19–20.
58. Hochgeschwender, K., Zirngiebl, E. 1979. Natriumhydroxid. In: Bartholomé, E., Biekert, E., Hellmann, H., Ley, H., Weigert, W. M., Weise, E., Eds. *Ullmanns Encyklopädie der Technischen Chemie*. Vol. 17. 4th ed. Weinheim: Verlag Chemie GmbH, pp. 201–209.
59. Leddy, J. J., Jones, I. C., Lowry, B. S., Spillers, F. W., Wing, R. E., Binger, C. D. 1978. Alkali and Chlorine Products. In: Grayson, M., Ed. *Kirk-Othmer Encyclopedia of Chemical Technology*. Vol. 1. 3rd ed. New York: Wiley, pp. 799–865.
60. Minz, F. R. 1993. Sodium Hydroxide. In: Elvers, B., Hawkins, S., Russey, W., Schulz, G., Eds. *Ullmann's Encyclopedia of Industrial Chemistry*. Vol. A24. 5th ed. Weinheim: VCH Verlagsgesellschaft mbH, pp. 345–354.
61. Schmittinger, P., Curlin, L. C., Asawa, T., Kotowski, S., Beer, H. B., Greenberg, A. M., Zelfel, E., Breitstadt, R. 1986. Chlorine. In: Gerhartz, W., Yamamoto, Y. S., Campbell, T., Pfefferkorn, R., Rounsaville, J. F., Eds. *Ullmann's Encyclopedia of Industrial Chemistry*. Vol. A6. 5th ed. Weinheim: VCH Verlagsgesellschaft mbH, pp. 399–481.
62. Gronchi, P., Principi, M., Pozzi, S. 2000. Renewed interest in the sodium metasilicate as soap and detergent builder. In: *5th World Surfactants Congress, Firenze, Italy, 29 May–2 June 2000*. Brussels: Comite Europeen des Agents de Surface et leurs Intermediaires Organiques, pp. 240–246.
63. Bertorelli, O. L., Mays, R. K., Williams, L. E., Zimmerman, H. F. 1975. Detergent composition employing alkali metal polysilicates. US3912649. Huber Corp.
64. Just, G. 1994. Process for the hydrothermal production of crystalline sodium disilicate. US5356607. Henkel KGAA.
65. Denkewicz, R. P., Borgsted, E. V. R. 1994. Functional Properties of Zeolite NaA/Silicate-Based Builder Systems. In: Cahn, A., Ed. *Proceedings of the Third World Conference on Detergents Global Perspectives*, New Horizons pp. 213–220.
66. Upadek, U., Kottwitz, B., Schreck, B. 1996. Zeolithe und neuartige Silicate als Waschmittelrohstoffe. *Tenside Surf. Det*. 33:385–392.
67. Pierce, R. 1975. Process for making granular hydrated alkali metal silicate. US3918921. Philadelphia Quartz Company.
68. Pierce, R. H. 1976. Compacted alkali metal silicate. US3931036. Philadelphia Quartz Company.
69. Steinreich, J. S. 1975. Production of high bulk density spray dried hydrous sodium silicate. US3875282. Stauffer Chemical Co.
70. Dokter, W. H., De Koning, H. J. M. 2001. Compacted sodium silicate. US6225280. Akzo-PQ Silica Vof.
71. Delwel, F., Osinga, T. J., Theunissen, J. P., Vrancken, J. M. 1994. Granular alkali metal silicate production. US5340559. Unilever.
72. Vezzani, C. 1995. Method of making a granular product with a high specific weight. US5409643. Vomm Impianti E Processi S.r.l.
73. Le Roux, J., Boittiaux, P., Joubert, D., Kiefer, J. 1992. Alkaline metal silicate based builder for detergent compositions. EP0488868. Rhone Poulenc Chimie.
74. Boittiaux, P., Taquet, P., Joubert, D. 1993. Builder based on silicate and a mineral product. EP0561656. Rhone Poulenc Chimie.
75. Falbe, J., Regitz, M., Eds. 1998. *Römpp-Lexikon Chemie*. Vol. 4. 10th ed. Stuttgart, New York: Thieme, p. 2839.
76. O'Neil, M. J., Smith, A., Heckelmann, P. E., Obenchain, J. R., Gallipeau, J. A. R., D'Arecca, M. A., Eds. 2001. *The Merck Index*. 13th ed. Whitehouse Station, NJ: Merck & Co., Inc., pp. 1543–1554.
77. Curll, D. B. 1941. Manufacture of silicates. US2239880. Philadelphia Quartz Company.
78. De Liedekerke, M. R. F., Michaels, W. M. J., Theunissen, J. P. H. 1975. Verfahren zur Herstellung von wasserfreien Alkalimetasilikaten. DE2519265. Zinkwit Nederland B.V.
79. Baker, C. L., Holloway, P. W. 1965. Spherical particles of anhydrous sodium metasilicate and method of manufacture thereof. US3208822. Philadelphia Quartz Company.
80. Baker, C. L., Holloway, P. W. 1969. Alkali metal silicates. GB1149859. Unilever.
81. Vrisakis, G., Chastel, J. 1981. Granulation of sodium metasilicate. US4253849. Rhone-Poulenc Industries.
82. Godard, G., Joubert, D., Gagnaire, P. 1985. Sodium metasilicate particulates and detergent compositions comprised thereof. US4518516. Rhone-Poulenc Chimie De Base.

83. Albertshauser, F. 1932. Verfahren zur Herstellung löslicher, schüttbarer und lagerbeständiger Natrium-silicatverbindungen. DE592292. Henkel & Cie G.m.b.H.
84. Jelen, F. C. 1962. Improvements in or relating to Silicates. GB908803. Cowles Chemical Company.
85. Joubert, D., Parker, P. 1989. Use of metasilicate/silica combination granulate in detergent compositions for washing machines. US4844831. Rhone Poulenc Chimie.
86. Detering, J., Bertleff, W., Essig, M., Kistenmacher, A. 1999. Influence of Various Builder Systems on the Composition and Morphology of Textile Incrustations during Laundering. *Tenside*. 36:399–407.
87. Tokuyama Siltech Co., Ltd. 2002. Tokuyama: Tokuyama Siltech Co., Ltd.
88. Kahlenberg, V., Dörsam, G., Wendschuh-Josties, M., Fischer, R. X. 1999. The crystal structure of δ-Na2Si2O5. *J. Solid. State. Chem.* 146:380–386.
89. de Lucas, A., Rodriguez, L., Sánchez, P., Lobato, J. 2000. Synthesis of Crystalline Layered Sodium Silicate from Amorphous Silicate for Use in Detergents. *Ind. Eng. Chem. Res.* 39:1249–1255.
90. Bauer, H. 1998. Schwermetallbindevermögen Silikatischer Waschmittelbuilder. *SÖFW Journal.* 124:698–701.
91. Bauer, H., Schimmel, G., Jürges, P. 1999. The Evolution of Detergent Builders from Phosphates to Zeolites to Silicates. *Tenside Surf. Det.* 36:225–229.
92. Bauer, H., Holz, J., Schimmel, G. 2002. Granular alkali metal phyllosilicate compound. US6455491. Clariant GmbH.
93. Bauer, H. 2000. Raw Materials in Laundry Detergent Tablets. *SÖFW Journal* 126:12–16.
94. Clariant. 2000. *The Performance Builder for a Bright Future* [TM]*SKS-6*. Sulzbach/Ts, Germany: Clariant.
95. Schimmel, G., Kotzian, M., Panter, H., Tapper, A. 1993. Process for producing crystalline sodium silicates having a layered structure. US5236682. Hoechst Aktiengesellschaft.
96. Schimmel, G., Gradl, R., Schott, M. 1993. Process for the preparation of crystalline sodium silicates having a sheet structure. US5211930. Hoechst Aktiengesellschaft.
97. Tapper, A., Schimmel, G., Rieck, H. P., Noltner, G. 1996. Process for preparing granular sodium silicates. US5520860. Hoechst AG.
98. Fawer, M., Concannon, M., Rieber, W. 1999. Life Cycle Inventories for the Production of Sodium Silicates. *Int. J. LCA.* 4:207–212.
99. Schweiker, G. C. 1987. Sodium Silicates and Sodium Aluminosilicates, A Worldwide Update. In: Baldwin, A. R. *Proceedings of the 2nd World Conference on Detergents*. New Horizons: AOCS, pp. 63–68.
100. Berth, P. 1978. Recent Develoments in the Field of Inorganic Builders. *J. Am. Oil Chem. Soc.* 55:52–57.
101. Berth, P. 1978. Neuere Entwicklungen bei anorganischen Buildern. *Tenside Det.* 15:176–180.
102. Condea Augusta S. p. A. 1998. Condea Vegobond AF, Technical Sheet 801, May 98. Milan: Condea Augusta S. p. A.
103. Ettlinger, M., Ferch, H. 1979. HAB A 40 für Waschmittel. *Seifen-Öle-Fette-Wachse.* 105:131–135, 160.
104. Christophilante, P., Worms, K. H. 1983. Synthetische Zeolithe. In: Winnacker, K., Küchler, L., Eds. *Chemische Technologie*. Vol. 3. 4th ed. München: Carl Hanser Verlag, pp. 63–75.
105. Zatta, A., Clerici, R., Faccetti, E., Mattioli, P. D., Rabaioli, M. R., Radici, P., Aiello, R., Crea, F. 1997. Zeolite AX—A New Zeolite Builder for Detergents. *XXVII Journadas del Comite Espanol de la Detergencia*, 27:71–82.
106. Breck, D. W. 1974. *Zeolite Molecular Sieves*. New York: Wiley.
107. O'Neil, M. J., Smith, A., Heckelmann, P. E., Obenchain, J. R., Gallipeau, J. A. R., D'Arecca, M. A., Eds. 2001. *The Merck Index*. 13th ed. Whitehouse Station, NJ: Merck & Co, Inc., p. 1808.
108. Vaughan, D. E. W. 1988. The Synthesis and Manufacture of Zeolites. *Chem. Eng. Prog.* 8:25–31.
109. Hauthal, H. G. 1996. Detergent Zeolites in an Ecobalance Spotlight. *SÖFW-Journal.* 122:899–911.
110. Puppe, L. 1979. Molekularsiebe. In: Bartholomé, E., Biekert, E., Hellmann, H., Ley, H., Weigert, W. M., Weise, E., Eds. *Ullmanns Encyklopädie der Technischen Chemie*. Vol. 17. 4th ed. Weinheim: Verlag Chemie GmbH, pp. 9–18.
111. Falbe, J., Regitz, M., Eds. 1999. *Römpp-Lexikon Chemie*. Vol. 6. 10th ed. Stuttgart, New York: Thieme, pp. 5053–5055.
112. Schwochow, F., Puppe, L. 1975. Zeolites. Production, Structure, Application. *Angewandte Chemie.* 87(18):659–667.

Production of Silicates and Zeolites   **417**

113. Frilette, V. J., Kerr, G. T. 1963. Process form making crystalline zeolites. US3071434. Socony Mobil Oil Co.
114. Michel, M., Papee, D. 1969. Method for the preparation of 4 angstrom unit zeolites. US3433588. Produits Chimiques Pechiney-Saint-Gobain.
115. ZEODET. 2000. *Zeolites for Detergents*. Brussels: ZEODET Association of Detergent Zeolite Producers.
116. Breck, D. W., Anderson, R. A. 1981. Molecular Sieves. In: Grayson, M. *Kirk-Othmer Encyclopedia of Chemical Technology*. Vol. 15. 3rd ed. New York: Wiley, pp. 638–669.
117. Strack, H., Roebke, W., Kneitel, D., Parr, E. 1978. Verfahren zur Herstellung eines kristallinen Zeolith-pulvers des Typs A. DE2660722. Degussa AG.
118. Christophliemk, P., Wust, W., Koch, O., Carduck, F. J., Peters, B. W., Vogler, R. 1981. Process for the preparation of very fine zeolitic sodium aluminium silicates. EP0037018. Degussa, Henkel KGAA.
119. Fawer, M. 1996. *Life Cycle Inventory for the Production of Zeolite A for Detergents*. St. Gallen: EMPA.
120. Flanigen, E. M. 1984. Molecular Sieve Zeolite Technology: The first twenty-five years. In: Ramoa Ribeiro, F., Ed., *Zeolites: Science and Technology (NATO ASI Series, Series E: Applied Sciences)*. 1st ed. Berlin: Springer, pp. 3–34.
121. HERA. 2004. HERA Human & Environmental Risk Assessment on Ingredients of European Household Cleaning Products Zeolite A. Version 3.0. http://www.heraproject.com (accessed December 26, 2004).
122. Jakobi, G., Löhr, A., Schwuger, M. J., Jung, D., Fischer, W. K., Gerike, P., Künstler, K. Detergents. 1987. In: Gerhartz, W., Yamamoto, Y. S., Kaudy, L., Pfefferkorn, R., Rounsaville, J. F., Eds. *Ullmann's Encyclopedia of Industrial Chemistry*. Vol. A8. 5th ed. Weinheim: VCH Verlagsgesellschaft mbH, pp. 315–448.
123. Kleinschmit, P. 1995. Zeolites. In: Thompson, R., Ed. *Industrial Inorganic Chemicals: Production and Uses*. Cambridge: Royal Society of Chemistry. pp. 327–349.
124. Nishi, K., Thompson, R. W. 2002. Synthesis of Classical Zeolites. In: Schueth, F., Sing, K. S. W., Weitkamp, J., Eds. *Handbook of Porous Solids*. Vol. 2. Weinheim: Wiley-VCH Verlag GmbH Co. KGaA, pp. 736–814.
125. Roland, E. 1989. Industrial Production of Zeolites. In: Karge, H. G., Weitkamp, J., Eds. *Zeolites as Catalysts, Sorbents, and Detergents Builders. Studies in Surface Science and Catalysis*. Vol. 46. Amsterdam: Elsevier. pp. 645–659.
126. Roland, E., Kleinschmidt, P. 1996. Zeolites. In: Elvers, B., Hawkins, S., Eds. *Ullmann's Encyclopedia of Industrial Chemistry*. Vol. A28. 5th ed. Weinheim: VCH Verlagsgesellschaft mbH, pp. 475–504.
127. Milton, R. M. 1959. Molecular sieve adsorbents. US2882243. Union Carbide Corporation.
128. Roebke, W., Kneitel, D., Parr, E. 1980. Process for the production of aluminum silicates. US4222995. Deutsche Gold- und Silber-Scheideanstalt vormals Roessler, Henkel & Cie.
129. Strack, H., Roebke, W., Kneitel, D., Parr, E. 1981. Crystalline zeolite powder of type A (IV). US4303628. Degussa & Henkel Kommanditgesellschaft auf Aktien.
130. Strack, H., Roebke, W., Kneitel, D., Parr, E. 1981. Crystalline zeolite powder of type A (I). US4303629. Degussa & Henkel Kommanditgesellschaft auf Aktien.
131. Corkill, J. M., Madison, B. L., Burns, M. E. 1986. Detergent compositions containing sodium aluminosilicate builders. US4605509. The Procter & Gamble Company.
132. Parr, E., Kneitel, D., Roebke, W. 1976. Process for the production of grit-free zeolitic molecular sieves. GB1517323. Henkel & Cie GmbH, Degussa.
133. Roebke, W., Kneitel, D., Parr, E. 1978. Process for the production of crystalline zeolitic molecular sieves of type A. US4073867. Deutsche Gold- und Silber-Scheideanstalt vormals Roessler, Henkel u. Cie GmbH.
134. Strack, H., Roebke, W., Kneitel, D., Parr, E. 1981. Crystalline zeolite powder of type A (V). US4303626. Degussa & Henkel Kommanditgesellschaft auf Aktien.
135. Strack, H., Roebke, W., Kneitel, D., Parr, E. 1981. Crystalline zeolite powder of type A (III). US4305916. Degussa & Henkel Kommanditgesellschaft auf Aktien.
136. Strack, H., Roebke, W., Kneitel, D., Parr, E. 1985. Process for preparing zeolite powder of type A (IV). US4551322. Degussa Aktiengesellschaft, Henkel Kommanditgesellschaft auf Aktien.
137. Wolf, F., Bergk, K. H. 1982. Fortschritte auf dem Gebiet der Molekularsiebe: Zur kontinuierlichen Herstellung von Zeolith A. *Swiss Chem.* 4(3a):61–74.

138. Weber, H. 1962. Process for the production of sodium zeolite A. US3058805. Bayer AG.
139. Kettinger, F. R., Laudone, J. A., Pierce, R. H. 1979. Preparing zeolite NaA. US4150100. PQ Corporation.
140. Feijen, E. J. P., Martens, J. A., Jacobs, P. A. 1999. Hydrothermal Zeolite Synthesis. In: Ertl, G., Knoezinger, H., Weitkamp, J., Eds. *Preparation of Solid Catalysts*. Weinheim: Wiley-VCH Verlag GmbH, pp. 262–284.
141. Strack, H., Roebke, W., Kneitel, D., Parr, E. 1981. Crystalline zeolite powder of type A (VI). US4303627. Degussa & Henkel Kommanditgesellschaft auf Aktien.
142. Fawer, M., Postlethwaite, D., Klüppel, H. 1998. Life Cycle Inventory for the Production of Zeolite A for Detergents. *J. Int. J. LCA.* 3(2):71–74.
143. INEOS Silicas. 2002. Warrington: INEOS Silicas.
144. Coope, B. M. 1989. Synthetic Silicas & Silicon Chemicals. *Ind. Miner. (London).* 298:43–55.
145. Griffith, J. 1987. Silica—Is the Choice Crystal Clear? *Ind. Miner. (London).* 235:25–43.
146. Lange, J. 1980. *Rohstoffe der Glasindustrie.* Leipzig: Dtsch. Verlag Grundstoffind.
147. Pincus, A. G., Davies, D. H., Eds. 1983. *Major Ingredients. Raw Materials in the Glass Industry*, Part 1. New York: Ashlee Publications.
148. Falbe, J., Regitz, M., Eds. *Römpp-Lexikon Chemie.* Vol. 5. 10th ed. Stuttgart, New York: Thieme, p. 3673.
149. Söll, J. 1999. Hochreiner Quarzsand für beste Glasqualität. *Glas Ingenieur.* 5:2–4.
150. Chemlink consultants. 1997. Alumina, Aluminium chemicals & zeolites. http://www.chemlink.com. au/alumina.htm (assessed December 26, 2004).
151. Anderson, W. A., Haupin, W. E. 1978. Aluminium and Aluminium Alloys. In: Grayson, M. *Kirk-Othmer Encyclopedia of Chemical Technology.* Vol. 2. 3rd ed. New York: Wiley, pp. 129–188.
152. Helmboldt, O., Hudson, L. K., Stark, H., Danner, M. 1985. Aluminium Compounds, Inorganic. In: Gerhartz, W., Yamamoto, Y. S., Campbell, F. T., Pfefferkorn, R., Rounsaville, J. F., Eds. *Ullmann's Encyclopedia of Industrial Chemistry.* Vol. A1. 5th ed. Weinheim: VCH Verlagsgesellschaft mbH, pp. 527–541.
153. Hudson, L. K. 1985. Aluminium Oxide. In: Gerhartz, W., Yamamoto, Y. S., Campbell, F. T., Pfefferkorn, R., Rounsaville, J. F., Eds. *Ullmann's Encyclopedia of Industrial Chemistry.* Vol. A1. 5th ed. Weinheim: VCH Verlagsgesellschaft mbH, pp. 557–594.
154. Falbe, J., Regitz, M., Eds. 1996. *Römpp-Lexikon Chemie.* Vol. 1. 10th ed. Stuttgart, New York: Thieme, p. 371.
155. Frigge, W. 1953. Oxide und Hydroxyde des Aluminiums. In: Foerst, W., Ed. *Ullmanns Encyklopädie der Technischen Chemie.* Vol. 3. 3th ed. München-Berlin: Urban & Schwarzenberg, pp. 368–407.
156. Kleinschmit, P. 1995. Aluminium Hydroxide. In: Thompson, R., Ed. *Industrial Inorganic Chemicals: Production and Uses.* Cambridge: Royal Society of Chemistry, pp. 278–280.
157. MacZura, G., Goodboy, K. P., Koenig, J. J. 1978. Aluminium Oxide (Alumina). In: Grayson, M. *Kirk-Othmer Encyclopedia of Chemical Technology.* Vol. 2. 3rd ed. New York: Wiley, pp. 218–244.
158. Falbe, J., Regitz, M., Eds. 1996. *Römpp-Lexikon Chemie.* Vol. 1. 10th ed. Stuttgart, New York: Thieme, p. 139.
159. Schepers, B., Neuwinger, H. D., Sroka, R. 1974. Aluminiumoxid. In: Bartholomé, E., Biekert, E., Hellmann, H., Ley, H., Eds. *Ullmanns Encyklopädie der Technischen Chemie.* Vol. 7. 4th ed. Weinheim: Verlag Chemie GmbH, pp. 293–332.
160. Busler, W. R. 1978. Aluminates. In: Grayson, M. *Kirk-Othmer Encyclopedia of Chemical Technology.* Vol. 2. 3rd ed. New York: Wiley, pp. 195–202.
161. Heck, W., Neuwinger, H. D., Schepers, B. 1974. Aluminiumverbindungen, Anorganische. In: Bartholomé, E., Biekert, E., Hellmann, H., Ley, H., Eds. *Ullmanns Encyklopädie der Technischen Chemie.* Vol. 7. 4th ed. Weinheim: Verlag Chemie GmbH, pp. 333–342.
162. Falbe, J., M. Regitz, Eds. 1998. *Römpp-Lexikon Chemie.* Vol. 4. 10th ed. Stuttgart, New York: Thieme, p. 2820.
163. Sittig, M. 1968. *Inorganic Chemical and Metallurgical Process Encyclopedia.* London: Noyes Development Corporation, p. 649.
164. Ashley, K. D., Sanborn, W. E. 1956. Composition comprising stable solid sodium aluminate and method of manufacturing the same. US2734796. American Cyanamid Co.
165. Kleinschmit, P. 1995. Sodium Aluminate. In: Thompson, R., Ed. *Industrial Inorganic Chemicals: Production and Uses.* Cambridge: Royal Society of Chemistry, p. 315.
166. Milton, R. M. 1959. Molecular sieve adsorbents. US2882244. Union Carbide Corporation.

Production of Silicates and Zeolites

167. Sittig, M. 1968. *Inorganic Chemical and Metallurgical Process Encyclopedia*. London: Noyes Development Corporation, p. 856.
168. Brown, G. T, Osinga, T. J., Parkington, M. J., Steel, A. T. 1990. Zeolite P, process for its preparation and its use in detergent compositions. EP0384070. Unilever.
169. Araya, A. 1993. Aluminosilicates. US5362466. Crosfield Joseph & Sons.
170. Adams, C. J., Araya, A., Carr, S. W., Chapple, A. P., Franklin, K. R., Graham, P., Minihan, A. R., Osinga, T. J., Stuart, J. A. 1997. Zeolite MAP. The New Detergent Zeolite. In: Chon, H., Ihm, S. K., Uh, Y. S., Eds. *Progress in Zeolite and Microporous Materials. Studies in Surface Science and Catalysis.* Vol. 105. Amsterdam: Elsevier Science BV, pp. 1667–1674.
171. Crosfield is Zealous about Zeolites. 1994. *European Chemical News*. July, 18 1994, 61:22.
172. Adams, C. J., Araya, A., Graham, P., Hight, A. 1996. Use of aluminosilicates of the zeolite p type as low temperature calcium binders. US5560829. Unilever.
173. Condea Augusta S. p. A. 1997. *Vegobond AX, Technical Sheet 803, April 97*. Milan: Condea Augusta S. p. A.
174. Aiello, R., Crea, F., Radici, P., Zatta, A., Mattioli, P. D., Rabaiolo, M. R. 1998. Microporous crystalline material, a process for its preparation and its use in detergent compositions. EP0816291. Condea Augusta S. p. A.
175. Mobil Oil Corporation. 1980. Manufacture of low-silica faujasites. GB1580928. Mobil Oil Corporation.

23 Production of Inorganic and Organic Bleaching Ingredients

Noel S. Boulos

CONTENTS

23.1 Introduction 421
23.2 Active Chlorine 422
    23.2.1 Sodium Hypochlorite 422
    23.2.2 Calcium Hypochlorite 423
    23.2.3 Organic Chlorinated Compounds 423
23.3 Active Oxygen 424
    23.3.1 Hydrogen Peroxide 424
    23.3.2 Persalts 426
        23.3.2.1 Sodium Perborate 426
        23.3.2.2 Sodium Percarbonate 428
    23.3.3 Percarboxylic Acids 430
        23.3.3.1 Reaction with Activators 430
        23.3.3.2 Equilibrium Peracids 432
        23.3.3.3 Preformed Peracids 433
    23.3.4 Inorganic Peracids 433
References 433

23.1 INTRODUCTION

A bleaching agent is an ingredient that can lighten the color of a substrate. The reaction involves destruction or modification of the chromophoric groups that impart the color. It could also involve breaking the color bodies into smaller entities that are more soluble and can be more easily removed in the bleaching process.

In the cleaning industry, bleaching is performed with oxidizing agents such as chlorine derivatives and peroxygen products. However, before the turn of the twentieth century, bleaching was mostly done by subjecting clothes to sunlight after washing. Oxidizing bleaching agents can also be powerful disinfectants. This chapter describes the production of the two main types of bleaches used in detergents and other cleaning products—compounds generating active chlorine or active oxygen.

23.2 ACTIVE CHLORINE

Active chlorine compounds are those that contain or are able to release the hypochlorous ion OCl^- or $MOCl$ in solution, where M is a common metal such as Na^+ and K^+. The most common products are as follows:

- *Inorganic* such as alkali metal or alkaline earth metal hypochlorite (e.g., NaOCl and $Ca(OCl)_2$).
- *Organic* containing a N-chloro moiety, which releases OCl^- on hydrolysis.

The concentration of the active ingredient is usually referred to as active or available chlorine. This relates to the amount of chlorine formed on reaction with acids, although under alkaline conditions chlorine, as such, is not released, but dissociation occurs to form the hypochlorous ion (OCl^-).

23.2.1 Sodium Hypochlorite

Sodium hypochlorite solutions [7681-52-9] have been used in households in southwestern Europe and the United States since 1869 and 1918, respectively [1]. They contain 4–6% of the active ingredient and their formulation has changed little since then. Most of the industrially produced sodium hypochlorite solutions are obtained from the dilute chlorine streams of chlor-alkali plants by reaction with caustic soda [2–4].

$$Cl_2 + 2NaOH \longrightarrow NaOCl + NaCl + H_2O \tag{23.1}$$

Production of sodium hypochlorite

A typical production process involves a continuous reaction between chlorine and alkali, resulting in concentrations of available chlorine of up to 15%. Chlorine gas is diluted with air and fed into a chlorination column packed with Raschig rings. Caustic soda is diluted with water to the desired concentration in a tank and is circulated through a heat exchanger, the chlorination column, and the tank. The resulting mixture of caustic soda and sodium hypochlorite is circulated until the desired concentration of hypochlorite is reached. A valve is then opened to withdraw the hypochlorite solution, and the process is repeated by adding more caustic soda solution to the tank. To reduce the decomposition of sodium hypochlorite, temperature must be controlled in the range 30–35°C (Figure 23.1) [5].

On-site electrolytic generation of dilute hypochlorite solutions is gaining increasing popularity. It is based on the electrolysis of sodium chloride into chlorine, which then hydrolyzes into hypochlorous acid. The latter dissociates into hypochlorite and chloride. Dilute hypochlorite solutions of up to 10% available chlorine can be obtained from these units. On-site generation is suitable for drinking water treatment and various environmental applications. This process is rarely used in industrial laundries.

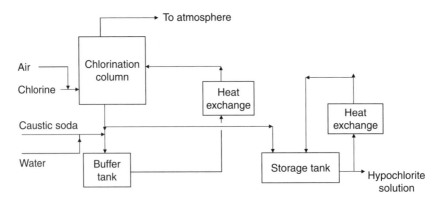

FIGURE 23.1 Flow diagram for the production of sodium hypochlorite. (Courtesy of Solvay Chemicals International, Brussels, Belgium.)

Production of Inorganic and Organic Bleaching Ingredients

23.2.2 CALCIUM HYPOCHLORITE

The main industrial product is the most stable dihydrate form $Ca(OCl)_2 \cdot 2H_2O$ [775854-3], which contains up to 16% water and 65% available chlorine. It also contains sodium chloride and small amounts of calcium hydroxide, calcium chloride, calcium chlorate, and calcium carbonate. The main use of calcium hypochlorite in the cleaning industry is not in laundry but in toilet bowl cleaners, which were introduced in the United States during the late 1970s [6]. It is also sometimes used for bleaching in commercial laundries, in areas where chlorine is not available [7].

The product is manufactured by passing dilute chlorine gas through columns of high-quality hydrated lime containing low levels of impurities [6].

$$2Cl_2 + 2Ca(OH)_2 \longrightarrow Ca(OCl)_2 \cdot 2H_2O + CaCl_2 \tag{23.2}$$

$$\underset{\text{Chlorine}}{} \quad \underset{\text{Calcium}}{} \quad \underset{\text{Calcium}}{} \quad \underset{\text{Calcium}}{}$$
$$\underset{\text{hydroxide}}{} \quad \underset{\text{hypochlorite}}{} \quad \underset{\text{chloride}}{}$$

Production of calcium hypochlorite

The process is designed to minimize the amounts of unwanted salts. It is most important to reduce the level of the coproduct $CaCl_2$ as it is hygroscopic and hinders drying while negatively affecting product stability. Its presence also creates filtration problems as it prevents the formation of large crystals [6].

In the Olin process, chlorination of a slurry of hydrated lime in sodium hypochlorite leads to the crystallization of the triple salt $Ca(OCl)_2 \cdot NaOCl \cdot NaCl \cdot 12H_2O$. This product is filtered and the filter cake is mixed with a chlorinated lime slurry. The molar concentration of the NaOCl in the triple salt is equivalent to that of $CaCl_2$ in the chlorinated lime. Heating this mixture leads to the precipitation of neutral calcium hypochlorite dihydrate crystals, which are filtered and dried [6].

The PPG process uses hypochlorous acid (HOCl) for the production of calcium hypochlorite. It consists of a rotating tubular reactor in which carbon dioxide containing ~10% chlorine and saturated with water vapor at 20–30°C runs counterflow to $Na_2CO_3 \cdot H_2O$. The HOCl gas containing 1–2% Cl_2O is produced, which is scrubbed in water generating a 10–15% solution of HOCl. Lime slurry is then added producing a 15–20% solution of $Ca(OCl)_2$. Spray drying this solution generates a product with an intermediate water content, which can be mixed with salt, compacted, granulated, and dried further. Excess Cl_2 and most of the CO_2 introduced to the process are recycled [6].

23.2.3 ORGANIC CHLORINATED COMPOUNDS

These are solid products derived from cyanuric acid; they slowly release hypochlorite ions in solution. There are three commercial N-chloroisocyanuric acid derivatives: sodium dichloroiso-cyanurate [51580-86-0], potassium dichloroisocyanurate [2244-21-5], and sodium trichloroisocy-anurate [87-90-1]. They are mainly used as household bleaches and in industrial and institutional cleaners.

Cyanuric acid is first produced by heating molten urea at temperatures between 200 and 300°C for several hours. Ammonia is obtained as a by-product.

$$\tag{23.3}$$

Urea   Cyanuric acid

Production of cyanuric acid

N-chloroisocyanuric acid derivatives are produced in a continuous process involving the reaction of chlorine with cyanuric acid and a solution of alkali at 0–15°C. The reaction conditions should be carefully controlled to avoid the formation of the explosive NCl_3 [8,9].

$$\text{Cyanuric acid} + Cl_2 + NaOH \longrightarrow \text{Sodium dichloroisocyanurate} \tag{23.4}$$

Production of dichloroisocyanurate

23.3 ACTIVE OXYGEN

All peroxygen systems contain the peroxide bond –O–O–. The simplest peroxide is hydrogen peroxide (H_2O_2). However, the neutral solution is a fairly mild oxidizing agent. Most of the chemistry of H_2O_2 in cleaning and other applications is based on its activation.

The simplest form of activation is to increase the pH of H_2O_2, thus allowing the formation of the perhydroxyl anion HOO^-.

$$H_2O_2 + HO^- \longrightarrow HOO^- + H_2O \tag{23.5}$$

This is the basis for the use of the solid forms of H_2O_2, that is, persalts such as sodium perborate (PBS) and sodium percarbonate (PCS).

If greater oxidizing/bleaching power is required, this can be achieved by converting H_2O_2 into stronger oxidants. These systems include

- Percarboxylic acids such as peracetic acid (PAA)
- Inorganic peracids or their salts such as potassium monopersulfate (KMPS)

The following sections describe the manufacturing process of the components used in different peroxygen bleaching systems.

23.3.1 HYDROGEN PEROXIDE

Hydrogen peroxide solutions [7722-84-1] were first prepared in 1818 by reacting sulfuric acid with barium peroxide. This produced only very dilute H_2O_2 solutions (3–6% w/w). With the introduction of the electrolytic process in 1908, higher concentrations could be made. The process involved the hydrolysis of peroxydisulfuric acid or its salts. The sulfuric acid formed in the process was recycled to make more peroxydisulfuric acid. Up to ~28% w/w H_2O_2 could be obtained by this process. In the mid-1950s, Shell developed the "isopropanol process," whereby isopropanol was oxidized by air to give H_2O_2 and acetone as by-products. This was used by Shell until 1980.

Today, H_2O_2 is almost exclusively produced by "anthraquinone autooxidation," also called the AO process. The reaction is composed of two steps:

Hydrogenation. Alkylanthraquinones (AQ) are catalytically reduced by hydrogen to the corresponding alkylanthrahydroquinone (AQH). The alkyl group can be ethyl [10], or t-butyl [11], and catalysis include Raney nickel or palladium. The catalyst can be suspended, on a fixed bed, or supported. If the catalyst is suspended or supported, a filtration step is included to retain the catalyst, which is then returned to the hydrogenator.

- *Oxidation.* Alkylanthrahydroquinone (AQH) is oxidized by air to yield H_2O_2 and the original AQ.

Production of Inorganic and Organic Bleaching Ingredients

$$(23.6)$$

Hydrogenation

$$(23.7)$$

Oxidation

Production of hydrogen peroxide: the autoxidation process

The net reaction is the generation of H_2O_2 from hydrogen and oxygen. These reactions take place in the so-called working solution, which is circulated through several stages. The main components of the working solution are as follows:

- AQ acts as the oxygen carrier and is present at ~10 to 20% w/w. The selected anthraquinone must have the appropriate solubility profile in the quinone and hydroquinone form and should be resistant to oxidation by H_2O_2.
- The "quinone solvent" allows the anthraquinone derivative to remain in solution. It is usually composed of aromatics with boiling points of 160–240°C. Examples include methylnaphthalene [12], trimethylbenzene [13], and polyalkylatedbenzenes [14].
- The "hydroquinone solvent" maintains the hydroquinone derivative in solution. This is a polar solvent such as an alcohol or ester. Examples include nonyl alcohols [12], N,N-dialkylcarbamates [15], and alkylcyclohexanol esters [16].

The solvents should have a good solubility profile for both the quinone and hydroquinone, and should be stable in both the oxidizer and hydrogenator. They must also be easy to separate from H_2O_2, that is, have low solubility in water and H_2O_2 solution, and have low volatility.

After the oxidation stage, the working solution is sent to the "extraction towers" where H_2O_2 is extracted by a countercurrent flow of water. After separation of the two phases, each phase is subjected to purification stages. The working solution is then adjusted to a specific water concentration and returned to the hydrogenator, whereas the crude H_2O_2 solution (~15 to 40% w/w) is subjected to fractional distillation. The actual chemistry of the process is quite complicated due to the formation of side products during these steps. These by-products can be converted into active quinones by passing the working solution through a regeneration step. Figure 23.2 illustrates the process [17].

Theoretically, the only feedstock for this reaction is hydrogen since the catalyst and the working solution are recycled, and air is used for oxidation. In reality, some of the solvent and anthraquinone are lost in the recycle and these have to be replenished on a regular basis.

On distillation, ~50 to 70% H_2O_2 is obtained. This is then used to generate the various commercial grades which include 70%, 50%, and 35% solutions. Stabilizers are added to all grades, the identity and level of which vary depending on the application and are considered proprietary information by the peroxide manufacturers. There are grades specifically designed to maintain stability on dilution. These are sold to the detergent industry for dilution to 3–6% concentration. Grades with lower additives are sold to the industrial and institutional industry, since the concentrated form is directly injected in the laundry machines by automated systems.

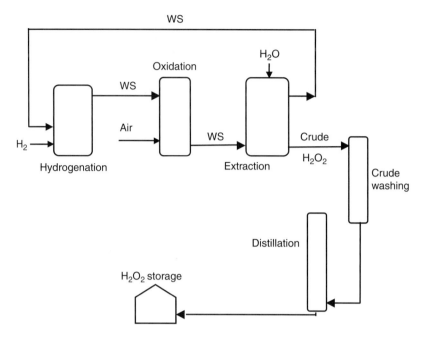

FIGURE 23.2 Hydrogen peroxide process flow diagram. WS = Working solution. (Courtesy of Solvay Chemicals, Inc., Houston, Texas.)

Other H_2O_2 manufacturing technologies are being developed for the on-site generation of dilute H_2O_2 for specific large volume applications such as pulp bleaching. These processes involve the generation of H_2O_2 by an oxygen reduction electrolytic technology or direct combination of hydrogen and oxygen.

23.3.2 PERSALTS

Persalts is a term that is used to refer to solid H_2O_2 carriers, that is, those products that release H_2O_2 on dissolution in water. These include PBS and PCS.

These products are mostly used as color safe bleach in dry bleaches and laundry detergents. They are also gaining popularity in automatic dishwasher detergents as a replacement for chlorinated products. Owing to its stability, PBS was the bleaching product of choice in laundry detergents. In the 1990s, PCS has replaced PBS in many laundry applications especially in Japan, Europe, and the United States.

23.3.2.1 Sodium Perborate

There are two grades of PBS that differ in the amount of water of crystallization present in the molecule. These are sodium perborate tetrahydrate (PBS4), and sodium perborate monohydrate (PBS1).

23.3.2.1.1 Sodium Perborate Tetrahydrate

This is the first solid peroxygen compound used as a bleach. Its nomenclature is confusing as it is not an accurate representation of its structure. When the product was first prepared, its molecular structure was reported as $NaBO_3 \cdot 4H_2O$, hence the nomenclature—tetrahydrate. It has only one oxygen useful for oxidation, called active or available oxygen (AvOx), therefore, the formula was also written as $NaBO_2 \cdot H_2O_2 \cdot 3H_2O$. The crystal structure has been elucidated in the 1960s and the product was found to be made of two molecules linked by two peroxide bonds in a six-membered ring.

$$\begin{bmatrix} HO & O-O & OH \\ & B & B & \\ HO & O-O & OH \end{bmatrix}^{2-} 2Na^+ \cdot 6H_2O$$

Structure of sodium perborate tetrahydrate
($NaBO_3 \cdot 4H_2O$)

Therefore, the product is also called sodium peroxoborate hexahydrate. The IUPAC name is disodium tetrahydroxo-di-μ-peroxo-diborate (III) hexahydrate [10486-00-7].

The commercial product is still called sodium perborate tetrahydrate and the molecular weight is reported as 150 referring to the molecular structure $NaBO_3 \cdot 4H_2O$ and not to the cyclic structure. The available oxygen of the commercial product is ~10%.

The raw materials used for the manufacturing process are sodium tetraborate (borax—$Na_2B_4O_7$), sodium hydroxide solution (NaOH), and hydrogen peroxide solution (H_2O_2). The basis of the process is as follows:

$$Na_2B_4O_7(s) + 2NaOH(l) \longrightarrow 4NaBO_2(l) + H_2O_2 \quad (23.8)$$
Borax $\quad\quad\quad\quad\quad\quad\quad\quad$ Na metaborate

$$NaBO_2(l) + H_2O_2 + 3H_2O \longrightarrow NaBO_3 \cdot 4H_2O(s) \quad (23.9)$$
Na metaborate $\quad\quad\quad\quad\quad\quad$ Na perborate tetrahydrate

Production of sodium perborate tetrahydrate

There are two forms of borate that can be used: the naturally occurring mineral (rasorite or tincal) and the purified borax (borax pentahydrate—5 mol borax—$Na_2B_4O_7 \cdot 5H_2O$).

Borax is dissolved in sodium hydroxide solution to make sodium metaborate ($NaBO_2$) solution. When the native mineral is used, this step is followed by decantation or filtration. The $NaBO_2$ solution is then mixed with H_2O_2 in a cooled reactor under carefully controlled conditions to produce attrition-resistant crystals. This leads to the crystallization of PBS4. Stabilizers are added either before or after the crystallization stage to improve storage and in-pack stability. Possible stabilizers include magnesium sulfate and alkali or alkaline earth metal silicate [18].

The mother liquor is recycled by mixing it with more boron-containing raw material and sodium hydroxide. The excess can be discharged after pH adjustment and reduction of H_2O_2 content to appropriate levels. Figure 23.3 illustrates the process.

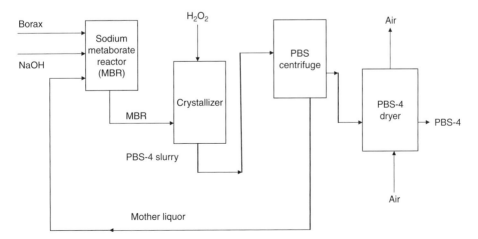

FIGURE 23.3 PBS-4 production flow diagram. (Courtesy of Solvay Chemicals, Inc., Houston, Texas.)

23.3.2.1.2 Sodium Perborate Monohydrate

As for PBS4, the nomenclature of PBS1 is not accurate and is due to the molecular structure historically drawn as $NaBO_3 \cdot H_2O$. It is believed that it has the same cyclic peroxodiborate ring as PBS4 and, as such, does not have any water of hydration [10332-33-9]. Its molecular weight (100) is also reported relative to its molecular structure. The available oxygen of the commercial product is ~15%.

$$\left[\begin{array}{ccc} HO & O-O & OH \\ & B \quad\quad B & \\ HO & O-O & OH \end{array} \right]^{2-} \quad 2Na^+$$

Structure of sodium perborate monohydrate
($NaBO_3 \cdot H_2O$)

PBS1 is obtained by further drying PBS4 to remove its water of hydration in a fluid bed at an inlet air temperature of ~180 to 210°C [19]. In practice, lower temperatures may be used. A cooler exhaust temperature of at least 60°C may be used if the relative humidity is kept at ~10 to 40% [20]. Since water of hydration is removed from inside the particle, the product has the potential for high attrition. However, all PBS1 manufacturers have developed processes that reduce this tendency, and the product in fact has excellent resistance to attrition and can be subjected to pneumatic transport.

23.3.2.2 Sodium Percarbonate

PCS [15630-89-4] is another granular oxygen bleach that has recently gained greater market share at the expense of PBS1. Like the latter, its nomenclature is also inaccurate, as the product does not contain the C–O–O–C structure. It contains H_2O_2 and sodium carbonate in the form of the adduct sodium carbonate perhydrate (also called sodium carbonate peroxyhydrate). Its stoichiometry is $2Na_2CO_3 \cdot 3H_2O_2$. PCS is made by the reaction of soda ash with H_2O_2 as follows:

$$2Na_2CO_3 \ + \ 3H_2O_2 \ \longrightarrow \ 2Na_2CO_3 \cdot 3H_2O_2 \tag{23.10}$$
$$\text{Soda ash} \quad \text{Hydrogen peroxide} \quad\quad \text{Sodium percarbonate}$$

Production of sodium percarbonate

PCS can be made by several processes: crystallization, fluid bed reaction, and direct route.

Crystallization process. This is similar to the one used for PBS4 manufacture (Section 23.2.1.1). Solutions of soda ash and H_2O_2 at the appropriate ratio are mixed in a cooled reactor. Owing to the greater solubility of PCS versus PBS4, a salting-out agent is usually added such as sodium chloride [21,22–24]. Other additives can include stabilizers such as silicates or phosphonates [25] and crystal habit modifiers such as polyacrylates and polyphosphates [26–30]. The crystals are separated from the mother liquor by centrifugation and dried in a fluidized-bed dryer. The mother liquor can be recycled by dissolving more soda ash and feeding the solution back to the crystallizer [22–24]. Process conditions can be manipulated to generate a product with various levels of AvOx and different mean particle size. Figure 23.4 illustrates the process.

Fluid bed process. In this case, solutions of soda ash and H_2O_2 are injected into a fluid bed containing PCS fluidized with hot air [31,32]. Alternatively, spraying and drying can occur in two steps [33]. The dry granules are then cooled in a fluid bed and stored for packaging. Stabilizers are added to the raw material solutions to improve the storage and in-pack stability of the product during its life cycle. Process conditions can be modified to generate products containing different levels of AvOx and mean particle size. Figure 23.5 illustrates the process.

Production of Inorganic and Organic Bleaching Ingredients

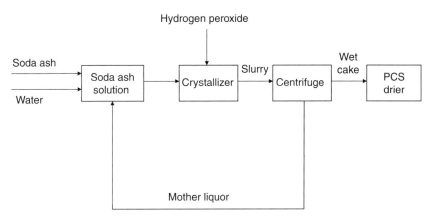

FIGURE 23.4 Crystallizer process for sodium percarbonate production. (Courtesy of Solvay Chemicals, Inc., Houston, Texas.)

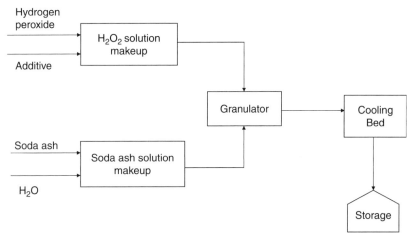

FIGURE 23.5 Fluid bed process for sodium percarbonate production. (Courtesy of Solvay Chemicals, Inc., Houston, Texas.)

Direct processes. These processes involve spraying a concentrated solution of H_2O_2 (50–80%) onto anhydrous or hydrated sodium carbonate granules (75–90% Na_2CO_3). This is done either in a blender [34] or a fluid bed reactor [35]. These processes suffer from extended reaction times as well as limitation on purity and particle size based on the soda ash being used. One process involves the addition of hydroxyalkylidenediphosphonate to enhance stability and assist in dehydrating the sodium carbonate monohydrate formed during the process [36,37]. These processes do not seem to achieve the high AvOx levels obtained by the other production techniques, possibly because of a greater potential for H_2O_2 decomposition during production.

Coating of PCS. Although the storage stability of PCS is high, it is more susceptible to decomposition by moisture than PBS4 or PBS1. This places a limitation on its use as a bleaching agent in zeolite-containing detergents. To allow the incorporation of PCS in these high moisture formulations, manufacturers have developed various coating processes. The coating creates a physical barrier between the product and its surroundings in the detergent formulation and reduces its decomposition rate. Several coatings have been developed and commercialized. Examples include borates [38,39], borate/silicate [40], sodium silicate

23.3.3 PERCARBOXYLIC ACIDS

Percarboxylic acids are stronger oxidizing agents than hydrogen peroxide or persalts. Their general structure is $R–CO_3H$, where R = an organic moiety.

In cleaning applications, they can be produced by three main routes:

- Generated *in situ* by the reaction of H_2O_2 with an activator
- Used as the equilibrium mixture of the raw materials with the peracid
- Preformed, extracted peracids

23.3.3.1 Reaction with Activators

In the past few years, the wash temperatures have dramatically dropped due to a combination of factors including governmental energy-saving campaigns and more. This has lead to a reduction in the bleaching efficiency of persalts.

This was compensated for, in the late 1970s, by the introduction of activators in detergent formulations. These are organic compounds that can be perhydrolyzed in the wash with the generation of a percarboxylic acid, which is a more effective bleaching agent than persalts at the lower washing temperatures. An activator contains at least one acyl group (RCO^-) and a leaving group (L). It reacts with the perhydroxyl anion generated from persalts in the wash as follows:

$$H_2O_2 + OH^- \longrightarrow HOO^- \qquad (23.11)$$
$$\text{(High pH)} \qquad \text{Perhydroxyl anion}$$

$$HOO^- + R–CO–L \longrightarrow R–CO–OOH + L^- \qquad (23.12)$$
$$\text{Activator} \qquad \text{Peracid} \qquad \text{Leaving group}$$

Perhydrolysis of bleach activator

Four hydrophilic activators were originally developed in Europe. They all generated PAA in the wash. These are tetraacetylglycoluril (TAGU), diacetyldioxohexahydrotriazine (DADHT), glucose pentaacetate, and tetraacetylethylenediamine (TAED). Today, the main commercial activator that produces PAA is TAED. The bleaching performance, chemical stability, moderate cost, and superior biodegradability were the major reasons for the success of TAED in Europe.

$$(CH_3CO)_2N–CH_2–CH_2–N(COCH_3)_2$$
Tetraacetylethylenediamine (TAED)

The relatively dilute and cool wash liquors in the Americas have presented a challenge to the industry. A more hydrophobic activator was finally introduced in the United States market during 1988. This is sodium nonanoyloxybenzenesulfonate (NOBS), which, on perhydrolysis in the wash, generates pernonanoic acid. Despite a proliferation of patents on new activators, new molecules have not been commercialized to date.

23.3.3.1.1 Tetraacetylethylenediamine

TAED is prepared by acetylating ethylenediamine ($H_2N–CH_2–CH_2–NH_2$) in a two-stage continuous process [46,47]. The first involves the reaction of ethylenediamine with acetic acid (AcOH) to generate diacetylethylenediamine (DAED) via the intermediate stage of a double salt. Water formed

Production of Inorganic and Organic Bleaching Ingredients

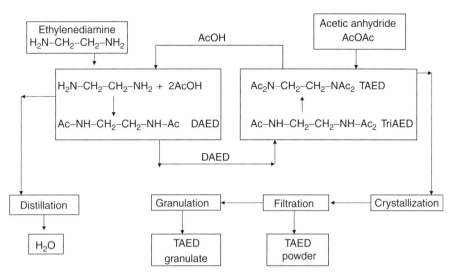

FIGURE 23.6 Flow diagram for production of TAED. (Reproduced from Peractive the clean and clever way of bleaching. A commercial brochure by Clariant GmbH, Basel, Switzerland.)

during the reaction, contaminated with traces of acetic acid, is distilled and disposed of. In the second stage, DAED is reacted with an excess of acetic anhydride (Ac_2O). This forms triacetylethylenediamine (TriAED), which is then converted to TAED. The acetic acid formed in this reaction is recycled to the first stage of the process [46] (Figure 23.6).

The product is purified, filtered, and dried. This gives colorless crystals of TAED of >99% purity. The product, as obtained, has a relatively small particle size and, therefore, a large surface area. This reduces its stability on contact with detergent ingredients. Therefore, the product is further granulated to reduce its contact with incompatible formulation ingredients. Granulation has an effect not only on storage stability but also reduces TAED dissolution rate, which prevents unwanted hydrolysis of the amide bond in the alkaline wash liquor.

Granulation ingredients used include carboxymethylcellulose [48] and surfactants [49]. Stabilizers can also be added in some grades to enhance compatibility with persalts in the detergent formulation.

A suspension of TAED in a structured liquid has been developed for use in a two-component system with alkaline H_2O_2 [50].

23.3.3.1.2 Sodium Nonanoyloxybenzenesulfonate

This is the activator of choice in North America. It is superior to TAED under U.S. washing conditions, that is, low temperatures (30–40°C) and low detergent concentrations. Above 40°C, bleaching performance of NOBS and TAED is equivalent.

$$C_8H_{17}-CO_2-C_6H_4-SO_3^- \; Na^+$$
Sodium nonanoyloxybenzenesulfonate (NOBS)

Several patents describe the production of NOBS starting with phenylsulfonic acid, which is converted to the sodium salt (SPS).

- SPS can be transesterified with phenol nonanoate, which can itself be made by catalyzed high-temperature reaction of phenol, either with alkene and carbon monoxide, or preferably with alcohol [51].

$$C_8H_{17}\text{–}CO_2\text{–}C_6H_5 + HO\text{–}C_6H_4\text{–}SO_3^-\ Na^+ \longrightarrow C_8H_{17}\text{–}CO_2\text{–}C_6H_4\text{–}SO_3^-\ Na^+ + C_6H_5\text{–}OH$$

Phenyl nonanoate   Na phenolsulphonate   NOBS   Phenol   (23.13)

Production of NOBS from phenyl nonanoate

- SPS can be reacted with nonanoic anhydride in the presence of a catalyst at 80–120°C in a polar aprotic solvent such as dimethyl formamide or dimethyl sulfoxide. Nonanoic anhydride is obtained from the reaction of the corresponding carboxylic acid and acetic anhydride at 150°C for 8 h followed by the removal of excess acetic acid/acetic anhydride by vacuum distillation [52,53].

$$C_8H_{17}\text{–}CO_2H + (CH_3CO)_2O \longrightarrow (C_8H_{17}CO)_2O + CH_3CO_2H$$

Nonanoic acid   Acetic anhydride   Nonanoic anhydride   Acetic acid   (23.14)

$$HO\text{–}C_6H_4\text{–}SO_3^-\ Na^+ + (C_8H_{17}CO)_2O \longrightarrow C_8H_{17}\text{–}CO_2\text{–}C_6H_4\text{–}SO_3^-\ Na^+ + C_8H_{17}CO_2H$$

Na phenolsulfonate   Nonanoic anhydride   NOBS   Nonanoic acid

Production of NOBS from nonanoic anhydride   (23.15)

- SPS is reacted with nonanoyl chloride by suspending it in a large excess of nonanoyl chloride at 120°C for 75 min followed by filtration, washing, and drying. Alternatively, only a small excess of nonanoyl chloride is absorbed onto the SPS in a fluid bed at 120–140°C for 45 min, resulting in a powder of the final product [54].

$$C_8H_{17}\text{–}COCl + HO\text{–}C_6H_4\text{–}SO_3^-\ Na^+ \longrightarrow C_8H_{17}\text{–}CO_2\text{–}C_6H_4\text{–}SO_3^-\ Na^+ + HCl$$

Nonanoyl chloride   Na phenolsulphonate   NOBS   Hydrochloric acid

Production of NOBS from nonanoyl chloride   (23.16)

23.3.3.2   Equilibrium Peracids

This refers to mixtures containing a peracid, the corresponding carboxylic acid, hydrogen peroxide and water. These are produced by loading a reactor with the carboxylic acid and water to achieve the desired concentration, and optionally, sulfuric acid to speed up the reaction. Hydrogen peroxide is then added and the mixture is stirred and left to equilibrate. Equilibration time varies depending on the desired peracid concentration, ratios of the raw materials, and process conditions. The concentration and molar ratio of the acid and hydrogen peroxide determine the position of the equilibrium.

The most common commercial equilibrium peracid is PAA. Its preparation is as follows:

$$CH_3CO_2H + H_2O_2 \xrightarrow{\ H^+\ } CH_3CO_3H + H_2O$$

Acetic acid   Hydrogen peroxide   Peracetic acid   Water   (23.17)

Production of peracetic acid

Typical grades used in the industrial and institutional industry are those containing 5 or 1% PAA with various concentrations of H_2O_2 depending on the manufacturer. Recently, a mixture of PAA and octanoic acid has been commercialized as a sanitizer.

Production of Inorganic and Organic Bleaching Ingredients

23.3.3.3 Preformed Peracids

In contrast to equilibrium peracids, preformed peracids refer to percarboxylic acids that were isolated from the reaction mixture and purified for use in formulations. The development of these peracids offers several challenges due to their susceptibility to alkaline hydrolysis, risk of spontaneous combustion, possible spotting of colored fabric, and poor cost/performance.

For these reasons, many products were developed in the past but were never commercialized. In the late 1980s, diperoxydodecanedioic acid (DPDDA) was test marketed as a laundry bleach. Magnesium monoperoxyphthalate (MMPP) was also tested in the 1990s as a bleach in detergent formulations. Today, the only peracid that has been commercialized on a limited scale is ε-phthalimidoperoxycaproic acid or PAP (trade name Eureco).

Production of ε-phthalimidoperoxycaproic acid

(PAP — Eureco)

Eureco is produced by reacting the corresponding carboxylic acid with hydrogen peroxide in a halogenated solvent in the presence of a strong acid catalyst. The carboxylic acid is dissolved in methylene chloride or chloroform and continuously reacted with hydrogen peroxide at a temperature range 10–35°C in the presence of a strong acid. The organic phase is then separated from the aqueous phase, the percarboxylic acid is recovered from the organic phase, and the solvent is recycled [55].

23.3.4 INORGANIC PERACIDS

The main inorganic peracid commercially used in cleaning products is peroxymonosulfuric acid. In the detergent industry, it is available as the triple salt of potassium [37222-66-5]. In the old nomenclature, the formula was written as if it contained three salts—hence the nomenclature $2KHSO_5 \cdot KHSO_4 \cdot K_2SO_4$. The formula should be written in a more general form as $K_5(HSO_5)_2(HSO_4)(SO_4)$ [56]. The product is now more commonly referred to as KMPS. The active oxygen content of the commercial product is 4.0–4.5%. In cleaning formulations, KMPS is introduced in dishwashing detergents and toilet bowl cleaners, but its main use is in denture cleaners.

The commercial product is made in two steps as follows:

- Mixing oleum with concentrated H_2O_2 at cold temperatures generates Caro's acid. This is an equilibrium mixture of peroxomonosulfuric acid and sulfuric acid with a small amount of H_2O_2. It is named Caro's acid after Henrich Caro.
- Partial neutralization of Caro's acid with potassium hydroxide is performed under controlled temperature conditions to avoid loss of available oxygen [57].

REFERENCES

1. Smith, W.L. Human and Environmental Safety of Hypochlorite. In: *Proceedings of the 3rd World Conference in Detergents: Global Perspectives*, Montreux, Switzerland, Sep. 26–30, 1994, pp. 183–92.
2. Murgatroyd's Salt and Chemical Co., GB 984 378, 1962.
3. Stanton, R. E., US 3 222 269, 1962.
4. Milwidsky, B. M. NaOCl. Manufacture of Sodium Hypochlorite Solution. *Manuf. Chem.* 10, 400–404, 1962.
5. Courtesy of Solvay Chemicals International, Brussels, Belgium.

6. Linak, E., Yokose, K. Hypochlorite Bleaches. In: *CEH Marketing Research Report*, ed. California: SRI International, 2003, pp. 508. 2000A–508.2002V.
7. Farr, J.P., Smith, W.L., Steichen, D.S.. Bleaching agents. In: Kroschwitz, J.L., Howe-Grant, M., ed. *Kirk-Othmer, Encyclopedia of Chemical Technology*. Vol. 4. 4th ed. New York: Wiley, 1992, pp. 271–300.
8. Nachbaur, E., Gottardi, W., *Monatsh. Chem.* 97, 115–120, 1966.
9. Shikoku Chem. Ind., FR 1 571 705, *Chem. Abstr.* 72, 1970, 121588, 1969.
10. I.G. Farbenindustrie AG, US 2 158 525, 1936.
11. E.I du Pont de Nemours & Co., US 2 689 169, 1949.
12. E.I. du Pont de Nemours & Co., DE 888 840, 1950.
13. Laporte Chemicals Ltd., DE 953 790, 1953.
14. Degussa, DE 1 261 838, 1963.
15. Kemira, US 114 701, 1990.
16. Laporte Chemicals Ltd., DE 933 088, 1953.
17. Courtesy of Solvay Chemicals, Inc., Houston, Texas.
18. Solvay, DE-AS 1 138 381, 1960.
19. Solvay, DE 1 930 286, 1969.
20. Peroxid-Chemie, DE 2 258 319, 1972.
21. Allied Chemical, US 2 986 448, 1958.
22. DuPont, US 2 541 733, 1947.
23. Kao Soap, US 4 025 609, 1976.
24. Mitsubishi Gas, US 4 409 197, 1983.
25. Peroxid-Chemie, GB 1 278 062, 1980.
26. Solvay & Cie, DE-OS 1 957 825, 1969.
27. Laporte, US 2 448 058, 1944.
28. Degussa, DE-OS 2 800 760, 1978.
29. Peroxid-Chemie, DE 2 328 803, 1973.
30. Kemira, WO 9405594A, 1994.
31. Solvay, DE-OS 2 060 971, 1970.
32. Interox, DE-OS 2 733 935, 1977.
33. Solvay, DE 2 250 720, 1972.
34. Dupont, DE-OS 2 344 017, 1973.
35. Ugine, DE-OS 2 434 531, 1974.
36. FMC, EP 429 321, 1991.
37. FMC Corp., EPA 363 852 A1, 1990.
38. Interox, DE-OS 2 800 916, 1978.
39. Kao, DE 3 321 082, 1982.
40. Kao, GB 2 123 044B, 1986.
41. Peroxid-Chemie, DE-OS 2 700 797, 1977.
42. Degussa, DE 2 651 442, 1976.
43. Interox, GB 1 538 893, 1979.
44. Degussa, DE 1 954 4293, 1995.
45. Interox, EP 279 282, 1987.
46. Peractive the clean and clever way of bleaching. A commercial brochure by Clariant GmbH Basel, Switzerland.
47. Reinhardt, G., Schuler, W., Quack, J.M. TAED Manufacture, Effects and Environmental Properties. In: *Proceedings of the 20th CED/AID meeting (Comunicaciones presentadas a las Jornadas del Comite Espanol de la Detergencia)*, Barcelona, March 8–10, 1989.
48. Hoeschst Aktiengesellschaft, US 5 100 576, 1992.
49. Warwick International Limited, US 4 921 631, 1990.
50. Warwick International Group Limited, US 6 080 710, 2000.
51. Ethyl Corp, EP 125 641, 1987.
52. Rhone Poulenc, EP 153 223, 1986.
53. Procter & Gamble, EP 105 673, 1988.
54. Monsanto, EP 148 148 1985.
55. Ausimont, S.p.A., US 5 208 340, 1993.
56. Comyns, A.E., 1996. Peroxides and Peroxide Compounds. In: Kroschwitz, J.L., Howe-Grant, M., ed. *Kirk-Othmer, Encyclopedia of Chemical Technology*, Vol. 18. 4th ed., New York: Wiley, pp. 202–229.
57. E.I. Du Pont De Nemours & Co., EPA 149 329, 1992.

24 Inorganic Bleaches: Production of Hypochlorite

William L. Smith

CONTENTS

24.1 Hypochlorite Technology Development..436
24.2 Raw Materials 437
    24.2.1 Brine..437
    24.2.2 Chlorine 438
    24.2.3 Sodium Hydroxide..439
    24.2.4 Potassium Hydroxide..440
    24.2.5 Lithium Hydroxide 441
    24.2.6 Lime and Hydrated Lime..441
    24.2.7 Crystalline Trisodium Phosphate..441
24.3 Chemistry..441
    24.3.1 Concentration Terms 441
    24.3.2 Equilibria...442
    24.3.3 Chlorination 443
    24.3.4 Decomposition of Solutions 444
    24.3.5 Decomposition of Solid Hypochlorites 445
    24.3.6 Metal Removal 446
    24.3.7 Temperature Control..446
    24.3.8 Measurement of Chlorination 447
24.4 Process Technology 447
    24.4.1 Sodium Hypochlorite 447
        24.4.1.1 Typical Production...447
        24.4.1.2 Low-Salt (High-Strength) Sodium Hypochlorite 449
        24.4.1.3 Solid Sodium Hypochlorite 451
        24.4.1.4 Electrolytic Hypochlorite Production.................................451
    24.4.2 Potassium Hypochlorite 452
    24.4.3 Hypochlorous Acid..452
    24.4.4 Chlorinated Trisodium Phosphate...454
    24.4.5 Lithium Hypochlorite..454
    24.4.6 Calcium Hypochlorite 455
        24.4.6.1 Bleach Liquor 455
        24.4.6.2 Hemibasic Calcium Hypochlorite 456
        24.4.6.3 Dibasic Calcium Hypochlorite 456
        24.4.6.4 Calcium Hypochlorite 456
        24.4.6.5 Bleaching Powder..461
    24.4.7 Magnesium Hypochlorite 461

24.5	Safety	462
24.6	Materials of Construction	463
24.7	Relative Economics	464
24.8	Future Trends	465
References		465

24.1 HYPOCHLORITE TECHNOLOGY DEVELOPMENT

This chapter reviews the production of sodium, potassium, calcium, and lithium salts of hypochlorite as well as a crystalline complex of hydrated trisodium orthophosphate and sodium hypochlorite known as chlorinated trisodium phosphate. These inorganic hypochlorite salts are used in detergent and cleaning products to kill all types of pathogens, decolorize stains, and break down and remove many kinds of soil. They are also widely used for water and wastewater disinfection and a variety of industrial uses.

Hypochlorites are usually made by the same reaction that was used in 1785 to make the first hypochlorite compound. Chlorine gas was bubbled into an aqueous solution of caustic potash (potassium hydroxide) to make potassium hypochlorite, KOCl. Sodium hypochlorite, NaOCl, was not made until 1820 when caustic soda (sodium hydroxide) was used instead of caustic potash.[1] Today, sodium hypochlorite is the major hypochlorite compound produced, and it is the hypochlorite compound that is most often used for cleaning. Potassium hypochlorite is only a minor product that is used in a limited number of cleaning products to avoid the precipitation of certain components. These compounds are almost always made and used as aqueous solutions that are called bleach. Usually the solution contains ~5–6% or ~12–15% sodium hypochlorite, an equal molar amount of sodium chloride, and 0.01–1% sodium hydroxide.

Sodium hypochlorite did not dominate hypochlorite markets until about 1930. Electrolytic production of chlorine did not begin until 1890, and liquid chlorine could not be shipped before 1909.[2] Before this, bleaching powder was essentially the only form of hypochlorite, or even chlorine, that was marketed.[3]

Bleaching powder was first made in 1790 by passing chlorine gas over finely divided, slightly moist, hydrate of lime (calcium hydroxide). This process is still used, but many improvements have been made, especially to control temperature, remove water, and improve mixing. Bleaching powder is a variable mixture of calcium hypochlorite, $Ca(OCl)_2$; dibasic calcium hypochlorite, $Ca(OCl)_2 \cdot 2Ca(OH)_2$; dibasic calcium chloride, $CaCl_2 \cdot 2Ca(OH)_2$; and small amounts of other calcium salts.[4] A fairly large portion of it does not dissolve in water. Bleaching powder also degrades notably during storage. Beginning about 1920, tropical bleach and stabilized tropical bleach were made from bleaching powder by adding up to 20% of calcium oxide to improve storage stability. However, this also increases the amount of insoluble matter.

An improved type of bleaching powder, hemibasic calcium hypochlorite, $Ca(OCl)_2 \cdot 0.5$ $Ca(OH)_2$, was introduced in Germany in the 1920s.[2] It is still made in Asia by chlorinating lime. The current product contains ~76% hemibasic calcium hypochlorite, 20% lime, and other calcium compounds. It has more insoluble matter, but it is cheaper than other types of calcium hypochlorite. It is more stable and contains more hypochlorite than bleaching powder or tropical bleach.[5] In 1928, calcium hypochlorite, $Ca(OCl)_2$, was introduced to replace bleaching powder in the United States. Its stability and solubility are much better than hemibasic calcium hypochlorite and bleaching powder. The product contained 70–75% calcium hypochlorite, about 1% water, sodium chloride, and small amounts of other calcium salts. In 1979, the calcium hypochlorite was reduced to ~65% and the water increased to 6–12% to reduce the risk of self-sustained decomposition due to organic impurities or ignition.[6–10] Dibasic magnesium hypochlorite, $Mg(OCl)_2 \cdot 2Mg(OH)_2$, was also developed,[11] and it may be produced in Asia. It is safer than calcium hypochlorite because its thermal stability is greater and its decomposition is endothermic rather than exothermic.

Inorganic Bleaches: Production of Hypochlorite

Because it forms insoluble calcium salts with water hardness, soaps, and anionic detergents, the use of calcium hypochlorite in cleaning products has been largely limited to automatic toilet bowl cleaners. It is mostly used to disinfect water and sewage. Calcium hypochlorite was used in commercial laundries and textile mills when its better stability or higher hypochlorite content made it more economical than sodium hypochlorite solutions. However, it was usually converted to sodium hypochlorite before use by mixing it with a solution of sodium carbonate and then removing the calcium carbonate that precipitates. Or, it was dissolved in the presence of enough sodium tripolyphosphate or some other calcium chelating agent to prevent the precipitation of calcium salts. Calcium hypochlorite solutions have been used by industry as they are cheaper to make than sodium hypochlorite. They are made on-site by chlorinating suspensions of lime or hydrated lime. These solutions, known as bleach liquor, have been mainly used by the pulp and paper industry. However, this industry uses much less hypochlorite than in the past.

Chlorinated trisodium phosphate was introduced in 1928, about the same time as calcium hypochlorite. It is a crystalline complex of hydrated trisodium orthophosphate and sodium hypochlorite with an approximate formula of $(Na_3PO_4 \cdot 11H_2O)_4 \cdot NaOCl$. It also contains other phosphate salts and sodium chloride.[12] Chlorinated trisodium phosphate is still used in commercial laundries and disinfectant cleaners. In other products, such as powdered abrasive cleansers and automatic dishwash detergents, it has been largely replaced by chlorinated isocyanurates to reduce cost, improve performance, or comply with phosphate restrictions.

A solid product with 30% lithium hypochlorite, LiOCl, was introduced in 1961.[13] This was shortly after the U.S. Atomic Energy Commission stopped buying large amounts of lithium in 1960. Because it is relatively expensive, the use of lithium hypochlorite in cleaning products is essentially limited to institutional laundries. It is mostly used as a shocking agent for swimming pool disinfection.

Hypochlorite is formed by the *in situ* oxidation of chloride ions by peroxymonosulfuric acid and its salts.[14–16] Ketones such as acetone[17,18] and aromatic diols[19] catalyze the reaction. Bromides can be used in place of chlorides to form hypobromite,[20] and such combinations are used to disinfect spas and hot tubs.

Hypochlorite is also formed by the hydrolysis of solid chloramines. Chlorinated isocyanurates are the most commonly used. Halogenated dialkylhydantoins are used in disinfectant cleaners and institutional laundries, but they dissolve too slowly for most household cleaning products. Small amounts of several other chloramines are used as disinfectants and several others have been unsuccessfully marketed.[21,22]

Point-of-use electrolytic generation of hypochlorite from salt solutions or seawater has been used since 1900.[2,23] It is mostly used to disinfect drinking water, cooling water, swimming pools, and sewage. It is also used in commercial laundries and in health care. Since about 1980, electrolytic generators have been used in appliances[24–26] and devices that dispense dilute solutions for cleaning, disinfection, and hygiene.[23]

24.2 RAW MATERIALS

24.2.1 BRINE

Brine, a solution of sodium chloride, NaCl, in water, is the basic raw material for making inorganic hypochlorite. The electrolysis of brine produces chlorine, Cl_2, and sodium hydroxide, NaOH, which are used to make sodium hypochlorite. The chlorine is also used to make other hypochlorite salts. Dilute solutions of sodium hypochlorite or hypochlorous acid are also made directly by the electrolysis of brine.

Salt is abundant and widely distributed. The worldwide production of salt in 2006 was 240 million tons. This rate of production can continue indefinitely. About 19% was made in the United States, of which 39% was rock salt and 46% was salt in brine. However, 90% of the salt used as a feedstock was salt in brine.[27]

Rock salt is obtained from mines and it is typically 95% sodium chloride. Alternatively, water is pumped into salt deposits to form brine underground. This brine is then removed through a brine well. After purification, the water is removed to make mechanically evaporated salt that is 99.99% sodium chloride. Controlled evaporation and fractional crystallization of seawater yields solar salt that is 99.7% sodium chloride.[28]

Chlorine can be made from all the three kinds of salt. Raw brine is obtained from a brine well or by heating salt with water. To prevent calcium or magnesium salts from depositing on electrodes or clogging membranes in electrolytic cells, sodium carbonate and sodium hydroxide are added to precipitate calcium carbonate and magnesium hydroxide. Iron and some other metals also form precipitates. Ammonia and other nitrogen compounds that may form nitrogen trichloride during electrolysis are removed by adding sodium hypochlorite. This also removes sulfide.[29] If needed, calcium chloride is added to precipitate sulfates. After filtration, the brine typically contains <4 mg kg^{-1} calcium and 0.5 mg kg^{-1} magnesium. This is suitable for diaphragm and mercury cells. If the brine will be used in a membrane cell, it is passed through a series of ion exchange columns to reduce calcium and magnesium concentrations below 20 μg kg^{-1}.[30] Ion exchange resins may also be used to remove metals that can form hydrogen in mercury cells.[31] Finally, hydrochloric acid is added to reduce the pH to ~3.5 to minimize the formation of oxygen and chlorate during electrolysis.

24.2.2 CHLORINE

The global consumption of chlorine in 2005 was ~50 million tons.[32] About 39% of the production capacity was in Asia, 25% in North America, and 27% in Europe. The capacity is expanding in China, followed by the Middle East. Older plants were located close to where the chlorine was used. Recently, larger plants have been built to reduce costs, and these are often located where energy is cheaper. Transportation has been considered less important, but this may change as shipping requirements become more stringent.[33]

Worldwide, 94% of the chlorine is made by the electrolysis of brine. Sodium hydroxide and hydrogen are produced as by-products. A few plants produce potassium hydroxide as well as, or instead of, sodium hydroxide by the electrolysis of potassium chloride solutions. About 3% of the chlorine is made by other processes.[30] These include electrolysis of hydrochloric acid; electrolysis of molten sodium or magnesium chloride, which, respectively, produce sodium or magnesium metal as well as chlorine[30,34]; nitric acid oxidation of potassium chloride to make nitrosyl chloride, which is further oxidized by oxygen to make potassium nitrate,[35] and the oxidation of hydrochloric acid directly with oxygen or air using a catalyst, or indirectly through the formation and subsequent oxidation of metal chlorides.[30,36]

Three types of divided electrical cells are used to make chlorine: mercury, diaphragm, and membrane cells. Diaphragm and mercury cells were both developed in the 1890s. Initially, diaphragm cells were preferred for their simpler operation. Mercury cells became popular after 1940 because they produce higher purity sodium hydroxide solutions that do not need to be concentrated.[37,38] To eliminate mercury emissions, mercury cells are no longer built, and older units are being replaced. Most new cells use membrane technology that was first used in 1975.[2] Membrane cells produce higher-purity sodium hydroxide and use less energy than diaphragm cells. The share of chlorine made in mercury cells has declined in the United States from a peak of 27% in 1970 to 10% in 2004; in Canada from 64% in 1966 to 2.7% in 1996; and in western Europe from 65% in 1995 to 55% in 2000.[39–41] Mercury cells are no longer used in Japan, where all chlorine is made in membrane cells.[41,42] Worldwide chlorine capacity in 2000 consisted of 19% mercury, 33% membrane, and 42% diaphragm cells.[30]

In a diaphragm cell, saturated brine with at least 320 g L^{-1} of sodium chloride flows through a porous diaphragm from the anode into the cathode chamber. The solution leaving the cathode chamber contains 10–12% sodium hydroxide and 13–16% sodium chloride. Hydrogen gas and

Inorganic Bleaches: Production of Hypochlorite

chlorine gas are removed from the cathode and the anode chambers, respectively, according to the following reactions:

$$2H_2O + 2e^- \rightarrow H_2 \text{ (gas)} + 2^-OH$$

$$2Cl^- \rightarrow Cl_2 \text{ (gas)} + 2e^-$$

The diaphragm prevents hydroxide and chlorine from reacting. Older diaphragms are asbestos fibers that have been vacuum-deposited on a hollow steel cathode. These were followed by diaphragms made of fibrous polytetrafluoroethylene (PTFE), with at least 75% asbestos, and then by diaphragms made of PTFE with embedded zirconium oxide or other inorganic materials.[2,30,37]

Membrane cells have two thin chambers separated by a cation-permeable membrane that allows sodium, but not anions such as chloride and hydroxide to pass. The anode and cathode are separated only by the thickness of the membrane. The membrane typically consists of a PTFE support, which is sandwiched between layers of perfluorosulfonate and perfluorocarboxylate polymers. Brine with $\sim$300 g L^{-1} of sodium chloride is circulated through the anode chamber, from which chlorine gas is obtained. Water is pumped into the cathode chamber and exits containing 31–35% sodium hydroxide and <50 mg kg^{-1} of sodium chloride. Hydrogen gas is also obtained from the cathode chamber.[2,30,43]

There are two varieties of diaphragm and membrane cells. They are equally used. Monopolar cells are connected in parallel using external electrical connections. Bipolar cells are connected in series, with the anode of a cell directly connected to the cathode of the adjacent cell. This makes bipolar cells more compact than monopolar cells.[30]

In mercury cells, the cathode is a stream of mercury flowing along the bottom of the electrolyzer. The anodes are suspended closely above and parallel to the mercury. Brine with $\sim$25.5% sodium chloride flows between the mercury and the anode. As in other cells, chlorine is generated at the anode, but the sodium ions dissolve in the mercury to form an amalgam with 0.25–0.5% sodium:

$$Na^+ + (Hg) + e^- \rightarrow Na(Hg)$$

The brine is recycled and the amalgam flows out of the electrolyzer into a graphite-filled decomposer where the graphite is the cathode and the amalgam becomes the anode. Purified water is added to react with the amalgam to form hydrogen and a 50% solution of sodium hydroxide. The following reactions occur at the anode and the cathode, respectively:

$$Na(Hg) \rightarrow Na^+ + (Hg) + e^-$$

$$2H_2O + 2e^- \rightarrow H_2 \text{ (gas)} + 2^-OH$$

The mercury is pumped back into the electrolyzer.[2,30,38]

With all cells, the exiting chlorine is cooled and brine mist is removed. The water content is reduced below 40 mg kg^{-1} on contact with a countercurrent of sulfuric acid in a multistage drier. Acid mist, hydrogen, carbon dioxide, and air are then removed. Chlorine gas is either used directly, compressed or liquefied.

24.2.3 SODIUM HYDROXIDE

The worldwide production of sodium hydroxide (caustic soda) in 2005 was $\sim$54 million tons.[32] More than 98% of the sodium hydroxide is made along with chlorine by the electrolysis of brine. The remainder is made by reacting soda ash with lime.[30] Most sodium hydroxide is sold as a 50% solution. However, it is usually sold in the form of Na_2O, where 77.5 g of Na_2O is the same as 100 g

440   Handbook of Detergents/Part F: Production

of NaOH. Of the sodium hydroxide in the United States, <1.5% is made into a 73% solution and <1.5% is solidified.[39] Although solid sodium hydroxide is cheaper to ship than the solutions, it is more expensive to make, is only available in drums and bags, and must be dissolved before it can be used. The 73% solution is shipped and stored above its freezing point of 62°C, which requires special handling.[44]

The heat of dilution of sodium hydroxide decreases from 4.9 kcal mol^{-1} for a 50% solution to -0.033 kcal mol^{-1} as the concentration decreases to 20%.[45] Thus, sodium hydroxide may be diluted to 20% when received so that it cools during storage to reduce the cooling needed while making hypochlorite. Dilution also decreases corrosion of iron and steel, decreases the freezing point from +12°C to -26°C, and may improve purity by precipitating iron and hardness salts.

As discussed in Section 24.2.2., sodium hydroxide is made in three kinds of electrical cells. The effluent from diaphragm cells contains 10–12% sodium hydroxide, 13–16% sodium chloride, 0.25% sodium sulfate, and 0.15% sodium chlorate. The water is removed in a multistage evaporator that is usually made of nickel. Sodium borohydride or sodium sulfite may be added to remove chlorate, which dissolves nickel by oxidation. This reduces corrosion and lowers the nickel concentration in the finished product.[46] Hydrazine can also be used.[47] Since the solubility of sodium chloride in 50% sodium hydroxide is 1–1.5%, most of the salt precipitates during evaporation and can be recycled into brine. Sodium sulfate also crystallizes in the later stages of evaporation.[48]

A special purified grade of the diaphragm cell sodium hydroxide is made by extracting the 50% solution with liquid ammonia to remove chloride, chlorate, and carbonate.[48] Metal ions, such as iron, nickel, cobalt, and copper, can be removed by electroplating them onto porous cathodes.[49] Electrolysis also removes some anions such as sulfide.[50] Iron and nickel are also removed by filtration using magnesium oxide filter aid after oxidation with hypochlorite.[51]

The concentration of sodium hydroxide in membrane cell effluents depends on the membrane.[30] They typically contain 30–40% sodium hydroxide, <50 mg kg^{-1} of sodium chloride, and only trace amounts of other salts. They are concentrated by evaporation, but there are no salts to dissolve metals, therefore the product does not become contaminated with nickel and other metals. Also, the salts do not separate and additional purification is usually not needed.

In a mercury cell, the brine is removed before sodium hydroxide is formed, thus the resulting solution typically contains <50 mg kg^{-1} of sodium chloride and even lower concentrations of other salts. The amount of water added to the decomposer is usually adjusted to make 50% sodium hydroxide, or rarely 73% sodium hydroxide. These solutions are simply cooled and filtered without the need for evaporation. This avoids contamination by nickel and other metals, which occurs during the concentration of sodium hydroxide from diaphragm cells. Because rayon production requires low concentrations of metals and salt, mercury cell sodium hydroxide is also called rayon grade. However, mercury cell sodium hydroxide typically contains 10–50 μg kg^{-1} of mercury. Some of this mercury ends up in hypochlorite products, although the concentration is reduced by dilution and processing operations such as filtration.

Sodium hydroxide is also made by mixing a 10–14% sodium carbonate solution with lime:

$$Ca(OH)_2 + Na_2CO_3 \rightarrow CaCO_3 + 2NaOH$$

This reaction does not go to completion, therefore only 90% of the stoichiometric amount of lime is added. The calcium carbonate is removed, calcined to reform lime, and recycled. The dilute sodium hydroxide solution is concentrated by evaporation. This causes most of the residual sodium carbonate to precipitate so that it can be removed.[52]

24.2.4 POTASSIUM HYDROXIDE

About 1.4 million tons of potassium hydroxide (caustic potash) were made globally in 2001.[30] Potassium hydroxide is made the same way as sodium hydroxide, by electrolysis of potassium chloride

Inorganic Bleaches: Production of Hypochlorite 441

instead of sodium chloride. Potassium hydroxide is typically sold as a 45–50% solution. Such solutions are obtained directly from mercury cells. These may be solidified by evaporation. Membrane cell effluents contain ~30% potassium hydroxide, from which water is removed by evaporation to produce 45–50% solutions.[53] In diaphragm cells, the cell effluent contains 10–15% potassium hydroxide and ~10% potassium chloride. Most of the potassium chloride crystallizes during concentration so that the final 50% potassium hydroxide solution contains only ~0.6% potassium chloride.[54]

24.2.5 LITHIUM HYDROXIDE

The worldwide production of lithium in all forms was 21,100 t in 2006 (see Ref. 27, p. 96). Lithium sulfate solutions are obtained by treating certain ores with sulfuric acid. Lithium is also produced from certain brines that have a low magnesium content. In both cases, adding carbonate precipitates lithium carbonate. Mixing a lithium carbonate slurry with calcium hydroxide produces lithium hydroxide, and lithium hydroxide monohydrate is crystallized from the supernatant solution. Lithium hydroxide is the least soluble alkali metal hydroxide, with a maximum concentration of $LiOH \cdot H_2O$ of ~19% at room temperature.[55]

24.2.6 LIME AND HYDRATED LIME

Lime, calcium oxide, is made by the thermal decomposition of limestone, marble, chalk, dolomite, oyster shells, and other sources of calcium carbonate. In 2006, 130 million tons of lime was made worldwide (see Ref. 27, p. 94). 100% calcium oxide is available, but most lime contains impurities such as magnesium oxide, silica, ferric oxide, and aluminum oxide.[56] Lime with low magnesium content is preferred for making calcium hypochlorite solutions since magnesium hydroxide interferes with the settling of the sludge.[57] Transition metals, especially iron, manganese, nickel, copper, and cobalt also need to be avoided since these metals catalyze the decomposition of hypochlorite compounds. In addition, lime used for making bleaching powder needs to be completely calcined to avoid carbonate salts, since these cause the loss of hypochlorite during storage.[4]

Hydrated lime, $Ca(OH)_2$, is made by hydrating lime with water in a process called slaking. Calcium hydroxide is a strong base, but only 0.219 g dissolves in 100 g of water (0.22%). Calcium hydroxide is often used because it is cheaper than sodium hydroxide.

24.2.7 CRYSTALLINE TRISODIUM PHOSPHATE

Crystalline trisodium phosphate has a variable formula, $(Na_3PO_4 \cdot 12H_2O)_x \cdot NaOH$, in which x ranges between 4 and 7. It is crystallized below 60°C from an aqueous solution of phosphoric acid that has been neutralized with a slight excess of sodium hydroxide. The crystals are isolated by centrifugation and dried at ~40°C to minimize dehydration. World production of phosphoric acid in 1993 was 33 million tons, of which over 90% is used in agriculture.[12]

24.3 CHEMISTRY

24.3.1 CONCENTRATION TERMS

Available chlorine is the amount of chlorine needed to make the hypochlorite in a solution or a solid. It is the amount of hypochlorite compound multiplied by the number of hypochlorite groups per molecule divided by its molecular weight times the molecular weight of chlorine. The last three terms are combined to derive the theoretical available chlorine factors as shown in Table 24.1. These factors multiplied by the amount of hypochlorite compound equals available chlorine. Available chlorine is usually measured by iodometric titration.[58,59] It is a convenient way of expressing the total activity of mixtures without knowing the concentrations of the components.

TABLE 24.1
Available Chlorine Factors

Compound	Factor
HOCl	1.3516
NaOCl	0.9525
KOCl	0.7830
LiOCl	1.2143
$Ca(OCl)_2$	0.9918
$Ca(OCl)_2 \cdot 2H_2O$	0.7922
$Ca(OCl)_2 \cdot 0.5Ca(OH)_2$	0.7877
$Ca(OCl)_2 \cdot 2Ca(OH)_2$	0.4870
$(Na_3PO_4 \cdot 11H_2O)4 \cdot NaOCl$	0.0466

TABLE 24.2
Equilibria for Hypochlorite Solutions

Equilibrium Reaction	Constant Equation	Constant Value at 25°C	Ionic Strength	Reference
Acid dissociation				
$HOCl \leftrightarrow H^+ + {}^-OCl$	$K = [H^+] [{}^-OCl]/[HOCl]$	3.98×10^{-8} M	1 M	60
Hydrolysis				
$Cl_2 + H_2O \leftrightarrow H^+ + Cl^- + HOCl$	$K = [HOCl] [Cl^-] [H^+]/[Cl_2]$	0.00104 M^2	0.5 M	63
$Cl_2O + H_2O \leftrightarrow 2HOCl$	$K = [HOCl]^2/[Cl_2O]$	86.6 M		64
Trihalide formation				
$Cl_2 + Cl \leftrightarrow Cl_3^-$	$K = [Cl_3^-]/[Cl_2] [Cl^-]$	0.18 M^{-1}	0.5–1 M	68
Henry's Law				
$HOCl \leftrightarrow HOCl$ (g)	$H = P_{HOCl}/[HOCl]$	0.00174 atm M^{-1}	1 M	68, 69[a]
$Cl_2O \leftrightarrow Cl_2O$ (g)	$H = P_{Cl_2O}/[Cl_2O]$	0.32 atm M^{-1}		72[b]
$Cl_2 \leftrightarrow Cl_2$ (g)	$H = P_{Cl_2}/[Cl_2]$	17.9 atm M^{-1}	1.1 M	73

[a] Calculated from the equation: $\ln(H) = 11.04 - 5196/T + 0.03998$ I, where I is the ionic strength in molarity and T is the temperature in degree kelvin. This equation was obtained by linear regression of the data from Imakawa, H., Chemical reactions in the chlorate manufacturing electrolytic cell (part 1); the vapour pressure of hypochlorous acid on its aqueous solution, *J. Electrochem. Soc. Jpn.*, 18, 382, 1950 and Imagawa, H., Studies on chemical reactions of the chlorate cell (part 2); the vapour pressure of hypochlorous acid on its mixed aqueous solution with sodium chlorate, *J. Electrochem. Soc. Jpn.*, 19, 271, 1951.

[b] Extrapolated using $\partial P/\partial T$ from Wojtowicz, J.A. In: Howe-Grant, M., Ed. *Kirk Othmer Encyclopedia of Chemical Technology*, Vol. 5, 4th ed., Wiley, New York, 1993, p. 932 and the equation: $\log(H2) = \log(H1) + \partial P/\partial T (1/T1 - 1/T2)$.

Trade percent is the grams of available chlorine in 100 mL of a sodium hypochlorite solution. This equals the weight percent of available chlorine times density. Grams per liter (gpl or g L^{-1}) is the grams of available chlorine in 1 L of solution. Trade percent and grams per liter are not affected by the density and thus do not vary with the concentration of other components such as sodium hydroxide.

24.3.2 Equilibria

The equilibria for hypochlorite solutions are shown in Table 24.2. The equilibrium constants have been determined for most of these equilibria as a function of temperature at low and moderate ionic strengths.[60-73] The effect of ionic strength has not been well determined, therefore calculations

Inorganic Bleaches: Production of Hypochlorite

become more uncertain with increasing deviations in ionic strength. Neglecting the effects of ionic strength, the ratio of hypochlorite ion to hypochlorous acid as a function of pH does not depend on the available chlorine concentration. Above pH 9.5, more than 99% of the available chlorine will be present as hypochlorite ions. The ratio of hypochlorous acid to hypochlorite ion increases with decreasing pH until pH 5.5, below which <1% of the available chlorine will be hypochlorite ions. Below pH 6, chlorine may be present. Its amount increases with decreasing pH and increasing total available chlorine. With 0.1% available chlorine, chlorine begins to appear at about pH 4 and becomes dominant about pH 2.5, whereas with 10% available chlorine, chlorine appears at about pH 6, and becomes dominant at about pH 4.5. Measurements as well as calculations show that hypochlorous acid, not chlorine, is responsible for the odor of uncontaminated sodium hypochlorite solutions.[74,75]

24.3.3 Chlorination

The reaction of chlorine with water is assisted by a base through an unstable intermediate, $HOCl_2^-$, which rapidly loses chloride according to the following reactions:[76-78]

$$Cl_2 + H_2O + A^- \leftrightarrow HOCl_2^- + HA$$

$$HOCl_2^- \rightarrow HOCl + Cl^-$$

or through an intermediate $[^-A \cdots H^+ \cdots ^-O(H) \cdots ^+Cl \cdots Cl^-]$ that forms by the transfer of Cl^+ from Cl_2 to the oxygen atom of H_2O at the same time H^+ from H_2O is transferred to A^- according to the following equation:[63]

$$Cl_2 + H_2O + A^- \leftrightarrow HOCl + HA + Cl^-$$

Both cases have the same kinetics in which

$$-\frac{dCl_2}{dt} = k[Cl_2][A-]$$

The rate constant increases with increasing basicity of A^-. At 15°C and an ionic strength of 0.50 M, the self-catalyzed second-order rate constant for water is 0.16 $M^{-1}s^{-1}$. The calculated rate constants for catalysis by hypochlorite and carbonate are 3.6×10^4 and 2.1×10^6 $M^{-1}s^{-1}$, respectively, using 7.633[61] and 10.431[79] for their respective pK_a. The extrapolated rate constant for reaction with hydroxide at 15°C and an ionic strength of 0.50 M is 4.1×10^9 $M^{-1}s^{-1}$.[63] However, hydroxide may react directly with chlorine by the following reaction, which is kinetically indistinguishable from the other two schemes:

$$Cl_2 + {}^-OH \leftrightarrow Cl^- + HOCl$$

On using a laminar liquid jet absorber, the second-order rate constant for this reaction at 20°C was found to be 1.4×10^9 $M^{-1}s^{-1}$ with hydroxide and 2.8×10^3 $M^{-1}s^{-1}$ with bicarbonate,[80] which agree well with the rates calculated earlier. Injecting chlorine diluted in helium at low pressure into unbuffered hydroxide at pH 8–11 to avoid depletion of hydroxide at the interface yields a similar second-order rate constant of 6×10^8 $M^{-1}s^{-1}$ at 20°C.[81]

In most cases during chlorination, hypochlorous acid is an intermediate, since it reacts rapidly with base to form hypochlorite as follows:

$$HOCl + {}^-OH \leftrightarrow {}^-OCl + H_2O$$

This proton-transfer reaction is much faster than the formation of hypochlorous acid, but the equilibrium constant is significantly less. Thus, the formation of hypochlorous acid and its subsequent conversion to hypochlorite occur in two reaction planes.[82] Hypochlorous acid can be obtained as a gas from the first reaction plane. High yields can be obtained from spray columns with high rates of gas mass transfer and low rates of liquid mass transfer.[83] Particulate bases such as calcium hydroxide can also be used to increase the yield of hypochlorous acid.[84] Chlorinating solutions of weak bases such as bicarbonate also yield hypochlorous acid.[80]

24.3.4 DECOMPOSITION OF SOLUTIONS

Solutions may lose hypochlorite during manufacture and storage by disproportionation, metal-catalyzed decomposition, and oxidation of adventitious compounds. These reactions must be managed during manufacture to maximize yields, minimize by-products, and maximize shelf life and product quality.

Hypochlorite and hypochlorous acid solutions are inherently unstable and they disproportionate into chloride and chlorate according to the following overall reaction (see Ref. 76, p. 385):

$$3MOCl \rightarrow 2MCl + MClO_3 \qquad M = H, Na, Li, K, Ca, etc.$$

The rate increases with the concentration of hypochlorite, temperature, and ionic strength.[85–89] The solutions are most stable above pH 11, where the disproportionation rate is independent of pH. In this region, disproportionation is a two-step process. Since the first step is rate limiting, the rate is second order in hypochlorite:

$$2NaOCl \rightarrow NaClO_2 + NaCl$$

$$NaOCl + NaClO_2 \rightarrow NaClO_3 + NaCl$$

About 5% of the disproportionation above pH 11 forms oxygen instead of chlorate by the following second-order reaction:[86,90,91]

$$2NaOCl \rightarrow O_2 + 2NaCl$$

Between pH 5 and 9, disproportionation also occurs at a much faster route that does not involve chlorite as an intermediate. The overall reaction is third order in available chlorine:[60]

$$2HOCl + NaOCl \rightarrow NaClO_3 + 2HCl$$

Because the rate depends on both hypochlorous acid and hypochlorite, it depends on pH. The maximum rate is at pH 6.9. The pH also decreases as the reaction progresses. Increasing hypochlorite concentration increases the rates of the preceeding reactions more than that predicted by their reaction orders, since the increased ionic strength also increases the rate constants.

Below pH 5, hypochlorous acid disproportionates to chloride and chlorate faster than hypochlorite above pH 11, but slower than between pH 5–9. In addition, the chloride that forms rapidly reacts with hypochlorous acid to generate chlorine, and chlorine becomes the major product of the decomposition of hypochlorous acid.

Figure 24.1 shows the instantaneous disproportionation rate as a function of pH and temperature. During hypochlorite production, adequate agitation and an excess of hydroxide is usually maintained to prevent localized areas of low pH where hypochlorite rapidly disproportionates. The figure also shows how adequate cooling limits decomposition by disproportionation.

During chlorination, oxidizable impurities consume some chlorine. The initial oxidation of many compounds is rapid, but these may be followed by slower secondary reactions[92] that reduce

Inorganic Bleaches: Production of Hypochlorite

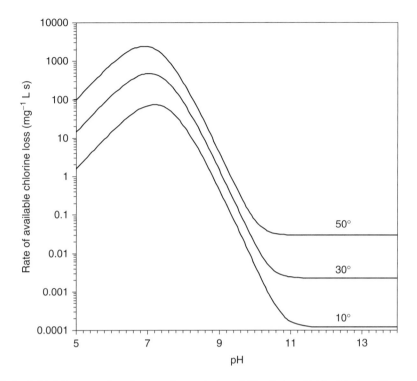

FIGURE 24.1 Initial disproportionation rate for a solution of 65 g L^{-1} available chlorine, C, as a function of pH and temperature calculated by the equation: $-dC/dt = k_1C^2/(1 + [H^+]/K_a)^2 + k_2C^3/(1 + K_a/[H^+])^2 (1 + [H^+]/K_a)$. (Derived from Adams, L.C. et al., Hypochlorous acid decomposition in the pH 5–8 region, *Inorg. Chem.*, 31, 3534, 1992; Gordon, G. et al., Predicting liquid bleach decomposition, *J. AWWA*, 89(4), 142, 1997.)

hypochlorite after packaging. The oxidation of 1 mole of an impurity may require several moles of chlorine.[93]

A few transition metals catalytically destroy hypochlorite. Hypochlorite oxidizes these metals to form intermediates that oxidize water to liberate oxygen. This returns the metal to its original oxidation state, thus it is again oxidized by hypochlorite. The overall reaction is as follows:[88,94]

$$2MOCl \rightarrow 2MCl + O_2 \qquad M = H, Na, Li, K, Ca, etc.$$

The rate is determined by the formation of the intermediate.[95] Nickel is the most effective catalyst followed by cobalt and then copper. Iridium is also a catalyst,[96] but manganese[94] and iron[57,88] are not. Dissolved metals have the greatest effect, followed by precipitated metal compounds and metal particles. The effect of particles increases with increasing surface area.

Hypochlorite is also decomposed by ultraviolet (UV) light to form oxygen, chloride, and chlorate through the formation of radicals,[97,98] but this is avoided by storing products in opaque containers.

24.3.5 Decomposition of Solid Hypochlorites

Solid calcium hypochlorite and bleaching powder decompose by the following reactions:[4]

$$3Ca(OCl)_2 \rightarrow 2CaCl_2 + Ca(ClO_3)_2$$

$$Ca(OCl)_2 \rightarrow CaCl_2 + O_2$$

Handbook of Detergents/Part F: Production

$$Ca(OCl)_2 + CaCl_2 \rightarrow 2CaO + 2Cl_2$$

$$Ca(OCl)_2 + CaCl_2 + 2CO_2 \rightarrow 2CaCO_3 + 2Cl_2$$

The first reaction predominates if the product contains a large amount of water (~18%). This reaction is analogous to the disproportionation of aqueous hypochlorite. However, disproportionation is much slower in solid calcium hypochlorite than in solution.[88] Under dry conditions, the second reaction predominates. It is catalyzed by transition metals including iron and manganese. It may occur explosively ~150°C. Thus, calcium hypochlorite products usually contain some water or an additive such as magnesium sulfate heptahydrate.[99] The third reaction is the reverse of chlorination. The fourth reaction is due to the adsorption of carbon dioxide from air or the release of carbon dioxide from carbonate salt impurities. It is accelerated by water and temperature. The first reaction accounts for ~70%, and the second reaction ~30%, of the decomposition of solid calcium hypochlorite made in the United States and stored in sealed containers at 25°C.[88]

24.3.6 METAL REMOVAL

Sufficient amounts of nickel and copper to decrease stability and release oxygen to deform or rupture storage containers is usually acquired from water, sodium hydroxide solutions, or contact with metal during hypochlorite manufacture. Metals such as calcium and iron may precipitate after packaging to make the product cloudy or form undesirable residues.[100] Iron or manganese may also turn the solution red, pink, or purple.[57]

For the hypochlorite that is produced on-site when needed, metals may not be a problem. For other uses, venting containers for storage and transportation may be sufficient. Steps can also be taken to avoid metals. Municipal water may be treated by ion exchange to remove transition metals and water hardness. Sodium hydroxide can be used directly from mercury or membrane cells without evaporation, or it can be treated to reduce metals as described in Section 24.2.3. Titanium or nonmetallic equipment, tanks, and piping can be used to avoid contact with offensive metals. However, additional purification is usually required to package hypochlorite solutions in nonventing bottles or pouches and to maximize shelf life.

A portion of the metals usually precipitate due to changes in pH, temperature, and concentrations that occur while hypochlorite is made. These are removed by filtration through a 0.5–1 μm filter,[88] or through filter aids such as diatomaceous earth or perlite deposited on a filter screen[101] or similar support.[57] Precipitation aids may also be used. Calcium carbonate precipitates on addition of calcium chloride and, if needed, sodium carbonate.[90,100] Iron chloride is added to coprecipitate iron hydroxide with hydroxides of nickel and copper.[102] Alternatively, sodium salts of anions such as hydroxide, carbonate, silicate, and oxalate that precipitate the problem metals can be added before filtration.[103,104]

Complexing agents can be added that stabilize dissolved metal catalysts in their higher oxidation states, therefore they no longer oxidize water and decompose hypochlorite. Periodate (IO_6^{5-}) and tellurate ($HTeO_6^{5-}$) are the most effective, but their salts and complexes have limited solubility that can cause cloudiness.[90,105–107] Iodide or iodate are equivalent since hypochlorite rapidly oxidizes them to periodate.[108] Polyphosphonic acids,[109,110] other organic chelating agents,[111–114] or smectic clays[115] may also be used. Other organic additives sacrificially reduce oxidized metal catalysts to prevent the oxidation of water or react with the oxygen formed from oxidation of water to prevent containers from bulging.[107,116–119]

24.3.7 TEMPERATURE CONTROL

Reacting chlorine gas with hydroxide generates 24.7 kcal mol^{-1} of heat. If liquid chlorine is used, its heat of vaporization adsorbs 3.9 kcal mol^{-1}.[5] Diluting or reacting bases with water may also

Inorganic Bleaches: Production of Hypochlorite **447**

release heat. Diluting 50% sodium hydroxide liberates 4.9 kcal mol^{-1},[45] but this heat is avoided by previously diluting to 20% or less sodium hydroxide and cooling. To minimize chlorate formation, the temperature should be kept below 35°C when making up to 7% sodium hypochlorite. With higher concentrations, the temperature should be kept below ~25°C. However, the temperature should be high enough to maintain the desired rate of chlorine absorption and prevent undesired precipitation of salt or solid forms of hypochlorite. Also greenish-yellow crystals of a chlorine clathrate in ice may form below 9.6°C. It is called chlorine ice or chlorine octahydrate, although the most likely structure is $Cl_2 \cdot 7.3H_2O$.[120,121] In a batch tank, these crystals can float to the surface and release chlorine gas.

24.3.8 MEASUREMENT OF CHLORINATION

The reaction of chlorine and alkali is an acid–base reaction that can be monitored by pH. However, ionic strength and sodium ion concentration are normally high, which cause large deviations in measured pH and make calibration difficult. In addition, pH electrodes respond slowly to changes in pH, and they do not last long in highly alkaline solutions.

Chlorination is usually measured by oxidation–reduction potential (ORP). The sensing electrode directly contacts the solution, therefore it responds much faster than pH electrodes. It is also much more mechanically robust. ORP measures the reduction of hypochlorite by the following reaction:

$$^-OCl + 2e^- + H_2O \rightarrow Cl^- + 2^-OH \qquad E° = 0.841 \text{ V}$$

Combining the Nernst equation for the preceding reaction with the dissociation of water shows that the reaction potential depends on pH as follows:

$$E = E° + \left(\frac{RT}{2F} \right) \ln \frac{[^-OCl]}{[Cl^-]} K_{H_2O}^2 + \left(\frac{RT}{F} \right) \ln [H^+]$$

The potential measured by the ORP electrode is E plus an electrode potential that depends on the installation and condition of the electrode.

24.4 PROCESS TECHNOLOGY

24.4.1 SODIUM HYPOCHLORITE

24.4.1.1 Typical Production

Sodium hypochlorite is almost always made as an aqueous solution by the chlorination of sodium hydroxide solutions as follows:

$$Cl_2 + 2NaOH \rightarrow NaOCl + NaCl + H_2O$$

Solutions of other bases such as sodium carbonate[57] and trisodium phosphate[122] may also be chlorinated to make sodium hypochlorite. Either chlorine gas or liquid chlorine can be used. Liquid chlorine is usually used in plants that do not generate their own chlorine. To allow for decomposition during shipping and storage, the concentration of hypochlorite is made slightly higher than the label strength. The excess depends on the concentration of the solution and the expected temperature and duration of storage.

Batch processes typically cost less to build, but they may cost more to operate than continuous processes. However, batch processes allow for more careful control of excess alkalinity and better removal of metals. The typical batch chlorinator is a large covered tank. The tank is oversized to accommodate expansion that occurs as the sodium hydroxide is consumed and as the temperature rises. The expansion ranges from 5 to 25%, depending on the concentration of hypochlorite that is produced.[57] The tank is filled with water and the amount of sodium hydroxide needed to make the target concentration of hypochlorite. Liquid chlorine is introduced through a pipe that has a series of holes to distribute the chlorine near the bottom of the tank. Alternatively, chlorine gas is introduced through coarse-grained ceramic diffusers. The flow rate is adjusted so that the chlorine is completely adsorbed within the depth of the solution. Air may be introduced underneath the chlorine to provide agitation to prevent localized overchlorination, which causes chlorate to form, prevent liquid chlorine from forming ice, and improve ORP measurement response. Agitation may also be supplied by recirculation, especially if a heat exchanger is used. Chlorine is usually added until only a small amount of hydroxide remains. Batch processes can be controlled by simply measuring the amounts of raw materials used.

In the United States, most sodium hypochlorite is made by continuous processes. These processes generally use smaller reaction vessels that have much larger production capacities than batch processes. They also allow better temperature control. Chlorine can be added in multiple stages with cooling in between. In addition to using heat exchangers and refrigerated reaction vessels, the solution can be cooled by spraying it into a vacuum, whereby some of the water absorbs heat to vaporize. Chlorine addition is controlled by the ORP and the temperature that are continually measured. The temperature may also be used to control the amount of hypochlorite solution that is cooled and recycled into the reactor. Conventionally, only 92–94% of the hydroxide is reacted to ensure that excess chlorine is not added. Higher conversion can be achieved using the derivative of the ORP with respect to time to control the rate of chlorine and hydroxide addition better.[123] Conversion is also improved by increasing the amount of product that is recycled. This also minimizes the temperature increase, which improves heat exchanger performance.

A common continuous process is to inject liquid chlorine into a reactor that is filled with recirculating hypochlorite solution. Sodium hydroxide is mixed with the hypochlorite solution just before it enters the reactor near the point where the chlorine is introduced. The reactor is packed with balls, rings, saddles, or some other shaped filler to promote mixing. The effluent passes through a heat exchanger, an averaging tank, and then a splitter that recycles some and discharges the rest to a storage tank. An example process is shown in Figure 24.2.

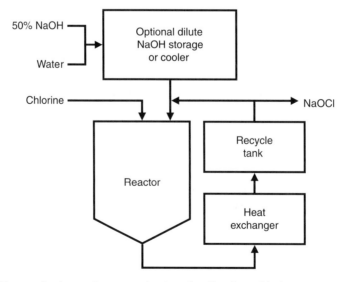

FIGURE 24.2 Diagram for the continuous production of sodium hypochlorite.

Inorganic Bleaches: Production of Hypochlorite

A nonflooded packed tower reactor is usually used to make hypochlorite from chlorine gas. Chlorine gas is injected from the bottom, while recirculating hypochlorite solution is sprayed from the top of the tower. Sodium hydroxide is mixed with the hypochlorite solution just before it enters the reactor. The effluent is withdrawn from the bottom of the reactor at a rate that prevents the reactor from filling completely with liquid. It is then cooled and recycled or sent to the storage tank. A less common implementation is to fill the packed tower from the bottom with recirculating hypochlorite solution to which fresh hydroxide has been added. Chlorine gas is also injected at the bottom of the tower into or near the stream of hypochlorite. The effluent is withdrawn from the top of the tower and treated as before. Chlorine gas and recirculating hypochlorite solution can also be concurrently injected at the top of the packed tower. The effluent is withdrawn from the bottom of the reactor and treated as before. Hydroxide is added to the recirculating hypochlorite before injection or in the recycle tank. Concurrent mixing avoids hydroxide depletion and the resulting chlorate formation better than countercurrent mixing.[124]

Streams of liquid chlorine and hydroxide can also be mixed directly. In a simple design, sodium hydroxide is mixed with cooled, recycled reaction mixture and then with liquid chlorine in a tubular reactor that empties into an averaging tank. The ORP and temperature are used to control the rates of addition and the amount of hypochlorite solution that is circulated through the heat exchanger and recycled.[125] More intense mixing is used to reduce or avoid recycling the reaction mixture. Hydroxide solution and liquid chlorine can be tangentially introduced from diametrically opposed locations into a reaction chamber so that they swirl upward through the chamber with laminar flow.[126] Chlorine gas and an aqueous solution of sodium hydroxide can be combined in an in-line venturi jet mixer and passed through a multistage turbine pump that creates intensive turbulence and a pressure >1 atm to make a solution of sodium hypochlorite or hypochlorous acid that is discharged to a storage tank.[127]

24.4.1.2 Low-Salt (High-Strength) Sodium Hypochlorite

Most sodium hypochlorite is made at a maximum concentration of 16% to avoid precipitating sodium chloride. Sodium chloride precipitates at 10°C when more than 18.5% sodium hydroxide is chlorinated to produce concentrations of sodium hypochlorite >17.2%.[128] As the temperature decreases, sodium chloride precipitates from solutions with less sodium hypochlorite.[129] The abrasive suspension of precipitated salt requires equipment and handling, but it can be recycled to brine for an electrolysis cell. Reducing salt decreases ionic strength and the disproportionation rate of hypochlorite. It may also increase the solubility of other ingredients in cleaning formulations. One product contains 15% sodium hypochlorite and 3.5%, instead of 12%, sodium chloride.[130] It is made by chlorinating sodium hydroxide to make a concentrated solution of sodium hypochlorite, removing the salt, and diluting with water. A product with 15% sodium hypochlorite and 0.5% sodium chloride is made by reacting hypochlorous acid with sodium hydroxide.[131]

Concentrated hypochlorite solutions can be made by chlorinating in stages. In a continuous process (Figure 24.3), liquid chlorine and sodium hydroxide solution are mixed in a reactor to make a solution with not more than 15% sodium hypochlorite and 4.5% sodium hydroxide so that sodium chloride does not precipitate as it forms. The solution is cooled in a heat exchanger. The output is sent to a crystallizer with an external heat exchanger and a mechanical agitator. Additional sodium hydroxide and chlorine are added to produce large sodium chloride crystals that are easily filtered. The product contains 25% sodium hypochlorite, 9.5% sodium chloride, and 0.2–0.8% sodium hydroxide.[132] In a similar process, water can be added to make a solution with 13% sodium hypochlorite and 3.5% sodium chloride.[133]

Reactors have also been designed to separate salt from hypochlorite solutions. The conical base of a vertical tower has a smaller diameter than the upper part. Liquid chlorine and sodium hydroxide solution are injected into the lower section. Hypochlorite is removed from the upper part of the tower, some of which is recycled into the lower section. The recycle and injection rates are

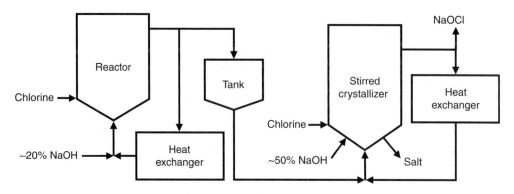

FIGURE 24.3 Process diagram for high-strength, low-salt sodium hypochlorite. (Redrawn from Dugua, J., U.S. Patent 4,780,303, 1988.)

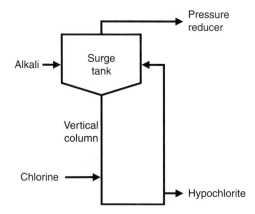

FIGURE 24.4 Evaporatively cooled chlorinator used for chlorinating slurries. (Redrawn from Nicolaisen, B.H., U.S. Patent 3,241,912, 1966.)

controlled to keep the sodium chloride crystals fluidized in the bottom part of the reactor, and the crystals are removed from the bottom of the tower.[134] Alternatively, the sodium chloride crystals may be kept in suspension throughout the reactor and recycle stream.[135] Another way to prevent salt clogging is to form a thin film of sodium hydroxide solution on the wall of a reactor below its introduction port and inject chlorine gas from a point below this port that is separated from the reactor wall. Air is also injected upward into the flowing hydroxide.[136] The reactor may also be designed to avoid clogging as shown in Figure 24.4. This device avoids cooled surfaces, such as those encountered with heat exchangers, on which components of the aqueous mixture can crystallize. The pressure in the tank is reduced to ~0.05 atm, thus water evaporates to cool the system. The vertical column is high enough so that the pressure exerted by the aqueous mixture is about 1 atm at the point where chlorine is added. The mixture is rapidly circulated to maintain the suspension, minimize temperature rise, and prevent hypochlorite decomposition.[137]

Continuous removal of salt allows higher concentrations of hypochlorite to be made without precipitating sodium hypochlorite pentahydrate. In a continuous process, sodium hydroxide and chlorine are continuously supplied to a cooled reaction tank, in which the degree of chlorination is maintained above 80%. The tank solution is circulated through a centrifugal separator to remove salt and produce a solution with 40.7% sodium hypochlorite, 2.4% sodium chloride, and 1.2% sodium hydroxide.[138]

Inorganic Bleaches: Production of Hypochlorite

Low-salt sodium hypochlorite can also be made by chlorinating a slurry of calcium hydroxide. The precipitated calcium hypochlorite compounds are removed by filtration and mixed with a solution of sodium hydroxide. The solids are removed by filtration. The resulting solution contains ~15% sodium hypochlorite and 1.3% sodium chloride.[139]

24.4.1.3 Solid Sodium Hypochlorite

A slurry of sodium hypochlorite pentahydrate crystals, $NaOCl \cdot 5H_2O$, in an aqueous solution of sodium hypochlorite has been made.[131] The total sodium hypochlorite concentration is >35% and the sodium chloride concentration is <2%. Because the crystals do not disproportionate, the slurry has better stability than a solution with an equal concentration of sodium hypochlorite. However, the slurry needs to be kept below ~20°C to keep the crystals from melting. The slurry is made by mixing a solution of at least 35% hypochlorous acid with sodium hydroxide below 25°C.[140]

None of the solid forms of sodium hypochlorite have sufficient stability for commercial use. However, hydrates of sodium hypochlorite can be separated from sodium chloride by fractional crystallization and then dissolved to make sodium hypochlorite solutions with low salt. Salt precipitates as a solution of 50% sodium hydroxide is chlorinated at 25–30°C. A solution with 32% of sodium hypochlorite and ~6% sodium chloride is separated from the salt. This solution is cooled to 10–22°C to crystallize $NaOCl \cdot 5H_2O$. The crystals are then dissolved to make a solution of 13% sodium hypochlorite and 0.1–2% sodium chloride.[5,141]

Crystals of $NaOCl \cdot 2.5H_2O$ can also be made. A 35–40% sodium hypochlorite solution is made by chlorinating sodium hydroxide. The precipitated salt is removed by filtration. Additional sodium hydroxide is added so that the solution contains at least 15% sodium hydroxide and 15% sodium hypochlorite. The solution is cooled to ~0°C when $NaOCl \cdot 2.5H_2O$ crystallizes. The crystals are removed by filtration, and the mother liquor is recycled into the process.[142] Solid mixtures of sodium chloride and sodium hypochlorite are made by reacting a fine spray of sodium hydroxide with chlorine gas. The stability is improved by melting the solid and adding an anhydrous salt that forms a stable hydrate such as sodium metaborate, trisodium phosphate,[143] or sodium carbonate.[144] Solid sodium hypochlorite is also made by vacuum evaporation of concentrated sodium hypochlorite solution that is essentially free of chloride ions.[145]

24.4.1.4 Electrolytic Hypochlorite Production

Integrated units are available that combine the outputs from the anode and the cathode chambers of conventional membrane cells to make 3–7% sodium hypochlorite. Higher concentrations are possible, but the units do not handle precipitated salt. The brine may be similar to that used by conventional chlorine cells (300 g L^{-1} salt) or it may be more dilute. The entire anode effluent may be mixed with the cathode effluent without separating chlorine or caustic.[146,147] In nonconventional cells with alkaline tolerant anodes, cathode effluent may be recycled to the anode,[148] or a mixture of sodium chloride and sodium hydroxide may be fed into the anode.[149] The concentration of sodium hypochlorite is determined by the concentration of brine, the relative volumes of effluent from the anode and the cathode, the flow rate, and the current. The units are usually sized for point-of-use operation. Conventional cells may also be used to separate chlorine gas from the anode effluent for point of use.[150]

Most hypochlorite generators electrolyze brine in undivided cells. The chlorine and hydroxide react to form hypochlorite directly in the cell. A typical unit uses 30 g L^{-1} sodium chloride to make 8 g L^{-1} sodium hypochlorite. Higher concentrations can be made only with nontypical coated electrodes that prevent the oxidation of hypochlorite to chlorate.[151] A maximum of 3 g L^{-1} sodium hypochlorite is made from seawater.[2] Excess salt increases cell efficiency and suppresses hypochlorite oxidation. The sodium hypochlorite is usually stored in a tank from which it is metered, and hydrogen is allowed to escape to air. In some cases, especially with swimming pools, 3–6 g L^{-1}

salt is added to the water being treated so that it can be circulated directly through the cell. Common cell designs include concentric tubes in which the outer and inner tubes are the electrodes and cylinders or rectangular boxes, which contain a series of flat, parallel electrodes. Cells are often cooled using a heat exchanger to increase cell efficiency and reduce chlorate formation. A series of cells may be used so that the solution can be cooled after it exits each cell. Cooling is done using heat exchangers[152] or by adding cold water.[148] Although these cells have no membranes to clog, scale forms on the electrodes. Some units reverse electrode polarity to remove scale, but all units require periodic cleaning and electrode replacement.

Electrolytic cells are also used in a variety of devices[23,153–160] and appliances[161–171] to produce dilute solutions of electrolyzed water for cleaning and sanitization. Miniature, battery-powered cells are used to generate hypochlorite in handheld sprayers[172–176] and small portable water disinfection devices.[177–180] The brine may be saturated or have 1–5 g L^{-1} of salt. An acidic solution of chlorine, hypochlorous acid, and possibly chlorine dioxide is obtained from the anode of a divided cell. It typically has 10–100 mg L^{-1} of available chlorine and pH values of 2–4. Its stability is poor, and the volatile oxidants are rapidly lost from open solutions. A neutral solution of hypochlorous acid and sodium hypochlorite is dispensed from undivided cells or by combining effluents from the anode and cathode. It typically has 80–100 mg L^{-1} of available chlorine and pH 5–8.[23] Devices to make 500–1000 mg L^{-1} of available chlorine are also available.[181] Neutral solutions made using larger versions of these devices are bottled and sold in some regions.

24.4.2 POTASSIUM HYPOCHLORITE

Potassium hypochlorite is made in the same manner as sodium hypochlorite, by substituting potassium hydroxide for sodium hydroxide. Solutions with up to 40% potassium hypochlorite and only small amounts of potassium chloride have been made.[182] These are made by mixing a solution with at least 35% hypochlorous acid and an aqueous slurry with at least 35% potassium hydroxide at a temperature below 45°C.[183] Unlike sodium hypochlorite, potassium hypochlorite remains completely in solution well below 0°C.[184,185]

24.4.3 HYPOCHLOROUS ACID

Chloride-free solutions that contain up to 50% hypochlorous acid have sufficient stability at 0°C to be sold for industrial use.[186] However, chloride reacts with hypochlorous acid to form chlorine. Thus, hypochlorous acid solutions made by chlorinating water and an alkaline compound are not stable enough to be sold. But, they may be prepared on-site by adding enough chlorine to reduce the pH to convert hypochlorite to hypochlorous acid. Hypochlorous acid may then be extracted into an organic solvent. Since hypochlorous acid is volatile, it can also be separated from the chlorination mixture as a gas and then condensed or adsorbed in water to make a chloride-free solution.

As shown in Figure 24.5, 50% sodium hydroxide is sprayed into a reactor to form droplets that react with recirculating chlorine gas. The resulting heat vaporizes the water to obtain a mixture of hypochlorous acid, chlorine, dichlorine monoxide, water vapor, and solid sodium chloride. The solid falls to the bottom of the reactor where it is removed. The gas is chilled to separate a solution of hypochlorous acid from the chlorine that is recycled.[187–194] Alternatively, the hypochlorous acid can be absorbed into water instead of being condensed.

In another process, acid is added to reduce the pH of a sodium hypochlorite solution to less than 5.5. The solution is sprayed into a stream of inert gas to make fine droplets or distilled. The vapor stream is adsorbed into water to make a hypochlorous acid solution with low salt.[195] Solutions of hypochlorous acid or sodium hypochlorite can also be stripped in a column against a current of chlorine, steam, and air at 95–100°C. The exit gases are condensed to form hypochlorous acid solutions that are free of chloride.[5] Similarly, a hydroxide solution is circulated downward through a stripping column against a stream of chlorine and an inert gas above 60°C. The effluent gas is condensed to form a salt-free solution of hypochlorous acid.[196] Hypochlorous acid can also be made

Inorganic Bleaches: Production of Hypochlorite

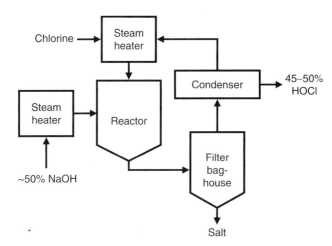

FIGURE 24.5 Process for making hypochlorous acid from sprayed sodium hydroxide droplets and chlorine gas. (Redrawn from Hilliard, G.E., Melton, J.K., and Helmstetter, D.A., U.S. Patent 5,116,594, 1992.)

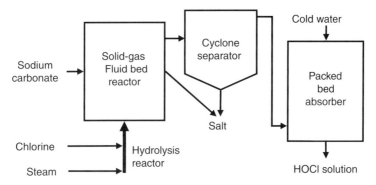

FIGURE 24.6 Production of hypochlorous acid from chlorine gas and steam. (Based on Yant, R.E. and Galluch, R.J., U.S. Patent 4,584,178, 1986.)

by passing chlorine over an agitated hydroxide solution. The exit gases are adsorbed into water to make a hypochlorous acid solution.[197] Efficiency can be increased using a rotating packed bed to mix a hot caustic solution with a countercurrent stream of chlorine. Hypochlorous acid desorbs into the chlorine stream and is absorbed into water after the gas mixture exits the device.[198]

Another approach is to produce hypochlorous acid as a gas by reacting chlorine with water vapor. In the past, chlorine and water vapor were passed over solid sodium carbonate in a tower or a rotating tubular reactor and the exit gases were dissolved in water to make a salt-free solution of hypochlorous acid.[199–201] This has been replaced by reacting chlorine with steam[202–204] and passing the resulting gases through a fluidized bed of sodium carbonate to convert hydrochloric acid to sodium chloride. The exit gases are absorbed in water to make a solution of hypochlorous acid as shown in Figure 24.6.

Hypochlorous acid for immediate use can be made by chlorinating a suspension of powdered limestone in water according to the following reaction:

$$2Cl_2 + 2H_2O + 2CaCO_3 + 2HOCl + CaCl_2 + Ca(HCO_3)_2$$

The suspended limestone is continuously introduced at the top of a tower, which has a series of perforated shelves. The suspension flows downward over the shelves while a stream of chlorine gas flows upward through the perforations with enough velocity to cause turbulent mixing. A solution of hypochlorous acid is removed from the bottom of the tower.[205] Sodium carbonate can also be used. Chlorine reacts with water to first neutralize sodium carbonate and then to neutralize the resulting sodium hypochlorite to form hypochlorous acid as follows:

$$Cl_2 + H_2O + 2Na_2CO_3 \rightarrow NaOCl + NaCl + 2NaHCO_3$$

$$Cl_2 + H_2O + NaOCl \rightarrow 2HOCl + NaCl$$

Adding more chlorine neutralizes bicarbonate, which releases carbon dioxide:

$$Cl_2 + H_2O + NaHCO_3 \rightarrow HOCl + NaCl + H_2CO_3$$

$$H_2CO_3 \rightarrow H_2O + CO_2$$

At this pH, the hypochlorous rapidly disproportionates to chlorate, which becomes the major product. Thus, chlorination is stopped when carbon dioxide first begins to evolve.[57]

24.4.4 CHLORINATED TRISODIUM PHOSPHATE

In the original process, sodium hypochlorite was added to a solution of sodium phosphate and chlorinated trisodium phosphate was then separated by crystallization.[206] The process was improved by adding a solution that contains 14–16% sodium hypochlorite and 2–4% sodium hydroxide to a hot, concentrated solution of sodium phosphate in a jacketed, heavy-duty mixer. The phosphate solution has a Na/P mole ratio of 2.60–2.85 and contains 27–28% Na_2O. It can be made by neutralizing phosphoric acid with sodium hydroxide, in which case the heat of neutralization is sufficient to heat the solution. Alternatively, partially hydrated disodium phosphate and sodium hydroxide, or a mixture of di- and trisodium phosphates can be melted. The solutions are mixed at 70–80°C, which is just above the melting point of the chlorinated trisodium phosphate, which ranges between 65 and 70°C. After mixing, just long enough to form a complete solution, water at room temperature is passed through the jacket to cool the mixture, which causes it to solidify. It is then air dried and granulated.[207,208] However, ~20% of the available chlorine is lost during drying. This is reduced by adding 0.5–2% of a water-soluble silicate to the reaction mixture.[209] The available chlorine loss is further reduced by evaporative cooling of the mixture under reduced pressure and drying the product with refrigerated air. Evaporative cooling also produces fine crystals that do not need to be granulated.[208] The formula, $(Na_3PO_4 \cdot 11H_2O)_4 \cdot NaOCl$, contains 4.89% sodium hypochlorite, a Na/P mole ratio of 3.25 and 52.05% water. However, the typical product contains 3.4–4.2% sodium hypochlorite and the Na/P mole ratio is 3.15–3.35. Sodium chloride and disodium phosphate dihydrate are also present.[12]

24.4.5 LITHIUM HYPOCHLORITE

The original product sold in the United States contains ~30% lithium hypochlorite (35% available chlorine), 34% sodium chloride, 20% of potassium and sodium sulfates, 3% lithium chloride, 3% lithium chlorate, 2% lithium hydroxide, 1% lithium carbonate, and the balance is water.[13] It is made from lithium sulfate that is extracted into water from a lithium aluminum silicate ore after it is treated with sulfuric acid. The resulting solution also contains sodium and potassium sulfates. It is neutralized with calcium carbonate to pH 6, treated to remove calcium and magnesium, filtered, and concentrated. Sodium hydroxide is added to convert lithium carbonate to lithium hydroxide. The solution is cooled to 0°C and the resulting sodium carbonate decahydrate crystals are removed by filtration. Slightly more sodium hydroxide than the molar equivalent of lithium hydroxide is then

Inorganic Bleaches: Production of Hypochlorite

added. Chlorine gas is added next to form lithium hypochlorite and sodium chloride according to the following equation:

$$LiOH + NaOH + Cl_2 \rightarrow LiOCl + NaCl + H_2O$$

The solution is then spray dried to produce a white powder.[210]

A granular solid that contains 77% lithium hypochlorite, 4–7% lithium chloride, and small amounts of lithium chlorate and lithium hydroxide has also been made.[211] A solution of 35–50% hypochlorous acid is added to an aqueous slurry of lithium hydroxide at 5–10°C with agitation. As the reaction proceeds, additional lithium hydroxide is added. More hypochlorous acid is also added until the concentration of lithium hydroxide drops below 1%. The resulting solution contains 25–35% lithium hypochlorite. It can be cooled between -5 and -15°C to produce crystals of lithium hypochlorite monohydrate, which are filtered and dried.[212] Alternatively, water is evaporated from the solution under reduced pressure to produce a slurry of lithium hypochlorite, and the solid is removed and dried.[213] In either case, the supernatant is recycled into the process to conserve lithium. The lithium hypochlorite solutions can also be used as made or they can be dried directly to make a less-pure solid with a lower concentration of lithium hypochlorite.

Lithium hydroxide solutions may also be chlorinated like sodium hydroxide. Chlorinating a saturated solution of lithium hydroxide produces a solution with ~10% lithium hypochlorite and equimolar lithium chloride. Chlorinating a mixture of lithium hydroxide and either potassium or sodium hydroxide preferentially forms lithium hypochlorite and sodium or potassium chloride. If a suspension containing 15–20% lithium hydroxide is chlorinated to make 25–35% lithium hypochlorite, much of the lithium hypochlorite crystallizes as the monohydrate at 20°C. If greater concentrations of lithium hypochlorite are made, a mixture of lithium hypochlorite monohydrate and lithium chloride is obtained.[214]

Lithium hypochlorite solutions can also be made by adding lithium carbonate or sulfate to a solution of calcium or magnesium hypochlorite and removing the precipitated calcium or magnesium salts.[143]

24.4.6 CALCIUM HYPOCHLORITE

24.4.6.1 Bleach Liquor

Calcium hypochlorite solutions are made by suspending lime or hydrated lime in water and adding chlorine. However, reacting lime with water generates lots of heat, and this heat must be allowed to dissipate before chlorination. The maximum concentration is ~85 g L^{-1} of available chlorine, above which dibasic calcium hypochlorite begins to precipitate.

Chlorination can be done in a batch or continuous reactor similar to those used to make sodium hypochlorite, except that agitation is needed to maintain the suspension. In a batch reactor, agitation can be achieved using a mechanical stirrer, or, more efficiently, by circulation. A centrifugal pump is used to pull the solution from the bottom of a conical tank and then reinject it tangentially to create a strong vortex. Alternatively, the solution passes through a packed tower before it returns to the tank. The packing in the tower creates shear that breaks the particles apart before the chlorine is injected at the bottom of the tower.[57] The reaction is as follows:

$$2Cl_2 + 2Ca(OH)_2 \rightarrow Ca(OCl)_2 + CaCl_2 + 2H_2O$$

A slight excess of lime is used to maintain alkalinity. After chlorination, the insoluble impurities from the lime such as silica and calcium carbonate are allowed to settle and the clear solution of calcium hypochlorite and calcium chloride is used.

In a typical continuous process, lime is hydrated in a tank that has a mechanical stirrer. The solution passes through a centrifugal classifier to remove and return unreacted lime to the mixing tank. The lime solution goes to a storage tank. The lime solution is then injected into a stream of chlorine gas as it enters a reactor with baffles that provide mixing. The exiting bleach goes through another centrifugal separator to remove particles larger than 50 μm. This produces a solution with ~0.2% suspended solids that can be used for pulp bleaching.[215]

24.4.6.2 Hemibasic Calcium Hypochlorite

Hemibasic calcium hypochlorite, $Ca(OCl)_2 \cdot 0.5Ca(OH)_2$, can be precipitated from a solution of calcium hypochlorite at 40–80°C.[143] These solutions are commercially made by chlorinating a slurry of hydrated lime.[5] The precipitate is removed by filtration and dried. Although it may still be made in Asia, it has not been made in the United States or Germany since 1955.[216]

24.4.6.3 Dibasic Calcium Hypochlorite

Dibasic calcium hypochlorite, $Ca(OCl)_2 \cdot 2Ca(OH)_2$, is commonly used as an intermediate in making calcium hypochlorite, $Ca(OCl)_2$. It has never been successfully marketed as a product because it dissolves much too slowly.[216] It is made by chlorinating a slurry of lime so that the optimum amount of unreacted calcium hydroxide remains to foster its crystallization.[217] Adding sodium chloride to a chlorinated lime solution causes dibasic calcium hypochlorite to crystallize. Dibasic calcium hypochlorite also precipitates by adding lime to saturated solutions of calcium hypochlorite as shown in the following reaction:

$$Ca(OCl)_2 + 2Ca(OH)_2 \rightarrow Ca(OCl)_2 \cdot 2Ca(OH)_2$$

This is often done to recycle filtrates and other solutions produced during the manufacture of calcium hypochlorite.

With poorer-quality lime, a gelatinous precipitate can be separated from the dibasic calcium hypochlorite crystals in a vertical classifier, flotation cell, hydrocyclone, or centrifugal classifier.[218] When the magnesium content of the lime is high, calcium hypochlorite crystallizes better if the chlorination is stopped after most of the calcium hydroxide has been chlorinated but before magnesium hypochlorite begins to form.[219]

24.4.6.4 Calcium Hypochlorite

Although calcium hypochlorite dihydrate, $Ca(OCl)_2 \cdot 2H_2O$, can be made by chlorinating a suspension of lime, it is difficult to filter, partly due to the presence of calcium chloride. Calcium chloride also reduces the stability of the dry product.[143] In addition, a large amount of calcium hypochlorite remains dissolved in the filtrate. Thus, in commercial processes, the formation of calcium chloride is avoided, it is converted to something else, or it is separated from the calcium hypochlorite. They recycle the effluents to minimize waste.

Sodium hypochlorite is commonly added to convert calcium chloride to calcium hypochlorite as follows:[5]

$$CaCl_2 + 2NaOCl \rightarrow Ca(OCl)_2 + 2NaCl$$

With concentrated calcium chloride solutions, sodium chloride precipitates after part of the sodium hypochlorite is added. After filtration, the remainder of the sodium hypochlorite is added. Calcium hypochlorite dihydrate then crystallizes. It is separated from the solution and dried.[220] Chlorinating a slurry of lime in a sodium hypochlorite solution prevents the formation of calcium chloride and allows large crystals of calcium hypochlorite dihydrate to form. These are readily separated and dried.[221,222]

Inorganic Bleaches: Production of Hypochlorite

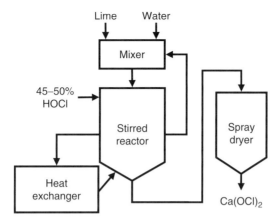

FIGURE 24.7 Diagram to make high-strength calcium hypochlorite from salt-free hypochlorous acid. Hypochlorous acid is made according to Figure 24.5. (Based on Hilliard, G.E. et al., World Patent Appl. 92 21,610, 1992.)

Calcium chloride is avoided by reacting hypochlorous acid with hydrated lime as follows:

$$Ca(OH)_2 + 2HOCl \rightarrow Ca(OCl)_2 + 2H_2O$$

In a process used from 1937 to 1984,[5] a solution of 10–15% hypochlorous acid, which contains very little chloride, is mixed with a slurry of hydrated lime at pH 10.0–10.5 to form a 15–20% solution of calcium hypochlorite.[199] Sodium hypochlorite may be added to convert calcium chloride to calcium hypochlorite.[223] The solution is spray dried, and the resulting solid is granulated and vacuum dried. Figure 24.7 depicts a current process, which uses salt-free 45–50% hypochlorous acid to make a slurry of hydrated calcium hypochlorite that is spray dried to make calcium hypochlorite with up to 82% available chlorine.[224–228] The product may also be agglomerated.[229]

Calcium chloride is also avoided by chlorinating a mixture of calcium hydroxide and sodium hydroxide since calcium hypochlorite and sodium chloride are preferentially formed.

$$Ca(OH)_2 + 2NaOH + 2Cl_2 \rightarrow Ca(OCl)_2 \cdot 2H_2O + 2NaCl$$

Chlorinating a slurry of lime in a sodium hydroxide solution forms crystals of calcium hypochlorite dihydrate, which can be filtered and dried.[230] The precipitation of basic compounds is avoided by incremental addition of the calcium hydroxide to the sodium hydroxide solution during chlorination. In a batch process, all of the sodium hydroxide and 10–80% of the calcium hydroxide are introduced into a stirred reactor. Chlorine is then added. When the chlorination is nearly complete, the remainder of the calcium hydroxide is added without interrupting the flow of chlorine. The solids are separated and dried to from $Ca(OCl)_2$.[231] In a continuous process, the sodium hydroxide solution is fed into one end of a long tubular reactor, and the calcium hydroxide slurry and chlorine are separately injected at a number of sites along the length of a jacketed reactor. The calcium hypochlorite is removed from the end of the reactor and dried.[232] When impurities in the lime interfere with the crystallization of calcium hypochlorite, potassium hydroxide is substituted for up to 25% of the sodium hydroxide.[233]

In a process used since 1984,[5] a mixture of calcium hypochlorite dihydrate and larger sodium chloride crystals are formed by chlorination and then separated based on size and density.[234–236] Solutions of calcium hypochlorite and sodium chloride recycled from the process are mixed with lime to make a slurry of dibasic calcium hypochlorite and unreacted lime. Gravity sedimentation

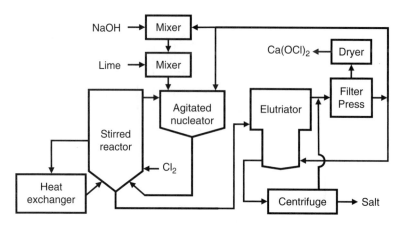

FIGURE 24.8 Continuous production of calcium hypochlorite by chlorination of mixed hydroxides and separation from salt. (Redrawn from Loehr, C.E. et al., U.S. Patent 4,390,512, 1983.)

may be used to concentrate this slurry before it is transferred to a tank where concentrated sodium hydroxide and additional recycled liquors are added. Alternatively, as shown in Figure 24.8, sodium hydroxide may be mixed with the recycled process solutions used to disperse the lime, and the slurry may be directly added to the tank. After some seed crystals of sodium chloride and calcium hypochlorite form, this mixture is transferred to a stirred reaction vessel. Chlorine is then added to form crystals of calcium hypochlorite and sodium chloride. The resulting slurry is introduced into the middle of a vertical elutriator or classifier. Recycled mother liquor is introduced at the bottom of the elutriator so that it flows upward and removes calcium hypochlorite crystals at the top. This slurry is pressure filtered. The filter cake contains ~50% calcium hypochlorite and 10% sodium chloride. It is dried with hot air to make a product containing at least 65% calcium hypochlorite, 2–12% water, and sodium chloride. Meanwhile, the denser sodium chloride crystals fall to the bottom of the classifier, where they are removed. This slurry can be centrifuged, the solids resuspended in the filtrate from the overflow, and then centrifuged again to separate washed sodium chloride. The supernatant liquids from these steps are recycled into the process. The crystals may also be separated using two elutriators in series to produce a moist calcium hypochlorite filter cake with <1% sodium chloride.[237] In a similar process, the sodium chloride and calcium hypochlorite crystals are separated in a flotation cell in which the smaller calcium hypochlorite crystals float to the top with the aid of air and a flotation agent.[238] A centrifugal separator, hydrocyclone, sieve, or classification unit may also be used to separate the crystals.[239–241]

The process can be simplified by chlorinating a mixture of calcium chloride and hydrated lime in the presence of prismatic seed crystals of calcium hypochlorite dihydrate. The resulting slurry is easy to filter and calcium hypochlorite dihydrate, which is not contaminated by sodium chloride, can be removed. The mother liquor is treated with hydrochloric acid or a metal oxide to convert dissolved calcium hypochlorite to calcium chloride. It is then recycled into the process. The calcium chloride aids the dispersion of lime. The seed crystals are made by chlorinating a mixture of calcium hydroxide and sodium hydroxide in the presence of citric acid.[242]

Other commercial processes make, or have made, calcium hypochlorite dihydrate by chlorinating a slurry of dibasic calcium hypochlorite.[243–246] The resulting calcium hypochlorite dihydrate is separated by filtration and dried. Lime is added to the filtrate to precipitate dibasic calcium hypochlorite, which is used as the starting material. The filtrate from this step can also be reused by adding sodium hypochlorite or sodium hydroxide and chlorine to convert calcium chloride into calcium hypochlorite and precipitate sodium chloride. The solution is then filtered and recycled into the process. In a process implemented in 1983,[5] a mixture of dibasic calcium hypochlorite, lime,

Inorganic Bleaches: Production of Hypochlorite

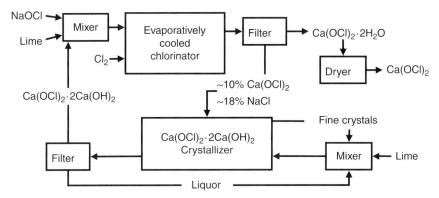

FIGURE 24.9 Continuous production of calcium hypochlorite from lime and dibasic calcium hypochlorite. (Redrawn from Sakowski, W.J., U.S. Patent 4,399,117, 1983.)

and high-strength, low-salt sodium hypochlorite are chlorinated to make calcium hypochlorite dihydrate. After filtering and drying, the product contains 65–75% calcium hypochlorite and 4–8% water and sodium chloride. The slurry of dibasic calcium hypochlorite is made by mixing lime with a recycled calcium hypochlorite solution as shown in Figure 24.9. Additional steps can be added to remove impurities from the low-quality lime or precipitated calcium salts.[247–249]

Instead of dibasic calcium hypochlorite, a triple salt was formed in a process that was used between 1928 and 1983.[5] A slurry of lime in concentrated sodium hypochlorite is chlorinated. After cooling to −15°C, a triple salt, $Ca(OCl)_2 \cdot NaOCl \cdot NaCl \cdot 12H_2O$, is easily removed by filtration. The triple salt is mixed with a slurry of chlorinated lime that contains enough calcium chloride to consume all of the sodium hypochlorite in the triple salt. This makes calcium hypochlorite dihydrate crystals that are easy to filter by the following reaction:

$$2Ca(OCl)_2 \cdot NaOCl \cdot NaCl \cdot 12H_2O + Ca(OCl)_2 + CaCl_2 \rightarrow 4Ca(OCl)_2 \cdot 2H_2O + 4NaCl + 16H_2O$$

Most of the sodium chloride remains in the filtrate and the dried product contains ~70% available chlorine.[250]

The precipitation of basic calcium hypochlorites can be used to remove calcium chloride. The excess calcium that is not associated with calcium hypochlorite can be precipitated by adding sulfate or carbonate, but this is less effective, and does not remove other impurities that remain in the calcium hypochlorite after the solution is evaporated. In either case, the conversion of lime to calcium hypochlorite is reduced by the amount of calcium salts that are removed from the process. In an earlier process, a concentrated lime slurry is chlorinated at 40–45°C to form crystals of hemibasic calcium hypochlorite. These are removed by filtration and suspended in a slurry of chlorinated lime. Chlorine is then added to make calcium hypochlorite dihydrate crystals that are removed by filtration. The filtrate is mixed with lime to precipitate dibasic calcium hypochlorite. The dibasic calcium hypochlorite is removed by filtration, suspended in recycled liquor, and chlorinated to form crystals of calcium hypochlorite dihydrate. Throughout the process, seed crystals are used and the filtrates are recycled.[251–254] Alternatively, slow chlorination at 15°C is used to directly produce calcium hypochlorite dihydrate instead of the hemibasic compound in the first step. Dibasic calcium hypochlorite is made from the filtrate and processed as before.[143]

The process is simplified by only isolating one type of intermediate. A lime slurry is chlorinated to produce a solution with equimolar amounts of calcium chloride and calcium hypochlorite. To remove insoluble impurities after filtration, sodium chloride is added to cause dibasic calcium hypochlorite to crystallize. The mixture is passed through a vertical classifier to separate a slurry of dibasic calcium hypochlorite. The overflow from the classifier is recycled into the beginning

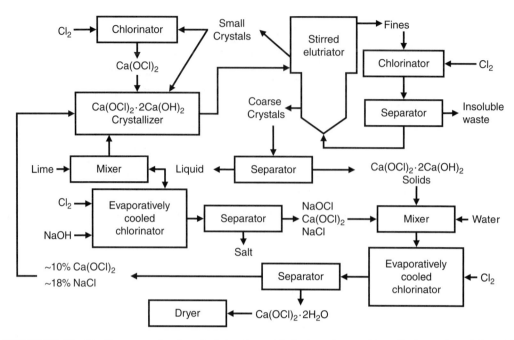

FIGURE 24.10 Continuous production of calcium hypochlorite from impure lime. (Redrawn from Sakowski, W.J., Bajaj, M.C., and Duncan, B.L., U.S. Patent 4,468,377, 1984.)

of the process. The dibasic calcium hypochlorite slurry is chlorinated to form a slurry of calcium hypochlorite dihydrate that is removed using a centrifuge. Concentrated sodium hypochlorite with a low chloride content is mixed with the solid from the centrifuge in a pug mill to reduce the amount of calcium chloride. The solid is then dried. The liquid from the centrifuge, which is saturated with calcium hypochlorite, is sent to the crystallizer.[255–259]

Intermediates may also be used to remove impurities from low-quality lime. Figure 24.10 shows the precipitation of dibasic calcium hypochlorite from mixing lime with recycled solutions of calcium hypochlorite and sodium chloride. The crystals are separated and purified by elutriation. Sodium hypochlorite is added and the mixture is chlorinated to make calcium hypochlorite dihydrate. This product is filtered and dried.[260] For further purification, the dibasic calcium hypochlorite crystals may be reacted with chlorine or a solution of calcium hypochlorite to precipitate hemibasic calcium hypochlorite. These crystals are separated, mixed with water, and chlorinated. The resulting calcium hypochlorite dihydrate is separated and dried. Sodium hydroxide may be added before chlorination, or sodium hypochlorite may be mixed with the filter cake to convert calcium chloride to calcium hypochlorite.[261–263]

Most calcium hypochlorite processes produce a filter cake of calcium hypochlorite dihydrate that contains 30–50% water. Zinc dust or salts,[264] carboxylic acids and their salts such as citric acid, or sugars such as glucose and galactose[265] can be used to modify crystal growth to make the crystals easier to filter and reduce water retention. The crystals are dehydrated using air or vacuum to form granular products that are mostly anhydrous calcium hypochlorite, $Ca(OCl)_2$. On air drying, some calcium carbonate forms by reaction with carbon dioxide. Some decomposition also occurs to form calcium chlorate and calcium chloride.[5] This decomposition is reduced by adding a concentrated solution of sodium hypochlorite and sodium hydroxide to the wet calcium hypochlorite to make a small amount of hemibasic calcium hypochlorite.[266] Drying can be aided by spraying the slurry of calcium hypochlorite, that contains ~55% water, into a fluidized bed of granular anhydrous calcium hypochlorite to produce a solid with ~20% water. This solid is then further dried using air or vacuum.[267]

Inorganic Bleaches: Production of Hypochlorite

24.4.6.5 Bleaching Powder

Bleaching powder may still be made by treating hydrated lime with chlorine gas in an enclosed chamber. Periodic mixing is required to expose all of the surfaces to chlorine. More uniform composition is obtained by passing a countercurrent flow of chlorine gas over hydrated lime that is moving along a nearly horizontal, rotating cylinder, or along a rotating screw in a vertical tower. A shower of finely powdered hydrated lime can also be projected down a high tower through a stream of chlorine gas.

Initially, chlorination produces a mixture of dibasic calcium hypochlorite, $Ca(OCl)_2 \cdot 2Ca(OH)_2$, and monobasic calcium chloride dihydrate, $CaCl_2 \cdot 2Ca(OH)_2 \cdot 2H_2O$, by the following reaction:

$$5Ca(OH)_2 + 2Cl_2 \rightarrow Ca(OCl)_2 \cdot 2Ca(OH)_2 + CaCl_2 \cdot Ca(OH)_2 \cdot 2H_2O$$

Further chlorination converts the dibasic calcium hypochlorite to a mixed crystal that is predominately calcium hypochlorite, $Ca(OCl)_2$. Some dibasic calcium hypochlorite remains as the minor component. The monobasic calcium chloride remains unchanged during this second stage of chlorination, and bleaching powder is typically a mixture of this compound and the calcium hypochlorite mixed crystal.[4,30]

The hydrated lime should contain 0.5–7% excess water, ideally, 4%. Additional water forms during chlorination, which is mostly removed by the stream of chlorine gas and the heat of reaction. However, ~11% water typically remains in the product. To improve product stability, this water is removed during the last stages of chlorination.[4] In a typical process, hydrated lime is chlorinated in a jacketed reactor with an agitator at 20–25°C and an absolute pressure of 40–90 mm Hg. After the product contains 28–30% available chlorine and 40% water, the second stage is carried out at 50–55°C and an absolute pressure of 30–50 mm Hg. The finished product has 36.8% calcium hypochlorite, 31.7% calcium chloride, 26.2% calcium hydroxide, 3% calcium carbonate, 0.2% calcium chlorate, and 0.9% water.[5] Additional drying further improves product stability. This is done by heating to 100°C, heating under reduced pressure, or briefly passing hot air or other gases through the bleaching powder. Adding up to one part of calcium oxide to five parts of bleaching powder to convert the free water to calcium hydroxide is another common way of stabilizing bleaching powder.

A more stable product is obtained when lime is hydrated with sodium hydroxide solution instead of water. In addition, the product may contain up to 60% available chlorine:

$$Ca(OH)_2 + 2NaOH + 2Cl_2 \rightarrow Ca(OCl)_2 + 2NaCl + H_2O$$

Stability is also improved by mixing bleaching powder with a strong sodium hypochlorite solution or chlorinating a mixture of lime and sodium hydroxide.[268] Or, by reacting lime[269] or hydrated lime[270] with dichlorine monoxide to produce calcium hypochlorite without forming calcium chloride:

$$CaO + Cl_2O \rightarrow Ca(OCl)_2$$

$$Ca(OH)_2 + Cl_2O \rightarrow Ca(OCl)_2 + H_2O$$

Dichlorine monoxide also reacts with calcium chloride to form calcium hypochlorite and chlorine.[271]

24.4.7 MAGNESIUM HYPOCHLORITE

Basic magnesium hypochlorite, $Mg(OCl)_2 \cdot Mg(OH)_2$, was first made more than a hundred years ago. It was made by adding chlorine to a suspension of magnesium oxide or hydroxide. It is also made by mixing concentrated aqueous solutions of calcium hypochlorite and magnesium chloride. After the mixture stands for 10 h, the precipitated magnesium hypochlorite is removed by filtration and rapidly dried in a vacuum.[143] It can also be made by the controlled addition of chlorine and a 20–35% magnesium chloride solution to a 10–35% sodium hydroxide solution to achieve pH 8.5–9.0. A solid with 35–44% available chlorine is then filtered and dried.[272]

Dibasic magnesium hypochlorite, $Mg(OCl)_2 \cdot 2Mg(OH)_2$, was not made until 1969. It can be made by adding a solution of sodium or calcium hypochlorite to a solution that contains an excess of magnesium chloride or nitrate. The solutions must be carefully combined to avoid the precipitation of magnesium hydroxide.[11] Alternatively, a solid magnesium salt such as magnesium chloride hexahydrate or magnesium nitrate hexahydrate is mixed with solid calcium hypochlorite for 16–24 h, during which time a thick slurry forms. Water is then added to dissolve soluble compounds, and the mixture is filtered. The solids are dried in a vacuum to give a product that contains 75–85% dibasic magnesium hypochlorite.[273] The product is easier to filter if a concentrated slurry of calcium hypochlorite is used instead of the solid compound. A chloride salt, which may be magnesium chloride, should also be included to accelerate the reaction to minimize the formation of chlorate.[274] In all three processes, the reaction is as follows:

$$3Mg^{++} + 4^-OCl + 2Cl^- + 2H_2O \rightarrow Mg(OCl)_2 \cdot 2Mg(OH)_2 + 2Cl_2$$

About 40% of the hypochlorite is converted to chlorine, which may be recycled into sodium or calcium hypochlorite. About 40% of the hypochlorite ends up in the product. The remainder is lost due to product solubility in the filtrate or disproportionation to chloride and chlorate.

24.5 SAFETY

Exposure hazards and required safety precautions depend on the type and concentration of hypochlorite produced and the raw materials that are used. Generally, hypochlorites are corrosive irritants. They do not normally cause subchronic or chronic health effects.[275,276] Owing to the lithium ion, lithium hypochlorite also has moderate acute oral toxicity and acute dermal toxicity.[277] With hypochlorite solutions, the consequences of exposure increase with concentration.[278] Exposure to products made for household use usually have no effect or cause only minor irritation, which rapidly heals itself without leaving long-term or residual effects.[279–281] However, prolonged exposure, for example, to saturated clothing, can cause burns. Exposure to the most common industrial strength solutions usually do not damage skin if they are washed off shortly after contact, but they can permanently damage eyes. More concentrated solutions, solid hypochlorites, and hypochlorous acid solutions can cause severe burns. Gas streams of hypochlorous acid or dichlorine monoxide are usually treated as if they were chlorine.[282] Generally, protective equipment should be used to prevent exposure of eyes, skin, and clothing to hypochlorites. Where needed, respiratory protection should be used to prevent exposure to mists, dusts, and gases. Safety showers and eyewash fountains need to be readily available.

Calcium hypochlorite is stable at normal temperature, but it sustains its own decomposition at 175°C with the release of oxygen gas. Lithium hypochlorite is decomposed by high temperatures with the release of oxygen, but it does not sustain its own decomposition. Calcium and lithium hypochlorite react vigorously with certain other materials such as other oxidizers, nitrogen compounds, and organic compounds, and support their combustion.[5] All hypochlorites can liberate chlorine gas when mixed with acids and chloramines when mixed with ammonia.

Chlorine is listed as a toxic and reactive highly hazardous chemical in Title 29 CFR Part 1910 of the U.S. Occupational Safety and Health Standards. The requirements for handling chlorine are well documented.[2,283–288] Protective equipment is required to prevent inhalation and contact with eyes, skin, and clothing. Chlorine is often handled in confined areas to prevent accidental releases. These areas require continuous monitoring to detect chlorine leaks, air scrubbers,

Inorganic Bleaches: Production of Hypochlorite

and automatic devices to shut off the source of chlorine. Emergency eyewash fountains, deluge showers, self-contained breathing apparatus, and kits to stop leaks must be readily available. An emergency preparedness plan is essential. Proper storage is also required. Many materials react violently with chlorine including aluminum, titanium, polypropylene, and most organic compounds. Chlorine does not burn, but it supports combustion of other materials, even in the absence of air or oxygen. The volume of liquid chlorine increases considerably with increasing temperature, which can cause containers and other equipment to rupture. Pipes that handle liquid chlorine need periodic expansion chambers and pressure relief valves. Liquid chlorine is also a freezing hazard since its boiling point is –34°C. In addition, one volume of liquid expands to 460 volumes of gas.[289]

Sodium, potassium, and calcium hydroxide solutions or dusts are corrosive. Short exposures can severely burn skin or permanently damage vision. Protective equipment is required to prevent inhalation of dust, mist or spray, exposure to eyes, skin, and clothing. Safety showers and eyewash fountains must be readily available. Hydroxides react violently with acids, many organic chemicals, and aluminum. They rapidly destroy leather, wool, aluminum, zinc, and tin. They may react violently with water causing boiling, spattering, or violent eruptions.[290]

24.6 MATERIALS OF CONSTRUCTION

Only a limited number of materials are suitable for use in a bleach plant. Hypochlorite needs to be protected from contacting nickel, cobalt, copper, and alloys that contain these metals since enough of these metals can be adsorbed from piping and equipment to decrease stability. In addition, hypochlorite corrodes these and many other metals such as aluminum, tin, zinc, iron, and stainless steel.

Titanium is the material of choice for reaction vessels, pressure piping, pumps, heat exchangers, and other equipment that contacts hypochlorite solutions. Steel pipes lined with polypropylene, chlorobutylene, or fluorocarbon polymers can also be used for pressure piping.[291] However, titanium or tantalum must be used for metal fittings. Polyvinyl chloride (PVC) is often used for low pressure piping at temperatures below 54°C. Chlorinated polyvinyl chloride (CPVC) may be used up to 100°C, but it is more brittle and more likely to have stress failures than PVC. Valve seats are usually a fluorocarbon elastomer or ethylene propylene rubber. Hypochlorite solutions are best stored in titanium or PVC-lined fiberglass-reinforced polyester (FRP) tanks. If the hypochlorite solution does not contain nickel, cobalt, or copper, it can be stored in an FRP or a black, heavy-walled, cross-linked polyethylene tank. The FRP must be an epoxy vinyl ester resin that is cured with dimethylaniline and benzoyl peroxide.[292] Cobalt curing agents cannot be used since the residual cobalt will cause failures. Chlorobutylene or ethylene propylene rubber–lined steel and concrete have also been used for reaction vessels and storage tanks.[57] Polyethylene and polypropylene are suitable for shipping containers. Silver, gold, and platinum are compatible with hypochlorite solutions and are often used for ORP electrodes.

Piping for liquid chlorine is usually carbon steel that is periodically monitored for corrosion using ultrasonic and radiographic measurements. Stainless steel tends to stress crack, embrittle, and fail. Tantalum is used for critical components, it is one of a few metals that withstands both hypochlorite and dry chlorine. Hastalloy C22 and C276 have been used for valve parts that may see both hypochlorite and dry chlorine. Titanium must not be used where it might contact dry chlorine since a vigorous reaction will ensue. Fluorinated hydrocarbon and fluorinated silicone lubricants can be used with chlorine. Vinyl ester resins, FRP, PVC, CPVC, polyethylene, polypropylene, and acrylonitrile butadiene styrene can be used with chlorine gas. PTFE, perfluoroalkoxy polymers, polyvinylidene difluoride, ethylene tetrafluoroethylene copolymer, and ethylene chlorotrifluoroethylene copolymer can be used with liquid or gaseous chlorine.[293] Chlorine is usually stored in its shipping container.

Most plastics, including PVC, steel, cast iron, stainless steel, and titanium are compatible with 50% sodium hydroxide solutions. Iron and steel are the most commonly used in bleach plants.

464   Handbook of Detergents/Part F: Production

Nickel, monel, and nickel–copper alloys have better corrosion resistance, but they are not used to avoid adding nickel and copper, which catalyze hypochlorite decomposition.[44]

24.7 RELATIVE ECONOMICS

As an industrial chemical, sodium hypochlorite is a low-cost commodity item. Its price varies according to market demand and fluctuations in the prices of chlorine and sodium hydroxide. In addition, production capacity is much greater than the amount produced. Based on available chlorine, sodium hypochlorite is the least costly form of hypochlorite and lithium hypochlorite is by far the most expensive.

In the United States more than 30 companies sell sodium hypochlorite, most of which do not sell chlorine. During 2005, 553 million gal of household and institutional bleach with label strength of 5.25–6.15% sodium hypochlorite and 310 million gal of industrial solutions with a minimum of 12.5% sodium hypochlorite were sold in the United States.[294] Most commercial products are made at 10–20% above their label strength to allow decomposition during storage and transportation. Also, a small portion of household bleach is made by diluting industrial solutions. Essentially, all of the household and institutional bleach and a small portion of the industrial solutions are used as bleaches, cleaners, sanitizers, and disinfectants. Overall, treating water, including swimming pools, is the largest use of sodium hypochlorite. It is also the fastest growing use, since many facilities are switching to sodium hypochlorite to avoid transporting and storing chlorine.

In Europe and Japan, most sodium hypochlorite is made by chlorine producers. The solutions are made with 12–25% sodium hypochlorite. In western Europe a large number of companies dilute, repackage, and distribute products. Within Europe, the amounts and usage patterns vary widely between countries. Overall, the largest use of sodium hypochlorite in Europe is for household and institutional bleaches, cleaners, and disinfectants.

Sodium hypochlorite is also produced in most other regions around the world, primarily for cleaning and disinfection. Because of limited stability and high shipping costs, most sodium hypochlorite is locally produced. In some locations, hypochlorite producers must make their own chlorine and caustic. Some new plants being built in established markets are also including membrane cells to make their own chlorine and caustic. This avoids the transportation of chlorine. Storage of chlorine can also be avoided by recombining chlorine and caustic as they are formed to make hypochlorite.

Because the plants are more expensive to build and more difficult to operate, calcium hypochlorite is produced in relatively few locations. It is also more stable and cheaper to ship than sodium hypochlorite, thus it can be exported. The 1995 world capacity of calcium hypochlorite with 65–75% available chlorine was about 181,000 t, of which 107,500 t was in the United States.[295] Japan is the next largest producer. The rest is made in South Africa, Canada, China, Brazil, and Korea.[5] The world capacity of hemibasic calcium hypochlorite, which has about 60% available chlorine, was 28,000 t in 1988. Japan and China are the largest producers, followed by Italy and then India.[5] Lesser amounts of bleaching powder may still be made, primarily in less developed countries. In addition, some pulp mills may still make bleach liquor for captive use. Throughout the world, most calcium hypochlorite is used for water treatment. In the United States and Europe, more than 80% is used to sanitize swimming pools. Most of the rest is used for other water treatment, but some is used in toilet bowl cleaners and disinfectants for agriculture and food processing. Calcium hypochlorite containing about 51,000 t of available chlorine was used in the United States in 1995.[295]

The capacity for lithium hypochlorite with 37–39% available chlorine is ~3,650 t in the United States.[5] More than 80% is used to sanitize swimming pools and hot tubs. The rest is used in dry institutional laundry bleaches.

Chlorinated trisodium phosphate is still used in sanitizing cleaners and powdered cleansers. About 34,000 t with an average of 3.65% available chlorine were used in the United States in 1987.[5]

Inorganic Bleaches: Production of Hypochlorite 465

24.8 FUTURE TRENDS

New inorganic hypochlorite bleaching agents are unlikely to be developed. The relative economics and use patterns of hypochlorites for water treatment, cleaning, and disinfection are also unlikely to change. However, the use of hypochlorites as a safer alternative to transportation and storage of chlorine may increase. In the United States, this may also cause closer affiliation of hypochlorite production with chlorine sources. The competition with point-of-use electrolytic generators will also increase as these become cheaper and more prevalent.

Manufacturing processes will continue to evolve with incremental improvements, with an emphasis on the continuous production of sodium and calcium hypochlorites. Improvements will also be made to improve purity and control by-products such as chlorate and bromate. Partly, this will involve changes to raw materials such as the eventual elimination of mercury cell caustic and the use of brine with minimal bromide. Improvements in equipment that combine the generation of chlorine and the preparation of sodium hypochlorite will also continue.

New bleaching agents or biocides that can truly replace hypochlorite as a household and institutional cleaner and disinfectant are also unlikely to be developed. There will continue to be a battle with other bleaching and cleaning agents for market share in laundry and household cleaners, and the performance of competitive products will probably improve. Still, hypochlorite is expected to maintain its performance and cost advantages. The use of hypochlorite for these purposes is expected to increase, with the greatest growth in developing markets.[296] Growth is expected to continue because of its effectiveness in combating diseases and the increasing needs for sanitation due to increasing crowding, use of child day care, home care of ill and immuno-compromised people, and resistance of pathogens to antibiotics. Research has shown that although hypochlorite used for cleaning and disinfecting forms small amounts of chlorinated organic by-products, there is no evidence that these harm people or the environment. The by-products are largely degraded by sewage treatment; the remnants do not increase toxicity of treated effluents; and highly toxic, persistent, or bioconcentratable compounds are not formed. Further, the amount of by-products formed is probably insignificant compared to other sources.[281,297–307]

Many industrial uses of hypochlorite will also continue unchanged or increase, particularly for treatment of drinking water, cooling water, and wastewater. Hypochlorite avoids transport and storage of chlorine in populated areas, but it is more expensive and forms chlorate during storage. However, concerns about the formation of chlorinated organic by-products will continue to limit or decrease the use of hypochlorite for some other industrial purposes.

The demand for calcium hypochlorite is expected to continue to grow for water treatment and swimming pool disinfection in the United States, South America, and Europe.[295,308] Lithium hypochlorite will continue with very limited use in spa and swimming pool treatment due to its high price and limited availability. Its price will remain high due to the economics of lithium.

REFERENCES

1. Higgins, S.H., *A History of Bleaching*, Longmans, Green & Co., London, 1924.
2. White, G.C., *Handbook of Chlorination and Alternative Disinfectants*, 3rd ed., Van Nostrand Reinhold, New York, 1992.
3. Sheltmire, W.H., Chlorinated bleaches and sanitizing agents, in *Chlorine: Its Manufacture, Properties and Uses*, Sconce, J.S., Ed., Robert E. Krieger Publishing Co., Huntington, NY, 1972, p. 512.
4. Addison, C.C., The oxides and oxyacids of chlorine: bleaching powder, in *Mellor's Comprehensive Treatise on Inorganic and Theoretical Chemistry*, Supplement II, Part I, Longmans, London, 1956, p. 563.
5. Wojtowicz, J.A., Chlorine oxygen acids and salts: dichlorine monoxide, hypochlorous acid, and hypochlorites, In: Howe-Grant, M., Ed., *Kirk Othmer Encyclopedia of Chemical Technology*, Vol. 5, 4th ed., Wiley, New York, 1993, p. 932.
6. Dychdala, G.R. and Cox, R.J., U.S. Patent 3,645,005, 1972.
7. Dychdala, G.R. and Sloane, R.S., U.S. Patent 3,793,216, 1974.

8. Faust, J.P., U.S. Patent 3,669,894, 1972.
9. Tatara, S. et al., U.S. Patent 4,053,429, 1977.
10. Murakami, T., Igawa, K. and Hiraga, Y., U.S. Patent 4,355,014, 1982.
11. Bishop, J.J. and Trotz, S.I., U.S. Patent 3,582,265, 1971.
12. Gard, D.R., Phosphoric acids and phosphates, In: Howe-Grant, M., Ed., *Kirk Othmer Encyclopedia of Chemical Technology*, Vol. 18, 4th ed., Wiley, New York, 1996, p. 669.
13. Ellestad, R.B., Lithium hypochlorite bleaches, *Soap and Chem. Specialties*, 37, 77, 1961.
14. Fortnum, D.H. et al., The kinetics of the oxidation of halide ions by monosubstituted peroxides, *J. Am. Chem. Soc.*, 82, 778, 1960.
15. *OXONE™ Monopersulfate Compound*, EI du Pont Nemours & Co., Wilmington, DE, 1984.
16. Francis, R.C. et al., Caroate delignification enhancement by halides, *Tappi J.*, 77, 135, 1994.
17. Adam, W., Curci, R. and Edwards, J.O., Dioxiranes: a new class of powerful oxidants, *Acc. Chem. Res.*, 22, 205, 1989.
18. Montgomery, R.E., U.S. Patent 3,822,114, 1974.
19. Casella, V.M. and Fong, R.A., U.S. Patent 4,613,332, 1986.
20. Gray, F.W., U.S. Patent 4,300,897, 1981.
21. Farr, J.P., Smith, W.L. and Steichen, D.S., Bleaching agents, In: Howe-Grant, M., Ed., *Kirk Othmer Encyclopedia of Chemical Technology*, Vol. 4, 4th ed., Wiley, New York, 1992, p. 271.
22. Wojtowicz, J.A., Chloramines and bromamines, In: Howe-Grant, M., Ed., *Kirk Othmer Encyclopedia of Chemical Technology*, Vol. 5, 4th ed., Wiley, New York, 1993, p. 911.
23. Al-Haq, M.I., Sugiyama, J. and Isobe, S., Applications of electrolyzed water in agriculture and food industries, *Food Sci. Technol. Res.*, 11(2), 135, 2005.
24. Sanyo Electric Co., Sanyo introduces the world's first zero-detergent electrolyzed water cleaning powered washing machine, News Release, June 22, 2001.
25. Sanyo Electric Co., Sanyo introduces five new products with a 'virus washer' system suppresses 99% airborne viruses without any chemicals, News Release, Aug 28, 2006.
26. Best of what's new grand award winner: mountain safety research MIOX purifier, *Popular Sci.*, Dec, 90 2003.
27. United States Geological Survey, *Mineral Commodity Summaries*, 2007, p. 136.
28. Bertram, B.M., Sodium compounds: sodium halides: sodium chloride, In: Howe-Grant, M., Ed., *Kirk Othmer Encyclopedia of Chemical Technology*, Vol. 22, 4th ed., Wiley, New York, 1997, p. 354.
29. Mucenieks, P.R., U.S. Patent 4,323,437, 1982.
30. Bommaraju, T.V. et al., Chlorine, In: Seidel, A., Ed., *Kirk Othmer Encyclopedia of Chemical Technology*, 5th ed., Vol. 6, Wiley, New York, 2004, p. 130.
31. *Pamphlet 67: Safety Guidelines for the Manufacture of Chlorine*, The Chlorine Institute, Washington, DC, 1981.
32. World chloralkali market projected to be strong in 2005. *Chem. Marketing Reporter*, Jan 24, 2005.
33. Hess, G., Making hazmat transport safer, *Chem. Eng. News*, 84(11), 60, 2006.
34. Berkey, F.M., Electrolysis of hydrochloric acid solutions, in *Chlorine: Its Manufacture, Properties and Uses*, Sconce, J.S., Ed., Robert E Krieger Publishing Co., Huntington, NY, 1972, p. 200.
35. Fogler, M.F., The salt process for chlorine manufacture, in *Chlorine: Its Manufacture, Properties and Uses*, Sconce, J.S., Ed., Robert E Krieger Publishing Co., Huntington, NY, 1972, p. 235.
36. Redniss, A., HCl oxidation processes, in *Chlorine: Its Manufacture, Properties and Uses*. Sconce, J.S., Ed., Robert E Krieger Publishing Co., Huntington, NY, 1972, p. 250.
37. Kircher, M.S., Electrolysis of brines in diaphragm cells, in *Chlorine: Its Manufacture, Properties and Uses*. Sconce, J.S., Ed., Robert E Krieger Publishing Co., Huntington, NY, 1972, p. 81.
38. MacMullin, R.B., Electrolysis of brines in mercury cells, in *Chlorine: Its Manufacture, Properties and Uses*. Sconce, J.S., Ed., Robert E Krieger Publishing Co., Huntington, NY, 1972, p. 127.
39. *Pamphlet 10: North American Chlor-Alkali Industry Plants and Production Data Report—1995*, The Chlorine Institute, Washington, DC, 1996.
40. Johnson, J., Where goes the missing mercury? *Chem. Eng. News*, 82(11), 31, 2004.
41. Short, P.L., Negotiating away mercury emissions, *Chem. Eng. News*, 79(11), 21, 2001.
42. Rudd, E.J. and Savinell, R.F., Report of the electrolytic industries for the year 1988, *J. Electrochem. Soc.*, 136, 449C, 1989.
43. *Chlorine and Caustic Soda Membrane Cell Process*, Cellchem Nobel Industries Sweden, Stockholm.
44. *Caustic Soda Handbook*, Occidental Chemical Corporation, Dallas, TX, 1992.
45. Bertetti, J.W. and McCabe, W.L., Sodium hydroxide solutions: heat of dilution at 68°F, *Ind. Eng. Chem.*, 28, 247, 1936.

Inorganic Bleaches: Production of Hypochlorite

46. Bommaraju, T.V., Hauck, W.V. and Lloyd, V.J., U.S. Patent 4,585,579, 1986.
47. Khare, G.P., U.S. Patent 4,282,178, 1981.
48. Curlin, L.C., Bommaraju, T.V. and Hansson, C.B., Alkali and chlorine products: chlorine and sodium hydroxide, In: Howe-Grant, M., Ed., *Kirk Othmer Encyclopedia of Chemical Technology*, Vol. 1, 5th ed., Wiley, New York, 1991, p. 938.
49. Otto, J.M., U.S. Patent 3,784456, 1974.
50. Dotson, R.L., U.S. Patent 4,278,527, 1981.
51. Bommaraju, T.V., Sodium hypochlorite its application and stability in bleaching, *Water Qual. Res. J. Canada*, 30, 339, 1995.
52. Rauh, F., Alkali and chlorine products: sodium carbonate, In: Howe-Grant, M., Ed., *Kirk Othmer Encyclopedia of Chemical Technology*, Vol. 1, 4th ed., Wiley, New York, 1991, p. 1025.
53. *Caustic Potash Handbook*, Occidental Chemical Corporation, Dallas, TX, 2000.
54. Freilich, M.B. and Petersen, R.L., Potassium compounds, In: Seidel, A., Ed., *Kirk Othmer Encyclopedia of Chemical Technology*, Vol. 19, 5th ed., Wiley, New York, 2006, p. 608.
55. Kamienski, C.W. et al., Lithium and lithium compounds, In: Seidel, A., Ed., *Kirk Othmer Encyclopedia of Chemical Technology*, Vol. 15, 5th ed., Wiley, New York, 2005, p. 120.
56. Petersen, R.L. and Freilich, M.B., Calcium compounds, In: Howe-Grant, M., Ed., *Kirk Othmer Encyclopedia of Chemical Technology*, Vol. 4, 4th ed., Wiley, New York, 1992, p. 787.
57. *Chlorine Bleach Solutions*, 2nd ed., Solvay Process Division Allied Chemical Corporation, New York, 1960.
58. Kolthoff, I.M. et al., *Quantitative Chemical Analysis*, 4th ed., The Macmillan Co., New York, 1969, pp. 851–852.
59. *Standard Methods for the Examination of Water and Wastewater*, 16th ed., Greenberg, A.E. et al., Eds., American Public Health Association, Washington, DC, 1985, pp. 294–315.
60. Adams, L.C. et al., Hypochlorous acid decomposition in the pH 5–8 region, *Inorg. Chem.*, 31, 3534, 1992.
61. Morris, J.C., The acid ionization constant of HOCl from 5 to 35°, *J. Phys. Chem.*, 70, 3798, 1966.
62. Connick, R.E. and Chia, Y.T., The hydrolysis of chlorine and its variation with temperature. *J. Am. Chem. Soc.*, 81, 1280, 1959.
63. Wang, T.X. and Margerum, D.W., Kinetics of reversible chlorine hydrolysis: temperature dependence and general-acid / base-assisted mechanisms, *Inorg. Chem.*, 33, 1050, 1994.
64. Roth, W.A., Zur Thermochemie des Chlors und der unterchlorigen Säure, *Z. Phys. Chem.*, A145, 289, 1929.
65. Zimmerman, G. and Strong, F.C., Equilibria and spectra of aqueous chlorine solutions, *J. Am. Chem. Soc.*, 79, 2063, 1957.
66. Wang, T.X. et al., Equilibrium, kinetic, and UV-spectral characteristics of aqueous bromine chloride, bromine, and chlorine species, *Inorg. Chem.*, 33, 5872, 1994.
67. Huthwelker, T. et al., Solubility of HOCl in water and aqueous H_2SO_4 to stratospheric temperatures, *J. Atmos. Chem.*, 21, 81, 1995.
68. Imakawa, H., Chemical reactions in the chlorate manufacturing electrolytic cell (part 1); the vapour pressure of hypochlorous acid on its aqueous solution, *J. Electrochem. Soc. Jpn.*, 18, 382, 1950.
69. Imagawa, H., Studies on chemical reactions of the chlorate cell (part 2); the vapour pressure of hypochlorous acid on its mixed aqueous solution with sodium chlorate, *J. Electrochem. Soc. Jpn.*, 19, 271, 1951.
70. Blatchley III, E.R. et al., Effective Henry's law constants for free chlorine and free bromine, *Water Res.*, 26, 99, 1992.
71. Holzwarth, G., Balmer, R.G. and Soni, L., The fate of chlorine and chloramines in cooling towers; Henry's law constants for flashoff. *Water Res.*, 18, 1421, 1984.
72. Roth, W.A., Notiz zur Thermochemie des Chlormonoxydes, *Z. Phys. Chem.*, A191, 248, 1942.
73. Ruiz-Ibanez, G. et al., Solubility and diffusivity of oxygen and chlorine in aqueous hydrogen peroxide solutions, *J. Chem. Eng. Data*, 36, 459, 1991.
74. Chlor Ergänzungsband, in *Gmelins Handbuch der Anorganischen Chemie*, Teil B, Lieferung 2, System-Nummer 6, Verlag Chemie, Weinheim, 1969, p. 379.
75. Addison, C.C., The oxides and oxyacids of chlorine: hypochlorous acid and the hypochlorites, in *Mellor's Comprehensive Treatise on Inorganic and Theoretical Chemistry*, Supplement II, Part I, Longmans, London, 1956, p. 544.
76. Eigen, M. and Kustin, K., The kinetics of halogen hydrolysis, *J. Am. Chem. Soc.*, 84, 1355, 1962.
77. Brian, P.L.T., Vivian, J.E. and Piazza, C., The effect of temperature on the rate of absorption of chlorine into water, *Chem. Eng. Sci.*, 21, 551, 1966.

78. Lifshitz, A. and Perlmutter-Hayman, B., The kinetics of the hydrolysis of chlorine. III. The reaction in the presence of various bases, and a discussion of the mechanism, *J. Am. Chem. Soc.* 84, 701, 1962.
79. Temperature dependence of selected equilibrium constants in aqueous solutions, in Lange's Handbook of Chemistry, 13th ed., Dean, J.A., Ed., McGraw-Hill, New York, 1985, pp. 5–63.
80. Ashour, S.S., Rinker, E.B. and Sandall, O.C., Absorption of chlorine into aqueous bicarbonate solutions and aqueous hydroxide solutions. *AIChE J.*, 42, 671, 1996.
81. Gershenzon, M. et al., Rate constant for the reaction of Cl_2(aq) with OH^-, *J. Phys. Chem.* A106(34), 7748, 2002.
82. Hitika, H. et al., Adsorption of chlorine into aqueous sodium hydroxide solutions, *Chem. Eng. J.*, 5, 77, 1973.
83. 85. Lahiri, R.N., Yadav, G.D. and Sharma, M.M., Absorption of chlorine in aqueous solutions of sodium hydroxide, *Chem. Eng. Sci.*, 38, 1119, 1983.
84. Mogal, M.M.B. and Yadav, G.D., Theoretical analysis of absorption of chlorine in aqueous slurries of calcium hydroxide: desorption of hypochlorous acid gas, *Can. J. Chem. Eng.*, 73, 693, 1995.
85. Lister, M.W., The decomposition of hypochlorous acid, *Can. J. Chem.*, 30, 879, 1952.
86. Adam, L.C. and Gordon, G., Hypochlorite ion decomposition: effects of temperature, ionic strength, and chloride ion, *Inorg. Chem.*, 38, 1299, 1999.
87. Lister, M.W., The decomposition of sodium hypochlorite: the uncatalyzed reaction, *Can. J. Chem.*, 33, 465, 1956.
88. Gordon, G. et al., Predicting liquid bleach decomposition, *J. AWWA*, 89(4), 142, 1997.
89. Gordon, G., Adam, L. and Bubnis, B., Minimizing chlorate ion formation, *J. AWWA*, 87(6), 97, 1995.
90. Lister, M.W. and Petterson, R.C., Oxygen evolution from sodium hypochlorite solutions, *Can. J. Chem.*, 40, 729, 1962.
91. Church, J.A., Kinetics of uncatalyzed and Cu(II)-catalyzed decomposition of sodium hypochlorite, *Ind. Eng. Chem. Res.*, 33, 239, 1994.
92. Onodera, S. et al., Chemical changes of organic compounds in chlorinated water IV. Gas chromatography and mass spectrometric identification and determination of some chlorinated and oxidation-reuptured compounds formed during the reactions of napthols and aqueous hypochlorite, *Eisei Kagaku*, 28, 146, 1982.
93. Aizawa, T. et al., Characteristics of the formation of total organic chlorides from organic compounds by aqueous chlorination, *Suishitsu Odaku Kenkyu*, 7, 36, 1984.
94. Lister, M.W., Decomposition of sodium hypochlorite: the catalyzed reaction, *Can. J. Chem.*, 34, 479, 1956.
95. Gray, E.T., Taylor, R.W. and Margerum, D.W., Kinetics and mechanisms of the copper-catalyzed decomposition of hypochlorite and hypobromite. Properties of a dimeric copper(III) hydroxide intermediate, *Inorg. Chem.*, 16, 3047, 1977.
96. Ayres, G.A. and Booth, M.H., Catalytic decomposition of hypochlorite solution by iridium compounds. I. The pH-time relationship, *J. Am. Chem. Soc.*, 77, 825, 1955.
97. Young, K.W. and Allmand, A.J., Experiments on the photolysis of aqueous solutions of chlorine, hypochlorous acid, and sodium hypochlorite, *Can. J. Res.*, 27B, 318, 1949.
98. Buxton, G.V. and Williams, R.J.M., Photochemical decompositions of aqueous solutions of oxyanions of chlorine and chlorine dioxide, *Proc. Chem. Soc.*, 141, 1962.
99. Mullins, R.M., U.S. Patent 6,638,446, 2003.
100. Skrypa, M.J., Baran, F.R. and Low, W.W., U.S. Patent 3,557,010, Jan 19, 1971.
101. Ikeda, E. et al., Japanese Patent Appl. 55 121,902, 1980.
102. Kihara, Y., Katada, T. and Okada, Y., Japanese Patent Appl. 78 102,894, 1978.
103. Marchesini, M. and Trigiante, G., European Patent Appl. 743,279, 1996.
104. Trigiante, G., European Patent Appl. 743,280, 1996.
105. Lister, M.W., The stability of some complexes of trivalent copper, *Can. J. Chem.*, 31, 638, 1953.
106. Gamlen, P.H., U.S. Patent 4,065,545, 1977.
107. Foxlee, J.C., U.S. Patent 4,474,677, 1984.
108. Ahmed, F.U., U.S. Patent 5,229,029, 1993.
109. Blum, H., German Patent Appl. 29 03 980, 1980.
110. Crutchfield, M.M. and Irani, R.R., U.S. Patent 3,297,578, 1967.
111. Agostini, F. and Caire, G., European Patent Appl. 653,482, 1995.
112. Douglass, M.L., U.S. Patent 5,380,458, 1995.
113. Rahman, R.-U. et al., Stabilization of soda bleach liquor, *Pak. J. Sci. Res.*, 32, 279, 1980.

Inorganic Bleaches: Production of Hypochlorite   **469**

114. Burton, C.D., U.S. Patent 4,898,681, 1990.
115. Ciullo, P.A. and Andersson, M., Natural smectic clays as bleach stabilizers, *HAPPI*, Dec, 102, 2000.
116. Takenaka, S., Japanese Patent Appl. 60 051,608 and 60 051,609, 1985.
117. Kandori, H, and Sugawara, H., Japanese Patent Appl. 62 089,800, 1987.
118. Izumi, Y. and Takechi, M., Japanese Patent Appl. 62 205,199, 1987.
119. Aoyanagi, M., Miyamoto, S. and Nakagawa, Y., Japanese Patent Appl. 63 110,298, 1988.
120. Laubusch, E.J., Physical and chemical properties of chlorine, in *Chlorine: Its Manufacture, Properties and Uses*, Sconce, J.S., Ed., Robert E Krieger Publishing Co., Huntington, NY, 1972, p. 21.
121. Cotton, F.A. and Wilkinson, G., *Advanced Inorganic Chemistry*, 4th ed., Wiley, New York, 1980, p. 227.
122. Weber, H.M., U.S. Patent 1,522,561, 1925.
123. Lineur, W., U.S. Patent 4,330,521, 1982.
124. Powell, D., U.S. Patent Appl. 2005/0169832, 2005.
125. Pavis, E.H., U.S. Patent 3,702,234, 1972.
126. King, A.S., U.S. Patent 4,010,244, 1977.
127. Yant, R.E. and Larson, P.A., U.S. Patent 4,744,956, 1988.
128. Markarov, S.Z. and Shcharkova, E.F., Solubility isotherm (10°C) of ternary systems of calcium and sodium hypochlorites and chlorides, *Russ. J. Inorg. Chem.*, 14, 1632, 1969.
129. Hamano, A., Hamado, K. and Kumamoto, K., Phase diagram of the sodium hypochlorite-sodium chloride-water system, *Nippon Kagaku Kaishi*, 7, 1066, 1977.
130. *HyPure® CCF Sodium Hypochlorite*, Olin Corp., Stamford, CT, 1996.
131. *HyPure® N Sodium Hypochlorite*, Olin Corp., Stamford, CT, 1991, 1995.
132. Dugua, J., U.S. Patent 4,780,303, 1988.
133. Kawamura, S., Ishida, S. and Tanabe, K., Japanese Patent Appl. 5,139,701, 1993.
134. Nesty, P., Dugua, J. and Thery, P., European Patent Appl. 527,083, 1993.
135. Verlaeten, J., U.S. Patent 4,428,918, 1984.
136. Fukuma, Y. and Fukushima, S., Japanese Patent Appl. 60 081,003, 1985.
137. Nicolaisen, B.H., U.S. Patent 3,241,912, 1966.
138. Tatara, S., Minakami, T. and Nishomiya, M., Japanese Patent Appl. 72 15,463, 1972.
139. Murakami, T. and Hiraga, Y., Japanese Patent Appl. 59 008,603, 1984.
140. Duncan, B.L. and Ness, R.C., U.S. Patent 5,194,238, 1993.
141. Tomotake, A., Tokuji, T. and Yoshimi, I., Japanese Patent Appl. 2000290003, 2000.
142. Brahm, J. and Demilie, P., European Patent Appl. 99,152, 1984.
143. Addison, C.C., The oxides and oxyacids of chlorine: the hypochlorites, in *Mellor's Comprehensive Treatise on Inorganic and Theoretical Chemistry*, Supplement II, Part I, Longmans, London, 1956, p. 556.
144. Japanese Patent Appl. 70 037,134, 1970.
145. Walsh, R.H. and Dietz, A., U.S. Patent 3,498,924, 1970.
146. Yamamoto, M., U.S. Patent 6,592,727, 2003.
147. Shinomiya, Y., Miyoshi, K. and Sudo, S., U.S. Patent 5,935,393, 1999.
148. Bess, J.W., Matousek, R.C. and Simmons, B., U.S. Patent 6,805,787, 2004.
149. Nakamura, S. and Fukuzuka, K., U.S. Patent Appl. 2003/0146108, 2003.
150. Rhees, R.C. et al., U.S. Patent 5,688,385, 1997.
151. Arimoto, O. and Kishi, T., U.S. Patent 5,622,613, 1997.
152. Ponzano, G.P., U.S. Patent 6,409,895, 2002.
153. Kaestner, E.A. and Spink, J., U.S. Patent 3,819,329, 1974.
154. Kumazawa, E., U.S. Patent 4,533,451,1985.
155. Kamitani, Y., Fujita, M. and Funabashi, T., U.S. Patent 5,728,274, 1998.
156. Yoshiya, O., Japanese Patent Appl. 10,263,542, 1998.
157. Eki, T. et al., U.S. Patent 5,846,390, 1998.
158. Fukuzuka, K. and Nakamra, S., U.S. Patent 6,231,747, 2001.
159. Kurokawa, K. et al., U.S. Patent 6,905,580, 2005.
160. Salerno, M., U.S. Patent Appl. 2006/0054510, 2006.
161. Hoehne, J. and Heinrich, H.-J., U.S. Patent 7,055,183, 2006.
162. Civanelli, C. and Vanetti, A., European Patent Appl. 146,184, 1985.
163. Porta, A. et al., U.S. Patent 4,560,455, 1985.
164. Sumidu, Y. et al., U.S. Patent 5,947,135, 1999.
165. Rikio, T. et al., Japanese Patent Appl. 2003214757, 2003.
166. Hiroaki, K. and Kazuo, T., Japanese Patent Appl. 2003172531, 2003.

167. Kurokawa, K. et al., U.S. Patent 6,827,849, 2004.
168. Keun, O.J., Japanese Patent Appl. 2003250721, 2003.
169. Takahiro, H. et al., Japanese Patent Appl. 2000126210, 2000.
170. Malchesky, P.S., U.S. Patent 5,932,171, 1999.
171. Tennakoon, C.L.K., Yalamanchili, R.C. and McGrew, E.I., U.S. Patent Appl. 2004/0213698, 2004.
172. Nakamura, S., Fukuzuka, K. and Miyashita, M., U.S. Patent 6,926,819, 2005.
173. Kasuya, S., U.S. Patent 6,632,336, 2003.
174. Kiyoteru, O. et al., Japanese Patent Appl. 2003181338, 2003.
175. Yuji, O., Noriyuki, K. and Kiyoteru, O., Japanese Patent Appl. 2004130263, 2004.
176. Herrington, R.E., U.S. Patent 7,008,523, 2006.
177. Herrington, R.E. et al., U.S. Patent 6,524,475, 2003.
178. Herrington, R.E. et al., U.S. Patent 6,261,464, 2001.
179. Herrington, R.E. and Hand, F., U.S. Patent 6,736,966, 2004.
180. Römer, H.G., European Patent Appl. 803,476, 1999.
181. Anderson, D. et al., U.S. Patent Appl. 2004/0055896, 2004.
182. *HyPure™ K. Potassium Hypochlorite*, Olin Chemicals, Stamford, CT, 1991.
183. Duncan, B.L. and Flowers, W.O., U.S. Patent 5,055,285, 1991.
184. Duncan, B.L., Density and viscosity of concentrated aqueous solutions of potassium hypochlorite, *J. Chem. Eng. Data.*, 39, 863, 1994.
185. Seidell, A., *Solubilities of Inorganic and Metal Organic Compounds*, D. Van Nostrand Co., New York, 1940, p. 1249.
186. *HyPure™ A Hypochlorous Acid*, Olin Chemicals, Stamford, CT, 1991.
187. Shaffer, J.H., Melton, J.K. and Borcz, J., U.S. Patent 5,322,677, 1994.
188. Melton, J.K., Hilliard, G.E. and Shaffer, J.H., U.S. Patent 5,116,593, 1992.
189. Melton, J.K. et al., U.S. Patent 5,037,627, 1991.
190. Brennan, J.P., Wojtowicz, J.A, and Campbell, P.H., U.S. Patent 4,146,578, 1979.
191. Melton, J.K., Hilliard, G.E. and Shaffer, J.H. U.S. Patent 5,270,019, 1993.
192. Hilliard, G.E., Melton, J.K. and Shaffer, J.H., U.S. Patent 5,213,771, 1993.
193. Hilliard, G.E. et al., U.S. Patent 5,106,591, 1992.
194. Hilliard, G.E., Melton, J.K. and Helmstetter, D.A., U.S. Patent 5,116,594, 1992.
195. Trent, D.L. et al., U.S. Patent 5,532,389, 1996.
196. Belgium Patent Appl. 1,007,467, 1995.
197. Wojtowicz, J.A. and Klanica, A.J., U.S. Patent 4,147,761, 1979.
198. Quardere, G.J. et al., U.S. Patent 6,048,513, 2000.
199. Muskat, I.E. and Cady, G.H., U.S. Patent 2,240,344, 1941.
200. Cady, G.H., U.S. Patent 2,157,524, 1939.
201. Cady, G.H., U.S. Patent 2,157,525, 1939.
202. Hoekje, H.H. and May, R.R., U.S. Patent 4,190,638, 1980.
203. Yant, R.E. and Galluch, R.J., U.S. Patent 4,584,178, 1986.
204. Yant, R.E. and Galluch, R.J., U.S. Patent 4,504,456, 1985.
205. Penard, H.F.L. and Risseeuw, J.I., U.S. Patent 4,017,592, 1977.
206. Mathias, L.D., U.S. Patent 1,555,474, 1925.
207. Alder, H., U.S. Patent 1,965,304, 1934.
208. Shen, C.Y., U.S. Patent 4,402,926, 1983.
209. Toy, A.D.F. and Bell, R.N., U.S. Patent 3,656,890, 1972.
210. Orazem, G.J., Ellestad, R.B. and Nelli, J.R., U.S. Patent 3,171,814, 1965.
211. *HyPure™ L Lithium Hypochlorite*. Olin Chemicals, Stamford, CT, 1991.
212. Duncan, B.L. et al., U.S. Patent 5,102,648, 1992.
213. Duncan, B.L., Carpenter, L.D. and Osborne, L.R., U.S. Patent 5,028,408, 1991.
214. Korzhenyak, I.G. and Furman, A.A., Investigation of equilibrium in the system LiClO-LiCl-H$_2$O, *J. Appl. Chem. USSR*, 50,736, 1977.
215. Partridge, H.deV., Pulp bleaching and purification, in *Chlorine: Its Manufacture, Properties and Uses*, Sconce, J.S., Ed., Robert E Krieger Publishing Co., Huntington, NY, 1972, p. 273.
216. Robson, H.L., Chlorine oxygen acids and salts: chlorine monoxide, hypochlorous acid, and hypochlorites, in *Kirk Othmer Encyclopedia of Chemical Technology*, Vol. 5, 2nd ed., Wiley, New York, 1964, p. 7.
217. Cherkasova, L.V., Mazanko, A.F. and Rabovskii, B.G., Soviet Union Patent Appl. 1,671,612, 1991.
218. Morgan, D.L., U.S. Patent 4,529,578, 1985.
219. Farmer Jr., D.A. and Wotowicz, J.A., U.S. Patent 4,108,792, 1978.

Inorganic Bleaches: Production of Hypochlorite

220. Brahm, J. et al., U.S. Patent 4,857,292, 1989.
221. Sakowski, W.J., U.S. Patent 3,895,099, 1975.
222. Sakowski, W.J., U.S. Patent 3,954,948, 1976.
223. Gleichert, R.D., U.S. Patent 3,134,641, 1964.
224. Hilliard, G.E. et al., World Patent Appl. 92 21,610, 1992.
225. Shaffer, J.H., Melton, J.K. and Hilliard, G.E., U.S. Patent 5,091,165, 1992.
226. Shaffer, J.H., Kurtz, W.L. and Hubbard, J.H., U.S. Patent 5,149,398, 1992.
227. Shaffer, J.H. and Kurtz, W.L., U.S. Patent 5,085,847, 1992.
228. Wojtowicz, J.A., U.S. Patent 4,416,864, 1983.
229. Bridges, W.G. et al., World Patent Appl. 91 17,951, 1991.
230. Tatara, S. et al., U.S. Patent 3,572,989, 1971.
231. Thibault, G. and Mayotte, D.C., U.S. Patent 4,397,832, 1983.
232. Tiedemann, H.H., Thibault, G. and Laberge, J.G.J., U.S. Patent 4,364,917, 1982.
233. Tiedemann, H.H., U.S. Patent 4,374,155, 1983.
234. Welch, C.N. et al., U.S. Patent 4,328,200, 1982.
235. Loehr, C.E. et al., U.S. Patent 4,390,512, 1983.
236. Stermole, D.A., Loehr, C.E. and Chun, D.S., U.S. Patent 4,428,919, 1984.
237. Chun, D.S., Stermole, D.A. and Loehr, C.E., U.S. Patent 4,448,759, 1984.
238. Hoffer, J.O. et al., U.S. Patent 4,258,024, 1981.
239. Miyashin, N. et al., U.S. Patent 3,950,499, 1976.
240. Nishomya, M. et al., Japanese Patent Appl. 60 141,605, 1985.
241. Tatara, S. et al., U.S. Patent 3,767,775, 1973.
242. Abe, Y. et al., U.S. Patent 6,309,621, 2001.
243. Kitchen, F.N., British Patent 378,847, 1932.
244. Meehan, F.T. and Kitchen, F.N., British Patent 404,627, 1934.
245. Nakaya, K. and Sato, K., U.S. Patent 4,521,397, 1985.
246. Kikuchi, M. and Murakami, T., Japanese Patent Appl. 52 093,694, 1977.
247. Sakowski, W.J., U.S. Patent 4,399,117, 1983.
248. Foster, C.A. and Shaffer, J.H., U.S. Patent 4,487,751, 1984.
249. Sakowski, W.J. and Duncan, B.L., U.S. Patent 4,367,209, 1983.
250. Sakowski, W.J., Carty, L.G. and Foster, C.A., U.S. Patent 4,335,090, 1982.
251. Ourisson, J., British Patent 187,009, 1938.
252. Ourisson, J., Camescasse, P. and Kastner, M., French Patent 858,057, 1940.
253. Ourisson, J., Camescasse, P. and Kastner, M., French Patent 862,483, 1941.
254. Vorburger, A.J., French Patent 1,019,027, 1953.
255. Sprauer, J.W., U.S. Patent 2,441,337, 1948.
256. Sprauer, J.W., U.S. Patent 2,469,901, 1949.
257. Sprauer, J.W., U.S. Patent 2,587,071, 1952.
258. Bruce, E.A., U.S. Patent 3,094,380, 1963.
259. Mohan Jr., J.C., U.S. Patent 3,030,177, 1962.
260. Sakowski, W.J., Bajaj, M.C. and Duncan, B.L., U.S. Patent 4,468,377, 1984.
261. Saeman, W.C., U.S. Patent 4,842,841, 1989.
262. Saeman, W.C., U.S. Patent 4,504,457, 1985.
263. Morgan, D.L., U.S. Patent 4,500,506, 1985.
264. Faust, J.P. and Robson, H.L., U.S. Patent 3,440,024, 1969.
265. Murakami, T., Kikuchi, M. and Igawa, K., U.S. Patent 4,248,848, 1981.
266. Sakowski, W.J., Shaffer, J.H. and Carty, L.G., U.S. Patent 4,337,236, 1982.
267. Saeman, W.C., U.S. Patent 3,969,546, 1976.
268. Wojtowicz, J.A., U.S. Patent 4,197,284, 1980.
269. Muskat, I.E. and Cady G.H., U.S. Patent 2,157,559, 1939.
270. Wojtowicz, J.A., U.S. Patent 4,105,565, 1978.
271. Muskat, I.E. and Cady G.H., U.S. Patent 2,225,923, 1940.
272. Nobuyuki, T. and Kyuichi, F., Japanese Patent Appl. 09,227,104, 1997.
273. Wojtowicz, J.A., U.S. Patent 4,071,605, 1978.
274. Wojtowicz, J.A., U.S. Patent 4,388,533, 1981.
275. *Patty's Industrial Hygiene and Toxicology*, Vol. 2F, chap. 43, Wiley, New York, 1996, pp. 4498–4500.
276. *R.E.D. Facts: Sodium and Calcium Hypochlorite Salts*, United States Environmental Protection Agency, Washington, DC, 1991.

472 Handbook of Detergents/Part F: Production

277. *R.E.D. Facts: Lithium Hypochlorite.* United States Environmental Protection Agency, Washington, DC, 1993.
278. Gosselin, R.E., Smith, R.P. and Hodge, H.C., *Clinical Toxicology of Commercial Products*, 5th ed., Section III Williams & Wilkins, Baltimore, 1984, pp. 202–205.
279. Racioppi, F. et al., Household bleaches based on sodium hypochlorite: review of acute toxicology and poison control center experience, *Fd. Chem. Toxic.*, 32, 845, 1994.
280. Lai, M.W. et al., 2005 Annual report of the American association of poison control centers' national poisoning and exposure database, *Clin. Toxicol.*, 44, 803, 2006.
281. Smith, W.L., Human and environmental safety of hypochlorite, in *Proceedings of the Third World Conference and Exhibition on Detergents: Global Perspectives*, Cahn, A., Ed., AOCS Press, Champaign, IL, 1994, p. 183.
282. Hypochlorous acid, *Dangerous Properties of Industrial Materials Report* 14, 31, 1994.
283. *Chlor-Alkali Technical and Safety Information*, The Chlorine Institute, Washington, DC, 1992.
284. *Chlorine Handbook*, Occidental Chemical Corporation, Dallas, TX, 1991.
285. Laubusch, E.J., Safe handling of chlorine, in *Chlorine: Its Manufacture, Properties and Uses*, Sconce, J.S., Ed., Robert E. Krieger Publishing Co., Huntington, NY, 1972, p. 46.
286. Chlorine, in *Documentation of Threshold Limit Values*, American Conference of Governmental Industrial Hygienists, 1987, pp. 117.1–117.3.
287. *Occupational Health Guideline for Chlorine.* National Institute for Occupational Safety and Health, 1978.
288. Ewers, U. et al., Chlorine, in *Handbook on Toxicity of Inorganic Compounds*, Seiler, H.G. and Sigel, H., Eds., Marcel Dekker, New York, 1988, p. 223.
289. *The Chlorine Manual*, 5th ed., The Chlorine Institute, Washington, DC, 1986.
290. *Pamphlet 87: Sodium Hydroxide Solution and Potassium Hydroxide Solution (Caustic): Tank Car Loading/Unloading, The Chlorine Institute, Washington, DC, 1993.*
291. *Pamphlet 96: Sodium Hypochlorite: Safety and Handling*, 2nd ed., The Chlorine Institute, Washington, DC, 2000.
292. *DERAKANE™ Epoxy Vinyl Ester Resins: Chemical Resistance and Engineering Guide.* Dow Chemical Co., 1992.
293. *Pamphlet 6: Piping Systems for Dry Chlorine*, The Chlorine Institute, Washington, DC, 1993.
294. Kirschner, M., Chemical profile: sodium hypochlorite, *Chem. Marketing Reporter*, Feb 27, 26, 2006.
295. Cal hypo market faces shortage, *Chem. Marketing Reporter*, Jun 24, 4, 1996.
296. Singletary, L., Chlorine bleach reigns in households despite industry changes, *Chem. Marketing Reporter*, Sep 21, 29, 1992.
297. Ong, S.K. et al., Toxicity and bioconcentration potential of adsorbable organic halides from bleached laundering in municipal wastewater, *Environ. Toxicol. Chem.*, 15, 138, 1996.
298. McCulloch, A., Chloroform in the environment: occurrence, sources, sinks and effects, *Chemosphere*, 50, 1291, 2003.
299. Braida, W. et al., Fate of adsorbable organic halides (AOX) from bleached laundering in septic tank systems, *Environ. Toxicol. Chem*, 17, 398–403, 1998.
300. Crebelli, R. et al., Genotoxicity of the disinfection by-products resulting from peracetic acid or hypochlorite disinfected sewage wastewater, *Wat. Res.*, 39, 1105, 2005.
301. Johnson, I., Pickup, J.A. and van Wijk, D., A perspective on the environmental risk of halogenated by-products from uses of hypochlorite using a whole effluent toxicity based approach, *Environ. Toxicol. Chem.*, 25, 1171, 2006.
302. Schowanek, D. et al., Quantitative in-situ monitoring of organohalogen compounds in domestic sewage resulting from the use of hypochlorite bleach, *Wat. Res.*, 30, 2193, 1996.
303. Rappe, C. and Andersson, R., Analyses of PCDDs and PCDFs in wastewater from dish washers and washing machines, *Organohalogens*, 9, 191, 1992.
304. Rappe, C. et al., Levels of polychlorinated dioxins and dibenzofurans in commercial detergents and related products, *Chemosphere*, 21, 43, 1990.
305. Horstmann, M. and McLachlan, M.S., Polychlorierete Dibenzo-p-Dioxine und Dibenzofurane in Textilien, *Melliand Textilberichte*, 1–2, 66, 1995.
306. Horstmann, M. et al., An Investigation of PCDD/F formation during textile production and finishing, *Organohalogen Compd.*, 11, 417, 1993.
307. Klororganiska föreningar från disk- och blekmedel? En försöksstudie, Rapport 4009, Naturvårdsverket, Solna, Sweden, 1992.
308. Westervelt, R., Looking for a big splash as summer heats up, *Chem. Week*, Jul 23, 40, 1997.

25 Production of Key Ingredients of Detergent Personal Care Products

Louis Ho Tan Tai and Veronique Nardello-Rataj

CONTENTS

25.1 Introduction 474
25.2 Fats/Oils Used in Making Soaps 474
    25.2.1 Preparation of Raw Materials 476
25.3 Fatty Acids: Used in Soaps and Toilet Bars 477
25.4 Surfactants 478
    25.4.1 Anionic Surfactants 478
        25.4.1.1 Production of Fatty Alcohols..478
        25.4.1.2 Alkyl Ether Sulfates or Fatty Alcohol
                  Ether Sulfates...479
        25.4.1.3 Primary Alcohol Sulfates 479
        25.4.1.4 Sulfosuccinates 480
        25.4.1.5 Alkyl Isethionates 480
        25.4.1.6 Linear Alkylbenzene Sulfonates 480
    25.4.2 Nonionic Surfactants: Used for Shampoos 481
        25.4.2.1 Fatty Alcohol Polyethylene Glycol Ether 481
        25.4.2.2 Alkyl Polyglycosides 481
    25.4.3 Amphoteric Surfactants 482
    25.4.4 Quaternary Ammonium Surfactants 483
25.5 Silicones: Used as Softening Agents for Shower Gels, Bath Foams,
    and Shampoos 483
25.6 Cationic Polymers: Used as Aid Deposit of Silicones in Shampoos 484
25.7 Glycerol: Used as a Humectant in Toothpastes 485
25.8 Sorbitol: Used as a Humectant in Toothpastes 485
25.9 Precipitated Calcium Carbonates: Used as an Abrasive in Toothpastes 486
25.10 Amorphous Silica: Used as an Abrasive in Toothpastes 486
25.11 Antimicrobial Agent in Toothpastes 487
25.12 Fluorides: Used as Anticarie Agents in Toothpastes 488
Acknowledgments...488
References...488

25.1 INTRODUCTION

A wide range of raw materials is used in the formulation of personal care products [1,2]. This chapter deals only with the most important ingredients contained in

- Skin care products: Soaps, toilet bars, shower gels, bath foams
- Hair care products: Shampoos and conditioners
- Oral care products: Toothpastes

Several ingredients are common among the aforementioned formulations. Therefore, only the production of the most important ones are discussed, with the indications in which formulations they are used. The synthesis of the following ingredients are described:

- Fats/oils
- Fatty acids
- Surfactants such as anionics, nonionics, amphoterics, and cationics
- Silicones
- Cationic polymers
- Glycerol
- Sorbitol
- Precipitated calcium carbonate
- Amorphous silica
- Zinc salts
- Fluorides

25.2 FATS/OILS USED IN MAKING SOAPS

Fatty matter and oils, whether animal or vegetable, are made up of triglycerides. The main raw materials used in soap making include tallow from beef and sheep and coconut oil. The fruit of the coconut tree is much larger than the fruit of the palm tree. The coconut is green, which becomes brown on drying. It is the nut of the fruit that is used to make oil (Figure 25.1). Areas of cultivation are the Solomon Islands and the Philippines.

There are other raw materials that are used:

- Palm oil, which is obtained from the skin of the palm fruit
- Palm kernel oil, which is obtained from the kernel of the palm fruit

Areas of cultivation include Colombia, the Ivory Coast, Ghana, the Congo, Thailand, and Malaysia (Figure 25.2).

In Europe and the United States, a mixture of tallow and coconut oil is generally used. Palm oil and palm kernel oil are used more frequently in the producing areas such as Africa and southeast Asia. Each fat molecule is made up of complex mixtures of natural fatty acids of different chain lengths from C_4 (butyric acid) to C_{22} (erucic acid) (Table 25.1). The chains of fatty acids, which are derived from biosynthesis, are built from two carbon units, and cis double bonds are inserted by desaturase enzymes at specific positions resulting in even-chain-length fatty acids. The most common acids are C_{16} and C_{18}.

Table 25.1 shows that coconut and palm kernel oil are rich in C_{12} along with few long chains, whereas tallow and palm oil do not contain C_{12} chains but a mixture of longer saturated and unsaturated chains. Coconut and palm kernel oils, which are sources of medium-chain fatty acids are referred to as lauric oils.

Production of Key Ingredients of Detergent Personal Care Products 475

FIGURE 25.1 Photograph of the coconut tree and its coconuts.

FIGURE 25.2 Photograph of the palm tree and its fruit.

Soaps are generally made from the following mixtures:

- Tallow/coconut
- Tallow/palm kernel oil
- Palm oil/coconut
- Palm oil/palm kernel oil

TABLE 25.1
Average Composition (wt%) of the Major Fats Used in Soaps

Chain Length	$<C_{10}$	C_{10}	C_{12}	C_{14}	C_{16}	$C_{16:1}$	C_{18}	$C_{18:1}$[a]	$C_{18:2}$[b]	$C_{18:3}$[c]	C_{20}
Common Name	—	Capric	Lauric	Myristic	Palmitic	Palmitoleic	Stearic	Oleic	Linoleic	Linolenic	Arachidic
Coconut Oil	9	6	47	18	9	1	2	5	2	1	—
Palm Kernel Oil	5	3	50	16	8	—	2.5	14	2	—	Traces
Tallow	—	—	—	3	25	2	22	45	3	0.5	—
Palm Oil	—	—	Traces	1	47	—	4	38	9	0.5	0.5

[a] C18:1 refers to an octadecenoic acid chain with 18 carbons and 1 double bond.

[b] C18:2 means that there are two double bonds in the octadecenoyl chain.

[c] C18:3 means that there are three double bonds in the octadecenoyl chain.

TABLE 25.2
Fatty Acids for an 80/20 Mixture of Palm Oil and Coconut Oil

Chain Length	$<C_{10}$	C_{10}	C_{12}	C_{14}	C_{16}	C_{18}	$C_{18:1}$	$C_{18:2}$
Percent	2	1.3	9.5	4.4	36.6	4.6	32.5	9
	Short-chain saturated fatty acids			Long-chain saturated fatty acids			Long-chain unsaturated fatty acids	

with 10–40% of coconut or palm kernel oil (usually ≈20%) and 60–90% of tallow or palm oil. Taking as an example an 80/20 mix of palm oil and coconut, the soap produced will have a mixture of fatty acids split approximately as shown in Table 25.2.

The choice of a specific mixture will affect the quality of the final soap. [For example, short-chain fatty acids are more soluble; they yield soaps, which despite giving out more foam are also more irritating for the skin and wear out faster.] A compromise must be found to get a mixture that satisfies the desired performances and cost criteria [3].

25.2.1 PREPARATION OF RAW MATERIALS

Before being used in making soaps, the untreated raw materials go through the twin processes of bleaching and deodorizing.

Bleaching. The first step in bleaching oils is vacuum drying at a high temperature (90°C). The vacuum helps to avoid oxidation and consequent deterioration. Once the water has been removed, bleaching earth is added in the form of natural clay called montmorillonite. Its granulometry provides a very large exchange surface of ~150–300 m²/g. When this clay is exposed to dilute acid, either before drying or during bleaching of the oil, some of its aluminum atoms are dissolved, leaving "holes" in the structure. Impurities such as dust, color, and various odors in the oil are absorbed into the holes. This stage is completed by adding 5% of bleaching earth to the dried oil. The mixture is agitated for 15–30 min at

90°C. After slight cooling to 70°C, which reduces the solubility of certain contaminants, the mixture is pumped through filter presses.

Deodorizing. To obtain a perfect oil of almost edible quality, the next operation uses steam under vacuum to remove all odors that could give the finished product an unpleasant smell (rancidity). The pure oil is now ready to be transformed into soap.

25.3 FATTY ACIDS: USED IN SOAPS AND TOILET BARS

Fatty acids are obtained industrially by heating water and fats directly at a high temperature (250°C) under strong pressure (50 atm). Manufacture of soaps requires three steps including preparation and distillation of fatty acids followed by their neutralization.

$$\begin{array}{c} R-C(=O)-O-CH_2 \\ R-C(=O)-O-CH \\ R-C(=O)-O-CH_2 \end{array} + 3H_2O \xrightarrow[50\ \text{atm}]{250°C} \begin{array}{c} HO-CH_2 \\ HO-CH \\ HO-CH_2 \end{array} + 3\ R-C(=O)-OH$$

Fatty matter → Glycerol + Fatty acid

Under these conditions, water is moderately soluble in the oil phase and stepwise hydrolysis of the triglycerides proceeds without a catalyst. The reaction takes place using an excess of water, which removes the glycerol formed, that is, glycerin, from the insoluble fatty acids.

During the process, water is pumped in at the top of a separator column and fat is introduced at the bottom. In the lower part of the column, the lower-density fatty matter rises, meeting the water/glycerine mixture near the top of the column. Conversely, in the upper part, the higher-density water drops through the rising fats (Figure 25.3).

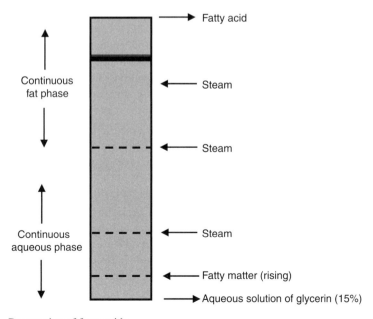

FIGURE 25.3 Preparation of fatty acids.

Steam is injected at different levels to maintain the temperature and disperse the fats in the water. The fatty acids are then subjected to reduced pressure, the water they still contain boils off and, thus, separate easily from the 15% of glycerin (glycerol recovery).

The fatty acids are then purified by distillation or use by the soap maker. They are then neutralized with caustic soda. Fatty acids are distilled in the two following steps:

- A predistillation removes volatiles/odorants (<5%).
- The main distillation eliminates unsaponified fatty matter as well as degraded or polymerized fatty acids or polymers (≥5%).

Historically, soaps were produced by alkaline hydrolysis, that is, *saponification*, of oils and fats. This process, which uses less energy since only 100°C is required at atmospheric pressure, can still be preferred for heat-sensitive fatty acids. The fatty acids formed are immediately converted into the sodium carboxylates, that is, the soaps, since the fats are boiled directly in a caustic soda solution.

| Fatty matter | Sodium hydroxide | Glycerol | Soap |

The problem with this process is the difficulty in obtaining a homogeneous mix since as soon as the fat reacts with the caustic soda, the soap formed thickens the mixture [3].

25.4 SURFACTANTS

Surfactants are "the main components in detergent personal care formulations" since they possess several properties such as detergency, foaming, solubilizing, emulsifying, dispersing, conditioning, and wetting. In a wide range of formulations, combinations of surfactants are preferred. They can be derived from either petroleum or vegetal and animal sources. The main nonfood use of oils and fats, for example, is the production of surfactants. Actually, the carboxylic groups can be converted into various other hydrophilic groups by more or less simple reactions leading to anionic, cationic, nonionic, and amphoteric surfactants [4].

25.4.1 ANIONIC SURFACTANTS

25.4.1.1 Production of Fatty Alcohols

Long-chain alcohols are produced both from oleo- and petrochemical feedstocks. The main oils and fats, which provide straight, even, and saturated or unsaturated chains, are coconut and palm kernel oil for C_{12}–C_{14} fatty alcohols and tallow and palm oil for C_{16}–C_{18} alcohols. However, crude oil is used for the production of the synthetic fatty alcohol chains leading to mixtures of branched and odd alkyl chains.

Fatty alcohols can be synthesized from various processes. They can result from a catalytic hydrogenation of fatty acids and methyl esters. They can also be obtained from the Ziegler process in which ethylene $H_2C=CH_2$ reacts with triethyl aluminum to form an alkyl aluminum that is then oxidized into an aluminum alcoholate, which is finally hydrolyzed under acidic conditions

Production of Key Ingredients of Detergent Personal Care Products 479

to provide the fatty alcohols. In the Oxo process, carbon monoxide and hydrogen react in a first step with an olefin R–CH=CH$_2$, providing an aldehyde R–CH$_2$–CH$_2$–CHO, which is reduced in a second step into the corresponding fatty alcohol. The Oxo process involves some branched and secondary alcohols, leading to a mixture of odd- and even-chain alcohols.

Fatty alcohols are an important base stock for the production of various surfactants.

25.4.1.2 Alkyl Ether Sulfates or Fatty Alcohol Ether Sulfates

Alkyl ether sulfates (AES), which are the sulfuric acid monoesters of alkoxylated alcohols, are the most widely used surfactants in detergent personal care products. They are mainly encountered in liquid formulations such as shower gels, bath foams, and shampoos. Their chemical formula is as follows:

$$R-O-(CH_2-CH_2-O)_n-SO_3^-$$

Alkyl ether sulfates

The alkyl moiety usually consists of a mixture of C_{12}–C_{16} chains. Lauryl ether sulfate (LES) or sodium laureth sulfate (according to the cosmetic, toiletry, and fragrance [CTFA] nomenclature), with mainly lauryl chain, is commonly used in personal care products. The degree of ethoxylation is usually equal to 2 or 3, and sodium, ammonium, or magnesium salts can be encountered in formulations.

Alkyl ether sulfates differ from alkyl sulfates (see Section 25.4.1.3) in their ether glycol units between the carbon chain and the sulfate group:

$$R-O-SO_3Na \qquad \text{Alkyl sulfates}$$

$$R-O-(CH_2-CH_2-O)_n-SO_3Na \qquad \text{Alkyl ether sulfates}$$

Formerly, chlorosulfonic acid (ClSO$_2$OH) or sulfamic acid (NH$_2$SO$_3$H) were used in a batch process. Nowadays, AES is produced in two stages as follows:

1. Addition of ethylene oxide molecules to the fatty alcohols leading to ethoxylated fatty alcohols.
2. Continuous sulfation of the ethoxylated fatty alcohols by a mixture of air/sulfur trioxide in a thin-film reactor (as for alkyl sulfonates), which is immediately (to avoid autohydrolysis) followed by neutralization typically with sodium hydroxide but also with ammonia or alkylamines to give the corresponding salts. The chemical reaction is as follows:

$$R-O-(CH_2-CH_2-O)_n-H \ + \ SO_3 \longrightarrow R-O-(CH_2-CH_2-O)_n-SO_3H$$

Alkyl ether sulfonic acid

25.4.1.3 Primary Alcohol Sulfates

Alkyl sulfates or primary alcohol sulfates (PAS), which were the first synthetic surfactants used in personal care in the United States, are mainly used in toothpaste formulations. As for the AES, the alkyl moiety is typically a mixture of C_{12}–C_{16} chains. According to the application, one can find sodium, ammonium, magnesium, or triethanolamine salts of fatty alcohol sulfuric acid. They are obtained by the sulfation of a fatty (natural or synthetic) primary alcohol with a mixture of air/SO$_3$ according to the following reaction:

$$R-OH + SO_3 \longrightarrow R-O-SO_3H$$

Fatty alcohol sulfuric acid

480   Handbook of Detergents/Part F: Production

It should be noted that continuous sulfation is widely used today, with the mixture of SO_3/air in reactors such as Chemithon, Allied, or Ballestra. Moreover, alcohol sulfates are preferably used in alkaline formulations because of their low stability in acid medium. However, they are completely biodegradable and compete with petrochemical-derived linear alkylbenzene sulfonates (LAS).

25.4.1.4 Sulfosuccinates

Sulfosuccinates are the hemiesters of succinic acid with two anionic groups: carboxylic acid and sulfonic. They have the following chemical formula:

$$R\text{—}O\text{—}\underset{\underset{O}{\|}}{C}\text{—}\underset{\underset{SO_3H}{|}}{CH}\text{—}CH_2\text{—}COOH$$

Sulfosuccinic acid

They are prepared by an equimolar reaction of maleic anhydride with fatty alcohol to produce the monoester first. In the second step, the latter reacts with sodium bisulfite providing the sulfosuccinic acid.

Maleic anhydride   Sulfosuccinic acid

Typically, the alkyl chain bears 12–14 carbon atoms. The sodium salts are widely used in hair and skin personal care products at pH 5–7.

25.4.1.5 Alkyl Isethionates

Alkyl isethionates such as *sodium cocoyl isethionate* are used mainly in creams, toilet bars, shower gels, bath foams, and baby products. These derivatives are obtained from sodium 2-hydroxyethanesulfonate, which can react either by esterification with fatty acids or condensation with fatty acid chlorides as follows:

$$R\text{—}\underset{\underset{O}{\|}}{C}\text{—}Cl \; + \; HO\text{—}CH_2\text{—}CH_2\text{—}SO_3Na \longrightarrow R\text{—}\underset{\underset{O}{\|}}{C}\text{—}CH_2\text{—}CH_2\text{—}SO_3Na + HCl$$

Sodium 2-hydroxy-ethanesulfonate   Alkyl isethionates

They are sold under the brand HeO S 3390-2 (Hoechst) or Fenipon AC (G.A.F.). They have the same properties as the sulfosuccinates in terms of mildness and stability toward hydrolysis (maximum of stability in the pH range 5–7).

25.4.1.6 Linear Alkylbenzene Sulfonates

LAS have very good foaming and detergency properties. However, they have a tendency to leave a dry feel on skin and hair. Hence, they are not widely used in cosmetic cleansing products. They can

Production of Key Ingredients of Detergent Personal Care Products 481

be found in toilet bars and cheap toothpaste formulations to reduce the cost of raw materials. Most of time, the sodium salt is used. They have the following chemical structure:

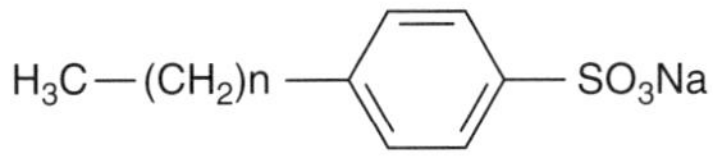

$$H_3C-(CH_2)n \longrightarrow \bigcirc \longrightarrow SO_3Na$$

Linear ABS

They are prepared by sulfonation of alkyl aromatic hydrocarbons, which are obtained by alkylation of benzene via a Friedel–Crafts reaction. There are two processes to produce the alkylates:

1. Alkylation with chlorinated *n*-paraffins according to the following reaction:

$$C_nH_{2n+1}Cl + \bigcirc \longrightarrow C_nH_{2n+1}\bigcirc + HCl$$

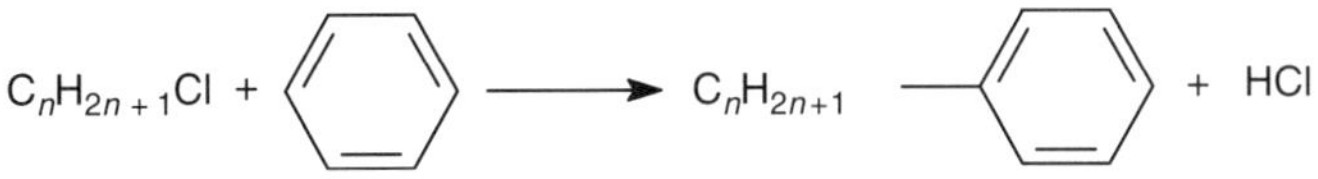

2. Alkylation with linear olefins according to the following reaction:

$$C_nH_{2n} + \bigcirc \longrightarrow C_nH_{2n+1}\bigcirc$$

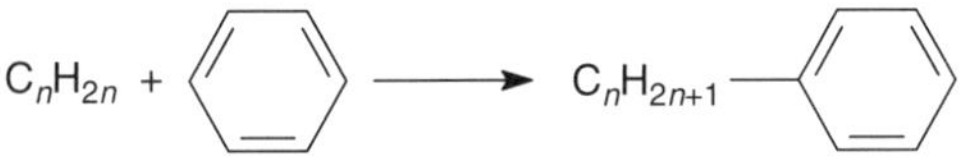

It should be noted that the double bond can be either at the end or inside the carbon chain; isomeric forms are obtained with the phenyl group in positions 1, 2, or 3.

Sulfonation of the resulting alkylaryl hydrocarbon is generally accomplished with a mixture of air/SO_3. It is then followed by neutralization with different alkaline components to produce the required salt of LAS, generally, sodium salts.

25.4.2 NONIONIC SURFACTANTS: USED FOR SHAMPOOS

25.4.2.1 Fatty Alcohol Polyethylene Glycol Ether

Among commercial nonionic surfactants, those made from fatty alcohols with ethylene oxide are the most commonly used. Ethoxylation offers the production of a wide range of nonionic surfactants as the hydrophobic part, and the ethylene oxide number can be easily adjusted according to the desired properties. The chemical reaction to convert a fatty alcohol into a nonionic ethoxylated surfactant uses ethylene oxide under pressure (typically 2–8 bars) and heat (typically 120–200°C). Actually, fatty alcohols have a hydroxyl group that can react further with ethylene oxide providing polyoxyethylene compounds with a range of molecular weights.

The reaction is often catalyzed by alkalis such as potassium or sodium hydroxide. The average number of ethylene oxide groups depends on the reaction conditions. However, commercial ethoxylated alcohols have a broad range of ethylene oxide units along with some amount of unreacted fatty alcohols. Narrow range ethoxylated fatty alcohols can be obtained by using specific catalysts.

25.4.2.2 Alkyl Polyglycosides

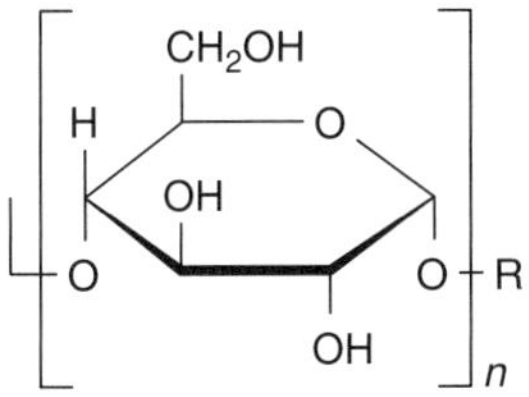

Alkyl polyglycosides (APG) are based on carbohydrates. They are generally complex mixtures because of the C-chain distribution of the alkyl moiety and the variation in the degree of glycosidation. Typically, APG have C_8–C_{16} alkyl chains with a mean value degree of glycosidation (n) of 1–3. They are prepared by glycosidation of starch or monomer glucose with fatty alcohols. This can be done either directly with acid catalysts or by transglycosidation using butylpolyglycosides as intermediates in a two-step process. These processes are proton catalyzed and carried out at 120–140°C under pressure and with excess fatty alcohol. In a separate step, the excess fatty alcohol is removed by vacuum distillation and recycled. APG production is ~100,000 t/year. These products, which are mild to the skin and easily biodegradable, are used in shampoos and shower gels.

25.4.3 AMPHOTERIC SURFACTANTS

Amphoteric surfactants are encountered in *shower gels, bath foams,* and *shampoos.* These surfactants behave like cationic surfactants at low pH and anionic surfactants at high pH. At medium pH, they carry both positive and negative charges leading to a bipolar ion structure. Unlike amphoterics, zwitterionic surfactants maintain a bipolar structure over a large range of pH. In this group of products, *alkyl amido propyl betaines* are the most commonly used. Their chemical formula is as follows:

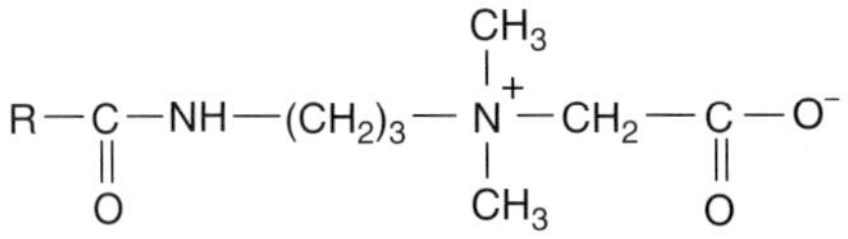

Alkyl amido propyl betaines

The most important alkyl amido betaine used in personal care products such as shampoos, bath foams is by far *coco amido propyl betaine* (CAPB), which has a lauryl alkyl chain. The synthesis of alkyl amido betaines involves, in the first step, the formation of the amide and in the second, the condensation of the alkyl dimethylamine with sodium chloroacetate:

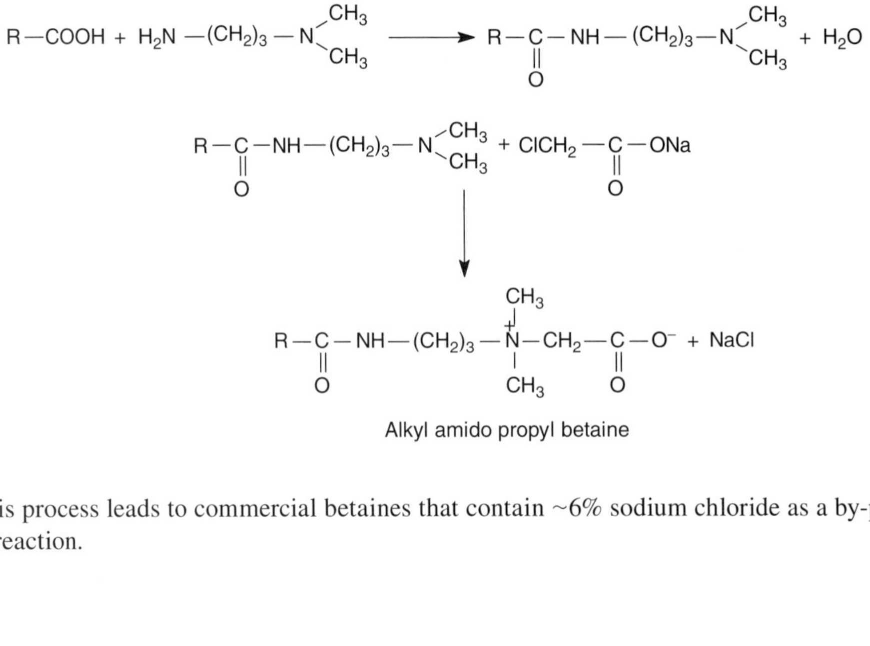

Alkyl amido propyl betaine

This process leads to commercial betaines that contain ~6% sodium chloride as a by-product of the reaction.

Production of Key Ingredients of Detergent Personal Care Products **483**

25.4.4 QUATERNARY AMMONIUM SURFACTANTS

Cationic surfactants and in particular *quaternary ammonium surfactants* are mainly used in conditioning products for hair. They exhibit poor cleansing properties but they have good antimicrobial and conditioning effects.

Single-chain quaternary ammonium surfactants are prepared from the reaction of fatty alcohols or fatty acids with a secondary amine. The resulting tertiary amine then reacts with methyl chloride to provide quaternary ammonium chlorides. The most important is *cetyl trimethyl ammonium chloride* (CTAC), which has excellent substantivity and conditioning properties.

In a similar way, double-chain quaternary ammonium surfactants are obtained from a dialkylamine, which is first made from either fatty alcohols or fatty acids. With fatty alcohols, we obtain the following reaction:

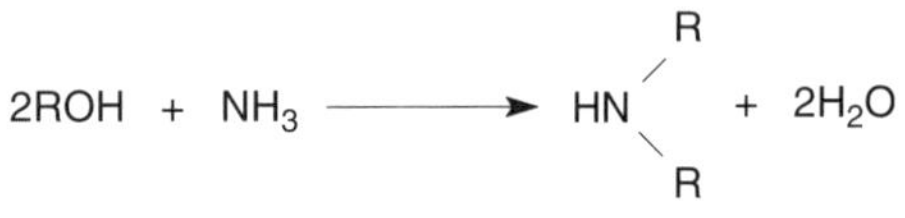

$$2ROH \;+\; NH_3 \longrightarrow HN\!\!\begin{smallmatrix}R\\[2pt]\\R\end{smallmatrix} \;+\; 2H_2O$$

Dialkylamine

Quaternization then takes place with methyl chloride in the presence of 50% sodium hydroxide to neutralize the hydrochloric acid that is formed:

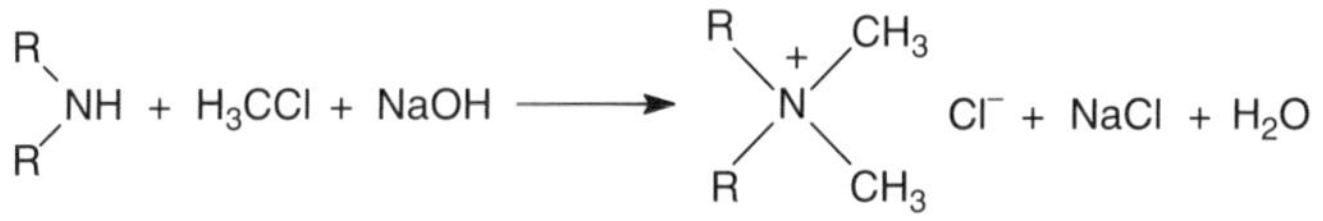

$$\begin{smallmatrix}R\\[2pt]\\R\end{smallmatrix}\!\!NH \;+\; H_3CCl \;+\; NaOH \longrightarrow \begin{smallmatrix}R\\[2pt]\\R\end{smallmatrix}\!\!N^{+}\!\!\begin{smallmatrix}CH_3\\[2pt]\\CH_3\end{smallmatrix} \; Cl^- \;+\; NaCl \;+\; H_2O$$

Dialkyldimethylammonium chloride

Distearyldimethylammonium chloride or dihydrogenated tallow dimethylammonium chloride, also known as *Quaternium-18*, exhibits strong conditioning properties. However, these compounds exhibit very poor biodegradability.

25.5 SILICONES: USED AS SOFTENING AGENTS FOR SHOWER GELS, BATH FOAMS, AND SHAMPOOS

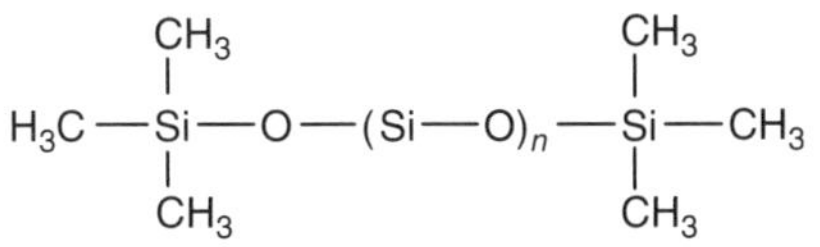

$$H_3C-\underset{\underset{CH_3}{|}}{\overset{\overset{CH_3}{|}}{Si}}-O-(Si-O)_n-\underset{\underset{CH_3}{|}}{\overset{\overset{CH_3}{|}}{Si}}-CH_3$$

PDMS

Silicones are synthetic polymers containing a Si–O backbone. The most common example of silicones is *polydimethylsiloxane* (PDMS) or *dimethicone*, which is the basic building block polymer of the silicone industry. Modification of the molecular weight (above 200,000) of the substituents and structure leads to numerous families of silicones.

Silicone materials are contained in 76% of hair conditioners, 66% of hair care, and 55% of hair shampoos. Concerning skin care, they are present in 65% of skin care, 62% of facial care, and 54% of hands and body formulations.

Silicones are made from silicon and methyl chloride in a process known as the *direct reaction* or *direct process*. This reaction yields methyl chlorosilanes. They are distilled for purification and the dimethyldichlorosilane is hydrolyzed to give PDMS. This product can be formulated into thousands of different products, which are sold to every major industrial segment.

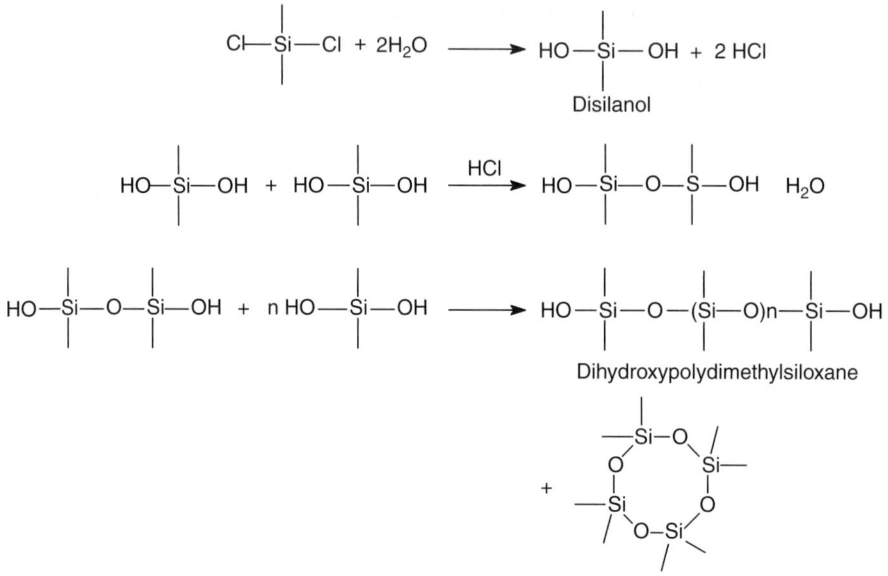

$$Cl-Si-Cl + 2H_2O \longrightarrow HO-Si-OH + 2\ HCl$$

Disilanol

$$HO-Si-OH + HO-Si-OH \xrightarrow{HCl} HO-Si-O-S-OH \quad H_2O$$

$$HO-Si-O-Si-OH + n\ HO-Si-OH \longrightarrow HO-Si-O-(Si-O)n-Si-OH$$

Dihydroxypolydimethylsiloxane

Cyclic D_4

Cyclic D_4 leads to dihydroxypolydimethylsiloxane in the presence of potassium hydroxide, whereas in the presence of $Me_3SiOSiMe_3$, it gives PDMS under similar conditions. The ratio $D_4/Me_3SiOSiMe_3$ determines the average molecular weight of the polymer.

Some silicones are volatile. They are cyclic and are called *cyclomethicones*. D_4, D_5, or D_6 are, respectively, cyclotetrasiloxane with four units of Me_2SiO, cyclopentasiloxane with five units of Me_2SiO, and cyclohexasiloxane with six units of Me_2SiO [5–10,11].

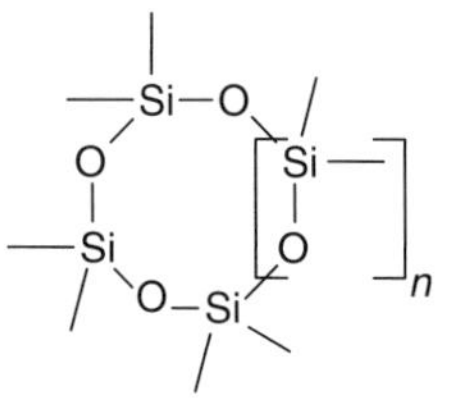

Cyclomethicones

25.6 CATIONIC POLYMERS: USED AS AID DEPOSIT OF SILICONES IN SHAMPOOS

Cationic polymers include homopolymers and cationic copolymers obtained by the polymerization of a vinylic monomer with a quaternary ammonium group or a quaternized amine with another soluble monomer in water such as acrylamide and methacrylamide. The most frequently used polymers are those derived from gum guar such as guar hydroxyl propyl triamonium chloride. They are sold under the trade names Jaguar C13S, C17, etc. [12–16].

$$G-O-CH_2-CH-CH_2-N^+(CH_3)_3Cl^-$$
$$OH$$

Production of Key Ingredients of Detergent Personal Care Products485

25.7GLYCEROL: USED AS A HUMECTANT IN TOOTHPASTES

$$CH_2—OH$$
$$|$$
$$CH—OH$$
$$|$$
$$CH_2—OH$$

Glycerol, which is commercially well known as glycerin or glycerine, is a sugar alcohol. The name "glycerol" was derived from the Greek word *glykys*, which means sweet. Glycerol has a wide range of applications and is the humectant used in greatest bulk quantity in toothpaste formulations.

As a principal component of all fats and oils in the form of its esters (called triglycerides), nearly all commercial glycerol was produced as a by-product in the manufacture of soap or from the hydrolysis of fats and oils until after World War II. Synthetic glycerol was then prepared from propylene. Crude glycerol is purified to make various marketed grades such as high gravity, dynamite, yellow distilled, United States Pharmacopeia (USP), and Food Chemicals Codex (FCC). USP grade is suitable for personal care products.

Glycerol can also be obtained by a transesterification process of oils and fats rich in triglycerides in a mixture of methanol and an alkali such as potassium or sodium hydroxide, resulting in a methyl ester and glycerin. This latter is a mixture of glycerol; methanol; water; inorganic salts; free fatty acids; unreacted mono-, di-, and triglycerides; methyl esters; and a variety of other matter organic nonglycerol in varying quantities. The methanol is typically stripped and reused, providing, after neutralization, what is known as crude glycerin. In raw form, this crude glycerin has high salt and free fatty acid content and substantial color (yellow to dark brown). Consequently, crude glycerin has few direct uses due to the presence of the salts and other species. Fat, soap, and other organic impurities can be separated and removed by filtration or centrifugation. Final purification is typically completed using vacuum distillation followed by activated carbon bleaching for large operations or ion exchange followed by flash drying to remove water for smaller capacity plants. Vacuum distillation is very expensive, and since energy consumption cannot always be carried out continuously, it is accompanied by considerable losses of glycerol. To separate glycerol from higher boiling point impurities, the mixture needs to be additionally subjected to severe thermal stresses, which further result in additional losses of glycerol and creates more decomposition products. Because of the high salt content, ion exchange is not economically practical, unless it is used to polish a diluted low salt content glycerol-in-water solution.

There is also a microbial production of glycerol. Not all organisms have a natural capacity to synthesize glycerol but the biological production of glycerol is known in some species of bacteria, algae, and yeast. The *Bacillus licheniformis* and *Lactobacillus lycopersica* bacteria synthesize glycerol. Glycerol production is found in the halotolerant algae *Dunaliella* sp. and *Asteromonas gracilis*. Various osmotolerant yeast also synthesize glycerol as a protective measure. Most strains of *Saccharomyces* produce some glycerol during alcoholic fermentation. Recently, glycerol was produced commercially with *Saccharomyces* cultures to which steering reagents were added such as sulfites or alkalis. Through the formation of an inactive complex, the steering agents block or inhibit the conversion of acetaldehyde to ethanol; thus, excess reducing equivalents (NADH) are available to or "steered" toward dihydroxyacetone phosphate (DHAP) for reduction to produce glycerol [11,17,18].

25.8SORBITOL: USED AS A HUMECTANT IN TOOTHPASTES

Sorbitol is a hexitol, which was originally found in the rowan fruit. Despite its natural abundance, it was not possible to carry out an industrial extraction of sorbitol from plants mainly owing to its high water solubility. Thus, the catalytic hydrogenation of D-glucose was patented in 1925 [19] and the industrial production of sorbitol began in the mid-1930s in the United States. D-sorbitol is obtained from high-pressure catalytic hydrogenation of D-glucose in the presence of Raney nickel catalyst.

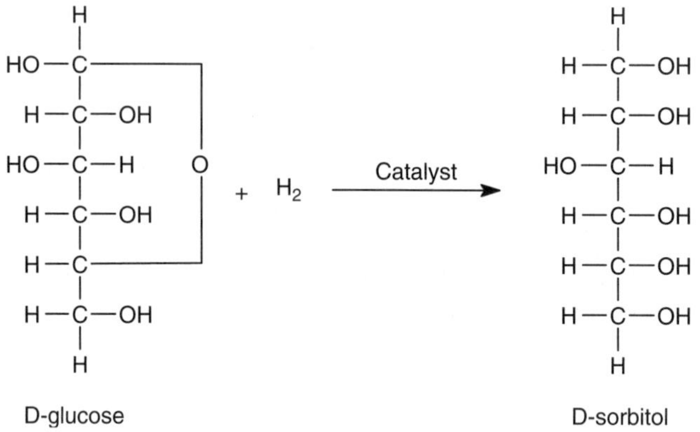

$$(SiO_2)_x(Na_2O)_y \cdot nH_2O + yH_2SO_4 \rightarrow xSiO_2 + yNa_2SO_4 + (y + n)H_2O$$

This technique is the basis of all industrial continuous and discontinuous processes and is typically carried out at temperatures of 100–150°C and pressures of 30–100 bars. The manufacture of D-sorbitol requires a raw material with a very high D-glucose content. For this, it is possible to use either sucrose, which after inversion leads to an equimolar mixture of D-glucose, but the use of this raw material leads to a mixture containing 75% D-sorbitol and 25% D-mannitol. The second method starts with D-fructose, or starch, the total hydrolysis of which provides pure D-glucose, and hence, pure D-sorbitol. In addition to glycerol, sorbitol is extensively used throughout the industry. Figure 25.4 describes the different steps in the manufacture of liquid and powder sorbitol starting from glucose [2,11,17–19].

25.9 PRECIPITATED CALCIUM CARBONATES: USED AS AN ABRASIVE IN TOOTHPASTES

Precipitated calcium carbonates $CaCO_3$ (chalk) are additives used in numerous applications. They are one of the most commonly used dental abrasives for toothpaste formulations. They are synthetic products from the decomposition and resynthesis of lime after purification.

Commercial precipitated calcium carbonates are available in variable particle sizes and crystalline forms, depending on manufacturing conditions. For example, some precipitated calcium carbonates are coated with stearin, thus exhibiting particular rheological and hydrophobic properties of toothpastes.

They are sold under the trademark "Socal and Winnofil" by the Solvay company. It is noteworthy that precipitated calcium carbonates are completely natural, and intermediate products can be recycled [2,11,17].

25.10 AMORPHOUS SILICA: USED AS AN ABRASIVE IN TOOTHPASTES

Precipitated silica (silica P) abrasive SiO_2 have been introduced in toothpaste formulation in the 1970s. They are synthetic, noncrystalline, or amorphous silica. They represent ~80% of the world production of synthetic amorphous silica. Variation in the particle size allows the control of the abrasivity and the thickness of the product. Amorphous silica is obtained by reaction of a sodium silicate solution at pH >7 with an acid such as H_2SO_4 or HCl, according to the following reaction:

$$(SiO_2)_x(Na_2O)_y \cdot nH_2O + yH_2SO_4 \rightarrow xSiO_2 + yNa_2SO_4 + (y + n)H_2O$$

Production of Key Ingredients of Detergent Personal Care Products

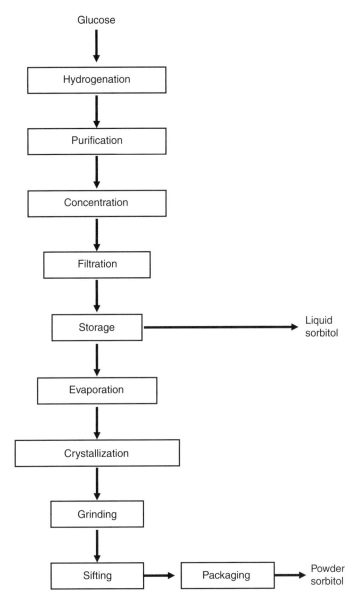

FIGURE 25.4 Production of sorbitol from glucose. The information contained in this document comes from Roquette Frères, the worldwide leading company for polyols such as sorbitol and isosorbide.

Sodium silicate is prepared either by alkaline fusion of natural sand in the presence of Na_2CO_3 at 1050–1100°C or by reaction of sand with sodium hydroxide at 180–220°C [2,11,17].

25.11 ANTIMICROBIAL AGENT IN TOOTHPASTES

Compounds with antimicrobial properties are important in toothpaste formulations to retard plaque formation and prevent gingivitis. Zinc salts, and particularly zinc citrate, are the most widely used antibacterials in toothpaste formulations. The sources of the zinc ion are zinc chloride, zinc sulfate, or zinc thiocyanate.

Finely divided zinc citrate is produced by reacting with a stoichiometric amount of citric acid (derived from corn) with zinc oxide, ZnO, under vigorous agitation such that the particles (<50 μm) of zinc citrate precipitate. Zinc citrate precipitates and is then separated, purified, and dried. Any remaining citric acid is removed by washing zinc citrate with water [21].

The effectiveness of the antimicrobial activity can be enhanced by combining zinc citrate with triclosan [22–24]. Triclosan, or 2,4,4'-trichloro-2-hydroxy diphenyl ether, is an aromatic, trichlorinated synthetic compound. It can be obtained by reaction of 2,4-dichlorophenol with 2,5-dichloronitrobenzene in the presence of alkali and reduction of the 2,4,4'-trichloro-2'-nitrodiphenyl ether into 2,4,4'-trichloro-2'-aminodiphenyl ether. This is diazotized with sodium nitrite with excess sulfuric acid. After hydrolization, 2,4,4'-trichloro-2'-hydroxydiphenyl ether is extracted with xylene and purified [20–25].

25.12 FLUORIDES: USED AS ANTICARIE AGENTS IN TOOTHPASTES

Fluorides are used as anticarie agents in toothpastes since they reduce decay by increasing the strength of teeth. Sodium fluoride, NaF, is the most commonly used fluoride. However, the literature mentions several other sources of fluoride ions such as potassium fluoride, lithium fluoride, aluminum fluoride, zinc fluoride, sodium monofluorophosphate, acidic phosphate fluoride, ammonium fluoride, titanium tetrafluoride, and amine fluoride.

In practice, the compounds that provide the fluoride ions are sodium fluoride, sodium monofluorophosphate (Na_2FPO_3), and sometimes certain amine fluorides.

Sodium monofluorophosphate is prepared by heating a difluorophosphate solution in dilute sodium hydroxide according to the following reaction:

$$PO_2F_2^{2-}{}_{(aq)} + 2NaOH_{(aq)} \rightarrow Na_2PO_3F_{(aq)} + 2HF_{(aq)}$$

One way to make sodium fluoride is to react the hydroxide with hydrofluoric acid. The resulting salt can then be purified by recrystallization [2,11,17–20,26].

$$NaOH_{(aq)} + HF_{(aq)} \rightarrow NaF_{(aq)} + H_2O_{(l)}$$

ACKNOWLEDGMENTS

The authors are grateful to AOCS Press for permission to reuse extracts from Ref. 2.

REFERENCES

1. *Chemistry and Technology of the Cosmetics and Toiletries Industry*, eds. D. F. Williams and W. H. Schmitt, 2nd Edition, Klomer Academic Publishers, 1996.
2. Ho Tan Taï, L., *Formulating Detergents and Personal Care Products. A Complete Guide to Product Development*, AOCS Press, 2000.
3. Bailey's Industrial Oil & Fat Products. In *Edible Oil & Fat Products: Chemistry, Properties and Health Effects*, Vol. 1, 6th Edition, Ed. F. Shahidi, Wiley-Interscience, A John Wiley & sons, Inc., New York, 2005.
4. *Surfactant in Cosmetics*. Surfactant Science Series, Vol 68, Eds. M. M. Rieger and L. D. Rhein, Marcel Dekker, New York, 1997.
5. Berthiaume, M. D., Silicones. In *Chemistry and Manufacture of Cosmetics*, Vol. 3, 3rd Edition, Ed. M. L. Schlossman, 2002, pp. 827–840.
6. McCarthy, J. P., G. H. Greene, A. G. Anthony, WO 9200303, 1992.
7. Chandra, G., G. S. Kohl, J. A. M. Tassof, Dow Corning, US 4559227, 1985.
8. O'Lenick, A. J., Siltech, US 5210133, 1993.
9. Kawamoto, H., S. Ichinohe, K. Tsubone, Shin-Etsu Chemical, EP 423696, 1991.

10. O'Lenick, A. J., J. K. Parkison, Siltech, US 5091493, 1992.
11. *Poucher's Perfumes, Cosmetics and Soaps*. In Cosmetics, Vol. 3, 9th Edition, W. A. Poucher, Ed. H. Butler, 1993.
12. Chowdhary, M. S., F. Robinson, Rhône-Poulenc, WO 9818828, 1998.
13. De La Mettrie, R., L'Oréal, WO 9901105, 1999.
14. Gallagher, P., T. Kreu-Nopakun, A. M. Murray, Unilever, WO 9953889, 1999.
15. Cottrell, I. W., G. T. Martino, K. A. Fewkes, US 2001051143, 2001.
16. Gati, S., L. Moldovanyi, *SOFW*, 111(17), 529–532, 1989.
17. Collins, M. A., J. M. Duckenfield, Colgate-Palmolive, EP 549287, 1993.
18. Nair, R. V., M. S. Payne, D. E. Trimbur, F. Valle, Du Pont de Nemours, WO 9928480, 1999.
19. Riley, P., Unilever, EP 0740932, 1996.
20. Asano, A., M. C. Gaffar, Johnson & Johnson, EP 0162574, 1985.
21. Hoyles, R., D. MacPherson, Unilever, GB 1373002, 1974.
22. Bhargava, H. N., P. A. Leonard, *Am. J. Infection Control*, 24(3), 209–218, 1996.
23. DeSalva, S. J., B. M. Kong, Y. J. Lin, *Am. J. Dent.* 2, 185–196, 1989.
24. Svatun, B., C. A. Saxton, G. Rolla, *Scand. J. Dent. Res.* 98(4), 301–304, 1990.
25. Varalwar, S., V. Satyanarayana, D. S. Chandra, IN 183696, 2000.
26. I. G. Farben Industrie A. G., DRP 544.666 & 554.074, 1925.

26 Production of Solvents for Detergent Industry

Rakesh Kumar Khandal, Sapana Kaushik, Geetha Seshadri, and Dhriti Khandal

CONTENTS

26.1 Introduction 492
26.2 Surfaces, Phases, Detergents, and Solvents 493
26.3 Classification of Detergents 494
    26.3.1 Ionic Detergents 494
        26.3.1.1 Anionic Detergents 494
        26.3.1.2 Cationic Detergents 495
    26.3.2 Nonionic Detergents 495
    26.3.3 Zwitterionic Detergents 495
26.4 Composition of Detergents 496
    26.4.1 Detergent Formulations 496
    26.4.2 Components of a Detergent 496
    26.4.3 Role of Solvents Influencing Detergency 496
    26.4.4 Different Types of Detergents 497
        26.4.4.1 Household Detergents 497
        26.4.4.2 Detergents in the Developing Countries 498
26.5 Role of Solvents in Determining the Behavior
    of Detergents in Water 499
26.6 Role of Solvents in Enhancing the Performance
    of Detergents 502
26.7 Solvent Groups and Properties 506
    26.7.1 Solvent Characteristics 507
    26.7.2 Solvents for the Detergent Industry 507
        26.7.2.1 Detergent Range Alcohols 507
        26.7.2.2 Isopropyl Alcohol (2-Propanol) 512
        26.7.2.3 Ethanol 513
        26.7.2.4 Alkanolamines 516
        26.7.2.5 Glycols 518
        26.7.2.6 Glycol Ethers and Esters 522
        26.7.2.7 Green Solvents 526
        26.7.2.8 Ionic Liquids 528
26.8 Conclusions 528
Acknowledgments 529

26.1 INTRODUCTION

Detergents are a unique class of compounds with special structural attributes. Detergents are amphipathic substances consisting of materials with structures having both hydrophobic and hydrophilic groups in the same molecule. For the purpose of understanding the behavior of detergents, the structure of a detergent molecule is described as a "head" attached to a "tail." Basically, for simplicity, the polar or hydrophilic group of the detergent is symbolized as a "head," whereas the nonpolar or hydrophobic group is designated as a tail. As the chemical nature of "head" is completely different from that of the "tail," the presence of the two completely diverse groups attached to each other in the same molecule is bound to exhibit unique properties.

Thus, the detergents are said to have dual functionality, that is, while one group (hydrophilic) can remain (or would prefer to remain) associated with the polar media, the other group (hydrophobic) can remain attached to the nonpolar phase, at the same time. In other words, the detergent molecules have the capacity of reorienting themselves as per the nature of phases they are encountered with. As all systems (domestic as well as industrial and natural as well as synthetic) consist of more than one phase, the presence of detergents play a significant role in stabilization as well as destabilization of the systems. The mechanism by which detergents perform is quite simple and basic as it involves the reorientation of detergent molecules in such a way that they become active at the surfaces or, to say more appropriately, at the interfaces. It is due to this reason that the detergents are also called as surfactants or surface-active substances. Further, depending on the types of phases involved in any system, detergents have been given various names—all of them specifying the performance for which the detergents are used.

Since most of the properties of detergents have been well understood, designing new types of detergents as per the requirements and with unique structural attributes has become much easier than before. It basically involves having a molecule with varying size, shape, and chemistry to start with, and then building on this is the complementary group having again varying size, shape, and chemistry.

Therefore, the starting materials can either be a hydrophobic molecule with sites that can be used to attach the hydrophilic groups or they can be hydrophilic molecules on which hydrophobes can be attached to design the detergents. It is obvious, then, that the chemical nature, size, and shape of both the hydrophilic and hydrophobic groups would play a key role in determining the performance characteristics of a detergent. Moreover, the balance between hydrophilic and hydrophobic groups of detergent molecules would be the key factor to decide about the applicability of a detergent for a given system.

The starting materials for detergents are also called as chain starters. Since in most of the cases, the chain starters are hydrophobes, they are also referred to as *solvents* for the detergents. The chain starters of the hydrophilic type are not quite common and hence, when the chain starters are referred to as *solvents* in this text, they must be understood as solvents of the hydrophobic type. The other type of solvents used in detergent industry are those with high degree of solvency and capacity to clean the dirt, soil, and so on from the surface. The formulations of detergents often, therefore, consist of various types of solvents to achieve the desired effects and performance. This chapter deals with various aspects of solvents used in the detergent industry. The subject has been discussed under the following sections:

1. Surfaces, phases, detergents, and solvents
2. Classification of detergents
3. Composition of detergents
4. Role of solvents in determining the behavior of detergents in water
5. Role of solvents in enhancing the performance of detergents
6. Solvent groups and properties
7. Conclusions

Production of Solvents for Detergent Industry 493

26.2 SURFACES, PHASES, DETERGENTS, AND SOLVENTS

A brief discussion about the surfaces and interfaces would be useful before introducing the subject of detergents because of the fact that the developments in the field of detergents are driven by the multiphase systems involving various types of surfaces.

Basically, the whole material world is divided into two main categories, that is, water loving or water hating. The surfaces that are water loving are oil hating and the ones that are water hating are oil loving, by nature. Further, all the systems, industrial or household, bio or nonbio involve more than one type of surface. The presence of interfaces, therefore, is a very common experience. Often, when the interfaces involve completely different and diverse surfaces, there rises the need of substances that can help in keeping the different types of surfaces in contact with each other without being separated. This is also called stabilization of the multiphase systems.

At certain times, there arises the need of separating the surfaces from each other. The separation of surfaces in the stabilized system is also referred to as destabilization of the system. The stabilization as well as destabilization depends on the interfacial tension. While the reduction of interfacial tension brings the two surfaces close to each other, the increase in the interfacial tension causes the separation of the surfaces. In other words, it is all about surfaces and modification of surfaces.

As such, there is no system with only one surface—the material surfaces are always in equilibrium with air or any other surface. Let us take the example of water in a container. Here, water (liquid phase) surface is in equilibrium with the surface of the container (solid phase) on one side and with the air (gas phase) on the other. If any other liquid that is not quite miscible in water is added to this system, further interfaces involving different phases are created in the system. Such systems are not only characterized as multiinterface systems but they are also called as multiphase systems. To homogenize the non-water-loving liquid substance with water, it is essential that the interfacial tension between water and the liquid substance is reduced. This is possible and achieved by incorporation of substances, which can act as the bridge between the two surfaces of opposite nature, although the surfaces are of the same phase. Such substances are called as surfactants or detergents.

Thus, the incorporation of solvents in detergent formulations would help in the homogenization of the complex formulations on the one hand and would aid in the performance of the formulation once used for a given purpose of cleaning surfaces on the other. In other words, if surface-active substances can modify interfaces, then solvents play a key role in enhancing the effect of surfactants in more than one way.

Therefore, surfactants or detergents are substances that have the unique attributes of modifying the interfaces as well as stabilizing or destabilizing multiphase systems. In the cleaning industry, for example, detergents play the role of removing the dirt or soil (liquid- or solid phase) from the surfaces (solid phase). The cleaning is achieved by using water as the cleaning medium. The complete cleaning operation involves various inputs that are in one way or the other derived from solvents as described in the following:

1. Detergents detach the soil or dirt from the surface. For ensuring that the detergent molecules are delivered to the interface, detergents are formulated accordingly. Often, detergent formulations are based on an aqueous medium, and to develop a stable detergent formulation, one needs different solvents so that the detergent formulations are of a homogeneous single phase.
2. For detachment of soil from the surface, it is essential that the detergent molecules penetrate into the vulnerable parts of the surface. This is also achieved by incorporation of appropriate solvents in the formulation.
3. The detached soil is removed from the surface by solubilizing or dispersing the soil in the cleaning medium, that is, water. For this again, suitable solvents are used in the detergent formulations.

26.3 CLASSIFICATION OF DETERGENTS

A large number of detergents with various combinations of hydrophobic and hydrophilic groups are now commercially available. The new detergents can, thus, be designed by creating combinations of hydrophilic and hydrophobic groups in the following ways:

1. Taking a particular hydrophobe and attaching it with various types and sizes of hydrophiles
2. Starting with a hydrophile and then attaching different types and sizes of hydrophobes
3. Making a combination of hydrophobes to which one or more types and sizes of hydrophiles are attached
4. Synergistic combination of various types of detergents for unique behavior

Based on the similarities in their behavior and characteristics of various types of detergents, it is essential that they are classified, and different types of detergents are categorized based on certain commonalities. Based on the nature of the hydrophilic head group, they can be broadly classified as ionic, nonionic, and zwitterionic detergents. Some of the examples of these surfactants are given in Table 26.1.

26.3.1 IONIC DETERGENTS

Ionic detergents contain a head group with a net charge. They can be either negatively (anionic) or positively charged (cationic).

26.3.1.1 Anionic Detergents

Anionic detergents are basically the negatively charged molecules. The detergency of the anionic detergents is vested in the anion. The anionic detergents are produced in their acidic form initially, but before their use they are neutralized with an alkali or materials that are basic in nature to produce full detergency in the product. For example, sodium dodecyl sulfate (SDS), which contains the

TABLE 26.1
General Formulations for Liquid Heavy-Duty Detergents

	Composition (%)					
	Western Europe		Japan		United States	
Ingredients	With Builders	Without Builders	With Builders	Without Builders	With Builders	Without Builders
Anionic surfactants	5–7	10–15	5–15	–	5–17	0–10
		10–15	10–20	–	0–14	
Nonionic surfactants	2–5	10–15	4–10	10–35	5–11	15–35
Suds-controlling agents	1–2	3–5	–	–	–	–
Foam boosters	0–2	–	–	–	–	–
Enzymes	0.3–0.5	0.6–0.8	0.1–0.5	0.2–0.8	0–1.6	0–2.3
Builders	20–25	0–3	3–7	–	6–12	–
Formulations aids	3–6	6–12	10–15	5–15	7–14	5–12
Optical brighteners	0.15–0.25	0.15–0.25	0.1–0.3	0.1–0.3	0.1–0.25	0.1–0.25
Stabilizers	–	1–3	1–3	1–5	–	–
Fabric softeners	–	–	–	–	0–2	0
Fragrances	+	+	+	+	+	+
Dyes	+	+	+	+	+	+
Water	Balance	Balance	Balance	Balance	Balance	Balance

Production of Solvents for Detergent Industry

negatively charged sulfate group, is first produced as the dodecyl sulfonic acid and then neutralized with caustic to produce SDS.

The commonly known and commercially available detergents are based mainly on the following anionic groups: (a) sulfates, (b) sulfonates, (c) carboxylates, (d) phosphates, and (e) sulfosuccinates.

26.3.1.2 Cationic Detergents

Detergents having the cationic hydrophilic group or positively charged moiety are called as cationic detergents. The detergency is due to the cation, which can be a substantially sized molecule. Most of the cationic detergents are basically quaternary ammonium compounds. For the quaternization, strong acids, such as hydrochloric acid, is used as the "neutralizing" agent leaving chloride ions as the counteranion, although, in essence, no neutralization takes place during the manufacturing process. For example, cetyl trimethyl ammonium bromide (CTAB), which carries the positively charged trimethylammonium group, is a cationic detergent with bromide ions as the counteranions. This is the result of using hydrobromic acid for quaternization.

Cationic detergents are known not for their detergency, but mainly for their germicidal behavior or antistatic behavior. The cationic detergents get easily adsorbed on the surfaces, especially the ones that carry the negative charge, and produce the softening and antistatic effect. As a result, cationic detergents are used mainly as germicides as well as conditioners in shampoos, fabric softeners, antistatic agents for surfaces, and specific emulsifiers.

For the production of cationic detergents, the major hydrophobes (solvents), used as the starting material, are amines, amides, and so on. The straight chain or cyclic cationic detergents can be designed based on the understanding about the nature of surfaces to be treated with the cationic besides the activity of various chemical groups.

26.3.2 Nonionic Detergents

The detergent molecules having no charge (anionic or cationic) are called as nonionic detergents. Here, not only the hydrophobes or the chain starter (solvent), like in all other cases, is the nonionic chemical but the hydrophile group attached to it is also of nonionic nature. When dissolved in water, therefore, unlike the ionic detergents, nonionics do not form ions. The nonionic detergents can be produced in the following manner:

1. Taking the chain starter possessing the groups to which the hydrophiles can be attached.
2. Taking a chain starter possessing the groups to which another hydrophobe can be attached and this results in a block of hydrophobe consisting of more than one chain starter. The hydrophile can then be attached to this block to produce the detergent.
3. Taking a hydrophobe and converting it into a bigger hydrophobe by reacting with a small but multifunctional hydrophile. The hydrophile can then be attached to this bigger hydrophobe for obtaining a detergent that has a hydrophobe as well as a hydrophile of varying size and shape.

26.3.3 Zwitterionic Detergents

This is a type of detergent where the anionic as well as the cationic groups are both present in the same molecule. Like nonionic detergents, the zwitterionic detergents do not possess a net charge. They lack conductivity, electrophoretic mobility, and do not bind to ion exchange resins.

From the preceding discussion, it is quite evident that the detergent actives are designed on the solvents (hydrophobes) of various types having one common feature necessary for detergency and that is an alkyl or aryl alkyl chain with the functional group or sites for derivatization to attach the hydrophilic groups. Thus, the selection of a particular hydrophobe at the design stage would determine the performance of detergents.

26.4 COMPOSITION OF DETERGENTS

26.4.1 DETERGENT FORMULATIONS

Many times, the terms *detergents* and *surfactants* are used interchangeably. Surfactants are a class of substances including different types and categories of products, whereas detergents are generally referred to those products that are used for the purpose of cleaning and so on. Both in industries as well as households. It is in this context that various formulations containing different content of surfactants are generally called detergents. When the formulation consists of only the surfactants, then the detergents and surfactants are the same. Incidentally, the detergents of any grade would exhibit surface-active properties. In other words, all detergents can be called as surfactants but all surfactants are not detergents.

26.4.2 COMPONENTS OF A DETERGENT

1. *Surfactants.* Surfactants are the main active components of detergents. Various types of detergents, as Section 26.3, can be taken as the surfactants used for formulating detergents.
2. *Builders.* Builders are substances that augment the detersive effects of surfactants. Most important is their ability to remove hardness (i.e., soften the water) from the wash liquor and, thus, prevent them from interacting with the actives of detergents, that is, surfactants.

   The hardness of water is more crucial for detergents based on the anionic surfactants and in such cases, the use of builders becomes important to prevent any reduction in detergency. In addition, they can exert a suspending (antiredeposition) effect and keep the detached soil from depositing on the fabric.
3. *Organic additives.* Certain nonsurfactant organic additives are also added to improve cleaning performance as well as for other desirable properties. Such additives are usually present in low percentages and serve one or more of the following specific functions: antiredeposition of soil from the detergent bath onto the substrate, increased whiteness or appearance of cleanliness, and enhanced cleaning effect on specific types of solid surfaces and stains.
4. *Solvents.* The detergents are designed for special effects including the enhanced detergency by incorporation of various types of solvents. For taking care of the various aspects related to the applications of detergents, formulations using certain types of solvents are designed keeping in view the factors that influence detergency. Moreover, solvents help in the inhibition of foaming power and foaming stability, besides providing stability to the physical form of the detergent composition.

26.4.3 ROLE OF SOLVENTS INFLUENCING DETERGENCY

The surfactants are surface specific, whereas solvents are solute specific; solvents do not act at the interface, unlike the surfactants. The solutes are dissolved in solvents by the mechanism of solvation of solute molecules in the bulk of the solvent. The end result is always a single phase with no interfaces because the solute exists in the molecular size in the solvents.

Sometimes, for dissolving the not-quite-soluble solute into a given solvent, one needs to use cosolvents. In the case of detergent formulations based on either the aqueous or nonaqueous phase, one needs to add solvents to ensure that the components of the formulations dissolve in the dispersing medium.

Once the detergent is used, the soil on the surface needs to be detached from the surface, dissolved, or dispersed in the cleaning medium, and then has to be removed from the system.

Production of Solvents for Detergent Industry 497

The presence of solvents in the detergent formulations brings several positive influences on the performance of detergents. The following major influences are the notable ones:

1. Enhancement of surfactant activity
2. Increased resistance to water hardness and builders
3. Antiredeposition effect
4. Solubilization of liquid soil
5. Dispersion of solid soil by diffusion

Here, it may be noted that despite the fact that there are a large number of detergent formulations already existing, there has always been a need for a new and improved one. For example, having a detergent formulation that can efficiently remove the soil from the surfaces at ambient temperature, without any mechanical action, without foaming, with minimum amount of water required for cleaning, and leaving no burden on the environment is always a desirable proposition. The challenging criteria of performance, increasing concern for environment and ecology, and increasing awareness about the quality and safety besides the customer's preference for the value for money are the major driving forces for the development of a new detergent formulation. The development leading to the extended range of solvents for reformulation of detergents provides the choice to the development chemists for meeting the desired effects.

26.4.4 DIFFERENT TYPES OF DETERGENTS

26.4.4.1 Household Detergents

Detergents currently on the market in various parts of the world can be classified into the following groups from a detergency standpoint:

- Heavy-duty or all-purpose detergents
- Specialty detergents
- Laundry aids
- Aftertreatment aids

Heavy-duty detergents. Heavy-duty detergents include those detergent products suited to all types of washing and cleaning at all wash temperatures. They are offered in both powder and liquid forms and huge differences are found from one formulation to another. The powdered formulations generally do not contain solvents, and hence, only the liquid detergents are discussed here.

Liquid heavy-duty detergents. These products are distinctive because of their relatively high surfactant content (up to ca. 40%). For reasons of solubility and stability, they seldom contain builders and are generally devoid of bleaching agents. General formulations for liquid heavy-duty detergents are given in Table 26.1. Their effectiveness is concentrated on the removal of grease and greasy soil, especially at wash temperatures <60°C and doses of 6–12 g/L (depending on water hardness and the amount of soil to be removed). The unique feature of these formulations is the fact that up to 10% by weight of the formulation constitute solvents of various types.

Specialty detergents. Specialty detergents play a relatively minor role in the United States, but they are quite important in Western Europe.

Specialty detergents are products developed for use with specific types of household fabrics. Such detergents are generally employed with washing machines and usually require the use of special cycles (e.g., a wool cycle to prevent felting, a colored wash cycle to eliminate dye transfer, or a curtain cycle to prevent wrinkling).

TABLE 26.2
General Formulations of West European Liquid Specialty Detergents

	Composition (%)		
	Detergents for Woolens		
Ingredients	With Incorporated Softeners	Without Incorporated Softeners	Detergents for Curtains
Anionic surfactants (alkylbenzene sulfonates, fatty alcohol ethylene glycol ether sulfates)	–	10–30	0–8
Nonionic surfactants (fatty alcohol poly(ethylene glycol) ethers, fatty acid amides)	20–30	2–5	15–30
Cationic surfactants (dialkyldimethyl ammonium chloride)	1–5		–
Solvents (ethanol, propylene glycol, etc.)	0–10	0–10	0–5
Toluenesulfonates, xylenesulfonates, cumenesulfonates	–	0–3	–
Builders (potassium diphosphate, sodium citrate)	–	0–15	2–5
Optical brighteners	–	–	+
Fragrances, dyes	+	+	+
Water	60–70	60–80	65–75

Note: Spray stain removers are composed largely of mixtures of solvents and surfactants.

Many liquid detergents are intended for manual use; however, specialty liquid products that are intended for machine washing applications, including detergents for wool and curtains, also exist. Liquid detergents for wool may be free of anionic surfactants, in which case they usually contain mixtures of cationic and nonionic materials. The general formulations of west European liquid specialty detergents are given in Table 26.2.

The specialty liquid detergents are designed using different types of solvents, which provide the synergistic effect, in combination with the actives of detergents.

Detergents for large-scale institutional use generally differ to the extent that they must be designed to meet the special circumstances associated with laundry on an industrial scale. Commercial laundries, in contrast to those in the household, normally have soft water at their disposal, usually obtained from softening systems that employ softening filters.

26.4.4.2 Detergents in the Developing Countries

The existing detergent formulations and the development of new ones in the developing countries are quite different from the ones being used in the developed world. In developing countries, the market for detergents is quite complex, just like in the case of other consumer products. This is mainly due to the fact that there exists several sections of societies with wide gaps in their purchasing power. The market demand is divided into various segments such as low income, lower medium income, medium income, higher medium income, and high income. Accordingly, the manufacturers introduce the detergent products in the market catering to the requirements of each segment. During the last few years, there has certainly been a shift from low- to high-income group but the situation cannot be compared with what it is like in the developed countries. In fact, all ranges of detergents are available in the market keeping in view the purchasing power of the people. Mostly, the detergents are formulated with a very low content of detergent actives and high amount of fillers. The emphasis

Production of Solvents for Detergent Industry 499

is still on the foaming power of the formulations as the general perception of performance being driven by foaming power still persists in developing countries. The sizeable market segment catering to the middle- and higher middle-income group of the society is the one that has been serviced with innovative formulations of detergents. Some of the low-end formulations of developed countries, with certain modifications such as the formulations with locally available resources, are being introduced for this segment. For the high-income group, the products such as heavy-duty detergents and specialty detergents are also available with the modifications suiting to the local needs. The increasing mechanization in urban areas presents opportunities for manufacturers to introduce high active detergents in recent times.

In developing countries, the detergent formulations are largely of solid type; very small fraction of the market is being serviced by the liquid detergents. Further, the solid formulations are largely in the form of cakes as the large section of the society uses detergents for outdoor purposes, and the availability of appliances for using powder formulations is an issue. Another very interesting aspect of detergent usage by the low-income group is the fact that the detergent liquor is used for several washings: first, the high-value fabric is washed with the detergent liquor and then other fabrics are cleaned with the same solution. Even while throwing the solution after the last cleaning, an attempt is made to clean the floor to take full advantage of the product. This may sound quite surprising for those who are not familiar with this type of market but this is an important aspect that the formulator must keep in mind while developing the product.

26.5 ROLE OF SOLVENTS IN DETERMINING THE BEHAVIOR OF DETERGENTS IN WATER

As the molecules of detergents consist of two different types of moieties (referred to as head and tail) present in their structure, the behavior of the detergents once dissolved in water would depend largely on the nature, type, size, and shape of moieties. Since the detergents are surface-active substances, their behavior in water provides a lot of insight into the various possibilities of their exploitation for commercial purposes in various applications.

The behavior of the solution of detergents in water can be understood from the fact that on the one hand, the hydrophilic moiety (or head) prefers to stay with water, and on the other, the hydrophobic moiety (or tail) would tend to move away from the bulk. In such a situation, the detergent molecules will have to orient in such a manner that the head remains with the bulk while the tail remains away from it.

Section 26.5, all the systems that we encounter consist of interfaces; the solution of detergents in water would also involve the interface between the water and air. For the detergent molecules dissolved in water, their tail part can remain in the interface in contact with air rather than being with water. However, this can be possible only up to the point till the detergent molecules saturate the interface. Once the concentration of detergents in water is increased beyond the level so that the interface gets saturated, the hydrophobic moiety gets constrained, and it is at this stage that the hydrophobic moieties reorganize and reorient themselves further. The two stages of solution of detergents in water can be visualized in Figures 26.1a and 26.1b. The stage I (AB_1) till the point of saturation of detergent at the water/air interface helps in the reduction of the interfacial tension (for interface with air, it is always referred to surface tension), whereas the stage II (B_1D) beyond the saturation point at interface, at higher concentration of detergent, results in the formation of aggregates (known as micelles) of detergents in the bulk. Thus, the increase in concentration of detergent in water, to start with, brings a reduction in surface tension but only up to a point beyond which there is no further change in surface tension. The concentration at which the minimum surface tension is achieved is called as the critical micelle concentration (CMC) because beyond this concentration, the detergent molecules exist as aggregates or micelles. From this, it is quite clear that the hydrophobic reorientation or hydrophobic interactions play a key role in determining the behavior of detergents dissolved in water.

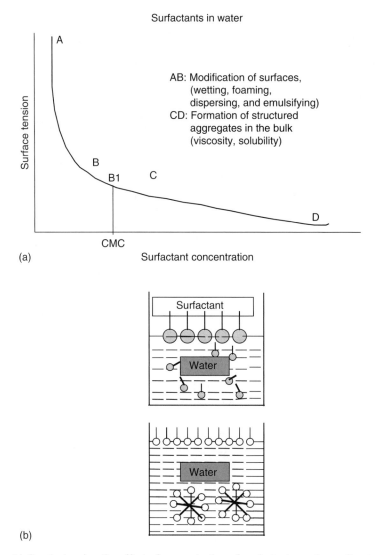

FIGURE 26.1 (a) Graph showing the effect of concentration of surfactants on the surface tension of water. AB represents the higher surfactant concentration at interface than in the bulk and CD represents the formation of structured aggregates in the bulk. (b) Formation of micelles during addition of surfactant in water.

From the preceding discussion, the following points are quite clear:

1. The addition of detergents to water results in the reduction of surface tension with the increase in the concentration of detergent. This effect is ascribed to the fact that the detergent molecules prefer to stay at the water/air interface instead of being in bulk. It can thus be said that the reduction in surface tension is due to the adsorption of detergent at the interface.

   When one is looking for a detergent that wets fast or, in other words, adsorbs at the surface that is meant to be cleaned, the hydrophobic moiety has to be chosen such that the micellization takes place at very low concentrations. Generally, among the solvents, such as the fatty alcohols of chain length C_{12}–C_{18}, alkyl chains of C_{12}, and those with unsaturation in the chain, would bring such an effect.

Production of Solvents for Detergent Industry

2. The reduction in the surface tension is observed only up to a certain concentration beyond which there is hardly any change in the surface tension, no matter how much the concentration of the detergent is raised. The concentration at which the minimum surface tension is achieved is called as the CMC. It is called critical because it represents the change from adsorption at the interface to aggregation in the bulk of the detergent molecules in solution. At CMC, adsorption ceases to exist while the aggregation of molecules begins. Thus, CMC indicates the concentration of detergent in water that is needed to form the micelles.

3. As the adsorption at the interface brings reduction of surface tension, the formation of micelles is expected to cause changes in the bulk properties, for example, viscosity of the solution of detergent in water.

Since the behavior of the detergent in water is determined by the way the detergent molecules reorient themselves at the interface and the way they regroup in the bulk, the role played by the factors concerning the hydrophobic as well as the hydrophilic portions of the detergent molecules becomes highly significant, the hydrophobic interactions being the key (Figure 26.2 shows the hydrophobic interactions). Based on the adsorption data and the bulk properties of the solution of detergents in water, one can, therefore, assess the suitability of a given detergent for a particular application.

The CMC of a detergent depends on the nature, size, and shape of the hydrophile as well as the hydrophobe. This can be understood from Figure 26.3. Here, the variations in the hydrophobic part as well as hydrophilic part can be made to design the detergents with the desired adsorption or CMC.

For the purpose of formation of micelles of desired size and shape, one needs to use a combination of solvents. To start with, having the chain initiators balancing the hydrophile group will help in the formation of micelles. Then, by choosing the right type of solvent or cosurfactant, the core of the micelles can be enlarged for the removal of the oily soils from the surfaces.

Moreover, the presence of solvents in the detergent always brings effects such as interfacial turbulence, diffusion, and ultralow surface tension. Thus, for cleaning efficiency of the detergents, solvents play an important role.

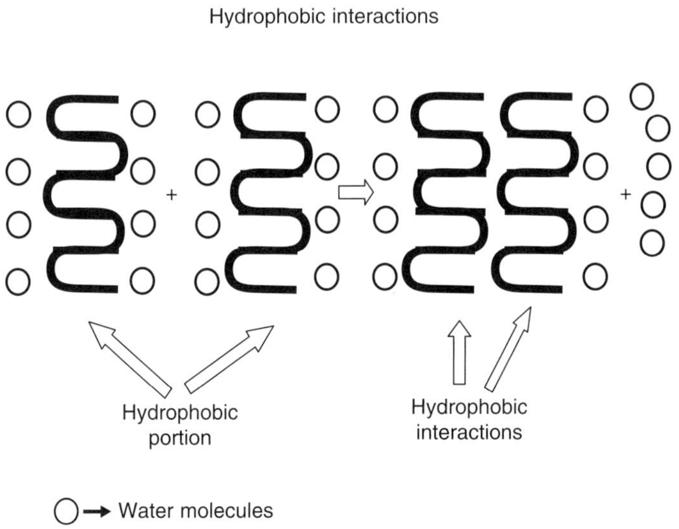

FIGURE 26.2 Interactions of hydrophobic molecules.

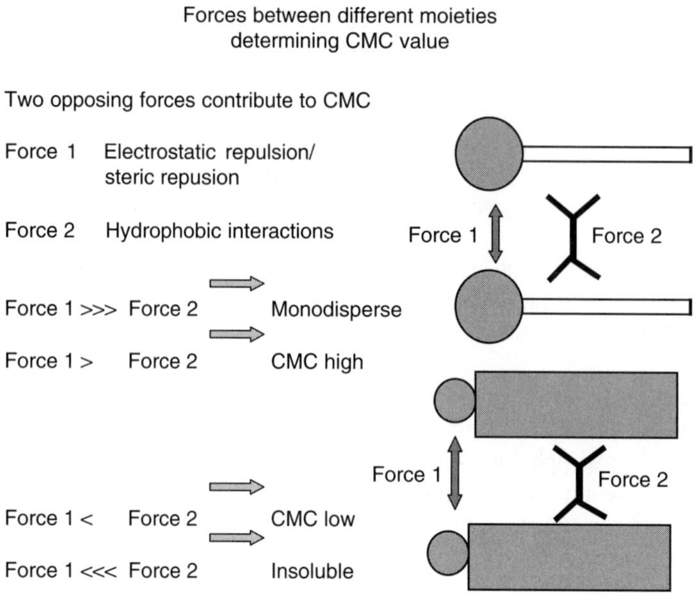

Forces between different moieties determining CMC value

Two opposing forces contribute to CMC

Force 1   Electrostatic repulsion/
          steric repusion

Force 2   Hydrophobic interactions

Force 1 >>> Force 2 → Monodisperse

Force 1 > Force 2 → CMC high

Force 1 < Force 2 → CMC low

Force 1 <<< Force 2 → Insoluble

FIGURE 26.3   Forces contributing to the formation of micelles.

26.6 ROLE OF SOLVENTS IN ENHANCING THE PERFORMANCE OF DETERGENTS

As evident from the Section 26.5, detergents play an important role in bridging the gap between two incompatible or completely opposite surfaces by bringing the interfacial tension down.

Adsorption at the interface and the aggregation in the bulk are the two important parameters determining the performance of the detergents. Incidentally, in all the applications of the detergents, water is involved at every stage of cleaning and hence the behavior of detergents in water comes handy while altering the performance of detergents for a desired performance.

Water can also be used as a medium in the concentrated formulations of detergents. But, like water, there can be other polar solvents or solvents with functionality that can be tried in the detergent formulations. To understand the role of solvents in determining the performance of detergents, it would be better if different steps involved in the detersive action are discussed. The stagewise depiction of the detergent action on the surfaces and the role of solvents in enhancing the performance are shown in Figure 26.4.

The first two stages of the cleaning process involve the adsorption of detergents at the interface. The adsorption depends on the hydrophobic chain length, nature of hydrophobe, and so on. For efficient adsorption, it is essential that the detergent molecules penetrate into the surfaces. For penetration, the solvents play an important role as the small solvent molecules with affinity for soil take the detergent molecules into the vulnerable parts of the soil surface with great ease.

Once the adsorption reaches its equilibrium, the next step in the cleaning operation involves the detachment of soil from the surface and the stabilization of the detached soil in water. Both these actions are determined by the degree of micellization of detergents. It is in the micelles that the oily soils get entrapped and removed from the soiled surfaces. Here, the solvents help in enhancing the soil removal by playing a role of cosurfactant as can be seen from Figure 26.5.

During this step, there is always a possibility of the soil redepositing on the surface, and the role played by the micelles as well as solvent are crucial for this.

The last step in the cleaning operation pertains to the rinsing of the surface making it free from detergents, and so on. Here again, the role played by solvents is crucial in the following two ways: ensuring the detergent gets washed with water and reducing the foam of the detergent.

Production of Solvents for Detergent Industry

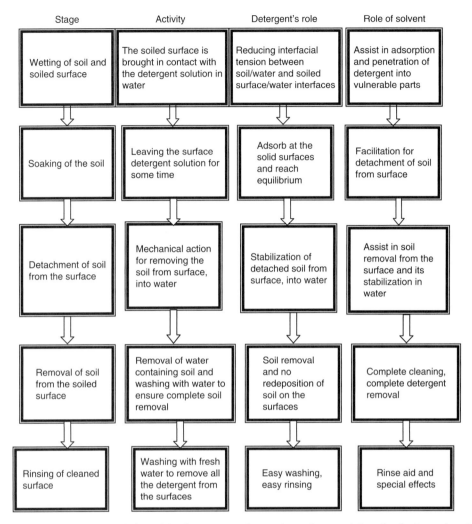

FIGURE 26.4 Stagewise depiction of the detergents action on the surfaces and the role of solvents in enhancing the performance.

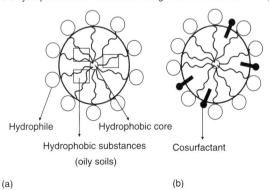

FIGURE 26.5 Effect of solvents as cosurfactants.

504   Handbook of Detergents/Part F: Production

The solvents play an important role in enhancing the performance of detergents. There are several types of solvents that are being tried for formulating the improved detergents. Depending on the structural attributes of detergents, detergent formulations are being prepared with special features.

The difference between the solvents and the surfactants (detergents) is quite a thin line; while the detergents homogenize the two opposing surfaces or phases into one by acting at the interfaces, solvents have the capacity to homogenize a given solute in the bulk by providing the sufficient energy (of solvation) to break down the particles of solute. Solvents, therefore, do not have preference for the interface like the detergents but they have strong affinity for the solutes. For the purpose of understanding the behavior of solvents and the mechanism by which the solvents perform, it would be better to analyze the system based on their polarity (Figure 26.6). On the basis of polarity, the systems can easily be categorized into two basic groups: polar and nonpolar. The polar systems will have a liking for the polar one and disliking for nonpolar ones and vice-versa. This basic difference is the reason behind the fact that there are two types of solvents, that is, polar and nonpolar. Because of the strong affinity of solvents for the solutes, the solute particles disintegrate into molecular sizes on contact with the solvent. For example, greasy and oily soil (dirt) when brought in contact with the hydrocarbons (nonpolar solvents) disintegrate and dissolve into the hydrocarbons. If one tries to clean the surfaces soiled with greasy or oily materials with water, the soil will not detach from the surface simply because the nonpolar soil will not be wet with the water. In the same way, the surfaces having the soil and dirt that disperse well with the aqueous medium will not get cleaned with the nonpolar solvents. For cleaning different types of soils, different types of detergents or their combinations can be tried. This is how several types of cleaning formulations both for dry- and wet cleaning are developed. The cleaning formulations consist of detergents and solvents and not just solvents. It is a fact that detergents have the capacity to detach the soils and dirt from different surfaces and they are used as the cleaning agents, but the part played by solvents

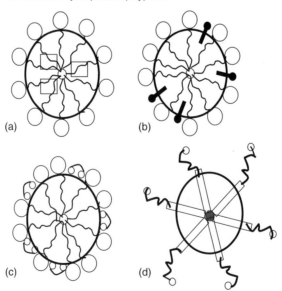

FIGURE 26.6 Micelle formation by hydrophobic interactions. (a) The hydrophobe gets dissolved in the hydrophobic core of the micelle. (b) The hydrophobe adjusts within the micelle so that the molecule orients as the micelle. (c) The hydrophobe gets coiled around the hydrophilic core of the micelle. (d) The tails encapsulating the hydrophobes.

Production of Solvents for Detergent Industry 505

in removing the soils from surfaces is no less significant. While the use of solvents alone serves only a limited purpose, the use of detergents alone also cannot provide the complete solution to the problems associated with the cleaning of surfaces. A combination of the solvents and surfactants can certainly provide a synergistic effect if used appropriately. The role played by the solvents in combination with the detergents can be understood from the fact that the following special effects are attributed to the presence of solvents in the detergents:

1. Interfacial turbulence
2. Diffusion
3. Ultralow interfacial tension

Interfacial turbulence. When two different (opposite in nature) surfaces are brought in contact with each other, there exists the interfacial tension causing the surfaces to remain away from each other. However, if one of the surfaces contains the detergent molecules at the interface, then the contact between the two surfaces would result in an affinity for the two surfaces. Furthermore, if the solvents are also present along with the detergents, then there exists an interfacial turbulence (Figure 26.7) mainly due to the unequal adsorption of detergent molecules, at different contact points of interface, aided by the presence of solvents. This interfacial turbulence facilitates the easy removal of oily soil from the surface. This concept has been in existence for years in formulations of several agrochemicals where the self-emulsifiable concentrate formulations are developed for applications of pesticides. In detergents, the same concept is applied while formulating for the complex soils and not so smooth surfaces.

Diffusion. Diffusion of oily soil materials in an aqueous medium due to the presence of detergents is the common mechanism by which the oily materials disperse in water through the reduction of interfacial tension. The phenomenon of diffusion is accelerated and enhanced by the presence of solvents in the detergents. Mass transfer due to the partitioning of the solvent along with the higher hydrophilic detergent from the oily surface in the water causes the diffusion of oily particles (Figure 26.8). Partitioning due to osmotic pressure arising out of differences in concentration of the oily soil material in two phases is responsible for diffusion.

Any hydrophobic soil on the surface gets dissolved in polar solvents such as isopropyl alcohol (IPA) and butanol. It then self-emulsifies due to the diffusion of the solvent taking

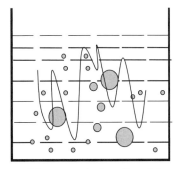

Because of the unequal adsorption of the detergent molecules on the interface aided by the presence of solvents, the soil particles break down into smaller droplets when brought in contact with the detergent solution in water.

FIGURE 26.7 Interfacial turbulence between solvent and detergents.

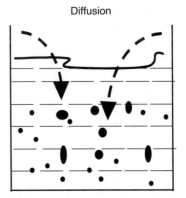

Because of the diffusion of highly hydrophilic material (solvent) from oil-to aqueous phase, the particles of hydrophobe are broken into smaller droplets resulting in their emulsification or dispersion in water.

FIGURE 26.8 Diffusion behavior of a hydrophile.

the hydrophobes along with it. This works more effectively when one uses solvents such as primary alcohols.

Ultralow interfacial tension. The surfactant concentration at the interface provides the necessary driving force (spreading pressure). But the presence of solvent along with the surfactant aids in increasing the spreading pressure resulting in lowering the interfacial tension to ultralow levels leading to self-emulsification of oily soils. In fact, the self-emulsification is just short of solubilization and can be explained by the following equation:

$$\gamma_1 = \gamma_{o/w} - \pi$$

$$\text{If } \pi \geq \gamma_{o/w}$$

$$\gamma_1 \leq 0$$

where

γ_1 = total interfacial tension
$\gamma_{o/w}$ = interfacial tension for oil and water (o/w) interface
π = spreading pressure

26.7 SOLVENT GROUPS AND PROPERTIES

Solvents are divided into 14 chemical groups, which include all commercial solvents used by various industries. The majority of commercial solvents consist of the following groups:

- Hydrocarbons (aliphatic and aromatic)—benzene and alkyl derivatives of benzene
- Halogenated hydrocarbons—chloroform
- Nitrogen-containing compounds (nitrates and nitriles)—ammonia, triethanol amine
- Organic sulfur compounds—carbon disulfide
- Monohydric alcohols—ethanol
- Polyhydric alcohols—glycols, glycerol, sorbitol, and so on
- Aryl alcohols—phenols and alkyl derivatives of phenols
- Aldehydes—formaldehyde, acetaldehyde, and so on
- Ethers—diethyl ether
- Ketones—acetone

Production of Solvents for Detergent Industry

In present day formulations, only a select few of the total range of solvents are being used mainly in the detergent industry. Majority of them include alcohols, polyhydric alcohols, glycol ethers, and esters. Most of these solvents are known for enhancing the performance of detergents.

26.7.1 Solvent Characteristics

The detergents were largely formulated using the hydrocarbon and chlorinated hydrocarbon types. With the rising concern for environment and ecology, such solvents have been replaced by the environment-friendly ones. The general characteristics of a solvent while considering them for potential applications are as follows:

1. Solvent power
2. Solubility parameters
3. Activity coefficient
4. Solvent–solute interactions

From the viewpoint of performance of detergents, the aforementioned characteristics of solvents help decide the appropriate solvent to be used for cleaning a given surface. The affinity of the solvent for a given type of soil plays a key role in the selection of the solvent.

26.7.2 Solvents for the Detergent Industry

The major solvents for the detergent industry are as follows:

1. Detergent range alcohols
2. IPA or propanol
3. Ethyl alcohol or ethanol
4. Alkanolamine
5. Glycols
6. Glycol ethers and esters
7. Green solvents
8. Ionic liquids

26.7.2.1 Detergent Range Alcohols

Although fatty alcohols are not used to formulate the detergents, they are the main building blocks for manufacturing a detergent and it would be worthwhile mentioning about them here.

Natural or synthetic detergent range alcohols are classified as follows:

1. *Middle cut.* Number of carbon atoms being 12 through 15.
2. *Heavy cut.* Number of carbon atoms being 16 through 18 corresponding to the high distillation fractions of fatty alcohols derived from vegetable oils such as coconut oil and palm oil.

The production of the desired cut can be increased by the proper choice of feedstock and processing conditions.

26.7.2.1.1 Specifications and Standards

The major portion of detergent range alcohols used commercially consists of mixtures of alcohols and a great variety of products that is available. Typical specifications for commercial linear detergent range alcohols are given in Table 26.3.

A small amount of detergent range alcohols are used as it is while most of it is used as derivatives such as poly(oxyalkylene) ethers or esters of acids such as sulfuric acid, phosphoric acid, and mono- and dicarboxylic acids.

TABLE 26.3

Typical Specifications for Commercial Linear Detergent Range Alcohols

Descriptive Name	Alcohol (% Minimum)	Hydro-Carbon (% Maximum)	Hydroxyl Value	Saponification Value (Maximum)	Acid Value (Maximum)	Iodine Value (Maximum)	Specific Gravity	Melting Point (°C)	Boiling Range (°C)	Color, APHA (Maximum)	Moisture (% Maximum)
Dodecanol (99% C_{12})	98.5	—	295–302	1.0	0.4	0.2	$0.832^{24/25}$	24	255–259	15	0.1
Dodecanol (67% C_{12})	98.5	1.0	280–290	0.7	0.1	1.0	$0.823^{35/25}$	22	272–299	10	0.1
Dodecanol-tridecanol	98.7	0.5	271–282	0.5	0.5	0.4	$0.83^{25/25}$	21–23	270–293	25	0.1
Hexadecanol	98.0	1.5	225–235	1.0	0.5	2.0	$0.814^{55/25}$	50	311–316	35	0.1
Hexadecanol-octadecanol (tallow)	96.0	1.5	204–215	3.0	1.0	3.0	$0.810^{65/25}$	53	316–332	35	0.1
Octadecanol	97.0	1.5	200–215	2.0	0.5	2.0	$0.811^{65/25}$	58	326–d	35	0.1
Octadecanol	—	—	200–210	1.0	0.2	93–96	$0.83–0.84^{40/25}$	4	—	—	—

Note: APHA, American Public Health Association.

TABLE 26.4
Composition of Detergent Range Commercial Alcohols

Descriptive Name	Representative Trade Names	Derived From	Approximate Composition (wt%, 100% Alcohol Basis)						
			C_{12}	C_{13}	C_{14}	C_{15}	C_{16}	C_{18}	C_{20}
Dodecanol	Alfol[a] 12	Ethylene	99		1				
	Epal[b] 1214	Ethylene	66		27		7		
	Alfol 1216	Ethylene	65		25		10		
	CO-1214	Coconut	67		26		7		
	Alfol 1218	Ethylene	40		30		20	10	
	Dehydag[c]	Coconut	72		27		1		
Tetradecanol	Alfol 14	Ethylene	1		99				
Tetradecanol-octadecanol	CO-1418	Coconut	12		43		22	23	
	Epal 1418	Ethylene			36		40	23	1
Dodecanol-tridecanol	Neodol[d] 23	Olefin	32	66	2				
	Neodol 25	Olefin	20	30	30	20			
Hexadecanol	CO-1695	Coconut			1		96	3	1
Hexadecanol-Octadecanol	TA-1618[e]	Tallow			4		28	67[f]	1
	Alfol 1618	Ethylene					61	37	2
Octadecanol	CO-1895	Coconut					2	97	1
Octadecanol	Dehydag HD	Natural Oils					4	94[g]	2
Octadecanol-Octadecanol	Dehydag 60/65	Natural Oils	1		4		26	68[g]	1

[a] Registered trademark for Continental Oil Company alcohols.

[b] Registered trademark for Ethyl Corporation alcohols.

[c] Registered trademark for Henkel alcohols.

[d] Registered trademark for Shell alcohols.

[e] Registered trademark for Procter & Gamble alcohols.

[f] Includes 1% C17 alcohol.

[g] Octadecanol.

The alcohols provide the starting material for all of the surfactant types such as nonionics, anionics, cationics, and zwitterionics. The fatty alcohol sulfates, such as sodium lauryl sulfate (SLS), $C_{12}H_{25}OSO_3Na$, are known for their cleaning ability besides being soft on the fabric and hands. Alkyl sulfates derived from C_{12} through C_{15} linear alcohols are widely used in consumer products, for instance, in toothpastes, hair shampoos, carpet shampoos, and light-duty household cleaners, whereas those derived from C_{16} and C_{18} linear alcohols are used in heavy-duty household detergents (Table 26.4). Detergents having good foam stability are prepared from mixtures of C_{14}, C_{16}, and C_{18} linear alcohols. Minor amounts of unsulfated alcohol left in the alkyl sulfate detergents serve as foam stabilizers, especially for products such as shampoos. The polyethoxylated alcohols, when sulfated and neutralized with a base such as sodium or ammonium hydroxide to give anionic surfactants, have wide applications as light-duty dishwashing detergents and as part of the surfactant system of heavy-duty household liquid and granular detergents. These and other more specialized surfactants have a wide variety of industrial and household applications.

The polyethoxylated alcohols in household use have an average ethylene oxide (EO) content of less than 3 to about 15 mol/mol of alcohol. Ethoxylation gives a broad range of products; for example, in a material with an average of three EO units, some of the molecules may have as many as 12 or 14 EO units and others exist as the free alcohol. This gives the effect of a mixed surfactant system. The polyethoxylated alcohols are also used directly as nonionic surfactants.

510   Handbook of Detergents/Part F: Production

26.7.2.1.2   Fatty Alcohols
26.7.2.1.2.1   Introduction
Fatty alcohols are aliphatic alcohols with chain lengths between C_6 and C_{22}:

$$CH_3(CH_2)_nCH_2OH \qquad (n = 4\text{--}20)$$

They are predominantly straight chained and monohydric, and can be saturated or have one or more double bonds. Alcohols with a carbon chain length above C_{22} are referred to as wax alcohols. Diols whose chain length exceeds C_8 are regarded as substituted fatty alcohols. The character of the fatty alcohols (primary or secondary, linear or branched chain, saturated or unsaturated) is determined by the manufacturing process and the raw materials used. Depending on the raw materials used, fatty alcohols are classified as natural or synthetic. Natural fatty alcohols are based on renewable resources such as fats, oils, and waxes of plant or animal origin, whereas synthetic fatty alcohols are produced from petrochemicals such as olefins and paraffins.

26.7.2.1.2.2   Production from Natural Sources
Two groups of natural raw materials are used for the production of fatty alcohols: (1) fats and oils of plant or animal origin, which contain fatty acids in the form of triglycerides that can be hydrogenated after suitable pretreatment to yield fatty alcohols and (2) wax esters from whale oil (sperm oil), from which the fatty alcohols are obtained by simple hydrolysis or reduction with sodium.

The commercial exploitation of sperm oil has led to the depletion of whale populations and is banned in some countries. Attention has, therefore, turned to the jojoba plant whose oil also consists of wax esters. Most fatty chemicals obtained from natural sources have chain lengths of C_{16}–C_{18}. The limited availability of compounds with 12–14 carbon atoms, which are important in surfactants, was one of the driving forces behind the development of petrochemical processes for the production of fatty alcohols. Higher alcohols, such as C_{20}–C_{22} alcohols, can be produced from rapeseed oils rich in erucic acid and fish oils. Unsaturated fatty alcohols may be manufactured in the presence of selective catalysts.

Hydrolysis of wax esters. The hydrolysis of wax esters is of limited importance today. It is carried out by heating sperm oil with concentrated sodium hydroxide at 300°C and distilling the alcohol from the sodium soap.

$$R^1 - \overset{\overset{\textstyle O}{\textstyle \|}}{C} - OR^2 + NaOH \longrightarrow R^1 - \overset{\overset{\textstyle O}{\textstyle \|}}{C} - ONa + R^2OH \qquad (26.1)$$

where
$$R^1 = C_{16}H_{30} \text{ (alkyl group of palmitic acid)}$$
$$R^2 = C_{18}H_{34} \text{ (alkyl group of oleic acid)}$$

The distillate consists of partially unsaturated C_{16}–C_{20} alcohols, which are hardened by catalytic hydrogenation to prevent autooxidation. Since sperm oil contains only 70% wax esters, the alcohol yield is ~35%.

Reduction of wax esters with sodium. The reduction of esters with sodium was first described by Bouveault and Blanc in 1902. Large-scale application of this process was achieved in 1928 (Dehydag).

$$R^1 - \overset{\overset{\textstyle O}{\textstyle \|}}{C} - OR^2 + 4Na + 2R^3OH \longrightarrow$$

$$R^1 - CH_2ONa + R^2ONa + 2R^3ONa \xrightarrow[-4NaOH]{+4H_2O} R^1 - CH_2OH + R^2OH + 2R^3OH \qquad (26.2)$$

Production of Solvents for Detergent Industry

where

$R^1 = C_{16}H_{30}$ (alkyl group of palmitic acid)
$R^2 = C_{18}H_{34}$ (alkyl group of oleic acid)
$R^3 = C_{20}H_{38}$ (alkyl group of eicosanoic acid)

Molten sodium is dispersed in an inert solvent and the carefully dried ester and alcohol are added. When the reaction is complete, the alkoxides are split by stirring in water and the alcohols are washed and distilled.

The reduction proceeds selectively without the production of hydrocarbons and isomerization or hydrogenation of double bonds. Extensive safety measures are required due to the large quantity of metallic sodium used. The process was used until the 1950s to produce unsaturated fatty alcohols, especially oleyl alcohol from sperm oil. These alcohols can now be produced by selective catalytic hydrogenation processes using cheap raw materials, and the sodium reduction process is of interest only in special cases.

Transesterification of triglycerides. This reaction is carried out continuously with alkaline catalysts. Transesterification is an equilibrium reaction and is shifted toward the desired ester by excess methanol or removal of glycerol.

Hydrogenation processes. Three large-scale hydrogenation processes are used commercially: (1) suspension hydrogenation, (2) gas-phase hydrogenation, and (3) trickle-bed hydrogenation. In all variants, hydrogenation is carried out with copper-containing, mixed-oxide catalysts at 200–300°C and 20–30 MPa.

If a stainless-steel reactor is used, this process can be applied to the direct hydrogenation of fatty acids. In this case, an acid-resistant catalyst is required, and catalyst consumption is increased.

26.7.2.1.2.3 Synthesis from Petrochemical Feedstocks

Ziegler Alcohol Processes. Two processes for the production of synthetic fatty alcohols are based on the work of Ziegler on organic aluminum compounds: the Alfol process, developed by Conoco; and Ethyl Corporation's Epal process. Fatty alcohols synthesized by these processes are structurally similar to natural fatty alcohols and are thus ideal substitutes for natural products.

Uses

- Fatty alcohols and their derivatives are used in synthetics, surfactants, oil additives, and cosmetics and have many specialty uses.
- Fatty alcohols are mainly employed as intermediates. In Western Europe, only 5% are used directly and 95% in the form of derivatives.
- Fatty alcohols orient themselves at phase interfaces and can, therefore, be used in emulsions and microemulsions. They impart body to cosmetic creams and lotions and solvency to industrial emulsions. They also serve as lubricants in polymer processing. The products obtained from fatty alcohols, EO, and propylene oxide (PO) are weak foaming surfactants.

26.7.2.1.3 Unsaturated Fatty Alcohols

Unsaturated fatty alcohols can only be obtained from natural sources; petrochemical processes for their manufacture do not exist.

26.7.2.1.3.1 Production

The first large-scale hydrogenation plant (Henkel) went into operation in the late 1950s. Previously, unsaturated fatty alcohols could be obtained only by hydrolysis of whale oil. The hydrogenation processes Section 26.7.2.1.2.2 are suitable for the large-scale production of unsaturated fatty alcohols.

The fixed-bed processes are preferred because of the mild reaction conditions. In suspension hydrogenation, the prolonged contact between the fatty alcohol and catalyst results in side reactions such as saturation of the double bond and formation of trans isomers, which leads to a higher solidification point and, hence, loss of quality. With polyunsaturated fatty acids, the formation of conjugated double bonds cannot be completely prevented.

Hydrogenation is generally carried out at temperatures of 250–280°C and pressures of 20–25 MPa. Catalysts include zinc oxide in conjunction with aluminum oxide, chromium oxide, or iron oxide, and possibly, other promoters; copper chromite whose activity has been reduced by the addition of cadmium compounds; and cadmium oxide on an alumina carrier. Selective hydrogenation can also be carried out in a homogenous phase with metallic soaps as catalysts.

26.7.2.1.3.2 Uses
Unsaturated fatty alcohols are used in detergents, cosmetic ointments and creams as plasticizers and antifoaming agents, and textile and leather processing. Oleyl alcohol is also used as an additive in petroleum and lubricating oils.

26.7.2.1.3.3 Economic Aspects
Prices of detergent range alcohols are strongly dependent on feedstock prices. The abrupt increase in petrochemical prices in 1973–1974 halted the trend of natural alcohols losing ground to synthetic alcohols derived from low-cost ethylene. It is quite clear that the prices of natural gas and crude oil are responsible for the price fluctuations for synthetic alcohols on the one hand and the production of renewable resources such as vegetable oils are responsible for the prices of fatty alcohols derived from vegetable oils on the other. However, things are not as simple as they appear to be because of the fact that several other factors such as social and political become important.

Hence, the future prices of detergent range alcohols cannot be accurately predicted. The major difference between synthetic alcohols and alcohols derived from vegetable source lies in the fact that alcohols derived from natural sources would only be of straight chain, whereas the ones derived from synthetic (petroleum source) can be designed with branched structure.

Synthetic alcohols can be substituted for natural alcohols except for applications that require the absence of chain branching or the minor amounts of secondary alcohols that are found in some synthetic alcohols, and for applications that are sensitive to minor impurities or odor characteristics. The lion's share of detergent range alcohol production at the present is by the synthetic processes. This situation is bound to change in the times to come mainly because of the fact that (a) synthetic alcohols would cost more and this would make the alcohols from natural sources cost competitive and (b) special performance features of natural alcohols.

26.7.2.2 Isopropyl Alcohol (2-Propanol)

26.7.2.2.1 Introduction
2-Propanol has been identified as a metabolic product of a variety of microorganisms. It is emitted in waste gases and wastewater from industrial sources, and may be removed by biological oxidation or reverse osmosis.

26.7.2.2.2 Production
2-Propanol is commonly manufactured from propene. Strong and weak acid processes, used previously and involving potentially hazardous intermediates and by-products, have largely been replaced by the catalytic hydration of propene today. The catalytic reduction of acetone is an alternative process.

2-propanol is produced from propene by two different processes:

- Indirect hydration
- Direct hydration

Production of Solvents for Detergent Industry   **513**

Indirect catalytic hydration has been replaced by the direct hydration process in countries such as Japan, United States, and Western Europe. Indirect catalytic hydrogenation of acetone involves the feeding of 88–93% sulfuric acid and propene gas into a reactor to produce a mixture of isopropyl and di-isopropyl sulfates, which are hydrolyzed with water to 2-propanol. Principal by-products are di-isopropyl ether and isopropyl oils consisting mainly of polypropylenes of high relative molecular mass. It has gradually been replaced by the weak-acid process, in which propene gas is absorbed in, and reacted with, 60% sulfuric acid. The resulting sulfates are hydrolyzed in a single-step process. 2-Propanol is stripped and refined from the condensate, which also contains di-isopropyl ether, acetone, and polymer oils of low relative molecular mass.

Removal of the compound by reverse osmosis (hyperfiltration) from wastewater is successful depending on the type of membrane used. Membranes such as cellulose acetate lead to 40–60% separation of 2-propanol, whereas cross-linked polyethyleneimine and aromatic polyamine membranes yield 80–90% separation of 2-propanol.

26.7.2.2.3   Uses

2-Propanol is mainly used as a solvent, and in pharmaceutical, household, and personal products. It is a low-cost solvent with many consumer and industrial applications. 2-Propanol also possesses cooling, antipyretic, rubefacient, cleansing, and antiseptic properties.

The other major uses of IPA include the following:

- Process solvent
- Coating and dye solvent
- Cleaning and drying agent
- Solvent in topically applied preparations
- Aerosol solvent

26.7.2.3   Ethanol

26.7.2.3.1   Introduction

Besides being used as an essential ingredient of alcoholic beverages, ethanol is used as a solvent, an antifreeze, and an intermediate in the synthesis of innumerable organic chemicals.

26.7.2.3.2   Production

The production of ethanol is done using the following processes.

26.7.2.3.2.1   Petrochemical Route

- Direct hydration of ethylene (Figure 26.9)
- Indirect hydration of ethylene (Figure 26.10)

26.7.2.3.2.2   Fermentation Route (Bioroute)

- Sugar crop fermentation
- Corn dry milling fermentation
- Corn wet milling fermentation
- Lignocellulose fermentation (Figure 26.11)

26.7.2.3.2.3   Direct Catalytic Hydration of Ethylene

Ethanol is synthetically produced from the catalytic hydration of ethylene. The common catalyst used today is phosphoric acid impregnated on an inert support, such as Celite diatomite. The reaction is carried out at high pressures and temperatures, typically 75 kg/cm^2 and 300°C. The reaction is near quantitative, with relatively minor side reactions producing ethers, aldehydes, ketones, and higher hydrocarbons (these minor products results from unwanted polymerization of ethylene). The major by-product is diethyl ether, which is usually recycled back to the reactor to form ethanol.

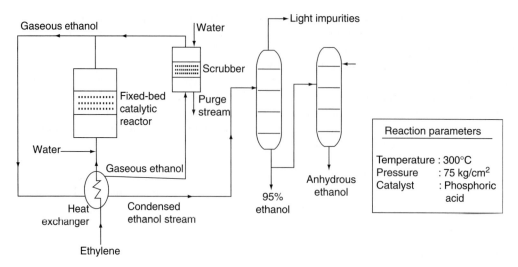

FIGURE 26.9 Direct hydration process for ethanol manufacture. (Direct hydration of ethylene for the production of ethanol is carried out at high pressure (75 kg/cm^2) and high temperature (300°C). The major by-product is diethyl ether, which is recycled back to the reactor to form ethanol.)

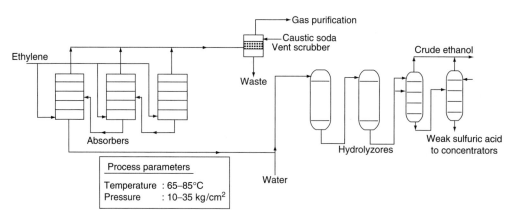

FIGURE 26.10 Indirect hydration process for ethanol manufacture. (Ethanol is produced from ethylene by using sulfuric acid in a three-step process. The process involves the formation of ethyl sulfate, which is hydrolyzed to ethanol.)

26.7.2.3.2.4 Indirect Hydration of Ethylene
The preparation of cthanol from ethylene by the use of sulfuric acid is a three-step process:

- Absorption of ethylene in concentrated sulfuric acid to form monoethyl sulfate (ethyl hydrogen sulfate) and diethyl sulfate
- Hydrolysis of ethyl sulfates to ethanol
- Reconcentration of the dilute sulfuric acid

None of the producers in the world have employed the indirect hydration of ethylene to manufacture ethanol since mid-1980s. The shift away from the indirect- to the direct route has been due to better yields, lesser by-products, and reduced quantity of pollutants.

Production of Solvents for Detergent Industry

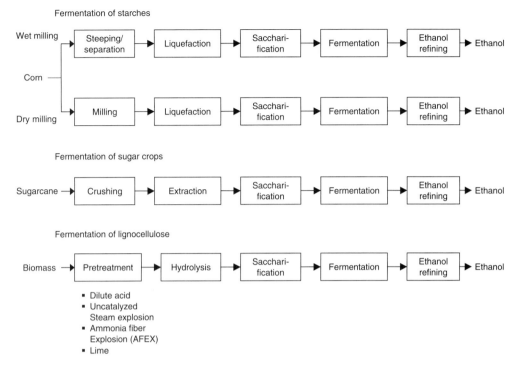

FIGURE 26.11 Industrial fermentation process for ethanol production. (Carbohydrates from sugar crops can be directly used for fermentation, whereas starch and other organic material must be converted into a fermentable form.)

26.7.2.3.2.5 Fermentation of Sugar Crops

Sugar crops include sugarcanes, sugar and fodder beets, and fruit crops. Although sugarcane is grown primarily for sucrose and molasses production, it is used as a raw material for ethanol production. It has a desirable composition for high ethanol yield. The fermentable carbohydrates from sugarcane may be directly utilized in the form of cane juice or in conjunction with a sugar factory from black strap molasses.

Cane juice extract is a green, sticky fluid, slightly more viscous than water, with an average sucrose content of 12 or 13%. It may then be evaporated to the desired concentration and used directly in the fermentation. A major disadvantage in the utilization of sugarcane juice is its lack of stability over an extended period of storage.

Black strap molasses is the noncrystallizable residue remaining after the sucrose has been crystallized from cane juice. This heavy viscous material is composed of sucrose, glucose, and fructose at a total carbohydrate concentration of 50–60%. Molasses may be easily stored for a long period of time and diluted to the required concentration before use.

26.7.2.3.2.6 Fermentation of Corn

Corn can be processed into ethanol using two main routes: dry- and wet milling. The major differences in unit operations are the initial treatment of corn (milling versus steeping) and the production of by-products (distiller's dried grains [DDGs] versus high fructose corn syrup [HFCS] and corn gluten feed). Each process has inherent advantages and disadvantages. The whole-kernel dry milling process is the simplest of the processes considered, and is generally the one recommended for new entrants into the market. Dry milling has certain advantages over wet milling:

- The process is simpler to operate than wet milling.
- Dry milling has lower capital and operating costs than wet milling.

Although dry milling produces a slate of by-products including DDGs, which are less valuable overall than the wet-milling process, it avoids the need for swing production and syrup integration, hence avoiding the need of the entrant to compete in the corn syrup and sweetener market.

Wet milling of corn is the conversion technology used when HFCS is desired as the main by-product of ethanol formation. HFCS is often used in conjunction with or as a substitute for sugar and other sweeteners in many food products, specifically soft drinks and baked goods.

The corn is not milled. It is first steeped in a solution of water and sulfur dioxide for 24–48 h. This loosens the germ and hull fibers. The germ is then removed from the kernel, and corn oil is extracted from the removed germ. The crude corn oil can then be processed in an edible oil plant.

The remaining germ oil from the corn oil extraction is combined with the hulls and fiber to produce corn gluten feed. The corn gluten feed is combined with the heavy spillage from the beer still and dried forming the corn gluten feed. The high protein fraction of the corn kernel is later separated out to produce corn gluten meal—a high-value animal feed made up of ~60% animal protein.

The remaining starch fraction is liquefied and fermented in a process similar to dry milling. In wet milling, often the clear, liquefied starch is split into two fractions: one fraction diverted to ethanol production and the other used for the production of HFCS or other sweeteners. Typically, HFCS enjoys a higher margin, and more starch is diverted to HFCS production than ethanol.

Hydrolyzed cornstarch is converted to dextrose (D-glucose), which is then partially isomerized into fructose. This moisture is further refined and concentrated for sale as HFCS.

The primary capital associated with the wet milling plant is the front end, where the corn oil, gluten feed, and gluten meal are separated out.

Based on the cost of production, both the dilute acid and enzymatic hydrolysis process of corn stover and the dry milling process of corn are equally competitive. However, capital requirement for the biomass-based process has been a major research and development (R&D) and engineering thrust of the biotechnology companies and research institutes.

Ethanol production by direct ethylene dehydration has recently become less competitive due to the high ethylene price, resulting in many shutdowns of ethylene hydration facilities. With the rising prices of petroleum crude, it is most likely that the bioroute would always be the preferred one for the production of ethanol.

26.7.2.3.3 Global Trend

In 2004, the total global ethanol production was at a historical high of 40.9 billion L (10.8 billion gal), contributed mainly by Brazil and the United States (and Canada) at 37 and 33% share of world production, respectively. On a global scale, synthetic ethanol plays a minor role, with less than 5% of overall ethanol output in 2003. More than 95% of ethanol came from agricultural crops and, given the strong interest in fuel ethanol production worldwide, this share can be expected to grow even higher. The world ethanol production is shown in Figure 26.12. Global ethanol market is expected to exceed the 120,000 million mark by the end of the year 2020, with a growth rate of above 6.5% from 2006. Contributing about half of the world bioethanol production, Brazil is the world's leading producer of the same, with the United States following next. The European Union, with a production of about 0.5 million tons, is estimated to be accountable for ~10% of the total bioethanol in the world.

26.7.2.4 Alkanolamines

26.7.2.4.1 Introduction

Alkanolamines are versatile reagents that can be used as starting points for the synthesis of a number of industrial chemicals, including a number of important heterocyclic compounds. The most important of these are the reaction products with long-chain fatty acids to produce neutral alkanolamine soaps used as emulsifying agents in foods, agricultural sprays, cleansers, cosmetics, and pharmaceuticals.

Production of Solvents for Detergent Industry

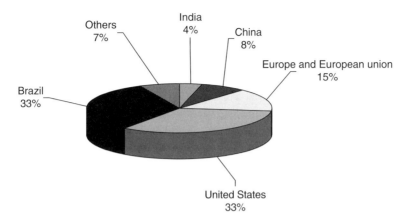

FIGURE 26.12 World ethanol production.

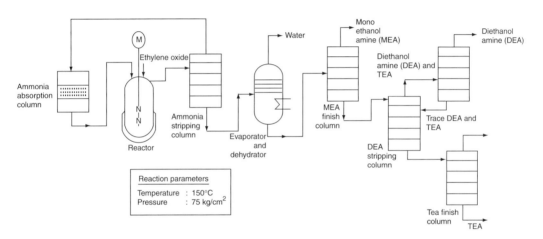

FIGURE 26.13 Flow diagram for the production of ethanolamines. (Ammonia concentration between 50 and 100% is fed into the reactor. The reaction takes place in the reactor where the pressure is maintained at 160 kg/cm² and the reaction temperature at 150°C. The reaction is highly exothermic.)

26.7.2.4.2 Production
26.7.2.4.2.1 Alkanolamines from Olefin Oxides and Ammonia
Alkanolamines are prepared from ammonia and EO, PO, or butylene oxide. The second series, the isopropanolamines, were first produced on a commercial scale in the United States in 1937. The three products obtained from the reaction of ammonia and PO are mono-, di-, and tri-isopropanolamine, $(CH_3CHOHCH_2)n\ NH_{3-n}$, where, $n = 1, 2$, or 3 (Figure 26.13).

The reactions are exothermic and are usually carried out at 50–100°C. The product ratio is generally controlled by the ratio of ammonia to alkylene oxide. A high ammonia-to-alkylene oxide ratio is used when mono- and dialkanolamines are desired. A recycle technique can be used when di- and triethanolamine are desired; excess monoalkanolamine is added to suppress its further formation. This product ratio flexibility is of great value when demands shift from one product to another. Some of the more commercially important substituted alkanolamines made this way are aminoethyl ethanolamine (from ethylenediamine and EO), dimethylethanolamine (from dimethylamine and EO), and diethylethanolamine (from diethylenamine and EO).

26.7.2.4.2.2 Uses

Surfactants: Ethanolamines are used widely as intermediates in the production of surfactants, which have become commercially important as detergents, textile and leather chemicals, and emulsifiers. Their uses range from drilling and cutting oils to medicinal soaps and high-quality toiletries.

They are, moreover, noncorrosive and can be used on virtually all textiles without damage. Ethanolamine soaps prepared from oleic-, stearic-, lauric-, or caprylic acid are constituents of many toiletries and medicinal soaps.

Ethanolamine soaps produced from fatty acids are among the most industrially important emulsifiers. They are used in cosmetics, polishes, shoe creams, car care products, drilling and cutting oils, and pharmaceutical ointments. Ethanolamine soaps combined with wax and resins are used as impregnating materials, protective coatings, and products for the care of textile and leather goods.

Ethanolamine soaps obtained from alkylarylsulfonic acids, preferably alkylbenzene sulfonic acids, or from alcohol sulfates are growing steadily in importance and dominate the market for household cleansers. The use of linear alkyl groups instead of branched chains in these products has resulted in greater biodegradability.

Fatty acid ethanolamides and products obtained by further ethoxylation have foam-stabilizing ability and are, therefore, important additives for detergents. Diethanolamides obtained from coconut fatty acids and oleic acids are used industrially.

In the manufacture of leather, ethanolamine-based chemicals are used for dressing, dyeing, and finishing. For paints and coatings, ethanolamines are employed both in production and in softeners and paint removers.

Large-scale economical production of ethanolamines from EO and ammonia in large, single-line plants has greatly promoted their use in many industrial sectors. About 50% of world production is monoethanolamine; 30–35%, diethanolamine; and 15–20%, triethanolamine. Alkanolamines are the potential solvents for detergent formulations with unique properties.

26.7.2.5 Glycols

Ethylene glycol and propylene glycol (PG) are the two most important glycols finding application in the detergent industry.

26.7.2.5.1 Ethylene glycol
26.7.2.5.1.1 Introduction
Ethylene glycol, 1,2-ethanediol, $HOCH_2 CH_2OH$, $M_w = 62.07$, usually called glycol, is the simplest diol. Previously, treatment of 1,2-dibromoethane with silver acetate yielded ethylene glycol diacetate, which was then hydrolyzed to ethylene glycol. Today, the complete production of ethylene glycol is based on the process technology of hydrolysis of EO. The worldwide capacity for the production of ethylene glycol via the hydrolysis of EO is estimated to be ca. 7×10^6 t/year.

26.7.2.5.1.2 Production
Ethylene oxide hydrolysis (current production method). Only one method is currently used for the industrial production of ethylene glycol. This method is based on the hydrolysis of EO, which is obtained by direct oxidation of ethylene with air or oxygen. Figure 26.14a shows a simplified scheme of a plant producing ethylene glycol. The EO is thermally hydrolyzed to ethylene glycol without a catalyst. The EO–water mixture is preheated to 200°C, whereby the EO is converted to ethylene glycol. Di-, tri-, tetra-, and polyethylene glycols are also produced, but with respectively decreasing yields. In practice, almost 90% of the EO can be converted to monoethylene glycol, the remaining

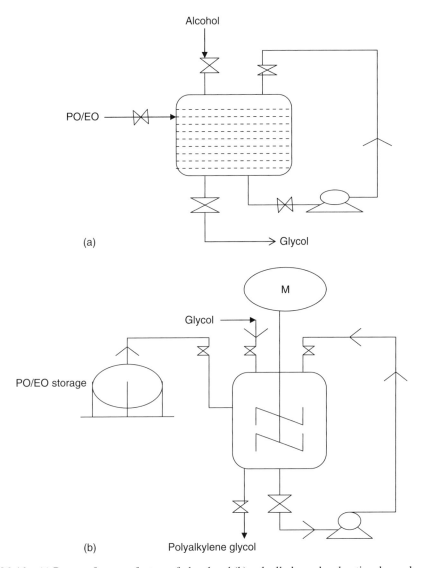

FIGURE 26.14 (a) Process for manufacture of glycol and (b) polyalkylene glycols: stirred-cum-loop reactor.

10% reacts to form higher homologues. The product mixture is purified by passing it through successive distillation columns with decreasing pressures. Water is first removed and returned to the reactor, the mono-, di-, and triethylene glycols are then separated by vacuum distillation. The yield of tetraethylene glycol is too low to warrant separate isolation. The heat liberated in the reactor is used to heat the distillation columns. A side stream must be provided to prevent the accumulation of secondary products, especially small amounts of aldehydes, which are produced during hydrolysis. The shape of the reactor affects the selectivity of the reaction. Plug-flow reactors are superior to both agitator-stirred tanks and column reactors.

The glycol production method is simple, but has some major drawbacks:

1. The selectivity of the first step—the production of EO—is low (80%).
2. The selectivity of EO hydrolysis is low—ca. 10% is converted to di- and triethylene glycol.
3. Energy consumption for the distillation of the large amount of excess water is high.

Therefore, much research has been carried out to improve this process. The search for a better silver catalyst is an objective for point 1 (EO). Points 2 and 3 must be considered together, as higher selectivity for EO hydrolysis automatically reduces the excess of water required.

Many catalysts have been described in the literature that are able to optimize selectivity or lower the reaction temperature and the required excess of water. Acids and bases are known to accelerate the reaction rate. Although the use of catalysts allowed the reaction temperature to be lowered, selectivity was not significantly enhanced. Furthermore, the catalyst needed to be separated and either fed back into the reaction mixture or replaced. As a result of these disadvantages, these types of catalysis have not proved to be of commercial use.

Problems such as product separation and catalyst feedback still need to be resolved, but this method for the selective synthesis of ethylene glycol from EO seems to be the most promising for industrial-scale application.

For production of polyethylene glycols, the process technology of alkoxylation involves the reaction between ethylene glycol and EO. The latest reactor design is such that the reactants circulate in a loop besides being stirred by the agitator at the same time. The stirred-cum-loop reactors (Figure 26.14b) not only help reduce the dissolved-free EO but it also ensures the narrow particle size distribution.

26.7.2.5.1.3 Uses
Ethylene glycol is used mainly as an antifreeze in automobile radiators and as a raw material for the manufacture of polyester fibers.

Because of the toxic nature of ethylene glycol as well as diethylene glycol, they are generally restricted. The higher glycols above the molecular weight of 200 are the preferred solvents for various applications, mainly in detergents. The new generation specialty detergents, mainly the liquids based on the nonaqueous medium are designed using polyethylene glycols.

26.7.2.5.2 Propylene Glycol
26.7.2.5.2.1 Introduction
1,2-Propylene glycol is a clear, viscous, colorless liquid that is practically odorless and has a slight characteristic taste. Although more volatile than ethylene glycol, PG is about three times as viscous at room temperature. It has a very low order of toxicity and, hence, becomes the most preferred solvent for detergents.

26.7.2.5.2.2 Production
It was first produced by the hydrolysis of PO. Today, the hydrolysis is carried out under pressure and at high temperature without a catalyst.

The proportion of products is controlled by the molar ratio of water to PO. Higher hydrolysis ratios increase not only the yield of PG but also the cost of purification. A ratio of 15 provides a product mix of 85% PG, 13% dipropylene glycol, and 1.5% higher adducts.

Although PG has a secondary hydroxyl group, its chemistry parallels that of ethylene glycol.

All commercial production of PG is by noncatalytic hydrolysis of PO carried out under high pressure and high temperature. A large excess of water is used in the conversion of PO into a mixture of mono-, di-, and tripropylene glycols. Typical product distribution is 90% PG and 10% coproducts. Hydration reactor conditions are 120–190°C at pressures up to 2170 kPA. After the hydration reaction is completed, excess water is removed in multieffect evaporators and drying towers, and the glycols are purified by high vacuum distillation.

26.7.2.5.2.3 Derivatives
The derivatives of PG are prepared by methods analogous to those for ethylene glycol derivatives. The base-catalyzed reaction of PO with alcohols gives predominantly primary monoalkyl ether,

Production of Solvents for Detergent Industry

$CH_3CHOHCH_2OR$, and small amounts of secondary ether, $CH_3CHORCH_2OH$. Acid catalysis increases the ratio of secondary to primary ethers. Monoalkyl ethers can also be prepared without catalysts under the proper conditions of temperature and pressure.

The monoalkyl ethers of PG are excellent solvents for a wide variety of organic materials. The lower alkyl ethers are completely miscible with water at room temperature and are soluble in some hydrocarbons.

The ethers and esters of PG are prepared from the glycol by conventional methods. PG yields isomeric mixtures of esters because it contains both primary and secondary hydroxyl groups. Monoesterification occurs usually at the primary hydroxyl group. Fatty acid esters, such as the dioleate and the monohydroxystearate, are used in ointments, drug creams, cosmetics, and surfactants.

26.7.2.5.2.4 Polypropylene glycols
Polyether polyols. The formation of polyether polyols is commercially the most important reaction of PO. A polyol is the product of reaction of an epoxide and compounds or initiators (e.g., glycols amines, acids, or water), which contain active hydrogen.

Poly(propylene glycol), the simplest PO-based polyol, is prepared by the base-catalyzed polymerization of PO with PG as the initiator. Such a polyol is commonly known as a polyol diol; a polyol triol results from the polymerization of PO initiated with glycerol. Other polyol triols may be obtained by initiating the reaction with trimethylolpropane, triethanolamine, and hexanetriols.

Polyols with a large number of terminal hydroxyl groups result when the reaction is initiated with compounds such as pentaerythritol, 2,2,6,6-tetrakis(hydroxymethyl) cyclohexanol, or with natural products, for example, sucrose, raffinose, or D-mannitol. Amines such as ethylenediamine, diethylenetriamine, and 2,4-diaminotoluene may also be used as the active hydrogen-containing initiator.

PO also reacts with active hydroxyl hydrogen derived from the ring opening of other compounds such as EO and tetrahydrofuran; thus, a copolymer polyol is obtained. Typically, polyols are obtained from base-catalyzed reactions with aqueous ammonia, sodium or potassium hydroxide, or lower alkyl tertiary amines such as trimethyl- and triethylamine. The reaction of PO with tetrahydrofuran is catalyzed by boron trifluoride etherate. The molecular weights of polyols prepared according to the reactions described earlier range from 200 to 7000.

High polymers. Poly(propylene oxide) polymers with molecular weights of 100,000 or more can be prepared with a catalyst that consists of $FeCl_3$ and approximately five equivalents of PO. The addition of small amounts of toluene 2,4- and 2,6-diisocyanates greatly increases the molecular weights of the polymers obtained. PO homopolymers can also be prepared with catalysts such as diethyl zinc and trialkyl aluminum compound.

The reaction, or hydration, of PO with water to produce PG is utilized commercially. Some dipropylene glycol, tripropylene glycol, and higher-order PGs are also produced. As the molar ratio of PO to water is increased, the proportion of higher-molecular-weight glycols in the product is increased. Usually, about 15–20 mol water per mole epoxide is used in the production of PG. The reaction is catalyzed by acids and bases; however, in the commercial processes, heat and pressure are applied without catalyst.

PO reacts with carbon dioxide to yield propylene carbonate, which, in turn, can be hydrolyzed to PG. The reaction is catalyzed by potassium iodide, tetraalkyl ammonium bromides, calcium bromide, or magnesium bromide.

Carboxylic acids and PO give a mixture that contains the monoesters of the primary and the secondary alcohol groups of PG. The monoesters may then react with additional acid to form the glycol diester. With a sufficient concentration of epoxide, the monoester may add PO to yield the ester of dipropylene glycol, higher poly(propylene glycols) and their esters, and the glycol diester. The esterification is catalyzed with sodium or potassium hydroxide, and anhydrous chromium

(III) tricarboxylate salts. Cyclic carboxylic acid anhydrides, for example, phthalic anhydride, give polyesters with PO. The reaction is catalyzed by diethyl zinc, lithium chloride, tertiary amines, and quaternary ammonium halides.

Alcohols or phenols and PO give monoethers of PG. These glycol ethers may then react further to produce di-, tri-, and poly(propylene glycol) ethers. As the ratio of alcohol to epoxide in the reaction mixture is increased, the molecular weight of the product tends to decrease, that is, the yield of PG ether increases relative to that of the di-, tri-, and poly(propylene glycol) ethers. A basic catalyst favors formation of secondary alcohols, whereas an acidic catalyst leads to a mixture of the primary and secondary alcohols.

26.7.2.5.2.5 Environmental Considerations

The glycols vary in biodegradability. The tests conducted with standard municipal inoculum, and other studies have shown that biodegradability can be greatly enhanced when using acclimated bacteria. For example, tripropylene glycol has shown 66% of theoretical oxygen demand at 20 days with an industrial seed. Thus, it is expected that all of the PGs will exhibit moderate to high biodegradability in a natural environment. The mono- and diethylene glycols are not considered safe.

26.7.2.5.2.6 Economic Aspects

World capacity for PGs in mid-2000 was 1.6×10^6 t. Production was concentrated in the United States (43%) and Western Europe (36%). Japan (5%) and the Republic of Korea (3%) constituted the next group. The remaining capacity was located in Mexico, India, Eastern Europe, Singapore, China, and Australia.

PG's total growth rate should drop from 5% to 2–2.5% per year. PG's growth in cosmetics and liquid detergents continues strongly, growing at the rate of 3–3.5%. The aircraft deicing market has matured. Unsaturated polyesters resins, polyethylene glycol's largest application, is eroding because of competition from dicyclopentadiene-based resin.

26.7.2.5.2.7 Global Trend

The overall glycol market approaches 15–16 million tons/year with an annual growth rate of 3–5% as the materials tend to be leading indicators of growth. Industrial market for PG is 1.2–1.4 million tons/year, and for ethylene glycol it is 12–13 million tons/year.

26.7.2.5.2.8 Uses

PG has a variety of applications. In the food industry, it is used as a solvent, humectant, and preservative in the manufacture of products that come in contact with food such as plasticizers for food wraps, as a solvent for food processing, and as a lubricant for food machinery. Glycols can be the preferred solvent for the cleaning formulations.

26.7.2.6 Glycol Ethers and Esters

26.7.2.6.1 Introduction

Most commonly known glycol ethers are a group of solvents based on alkyl ethers of ethylene glycol, also called Cellosolve. These solvents typically have higher boiling point, together with the favorable solvent properties of lower-molecular-weight ethers and alcohols. Acetates of glycols are also an important type of potent solvents.

Glycol ether solvents include

Monoalkyl ethylene glycol ethers:
- Ethylene glycol monomethyl ether (2-methoxyethanol): $CH_3OCH_2CH_2OH$
- Ethylene glycol monoethyl ether (2-ethoxyethanol): $CH_3CH_2OCH_2CH_2OH$
- Ethylene glycol monopropyl ether (2-propoxyethanol): $CH_3CH_2CH_2OCH_2CH_2OH$

Production of Solvents for Detergent Industry

- Ethylene glycol monoisopropyl ether (2-isopropoxyethanol): $(CH_3)_2CHOCH_2CH_2OH$
- Ethylene glycol monobutyl ether (2-butoxyethanol): $CH_3CH_2CH_2CH_2OCH_2CH_2OH$
- Ethylene glycol monophenyl ether (2-phenoxyethanol): $C_6H_5OCH_2CH_2OH$
- Ethylene glycol monobenzyl ether (2-benzyloxyethanol): $C_6H_5CH_2OCH_2CH_2OH$
- Diethylene glycol monoethyl ether (2-(2-ethoxyethoxy)ethanol, carbitol cellosolve): $CH_3CH_2OCH_2CH_2OCH_2CH_2OH$
- Diethylene glycol mono-n-butyl ether (2-(2-butoxyethoxy)ethanol): $CH_3CH_2CH_2CH_2OCH_2CH_2OCH_2CH_2OH$

Dialkyl ethylene glycol ethers:

- Ethylene glycol dimethyl ether (dimethoxyethane): $CH_3OCH_2CH_2OCH_3$
- Ethylene glycol diethyl ether (diethoxyethane): $CH_3CH_2OCH_2CH_2OCH_2CH_3$
- Ethylene glycol dibutyl ether (dibutoxyethane): $CH_3CH_2CH_2CH_2OCH_2CH_2OCH_2CH_2CH_2CH_3$

Esters:

- Ethylene glycol methyl ether acetate (2-methoxyethyl acetate): $CH_3OCH_2CH_2OCOCH_3$
- Ethylene glycol monoethyl ether acetate (2-ethoxyethyl acetate): $CH_3CH_2OCH_2CH_2OCOCH_3$
- Ethylene glycol monobutyl ether acetate (2-butoxyethyl acetate): $CH_3CH_2CH_2CH_2OCH_2CH_2OCOCH_3$

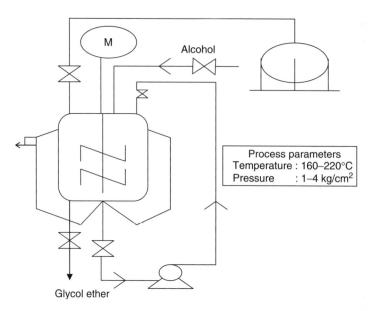

FIGURE 26.15 Process for manufacture of glycol ether: stirred-cum-loop reactor.

26.7.2.6.2 Production

Glycol ethers, consisting of a series of over 30 ethylene glycol (E-series) and PG (P-series) derivatives, are being produced by reacting an alkylene oxide (EO for E-series and PO for P-series) with an alcohol (Figure 26.15). During the reaction, monoglycol, diglycol, triglycol, and higher glycol ethers are produced. The reaction of an alkylene oxide with an alcohol may be carried out in liquid or vapor phases. The former is preferred and requires sufficient pressure to maintain the reactants in the liquid state.

Generally, the reaction proceeds with the opening of the epoxide ring leading to form an addition product with the alcohol as follows:

Reactions for ethylene and propylene ether

Production of Solvents for Detergent Industry

A homogeneous catalyst such as an alkali metal hydroxide or an alkali metal alcoholate is generally used. Anion exchange catalysts or acidic catalysts can also be used.

Although typically a tenfold molar excess of alcohol is used, the molar alkylene oxide selectivity to the monoglycol ether is ~75–85 %. Diglycol, triglycol, and higher glycols are formed as by-products.

PG ethers are of increasing interest, as the corresponding ethylene glycol ethers are reported to be more toxic and volatile. This has motivated a shift toward using PG analogues from ethylene glycol ethers. Glymes can be produced by a variety of methods as discussed in the following section. The most commonly used method for commercially producing glymes (Figure 26.15) involves the cleavage of epoxides in the presence of a low-molecular-weight ether and a Lewis acid catalyst. This reaction enables the insertion of oxacycloalkanes into chain-type ethers. One of the principle disadvantages of the cleavage reaction of epoxides is that it is not particularly selective. Insertion of a specific number of oxacycloalkane units is difficult to control. Therefore, the final product consists of a mixture of glymes. It is necessary to separate the reaction mixture by complex distillation techniques or other means for obtaining pure glymes, which adds time and complexity to the manufacturing process. Despite all these disadvantages, all the glycol ethers available today in the world are being manufactured by this process.

Glymes are synthesized using Williamson synthesis reaction that involves the cleavage of epoxides. In the Williamson synthesis, a monoalkyl polyalkylene glycol is treated with a base or an alkali metal, typically molten sodium, to form an alkoxide ion, which is then reacted with an alkyl halide such as methyl chloride to form the glyme. The by-products from the Williamson synthesis are hydrogen gas and salt.

Although the Williamson synthesis is one of the conventional methods of producing glymes on a commercial scale, the process presents several disadvantages: (a) use of costly and potentially hazardous starting materials, (b) generation of hydrogen gas as a by-product, (c) generation of salt as a major by-product, and (d) slow rate of reaction.

Another method of producing glymes that overcomes the disadvantages of both the Williamson synthesis and the cleavage reaction of epoxides comprises contacting a glycol with a monohydric alcohol in the presence of a polyperfluorosulfonic acid resin catalyst under conditions that are effective for the production of the glyme. The starting materials used in the method are commercially readily available, not expensive or hazardous, and the catalyst used in the reaction can be recovered, regenerated, and reused. It does not generate hydrogen gas or salt, proceeds at a relatively rapid rate, and produces a single glyme.

Preferably, a molar excess of the monohydric alcohol is used in the reaction. Typically, a molar excess of about 3–about 5 mol of monohydric alcohol is used in the reaction for every mole of glycol.

The glycol, the monohydric alcohol, and the polyperfluorosulfonic acid resin catalyst are combined in a suitable reactor vessel under agitation and heated. Owing to the high vapor pressures of the reactants and the products formed during the reaction, the reactor vessel must be capable of handling pressures as high as 75 kg/cm^2. A conventional autoclave is a preferred reactor vessel for use in accordance with the invention.

An elevated reaction temperature allows the catalyst to partially dissolve in the reaction mixture, thus providing semihomogeneous catalysis conditions. For the production of glymes, a reaction temperature in the range 100°C–300°C, preferably between 160°C and 220°C is suitable. Once the reaction has been completed, the reactor contents are cooled to ambient temperatures and the reactor contents are separated by distillation. Section 26.7.2.6.2, in addition to the formation of the desired products such as ethyl glyme, ethyl diglyme, triglyme, butyl diglyme, and tetraglyme, the method also produces a quantity of the corresponding intermediate monoalkyl ether. These materials can easily be separated and recovered by conventional distillation to be recycled in the process to achieve further conversion. In the case of diglyme production, 1,4-dioxane is formed as a coproduct.

By-products from the reaction include water and dialkyl ethers, which are also easily separated and recovered by distillation. Dialkyl ethers such as dimethylether are high-value products, and can be used in a variety of applications. The polyperfluorosulfonic acid resin catalyst can also be recovered in concentrated form in the glycol bottoms for reuse in accordance with known methods. Once the activity of the polyperfluorosulfonic acid declines, the material can be regenerated by treatment with a strong mineral acid (i.e., nitric acid) to restore the proton sites on the resin.

The starting materials used in the reaction, being glycols and monohydric alcohols, are generally inexpensive and readily available. These starting materials do not present significant toxicity and handling problems, especially when compared to the starting materials used in the Williamson synthesis. Here, it may be noted that although this method presents several advantages, it is generally preferred only for the production of alkyl ethers. Owing to the fact that the main raw material for this process is glycol, which is produced from alkylene oxide, it is always cost effective and logical to produce the monoalkyl glycol ethers using alkylene oxide route.

26.7.2.6.3 Uses

Because of their physical properties, the largest end use of E-series glycol ethers is as solvents, as shown in the Figure 26.16. The next largest volume is for derivative products such as the glycol ether acetates, which are also used principally as solvents. Monoglycol ethers are also used as intermediates to produce diglycol and triglycol ethers.

Brake fluids are the third largest application of the E-series. The high boiling diethylene glycol and, principally, triethylene glycol and higher ethers are used in this application. The last significant application is in the jet fuel deicers.

The market for P-series glycol ethers is also focused on solvent uses. End uses are similar to those for the E-series solvents, however, with a greater focus on use in cleaners. Production of intermediates accounts for most of the remaining demand. The PG ethers are used as intermediates almost exclusively in glycol ether acetate production. The acetates are being considered as environment-friendly solvents.

Glycol ethers, which are also commonly known as glymes, are used as aprotic solvents in a variety of applications.

26.7.2.7 Green Solvents

Efficient cleaning methods require the use of solvents along with the detergents; thus, safety and consideration of the environment are of prime importance when it comes to the selection of solvents for this purpose. There are significant regulation reforms occurring with all the products being used especially for household applications due to the safety and health issues related to the uses on the one hand and protection of environment and ecology on their disposal on the other.

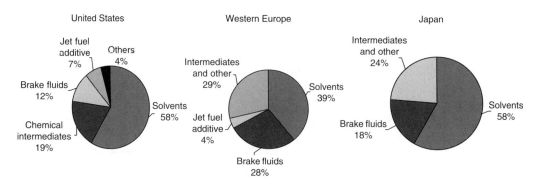

FIGURE 26.16 Principal applications of E-series glycol ethers, 2001.

Production of Solvents for Detergent Industry

Toxicity, flammability, material incompatibility, and biodegradability are all important safety factors, which must be considered in choosing the proper cleaning or surface treating systems. Environmental factors that must be considered are volatile emissions or volatile organic compounds (VOC) and waste handling, storage, and disposal. The formulator or end user has limited choices: either to use newer solvents with minimum VOCs or to reduce the use of solvents if it is impossible to eliminate them completely by adopting more efficient cleaning processes.

The environment-friendly solvents being developed in recent times are also called as *green solvents*.

Green solvents or biosolvents is the term used for solvents that are derived from the processing of agricultural crops. Such solvents are safer and more environment friendly.

The use of petrochemical solvents has been the key to the majority of chemical processes but not without severe implications on the environment. The Montreal Protocol identified the need to reevaluate all chemical processes with regard to the use of VOCs and the impact these VOCs may have on the environment. Green solvents are being developed as a more environmental-friendly alternative to petrochemical solvents.

Ethyl lactate is a green solvent derived from processing corn. Ethyl lactate is the ester of lactic acid. Lactate esters solvents are commonly used solvents in the paints and coatings industry and have numerous attractive advantages including being 100% biodegradable, easy to recycle, noncorrosive, noncarcinogenic, and nonozone depleting.

Ethyl lactate is a particularly attractive solvent for the coatings industry as a result of its high solvency power, high boiling point, low vapor pressure, and low surface tension. Other applications of ethyl lactate include being an excellent cleaner for the polyurethane industry. Ethyl lactate has a high solvency power, which means it has the ability to dissolve a wide range of polyurethane resins. The excellent cleaning power of ethyl lactate also means it can be used to clean a variety of metal surfaces, efficiently removing greases, oils, adhesives, and solid fuels. The use of ethyl lactate is highly variable as it has eliminated the use of chlorinated solvents.

Figure 26.17 shows the synthetic route to products made from ethyl lactate.

Unlike other solvents, which can damage the ozone layer or pollute groundwater, ethyl lactate is so benign that the U.S. Food and Drug Administration approved its use even in food products

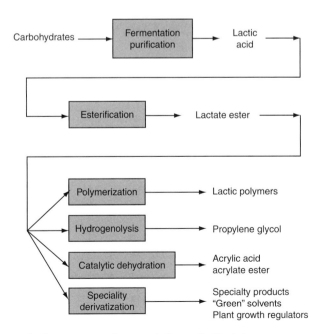

FIGURE 26.17 The synthetic route to products made from ethyl lactate.

long ago. Although ethyl lactate has been around for years, the cost of producing it has been too high to allow it to compete economically with lower-priced chemical solvents. Ethyl lactate normally sells at a much higher price than the conventional solvents. The efforts have been on to develop a new process technology that can lower the market price of such solvents considerably and make them cost competitive with the conventional solvents.

There are basically two important challenges in this regard: (a) process for manufacturing lactic acid with greater efficiency and (b) process that can convert the lactic acid to lactate with greater efficiency. For the second challenge, the key technology that will significantly reduce the market price is a new patented purification-separation system. It uses a membrane to selectively remove excess water and bring the reaction to nearly 100% completion with higher purity and lower processing costs.

A recent study analyzes the U.S. solvents industry, presenting historical data for 1992, 1997, and 2002 and forecasts to 2007 and 2012 by product (e.g., alcohols, hydrocarbons, ethers, ketones, esters, chlorinated solvents, PG, terpenes, butanediol, vegetable oils, tetrahydrofuran, and hydrogen peroxide); by function (e.g., vehicle/carrier/thinner, antifreeze and deicers, cleaners, and extraction agents); and by market. The market for green solvents would grow at rates faster than ever before the gainers would be the environment-friendly solvents, especially for applications in detergents.

26.7.2.8 Ionic Liquids

Recently, however, a new class of solvents has emerged—ionic liquids. These solvents are often fluid at room temperature, and consist entirely of ionic species. They have many fascinating properties, which make them of fundamental interest to all chemists as both the thermodynamics and kinetics of reactions carried out in ionic liquids are different from those in conventional molecular solvents. As they are made up of at least two components, which can be varied (the anion and cation), the solvents can be designed with a particular end use in mind, or to possess a particular set of properties. Hence, the term *designer solvents* has come into common use. Ionic liquids come in two main categories, namely, simple salts (made of a single anion and cation) and binary ionic liquids (salts where an equilibrium is involved). For example, $[EtNH_3][NO_3]$ is a simple salt, whereas mixtures of aluminum (III) chloride and 1,3-dialkylimidazolium chlorides (a binary ionic liquid system) contain several different ionic species, and their melting point and properties depend on the mole fractions of the aluminum(III) chloride and 1,3-dialkylimidazolium chloride present.

Ionic liquids can be considered as the potential solvents meeting the requirements of environment protection and safety for their usage in detergents.

26.8 CONCLUSIONS

Detergents are surface-active agents, having preference for the interface. The preferential adsorption at the interface at the lower concentration and formation of aggregates (micelles) in the bulk at the higher concentration are the two unique features of detergents. The structural attributes of detergents are responsible for their wide applications in different fields. Although detergents are often considered as synonyms of surfactants, the fact is that the surfactants used for cleansing different surfaces are classified as detergents. The performance of detergents is determined by their ability to form micelles. Generally, the assessment of quality as well as benchmarking of detergents is done by studying the behavior of detergents in water.

The most important factor that determines the behavior of detergents in water is the hydrophobic interactions. The hydrophobic part of detergents is basically the starting material and the hydrophilic portion is attached to obtain the detergents. Detergents are classified depending on the nature of the hydrophilic portion; ionic hydrophile means ionic detergent and nonionic hydrophile means nonionic detergents.

Production of Solvents for Detergent Industry

There is another category, that is, zwitterionic detergent, which exhibits both the anionic as well as cationic behavior in solution.

For a given application, detergents can be designed by choosing an appropriate hydrophobe and attaching to it a hydrophile of desired type, size, and shape.

In fact, it is the balance between the hydrophobe and hydrophile that plays the key role in deciding about the application of a detergent.

The hydrophobic portion of the detergent is almost always a long alkyl, aryl, or alkyl aryl chain with functional groups that can be derivatized by attaching hydrophile groups to obtain detergents. The anionic detergents are mostly sulfates or sulfonates and phosphates obtained by esterification with acids like sulfuric- and phosphoric acid or their anhydrides, respectively. The cationics are mostly quaternary ammonium compounds made from amines or amino derivatives. The nonionic detergents are mostly the EO derivatives with varying moles of EO with a fuel range beginning with oil- (low EO content) to water-soluble (high EO content) detergents. There is another category of nonionic detergents based on ester groups and they are generally of oil-soluble type. The sulfonate esters these days are the new trend in the detergent industry with better performance.

The performance of detergents can be modified and enhanced by the incorporation of solvents. Since solvents have the capacity to disintegrate and dissolve solutes, they are useful in dissolving or thereby removing soil or dirt from the surfaces. Solvents alone cannot work as the surface-active agents but they certainly can in combination with surfactants. One needs to understand the mechanism by which the synergism is achieved to design the detergent formulations for a given application.

With the increasing awareness about the health safety as well as about the environment and ecology, the market trends show that the biodegradable and environmentally safe products would be preferred in future. This would demand for the solvents that are not only safe but are also derived from the renewable resources. Even the chain starters (hydrophobes) for detergents would have to be based on the renewable resources. As a result, for example, the fatty alcohols derived from the vegetable oils would be preferred rather than the ones derived from the petroleum resources.

The trends in detergent industry, world over, clearly show that the quality and grades of detergents in demand in a country depend largely on the economic situation in that country. Similarly, the type of detergents being used gives the indication about the state of mechanization and automation in a country. In the developed world, the detergent formulations consist of a high content of the active component, that is, surfactants with the significant amount of builders but very little amount of fillers. Here, the emphasis is on the products meant for cleaning at all temperatures with minimum energy required besides the minimum demand for water. Therefore, the products in developed countries are foamless and highly efficient. In contrast, the products being used in developing as well as underdeveloped countries are composed of the sizeable quantity of fillers and much less quantity of detergent actives. The emphasis is more on the foaming behavior of the detergents. This has to change with time and with the changing times, the formulators will require, new types of detergents with high efficiency but with low cost. With a better understanding of the behavior of detergents in solution, it would not be a difficult task for the scientists to design such detergent actives in times to come.

ACKNOWLEDGMENTS

The authors wish to express their sincere thanks to the team of Meenu Kapoor, Gouri Shanker Jha, Mukti Tyagi, Gunjan Suri, Pranshu Chhabra, Neeraj Verma, R.K. Bhatt, and Surinder Negi for their assistance in preparing the manuscript.

27 Production of Proteases and Other Detergent Enzymes

T. T. Hansen, H. Jørgensen, and M. Bundgaard-Nielsen

CONTENTS

27.1 Introduction 531
27.2 Fermentation 532
    27.2.1 Production Strains.. 532
    27.2.2 Process Conditions... 533
    27.2.3 Equipment.. 534
    27.2.4 Mode of Operation... 534
    27.2.5 Raw Materials.. 534
27.3 Recovery.. 536
    27.3.1 Recovery Process Design... 536
        27.3.1.1 Influence of the Production Organism 536
        27.3.1.2 Enzyme Characteristics 537
        27.3.1.3 Demand for Product Quality.. 537
        27.3.1.4 Type of Product 537
        27.3.1.5 The Environmental Impact of the Process............................... 537
    27.3.2 Recovery Process Unit Operations 538
        27.3.2.1 Pretreatment... 538
        27.3.2.2 Primary Separation... 539
        27.3.2.3 Concentration... 539
        27.3.2.4 Purification 540
27.4 Formulation 541
    27.4.1 History of Formulation of Detergent Enzymes.. 541
    27.4.2 Liquid Formulations of Enzymes 542
        27.4.2.1 Enzymatic Stability.. 542
        27.4.2.2 Physical Stability 543
        27.4.2.3 Microbial Stability... 543
    27.4.3 Solid Formulation of Enzymes 543
        27.4.3.1 High-Shear Granulation.. 543
        27.4.3.2 Fluidized Bed Spray Coating... 544
        27.4.3.3 Other Granulation Techniques... 545
References.. 545

27.1 INTRODUCTION

It is a little over 40 years ago that the modern use of enzymes in the detergent industry started; and since then, production of enzymes has undergone a dramatic change. It has evolved from being a small, curious spin-off from the pharmaceutical or chemical industry to the present large volume, efficient production process. Production of enzymes is an industrial biotechnology in large scale, and it has profited from all the new technologies discovered during this period. The most important of these technologies has probably been the use of genetic engineering or recombinant technology, which allows constructions of very efficient production strains that can produce the desired enzymes in high purity. The development has also been impelled by a considerable pressure from the detergent industry itself for getting better, more robust, and more cost-effective enzymes.

The enzymes, which have been most commonly used in the detergent industry are proteases, amylases, cellulases, and lipases. Recently, other types of enzymes such as pectate lyases and oxidoreductases have been introduced, either in detergents or related applications. In principle, the same production method apply to all the enzymes mentioned earlier, but special methods can be required when producing proteases, as they will start self-digestion (hydrolysis of the enzyme protein itself) to a certain degree, depending on the conditions. There is also a risk that unwanted protease side activity will decrease the yield of the other enzymes. Production of the other types of enzymes follows the same type of production methods as with proteases. However, depending on the type of enzyme class, special precautions might be necessary. For example, amylases have in general a higher tendency to spontaneously form enzyme crystals making the downstream recovery and purification process more complicated. Another example is fermentation of oxidoreductases, where extra addition of transition metals (iron or copper) is necessary as these are an essential part of the active site of these enzymes, making the fermentation media more complex. In spite of these differences, the basic principles of production for all the mentioned type of enzymes are the same. The production of enzymes can be divided into three subprocesses: Fermentation, recovery, and formulation of the final enzyme product. Each area is described in more detail in the following sections.

27.2 FERMENTATION

Today, the majority of enzymes used in the detergent industry are produced by cultivation of selected microorganisms in large submerse fermentation tanks. The production process and production strains have steadily been developed over the past 50 years and nowadays the fermentation of enzymes is based on the cutting-edge techniques within modern cell biology.

27.2.1 Production Strains

Initially, all new enzymes used in the detergent industry were found in microorganisms growing under extreme conditions, such as high pH and salinity, since it was believed that this kind of environment promoted growth of microorganisms that produced enzymes with the robustness needed for detergent applications. The main objective for the early screenings was to find enzymes with improved washing performance and at the same time it was required that the new product was produced by organisms that could secrete enzymes in high concentrations. One of the first microbial proteases on the detergent market to fulfill both these criteria was an alkaline protease from *Bacillus licheniformis*, Alcalase®, which was launched more than 40 years ago. This product is still in the market, but during the past 40 years the production strain has been considerably improved and the strain and fermentation process used today produces a significantly higher yield than the ones used in the 1960s.

With the development of the technology for making recombinant production strains, a variety of new possibilities for making better strains and enzymes have opened up. It is now possible to [1]

- Move the gene coding for an interesting enzyme from a wild-type organism to a high-yielding production strain (heterolog expression)
- Multiply the number of gene coding for the desired enzyme and thereby increase the productivity of the strain (homolog expression)

- Change individual amino acids in the enzyme and thereby improve the physical properties or performance of the enzyme (protein engineering)

All of the preceding techniques are extensively used for the production of detergent enzymes such that today the vast majority of industrial enzymes are produced by recombinant techniques. This is done in a limited number of optimized, well-known production hosts. Some of the most frequently used host organisms are the *Bacillus* species—*B. subtilis*, *B. licheniformis*, and *B. clausii*, which are used for production of proteases and amylases. The lipases and cellulases of fungal origin are produced by cultivation of the filamentous fungi *Aspergillus oryzae* and *Trichoderma reesei*.

Generally, genes derived from a prokaryotic organism are expressed in *Bacillus* strains, whereas the genes from eukaryotic sources are better expressed in filamentous fungi.

The production hosts are often further improved by either classical mutation or recombinant techniques to allow ease of production. The primary target for strain improvement is the removal of unwanted enzyme activities, such as extra cellular protease and amylase to improve product purity and stability.

27.2.2 Process Conditions

The classical process setup for a fermentation of detergent enzymes is the three-step process outlined in Figure 27.1.

A production batch is always initiated by seeding the selected microorganism onto a solid nutrient agar medium. On the agar, the cells will start to multiply and they will grow to eventually cover the whole surface of the media in the flask. The content of the seed flask is then used to inoculate the medium in the seed tank.

Both the seed and main tank fermentations are very similar; the former can be regarded as a smaller-scale version of the main fermentation tank. The main purpose of the seed stage is to start the fermentation process on a small scale and get more efficient utilization of the large fermentation tank. The size of the seed tank is dependant on the main tank, but usually the volume is in the range of 10% of the main tank.

Most of the enzyme is formed during the main tank fermentation process, which is carried out in large, aerated, and stirred fermentation tanks ranging from 40 m^3 to several hundred cubic meter in volume [2].

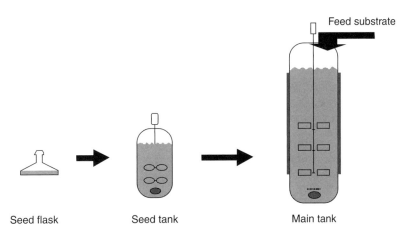

FIGURE 27.1 Classical process setup for an enzyme production process.

27.2.3 Equipment

The stainless steel–made fermentation tanks are designed to allow easy cleaning and proper sterilization of the substrate. The sterilization of the substrate is performed in the tank by heating with direct injection of steam. Every step in the process is carefully designed to protect the cultures from contamination with foreign wild-type organisms, since the intrusion of such organisms would ruin the whole batch.

The most challenging part in most fermentation processes is to supply the cells with sufficient oxygen. Most fermentation processes require a relatively high oxygen concentration in the broth, to obtain optimal enzyme productivity rates. Free oxygen has a relatively low solubility in aqueous systems and a continuous transfer of oxygen is required to maintain the oxidative metabolism of the culture [3,4]. This can be accomplished by introducing relatively large amounts of atmospheric air into the bottom of the fermentation tank. Normally, an aeration rate of 0.5–1.0 VVM (volume air/volume tank/minute) is applied in the process. This corresponds to an air-flow rate of 100 Nm^3/min for a middle-sized production scale fermentation tank.

The oxygen in the gas phase is transferred over the bubble surface and into the fermentation medium by simple diffusion. To improve the oxygen transport, the tanks are equipped with high-power agitation systems, which ensure that the medium is well mixed, and that air bubbles are broken up into smaller bubbles with a larger contact area. Many sophisticated mechanical systems exist to improve the oxygen transfer to the tank, but these are very often too complicated to operate and have gained little acceptance in the fermentation industry. Today, all major producers of industrial enzymes use traditional, stirred tank reactors for production. If even higher oxygen transfer rates are required, pure oxygen can be added to the inlet air to increase the oxygen concentration in the bubbles. However, this is an expensive way of increasing the oxygen transfer and is therefore used only in a limited number of fermentation processes.

27.2.4 Mode of Operation

The main tank fermentation process is carried out as a fed batch process, since this mode of operation is ideal for many microbial processes. The feed substrate is in most cases the carbon source, which is added continuously during the entire batch. This allows the process engineer to control the growth rate of the organism and thus to keep the culture at its optimal conditions for product formation.

The production batches are extensively monitored during the whole process. Often, the tanks are equipped with several on-line and at-line sensors (Figure 27.2).

During the production, measurements of pH, dissolved oxygen (pO_2), temperature, pressure, air flow, off-gas analysis, and batch weight are collected by the process computer system (Figure 27.3); and the process engineers can use these measurements to evaluate the development of the fermentation and interfere if something unexpected should occur.

The fermentation batch is harvested when the enzyme concentration is at its maximum, which will vary from one product to another. Often, the *Bacillus* processes producing protease have a relatively short fermentation time, 40–70 h, due to the problems with the autoproteolysis. For other products where the enzyme is more stable, it is beneficial to continue the process and some of these fermentations may run for as long as 10 days.

27.2.5 Raw Materials

The substrates used in the fermentation medium are typically agricultural products, and they are exclusively of either vegetable or inorganic origin. The main incentive obviously is cost and performance in the process, but other requirements such as ease of downstream processing are also important.

The composition of the substrate for a typical industrial fermentation medium is shown in Table 27.1.

Production of Proteases and Other Detergent Enzymes

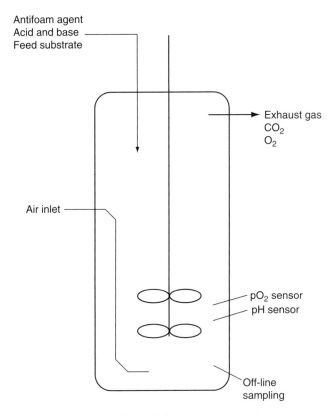

FIGURE 27.2 Configuration of a fermentation tank for enzyme production.

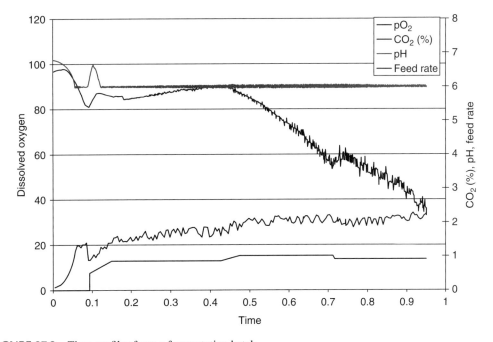

FIGURE 27.3 Time profiles from a fermentation batch.

TABLE 27.1
Overview of Substrate Composition for an Enzyme Fermentation Process

	Bacillus	**Aspergillus**
Nitrogen source	Soy protein, potato protein or $NH_3(g)$	Yeast extract or $NH_3(g)$
Carbon source	Sucrose or glucose	Glucose or dextrin
Minerals	K_2HPO_4, K_2SO_4, $MgSO_4$, $(NH_4)_2SO_4$	K_2HPO_4, K_2SO_4, $MgSO_4$, $(NH_4)_2SO_4$
Others	Trace metals, vitamins, antifoam agent	Trace metals, antifoam agent
Feed substrate	Glucose or sucrose	Dextrin or maltose syrup
	Nitrogen source	

The substrate requirements vary significantly from one type of organism to another [5] and amount of complex protein used as nitrogen source is highly strain dependent.

In the protease production in particular, it is beneficial to add significant amounts of protein, since this can reduce the autoproteolysis or self-digestion, which in many cases is the limiting factor in a protease production process.

Many recombinant strains are based on the gene regulation from an amylase, and some of these strains may have a need for dextrin as an inducer or activator of the promoter of the gene to obtain maximal production of the desired enzyme.

27.3 RECOVERY

The purpose of the recovery process, often referred to as the downstream process, is to separate the enzyme from the biomass and to produce a solution that contains the enzymes in a desired purity to be used for formulation of the final enzyme product. The production volumes and costs of enzyme products can be compared to fine chemicals, and thus the recovery processes need to be designed to handle large volumes in a fast and cost-effective way.

27.3.1 RECOVERY PROCESS DESIGN

The main factors influencing the design of the enzyme recovery process are as follows:

1. The properties of the production organism
2. The characteristics of the enzyme
3. The product quality demands
4. The type of product to be produced
5. The environmental impact of the process

27.3.1.1 Influence of the Production Organism

Almost all enzymes used in detergents are produced by submerse fermentation of a microorganism. Depending on the type of enzyme and microorganism, the enzyme can be secreted either extracellularly or intracellularly. Extracellular production is preferable, since the enzyme is secreted into the fermentation media allowing a simpler product recovery. In a few cases, the enzyme can only be produced and accumulate intracellularly. In these cases, the recovery of the enzyme requires that the cells are disrupted, either by homogenization (e.g., high pressure) or extracting of the enzyme by solvents.

The type of production organism (bacterium or fungus) also has great impact on the design of the process and even within the two types of producer hosts there are significant differences. Primarily, the differences in size of the microorganisms are an issue, but also the differences in the composition of the fermentation media, for example, well-defined media versus complex media, will influence the design of the processes.

Production of Proteases and Other Detergent Enzymes 537

27.3.1.2 Enzyme Characteristics

Most enzymes for detergent formulations, which are produced by genetically modified microorganisms, are monocomponents, that is, only one enzyme protein contributes to the overall activity. In a few cases, like cellulases, the enzyme product can be a multicomponent mixture of several enzyme proteins. In these cases, the recovery process must be designed to ensure that the ratio between the enzyme proteins important for the application is maintained throughout the recovery process.

The properties of the enzyme have an impact on the design of the conditions for the recovery process, especially characteristics like temperature and pH stability, solubility of the enzyme protein, and hydrophobicity. The enzyme activity as a function of pH and temperature also influences the choice of process conditions. When processing (e.g., proteases), it can be advantageous to select process conditions where the enzyme has a low activity, to avoid self-digestion (auto proteolysis) of the enzyme.

27.3.1.3 Demand for Product Quality

The product quality demands are highly dependent on the application of the enzyme. Detergent enzymes typically have requirements for their physical appearance (color, odor) and also for compatibility with the detergent. Impurities related to the process (e.g., the presence of unfermented sugar from the media) and microorganism (e.g., metabolites produced during the fermentation) can influence the quality of the product. Dealing with these issues typically requires addition of one or more purification steps in the process.

Typical requirements also include protein purity, germ count, and absence of other enzyme activities. The presence of other enzyme activities can interfere with the application or more commonly that the product stability is negatively affected, for example, by proteases present in the product, which degrade the enzyme of interest.

27.3.1.4 Type of Product

The result of the recovery process is an enzyme contained in a solution that serves as the starting point for the formulation of the final product. The recovery process therefore needs to take the different formulation requirements into account. An enzyme concentrate to be used for a liquid enzyme formulation needs to have an enzyme concentration high enough to allow for dilution by formulation chemicals. Furthermore, the demand for physical and enzymatic stability of the liquid enzyme product typically requires that the recovery process reduces compounds (salts) that can precipitate or degrade the enzyme (e.g., proteases). Enzyme concentrates to be used for a dry formulation (granulate) need to fill the requirements on activity and dry matter content. A low activity to dry matter ratio will make it difficult to obtain sufficiently high enzyme concentration.

27.3.1.5 The Environmental Impact of the Process

From the recovery process, waste is generated in the form of sludge, for example, used biomass and liquid waste (e.g., in the form of cleaning liquids). A significant cost of the recovery process is therefore the disposal of these materials and the design of the recovery process needs to consider this aspect. The environmental impact can be reduced by considering reuse of process liquids, selecting unit operations that have low impact, or producing valuable products from the waste. An example of the latest is the use of the utilized biomass as fertilizer on farmland or animal feed.

An example of reuse/recycling of process liquids is the use of permeates from the ultrafiltration for dilution during the first steps of the recovery process. The benefit of this approach is not only that generation of waste water is reduced, but also that the dry matter returned with the permeates can be expected to have a stabilizing effect on the enzyme (e.g., proteases).

27.3.2 Recovery Process Unit Operations

The type and sequence of unit operations for conversion of the fermentation broth into a liquid that can be used for the final formulation of the enzyme are selected to provide a product of the desired quality, and at the same time ensure a high enzyme yield and low production costs.

Since most industrial enzymes are produced extracellularly, only unit operations used for recovering these types of enzymes are considered. For intracellularly produced enzymes, additional steps must be included to harvest and open the cells (e.g., by a bead mill or high-pressure homogenizer). This complicates the process and increases the costs; hence this type of processing is avoided, if possible. A recovery process typically consists of a pretreatment, followed by a primary separation of the enzyme from the biomass. Later, the enzyme is concentrated by removal of water, and unwanted impurities are removed in a purification step. Figure 27.4 illustrates an example of a simplified recovery process.

27.3.2.1 Pretreatment

The fermentation broth obtained is often pretreated to improve its separation properties. For extracellularly produced enzymes, the complexity of the pretreatment varies, depending on the type of microorganism and product to be produced.

The pretreatment of a fungal broth is simple and typically consists of a dilution combined with a pH adjustment. The dilution is done to reduce the viscosity and thereby enhance the separation. Since the surface of the cells is negatively charged, the enzyme might bind to the surface if the pH of the process is higher than the isoelectrical point of the enzyme. In these cases, a pH adjustment is needed to release the product from the cells. Addition of salts can also be used to neutralize the surface charge of the microorganisms. Furthermore, the salts can be used as precipitation agents, for example, for removal of metabolites such as organic acids that have been formed during the fermentation.

For bacteria, the pretreatment of the broth typically consists of a coagulation followed by a flocculation [6]. In the coagulation process, salts (e.g., calcium, sodium, and aluminum) are added for the neutralization of the surface charges of the microorganism and other colloid materials.

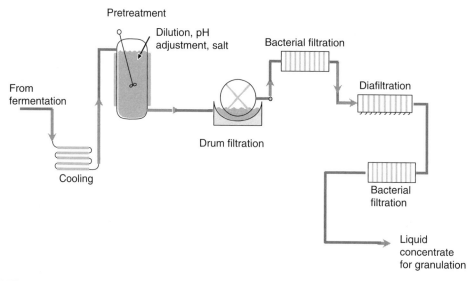

FIGURE 27.4 Example of a recovery process. In this case, the primary separation is a drum filtration and concentration step is done by ultrafiltration.

Production of Proteases and Other Detergent Enzymes

Then, a flocculating agent like a polyelectrolyte (e.g., a cationic polyacrylamide-based polymer) is added. Depending on the conditions and dosage, flocs (clusters) with different morphologies are obtained in this process. The optimal dosage and conditions are characterized by giving dense flocs and a clear supernatant, but the exact conditions also depend on the separation devices used in the primary separation.

27.3.2.2 Primary Separation

In the primary separation, the pretreated liquid containing the enzyme is separated from the biomass and other insoluble materials by filtration [7] or centrifugation [8]. The two technologies use different driving forces (pressure and gravity in case of filtration and centrifugation, respectively), but both of them can be used for bacteria as well as fungi.

Rotary drum vacuum filters are a widely used piece of equipment for this filtration step, due to its simple operation and low investment costs. The drum is typically precoated (e.g., with diatomaceous earth, silica, perlite) to improve the process performance [7]. Additional filter aid is often added to improve the filtrate quality; hence, the amount of sludge generated from this unit operation is relatively large. Several other filtration devices such as filter automates and cross-flow filtration devices are also used for the filtration. One of the benefits of the latter is that they can be operated without addition of filter aids and thus, the amount of generated sludge can be reduced.

The centrifugation step is an attractive, high-capacity unit operation, which is available in several different configurations such as disc stack centrifuges, decanters, and basket centrifuges. The centrifugation step often results in a liquid containing small amount of sludge and therefore, a fine centrifugation or filtration step is needed to ensure a complete removal of the microorganisms and produce a liquid with a satisfactory low turbidity.

The two main technologies mentioned earlier—drum filtration and centrifugation—can also be combined, for example, the centrifuges are responsible for the first separation and drums are used for removing the residual turbidity.

27.3.2.3 Concentration

The liquid obtained from the primary separation is a clear liquid, containing the enzyme of interest in a diluted form. The succeeding step is therefore to increase the concentration of the enzyme. Two technologies are widely used for this purpose, evaporation and ultrafiltration.

Evaporation is done in vacuum where water can be removed at conditions where the enzyme is stable (30–40°C). By evaporation, water and other volatile compounds are removed, whereas enzymes, salts, and other dry matters are retained. The liquid is typically concentrated up to 20–40% dry matter during the evaporation. When using evaporation for concentration of enzymes, it is important to ensure that the temperature of the heating surface is as low as possible, to avoid denaturation of the enzyme. The advantage of evaporation is that the dry matter, which typically stabilizes the enzymes, is retained. Thereby, savings on the formulation costs can be realized, but of course these savings should be weighed against the costs associated with the higher energy input.

Concentration using membranes is another widely used technology [9]. Membranes are available in a range of pore sizes, from nanofiltration membranes that only allow passage of water molecules to ultrafiltration membranes that allow passage of salts and lower molecular materials. For concentration of enzymes between 10 and 100 kDa, ultrafiltration is preferred. The ability of the membranes to retain the high-molecular molecules is described by the membranes' cut-off value, for example, a membrane with a cut-off value of 10 kDa will allow passage of molecules with a molecular weight of less than 10 kDa. The selected membranes typically only allow for less than 1% passage of the enzyme to the permeate. Ultrafiltration can be used to concentrate liquids to 20–25% dry matter. Concentration to higher dry matter is possible, but the viscosity of the liquid often becomes a limiting factor and flux (flow per membrane area) decreases significantly.

Ultrafiltration is attractive compared to evaporation due to the lower energy consumption, and compared to nanofiltration, the capacity (flow per membrane area) is significantly higher.

An ultrafiltration setup can also be used to wash out lower molecular impurities by diafiltration. In this mode of operation, water (tap water or deionized water) is added to the ultrafiltration concentrate and the liquid is concentrated to compensate for the dilution. With this approach, the content of low molecular impurities can be reduced.

Other concentration tools include precipitation with salts, polymers, or organic solvents but due to the high environmental impact (disposal or regeneration of salts and organic solvents), they are less attractive. Finally, concentration can be done by spray drying or lyophilization (freeze drying). Handling issues are observed for both methods due to formation of enzyme dust, but especially the latter is potentially beneficial for heat-sensitive enzymes.

27.3.2.4 Purification

The purification steps are aimed at addressing the issues related to undesirable effects of impurities, in the product or application. The technology used depends on the type of impurity to be removed. Some of the tools available are

- Precipitation
- Adsorption
- Inactivation
- Crystallization
- Chromatography
- Two-phase extraction

When precipitation [10] is used for purification, the enzyme is precipitated with salts (e.g., Na_2SO_4), an organic solvent (acetone), or isoprecipitation. This technique is primarily useful if the impurity is of nonprotein origin, since the precipitation typically is unspecific.

Adsorption, using activated carbon, is known to be useful for adsorption of process-related impurities like color, odor, or antifoam used in fermentation. A range of activated carbon types is available and screening is typically needed to find the best carbon for the specific problem. The main concern about including a carbon-adsorption step in a recovery process is the handling issues, especially related to poor filterability of activated carbon. Lately, filter plates/cassettes impregnated with activated carbon have overcome some of these problems.

Inactivation is applicable when the impurity is another enzyme activity, for example, a protease, which is labile at conditions where the main enzyme activity is stable. With this approach, the enzyme impurity is denatured at a specific pH and temperature and will afterward be removed by filtration.

Protein crystallization is a very complex process, much more complex than for small molecules (e.g., salts) that crystallize when the liquid is supersaturated. The crystallization process for proteins and enzymes in particular consists of at least two steps—nucleation and crystal growth. When developing a crystallization process, the conditions for both stages need to be optimized. Addition of salts, adjusting of pH, and addition of polymers and organic solvents are some of the techniques available to generate a supersaturated liquid and induce crystal formation (nucleation). The selecting of process conditions often optimizes the crystal growth. If conditions can be found where crystals are formed at a sufficiently high yield, the process is a very simple and cost-effective protein separation technique.

Chromatography is a widely used piece of tool for protein separation in the biopharmaceutical industry where very high purity is required. For industrial enzyme production, the technology in general is regarded as being too expensive, except for a few high-value products.

Aqueous two-phase extraction is an alternative recovery technology that has been investigated for many years in the industry and at universities. The standard two-phase system consists

Production of Proteases and Other Detergent Enzymes 541

of polyethylene glycol (PEG) and salts (e.g., potassium phosphate). The polymer is added to the enzyme concentrate and after mixing, the enzyme is isolated from the PEG-containing phase. One of the main hurdles for application of this type of process is the cost related to disposal or regeneration of the PEG and salts.

27.4 FORMULATION

After recovery and purification of the enzymes, the third important process step is to make a final formulation of the enzymes. The enzyme concentrate, whether it is present in an aqueous solution or in a dried form, has to be transformed into a stable form that fulfils the detergent producers' requirements. Obviously, the chosen formulation needs to be compatible with the detergent and a range of different quality parameters should be defined.

Table 27.2 states a list of typical quality parameters.

All formulations are a compromise between the parameters listed in Table 27.2 and the cost associated with the formulation of the final product. For example, the pH necessary for good microbial or physical stability of a liquid formulation may differ from the pH that gives optimum enzyme stability.

Although most efforts in developing new and even more efficient products have been focused on finding the right enzyme molecules for a given application, formulation can play a big role in the final use and success of the product. A new enzyme molecule with excellent performance can risk having limited use, because of instability of the protein during storage, unless the right formulation can be developed to remove these shortcomings.

Generally, the different formulations can be used for all types of enzymes. However, special care needs to be taken when working with proteases. Proteases will start self-digestion in an aqueous environment as well as degrading other types of enzymes if they are present. To minimize this problem, proteases should be kept for only a limited time in water or in systems with a low-water activity.

Two different types of formulations exist, a liquid and a solid formulation, reflecting the different types of detergent formulations. Specific issues are linked to the physical form of the formulation, thus they are treated separately.

27.4.1 HISTORY OF FORMULATION OF DETERGENT ENZYMES

Although enzymes have been used sporadically for more than 90 years in the detergent industry, the modern industrial application of enzymes started in 1962 with the introduction of Alcalase®. At that time, the formulation consisted of a simple drying, followed by standardization with NaCl. Other simple dry formulations followed, such as enzyme solutions sprayed on cores of sodium

---

TABLE 27.2
Quality Parameters for a Formulated Enzyme Product

Stability of Formulated Product	Stability in Detergent
Release rate	Solubility
Dust formation potential	Odor
Color	Particle size and distribution
Viscosity	Density
Flow ability	Loading of enzyme
Protein purity	Homogeneity

tripolyphosphate and prill formulations (i.e., particles containing enzyme powders suspended in a surfactant). The enzyme formulations generally were rather stable, as the proteases themselves were rather stable and detergents were phosphate-based without bleaching agents. Owing to this, the enzyme granulates were seldom coated.

In the early 1970s, the enthusiasm about enzymes suddenly cooled down, as a result of reported cases of in-plant safety problems [11]. The simple noncoated dry formulation gave rise to allergic reactions among factory workers, as any protein can do. The problem was solved rather quickly by developing a new and more robust type of dry formulation, based on extrusion and spheronisation, followed by coating of the granulate. Although the problem was quite rapidly solved, it took almost a decade before sales volume of detergent enzymes were back on their previous level. This incident clearly demonstrated how a suboptimal formulation can have a dramatic impact; and since then, high safety standards have always been a top priority for all formulation work with enzymes and today, all the existing products are perfectly safe.

Liquid formulation of enzymes on a larger scale was introduced during the 1970s. Although a range of different, rather sophisticated liquid formulations have been introduced during the following years, the bulk of liquid enzyme formulations have not changed much since the introduction. A typical liquid formulation primarily consists of water combined with polyols, sugars, salts, and other minor components such as antioxidants and fungicides.

The development of much more complex, efficient, and sophisticated detergents (e.g., by introducing bleaching systems) has increased the demand for robustness of the enzyme formulation. Furthermore, the popularity in recent years of heavy-duty liquid detergents has pressed for the development of stable liquid formulations of enzymes for this segment. Finally, the majority of the more sophisticated detergents today have more than one enzyme present—up to four have been described—and as proteases normally are always present, it can sometimes create challenges.

27.4.2 Liquid Formulations of Enzymes

Liquid formulation in this context means nonsolid formulations. The vast majority of liquid formulations are based on water with a varying degree of other components present, although liquid nonwater formulations of enzymes have been described. The most important of the issues listed in Table 27.2, when working with liquid formulations, is the stability of the formulated enzyme product as well as the final liquid detergent. A comprehensive overview of the stability issues of enzymes in detergents is given in Ref. 11. Stability can be divided into enzymatic, physical, and microbial stability.

27.4.2.1 Enzymatic Stability

Loss of enzyme-catalytic stability is a serious problem and it may be caused by many things. Denaturation of the protein will take place by the use of excessive heat and acidic or alkaline hydrolysis, proteolysis, oxidants, surfactants, solvents, etc. In a liquid detergent, relatively harsh conditions for a protein will occur, and thus it is important that the enzyme itself has been selected with a view to its robustness, either by choosing a native enzyme with sufficient stability in itself or, which is more common today, by protein engineering of the original enzyme protein, to improve the stability.

All enzymes need to be within a certain window of ionic strength and pH to maintain a high degree of stability, and thus the final formulation is often adjusted by salt and acid/alkali. The ions themselves will also in certain cases dramatically improve the stability of certain enzymes. The best-known example is Ca^{2+}, which stabilizes most α-amylases.

A special case is the problem with proteolysis. From a formulation point of view, this problem can be addressed in three ways: slowing the activity in the formulation, reversibly inhibit the reaction, or physical separation of the enzymes. The most common way to slow the activity of the protease in the formulation is to reduce the water activity, typically by addition of polyalcohols

Production of Proteases and Other Detergent Enzymes   **543**

such as glycerol, sugar, sorbitol, or propylene glycol. Another approach has been to make low water containing slurry formulations in nonionic detergents, where the protease is precipitated and therefore shows low activity. Inhibition of the proteolytic activity can also be obtained by addition of peptides, amino acids, and carboxylic acids such as formate. Addition of borate, which to some extent is converted into polyborate, is a very good reversible inhibitor of the protease and also used in some cases, but the widest use of this inhibitor type is in liquid detergents, where a borate complex with a polyol (e.g., propylene glycol) is routinely used as a way to avoid unwanted proteolysis. Lately, much more efficient inhibitor types based on 4-substituted-phenyl-boronic acid derivatives have been described [12]. Physical separation by microencapsulation of a protease has also been described [13].

27.4.2.2 Physical Stability

It is almost always desirable to have a homogenous enzyme formulation, that is, without precipitations. Unwanted crystallization in the liquid formulation of the enzyme can occur if the enzyme concentration, pH, ion strength, etc. are in the right range. This can be avoided by addition of polyols and adjustment of pH. Impurities remaining from the fermentation broth may also precipitate and can, to a certain degree, be handled by adjustment of the formulation. However, it is generally more optimal to solve these issues upstream in the production process, that is, in the fermentation or primary separation steps.

27.4.2.3 Microbial Stability

As enzymatic liquid formulations contain a high amount of water, there is a risk of microbial growth taking place, either during storage or during handling of the formulation at the detergent plants. This can be avoided by lowering the water activity and adding preservatives. The most common method is to add small amounts of tensides and up to 50% polyols as propylene glycol, which will both lower the water activity and have a certain preservative effect in itself.

27.4.3 Solid Formulation of Enzymes

Solid formulation of enzymes to be used in powdered detergents is nowadays made by granulation. Granulation is the generic term for a particle size enlargement process, but in the industrial enzymes production, the word is more broadly used as a term for a process resulting in granulates with a diameter from 300 to 1200 μm. A granulate is preferred to powders, especially to secure a dust-free product, that is, it should not contain particles that can become airborne and the particles must be strong and resilient, to avoid creating dust during handling of the enzyme in the detergent factories. Further technical advantages of granulates to powders are improved flow properties, reduced risk of segregation, improved product homogeneity, and better stability [14]. The final granulates should be the optimal compromise between the quality parameters listed in Table 27.2 and costs associated with the process and raw materials. Historically, many different processes have been used to make enzyme granulates; but today, the two dominating enzyme granulate types for the detergent industry are made by either a high-shear granulation process (T-granulate) or a fluidized bed spray coating process (Enzoguard) (see Figure 27.5). In some special cases, extrusion is used as well.

27.4.3.1 High-Shear Granulation

In high-shear granulation, an enzyme solution is mixed with fillers, typically Na_2SO_4, binders, cellulose fibres, etc., as well as an appropriate amount of water [15]. The mixing and granulation are carried out in drums, using plows and high-speed chopping blades. The resulting granulates are then sieved and finally, one or more layers of coating are laid on the granulates. The cellulose fibres are entangled in a strong network inside the granulates, which together with the highly intense

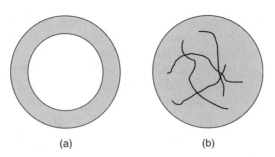

FIGURE 27.5 (a) Spray-coated granulate consisting of an inert core with the enzyme layer on the outside, (b) high-shear granulate with the enzyme evenly distributed inside. The enzyme layer and cellulose fibers are highlighted. One or several layers are further coated on the outside of the granulates, to obtain lower dust numbers and improved stability.

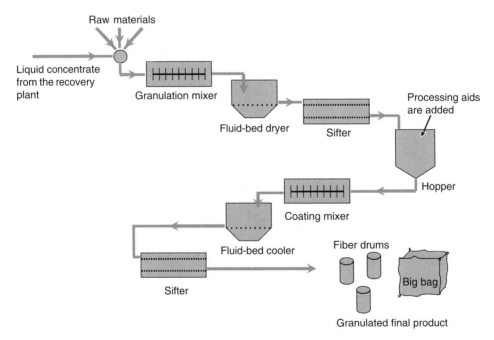

FIGURE 27.6 Example of a two-step continuous high-shear granulation process. In the first step, the raw granulate is made. In the second step, the granulate is coated in a coating mixer. The over- and undersize granulates are taking out in the sifters and recycled back into the process.

pummelling of the particles result in robust and tough granulates that are very resistant to physical breakage.

The process can be operated in both a batch- and continuous mode. In the latter case, the over- and undersized granulates are recycled in the process, making it a very efficient process for high-volume production (see Figure 27.6).

27.4.3.2 Fluidized Bed Spray Coating

In a spray-coating process, an inert soluble core composed of sugars, salts, polymers, or a combination of these are fluidized within a fluid bed and an enzyme solution is sprayed on the core to make a continuous film on the surface. The process can be carried out either in a top spray mode, that is, where the spray nozzles are placed in the top of the fluid bed chamber or as a bottom spray

Production of Proteases and Other Detergent Enzymes | **545**

mode where the nozzles are placed in the bottom. The latter is the preferred process today as, in general, it secures a more uniform coating but with a lower capacity, although. Then, the resulting enzyme coated cores are further coated with one or more layers of final coating, primarily to reduce dust levels as in the case of high-shear granulates.

The process is a batch process and as such, flexible to changes in formulations, but reworking of the granulates in the process is difficult.

27.4.3.3   Other Granulation Techniques

As stated earlier, high-shear granulation or spray coating are by far the two dominating techniques for enzyme granulate productions. In some cases, however, special requirements will make it advantageous to introduce another process. Extrusion is used in some cases where large uniform particles are desired, but it is a rather dusty open process, which is somewhat sensitive to formulation variation in moisture and enzyme feed composition, and also the reworking of off-specification products is difficult [16].

Prilling, where enzyme powders are suspended in molten nonionic surfactants and later spray cooled into particles, is no longer in use. It was a preferred process in the early days of industrial enzymes for detergents but now more strict demands to dust levels of enzyme granulates have made them obsolete. The term *prills* is still used although, as a general description of enzyme granulates, but it is technically not correct.

REFERENCES

1. Mountain, A., Ney, U.M., Schomburg, D. 1999. *Biotechnology. Recombinant Proteins, Monoclonal Antibodies, and Therapeutic Genes.* Vol. 5a, Weinheim: Wiley-VCH.
2. Ward, O.P. 1989. *Fermentation Biotechnology: Principles, Processes, Products.* Vol. 1, Milton Keynes: Open University Press.
3. Nielsen., J., Villadsen, J., Liden, G. 2002. *Bioreaction Engineering Principles.* 2nd ed., Dordrecht: Kluwer Academic Publishers.
4. Sahm, H. 1993. *Biotechnology, Biological Fundamental.* Vol. 1, 2nd edition, Weinheim: Wiley-VCH.
5. Stephanopoulos, G. 1993. *Biotechnology, Bioprocessing.* Vol. 3, 2nd edition, Weinheim: Wiley-VCH.
6. Wheelwright, S.M. 1991. *Protein Purification: Design and Scale Up of DownStream Processes.* New York: Wiley.
7. Dickenson, C. 1992. *Filters and Filtration Handbook.* 3rd edition, Oxford: Elsevier.
8. Leung, W.W-F. 1998. *Industrial Centrifugation Technology.* New York: McGraw-Hill.
9. Mulder, M. 1996. *Basic Principles of Membrane Technology.* 2nd edition, Dordrecht: Kluwer Academic Publishers.
10. Harrison, R.G. 1994. *Protein Purification Process Engineering.* New York: Marcel Dekker.
11. Crutzen, A., Douglass, M.L. 1999. Detergent enzymes: A challenge! In: Broze, G., ed., *Handbook of Detergents Part A.* New York: Marcel Dekker, pp. 639–690.
12. Nielsen, L.K., Deane-Wray, A. U.S. Patent 5972873 (to Novozymes), 1997.
13. Ness, J., Simonsen, O., Symes, K. 2003. Microcapsules for household products. In: Arshady, R., Boh, B., ed., *Microcapsule Patents and Products.* London: Citus Books, pp. 199–234.
14. Kirk, O. et al. 2004. Enzyme applications, industrial. In: *Kirk-Othmer Encyclopaedia of Chemical Technology.* New York: John Wiley & Sons, Inc.
15. Gormsen, E., Marcussen, E., Damhus, T. 1998. Enzymes. In: Showell, M.S., ed., *Powdered Detergents.* New York: Marcel Dekker, pp. 137–163.
16. Becker, T., Park, G., Gaertner, A.L. 1997. Formulation of detergent enzymes. In: Ee, J.H.v., Misset, O., Baas, E.J., eds., *Enzymes in Detergency.* New York: Marcel Dekker, pp. 299–325.

28 Chemistry, Production, and Application of Fluorescent Whitening Agents

Karla Ann Wilzer and Andress Kirsty Johnson

CONTENTS

28.1 Introduction and Background 547
28.2 Mechanism of Action of FWAs 548
28.3 Chemistry of FWAs and Production Schemes of Select FWAs 549
    28.3.1 Stilbene Derivatives 550
    28.3.2 Styryl Derivatives of Benzene and Biphenyl 553
    28.3.3 Coumarins 554
    28.3.4 Benzo[b]furans 556
    28.3.5 Benzoxazoles 556
    28.3.6 Physical Forms 556
28.4 Measurement of Whiteness 558
28.5 Detergent Applications 558
28.6 Conclusions 558
References 559

28.1 INTRODUCTION AND BACKGROUND

Although fluorescent whitening agents (FWAs) have only been developed and marketed on a large scale since the past 60 years, their scientific history dates back to more than 200 years. FWAs have also been referred to as opticals, optical dyes, optical bleaching agents, optical bleaches, optical brightening agents, optical brighteners, fluorescent bleaching agents, fluorescent brighteners, fluorescent whiteners, fluorescers, or just brighteners.

Their development arose from observations made within the bleaching industry. Toward the end of the eighteenth century, extracts of horse chestnut (*Aesculus hippocastanum*) were added to the normal bleaching solution to obtain increased whiteness. This was only one of the many additives used empirically by the industry. In 1780, Frischmann published his observations on the "double color (iridescence) of various wood tinctures," derived from, among other things, chestnut husks.[1] Five years later, Remmler succeeded in isolating this "Schillerstoff" (shimmer stuff or iridescence).[2] Kastner called the same substance *polychrome* and Dahlstrom called it *aesculin.*[3]

Despite this early work, however, no real advance in the application of fluorescence to textiles was made until the twentieth century. Initial research conducted on fluorescent dyes by Lagono in the 1920s encouraged Paul Krais to try and increase the white effects of cellulose textiles (viscose and linen) with a solution of 0.25 g/L of aesculin.[4] The result was a definite improvement in the fabric whiteness, explaining why the chestnut extracts used 150 years before (containing 6,7-dihydroxycoumarin, linked to sugar) had been so effective. In 1929, Krais published a paper on "A new black and a new white" in which he described the first application of optical whitening to textiles. His conclusion has a very modern ring—"It is actually possible by this method to make what has hitherto been the whitest white still whiter."[5]

In the following years, one company after another applied for patents for coumarin derivatives and for stilbene, benzoxazol, and benzimidazol derivatives. In 1935, Imperial Chemical Industries (ICI) applied for patents regarding the application of stilbene derivatives on cotton.[6] In 1940, Wendt of I.G. Farben industrie found that 4,4'-diaminostilbene-2,2'-disulfonic acid, a compound long used in the production of azo dyes, had the same effect of improving whiteness on bleached cellulosic fibers as the aesculin compound used by Krais.[7] In the following year, IG-Farben patented various stilbene derivatives for use on photographic papers, for improving the appearance of soap products, and for use in the bleaching liquors of the textile industry.[8] The postwar years have seen intensified research efforts by many companies and rapid market growth. More than 2000 patents for FWAs exist, there are several hundred commercial products, and approximately 100 producers and distributors. However, of these, about a half dozen are currently of commercial significance to the detergent industry.

28.2 MECHANISM OF ACTION OF FWAs

The word fluorescence may be familiar, but the mechanism is probably not so. Visible light forms only a fraction of the total energy radiated by the sun and out of this fraction, a colored object absorbs and reflects part of it. The part absorbed is transformed into heat, which is another form of energy. In most substances, ultraviolet (UV) light is also reemitted as heat, but certain substances, among them FWAs, reemit the invisible UV wavelengths as visible light anywhere between violet and blue-green (Figure 28.1).[9] Fluorescent whitening, therefore, is based on the addition of light, whereas the traditional bluing method achieves its white effect through the subtraction of light.[10] The additional reflected light also results in an increase in brightness.

The operation of detergent whitening, that is, bleaching or brightening, is concerned with imparting the highest possible whiteness to fabrics during the wash. In *chemical bleaching*, impurities are oxidized or reduced to colorless products. *Physical bleaching* involves the introduction of

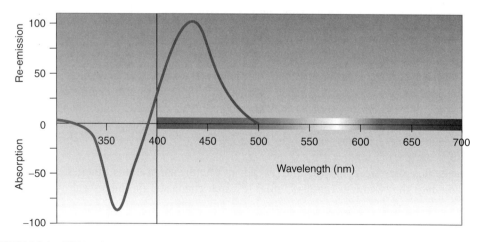

FIGURE 28.1 FWAs absorb UV light in the UV region of the electromagnetic spectrum and reemit in the short-wavelength (blue) region of the visible spectrum.

Chemistry, Production, and Application of FWAs

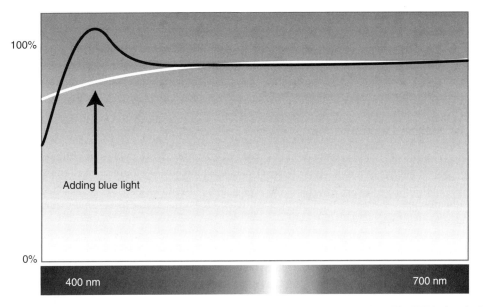

FIGURE 28.2 Lack of reflected blue light can be compensated by adding an FWA. The FWA absorbs UV light and converts the energy into blue light. The total amount of visible light increases and the cotton looks whiter.

a complementary color whereby the undesired color is made invisible to the eye in an optical manner. In bluing, the yellow cast of substrates such as textiles or paper is eliminated by means of blue or blue-violet dyes. Through color compensation, the treated product appears whiter to the eye.

With the aid of FWAs, optical compensation of the yellow cast may be obtained. White fabrics have a strong absorption band in the short-wavelength UV region, which extends into the visible region, depending mainly on the number and extension of conjugated systems in their component polymer molecules. The presence of impurities, degradation products, and dirt, all of which includes such extended conjugated systems, contributes to additional blue-light absorption, resulting in a yellowish-white shade. With FWAs this lost light is partially replaced, thus a high white is attained. As an example, bleached cotton is not pure white and FWAs and shading agents are added at the textile mill to enhance the base white of the fabric, resulting in total white of the fabric (Figure 28.2). During the wash cycle, these FWA and shading agents are removed and the rate is dependent on their wash-fastness properties. Thus they need to be replenished.

A fluorescent whitener should be optically colorless on the substrate and should not absorb in the visible part of the spectrum. In the application of FWAs, it is possible to replace the light lost through absorption, thereby attaining a neutral, complete white. Furthermore, through the use of excess whitener, still more UV radiation can be converted into visible light so that the whitest white is made more sparkling.

28.3 CHEMISTRY OF FWAs AND PRODUCTION SCHEMES OF SELECT FWAs

Many chemical compounds have been described in the literature as fluorescent, and since the 1950s intensive research has yielded a significant number of fluorescent compounds that provide a suitable whitening effect; however, only a small number of these compounds have found practical uses. Collectively, these materials are aromatic or heterocyclic compounds; many of them contain condensed ring systems. An important feature of these compounds is the presence of an uninterrupted chain of conjugated double bonds, the number of which is dependent on substituents as well as the

550   Handbook of Detergents/Part F: Production

planarity of the fluorescent part of the compound. Almost all of these compounds are derivatives of stilbene or 4,4′-diaminostilbene, biphenyl, 5-membered heterocycles such as triazoles, oxazoles, and imidazoles, or 6-membered heterocycles (e.g., coumarins, naphthalimide, and s-triazine).

Research has also been conducted in the area of polymeric FWAs for enhancing whitening of synthetic fabrics and blends. The approach with some of these polymers was to combine the effects of soil release polymers used for polyester with pendant fluorescent units.[11–13]

28.3.1 STILBENE DERIVATIVES

In the early twentieth century, patents were taken out on certain bis-(acylamino)-stilbene disulfonic acids for use as fluorescers in stage lighting effects produced by UV light. Their optical whitening properties were not recognized at that time, and it was some time later that a patent was filed for the production, as UV absorbers, of optical whiteners from the series of 4,4′-bis-(triazinylamino)-stilbene-2,2′-disulfonic acids.[14]

Today, a large percentage of commercial brighteners are bistriazinyl derivatives of 4,4′-dinitrostilbene-2,2′-disulfonic acid. The starting compound for producing 4,4′-diaminostilbene-2,2′-disulfonic acid is p-nitrotoluene (**1**, Scheme 28.1). This compound is sulfonated to give 4-nitrotoluene-2-sulfonic acid (**2**) and 2 moles of compound are then combined by oxidation to form, by means of 4,4′-dinitrodibenzyl-2,2′-disulfonic acid, 4,4′-dinitrostilbene-2,2′-disulfonic acid (**3**). The oxidation is carried out either with chlorine lye or by atmospheric oxygen in the presence of catalysts such as manganese salts.[15] The next step is the reduction of the dinitrostilbene disulfonic acid to 4,4′-diaminostilbene-2,2′-disulfonic acid (**4**). It can be carried out either by the Bechamp process using iron filings or with hydrogen in the presence of catalysts. The crude diaminostilbene disulfonic acid may undergo further refining operations to ensure a compound of maximum purity for the starting material.

The most familiar derivatives of diaminostilbene disulfonic acid are the 4,4′-bis-(triazinyl-amino)-stilbene disulfonic acids. Compounds of this type are produced by converting one molecule of 4,4′-diaminostilbene-2,2′-disulfonic acid with two molecules of cyanuric chloride (**5**) to yield compound **6** and then substituting the remaining labile halogen atoms with nonchromophoric substituents such as alkoxy or phenoxy groups, but generally by means of unsubstituted, substituted, aliphatic, or aromatic amine groups (examples are compounds **7** and **8**).

Compounds **9** and **10** represent FWAs of the bis-(triazinylamino)-stilbene disulfonic acid group. The choice of possible amines is very broad. It includes the ammonia group (or primary or secondary aliphatic amines), which can be further substituted, or it can be aromatic amines (or substituted aromatic amines or amino sulfonic acids). Heterocyclic amine groups such as morpholine will also give products with excellent properties.

An enormous number of different compounds have been prepared (**11–33**, see Table 28.1), but of all the possibilities mentioned in hundreds of patents, only a few are used in the industry. The different compounds used in industry have considerable effect on such properties as solubility and exhaustion rate at different temperatures and on the use of the products on a wide variety of materials. The range extends from readily water-soluble whiteners for paper to textile and soap and detergent whiteners to whiteners for polyamide fibers in the mass.

From the enormous amount of available material, only a few FWAs are discussed in this chapter. Table 28.1 shows some typical compounds of this class. The columns headed R and R′ show the different substituents introduced into the triazine ring. It must be stressed, however, that this table does not nearly show all the possible substituents. Compound **20** shows an example with halogen groups, whereas **21** and **22** are typical alkoxy or methoxy derivatives. In the following four compounds, the four substitutable halogen atoms have been replaced in all cases by four amine groups. Compounds **23–27** show materials with an ammonia group, alkylamino groups, and the ethanolamine group combined with aliphatic amines, but mainly with the aniline group. Compounds **18** and **28** also contain the aniline group with a wide variety of hydroxyalkylamine groups. Compound **29** shows

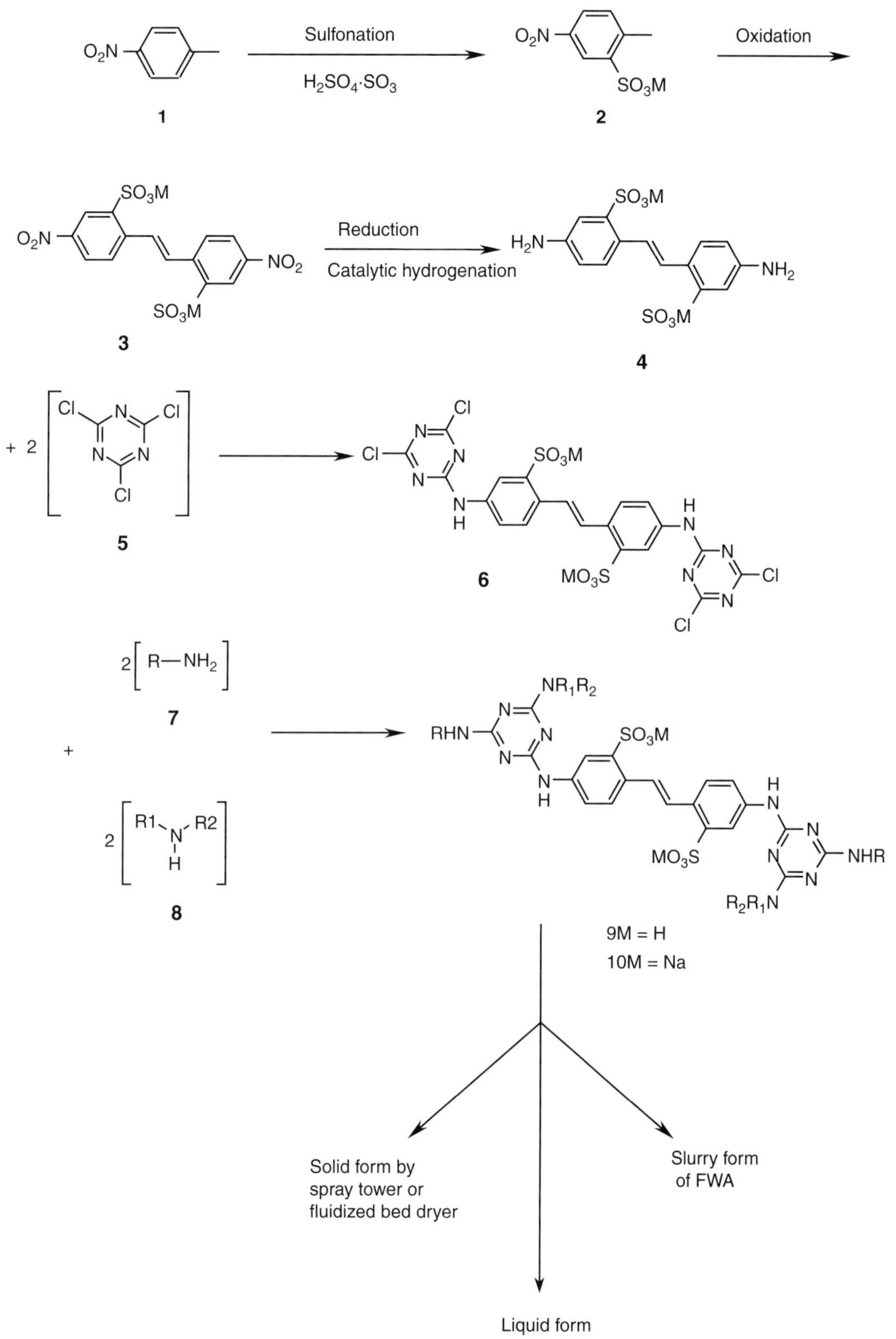

Chemistry, Production, and Application of FWAs

SCHEME 28.1 Production scheme for stilbene FWAs.

TABLE 28.1
Stilbene-Type FWAs

Compound	R	R′
11	$-NHC_6H_5$	$-OCH_2CH_2OH$
12	$-NHC_6H_5$	$-NHCH_3$
13	$-NHC_6H_5$	$-N(CH_3)CH_2CH_2OH$
14	$-NHC_6H_5$	$-N(CH_2CH_2OH)_2$
15	$-NH_2$	$-NH_2$
16	$-NHC_6H_5$	$-NHC_3H_6OCH_3$
17	(N-methylanilino-4-sulfonic acid)	$-N(CH_2CH_2OH)_2$
18	$-NHC_6H_5$	$-N(CH_2CH_2OH)_2$
19	(N-methylanilino-3,5-disulfonic acid)	$-N(CH_2CH_3)_2$
20	$-Cl$	$-N(CH_3)CH_2CH_2OH$
21	$-OCH_3$	$-NHCH_2CH(CH_3)OH$
22	$-OCH_3$	$-NHPh$
23	$-NH_2$	$-NHCH_2CH_2OH$
24	$-NH_2$	$-NHPh$
25	$-NHC_2H_5$	$-NHPh$
26	$-NHCH_2CH_2OH$	$-NHPh$
27	$-NHCH_2CH(CH_3)OH$	$-NHPh$
28	$-NHCH_2CH_2OH$	$-NHPh$
29	$-NHCH_2CH_2OCH_3$	$-NHPh$
30	$-N(CH_2CH_2)_2O$	$-NHPh$
31	$-NHPh$	$-NHPh$
32	$-N(CH_2CH_2OH)_2$	(N-methylanilino-3-sulfonic acid)
33	$-NHPh$	(N-methylanilino-4-sulfonic acid)

Chemistry, Production, and Application of FWAs

how alkoxyalkylene amine groups of various chain lengths can be used; and in heterocyclic compound **30**, the morpholino group has been introduced. Compound **31** is the tetraaniline derivative bis-(di-anilino-triazinylamino)-stilbene disulfonic acid. Compounds **32** and **33** illustrate possible combinations of aliphatic or aromatic amine groups with aniline sulfonic acid groups.

We cannot go into the details of production of optical whites on different materials and their behavior in application to liquid preparations, crystal lattices, crystal forms, etc., which represent a considerable and indeed often vital part of the properties of these products. In addition, it should be noted that within this class of products, not only the chemistry of the product itself but also the chemical composition and level of by-products from the manufacturing process are of significant importance to the performance as an FWA in a detergent application. It has been found that even if the chemistry of the major component is the same, the performance of FWA in a detergent application may show differences in a significant way, which is due to the by-products resulting from the manufacturing process. These by-products may result in a strong negative impact not only in fabric whitening, but also in detergent powder appearance and the smell of a detergent. The occurrence of these by-products can be avoided by the manufacturing process and processing technologies employed.

Asymmetric derivatives can be synthesized by 4-amino-4'-nitrostilbene-2,2'-disulfonic acid; however, their preparation is more expensive, and they show little advantage over the symmetrical compounds.[16] The principal effects of structural variations are changes in solubility, substrate affinity, acid fastness, etc. The bistriazinyl brighteners are employed principally on cellulosics such as cotton or paper. 2-(Stilben-4-yl)-naphthotriazoles (**34**, **35**) are prepared by diazotization of 4-aminostilbene-2-sulfonic acid or 4-amino-2-cyano-4'-chlorostilbene, coupling with an ortho-coupling naphthylamine derivative, and finally, oxidation to the triazole.

34	R1 = –SO$_3$H	R2 = –H
35	R1 = –CN	R2 = –Cl

28.3.2 STYRYL DERIVATIVES OF BENZENE AND BIPHENYL

Compounds based on the styryl group were prepared to lengthen the conjugated system of stilbene. 1,4-Di-(stryryl)-benzenes are obtained by the Horner modification of the Wittig reaction, for example, 1,4-bis-(chloromethyl)benzene is treated with 2 moles of triethyl phosphate and the resulting phosphonate reacts with 2 mol of o-cyanobenzaldehyde to yield compound **36**.[17] A strong brightening effect with a reddish cast is obtained on polyester fibers and is mainly used in the textile industry.

36

4,4'-Di-(styryl)-biphenyls are also obtained by the Horner–Wittig reaction of the phosphonate derived from 1,4-bis-(chloromethyl)-biphenyl (**37**) and triethyl phosphite (**38**) with benzaldehyde-2-sulfonic acid (**40**), giving the corresponding distyrylbiphenyl disodium salt (e.g., **41**, see Scheme 28.2).[18] This class of FWAs is widely used in detergent applications due to its unique whitening benefits

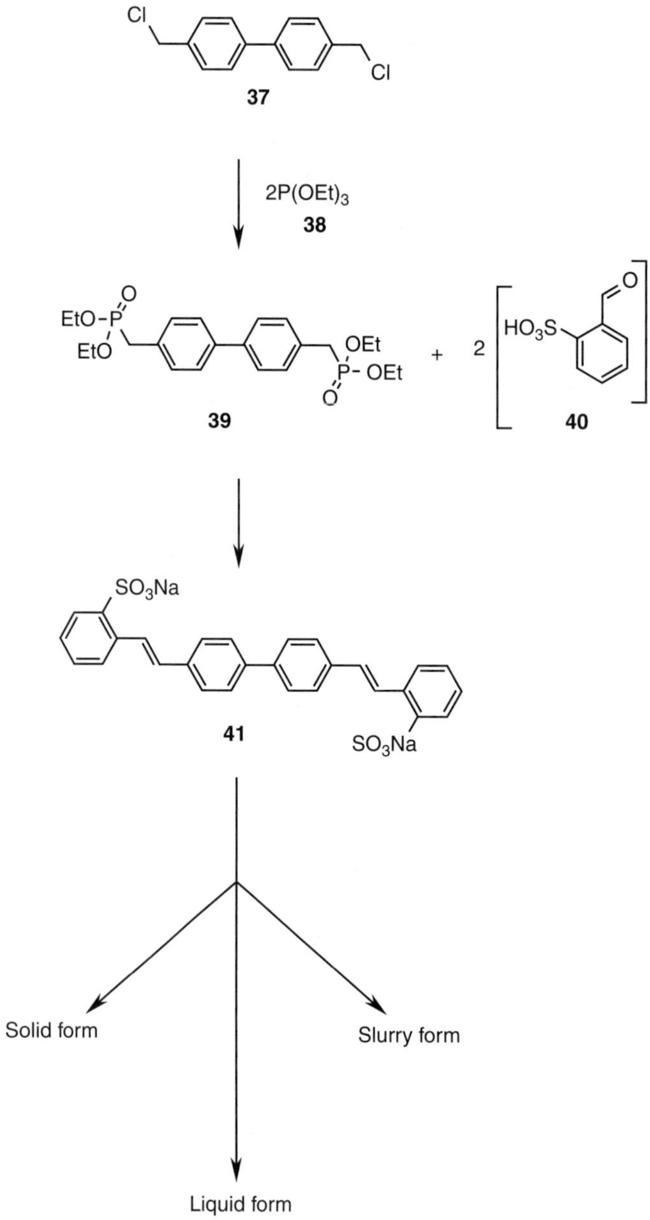

SCHEME 28.2 Production scheme for distyrylbiphenyl FWAs.

under various application conditions and detergent formulations. Thus, they are used in washing powders, liquids, bars, tablets, and sachets as well as fabric softening agents for brightening cellulosics to a high degree of whiteness with improved lightfastness as well as chlorine and oxygen bleach stability.[19]

28.3.3 COUMARINS

By treatment of flax with esculin, a glucoside of esculetin (**42**), a brightening effect is achieved; however, this effect is not fast to washing and light. The use of β-methylumbelliferone (**43**) and similar compounds as brighteners for textiles and soap has been patented.

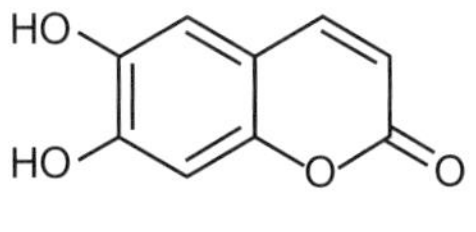

Chemistry, Production, and Application of FWAs

42

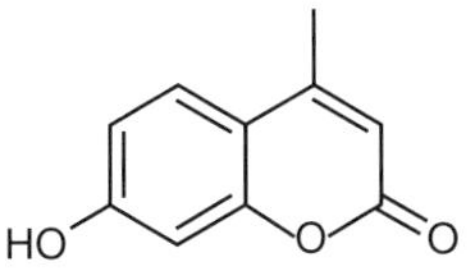

43

As an improvement over β-methylumbelliferone,[20–22] 4-methyl-7-aminocoumarin (**44**) and 7-dimethylamino-4-methylcoumarin (**45**)[23–25] were proposed. These compounds are used for brightening wool and nylon either in soap powders or detergents, or as salts under acid dyeing conditions. They are obtained by the Pechmann synthesis from appropriately substituted phenols (**46**) and β-ketocarboxylic acid esters (**47**) or nitriles in the presence of Lewis acid catalysts. An example for the synthesis of 4-methyl-7-diethylaminocoumarin, compound **48** is shown in Scheme 28.3.

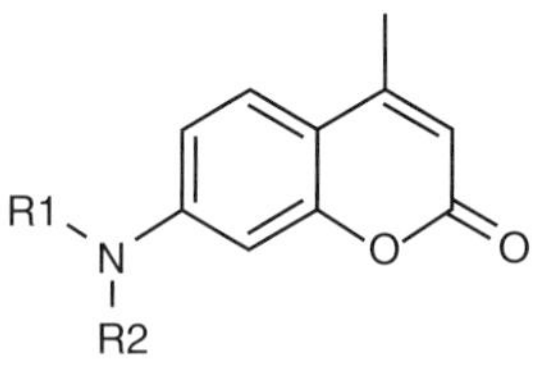

44 R1 = R2 = H
45 R1 = R2 = CH$_3$

A further development in the coumarin series is the use of derivatives of 3-phenyl-7-aminocoumarin (**49**) as building blocks for a series of light-stable brighteners for various plastics and synthetic fibers, and as quaternized compounds for brightening polyacrylonitrile.

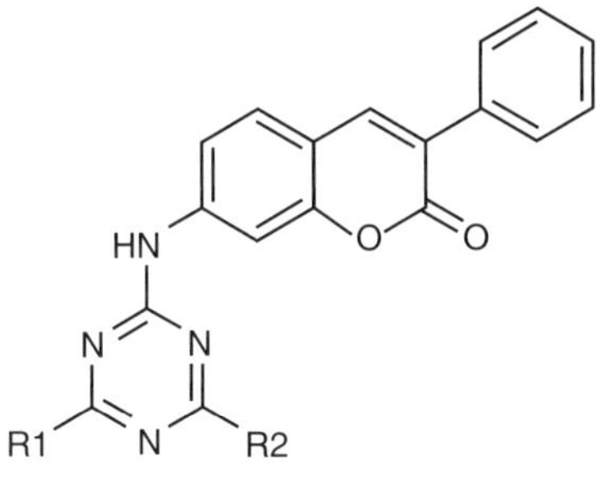

49 R1,R2 = Cl, amines

3-Phenyl-7-aminocoumarin is obtained by a Knoevenagel reaction of substituted salicylaldehydes with phenylacetic acid or benzyl cyanide. Further synthesis of the individual end products

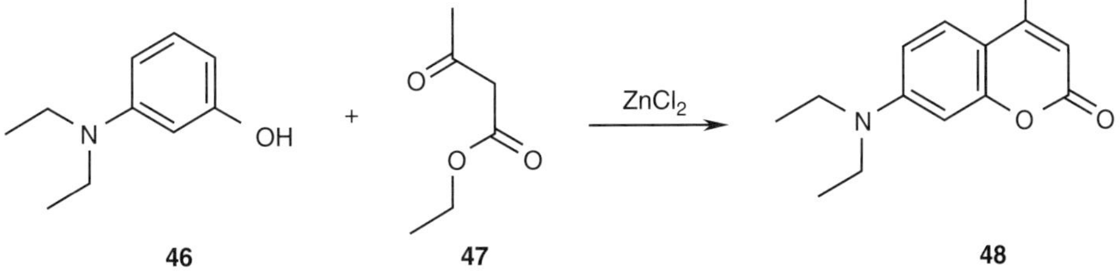

46

47

48

SCHEME 28.3 Production scheme for coumarin FWAs.

556   Handbook of Detergents/Part F: Production

is carried out by usual procedures. Other related substances are 3-phenyl-7-(azol-2-yl)coumarins (**50, 51**) and 3,7-bis-(azolyl)coumarins (**52, 53**).

28.3.4 BENZO[B]FURANS

Furans and benzo[b]furans are further building blocks of FWAs. They are used, for example, in combination with benzimidazoles and benzo[b]furans as biphenyl end groups. The 4,4'-bis (benzo[b]furan-2-yl)biphenyls can be regarded as distyrylbiphenyls that have their rings closed and are thus fixed in the *E*-configuration. Sulfonated derivatives are readily water-soluble and are used as lightfast brighteners for polyamides and cellulosic fibers.[26]

Compound **57** is produced by first reacting 4,4'-bis(chloromethyl)biphenyl (**54**) with salicylaldehyde (**55**) to give 4,4'-biphenyldiyl-bis(methylenoxy-2-benzaldehyde). The formyl groups of this compound are converted into phenyliminomethyl groups (**56**) with aniline. The double ring closure in dimethylformamide with potassium hydroxide is then carried out.[27] The final step is sulfonation (see Scheme 28.4). Compound **58** was a commercially available dibenzofuranylbiphenyl that was introduced in the market in 1997, but no longer is available. It possesses good lightfastness and outstanding stability to aggressive oxygen bleaching systems (e.g., peracids).

28.3.5 BENZOXAZOLES

Preparation of 2-(4-phenylstilben-4-yl)-benzoxazoles is done by anil synthesis from 2-(4-methylphenyl)-benzoxazoles and 4-biphenylcarboxaldehyde anil—an example is compound **59**. These FWAs were used in the past for brightening polyester fibers,[28] but are no longer widely used.

28.3.6 PHYSICAL FORMS

Spray-dried FWA powders were first introduced by European FWA producers in the 1970s.[29] These products are free-flowing, low-dusting beads. Investigation of alternative technologies

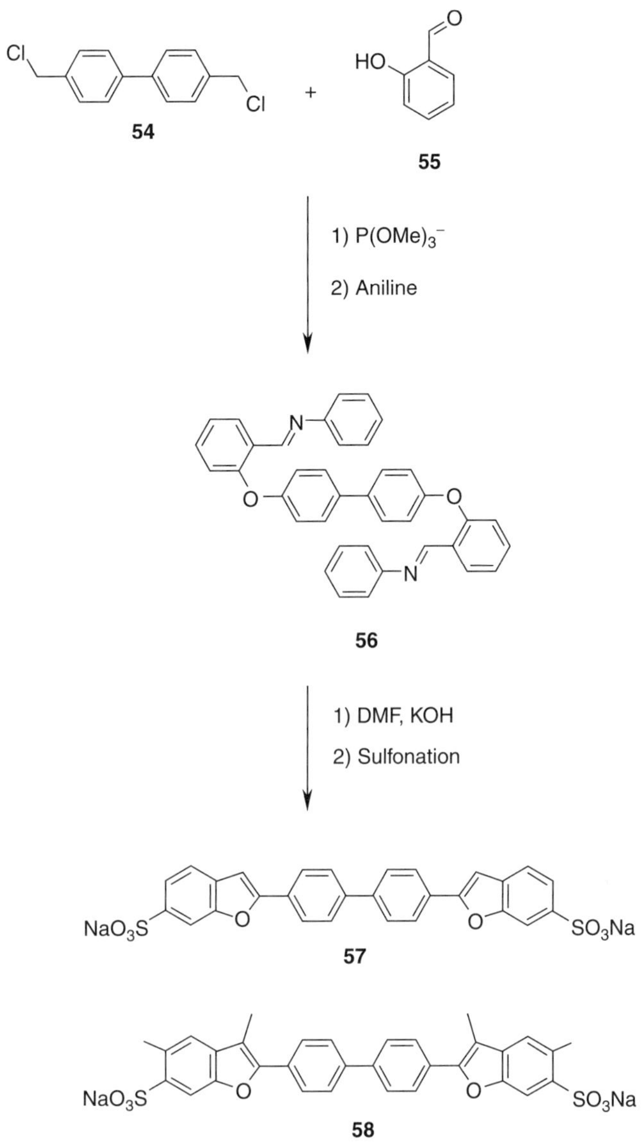

Chemistry, Production, and Application of FWAs

SCHEME 28.4 Production scheme for benzo[b]furans.

in the United States resulted in the introduction of free-flowing, low-dusting granular forms of FWAs. The granular form is produced by an agglomeration process with fluidized bed dryer. In some detergent manufacturing processes, automated solids metering devices are not available or the incorporation of powdered or agglomerated FWAs into liquid formulations cannot be readily accomplished because of insufficient shear in the available mixers. In such situations, the use of liquid forms of FWAs offers a practical solution. In the case of very soluble FWAs, stable solutions are available. Most of the FWAs, however, are relatively insoluble and although practical levels cannot be incorporated into solution, high-strength slurries can be produced. In the United States, Europe, and other countries, stabilized slurries were developed and became a popular form from the standpoint of ease of incorporation in both powder and liquid detergents.[30-32]

28.4 MEASUREMENT OF WHITENESS

There have been many approaches to express the visual white impression of the average human eye by means of physical parameters. Equations have been derived for this purpose and most of them use the International Commission on Illumination (CIE) color tristimulus to express the white appearance of a sample in numbers. It is important to note that it is necessary to measure the whole reflectance curve from the UV light through the visible range of red light and calculate the number for whiteness from this data. Owing to the fluorescence of a white sample, it is evident that the amount of UV light that is available to irradiate the sample has a major impact on the obtained numbers for the whiteness of the samples. Thus, a proper UV light calibration of the instrumentation is the key for accurate measurements. A difference among the various equations is how strongly they weigh the impact of various parameters, which results in preferences toward certain shades.[33-35] The white scale developed by Ganz and Griesser for D65 illumination and the $10°$ observer is the only equation which gives both the neutral white impression and its deviation toward a certain tint (tint value). The whiteness Ganz is expressed as $W(\text{Ganz}) = (D * Y) + (P * x) + (Q * y) + C$ and the tint value $\text{Tv} = (m * x) + (n * y) + k$. In the equations, Y, x, and y are based on the CIE tristimulus functions, whereas the variable parameters D, P, Q, C, m, n, and k are calibrated to a set of white standard fabrics.[36-39] The CIE whiteness ($Y - [800 * x] - [1700 * y] + 813.7$) is based on the Ganz/Griesser methodology, but with fixed parameters. The CIE Tv is equal to $(m * x) + (n * y) + k$. This simplifies the use of the method.[40-42] These two white scales are commonly used to assess the whiteness of fabrics.

28.5 DETERGENT APPLICATIONS

In the detergent industry, the chief function of FWAs is to increase the degree of whiteness and brightness of fabrics through the laundry process, and thus they also help to signal the cleaning effectiveness of the laundry product. FWAs used in the wash compensate for the reduction in whiteness, which occurs through normal wear as well as loss of textile FWAs during washing. Thus FWAs contribute not only to the aesthetic appeal of a garment, but also toward prolongation of the useful life of the textile material.

The most commonly used FWAs in laundry products today are shown in Table 28.2 and represent three chemistries—distyrylbiphenyl, coumarin, and stilbene. The selection of the FWA to be used in a specific type of laundry product will depend on several factors such as compatibility with the formulation, fabrics, product claims, laundry conditions, application, and manufacturing limitations. For example, compounds **60** and **62–65** are substantive to cellulosics and compound **61** is substantive to silk, wool, nylon, secondary acetate, and triacetate fibers. For bleach-based products (e.g., hydrogen peroxide) compound **60** is used as distyrylbiphenyl chemistry exhibits the required stability.

28.6 CONCLUSIONS

It is readily apparent that the palette of fluorescent whiteners commercially available to the detergent formulator remains limited. This limited range, however, is the result of a natural selection process by the detergent industry rather than a lack of new product development by the FWA manufacturers. As long as the available compounds are cost-effective and meet the technical requirements of the detergent industry, there is little incentive to develop new products. The tendency to use the current whiteners is reinforced by a long history of usage and the availability of significant data. The acceptance of a new whitener will be dependent on new technical requirements not met by the currently available products.

Chemistry, Production, and Application of FWAs

TABLE 28.2

Commonly Used FWAs for Laundry Detergents

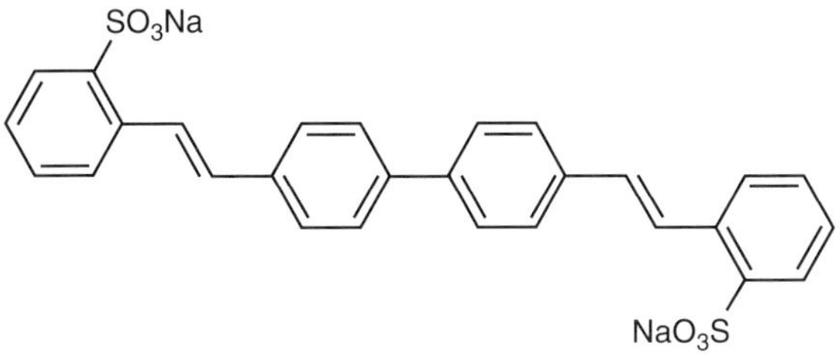

| Compound | R | R' |

60

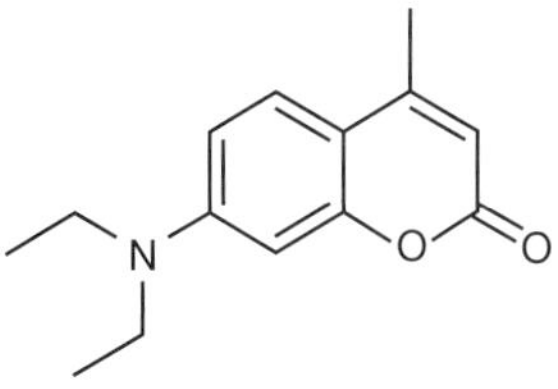

61

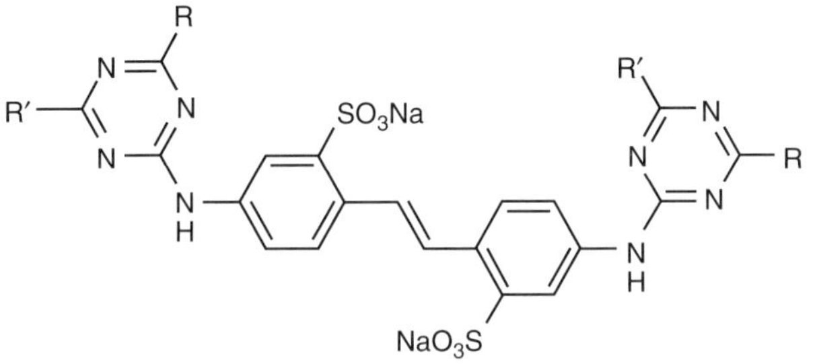

Compounds **62–65**

	R	R'
62	$-NHC_6H_5$	$-N(CH_2CH_2)_2O$
63	$-NHC_6H_5$	$-N(CH_2CH_2OH)Me$
64	$-NHC_6H_5$	$-N(CH_2CH_2OH)_2$
65	$-NHC_6H_5$	$-NHC_6H_5$

REFERENCES

1. Frischmann, *Crells Chem Jour.* 5 Teil, 1780.
2. Remmler, Göttlings Taschenbuch Für Scheidek, *J. Apoth.* 124, 1785.
3. Dahlstrom, B. *Jahresber. Chem.* 12, 274, 1883.
4. Krais, P., Ueber ein neues Schwarz und ein neues Weiss, *Meilliand Textilber.* 10, 468, 1929.
5. Zweidler, R., Einführung in die chemie der optischen Aufheller, *Textilveredlung* 4, 75, 1969.
6. Paine, C., Radley, J.A., Rendell, L.P., U.S. Patent 2,089,413, 1937.
7. Wendt, B., DE 752,677, 1940.
8. Petersen, S., Bayer, O., Wendt, B., DE 746,569, 1940.
9. Foerster, Th., Umwandlung der Anregungsenergie. 2. Internationales Farbensymposium, in *Optische Anregung von Organische Systemen*, ed., Jund, W., Verlag Chemie, Weinheim, 1966.
10. Brockes, A., Mechanism of whitening and quenching, in *Fluorescent Whitening Agents*, ed., Anliker, R., Muller, G., Georg Thieme, Stuttgart, p. 19, 1975.
11. Langer, M.E., Khorshahi, F., Aronson, M.P., U.S. Patent 5,039,782, 1991.
12. Rohrbaugh, R.H., Gosselink, E.P., U.S. Patent 5,834,412, 1998.
13. Rohrbaugh, R.H., Gosselink, E.P., World Patent WO 9723542, 1997.
14. Petersen, S., Organische Fluoreszenzfarbarbstoffe und ihre technische veriondung Über die optische Bleiche, *Angew. Chem.* 61, 17, 1949.
15. Lund, R.B. et al., U.S. Patent 4,952,725, 1990.
16. Siegrist, A., Hefti, H., Meyer, H., Schmidt, E., Fluorescent Whitening Agents 1973–1985, *Rev. Prog. Color.* 17, 79, 1987.

17. Stiltz, W., Pommer, H., Koenig, K., British Patent 920,988, 1960.
18. Maerker, A., *Org. Reactions* 14, 270, 1964.
19. Reinehr, D. et al., U.S. Patent 6,096,919, 2000.
20. Pechmann, H., Duisberg, C., Über die verbindungen der phenole mit acetessigäther, *Ber. Dtsch. Chem. Ges.* 16, 2119, 1883.
21. Shah, R., Bafna, S., ultraviolet absorption spectra of coumarins, *Indian J. Chem.* 1, 400, 1963.
22. Mellersch-Jackson, British Patent 472,473, 1937.
23. Ackermann, F., U.S. Patent 2,610,152, 1952.
24. Fleck, P., U.S. Patent 2,791,564, 1957.
25. Hausermann, H., U.S. Patent 2,929,822, 1960.
26. Eckhardt, C. et al., The changing face of fluorescent whitening agent requirements in the 1990s—A new approach, in *Proceedings of the 3rd World Conference on Detergents*, ed. Cahn, A., AOCS Press, Champaign IL, p. 193, 1993.
27. Schnizel, E., U.S. Patent 4,133,953, 1977.
28. Ackermann, F., Siegrist, A., U.S. Patent 2,875,089, 1959.
29. Findley, W., Fluorescent Whitening agents for modern detergents, *J. Am. Oil Chem. Soc.* 65(4), 679, 1988
30. Fringeli, W. et al., U.S. Patent 5,076,968, 1991.
31. Zelger, J., U.S. Patent 5,429,767, 1995.
32. Zelger, J., Schroeder, S., World Patent WO/2004/111330, 2004.
33. Hunter, R., Description and measurement of white surfaces, *J. Opt. Soc. Amer.* 48, 597, 1958.
34. Berger, A., Weissgradformeln und ihre, *Die Foabe* 8, 187, 1959.
35. Stensby, P., Optical brighteners and their evaluation, *Soap Chem. Special* 43, 80, 1967.
36. Ganz, E., Whiteness measurements, *J. Color Appearance* 1, 33, 1972.
37. Ganz, E., Griesser, R., Whiteness: Assessment of tint, *Appl. Optics* 20, 1395, 1981.
38. Griesser, R., Stand der instrumentellen Weissbewertung unter besonderer Berücksichtigung der Beleuchtung [Status of instrumental whiteness assessment with particular reference to the illumination], *Textilveredlung* 18, 157, 1983.
39. Griesser, R., Instrumental measurement of fluorescence and determination of whiteness: Review and advances, *Rev. Prog. Color.* 11, 25, 1981.
40. Brockes, A., The evaluation of whiteness, *CIE J.* 1, 38, 1982.
41. Ganz, E., Pauli, H.K.A., *Appl. Optics* 34, 2998, 1995.
42. Griesser, R., Assessment of Whiteness and tint of fluorescent substrates with good interinstrument correlation, *Color Res. Appl.* 19, 446, 1994.

29 Production of Gemini Surfactants

Bessie N. A. Mbadugha and Jason S. Keiper

CONTENTS

29.1 Introduction 561
29.2 Spacer 561
29.3 Cationic 563
29.4 Anionic 568
29.5 Heterogeminis and Amphoterics 569
29.6 Nonionic 570
29.7 Sugar Surfactants 571
29.8 Amino Acid–Based Geminis 572
29.9 Commercial Gemini Surfactants 572
29.10 Conclusion 575
References 575

29.1 INTRODUCTION

Since the early 1990's, gemini surfactants have been a wellspring for colloid scientists [1–5]. Although known before that time [6,7], publications by Zana and Menger focused attention on the multimeric molecules having the schematic form as shown in Figure 29.1 [8–10]. Conventional single-head group/single-chain surfactants are well known to associate into aggregates via self-assembly processes, whereas the gemini surfactant contains two or more individual surfactant units that are covalently associated. These new molecules have provided new challenges and opportunities for researchers inclined to physical, theoretical, or synthetic aspects of surfactants. In many cases, the properties of gemini surfactants are rather ordinary in comparison to their monomeric cousins. In some cases, however, unusual and potentially useful solution phenomena have been observed, where certain geminis prove to be a lot more than the sum of their parts. As a result, gemini surfactants remain promising candidates for commercial production in areas ranging from cosmetic formulation components to gene transfection agents. The quest to discover new geminis has led to a wide range of surfactant structures requiring varying degrees of synthetic sophistication. With the aim of highlighting the rich structural variety that has developed over the past 15 years, this chapter surveys routes to produce gemini surfactants, the different classes that have been reported in publications and patents, and some of the commercially produced geminis.

29.2 SPACER

Before delving into specific examples of gemini surfactants, it should be pointed out that the key feature differentiating them structurally from conventional surfactants is the linkage between individual amphiphilic moieties. The linkages are often referred to as spacers. Although the term

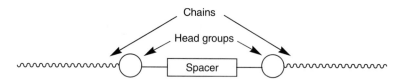

FIGURE 29.1 Schematic structure of a typical gemini surfactant.

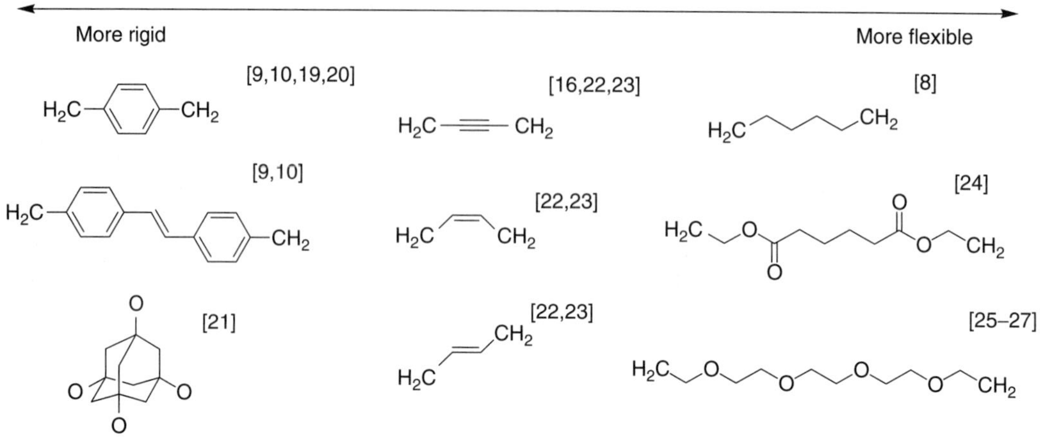

FIGURE 29.2 Examples of gemini spacers with varying rigidity or flexibility.

spacer may give a connotation of a group that spaces discrete units apart to restrict chain–chain interactions (an objective of Menger and Littau [9] while originally coining the term *gemini* for dimeric surfactants with rigid aromatic groups such as xylene and stilbene), it has become generalized to include flexible groups such as polymethylene or poly(oxyethylene) groups, with lower bending stiffness and greater freedom of rotation. Examples of spacers of varying rigidity are shown in Figure 29.2. The spacers may also differ in hydrophilicity or hydrophobicity, surface area and volume, and chemical functionality. Ultimately, the nature of the spacer has a profound impact on the surfactant's properties [11–15]. Any synthetic strategy to develop a gemini surfactant requires considerable thought regarding the nature of the spacer to incorporate. There have been many considerations and rationale for selecting particular gemini spacers. Reasons have included targeting unique micellar packing by using short spacers to bring ionic surfactant units into "unnatural" separation distances in solution [16], or utilizing "green" raw materials such as naturally occurring symmetrical spacers, for example, tartaric acid [17] or trehalose [18]. More specific examples will be reviewed in detail in the following text.

In determining a synthetic path, hand in hand with the "why" of a spacer is the "where and how" it will be incorporated into the gemini. Let us first consider where spacers are placed within the gemini structures. Examples are known wherein the surfactant units are directly linked at the head group or elsewhere on the molecule. Some scenarios are shown in simplified form in Figure 29.3, illustrating how individual surfactant units in a gemini can be linked at different levels. Also shown are amphiphiles that are close in form to, but not generally considered, gemini surfactants, which by convention are multimers of individual units, containing both a hydrophilic head group and hydrophobe. The nongeminis shown include surfactants with one head group and two hydrophobes, those with two head groups and one hydrophobe, and molecules known as bolaamphiphiles [19]—a surfactant class featuring head groups at either end of a chain, which also serves as a hydrophobe.

Production of Gemini Surfactants 563

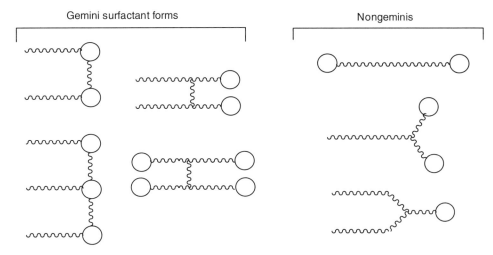

FIGURE 29.3 Schematic depictions of geminis versus nongeminis.

(It should be noted that certain geminis have spacers that may participate in hydrophobic association, such as bis-cationics that have flexible polymethylene spacer units long enough to presumably "bend" inward in a bilayer assembly [20].)

How the spacer group is incorporated to produce the gemini types, shown in Figure 29.3, is often the key to the synthetic strategy. From the benchtop to plant scales, conventional surfactants such as sulfonates or quats are produced, with a few exceptions, through well-established routes. With gemini surfactants, there have evolved a variety of ways to confront the synthetic wrinkle the spacer provides. Especially common are symmetrical synthons that are reactive at "alpha–omega" sites such as diols, dihalides, diamines, diesters, and bis-epoxides. These building blocks are usually commercially available fine or commodity chemicals, or readily accessible through standard transformations (e.g., dibromination of a diol by PBr_3 or N-bromosuccinimide). Many simple dimeric gemini surfactants thus involve reacting one equivalent of such a bis-reactive compound with two equivalents of another reagent to provide the core of, or complete, gemini surfactant. This is a typical route to produce cationic gemini surfactants, the first family we will survey.

29.3 CATIONIC

A multitude of reports have appeared on cationic, quaternary ammonium-based gemini surfactants. In particular, bis-quaternary surfactants known as m-s-m geminis [8] (where m = hydrophobe chain carbon number, and s = spacer methylene number) have proven popular due to their relatively straightforward preparation, accessibility to an extensive range of analogs for comparison studies [21–22], and unusual micellization behavior (rod- and wormlike structures) of short-spacer versions in solution [19,25]. The m-s-m surfactants are commonly produced in one step by either of the two pathways shown in Figure 29.4, where X = Cl or Br. The diamine route a is preferred where s = 2, as the alpha–omega dihalide route is reported to lead to incomplete quaternization [8]. This method has also been used to produce several other types of bis-quats, such as (1) dimethoxy-spaced geminis [26,27], (2) trimeric geminis, prepared by reacting alkylated polyamines with appropriate alkyl halides [28,29], and (3) asymmetric bis-quats from the sequential reaction of two different alkyl halides, followed by careful purification [30,31].

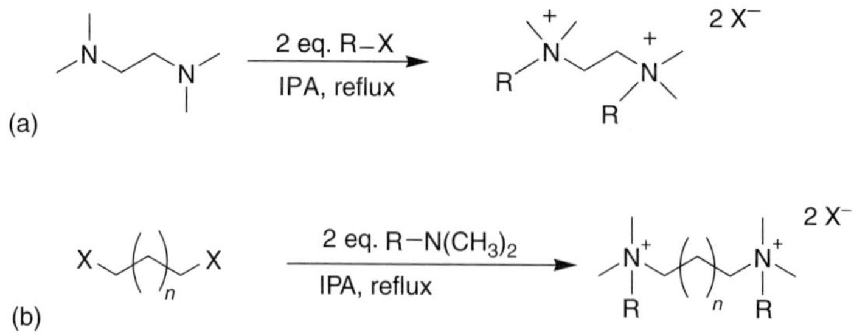

FIGURE 29.4 Two different pathways to *m-s-m*-type gemini surfactants. Path a is preferred for geminis, where *s* = 2; IPA—Isopropyl alcohol.

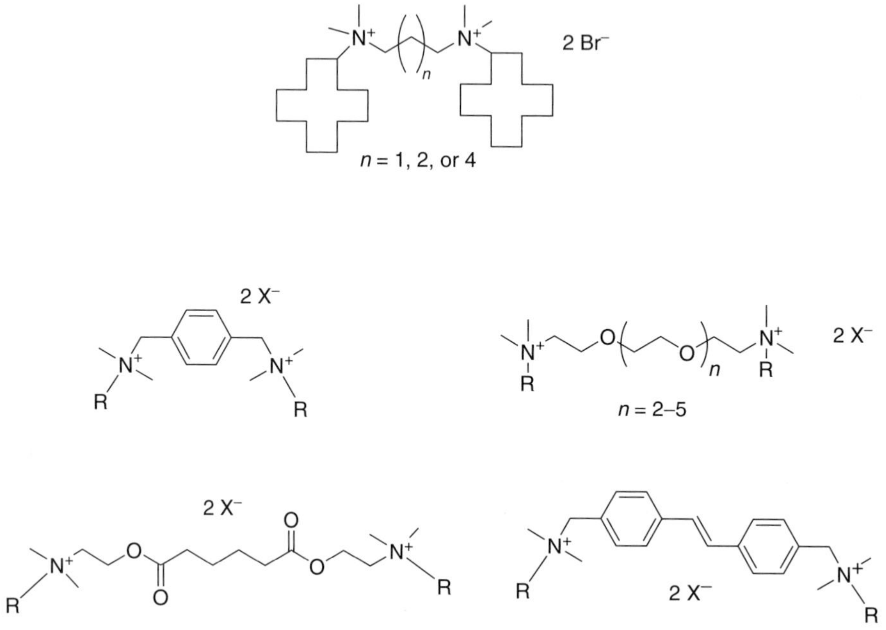

FIGURE 29.5 Examples of gemini surfactants produced by reacting dimethylalkylamines with alpha–omega dihalides.

The dihalide route b is preferred for longer *s* geminis, with successful syntheses reported for *m* chain lengths up to 22 carbons. Similar syntheses have been reported for other spacer types, including those containing xylene [9,10,32,33], stilbene [9,10], 2-butenyl [34,35], 2-butynyl [16,34,35], and polydeuterated spacers [36]. Bis-quaternary geminis having oligo(ethylene oxide)-based spacers have been prepared by first halogenating either end of a diol precursor followed by a reaction with an appropriate amine [37–39]. Structural examples of several surfactants produced by route b are shown in Figure 29.5.

Although bis-quat geminis derived from readily accessible diamines or dihaloalkanes are relatively straightforward to prepare, other gemini spacers require alternative and slightly more involved synthetic routes. For instance, to study decarboxylation and dephosphorylation reactions in cationic micelles, Brinchi et al. [40] synthesized a series of xylyl-spaced geminis similar in structure to the type shown in Figure 29.5, but having systematic variation of ring substitution and hydrophobic chain position (Figure 29.6). In another example, Eastoe et al. [41] reported a photodimerizable stilbenoid gemini bis-quat similar to the example shown in Figure 29.5, but having ester linkages

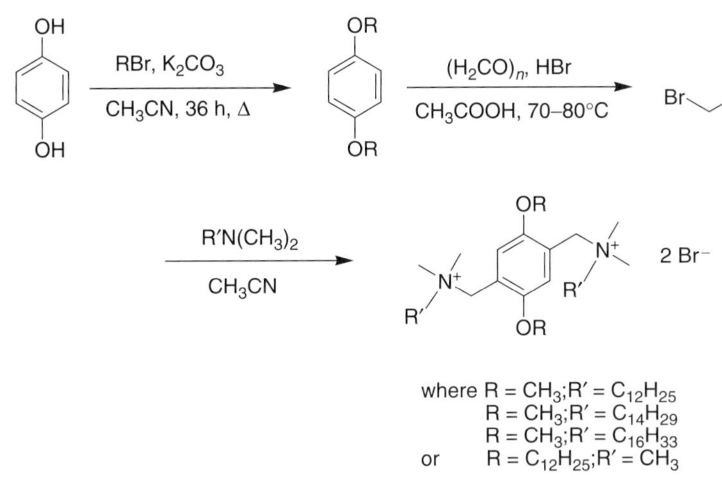

Production of Gemini Surfactants

FIGURE 29.6 Modified xylyl bis-quat geminis synthesized by Brinchi et al. (Reproduced from Brinchi, L., Germani, R., Goracci, L., Savelli, G., Bunton, C.A., *Langmuir*, 18, 7821–7825, 2002. With permission.)

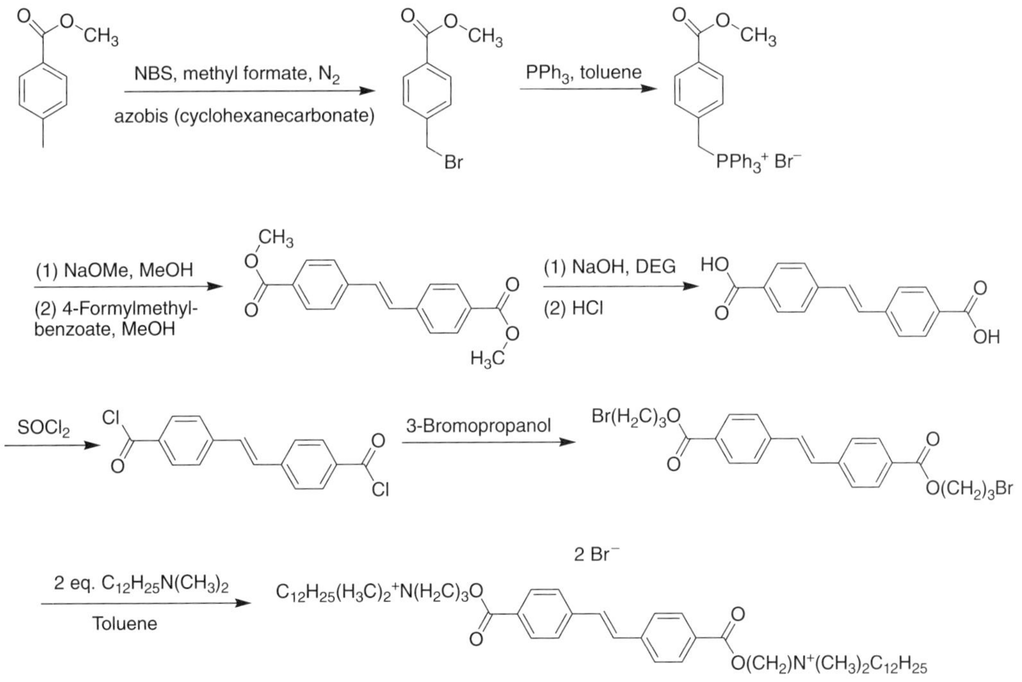

FIGURE 29.7 Stilbenoid bis-quat gemini (where NBS—N-bromosuccinimide; DEG—diethylene glycol) reported by Eastoe et al. (Reproduced from Eastoe, J., Sanchez Dominguez, M., Wyatt, P., Beeby, A., Heenan, R.K., *Langmuir*, 18, 7837–7844, 2002. With permission.)

between the spacer and the chains (Figure 29.7). The key to surfactant synthesis was the multistep construction of the dicarboxyl spacer core.

A variety of gemini bis-quats with hydroxylated spacers have been reported by Logan [42] and Kim et al. [43–45] through an alternative alkylation route using epichlorohydrin. This epichlorohydrin method has been reported for versions incorporating ester and amide functionalities in the hydrophobic chains; examples are shown in Figure 29.8. Ricci et al. [46] further reported a

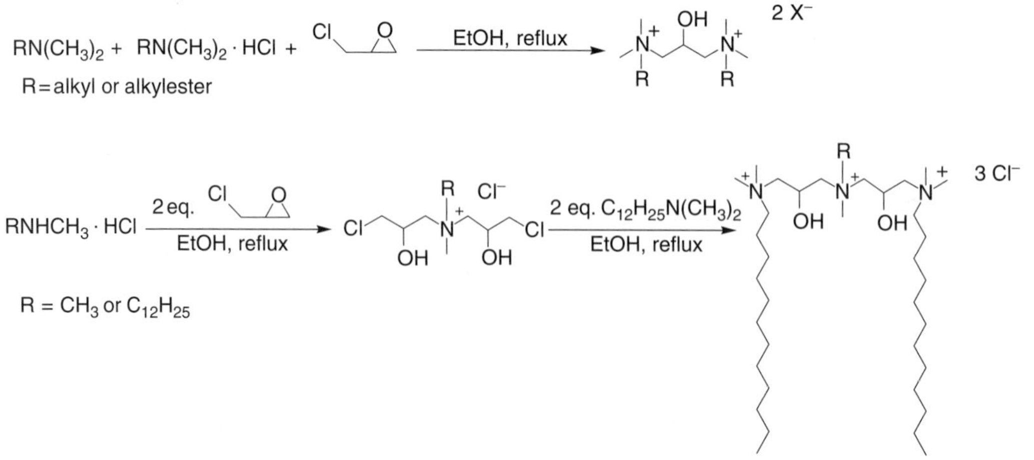

FIGURE 29.8 Examples of dimeric and trimeric gemini surfactants produced through use of epichlorohydrin. (Reproduced from Kim, T.-S., Kida, T., Nakatsuji, Y., Ikeda, I., *Langmuir*, 12, 6304–6308, 1996. With permission.)

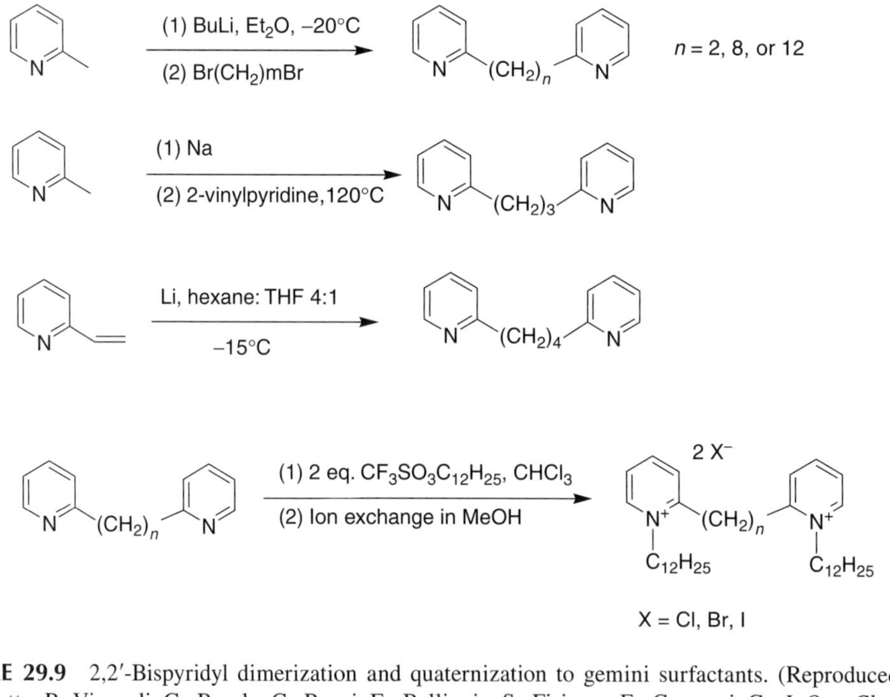

FIGURE 29.9 2,2'-Bispyridyl dimerization and quaternization to gemini surfactants. (Reproduced from Quagliotto, P., Viscardi, G., Barolo, C., Barni, E., Bellinvia, S., Fisicaro, E., Compari, C., *J. Org. Chem.*, 68, 7651–7660, 2003. With permission.)

modified, "micellar improved" synthesis, by carrying out the epichlorohydrin bis-quaternization in aqueous or hydroalcoholic media under mild conditions.

Nitrogen-containing aromatic groups can be incorporated into gemini bis-quats. A series of bis-2-pyridinium quats reported by Quagliotto [47] is shown in Figure 29.9, where individual pyridinyl reagents were dimerized before quaternization. Interestingly, the quaternization step proved difficult with conventional alkyl halides, requiring reaction with alkyl triflates followed by ion exchange to obtain the chlorides, bromides, and iodides. Other examples of bis-pyridinium geminis where

Production of Gemini Surfactants

the pyridine nitrogen was otherwise positioned were quaternized with alkyl halides (Figure 29.10) [47,48]. A series of bis-imidazolinium geminis were accessible via a two-step preparation shown in Figure 29.11 [49].

Cationic geminis have also been synthesized with compositions outside the "standard" hydrocarbon realm. *m-s-m* Geminis have been prepared with partially fluorinated hydrophobic chains [50,51] as well as thiol chain termini [52]. Geminis with a silicon-containing spacer were made starting from a dichlorosiloxane base, shown in Figure 29.12, used for templating mesoporous metal oxides [53,54]. To conclude, Figure 29.13 depicts the synthesis of polymerizable geminis,

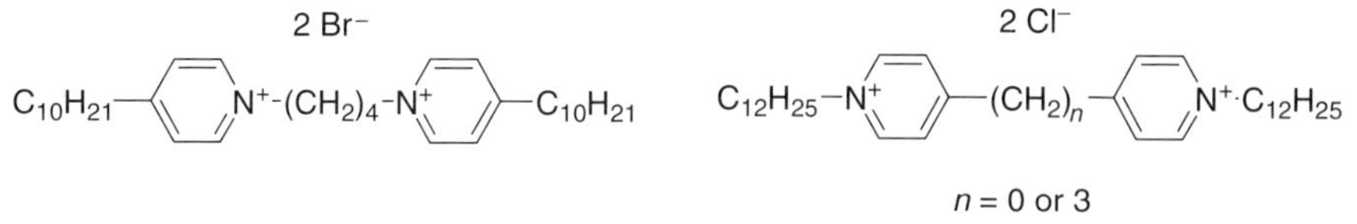

FIGURE 29.10 Bispyridinium gemini quats.

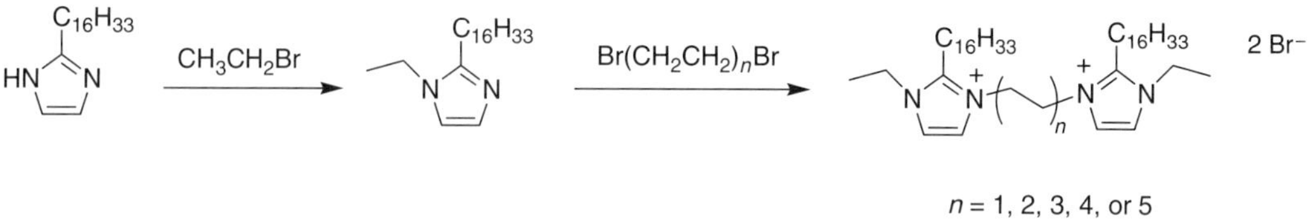

FIGURE 29.11 Synthesis of bis-imidizolinium geminis. (Adapted from Zhai, X., Zhang, L., Liu, M., *J. Phys. Chem. B*, 108, 7180–7185, 2004.)

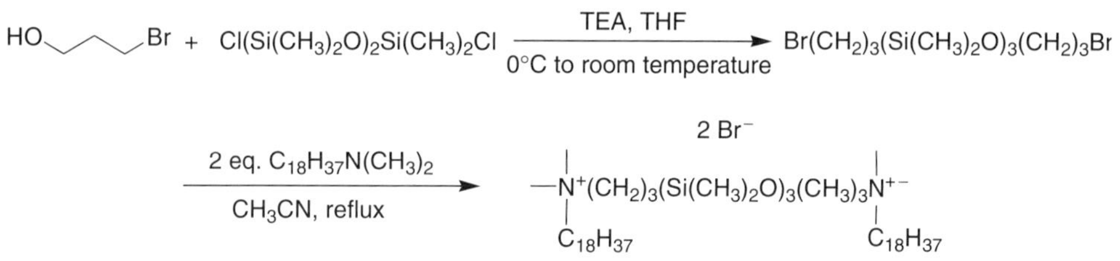

FIGURE 29.12 Synthesis of a siloxane-spacered bis-quat gemini surfactant. (Adapted from Lyu, Y.-Y., Yi, S.H., Shon, J.K., Chang, S., Pu, L.S., Lee, S.-Y., Yie, J.E., Char, K., Stucky, G.D., Kim, J.M., *J. Am. Chem. Soc.*, 126, 2310–2311, 2004.)

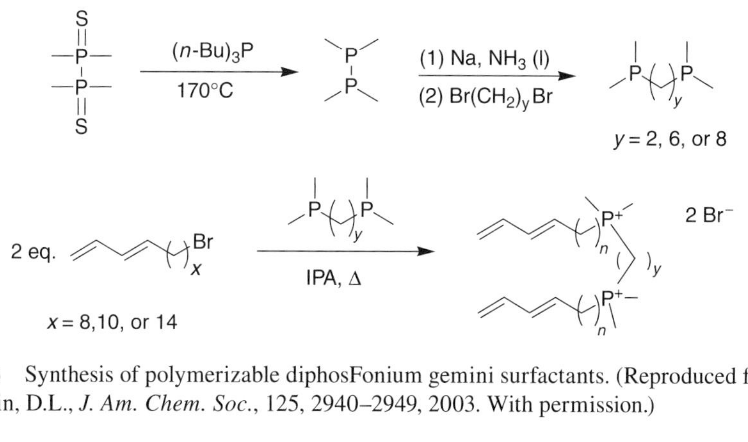

FIGURE 29.13 Synthesis of polymerizable diphosFonium gemini surfactants. (Reproduced from Pindzola, B.A., Jin, J., Gin, D.L., *J. Am. Chem. Soc.*, 125, 2940–2949, 2003. With permission.)

where the cationic units are phosphonium groups, reported by Pindzola et al. [55] and evaluated for their cross-linking and liquid crystalline behavior. As with many quaternary ammonium geminis, alpha–omega dibromoalkanes were used for creating the spacer for these phosphorous-containing geminis.

29.4 ANIONIC

Gemini surfactants bearing anionic charges have included phosphate esters, sulfates, sulfonates, and carboxylates containing a variety of spacer chemistries. Examples of phosphate ester geminis are shown in Figure 29.14. Straightforward preparations have been reported, wherein the ester spacer groups were established by either alkylation of a dihalide under basic conditions or phosphorylation of a diol [9,10,56]. Sommerdijk et al. [57] demonstrated the synthesis of bis-phosphate gemini analogs starting from three different stereoisomeric tetraols.

Epichlorohydrin played an important role in the preparation of sulfated geminis. A general strategy is shown in Figure 29.15, where the core can be, for example, a fatty amide [58] or a polymethylene group [59]. A similar epichlorohydrin-based strategy was used to prepare tetrasulfated geminis starting with a pentaerythritol core [60]. Acharya et al. [61] synthesized a bis-ester, bis-carboxylate gemini (disodium 2,3-didodecyl-1,2,3,4-butanetetracarboxylate), touted as the first gemini not to have a spacer, with the individual units linked directly at the chain. It is possible that this particular gemini may also be viewed as having a butylene spacer between the head groups, and Sommerdijk et al.'s [57] earlier published phosphate ester (Figure 29.15) is certainly of a similar format. Another set of carboxylate geminis was based on a diamino acid backbone, leading to a bis-aspartate structure (Figure 29.16) [62]. Further examples of amino acid geminis will be described in Section 29.8.

An interesting concept in gemini chemistry was that of a "pseudogemini" (Figure 29.17)—a metallosurfactant complex of an alkylsulfonatephosphine functionalized alkyphenol ethoxylate [63].

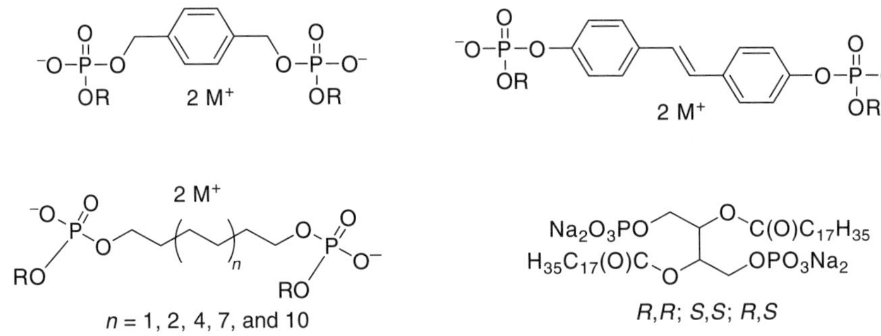

FIGURE 29.14 Phosphate ester gemini surfactants.

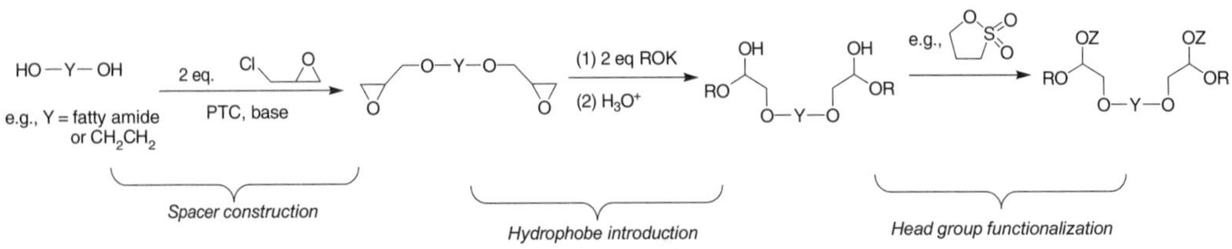

FIGURE 29.15 Synthetic strategy of sulfated or sulfonated geminis (where PTC—phase transfer catalyst). (Reproduced from Zhu, Y.-P., Masuyama, A., Kirito, Y., Okahata, M., *J. Am. Oil. Chem. Soc.*, 68, 539–543, 1991; Sumida, Y., Masuyama, A., Oki, T., Kida, T., Nakatsuji, Y., Ikeda, I., Nojima, M., *Langmuir*, 12, 3986–3990, 1996. With permission.)

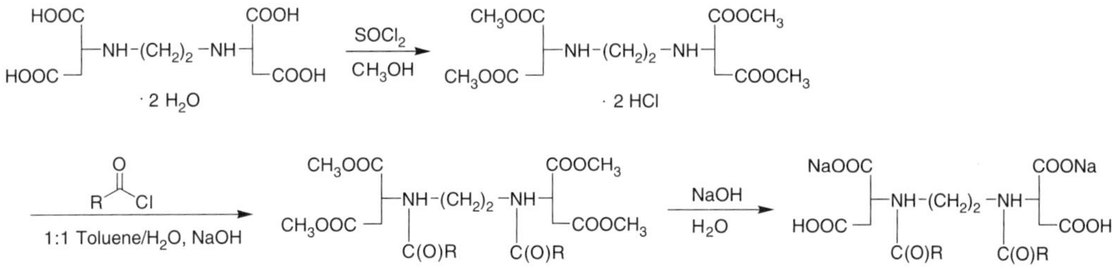

Production of Gemini Surfactants

FIGURE 29.16 Bis-aspartate gemini reported by Tsubone et al. (Reproduced from Tsubone, K., Ogawa, T., Mimura, K., *J. Surfactants Deterg.*, 6, 39–46, 2003. With permission.)

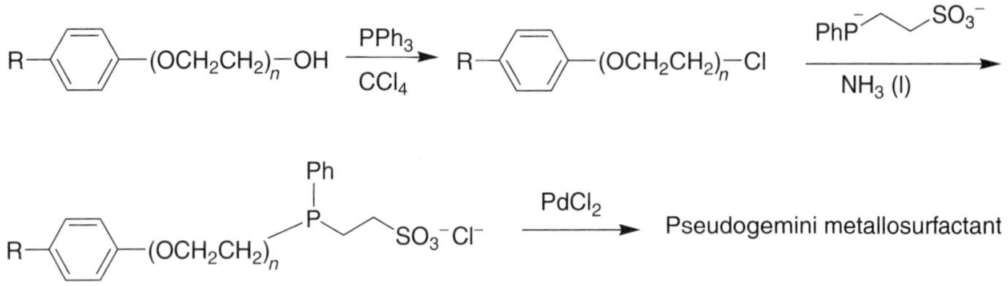

FIGURE 29.17 Pseudogemini metallosurfactant synthesis prepared by Valls et al. (Reproduced from Valls, E., Solsona, S.J., Mathieu, R., Comelles, F., Lopez-Iglesias, C., *Organometallics*, 21, 2473–2480, 2002. With permission.)

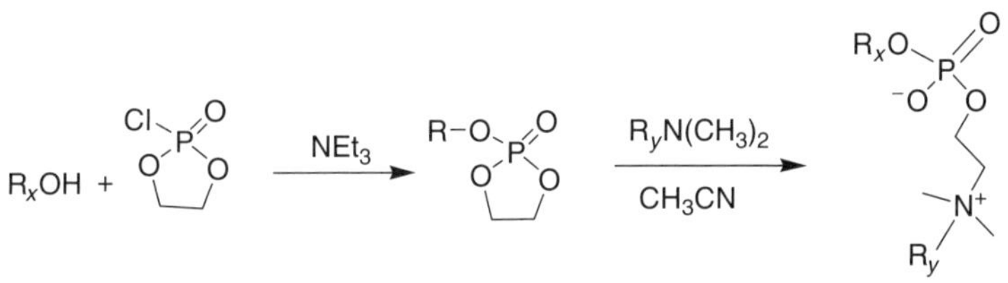

FIGURE 29.18 Two-step synthesis of zwitterionic geminis. (Reproduced from Peresypkin, A.V., Menger, F.M., *Org. Lett.*, 1, 1347–1350, 1999. With permission.)

As many gemini surfactants possess lower critical micelle concentrations (CMC) than their single-chained analogs [3], the pseudogemini complexes possessed CMC's lower than those of the individual ligand surfactants.

29.5 HETEROGEMINIS AND AMPHOTERICS

Gemini syntheses need not be excessively long or complex to afford surfactants with extremely interesting self-assembly properties. A prime example is the family of zwitterionic, neo-phosphocholine geminis developed by Peresypkin and Menger [64]. Accessible via a two-step synthetic sequence (Figure 29.18), dozens of neo-phosphocholine analogs (also referred to as *heterogeminis* [65,66]) have been prepared by simply varying the esterifying alcohol or quaternizing tertiary amine. Versions with branched [67] and fluorinated [68] chains have been reported. This has allowed for extensive physical studies probing the structure–property relationships of the geminis [66,69,70]. A nonionic/anionic "asymmetrical" heterogemini was reported by Renouf et al. [71], having one head group as a methyl-vapped poly(ethylene glycol) (PEG) and the other as a sulfonate. Another series of anionic/nonionic heterogemini surfactants was prepared using oleylnitrile as a functionalizable feedstock [72–74]. After epoxidation, reaction with methyl-capped PEG of molecular weight (MW)

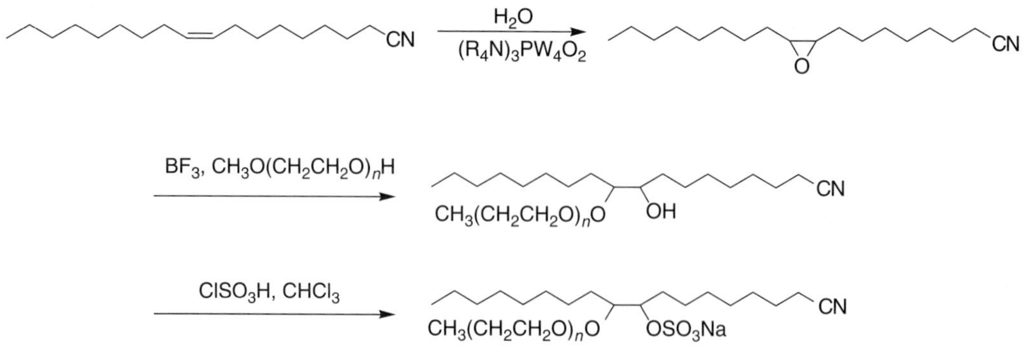

570

Handbook of Detergents/Part F: Production

FIGURE 29.19 Synthesis of nitrile heterogeminis having nonidentical head groups. (Reproduced from Alami, E., Holmberg, K., *J. Colloid Interface Sci.*, 239, 230–240, 2001, 73. Alami, E., Holmberg, K., Eastoe, J., *J. Colloid Interface Sci.*, 247, 447–455, 2002; 74. Alami, E., Abrahmsen-Alami, S., Eastoe, J., Heenan, R.K., *Langmuir*, 19, 18–23, 2003. With permission.)

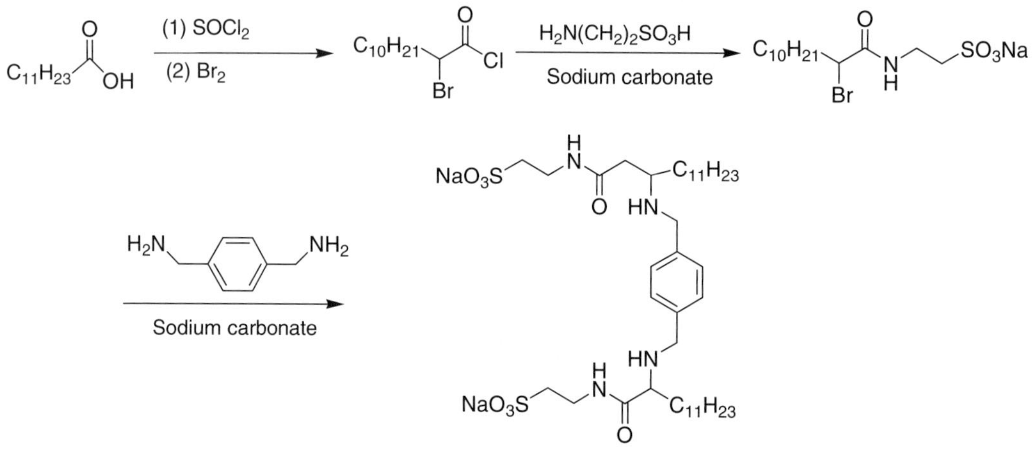

FIGURE 29.20 Synthesis of taurinate gemini. (Adapted from Tracy, D.J., Li, R., Yang, J., U.S. Patent No. 5,789,371.)

from 350 to 750 under Lewis acid catalysis yielded nonionic heterogeminis, while sulfated versions were accessible via reaction with chlorosulfonic acid (Figure 29.19).

Amphoteric surfactants are important in personal care applications and gemini versions have been synthesized for this purpose. Fischer et al. [75] prepared a bis-betaine gemini by first coupling dimethylaminoproylamine (DMAPA) with dimer acid (a feedstock often derived from tall oil), followed by a reaction with sodium monochloroacetate. Kwetkat [76] also synthesized gemini betaines through carboxymethylation of diamines. Tracy et al. [77] reported amphoteric geminis, including taurinates such as the one shown in Figure 29.20, with a variety of flexible and rigid spacers.

29.6 NONIONIC

Typically, nonionic surfactants are composed of hydrophiles composed of poly(oxyethylene) or polyhydroxyl groups. Examples of ethoxylated geminis were reported by FitzGerald et al. [78], who first constructed C-8 spacered diols, which were ethoxylated with up to 30 mol of ethylene oxide. In several previous examples, other epoxide-containing synthons were used in the preparation of a variety of gemini surfactants. Such synthons were also key to the production of a variety of nonionic gemini surfactants. Cognis/Henkel have reported nonionic geminis of the general structure

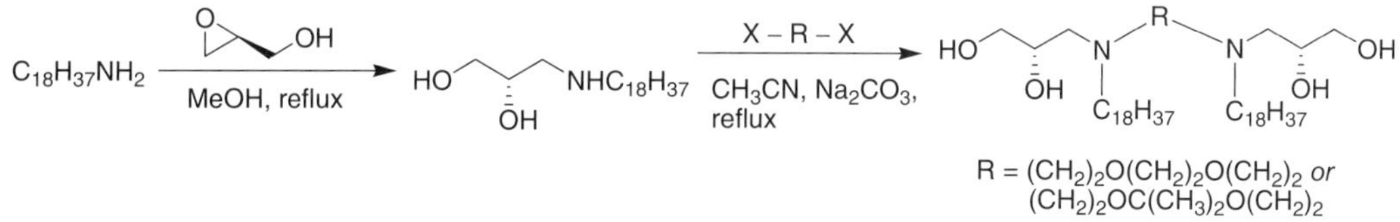

Production of Gemini Surfactants 571

FIGURE 29.21 Acid-sensitive nonionic geminis prepared by Asokan and Cho. (Reproduced from Asokan, A., Cho, M.J., *Bioconjugate Chem.*, 15, 1166–1173, 2004. With permission.)

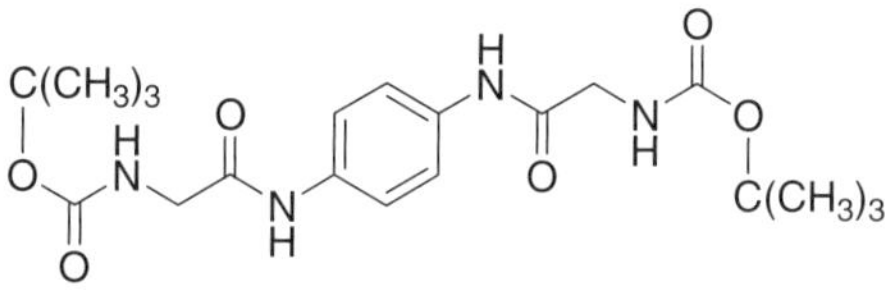

FIGURE 29.22 Krishnan et al.'s nonionic amide gemini. (Adapted from Krishnan, R.S.G., Thennarasu, S., Mandal, A.B., *J. Phys. Chem. B*, 108, 8806–8816, 2004.)

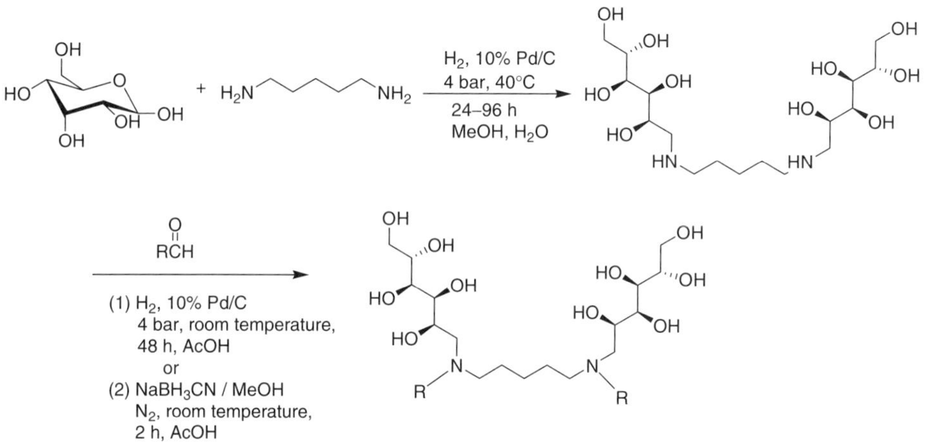

FIGURE 29.23 Bisglucamine geminis reported by Fielden et al. (Reproduced from Fielden, M.L., Perrin, C., Kremer, A., Bergsma, M., Stuart, M.C., Camilleri, P., Engberts, J.B.F.N., *Eur. J. Biochemi.*, 268, 1269–1279, 2001. With permission.)

R1–CHOH–CH$_2$-[OCH$_2$CH$_2$]x–O–CH$_2$–CHOH–R2, prepared by reacting 1,2 epoxyalkanes with polyols, which essentially constitute the hydrophilic spacer and head group portion of the surfactants [79–83]. Asokan and Cho [84] reported low hydrophile–lipophile balance (HLB), hydrolizable geminis prepared through initial reaction of *S*-glycidol with a fatty amine, followed by dimerization with an appropriate bis-halide (Figure 29.21). Another low HLB nonionic gemini was prepared by reacting a glycine-derived symmetrical anhydride with *p*-phenylene diamine (Figure 29.22) [85].

29.7 SUGAR SURFACTANTS

Carbohydrate-based surfactants are a continuing area of interest, given their green profile and perception. Several types of sugar-based geminis have been developed, both nonionic versions and ionically derivatized varieties. Engberts et al. prepared reduced-glucose amido and amino geminis with either hydrocarbon or polyether spacers [86–90]. An example synthesis is shown in Figure 29.23 [87]. In this sequence, the amino geminis are obtained through reductive alkylation procedure. Warwel et al. [91,92] took a different approach to obtain bis-glucamines, linking *N*-alkylglucamines via reaction

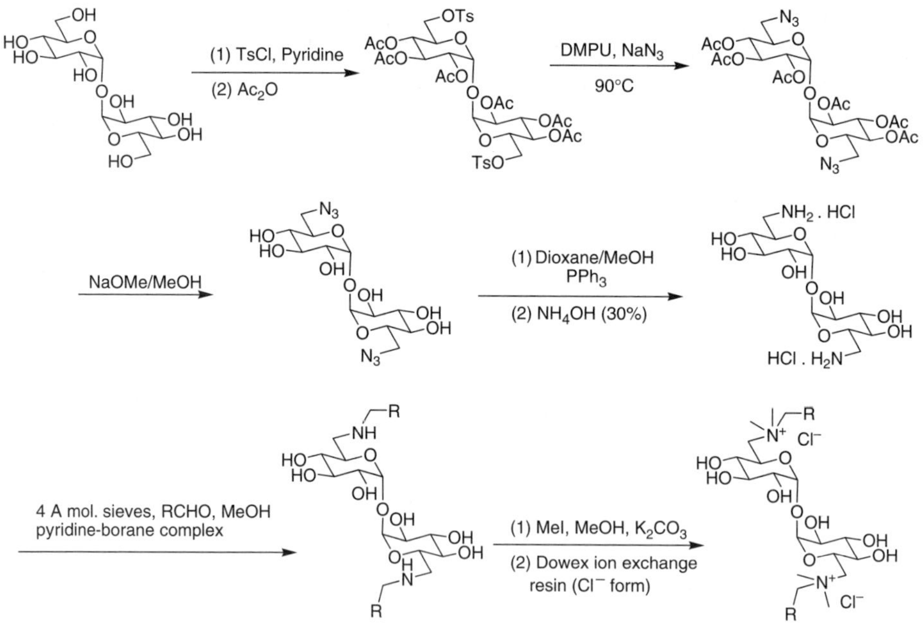

FIGURE 29.24 Synthesis of bis-quat trehalose geminis. (Reproduced from Menger, F.M., Mbadugha, B.N.A., *J. Am. Chem. Soc*, 123, 875–885, 2001. With permission.)

with bis-epoxides. In this manner, they were able to access a wide array of multimeric glucamine geminis with different spacer chemistries, including polyethers, polyhydroxyls, and polyaromatics. Other reduced-glucose geminis having siloxane hydrophobic chains [93] were reported.

Closed sugars have been incorporated into dimeric and trimeric gemini surfactants, where the sugar groups serve as the hydrophilic head groups. Gao et al. [94] utilized enzymes to regioselectively incorporate fatty acids onto a variety of mono- and disaccharides. By reacting bis-acyl chlorides as a linking group, Castro et al. [95–97] accessed multiple gemini alkyl glucosides. Menger and Mbadugha [18] took advantage of the symmetry of trehalose to produce geminis that were both nonionic and cationic in nature. The synthesis of the bis-quat versions are shown in Figure 29.24.

29.8 AMINO ACID–BASED GEMINIS

Figures 29.16 and 29.20 show examples of amino acid gemini surfactants, where the amino acid groups made up the head groups. This is one format for amino acid–based geminis, and similarly synthesized examples are known for beta-alaninate [98] and arginine [99]. Other types have been prepared, taking advantage of the symmetry of disulfide-containing amino acids; specific examples include geminis with spacers composed of a dimer of ornithin-extended cysteine [100] or cystine [101]. The most extensive reports of peptide-based gemini surfactants came from the European Network on Gemini Surfactants (ENGEMS) group, who pursued gemini surfactants as drug or genetic delivery agents [102–107]. An example of one of their surfactant syntheses, with polylysine groups as the head groups (AA) and a variety of hydrophobes is shown in Figure 29.25 [103].

29.9 COMMERCIAL GEMINI SURFACTANTS

To date, few of the many novel gemini surfactants developed have made their way out of the laboratory and onto the store shelf. Two families of long-existing commercial products that may be thought of as gemini surfactants are diamine-based block copolymers such as BASF's Tetronic® line and Air Products' various alkyne-containing surfactants. Diamine block copolymers, used as

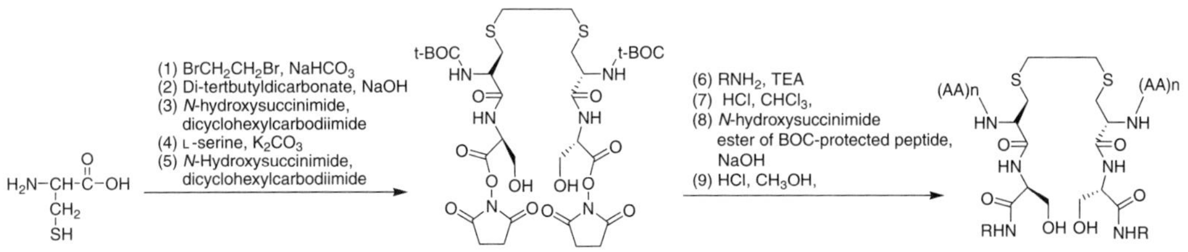

Production of Gemini Surfactants

FIGURE 29.25 Synthesis of peptide-based gemini surfactant. (Reproduced from McGregor, C., Perrin, C., Monck, M., Camilleri, P., Kirby, A.J., *J. Am. Chem. Soc.*, 123, 6215–6220, 2001. With permission.)

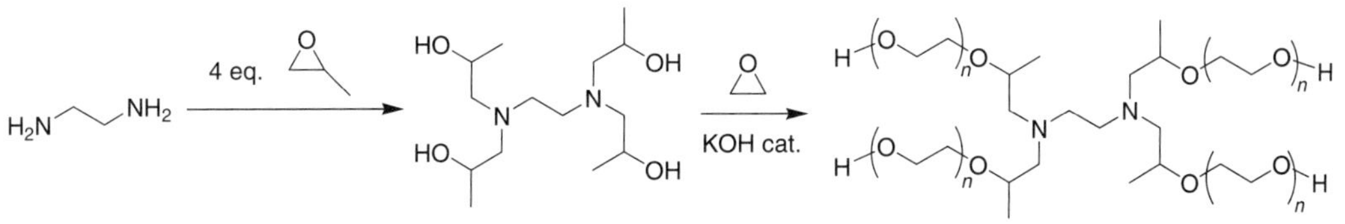

FIGURE 29.26 Example synthesis of a diamine alkoxylate block copolymer. (Reproduced from Nace, V.M., *Nonionic Surfactants: Polyoxyalkylene Block Copolymers*, Marcel Dekker, New York, 1996. With permission.)

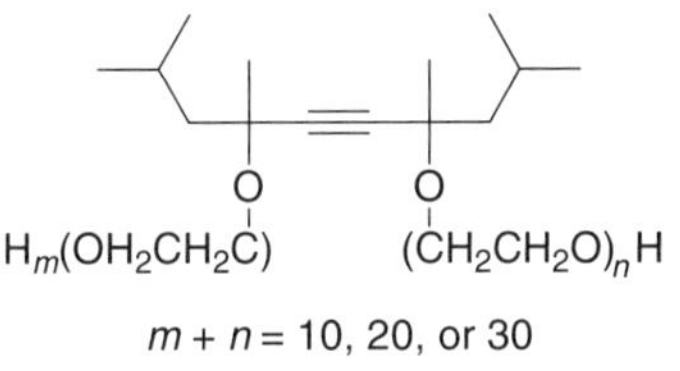

$m + n = 10, 20,$ or 30

FIGURE 29.27 Example structure of Air Products acetylenic diol nonionic geminis.

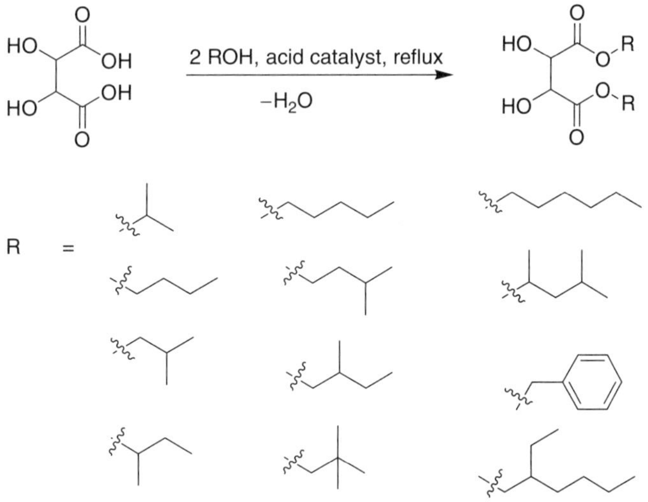

FIGURE 29.28 Dialkyl tartrate esters prepared via acid catalysis.

dispersants and emulsifiers in a broad range of applications, are conventionally manufactured by reacting at least four equivalents of propylene oxide, followed by ethoxylation under basic catalysis (Figure 29.26) [108]. The resulting molecules can be considered a "tetrafunctional" polymeric surfactant tethered at the hydrophobe termini. Air Products' acetylenic diol–based chemistries anchor what can be considered gemini-type structures, where ethoxylating the diols further produce nonionic surfactants such as the structure shown in Figure 29.27 [109].

Air Products has recently launched a class of products known as *EnviroGem*® surfactants, having excellent defoaming and wetting capacities. Air Products has described, in the patent literature, low HLB surfactants in this vein, including esters of dicarboxylic acids such as the tartaric diesters shown in Figure 29.28 [17]. The esters are produced via acid-catlyzed esterification, where the catalyst can be an acid ion exchange resin or *p*-toluenesulfonic acid (PTSA), and the reaction is further

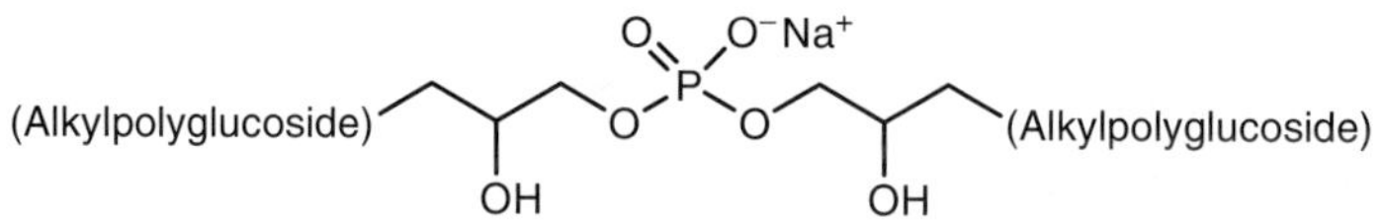

FIGURE 29.29 Structure of Suga-Phos surfactants produced Colonial Chemical. (Adapted from Sloane, C.S., Lassila, K.R., US Patent No. 6,423,376.)

driven by azeotropic water distillation with the reacting alcohol using a Dean–Stark trap. These surfactants were reported as potentially useful in applications such as inks, coatings, and adhesives.

Finally, Colonial Chemical has introduced anionic Suga®Phos surfactants, marketed as a gemini surfactant with mild properties for use in personal care applications [110]. The surfactants (Figure 29.29) are apparently produced by linking alkylpolyglucosides with a hydroxypropyl phosphate spacer via a chlorohydroxypropyl phosphate intermediate.

Given the amount of research carried out on gemini surfactants in the past 15 years, why have so few products been commercialized? Majority of the volume of manufactured surfactants are commodity in nature, with well-established supply chains, production assets, and processes already in place. It is unlikely that an upstart gemini could, for instance, easily break into the mass laundry detergent market. For successful entry into commercialization, a gemini would need to possess a favorable cost–performance profile as either a displacing or an enabling technology. To date, so far, it seems that geminis in actual applications have not shown enough performance benefit or profit potential to justify necessary expenses such as capital investment and product registration. Consider, for example, the widely examined m-s-m bis-quaternary geminis ($s = 2$ or 3) that have the potentially valuable quality to viscosify aqueous media—could they compete with currently used surfactants? Viscosification of water-based personal care products is readily accomplished with blends of low-cost conventional surfactants such as sodium laureth sulfate and cocoamidopropyl betaine. If gemini 12-2-12 dichloride could be manufactured via an hypothetical three-step process by (1) bis-amidation of ethylenediamine (EDA, $1.38/lb) with lauric acid ($0.54/lb) and (2) a "generic" reduction to the bis-amine, followed by exhaustive quaternization with methyl chloride ($0.37/lb), then based on March 2005 raw material prices [111; values in U.S. dollars], and conservatively assuming $0.10/lb overhead/disposal cost for each step, 100% active surfactant is estimated to approximately cost $0.93/lb*. Assuming the final product would be manufactured in a 50% active form in water, the cost "without" added margin would be in approximate parity with the commodity "viscosifying" surfactants. Unless the gemini can be used at a significantly lower level in a formulation than the existing surfactants, this does not represent a strong economic case for a displacing technology. A higher-end application might better tolerate the price point of the surfactant.

Thus, it can be concluded that since gemini surfactants differ from conventional surfactants by the presence of the connecting spacer, reducing the cost of production of this core component is critical. For 12-2-12, the spacer group constitutes about one-tenth of the total molecular weight of the final surfactant, and EDA is among the lower-priced spacer building blocks among those included in this chapter. Epichlorohydrin, a key raw material in many of the spacer groups described earlier, is fairly attractive pricewise ($1,500/t [111; values in U.S. dollars]), but is a carcinogen that requires special handling protocols, which add to the expense. Tartaric acid ($2.50/lb [111; values in U.S. dollars]), basis of the EnviroGem surfactants described earlier, represents about 30% of the overall molecular weight of the bis-2-ethylhexanol version. For the C-12 chained version of the sugar-based bis-quat shown in Figure 29.24, the disaccharide trehalose, a fine chemical estimated at over $5/lb, represents over 40% of the gemini molecular weight. Although such costs would seem to relegate geminis to applications in the pharmaceutical or electrochemical sectors, opportunities

---

* This rough process is "back of the envelope" and is not based on any publication or patent.

Production of Gemini Surfactants

clearly exist for process development and improvement or implementation of raw material alternatives that are not detrimental to performance. Based on the patented work since the 1990s, gemini surfactants have clearly been in the sights of industry—successful gemini product entries in the coming few years will tell if manufacturers have been investing in the necessary research and development efforts.

29.10 CONCLUSION

Gemini surfactants remain a vibrant field of research in the academic ranks, and the first products with the gemini "label" have now entered the market. Great opportunities are present for geminis in ongoing surfactant trends in areas such as green chemistry, nanotechnology and nanotemplating, and drug and genetic delivery systems. Advances in the conception of new gemini surfactants, exploration of fundamental properties, and development of application are necessary to build on the momentum of the past 15 years. Ultimately, however, the success of gemini surfactants certainly depends on the capability for chemical manufacturers to produce them for markets, which can absorb cost structures that are higher than the commodity surfactants, or determine processes that will bring their costs down without losing the properties that differentiate them from conventional surfactants. The synthetic work cited in this chapter is the foundation for such efforts, which will hopefully lead to a bright future for geminis.

REFERENCES

1. Rosen, M.J. *CHEMTECH* 1993, 30–33.
2. Zana, R. *Curr. Opin. Colloid Interface Sci.* 1996, *1*, 566–571.
3. Menger, F.M., Keiper, J.S. *Angew. Chem. Int. Ed. Engl.* 2000, *39*, 1906–1920.
4. Hait, S.K., Moulik, S.P. *Curr. Sci.* 2002, *82*, 1101–1111.
5. Zana, R., Xia, J. *Gemini surfactants: Synthesis, Interfacial and Solution-Phase Behavior, and Applications*, Marcel Dekker, New York, 2003.
6. Bunton, C.A., Robinson, L.A., Schaak, J., Stern, M.F. *J. Org. Chem.* 1971, *36*, 2346–2350.
7. Devinsky, F., Masarova, L., Lacko, I. *J. Colloid Interface Sci.* 1985, *105*, 235–239.
8. Zana, R., Benrraou, M., Rueff, R. *Langmuir* 1991, *7*, 1072–1075.
9. Menger, F.M., Littau, C.A. *J. Am. Chem. Soc.* 1991, *113*, 1451–1452.
10. Menger, F.M., Littau, C.A. *J. Am. Chem. Soc.* 1993, *115*, 10083–10090.
11. Zana, R. *J. Colloid Interface Sci.* 2002, *248*, 203–220.
12. Diamant, H., Andelman, D. *Langmuir* 1994, *10*, 2910–2916.
13. Diamant, H., Andelman, D. *Langmuir* 1995, *11*, 3605–3606.
14. De, S., Aswal, V.K., Goyal, P.S., Bhattacharya, S. *J. Phys. Chem.* 1996, *100*, 11664–11671.
15. Maiti, P.K., Chowdhury, D. *J. Chem. Phys.* 1998, *109*, 5126–5133.
16. Menger, F.M., Keiper, J.S., Azov, V. *Langmuir* 2000, *16*, 2062–2067.
17. Sloane, C.S., Lassila, K.R. US Patent No. 6,423,376.
18. Menger, F.M., Mbadugha, B.N.A. *J. Am. Chem. Soc.* 2001, *123*, 875–885.
19. Fuhrhop, J.-H., Wang, T. *Chem. Rev.* 2004, *104*, 2901–2938.
20. Danino, D., Talmon, Y., Zana, R. *Langmuir* 1995, *11*, 1448–1456.
21. Alami, E., Beinert, M.P., Zana, R. *Langmuir* 1993, *9*, 1465–1467.
22. Aswal, V.K., De, S., Goyal, P.S., Bhattacharya, S., Heenan, R.K. *Phys. Rev. E.* 1998, *57*, 776–783.
23. Li., Z.X., Dong, C.C., Thomas, R.K. *Langmuir* 1999, *15*, 4392–4396.
24. Menger, F.M., Keiper, J.S., Mbadugha, B.N.A., Caran, K.L., Romsted, L.S. *Langmuir* 2000, *16*, 9095–9098.
25. Buhler, E., Mendes, E., Boltenhagen, P., Munch, J.P., Zana, R., Candau, S.J. *Langmuir* 1997, *13*, 3096–3102.
26. Ryhanen, S.J., Pakkanen, A.L., Saeily, M.J., Bello, C., Mancini, G., Kinnunen, P.K.J. *J. Phys. Chem.* 2002, *106*, 11694–11697.
27. Caracciolo, G., Mancini, G., Bombelli, C., Luciani, P., Caminiti, R. *J. Phys. Chem. B* 2003, *107*, 12268–12274.

28. Mirviss, S.B., Steichen, D., Spellane, P.J., Cho, H.J. World Patent No. 00/39241.
29. In, M., Bec, V., Aguerre–Chariol, O., Zana, R. *Langmuir* 2000, *16*, 141–148.
30. Oda, R., Huc, I., Candau, S.J. *Chem. Commun.* 1997, 2105–2106.
31. Sikiric´, M., Sÿmit, I., Tusek-Bozic´, L., Tomasic´,V., Pucic´, I., Primozic, I., Filipovic´-Vincekovic, N. *Langmuir* 2003, *19*, 10044–10053.
32. Song, L.D., Rosen, M.J. *Langmuir* 1996, *12*, 1149–1153.
33. Rosen, M.J., Song, L.D. *J. Colloid Interface Sci.* 1996, *179*, 261–268.
34. Tatsumi, T., Zhang, W., Kida, T., Nakatsuji, Y., Miyake, K., Matsushima, K., Tanaka, M., Furuta, T., Ikeda, I. *J. Surfact. Deterg.* 2001, *4*, 271–277.
35. Tatsumi, T., Zhang, W., Nakatsuji, Y., Ono, D., Takeda, T., Ikeda, I. *J. Surfact. Deterg.* 2001, *4*, 279–285.
36. Li, Z.X., Dong, C.C., Wang, J.B., Thomas, R.K., Penfold, J. *Langmuir* 2002, *18*, 6614-6622.
37. De, S., Aswal, V.K., Goyal, P.S., Bhattacharya, S. *J. Phys. Chem. B* 1998, *102*, 6152–6160.
38. Dreja, M., *Chem. Commun.* 1998, *13*, 1371–1372.
39. Wettig, S.D., Li, X., Verrall, R.E. *Langmuir* 2003, *19*, 3666–3670.
40. Brinchi, L., Germani, R., Goracci, L., Savelli, G., Bunton, C.A. *Langmuir* 2002, *18*, 7821–7825.
41. Eastoe, J., Sanchez Dominguez, M., Wyatt, P., Beeby, A., Heenan, R.K. *Langmuir* 2002, *18*, 7837–7844.
42. Logan, R.B. U.S. Patent 4,734,277.
43. Kim, T.-S., Kida, T., Nakatsuji, Y., Ikeda, I. *Langmuir* 1996, *12*, 6304–6308.
44. Kim, T.-S., Hirao, T., Ikeda, I. *J. Am. Oil Chem. Soc.* 1996, *73*, 67–71.
45. Kim, T.-S., Kida, T., Nakatsuji, Y., Hirao, T., Ikeda, I. *J. Am. Oil Chem. Soc.* 1996, *73*, 907–911.
46. Ricci, C.G., Cabrera, M.I., Luna, J.A., Grau, R.J. *J. Surfact. Deterg.* 2003, *6*, 231–237.
47. Quagliotto, P., Viscardi, G., Barolo, C., Barni, E., Bellinvia, S., Fisicaro, E., Compari, C. *J. Org. Chem.* 2003, *68*, 7651–7660.
48. Buwalda, R.T., Engberts, J.B.F.N. *Langmuir* 2001, *17*, 1054–1059.
49. Zhai, X., Zhang, L., Liu, M. *J. Phys. Chem. B* 2004, *108*, 7180–7185.
50. Huc, I., Oda, R. *Chem. Commun.* 1999, *20*, 2025–2026.
51. Oda, R., Laguerre, M., Huc, I., Desbat, B. *Langmuir* 2002, *18*, 9659–9667.
52. Yokokawa, S., Tamada, K., Ito, E., Hara, M. *J. Phys. Chem. B.* 2003, *107*, 3544–3551.
53. Lyu, Y.-Y., Yi, S.H., Shon, J.K., Chang, S., Pu, L.S., Lee, S.-Y., Yie, J.E., Char, K., Stucky, G.D., Kim, J.M. *J. Am. Chem. Soc.* 2004, *126*, 2310–2311.
54. Lyu, Y.-Y., Chang, S., Kim, J.M., Park, J.G. U.S. Patent No. 2004/0138087.
55. Pindzola, B.A., Jin, J., Gin, D.L. *J. Am. Chem. Soc.* 2003, *125*, 2940–2949.
56. Duivenvoorde, F.L., Feiters, M.C., van der Gaast, S.J., Engberts, J.B.F.N. *Langmuir* 1997, *13*, 3737–3743.
57. Sommerdijk, N.A.J.M., Hoks, T.H.L., Synak, M., Feiters, M.C., Nolte, R.J.M., Zwanenburg, B. *J. Am. Chem. Soc.* 1997, *119*, 4338–4344.
58. Zhu, Y.-P., Masuyama, A., Kirito, Y., Okahata, M. *J. Am. Oil. Chem. Soc.* 1991, *68*, 539–543.
59. Sumida, Y., Masuyama, A., Oki, T., Kida, T., Nakatsuji, Y., Ikeda, I., Nojima, M. *Langmuir* 1996, *12*, 3986–3990.
60. Murguia, M.C., Grau, R.J. *Synlett* 2001, *8*, 1229–1232.
61. Acharya, D.P., Kuneida, H., Shiba, Y., Aratani, K. *J. Phys. Chem. B* 2004, *108*, 1790–1797.
62. Tsubone, K., Ogawa, T., Mimura, K. *J. Surfact. Deterg.* 2003, *6*, 39–46.
63. Valls, E., Solsona, S.J., Mathieu, R., Comelles, F., Lopez-Iglesias, C. *Organometallics* 2002, *21*, 2473–2480.
64. Peresypkin, A.V., Menger, F.M. *Org. Lett.* 1999, *1*, 1347–1350.
65. Seredyuk, V., Holmberg, K. *J. Colloid Interface Sci.* 2001, *241*, 524–526.
66. Seredyuk, V., Alami, E., Nyden, M., Holmberg, K., Peresypkin, A.V., Menger, F.M. *Langmuir* 2001, *17*, 5160–5165.
67. Menger, F.M., Seredyuk, V.A., Apkarian, R.P., Wright, E.R. *J. Am. Chem. Soc.* 2002, *124*, 12408–12409.
68. DeSimone, J.M., Keiper, J.S., Menger, F.M., Peresypkin, A.P., Clavel, C. World Patent No. 02/59243.
69. Menger, F.M., Peresypkin, A.V. *J. Am. Chem. Soc.* 2001, *123*, 5614–5615.
70. Menger, F.M., Peresypkin, A.V. *J. Am. Chem. Soc.* 2003, *125*, 5340–5345.
71. Renouf, P., Mioskowski, C., Lebeau, L. *Tetrahedron Lett.* 1998, *39*, 1357–1360.
72. Alami, E., Holmberg, K. *J. Colloid Interface Sci.* 2001, *239*, 230–240.
73. Alami, E., Holmberg, K., Eastoe, J. *J. Colloid Interface Sci.* 2002, *247*, 447–455.

Production of Gemini Surfactants

74. Alami, E., Abrahmsen-Alami, S., Eastoe, J., Heenan, R.K. *Langmuir* 2003, *19*, 18–23.
75. Fischer, P., Rehage, H., Gruening, B. *J. Phys. Chem. B* 2002, *106*, 11041–11046.
76. Kwetkat, K. U.S. Patent No. 6,034, 271.
77. Tracy, D.J., Li, R., Yang, J. U.S. Patent No. 5,789,371.
78. FitzGerald, P.A., Carr, M.C., Davey, T.W., Serelis, A.K., Such, C.H., Warr, G.G. *J. Colloid Interface. Sci.* 2004, *275*, 649–658.
79. Elsner, M., Weuthen, M., Raths, H.-C. U.S. Patent No. 6,666,217.
80. Kischkel, D., Rather, H.-C., Weuthen, M., Elsner, M. U.S. Patent No. 2003 / 78,182.
81. Rather, H.-C., Weuthen, M., Elsner, M. U.S. Patent No. 6,777,384.
82. Elsner, M., Weuthen, M., Raths, H.-C. U.S. Patent No. 6,794,345.
83. Elsner, M., Kischke, D., Weuthen, M. U.S. Patent No. 6,805, 141.
84. Asokan, A., Cho, M.J. *Bioconjugate Chem.* 2004, *15*, 1166–1173.
85. Krishnan, R.S.G., Thennarasu, S., Mandal, A.B. *J. Phys. Chem. B* 2004, *108*, 8806–8816.
86. Pestman, J.M., Terpstra, K.R., Stuart, M.C.A., Van Doren, H.A., Brisson, A., Kellogg, R.M., Engberts, J.B.F.N. *Langmuir* 1997, *13*, 6857–6860.
87. Fielden, M.L., Perrin, C., Kremer, A., Bergsma, M., Stuart, M.C., Camilleri, P., Engberts, J.B.F.N. *Eur. J. Biochem.* 2001, *268*, 1269–1279.
88. Bell, P.C., Bergsma, M., Dolbnya, I.P., Bras, W., Stuart, M.C.A., Rowan, A.E., Feiters, M.C., Engberts, J.B.F.N. *J.Am. Chem. Soc.* 2003, *125*, 1551–1558.
89. Johnsson, M., Wagenaar, A., Engberts, J.B.F.N. *J. Am. Chem. Soc.* 2003, *125*, 757–760.
90. Johnsson, M., Wagenaar, A., Stuart, M.C.A., Engberts, J.B.F.N. *Langmuir* 2003, *19*, 4609–4618.
91. Warwel, S., Bruse, F., Schier, H. *J. Surfact. Deterg.* 2004, *7*, 181–186.
92. Warwel, S., Bruse, F. *J. Surfact. Deterg.* 2004, *7*, 187–193.
93. Han, F., Zhang, G. *J. Surfact. Deterg.* 2004, *7*, 175–180.
94. Gao, C., Millqvist-Fureby, A., Whitcombe, M.J., Vulfson, E.N. *J. Surfact. Deterg.* 1999, *2*, 293–302.
95. Castro, M.J.L., Kovensky, J., Fernandez Grelli, A. *Tetrahedron Lett.* 1997, *38*, 3995–3998.
96. Castro, M.J.L., Kovensky, J., Cirelli, A.F. *Tetrahedron* 1999, *55*, 12711–12722.
97. Castro, M.J.L., Kovensky, J., Fernandez Grelli, A. *Langmuir* 2002, *18*, 2477–2482.
98. Tsubone, K., Arakawa, Y., Rosen, M.J. *J. Colloid Interface Sci.* 2003, *262*, 516–524.
99. Castillo, J.A., Pinazo, A., Carilla, J., Infante, M.R., Alsina, M.A., Hara, I., Clapes, P. *Langmuir* 2004, *20*, 3379–3387.
100. Dauty, E., Remy, J.-S., Blessing, T., Behr, J.-P. *J. Am. Chem. Soc.* 2001, *123*, 9227–9234.
101. Menger, F.M., Zhang, H., Caran, K.L., Seredyuk, V.A., Apkarian, R.P. *J. Am. Chem. Soc.* 2002, *124*, 1140–1141.
102. Jennings, K., Marshall, I., Birrell, H., Edwards, A., Haskins, N., Sodermann, O., Kirby, A.J., Camilleri, P. *Chem. Commun.* 1998, 1951–1952.
103. McGregor, C., Perrin, C., Monck, M., Camilleri, P., Kirby, A.J. *J. Am. Chem. Soc.* 2001, *123*, 6215–6220.
104. Jennings, K.H., Marshall, I.C.B., Wilkinson, M.J., Kremer, A., Kirby, A.J., Camilleri, P. *Langmuir* 2002, *18*, 2425–2429.
105. Camilleri, P., Feiters, M.C., Kirby, A.J., Ronsin, G.A.B., Nolte, R.J.M., Garcia, C.L. World Patent No. 03/82,809.
106. Camilleri, P., Kremer, A., Rice, S.Q.J.R. U.S. Patent No. 2004/43,939.
107. Camilleri, P., Kirby, A.J., Perrin, C., Ronsin, G.A.B., Guedat, P. U.S. Patent No. 2004/138,139.
108. Nace, V.M. *Nonionic Surfactants: Polyoxyalkylene Block Copolymers*, Marcel Dekker, New York, 1996.
109. Nieh, M.-P., Kumar, S.K., Fernando, R.H., Colby, R.H., Katsaras, J. *Langmuir*, 2004, *20*, 9061–9068.
110. O'Lenick, A.J., O'Lenick, K.A. U.S. Patent No. 6,627,612.
111. Chemical Market Reporter, March 21, 2005.
112. Menger, F.M., Migulin, V.A. *J. Org. Chem.* 1999, *64*, 8916–8921.
113. Pavlikova, M., Lacko, I., Devinsky, F., Mlynaereik, D. *Collect Czech. Chem. Commun.* 1995, *60*, 1213–1228.

Index

A

ABS. *See* branched-chain alkylbenzene (ABS)
acetals. *See* alkyl glucosides/glycosides
acid digester for methyl ester sulfonate, 209–210
acid *vs.* neutral bleaching of methyl ester sulfonate, 202–203
active chlorine in bleaching, 422–424
    calcium hypochlorite, 423
    introduction, 422
    organic chlorinated compounds, 423–424
    sodium hypochlorite, 422
active oxygen (peroxygen), 424–433
    hydrogen peroxide, 424–426
    inorganic peracids, 433
    introduction, 424
    percarboxylic acids, 430–433
        equilibrium peracids, 432
        preformed peracids, 433
        reaction with activators, 430–432
    persalts, 426–430
        sodium perborate, 426–428
        sodium percarbonate, 428–430
acyl/dilkyl ethylenediamines, 223–228
    amino propionates, 227
    amphoacetates, 15–16, 225–227
    carboxyamphoterics, 228, 229
    coco derivatives, 227–228
ADMA. *See* alkyl dimethyl amine (ADMA)
agglomeration of powdered detergents, 337–345
    defined, 338
    dense laundry process, 341
    dry neutralization, 344–345
    factors in, 338–339, 340–341
    generic description, 339–340
    high- and low-density agglomerates, 343–344
    introduction, 337–338
    mixer operation, 342–343
    process control strategies, 341–342
    regimes, 339
alcohols, 117–137. *See also* esters
    alcohol sulfates, 126–135
        alternative technology, 128–135
        introduction, 117–118
        natural sulfation, 126–128
        new technology, 128–129, 133–135
        oleo-derived, 133–135
        as source of future surfactants, 128–129
        synthetic sulfation, 126–128
    alkyl glucosides/glycosides, 70–71
    detergent range, 507–512
    fatty alcohol-based surfactants, 239–246
        alkylpolyglucosides, 241–242
        economic considerations, 512
        ethoxylation, 29
        fatty acid glucamides, 242–244
        fatty alkanolamides, 239–241
        historical perspective, 8–10

        introduction, 510
        natural C12-16 fatty alcohols, 28
        raw materials, 510–511
        sulfosuccinates, 244–245
        summary, 245–246
    isopropyl alcohol (2-propanol), 512–513
    oleo-derived, 118–120
        new technology, 133–135
        processes, 119–120
        raw materials, 118
    oleyl alcohol, 512
    petroleum-derived, 120–126, 129–133
        ethylene-derived processes, 122–124
        ethyl process, 124–125
        linear and branched, 120–126
        paraffin-derived processes, 125–126
        raw materials and olefin chemistry, 120–122
    polyvinyl alcohol for detergent pouches, 329, 360–362
    quaternary surfactants, 370–371
    sulfosuccinate production, 244–245
aliphatic alcohol. *See* fatty acids/fatty alcohols
alkaline hydrolysis process for soap production, 478
alkaline proteases, 532
n-alkanes, 146
alkanesulfonates, 139–157. *See also* secondary
        alkanesulfonate (SAS)
    introduction, 139–141
    lignosulfonates, 152–154
        consumption and uses, 154
        formation, 152–154
    petroleum sulfonates, 151–152
        consumption and uses, 152
        manufacture, 151–152
    primary alkanesulfonates, 142
alkanolamides
    DEA-based, 25
    fatty, 239–241
    monoalkanolamides, 25
    as nonionic surfactants, 25–26
alkanolamine esters (esterquats), 370–371
alkanolamines, 516–518
alkoxylated surfactants. *See also* ethoxylated surfactants
    overview, 26–29
    propoxylated surfactants, 27, 257
alkylamido betaines, 231–233
alkylamidopropyl betaines, 13, 482
N-alkylamino acids, 228, 230
alkylated phenol, 28
alkylated surfactants, 39–48
    dialkylation of phenols, 52
    future trends, 47
    homogenous *vs.* heterogeneous catalysts, 54
    n-olefins, 45–47
        α-olefins, 45
        applications, 45
        internal olefins, 45–46
        market demand for, 45

579

580 Index

alkylated surfactants (*contd.*)
  n-paraffins
    liquid-phase extraction, 47
    vapor-phase extraction, 47
  n-paraffins, 46–47
    Fisher-Tropsch process, 47
    linear alkylbenzene production, 46
  perfluroalkanesulfonamides, 310–313
  technology development, 47
alkylbenzenes, 39–44. *See also* linear alkylbenzene
    (LAB)
  alkylation, 41–44
    aluminum chloride (ALCl$_3$) alkylation, 42–43
    Detal process, 43
    hydrofloric alkylation, 41–42
    solid bed catalyst alkylation, 43
  branched alkylbenzenes, 39, 40, 47
  consumption, 39–40
  introduction, 39
  quality, 44–45
alkylbenzene sulfonate. *See* linear alkylbenzene
    sulfonate (LAS)
alkyl betaines, 13, 231–232
alkyl dimethyl amine (ADMA), 13
alkylene oxide
  propylene glycol ethers, 525–526
  safety and handling of, 267–269
alkyl ether sulfates, 479
alkyl fluoro fluids, 290
alkylglucamides, 242–244
alkyl glucosides/glycosides, 69–81
  chemistry, 71–77
    acetals, 71, 73
    degree of polymerization, 74–76
    D-glucose, 72–75
    hemiacetals, 71–73
    high performance liquid chromatography,
      76–77
    high temperature gas chromatography, 76–77
    lipophilicity (hydrophobicity), 74–75
    lipophobieity (hydrophilicity), 74–75
  defined, 69
  introduction, 69
  process and technology, 77–79
    direct (one-step) reaction, 77–78
    transacetalization (two-step) reaction, 77
  raw materials, 69–71
    alcohols, 70–71
    cornstarch, 69–70
  research and development, 79–80
  summary, 80
alkyl isethionates, 480
alkylnaphthalene sulfonates as hydrotropes, 247
alkylphenol ethoxylate (APE), 61–67
  di-nonylphenol, 62
  di-sec butylphenol, 63
  environmental considerations, 66–67
  future trends, 66–67
  introduction, 61–62
  p-dodecyl phenol, 62
  p-nonylphenol, 62
  p-tert-octylphenol, 62
  raw materials for, 63–65

    cumene, 63–64
    discussion, 65
    olefins, 65
    phenol, 63–65
alkylphenols (alkylphenolethoxylates), 49–67. *See also*
    alkylphenol ethoxylate (APE)
  defined, 49
  economics considerations, 59–61
    demand for alkylphenols, 60–61
    non-raw material costs, 60
    raw material costs, 59–60
  introduction, 49–50
  major applications, 50–52
    di-sec butylphenol, 52
    di-nonylphenol, 52
    4-dodecylphenol, 51
    introduction, 50–51
    4-nonylphenol, 51
    4-*tert*-octylphenol, 51
  nomenclature, 50
  processes, 52–59
    health and safety, 58–59
    physical properties, 58–59
    product isolation steps, 57–58
    reaction mechanism, 52–54
    reaction steps, 54–57
    reactor configuration, 54–57
  product research, 66–67
alkylpolyglucoside (APG), 29–30, 241–242
alkyl polyglycosides, 481–482
alkylsulfosuccinates, 244–245
all-block polyalkylene oxide block polymer
    subgroup, 254
all-heteric surfactants
  EO/PO block copolymers, 263–264
  polyalkylene oxide block polymer subgroup, 254
α-olefin sulfonate (AOS), 19, 102–109
aluminum chloride alkylation, 42–43
amide formation and purification, alkylamido betaines,
    232–233
amid ethoxylates, 29
amine oxides, 13–14, 15, 235
amino acid-based geminis, 572
amino propionates, 228
amorphous silica for toothpastes, 486–487
amphiphilic moieties' linkage (spacers), gemini, 562–563
amphoacetates (amphocarboxyglycinates), 15–16,
    225–227
amphopropionates, 15–16, 16
amphoteric surfactants, 13–16, 221–237
  acyl/dilkyl ethylenediamines, 223–228
    amino propionates, 228
    amphoacetates, 15–16, 225–227
    carboxyamphoterics, 228, 229
    coco derivatives, 227–228
  N-alkylamino acids, 229–230
  alkyl dimethyl amine, 13
  amine oxides, 13, 15, 235
  amphoacetates (amphocarboxyglycinates), 15–16
  amphopropionates, 15–16
  background, 222
  betaines, 13–15, 231–235
    alkylamido betaines, 231–233

Index

alkyl betaines, 13, 231–232
    imidazolinium betaines, 234
    sulfo betaines, 234–235
challenges facing, 16
characteristics, 223
classes, 13–16, 223, 224
conclusion, 236
dimethylaminopropylamine, 13
geminis, 569–570
introduction, 221
amylases, host organisms of, 533
anhydrous sodium metasilicate, 399–401
    direct fusion route, 400
    granulation of solution, 400–401
    introduction, 399–400
    solid-state reaction, 400
anhydrous sodium silicate lumps/cullets, 388–393
    introduction, 388–389
    open-hearth regenerative furnace process, 391–393
    quartz sand, 391
    raw materials, 390–391
    rotary furnace process, 393
    soda ash, 389–391
anionic surfactants. *See also* alcohols; sulfated
        surfactants; sulfonated surfactants
    classes and production, 16–21
        carboxylated surfactants, 21
        phosphated surfactants, 21
        sulfates, 17–21
        sulfonates, 17–21
    geminis, 568–569
    overview, 117–118
    perfluoroalkanesulfonyl fluorides, 310–311
    perfluorocarboxylate fluorides, 314–315
    personal care products, 478–479
    sulfate and sulfonate manufacturing processes, 140–141
    sulfosuccinates (sulfosuccinic acid esters), 244–245
anthraquinone autooxidation (AO) process, hydrogen
        peroxide, 424–426
antifoam compounds, silicone, 291–293
    background, 291
    for detergents, 292
    future trends, 299
    mechanism, 291–292
    process for making, 292–293
antimicrobial agents in toothpastes, 487–488
AO process, hydrogen peroxide. *See* anthraquinone
        autooxidation (AO) process
AOS. *See* α-olefin sulfonate (AOS)
APE. *See* alkylphenol ethoxylate (APE)
APG. *See* alkylpolyglucoside (APG)
asymmetric derivatives for fluorescent whitening, 553
AX-type zeolites, 412

B

BAB. *See* branched alkylbenzenes (BAB)
bacillus species as enzyme hosts, 532, 533
bacterial *vs.* fungal enzyme hosts
    overview, 533
    pretreatment considerations, 538
    recovery process design, 536

benzene. *See* alkylbenzenes
benzo[b]furans, 556
benzoxazoles, 556
betaines, 13–15, 231–235
    alkylamido betaines, 231–233
        amide formation and purification, 232–233
        caboxymethylation, 233
        minimizing contaminants, 233–234
    alkylamidopropyl betaines, 13, 482
    alkyl betaines, 13, 231–232
    imidazolinium betaines, 234
    phosphobetaines, 196–197
    sulfo betaines, 234–235
binder considerations in detergent agglomeration
        process, 340
biodiesel production, 12, 133
bioroute (fermentation)
    enzymes, 532–536
        equipment, 534
        mode of operation, 534
        process conditions, 533
        production strains, 532–533
        raw materials, 534–536
    ethanol production, 513, 515–516
    overview, 11
bistriazinyl derivatives for fluorescent whitening, 550
bisulfate, 19–20
bleaching, 421–434. *See also* active oxygen; fluorescent
        whitening agents (FWAs); hypochlorite
    active chlorine, 422–424
        calcium hypochlorite, 423
        introduction, 422
        organic chlorinated compounds, 423–424
        sodium hypochlorite, 422
    chemical *vs.* physical, 548–549
    introduction, 421
    methyl ester sulfonate, 202–203, 210–211
    soap raw materials, 476–477
bleaching powder, 461
bleach liquor, 455–456
block-heteric polyalkylene oxide block polymer
        subgroup, 254
block propylene oxide, 28
blown powder formulation, 332
branched alkylbenzenes (BAB), 39, 40, 47
branched and linear alcohols, 120–126
branched-chain alkylbenzene (ABS), 83–84
brine in hypochlorite, 437–438
builders, detergent, 375–384. *See also* zeolites
    carbonates (sodium carbonate/soda ash), 378–381
        introduction, 378
        natural soda ash, 378–380
        synthetic soda ash, 380–381
    defined, 496
    phosphates, 375–378
        introduction, 375–376
        pyrophosphate, 377–378
        sodium tripolyphosphate, 375–377
    polycarboxylates (polyacrylates), 381–384
        emulsion polymerization, 381–382
        introduction, 381
        processes, 382–384
        raw materials, 382

582 Index

C

cake strength of powdered detergents, 337
calcium hypochlorite, 423, 437, 455–462
  bleaching powder, 461
  bleach liquor, 455–456
  decomposition of, 445–446
  dibasic, 456
  economic considerations, 464, 465
  hemibasic, 456
  process considerations, 456–460
  safety considerations, 462
CAPB. *See* cocamidopropyl betaine (CAPB)
carbohydrate-based surfactants. *See also* alkyl
      glucosides/glycosides
  alkylpolyglucosides, 29–30, 241–242
  fermented sugar for ethanol, 515–516
  geminis, 571–572
carbonates (sodium carbonate/soda ash), 378–381
  introduction, 378
  precipitated calcium carbonate, 486
  production, 378–381
    natural soda ash, 378–380
    synthetic soda ash, 380–381
  silicate production, 398–399
  sodium percarbonate, 428–430
carboxyamphoterics, 228, 229
carboxylated surfactants, 21
carboxymethylation of alkylamido betaines, 233
cascade-type sulfonation reactors, 91–94
cationic polymers for silicone in shampoos, 484
cationic surfactants
  cationic polymers in shampoos, 484
  classification, 495
  geminis as, 563–568
  introduction, 21–24
  perfluoroalkanesulfonyl fluorides, 312–313
  perfluorocarboxylate fluorides, 316–318
  quaternary surfactants, 365–373
    alkanolamine esters (esterquat)-based cationics,
      370–371
    introduction, 365–366
    quaternization process, 371–372
    (tri)alkylamine-based cationics, 367–370
caustic potash (potassium hydroxide) in hypochlorite,
    440–441
caustic soda (sodium hydroxide)
  in hypochlorite, 439–440
  silicate production, 395–396
cellulases, host organisms of, 533
cetyl trimethyl ammonium chloride (CTAC), 483
chain extension polymerization, silicone compounds, 288
chain starters of detergents, 492
chemical *vs.* physical bleaching, 548–549
chlorinated trisodium phosphate, 437
chlorination, 443–444, 445
chlorine in hypochlorite, 438–439
chloroacetic acids, 233
chloro-glyceryl ethers, 160–162
chlorosulfonic acid (CSA), 17
chromatography
  EO/PO block polymer analysis, 264
  high performance liquid chromatography, 76–77
  high temperature gas chromatography, 76–77

CIE. *See* Commission on Illumination (CIE)
clay conversion process, zeolites, 409
cleavage of epoxides method for propylene glycol
    ethers, 525
CMC. *See* critical micelle concentration (CMC)
coal, alcohols from, 130–131
cocamidopropyl betaine (CAPB), 231–233
cocoamphodiacetates, 227–228
coco derivatives, acyl/dilkyl ethylenediamines, 227–228
Commission on Illumination (CIE), 558
compact powder detergents, 327–328
concentration phase of enzyme production, 539–540
continuous film reactor, 91, 92
cornstarch for alkyl glucosides/glycosides, 69–70
coumarins, 554–556
critical micelle concentration (CMC)
  amphoteric surfactants, 223
  defined, 499
  pseudogeminis, 569
crystalline-layered sodium disilicate, 402–404
crystalline trisodium phosphate in hypochlorite, 441
crystallization
  crystalline-layered sodium disilicate, 402–404
  crystalline trisodium phosphate, 441
  sodium metasilicate solution, 401–402
  sodium percarbonate, 428, 429
CSA. *See* chlorosulfonic acid (CSA)
CTAC. *See* cetyl trimethyl ammonium chloride (CTAC)
cumene for alkylphenol ethoxylates, 63–64
cuphea as oleo-derived detergent source, 133
cyanuric acid, organic chlorinated compounds, 423–424

D

DAP ester. *See* dialkyl phosphate (DAP) ester
DEA-based alkanolamides, 25
decomposition of solid hypochlorites, 445–446
deformability of powders in agglomeration, 338–339, 340
degree of polymerization (DP), 74–76
dense laundry process, 341
density of powdered detergents, 336
deodorizing of soap raw materials, 477
Detal process, 43
detergent range alcohols, 507–512
  fatty alcohols, 510–511
  specifications and standards, 507–509
  unsaturated fatty alcohols, 511–512
detergents, 323–363, 492–529. *See also* builders,
    detergent; heavy-duty liquid (HDL) detergents;
    powdered detergents; surfactants
  basic function, 493
  classification, 494–496
    ionic detergents, 494–495
    nonionic detergents, 495
    zwitterionic detergents, 495
  composition, 496–499
    for developing countries, 498–499
    household type, 497–498
    overview of components, 496
    solvents' role in, 496–497
  conclusions, 528–529
  processing, 323–363
    forms of product, 326–331

Index

introduction, 324–331
main surfactant (LAS), 326
unit dose technologies, 354–363
solvents' role in, 499–506
developing countries
detergent composition for, 498–499
syndet bar popularity in, 324, 329
dialkylation of phenols, 52
dialkyl esters of sulfosuccinic acid, 244–245
dialkyl ethylene glycol ether, 523
dialkyl phosphate (DAP) ester, 190, 194–196
diaminostilbene disulfonic acid for fluorescent
whitening, 550
dibasic calcium hypochlorite, 456
diffusion, solvents in detergents, 505, 506
dimethicone copolyol, 294–299
dimethicone (PDMS) (polydimethylsiloxane), 483
dimethiconol (silanol), 290
dimethylalkylamines, 367–370
dimethylaminopropylamine (DMAPA), 13
dimethyl ether (DME), 217
dimethyl sulfate (DMS), 216–217
direct catalytic hydration of ethylene for ethanol, 513–514
direct fusion route for anhydrous sodium metasilicate, 400
direct (one-step) reaction, alkyl glucoside/glycoside, 77–79
di-sec butylphenol (DSBP), 52, 63
disubstituted alkylphenol nomenclature, 50
diversified chemical producers, 3
DMAPA. *See* dimethylaminopropylamine (DMAPA)
DME. *See* dimethyl ether (DME)
DMS. *See* dimethyl sulfate (DMS)
dodecene (propylene tetramer), 65
4-dodecylphenol, 51
downstream process (recovery), enzymes, 536–541
process design, 536–537
unit operations, 538–541
DP. *See* degree of polymerization (DP)
dried silicate solutions, 396–398
cogranules, 398–399
granules, 397–398
introduction, 396–397
powders, 397
drum granulation process, anhydrous sodium
metasilicates, 400–401
dryer systems. *See also* spray drying
hydrous sodium polysilicate granules, 398
methyl ester sulfonate, 212–216
dry neutralization, powdered detergents, 344–345
dry- *vs.* wet-milling of corn for ethanol, 515–516
DSBP. *See* di-sec butylphenol (DSBP)
dual and multicompartment water-soluble pouches,
362–363

E

ECH. *See* epichlorohydrin (ECH)
economic considerations
alkanolamines, 518
alkylphenols, 59–61
ethanol, 516
fatty alcohol-based surfactants, 512
geminis, 574–575
hypochlorite, 464, 465

linear alkylbenzene sulfonate, 109–112
overview, 12
powder *vs.* liquid detergents, 324
propylene glycols, 522
silicone compounds, 299
electrochemical fluorination. *See* fluorinated surfactants
electrolysis. *See* hypochlorite
electrolytic sodium hypochlorite, 451–452
emulsions
polymerization of polycarboxylate, 381–382
semibatch polycarboxylate production, 384
silicone surfactant, 293, 297–298
environmental considerations
alkylphenol ethoxylates, 66–67
biodiesel production, 12, 133
enzyme production, 537
fluorinated surfactants, 318
glycols, 522, 526–527
linear alkylbenzene sulfonate, 129
oleo-derived detergent advantages, 271
sodium carbonate, 378
and surfactant technology, 129
enzymes, 531–545
fermentation, 532–536
equipment, 534
mode of operation, 534
process conditions, 533
production strains, 532–533
raw materials, 534–536
formulation, 541–545
history, 541–542
introduction, 541
liquid type, 542–543
solid type, 543–545
introduction, 532
recovery, 536–541
process design, 536–537
unit operations, 538–541
EO (ethylene oxide) block copolymers. *See* ethylene
oxide/propylene oxide (EO/PO) block
copolymers
epichlorohydrin (ECH), 16, 568
epoxidation reaction in glyceryl ether production,
160–163
equilibrium peracids, 432
equlibria in hypochlorite chemistry, 442–443
esteramines, 370–371
esterified surfactants, 29
esterquats (alkanolamine esters), 370–371
esters. *See also* methyl ester sulfonate (MES); phosphate
esters
ethylene glycol, 523
fatty acid methyl ester oxyethylation, 271–282
catalysts for, 273–275
introduction, 271–273
properties of, 281–282
synthesis of, 275–281
glycerol esters, 28
ethanol, 513–516
direct catalytic hydration of ethylene, 513–514
economic considerations, 516
fermentation route (bioroute), 513, 515–516
indirect hydration of ethylene, 514–515
petrochemical route, 513

ethanolamines, 517–518
ethoxylated surfactants. *See also* alkylphenols
    alkylphenol ethoxylate, 61–67
        di-nonylphenol, 62
        di-sec butylphenol, 63
        environmental considerations, 66–67
        future trends, 66–67
        introduction, 61–62
        p-dodecyl phenol, 62
        p-nonylphenol, 62
        p-tert-octylphenol, 62
        raw materials for, 63–65
    detergent range alcohols, 509
    ethylene oxide block copolymers, 256–262
    overview of process, 26–29
ethylene glycol, 518–520, 522–526
ethylene hydration methods for ethanol, 513–515
ethylene oligomerization of α-olefins, 102–109, 120–122
ethylene oxide hydrolysis, ethylene glycol production,
    518–520
ethylene oxide/propylene oxide (EO/PO) block
        copolymers, 253–270
    classification and applications, 255–256
    introduction, 253–254
    processes, 262–269
        all-heteric surfactant, 263–264
        analysis, 264–265
        pluronic surfactant, 263
        reactors, 264–267
        safety and alkylene oxide handling, 267–269
        synthesis, 262–263
        tetronic surfactant, 263
    reaction mechanisms and kinetics, 256–262
        catalyst formation, 257
        initiation, 257
        propagations, 258–259
        proton transfer, 259–262
ethyl lactate, 527–528
ethyl process, alcohols, 124–125
extracellular enzyme production, 536

F

fabric softeners, silicone fluids in, 289
falling film sulfonation reactor, 95–97, 208
FAME. *See* fatty acid methyl esters (FAME)
fatty acid glucamides, 242–244
fatty acid methyl esters (FAME), 271–284
    catalysts, 273–275
    introduction, 271–273
    properties, 281–282
    synthesis, 275–281
fatty acids/fatty alcohols, 239–246
    alkylpolyglucosides, 241–242
    economic considerations, 512
    ethanolamides, 518
    ethoxylation, 29
    fatty acid glucamides, 242–244
    fatty alkanolamides, 239–241
    historical perspective, 8–10
    hydrogenation processes, 511, 512
    introduction, 510

    natural C12-16 fatty alcohols, 28
    personal care products, 477–479, 481
    raw materials, 510–511
    sulfosuccinates, 244–245
    summary, 245–246
fatty alcohol ether sulfates (FES), 479
fatty alcohol polyethylene glycol ether, 481
fatty alkyl imidazolines, 228
feedstock producers, role of, 2–3
fermentation (bioroute)
    enzymes, 532–536
        equipment, 534
        mode of operation, 534
        process conditions, 533
        production strains, 532–533
        raw materials, 534–536
    ethanol production, 513, 515–516
    overview, 11
FES. *See* fatty alcohol ether sulfates (FES)
film sulfonation reactors
    continuous film, 91, 92
    falling-film, 95–97, 208
Fischer-Tropsch (FT) process
    alcohol from coal, 130–131
    n-paraffin alkylation, 47
fluidized bed process
    agglomeration, 342–343
    sodium percarbonate, 428, 429
    spray coating in enzyme formulation, 544–545
fluorescent whitening agents (FWAs), 547–560
    application to detergents, 558
    chemistry and production, 549–557
        benzo[b]furans, 556
        benzoxazoles, 556
        coumarins, 554–556
        physical forms, 556–557
        stilbene derivatives, 549–553
        styryl derivatives of benzene and biphenyl,
            553–554
    conclusions, 558–559
    introduction, 547–548
    mechanism of action, 548–549
    whiteness measurement, 557–558
fluorides in toothpastes, 488
fluorinated surfactants, 301–321
    electrochemical fluorination process, 303–307
    environmental considerations, 318
    introduction, 301–303
        nomenclature, 303
        properties of, 301–303
    perfluoroalkanesulfonyl fluorides, 307–313
        acids and metal salts of, 307–308
        alkylation of perfluroalkanesulfonamides,
            310–313
        anionic surfactants, 310–311
        cationic surfactants, 312–313
        halides/ammonia and amine reaction, 308–310
        introduction, 303
        nonionic surfactants, 311–312
        zwitterionic surfactants, 312–313
    perfluorocarboxylate fluorides, 313–318
        anionic surfactants, 314–315
        cationic surfactants, 316–318

Index

nonionic surfactants, 315–316
synthesis of, 314
zwitterionic surfactants, 316–318
perfluorooctanesulfonyl fluorides, 303–307, 308
fluoropolymers, uses for, 303
FMC processes, phosphate production, 377
foaming property, amphoteric surfactants, 223. *See also* antifoam compounds, silicone
Friedel-Crafts alkylation, 52
FT process. *See* Fischer-Tropsch (FT) process
fungal *vs.* bacterial enzyme hosts
overview, 533
pretreatment considerations, 538
recovery process design, 536
furans for fluorescent whitening, 556
FWAs. *See* fluorescent whitening agents (FWAs)

G

gas-to-liquids (GTL) technology, 131, 132–133
gemini surfactants, 561–577
amino acid-based, 572
anionic, 568–569
cationic, 563–568
commercial, 572–575
conclusion, 575
heterogeminis and amphoterics, 569–570
introduction, 561–562
nonionic, 570–571
spacers (amphiphilic moieties' linkage), 562–563
sugar surfactants, 571–572
glucose. *See* carbohydrate-based surfactants
D-glucose, alkyl glucosides/glycosides, 72–75
glycerol esters, 28
glycerol in toothpastes, 485
glyceryl ether sulfonates, 159–169
alternative process, 165–166
benefits and uses of, 166–168
chemistry, 163–164
conclusion, 168
defined, 159
glyceryl ethers, 159–161
glyceryl ether sulfonate, 163–165
glycidyl ethers, 161–163
introduction, 159
glycidal ethers, 161–163
glycols, 518–522
ethylene glycol, 518–520, 522–526
propylene glycol, 520–522
glymes (propylene glycol analogues), 525
green solvents, 526–528
GTL technology. *See* gas-to-liquids (GTL) technology

H

hard compressed tablets, 355–358
HDL detergents. *See* heavy-duty liquid (HDL) detergents
heavy cut detergent range alcohols, 507
heavy-duty detergents, defined, 497
heavy-duty liquid (HDL) detergents, 350–354
batch processing, 351–352

conclusion, 354
continuous operations, 352–353
defined, 497
general formulations, 494
and hydrotropes, 247, 248
introduction, 350–351
late product differentiation, 353
microbial growth issue, 326, 354
optimal process for market, 353
recycling considerations, 354
transformation requirements, 350–351
vs. powder detergents, 324
hemiacetals, alkyl glucosides/glycosides, 71–73
hemibasic calcium hypochlorite, 456
heteric-block polyalkylene oxide block polymer subgroup, 254
heterogeminis and amphoterics, 569–570
heterogeneous *vs.* homogenous alkylation catalysts, 54
HF alkylation. *See* hydrofloric (HF) alkylation
higher *n*-olefins for surfactants, 45–46
α-olefins, 45
applications, 45
internal olefins, 45–46
market demand for, 45
highly soluble alcohol sulfates (HSAS), 129–130
high-molecular-weight sulfonates, 151–154
lignosulfonates, 152–154
petroleum sulfonates, 151–152
high performance liquid chromatography (HPLC), 76–77
high-shear granulation, enzyme formulation, 544–545
high-strength (low-salt) sodium hypochlorite, 449–451
high temperature gas chromatography (HTGC), 76–77
HLB (Hydrophile-Lipophile Balance). *See* Hydrophile-Lipophile Balance (HLB)
Hoechst-Knapsack process, phosphate production, 376–377
homogenous *vs.* heterogeneous alkylation catalysts, 54
horizontal forming processes, water-soluble pouches, 361–362, 362
household-type detergent, composition of, 497–498
HPLC. *See* high performance liquid chromatography (HPLC)
HSAS. *See* highly soluble alcohol sulfates (HSAS)
HTGC. *See* high temperature gas chromatography (HTGC)
hydrated lime and lime in hypochlorite, 441
hydrated water soluble silicates, 396–398
cogranules, 398–399
granules, 397–398
introduction, 396–397
powders, 397
hydration processes in isopropyl alcohol production, 512–513
hydrofloric (HF) alkylation, 41–42
hydrogel processes for zeolite, 409–412
hydrogenation processes, fatty alcohol production, 511, 512
hydrogen peroxide in bleaching, 424–426
hydrolytic stability of fatty acid methyl esters, 281
hydrolyzate as silicone raw material, 287
Hydrophile-Lipophile Balance (HLB), 293–296
hydrophilic properties of detergents
alkyl glucosides/glycosides, 74–75
overview, 492–494
silicone compounds, 288–290

hydrophobic properties of detergents
  alkyl glucosides/glycosides, 74–75
  overview, 492–494
  silicone compounds, 288–290
    alkyl fluoro fluids, 290
    silanols, 290
    silicone fluids, 288–290
hydrosilylation, 297–298
hydrothermal dissolution of anhydrous lumps/cullets, 394–395
hydrothermal process from quartz sand and caustic soda, 395–396
hydrotropes, 247–251
  heavy-duty liquid detergents, 247, 248
  introduction, 247
  sulfonic acid salts, synthesis of, 247–251
hydrous sodium polysilicates, 396–398
  cogranules, 398–399
  granules, 397–398
  introduction, 396–397
  powders, 397
hydroxyl compounds in glyceryl ether production, 160–161
hydroxyl number in EO/PO block polymer analysis, 264
hypochlorite, 436–472
  chemistry, 441–447
    chlorination, 443–444, 447
    concentration terms, 441–442
    decomposition of solid hypochlorites, 445–446
    decomposition of solutions, 444–445
    equilibria, 442–443
    metal removal, 446
    temperature control, 446–447
  chlorinated trisodium phosphate, 437
  construction materials, 463–464
  economic considerations, 464
  future trends, 465
  process technology, 447–462
    calcium hypochlorite, 437, 455–462, 464, 465
    chlorinated trisodium phosphate, 454, 464
    hypochlorous acid, 452–454
    lithium hypochlorite, 454–455, 464
    magnesium hypochlorite, 461–462
    potassium hypochlorite, 436, 452
    sodium hypochlorite, 436, 447–452, 464
  raw materials, 437–441
    brine, 437–438
    chlorine, 438–439
    crystalline trisodium phosphate, 441
    lime and hydrated lime, 441
    lithium hydroxide, 436, 441
    potassium hydroxide (caustic potash), 440–441
    sodium hydroxide, 439–440
  safety considerations, 462–463
  technology development, 436–437
hypochlorous acid, 452–454

I

imidazolinium amphoteric surfactants, 225–228, 230, 234
indirect hydration of ethylene for ethanol, 514–515
interfacial turbulence, solvents in detergents, 505

ionic detergent classification, 494–495. *See also* anionic surfactants; cationic surfactants
ionic liquids, 528
isoelectric point, amphoteric surfactants, 223
isopropyl alcohol (2-propanol), 512–513

J

jatropha curcus as oleo-derived detergent source, 133

L

LAB. *See* linear alkylbenzene (LAB)
LAS. *See* linear alkylbenzene sulfonate (LAS)
laundry detergents. *See* detergents
light-duty liquid (LDL) detergents, 247
lignosulfonates, 152–154
lime and hydrated lime in hypochlorite, 441
linear alkylbenzene (LAB), 41–47, 83–86. *See also* linear alkylbenzene sulfonate (LAS)
  aluminum chloride ($ALCl_3$) alkylation, 42–43
  consumption of, 84
  Detal process, 43
  historical importance of, 83–84
  hydrofloric alkylation, 41–42
  production, 46
  quality of, 44–45
  solid bed catalyst alkylation, 43
  sources and specifications, 84–86
  technology development, 47
linear alkylbenzene sulfonate (LAS), 86–115
  economic considerations, 109–112
  environmental considerations, 129
  historical importance of, 83–84
  as hydrotropes, 247
  in laundry detergent manufacture, 326, 344–345
  personal care products, 480–481
  perspective development, 112–115
  in syndet bars, 330–331
  technology, 86–109
    cascade-type sulfonation reactors, 91–94
    continuous film reactor, 91, 92
    falling-film reactor, 95–97
    neutralization treatment, 97–102
    α-olefin sulfonate, 102–109
    sulfur trioxide-based sulfonation, 86–91
linear and branched alcohols, 120–126
lipases, host organisms of, 533
lipophilicity and lipophobeity in alkyl glucosides/glycosides, 74–75
liquid detergents. *See also* heavy-duty liquid (HDL) detergents
  light-duty, 247
  overview, 328–329
  pouches, 329, 360–363
liquid-phase extraction of *n*-paraffins, 47
liquid sulfur trioxide, 17–18
liquid type enzymes, 542–543
lithium hydroxide in hypochlorite, 436, 441
lithium hypochlorite, 454–455, 464
low-salt (high-strength) sodium hypochlorite, 449–451

Index

587

M

magnesium hypochlorite, 461–462

MAP. *See* zeolites

MAP esters. *See* monoalkyl phosphate (MAP) esters

M–D–M polymers, silicone compounds as, 288

MES. *See* methyl ester sulfonate (MES)

metal removal in hypochlorite chemistry, 446

metasilicate pentahydrate, 401–402

metasilicates of anhydrous sodium, 399–401
  direct fusion route, 400
  granulation of solution, 400–401
  introduction, 399–400
  solid-state reaction, 400

methanol recovery system for methyl ester sulfonate, 216

methyl esters, fatty acid. *See* fatty acid methyl esters (FAME)

methyl ester sulfonate (MES), 201–219
  acid *vs.* neutral bleaching of, 202–203
  background, 201–202
  chemical reactions and by-products, 206–207
  conclusions, 218
  feedstock specifications, 203–204, 205, 208
  overview, 19
  process, 207–217
    acid digester, 209–210
    basic sulfonation process, 208–209
    by-products and safety, 216–217
    conditions for, 207–208
    cooling system for dried product, 213–214
    dryer auxiliary systems, 214–216
    methanol recovery system, 216
    neutralizer system, 211–212
    stabilization and bleaching, 210–211
    tube dryer system, 212–216
  production, 204–206
    plant capacity, 205
    raw materials, 205–206
  vs. linear alkylbenzene/linear alkyl benzene sulfonate, 201

methyl-substituted alcohol sulfate surfactants, 129–130

micelle formation, amphoteric surfactants, 223

microalgae as biodiesel source, 133–134

microbial contamination issue
  enzyme formulation, 543
  liquid detergents, 326, 354
  toothpastes, 487–488

microbial fermentation to produce surfactants, 11

middle cut detergent range alcohols, 507

monoalkanolamides, 25

monoalkyl–dialkyl phosphate ester mixtures, 187–196

monoalkyl ethylene glycol ethers, 522–523

monoalkyl glycerin ethers, 165–166

monoalkyl phosphate (MAP) esters, 189–194
  hybrid processes, 190–193
  phosphorus oxychloride processes, 193–194
  polyphosphoric acid process, 189–190

mono esters of sulfosuccinic acid, 244, 245

monosubstituted alkylphenol nomenclature, 50

moulding processes, water-soluble pouches, 362–363

m-s-m geminis, 563–568

multicompartment water-soluble pouches, 362–363

multimeric molecules. *See* gemini surfactants

N

NaA-type zeolites, 404–411
  clay conversion process, 409
  continuous *vs.* batch-type processes, 409
  hydrogel process, 409–411
  introduction, 404–405
  principles of production, 406–409
  raw materials, 405–406
  requirements and resources, 411

natural alcohol sulfates. *See* soaps

natural C12-16 fatty alcohols, 28

natural soda ash, 378–380

natural sulfation, alcohols, 126–128

neo-phosphocholine geminis, 569–570

neutralization treatment, linear alkylbenzene sulfonate, 97–102

neutralizer system for methyl ester sulfonate, 211–212

NMR. *See* nuclear magnetic resonance (NMR)

NOBS. *See* sodium nonanoyloxybenzenesulfonate (NOBS)

nonene (propylene trimer), 65

nonether containing polyhydric alcohols, 29

nonionic surfactants. *See also* alkyl glucosides/glycosides
  alkanolamides, 25–26, 239–241
  defined, 495
  fatty acid methyl esters, 271–282
    catalysts, 273–275
    introduction, 271–273
    properties, 281–282
    synthesis, 275–281
  fluorinated surfactants as, 311–312, 315–316
  geminis as, 570–571
  polyalkylene oxide block polymers, 254–256

4-nonylphenol, 51

di-nonylphenol, 52, 62

nuclear magnetic resonance (NMR) in EO/PO block polymer analysis, 264

O

octylphenols, 50–51

olefins
  alkanolamines from olefin oxides and ammonia, 517–518
  alkylphenol ethoxylates, 65
  α-olefin sulfonate, 19, 102–109
  ethylene oligomerization of α-olefins, 102–109, 120–122
  n-olefins, 45–47
    applications, 45
    internal olefins, 45–46
    market demand for, 45
    α–olefins, 45
    technology development, 47
  α-olefins
    alkylation, 45
    ethylene oligomerization, 102–109, 120–122
    secondary alcohols from primary α-olefins, 28
  petroleum-derived alcohols, 120–122

oleo-derived surfactants. *See also* methyl ester sulfonate (MES); personal care products

oleo-derived surfactants (*contd.*)
  alcohols, 118–120, 133–135
  alkylpolyglucosides, 29–30, 241–242
  environmental considerations, 271
  fatty acid glucamides, 242–244
  fatty acid methyl esters, 271–282
    catalysts, 273–275
    introduction, 271–273
    properties, 281–282
    synthesis, 275–281
  fatty alkanolamides, 239–241
  historical perspective, 5, 8–9, 11–12
  overview, 5, 8–9, 10, 11–12
  quaternary surfactants, 365–373
  soaps as, 118
oleyl alcohol, 512
Olin process, calcium hypochlorite, 423
open-hearth regenerative furnace process, silicate
    production, 391–393
optical brightening agents. *See* fluorescent whitening
    agents (FWAs)
organic additives, defined, 496
organophosphate esters. *See* phosphate esters
oxo process, 124
oxyethylation of fatty acid methyl esters, 271–282
  catalysts for, 273–275
  introduction, 271–273
  properties of, 281–282
  synthesis of, 275–281

P

paraffins. *See also* secondary alkanesulfonate (SAS)
  alcohols, 125–126
  n-paraffins, 46–47
    Fischer-Tropsch process, 47
    linear alkylbenzene production, 45–46, 46
    liquid-phase extraction, 47
    vapor-phase extraction, 47
particle size distribution of powdered detergents, 337, 340
PAS. *See* primary alcohol sulfates (PAS); primary
    alkanesulfonates (PAS)
PDDP. *See* p-dodecyl phenol (PDDP)
PDMS. *See* polydimethylsiloxane (PDMS)
p-dodecyl phenol (PDDP), 62
PEG. *See* polyoxyethylene glycol (PEG)
peracids, 432–433
percarboxylic acids, 430–433
  equilibrium peracids, 432
  preformed peracids, 433
  reaction with activators, 430–432
perfluoroalkanesulfonamides, alkylation of, 310–313
perfluoroalkanesulfonyl fluorides, 307–313
  acids and metal salts of, 307–308
  alkylation of perfluroalkanesulfonamides, 310–313
  anionic surfactants, 310–311
  cationic and zwitterionic surfactants, 312–313
  halides/ammonia and amine reaction, 308–310
  introduction, 303
  nonionic surfactants, 311–312
perfluorocarboxylate (PFCA) fluorides, 313–318
  anionic surfactants, 314–315

cationic and zwitterionic surfactants, 316–318
  nonionic surfactants, 315–316
  synthesis of, 314
perfluorooctanesulfonyl fluorides, 303–307, 308
peroxygen (active oxygen), 424–433
  hydrogen peroxide, 424–426
  inorganic peracids, 433
  introduction, 424
  percarboxylic acids, 430–433
    equilibrium peracids, 432
    preformed peracids, 433
    reaction with activators, 430–432
  persalts, 426–430
    sodium perborate, 426–428
    sodium percarbonate, 428–430
persalts, 426–430
  sodium perborate, 426–428
    monohydrate, 428
    tetrahydrate, 426–427
  sodium percarbonate, 428–430
personal care products, 473–489
  amorphous silica, 486–487
  amphoteric surfactants in, 221
  antimicrobial agents in toothpastes, 487–488
  cationic polymers, 484
  fatty alkanolamides, 239–241
  fluorides in toothpastes, 488
  glycerol in toothpastes, 485
  introduction, 474
  precipitated calcium carbonates, 486
  silicones as softening agents, 289, 483–484
  soaps, fats/oils in, 474–477
    fatty acids used, 477–478
    raw materials, 474–477
  sorbitol as humectant in toothpastes, 485–486
  surfactants, 478–483
    alkyl ether sulfates or fatty alcohol ether sulfates,
      479
    alkyl isethionates, 480
    amphoteric, 482
    anionic, 478–479
    introduction, 478
    linear alkylbenzene sulfonates, 480–481
    nonionic, 481–482
    primary alcohol sulfates, 479–480
    quaternary ammonium, 483
    sulfosuccinates, 480
petroleum-derived surfactants. *See also* linear
    alkylbenzene sulfonate (LAS)
  alcohols, 120–126, 129–133
    ethylene-derived processes, 122–124
    ethyl process, 124–125
    linear and branched, 120–126
    paraffin-derived processes, 125–126
    raw materials and olefin chemistry, 120–122
  alkanesulfonates, 151–152
    consumption and uses, 152
    manufacture, 151–152
  feedstocks for fatty alcohols, 511
  historical perspective, 5–8, 9, 10
  petrochemical route for ethanol, 513
  raw materials, 12
  secondary alkanesulfonate, 146

Index

petroleum sulfonates, 151–152
PFCA fluorides. *See* perfluorocarboxylate (PFCA) fluorides
pH
    amphoteric surfactants, 221–222
    chlorination monitoring, 447
    liquid detergent stability, 350–351
phenols. *See also* alkylphenol ethoxylate (APE);
        alkylphenols
    alkylated phenol, 28
    defined, 50
    dialkylation of, 52
    di-nonylphenol, 52, 62
    di-sec butylphenol, 52, 63
    4-dodecylphenol, 51
    4-nonylphenol, 51
    4-*tert*-octylphenol, 51
    octylphenols, 50
    p-dodecyl phenol, 62
    p-nonylphenol, 62
    p-*tert*-octylphenol, 62
phosphated surfactants, 21. *See also* phosphate esters
    as builders, 375–378
    chlorinated trisodium phosphate, 437
    crystalline trisodium phosphate, 441
    overview, 21
    phosphation processes, 376–377
    pyrophosphate, 377–378
    sesquiphosphates, 187–189
    sodium tripolyphosphate, 375–377
phosphate esters, 183–199
    conclusion, 197–198
    defined, 183
    dialkyl phosphate ester, 190, 194–196
    geminis, 568
    introduction, 183–184
    monoalkyl phosphate esters, 189–194
        hybrid processes, 190–193
        phosphorus oxychloride processes, 193–194
        polyphosphoric acid process, 189–190
    phosphation reagents, 184–187
        phosphoric anhydride, 184–185
        phosphorus oxychloride, 187
        polyphosphoric acid, 185–186
    phosphobetaines, 196–197
    uses of, 183–184
phosphation reagents, 184–187
phosphobetaines, 196–197
phospholipids, 196–197
phosphoric anhydride, 184–185, 187–189
phosphorus oxychloride, 187, 193–194
physical *vs.* chemical bleaching, 548–549
pi (π) factor and sulfonate aromatic compounds, 249–250
pluronic surfactant, EO/PO block copolymers, 263
p-nonylphenol (PNP), 62
PNP. *See p*-nonylphenol (PNP)
p-*tert*-octyl phenol (PTOP), 62
polyacrylates (polycarboxylates), 381–384
    emulsion polymerization, 381–382
    introduction, 381
    processes, 382–384
    raw materials, 382
polyalklylene oxide block polymers, 254–256
polycarboxylates (polyacrylates), 381–384

    emulsion polymerization, 381–382
    introduction, 381
    processes, 382–384
    raw materials, 382
polydimethylsiloxane (PDMS) (dimethicone), 483
polyether polyols, 521–522
polyethylene glycol, 481, 520
polyhydric alcohol esters, 28
polymeric surfactants, 11
polymerization
    degree of polymerization, 74–76
    polycarboxylates, 381–382
    silicone compounds, 288, 289
polymers. *See* silicate production
polyoxyethylene glycol (PEG), 272
polyoxylalkylene block copolymers. *See* ethylene oxide/
        propylene oxide (EO/PO) block copolymers
polyphosphoric acid process, phosphate esters, 189–190
polypropylene glycols, 521–522
polysilicates, hydrous sodium, 394–396
    cogranules, 396–397
    granules, 395–396
    introduction, 394–395
    powders, 395
polyvinyl alcohol (PVA) for detergent pouches, 329,
        360–362
PO (propylene oxide) block copolymers. *See* ethylene
        oxide/propylene oxide (EO/PO) block
        copolymers
pore saturation of powders in agglomeration, 339
potassium hydroxide in hypochlorite, 440–441
potassium hypochlorite, 436, 452
pouches, liquid detergent, 329, 360–363
powdered detergents, 331–349
    agglomeration, 337–345
        defined, 338
        dense laundry process, 341
        dry neutralization, 344–345
        factors in, 338–339, 340–341
        generic description, 339–340
        high- and low-density agglomerates, 343–344
        introduction, 337–338
        mixer operation, 342–343
        process control strategies, 341–342
        regimes, 339
    economic considerations, 324
    finishing of dry laundry detergents, 343–349
        batch mixing, 346–347
        continuous mixing, 347–348
        dust control/explosion protection, 349
        liquid spray-ons, 348–349
        quality of product, 349
        transformations, 345–346
    overview, 327
    spray drying, 331, 334–337
        atomization, 334–335
        blown powder formulation, 332
        cooling and classification, 336
        diagram of process, 333
        drying, 335–336
        introduction, 331–332
        operation, 336
        powder properties, 336–337

590   Index

powdered detergents (*contd.*)
    pumping, 334
    slurry making, 333
    sodium tripolyphosphate, 333–334
PPG process, calcium hypochlorite, 423
precipitated calcium carbonates for toothpastes, 486
preformed peracids, 433
pretreatment phase of enzyme production, 538–539
primary alcohol sulfates (PAS), 479–480
primary alkanesulfonates (PAS), 142
primary separation phase of enzyme production, 538–539
2-propanol, 512–513
propoxylated surfactants, 27, 257
propylene glycol, 520–522
propylene glycol analogues (glymes), 525
propylene oxide (PO) block copolymers. *See* ethylene oxide/propylene oxide (EO/PO) block copolymers
propylene trimer (nonene), 65
proteases, host organisms of, 533
"pseudogemini," 568–569
PTOP. *See* p-*tert*-octylphenol (PTOP)
P-type zeolites, 411–412
punch sticking in compressed laundry tablets, 357, 359
purification phase of enzyme production, 540–541
PVA. *See* polyvinyl alcohol (PVA)
pyrosulfonic acid, 247, 249

Q

quartz sand, silicate production, 391, 395–396
quaternary ammonium, 483
quaternary surfactants, 365–373
    alkanolamine esters (esterquats), 370–371
    introduction, 365–366
    quaternization process, 371–372
    (tri)alkylamine-based cationics, 367–370
bis-quaternary surfactants (*m-s-m* geminis), 563–568

R

recovery (downstream process), enzyme, 536–541
    process design, 536–537
    unit operations, 538–541
recycling considerations
    dual and multicompartment water-soluble pouches, 363
    enzyme production, 537
    heavy-duty liquid detergents, 354
Rochow process for silicone, 286
rotary furnace process, silicate production, 393

S

salt and hypochlorite production, 437–438
saponification process for soap production, 478
secondary alcohols from primary α-olefins, 28
secondary alkanesulfonate (SAS) (paraffin sulfonates), 142–151
    n-alkanes, 146
    consumption, 151

historical considerations, 142
manufacture of, 143–151
    chemistry, 142–146
    industrial developments, 146–151
production, 151
sulfochlorination, 143–144, 147
    chlorine and sulfur dioxide, 143, 147
    sulfuryl chloride, 143–144
sulfoxidation, 144–146, 148–151
uses of, 151
sesquiphosphates, 187–189
shampoos. *See* personal care products
silanol (dimethiconol), 290
silicate production, 388–404
    anhydrous lumps/cullets, 388–393
        introduction, 388–389
        open-hearth regenerative furnace process, 391–393
        quartz sand, 391
        raw materials, 390–391
        rotary furnace process, 393
        soda ash, 389–391
    anhydrous sodium metasilicate, 399–401
        direct fusion route, 400
        granulation of solution, 400–401
        introduction, 399–400
        solid-state reaction, 400
    crystalline-layered sodium disilicate, 402–404
    hydrous sodium polysilicates, 396–398
        cogranules, 398–399
        granules, 397–398
        introduction, 396–397
        powders, 397
    introduction, 388
    requirements of resources, 404
    sodium carbonate cogranules, 398–399
    sodium metasilicate pentahydrate, 401–402
    sodium silicate solutions, 393–396
        hydrothermal dissolution of anhydrous lumps/cullets, 394–395
        hydrothermal process from quartz sand and caustic soda, 395–396
        introduction, 393–394
silicone compounds, 286–300
    antifoam compounds, 291–293
        background, 291
        for detergents, 292
        mechanism, 291–292
        process for making, 292–293
    background, 286
    conclusion, 299
    derivatives, 287–288
        construction, 288
        nomenclature, 287
    economic considerations, 299
    future trends, 299
    hydrophobic silicone compounds, 288–290
        alkyl fluoro fluids, 290
        silanols, 290
        silicone fluids, 288–290
    raw materials synthesis, 286
    Rochow process, 286
    silicone surfactants, 293–298

Index

background, 293–296
dimethicone copolyol, 294–299
as softening agents, 289, 483–484
silicone fluids, 288–290
single compartment, water-soluble pouches, 360–362
soaps
formulation of, 176
as oleo-derived surfactants, 118
personal care products, 474–477
fatty acids used, 477–478
raw materials, 474–477
vs. syndet bars, 172–173
sodium carbonate (carbonates/soda ash)
introduction, 378
precipitated calcium carbonate, 486
production, 378–381
natural soda ash, 378–380
synthetic soda ash, 380–381
silicate production, 398–399
sodium percarbonate, 428–430
sodium fluoride in toothpaste, 488
sodium hydroxide (caustic soda)
in hypochlorite, 439–440
silicate production, 395–396
sodium hypochlorite, 422, 436, 447–452
economic considerations, 464
electrolytic, 451–452
low-salt (high-strength), 449–451
solid, 451
typical production, 447–449
sodium nonanoyloxybenzenesulfonate (NOBS), 431–432
sodium perborate, 426–428
monohydrate, 428
tetrahydrate, 426–427
sodium percarbonate, 428–430
sodium silicates. *See* silicate production
sodium sulfite solution for methyl ester sulfonate, 214–216
sodium tripolyphosphate (STPP), 333–334, 375–377
soft compressed tablets, 359
softening of water. *See* zeolites
softergent formulations, silicone fluids in, 289
solid amides, 25–26
solid bed catalyst alkylation, 43
solid hypochlorites, decomposition of, 445–446
solid sodium hypochlorite, 451
solid-state reaction for anhydrous sodium metasilicate, 400
solid type enzymes, 543–545
solubility of powdered detergents, 337
Solvay process for synthetic soda ash, 380–381
solvents, 496–497, 506–528
alkanolamines, 516–518
characteristics, 507
defined, 496
detergent range alcohols, 507–512
fatty alcohols, 510–511
specifications and standards, 507–509
unsaturated fatty alcohols, 511–512
ethanol, 513–516
glycol ethers and esters, 522–526
glycols, 518–522
ethylene glycol, 518–520, 522–526
propylene glycol, 520–522
green solvents, 526–528

introduction, 492
ionic liquids, 528
isopropyl alcohol (2-propanol), 512–513
overview, 506–507
performance-enhancing role, 502–506
role in detergents, 499–506
sorbitol as humectant in toothpastes, 485–486
spacers (amphiphilic moieties' linkage), gemini, 562–563
specialty detergents, 497–498
spray drying, 331, 334–337
atomization, 334–335
blown powder formulation, 332
cooling and classification, 336
diagram of process, 333
drying, 335–336
hydrous sodium polysilicate granules, 397–398
introduction, 331–332
operation, 336
powder properties, 336–337
pumping, 334
slurry making, 333
sodium tripolyphosphate products, 333–334
stilbene derivatives, 549–553
STPP. *See* sodium tripolyphosphate (STPP)
Streckerization reaction, 161, 163
styryl derivatives of benzene and biphenyl, 553–554
sulfated surfactants
alcohol sulfates, 126–135
anionic surfactants, 17, 20
bisulfate, 19–20
continuous film reactor, 91, 92
dimethyl sulfate, 216–217
geminis, 568
manufacturing process, 140–141
overview, 17–21
personal care detergents, 479–480
sulfo betaines, 234–235
sulfochlorination, 142, 143–144, 147
chlorine and sulfur dioxide, 143, 147
sulfuryl chloride, 143–144
sulfonated surfactants
alkanesulfonates, 139–157
anionic surfactants, 17, 20
glyceryl ether sulfonates, 159–169
hydrotropes, 247–251
linear alkylbenzene sulfonate, 86–115
manufacturing process, 140–141, 146, 151–154
methyl ester sulfonate, 19, 202–219
α-olefin sulfonate, 19, 102–109
overview, 17–21
reactors, 91–97, 208–209
sulfonic acid salts, synthesis of, 247–251
sulfosuccinates (sulfosuccinic acid esters), 244–245, 480
sulfoxidation, 142, 144–146, 148–151
sulfur in oxidation state IV, 19
sulfur trioxide-based sulfonation, 86–91
sulfur trioxide in sulfation of alcohols, 127–128
sulfuryl chloride, 143–144
surfactancy property, amphoteric surfactants, 223
surfactants, overview, 1–35. *See also* detergents
classes and production, 12–30
amphoteric surfactants, 13–16
anionic surfactants, 16–21

592 Index

surfactants, overview (*contd.*)
    cationic surfactants, 21–24
    introduction, 12–13
    nonionic surfactants, 24–30
    specialty, 11
  competitive forces affecting production, 4–5
  construction issues, 31–35
  converter overview, 3
  defined, 496
  economic considerations, 12
  global consumption, 1–2, 3–4
  historical perspective, 5–11
  introduction, 1–4
  operational issues, 31–35
  raw materials, 11–12
  regulatory standards, 31–35
  survival strategies, 35
  vs. detergents, 496
  vs. solvents, 504
syndet toilet bars, 171–181
  demand for, 324
  in developing countries, 324, 329
  dual-function prerefining finishing line, 175
  formulas, 173, 176, 177
  introduction, 171–172
  overview, 329–331
  process, 174, 176–181
    conditioning, 181
    cutting, 181
    extruding, 179
    manufacture, 176–177, 178
    mixing, 177
    prerefining, 177
    refining, 177, 179
  vs. soaps, 172–173
synthetic fatty alcohols, 28
synthetic soda ash, 380–381
synthetic sulfation, alcohols, 126–128

T

tablet detergents
  hard compressed, 355–358
  overview, 328
  punch sticking problem, 357, 359
  soft compressed, 359
temperature control in hypochlorite chemistry,
    446–447
tetraacetylethylenediamine (TAED), 431–432
tetrasodium diphosphate (TSPP), 377–378
tetronic surfactant, EO/PO block copolymers, 263
toothpaste products
  amorphous silica, 486–487
  antimicrobial agents, 487–488
  fluorides in, 488
  glycerol for, 485
  precipitated calcium carbonates, 486
  sorbitol as humectant in, 485–486
transacetalization (two-step) alkyl glucoside reaction, 77
transesterification of triglycerides, fatty alcohol
    production, 511

(tri)alkylamine-based cationics, 367–370
bis-(triazinyl amino)-stilbene disulfonic acids for
    fluorescent whitening, 550
TSPP. *See* tetrasodium diphosphate (TSPP)
TTD. *See* Turbo Tube Dryer (TTD)
tube dryer systems, methyl ester sulfonate, 212–216
turbodryer process for hydrous sodium polysilicate
    granules, 398
Turbo Tube Dryer (TTD), 212–216

U

ultralow interfacial tension, solvents in detergents, 506
unit dose technologies, 354–363
  hard compressed tablets, 355–358
  introduction, 354–355
  soft compressed tablets, 359
  water-soluble pouches, 329, 360–363
    dual and multicompartment, 362–363
    overview, 329
    single compartment, 360–362
    summary and trends, 363
unsaturated fatty alcohols, 511–512

V

vapor-phase extraction of *n*-paraffins, 47
vegetable oil-based surfactants. *See* oleo-derived
    surfactants
vertical form fill seal (VFFS) liquid pouch process,
    360–361, 363
vertical forming processes, water-soluble pouches,
    360–361

W

water-soluble pouches, 329, 360–363
  dual and multicompartment, 362–363
  overview, 329
  single compartment, 360–362
  summary and trends, 363
wax esters, reduction of, fatty alcohol production,
    510–511
wet- *vs.* dry-milling of corn for ethanol, 515–516
whiteness measurement, fluorescent whitening agents,
    557–558
whitening agents. *See* fluorescent whitening agents
    (FWAs)
Williamson synthesis, 525

Z

zeolites, 404–412
  AX-type, 412
  NaA-type, 404–411
    clay conversion process, 409
    continuous *vs.* batch-type processes, 409

Index

zeolites (*contd.*)
    hydrogel process, 409–411
    introduction, 404–405
    principles of production, 406–409
    raw materials, 405–406
    requirements and resources, 411
    P-type (zeolite maximum aluminum P (MAP)/zeolite A24), 411–412

Ziegler alcohol processes, 511
zinc salts (zinc citrate) as antimicrobial agents, 487–488
zwitterionic surfactants
    defined, 495
    geminis, 569–570
    perfluoroalkanesulfonyl fluorides, 312–313
    perfluorocarboxylate fluorides, 316–318